T0319486

Structural Reliability

Structural Reliability

Approaches from Perspectives of Statistical Moments

Yan-Gang Zhao and Zhao-Hui Lu

Registered Offices
John Wiley & Sons, Inc., 111 River Street, Hoboken, NJ 07030, USA
John Wiley & Sons Ltd, The Atrium, Southern Gate, Chichester, West Sussex, PO19 8SQ, UK

Editorial Office
9600 Garsington Road, Oxford, OX4 2DQ, UK

For details of our global editorial offices, customer services, and more information about Wiley products visit us at www.wiley.com.

Wiley also publishes its books in a variety of electronic formats and by print-on-demand. Some content that appears in standard print versions of this book may not be available in other formats.

Limit of Liability/Disclaimer of Warranty
In view of ongoing research, equipment modifications, changes in governmental regulations, and the constant flow of information relating to the use of experimental reagents, equipment, and devices, the reader is urged to review and evaluate the information provided in the package insert or instructions for each chemical, piece of equipment, reagent, or device for, among other things, any changes in the instructions or indication of usage and for added warnings and precautions. While the publisher and author have used their best efforts in preparing this work, they make no representations or warranties with respect to the accuracy or completeness of the contents of this work and specifically disclaim all warranties, including without limitation any implied warranties of merchantability or fitness for a particular purpose. No warranty may be created or extended by sales representatives, written sales materials or promotional statements for this work. The fact that an organization, website, or product is referred to in this work as a citation and/or potential source of further information does not mean that the publisher and author endorse the information or services the organization, website, or product may provide or recommendations it may make. This work is sold with the understanding that the publisher is not engaged in rendering professional services. The advice and strategies contained herein may not be suitable for your situation. You should consult with a specialist where appropriate. Further, readers should be aware that websites listed in this work may have changed or disappeared between when this work was written and when it is read. Neither the publisher nor author shall be liable for any loss of profit or any other commercial damages, including but not limited to special, incidental, consequential, or other damages.

Library of Congress Cataloging-in-Publication Data

Names: Zhao, Yan-Gang, author. | Lu, Zhao-Hui, author.
Title: Structural reliability : approaches from perspectives of statistical
    moments / Yan-Gang Zhao and Zhao-Hui Lu.
Description: Hoboken, NJ, USA : Wiley-Blackwell, 2021. | Includes
    bibliographical references and index.
Identifiers: LCCN 2020041116 (print) | LCCN 2020041117 (ebook) | ISBN
    9781119620815 (cloth) | ISBN 9781119620693 (adobe pdf) | ISBN
    9781119620747 (epub)
Subjects: LCSH: Structural engineering–Statistical methods. | Reliability
    (Engineering)–Mathematics. | Moments method (Statistics)
Classification: LCC TA650 .Z53 2021 (print) | LCC TA650 (ebook) | DDC
    624.1–dc23
LC record available at https://lccn.loc.gov/2020041116
LC ebook record available at https://lccn.loc.gov/2020041117

Cover Design: Wiley
Cover Image: © LIU KAIYOU/Getty Images
Printed and bound by CPI Group (UK) Ltd, Croydon, CR0 4YY

C9781119620815_170321

Contents

Preface

This book provides a unified presentation of structural reliability approaches from the perspective of statistical moments. It outlines the unique framework of structural reliability theory centered around the methods of moment. The book is different from many of other books for structural reliability methods that generally more focus on the first-order reliability method (FORM).

The book is composed of the following 13 chapters.

In Chapter 1, the necessity and importance for dealing with the various uncertainties in structural design is emphasised, and different existing measures of safety are reviewed. After pointing out the limit of the traditional deterministic measure of structural safety, the importance of probabilistic measure of safety, i.e., the structural reliability theory is emphasised.

In Chapter 2, basic concepts of structural reliability such as performance function, failure probability, and reliability index are introduced through a fundamental case, which includes only two statistically independent random variables (i.e. a load effect variable S and a resistance variable R). Since an accurate integration for failure probability is almost impossible for many practical engineering problems, the Monte Carlo Simulation (MCS) is first introduced.

Chapter 3 presents the moment computation for performance functions. For some explicit functions, the first few moments can be easily computed from their definitions. For a good behaviour function with only one variable, point estimates with arbitrary numbers of points are developed. Using the point estimate method, the first few moments of a function of single variable can be quickly obtained with required accuracy. Two additional approximate methods are also introduced to obtain the first few moments of performance functions of multiple random variables.

In Chapter 4, direct methods of moment are presented for structural reliability, including the direct second-moment method (2M), the direct third-moment method (3M), and the direct fourth-moment method (4M). These methods are easy to implement for structural reliability analysis and have no issues associated with the design points that are necessary in FORM. No iteration or computation of derivatives is required. The applicable range is then determined for the 2M, 3M, and 4M reliability methods, and simple 3M and 4M reliability indices are suggested for structural reliability analysis in engineering.

Unlike the direct methods of moment in Chapter 4, where the first few moments of the original performance function are used, methods of moment based on the first-order and second-order approximation of the performance function are introduced in Chapter 5. Under the first order approximation of a performance function in standardised normal space, the first-order second-moment reliability index is equal to the shortest distance and therefore perpendicular to the hyper plane in the standardised normal space. While under the second-order approximation of performance function in standardised normal space, second-order second-moment method, second-order third-moment method, and the second-order fourth-moment method, are further developed.

In Chapter 6, reliability evaluation is discussed for the problems without using probability distributions of random variables. The information of the first few moments of the random variables will be used for structural reliability assessment instead of probability distributions.

Chapter 7 presents transformation of non-normal variables to independent normal variables, which are necessary tools in Chapters 3, 5, and 6. Normal tail transformation for a single random variable (or independent random variables) and Rosenblatt or Nataf transformation for correlated random variables are reviewed. The emphasis of this chapter is to introduce the so-called pseudo normal transformation methods for transformation of correlated non-normal random variables into independent standard normal random variables, which can be achieved even when the probability distributions of the basic random variables are unknown.

In Chapter 8, the concept of system reliability is introduced. In order to improve the narrow bounds of the failure probability of a series structural system, a point estimation method is introduced for calculating the joint probability of every pair of failure modes of the system. A moment-based method is presented for the system reliability assessment of series and non-series structures, with emphasis on series systems. The method directly calculates the reliability indices based on the first few moments of the system performance function. This method does not require the reliability analysis of individual failure modes, nor does it need the iterative computation of derivatives, any design points, or the mutual correlations among the failure modes.

Chapter 9 presents the application of the methods of moment for the determination of load and resistance factors. Derivative-based iteration, which is necessary in FORM, is not required in the method. For this reason, the methods of moments are easier for implementation. Although the obtained load and resistance factors are different from those by FORM, the target mean resistances are essentially the same for both methods.

In Chapter 10, the time-variant problems in structural reliability are discussed. Simulating stationary non-Gaussian vector process using the third-order polynomial normal transformation model is first described. The transformation is then applied to evaluate the first passage probability of stationary non-Gaussian structural responses in structural dynamical reliability. A moment-based approach is introduced for time-variant static reliability assessment, where both the resistance and load effects are time-dependent.

Two typical problems of hierarchical models encountered in structural reliability are discussed in Chapter 11, one is structural reliability analysis considering the uncertainties in distribution parameters, and another is the dynamic reliability considering uncertainties contained in input parameters. The application of FORM, methods of moment, and the point-estimate method to evaluate the overall probability of failure with consideration of both the two levels of uncertainties is introduced. The point-estimate method for evaluating the quantile of the conditional failure probability is also introduced.

In Chapter 12, the reliability evaluation using the first few linear moments (L-moments) of random variables is discussed. The chapter introduces the definition of the first four L-moments and the computation of the first four L-moments from the probability distributions and statistical data of random variables. The second- and third-order polynomial normal transformation techniques are then investigated using the first three and four L-moments, respectively, and FORM based on the transformation techniques using L-moments is demonstrated to be sufficiently accurate in structural reliability assessment.

Chapter13 presents the methods of moment combined with Box-Cox transformation, which may significantly abate the non-normality of the performance functions, for structural reliability analysis. A criterion for determining the Box-Cox transformation parameter is introduced, and the procedure of the methods with Box-Cox transformation for structural reliability is presented.

A large number of examples are provided with step-by-step procedures to better illustrate the concepts and methodology. Numerical simulations with more significant digits are also provided to track and calibrate these procedures. This book can be used as textbook for undergraduate and graduate student as well as reference for readers in the areas of structural reliability, reliability engineering and risk management.

Acknowledgements

This book has been a long time in the making. The first manuscript was finished in April 2006, and was only distributed to the graduate students who took part in the course with Yan-Gang Zhao in graduate school at Nagoya Institute of Technology. During the preparation and development of the material for this book, the authors are indebted in many ways to colleagues and students.

A large part of this book is composed of published papers on structural reliability co-authored with Prof. T. Ono, Nagoya Institute of Technology, Japan; the supervisor of Yan-Gang Zhao's doctorate degree, who had a profound effect on so much of the research life of the authors, and who has provided the authors with continuous encouragement, advice, and support on the research of structural reliability over the years. Also, Prof. Alfredo H.-S. Ang, University of California, Irvine, who provided helpful advice and fruitful discussions, not only during the author's visiting research at UCI, but also from continuous e-mail communications and support.

The authors are indebted to the current and former graduate students who had to endure imperfect and incomplete versions of the material; to Mr. C.H. Cai, Ms. F.W. Ge, Mr. D.Z. Hu, Ms. Y. Leng, Ms. P.P. Li, Dr. W.C. Pu, Dr. A. Rasooli, Ms. L. Ren, Dr. M. Sharfuddin, Ms. M.N. Tong, Dr. H.Z. Zhang, Dr. L.W. Zhang, Dr. X.Y. Zhang, and Dr. W.Q. Zhong, for numerous discussions and suggestions on the material. The fruitful discussions with Prof. T. Takada, University of Tokyo; Prof. Y. Mori, University of Nagoya; and Prof. H. Idota, Nagoya Institute of Technology, during the committee of limit state design, Architectural Institute of Japan are gratefully acknowledged. The fruitful discussions with Prof. A. Der Kiureghian, University of California, Berkeley and Prof. Dan M. Frangopol, Lehigh University are gratefully acknowledged.

Special thanks are given to Prof. J.R. Jiang, Institute of Engineering Mechanics, the supervisor of the master's degree of Yan-Gang Zhao, who lead the author to the area of structural reliability; to Dr. K. Ishii, Dr. M. Suzuki, and J. Yoshida, Shimizu Corporation, and Prof. C. Chen, San Francisco State University, who gave the authors much helpful advice on their research over the years.

Finally, a votes of thanks go to Mr. Sakthivel Kandaswamy and other editors at John Wiley & Sons for their patience and scrutiny in the editing of this book. The academic atmosphere conducive to research and innovative developments, Kanagawa University, Beijing University of Technology and Central South University, without which this book would not have been completed, and the supports by the added in Grant-in-Aid from ESCST, Japan and National Natural Science Foundation of China, China, are gratefully acknowledged.

Yan-Gang Zhao and Zhao-Hui Lu

1

Measures of Structural Safety

1.1 Introduction

Engineering structures are generally built in a severe environment. During the service life of some structures, there may be extreme loads caused by earthquakes, wind, snow, fire, etc. To ensure the safety of such structures, engineers do their best to predict the extreme loads and design the structures accordingly. Despite this, some structures may still not be able to withstand disasters because of factors such as overloading, poor quality of structural materials, poor workmanship, or human error. Nevertheless, the reason for structural failure is that there is a gap between our predictions of structural performance and actions and the reality; and this gap generally presents as uncertainties.

Along the development of engineering structural designs, decisions are often required to be made under the conditions of such uncertainties, in the sense that the consequence of a given decision cannot be determined with complete confidence. Therefore, decision making under conditions of uncertainties involves risks. How to quantitatively evaluate the risk and provide a measure of safety is an important problem in engineering.

1.2 Uncertainties in Structural Design

1.2.1 Uncertainties in the Properties of Structures and Their Environment

First, consider a simple beam with a rectangular section of $b \times h$ and a span of l. The mechanical model is illustrated in Figure 1.1a, with its moment diagram shown in Figure 1.1b.

The maximum moment caused by a load P is given as $Pl/4$, and the plastic moment is calculated as $\sigma_y bh^2/4$, where σ_y is the yield stress of the material. Assume that the beam is made of an ideal elastic-plastic material. When the load P becomes sufficiently large to make the maximum moment $Pl/4$ equal to the plastic moment $\sigma_y bh^2/4$, the beam will form a mechanism. The safety of the beam can be checked if the following equation is satisfied during the design of the beam:

$$\frac{Pl}{4} \le \frac{bh^2}{4}\sigma_y \tag{1.1}$$

Structural Reliability: Approaches from Perspectives of Statistical Moments, First Edition.
Yan-Gang Zhao and Zhao-Hui Lu.
© 2021 John Wiley & Sons Ltd. Published 2021 by John Wiley & Sons Ltd.

(a)

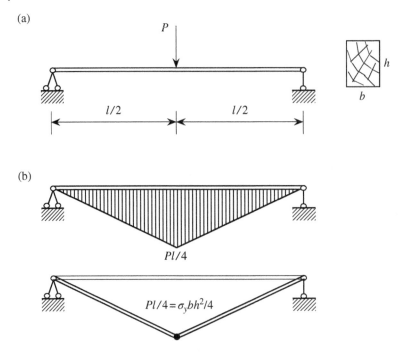

(b)

Figure 1.1 A simple beam. (a) Load bearing condition and cross section of the beam. (b) Bending moment diagram and the plastic moment.

If the quantities of P, l, b, h, and σ_y in Eq. (1.1) are known exactly, from the comparison of the two sides in Eq. (1.1), we can deterministically understand whether the beam is absolutely safe or not. However, the problem arises when we cannot know these quantities exactly. For example, if the beam is used as a bridge, people who step on the bridge may be a child with a weight of 20 kg, an adult with a weight of 70 kg, or in an extremely rare case, a Japanese sumo wrestler with a weight of 200 kg. Therefore, the load P is clearly an uncertain variable. This raises a similar question about the yield stress σ_y. For a specific material, one can obtain the yield stress from tests. However, different values of yield stress will be obtained from different specimens, and it is difficult, if not impossible to determine the yield exact stress of the material used for the beam. As for the parameters l, b, and h, they are generally considered to be known. However, in reality, they cannot be known exactly because of inevitable errors in the process of construction.

If we define $\sigma_y bh^2/4$ as resistance R and $Pl/4$ as load effect S, then Eq. (1.1) becomes

$$S \leq R \tag{1.2}$$

When we check the safety of the beam using Eq. (1.2) under deterministic conditions, if S is less than R, the beam will be absolutely safe. Since the quantities of P, l, b, h, and σ_y in Eq. (1.1) are uncertain, R and S are also uncertain. This uncertainty may lead to a possible reversal of the balance between R and S, and ultimately, the beam may collapse. For example, when $S = 100$ and $R = 150$, the bridge will be considered to be absolutely safe under deterministic conditions. However, since the values of R and S are uncertain, there is a possibility of $S = 140$ and $R = 130$; in this case, the beam may fail.

The example above indicates that not all the parameters in the design process can be exactly known and they generally contain uncertainties. Uncertainty refers to imprecise and incomplete information about the phenomenon being investigated. For example, if it is known that a structure will be subjected to an earthquake load, the exact magnitude of the earthquake is unknown. At least at the present stage, we cannot expect to know exactly when, where, and through what process an earthquake will occur. Furthermore, the duration, the peak ground acceleration, and the frequency of the earthquake waves are also uncertain. Figure 1.2 shows the time histories of ground motion records during the Chuetsu earthquake 2004, Japan (Li and Zhao 2005), and it has been observed that no earthquake histories are the same. Generally, uncertainty is significant in the load and environment. Uncertainties exist in almost all the aspects of a load, e.g. the live load varies every day and the wind load changes all the time.

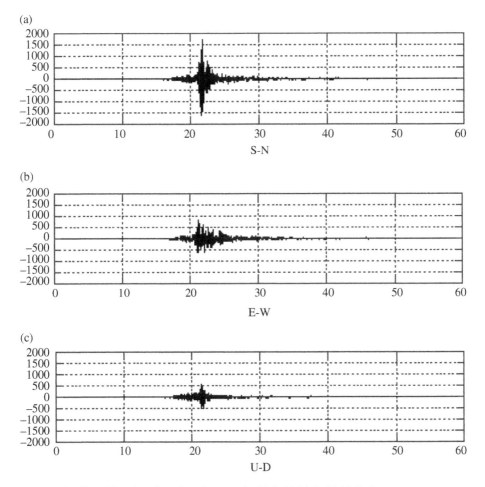

Figure 1.2 Time histories of earthquake records. (a) S–N. (b) E–W. (c) U–D.

Other than the load and environmental effects, there are also many uncertainties associated with structures themselves. It can be expected that the grade of concrete will be, for example, C30, but it is not known what the values of the compressive strength in a specific cross-section will be, unless a test is carried out for that section. Assume that 15 concrete cylinders are sampled from the concrete mixes used in construction and subsequently tested in compression with the following results (in MPa): 30.5, 29.8, 31.2, 30.8, 28.9, 31.6, 30.7, 30.9, 30.4, 30.5, 30.8, 30.3, 30.5, 31.1, and 30.9.

Based on these observations, the sample mean (average) is obtained as

$$\overline{f}_c = \frac{1}{15} \sum_{i=1}^{15} f_{ci} = 30.6 \tag{1.3}$$

where f_{ci} is the concrete compressive strength of the ith samples; and the $\overline{f}_c$ is the average of the concrete compressive strength.

The sample mean is, of course, an estimate of the true concrete strength (which remains unknown). The fact that the observed strength of the 15 different cylinders is significantly scattered gives rise to an uncertainty in the actual strength of any given section of any part of the structure. Furthermore, the strength of structural members may change with time, as most of them degrade in the corrosive environment.

Another type of uncertainty is human error. Human error is an inappropriate or undesirable human decision or behaviour that reduces or has the potential of reducing the effectiveness or safety of a system. Human error includes badly designed or faulty equipment, poor management practices, inaccurate or incomplete procedures, and inadequate or inappropriate training.

1.2.2 Sources and Types of Uncertainties

Uncertainties in engineering may be associated with physical phenomena that are inherently random, or with predictions and estimations of reality performed with incomplete or inadequate information. From this standpoint, uncertainties may be associated with the inherent variability of a physical process or with imperfections in the modelling of a physical process. That is, randomness in a physical process contributes to uncertainty because inherent errors of an imperfect prediction model cannot be entirely corrected deterministically. Furthermore, prediction or modelling errors may contain two components; a systematic component and a random component. In measurement theory, these are known as the systematic error and random error, respectively. From a practical standpoint, inherent variability is essentially a state of nature, and the resulting uncertainty may not be controlled or reduced, i.e. the uncertainty associated with inherent variability is something that we have to live with. The uncertainty associated with prediction or modelling errors may be reduced using more accurate models or the acquisition of additional data. In some literatures, the uncertainties above are also classified as aleatory and epistemic uncertainties (ISO 2394 2015; Ang and Tang 2006).

1.2.3   Treatment of Uncertainties

Since we cannot exactly predict the outcomes of uncertain phenomena, such uncertain phenomena are characterised by experimental observations that are invariably different from one to another (even if performed under apparently identical conditions). In other words, there is usually a range of measured or observed values. Furthermore, within this range, certain values may occur more frequently than others. The characteristics of such experimental data can be portrayed graphically in the form of a *histogram* or *frequency diagram*. Such examples are shown in Figure 1.3 for the residual stress of H-shaped steel (JIS-G-3192 1977) and the Young's modulus of SS400 steel (JIS-G-3101 1976; Ono et al. 1986). Therefore, the histogram or frequency diagram is a graphical description of the variability of experimental information.

For a specific set of experimental data, the corresponding histogram may be constructed as follows.

Figure 1.3   Examples of histograms. (a) Histogram of residual stress of H-shaped steel (JIS) (over a number of observations). (b) Histogram of the Young's modulus of SS400 steel (JIS) (over a number of observations).

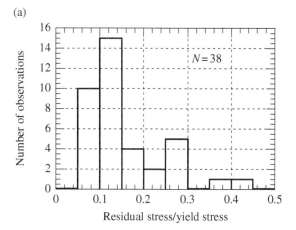

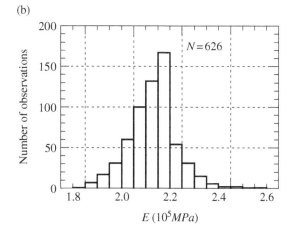

Table 1.1 Data of residual stress in H-shaped steel (residual stress/nominal yield stress).

Sample No.	Sample values	Sample No.	Sample values	Sample No.	Sample values	Sample No.	Sample values
1	0.109	11	0.102	21	0.102	31	0.15
2	0.143	12	0.101	22	0.131	32	0.403
3	0.143	13	0.127	23	0.086	33	0.277
4	0.1	14	0.091	24	0.106	34	0.266
5	0.139	15	0.106	25	0.09	35	0.277
6	0.076	16	0.098	26	0.082	36	0.266
7	0.082	17	0.184	27	0.117	37	0.202
8	0.143	18	0.184	28	0.194	38	0.394
9	0.096	19	0.072	29	0.275		
10	0.086	20	0.123	30	0.20		

From the observed range of experimental measurements, choose a range on the abscissa sufficient to include the largest and smallest observed values, and divide this range into convenient intervals. Then, count the number of observations within each interval, and draw vertical bars with heights representing the number of observations in the respective intervals. Alternatively, the heights of the bars may be expressed in terms of the fractions of the total number of observations in each interval. For example, consider the data of residual stress in H-shaped steel in Table 1.1. It can be seen that the data ranges from 0.072 to 0.403. Choosing a uniform interval of 0.05, between 0.05 and 0.45, the number of observations and the corresponding fraction of total observations within each interval are shown in Table 1.2.

Table 1.2 Data for histogram and a frequency diagram.

Interval	Number of observations	Fraction of total observations
0.05–0.1	10	0.263158
0.1–0.15	15	0.394737
0.15–0.2	4	0.105263
0.2–0.25	2	0.052632
0.25–0.3	5	0.131579
0.3–0.35	0	0
0.35–0.4	1	0.026316
0.4–0.45	1	0.026316
Total	38	1

Plotting the number of observations in the given interval, we can obtain a histogram of the residual stress as shown in Figure 1.3a, whereas, in terms of the fraction of the total observations, the histogram would be as shown in Figure 1.4a.

For comparing an empirical frequency distribution with a theoretical probability density function, the corresponding frequency diagram is required. This may be obtained from the histogram by simply dividing the ordinates of the histogram by its total area. In the case of the histogram in Figure 1.4, we can obtain the corresponding frequency diagram by dividing the ordinates in Figure 1.3a by $38 \times 0.05 = 1.9$, or alternatively, by dividing the ordinates in Figure 1.4a by $1 \times 0.05 = 0.05$. The result would be as shown in Figure 1.4b, which is the frequency diagram for the residual stress.

The histogram, or frequency diagram, provides a graphical picture of the relative frequency of various observations or measurements. For most engineering purposes, certain

(a)

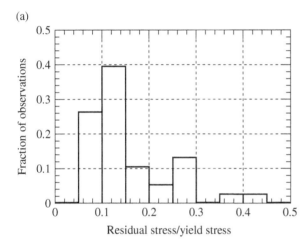

(b)

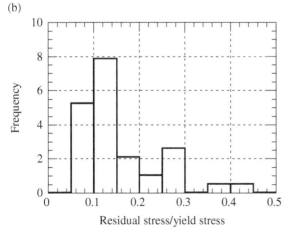

Figure 1.4 Histogram and frequency diagram of residual stress in H-shape steel. (a) Histogram for fraction of observations. (b) Frequency diagram.

aggregate quantities from the set of observations are more useful than the complete histogram. These include, in particular, the mean value (or average) and the measure of dispersion. Such quantities may be evaluated from a given histogram; statistically however, these are usually obtained in terms of the sample mean and the sample standard deviation, which are described in detail in Appendix A.3.

Clearly, if the recorded data of a variable exhibits scatter or dispersion, such as those illustrated in Figures 1.3 and 1.4, the value of the variable cannot be predicted with certainty. Such a variable is known as a random variable, and its value, or range of values, can be predicted only with an associated probability.

When two or more random variables are involved, the characteristics of one variable may depend on the value of the other variable. Pairs of observed data for the two variables, when plotted on a two-dimensional space, as shown in Figure 1.5, are characterised by scatter or dispersion in the data points, and these are called scattergrams. In view of such scatter, the value of one variable, given that of the other, cannot be predicted with certainty. The degree of predictability will depend on the degree of mutual dependency or the correlation between the variables.

When we deal with information such as that illustrated in Figures 1.3 through 1.5, which requires a probabilistic description, proper utilisation of this information will necessarily require concepts and methods of probability for engineering decision making. For example, if a design equation involves random variables, such as those described in Figures 1.3 through 1.5, quantitative analysis of the effects on the design and the formulation of the design will necessarily involve probabilistic concepts. In this book, basic theories of statistics and probability are included in the Appendix A. Alternatively, one may also refer to other textbooks.

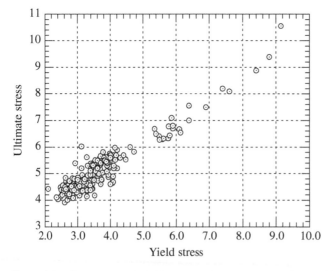

Figure 1.5 Scattergram of yield stress and ultimate stress for steel bars.

1.2.4   Design and Decision Making with Uncertainties

If engineering decisions must be made under conditions of uncertainties, how should designs be formulated, or how can decisions affecting a design be resolved? Presumably, one can consistently assume the worst scenarios and develop conservative designs. From the standpoint of system performance and safety, this approach may be suitable. However, the resulting design may be too costly as a consequence of compounded conservatism. On the other hand, a conservative design may not ensure the desired level of performance or safety. Therefore, decisions should be made as a trade-off between costs and benefits. The most desirable solution would be optimal in the sense of the minimum cost and/or maximum benefits. If the available information and/or the evaluation models contain uncertainties, the required trade-off analysis should include the effects of such uncertainties for any decision making.

1.3   Deterministic Measures of Safety

The traditional definition of safety is through a *factor of safety*, usually associated with the following allowable stress design formula:

$$\sigma \leq [\sigma] \tag{1.4}$$

where σ is the applied stress; and $[\sigma]$ is the allowable stress.

The allowable stress is usually defined in structural design codes. It is derived from the material strength expressed in ultimate stress σ_u (or yield stress σ_y) but is lowered through a multiplier F:

$$[\sigma] = \frac{\sigma_u}{F} \tag{1.5}$$

where F is the factor of safety or coefficient of safety. The greater the uncertainty about the strength of materials, the higher the coefficient of safety. One may also call it a factor of prudence or if one likes, of ignorance. The factor F is usually selected based on experimental observations, previous practical experience, economic, and perhaps political considerations. The selection of F is one of the responsibilities of code committees, and F generally varies with different load cases for different types of structures.

In the allowable stress design formula described above, the safety factor F is introduced to deal with the uncertainties contained in the applied stress and the ultimate strength. However, as a deterministic measure of safety, the factor F may not provide a quantitative measure of safety. If different members of the structure use different safety factors, then the safety of the entire structure may be difficult to evaluate. In contrast, a structure designed using a larger factor of safety F may not mean that it is safer than those designed with a smaller factor F. For example, the safety factor for steel structures is generally smaller than that for reinforced concrete (RC) structures. However, this does not mean that RC structures are safer than steel structures. Similarly, the safety factor under an earthquake load is larger than that under a wind load. It does not mean that the structure is safer under an earthquake load.

There are also some other disadvantages for the allowable stress design formula, e.g. since the factor *F* is usually selected on the basis of previous experience, it may be unsuitable for checking the safety of structures under loads that are difficult to predict. Furthermore, the allowable stress design code lacks formula invariance. An investigation has been performed in detail by Melchers (1987). Since it is not a quantitative measure of safety, it is obviously difficult to determine a trade-off between safety and economy.

1.4   Probabilistic Measure of Safety

In the probabilistic measure of safety, the quantification of uncertainty, and the evaluation of its effects on the performance and design of an engineering structure are conducted using concepts and methods of probability. This measure is also referred to as structural reliability theory.

When an engineering structure is loaded in some way, its response would depend on the type and magnitude of the load, and the strength and stiffness of the structure. Whether the response is considered satisfactory depends on the requirements that must be satisfied. Such requirements may include safety of the structure against collapse, damage limitations, deflections limitations, or any of a range of other criteria. Each of these requirements may be termed as a limit state. The violation of a limit state can then be defined as the attainment of an undesirable condition for the structure.

The study of structural reliability is concerned with the calculation and prediction of the probability of limit state violation for engineering structures at any stage during their service life. In particular, structural safety is concerned with the ultimate or safety limit states of a structure.

The probability of the occurrence of an event such as limit state violation is a numerical measure of the chance of its occurring. This measure may either be obtained from measurements of the long-term frequency of the occurrence of the event for generally similar structures, or may simply be a subjective estimation of the numerical value. In practice, it is usually not possible to observe other similar structures for a sufficiently long period of time, and a combination of subjective estimation and frequency observation for structural components and properties is usually used to predict the probability of limit state violation for the structure as a whole.

In probabilistic assessments, any uncertainty about a variable is explicitly taken into account. This is not the case in traditional methods to measure safety, such as the factor of safety. These are deterministic measures in that the variables describing the structure and its strength are assumed to take on known values for which no uncertainty is assumed.

1.5   Summary

In this chapter, the necessity of dealing with various uncertainties in engineering decision making is emphasised and different measures of safety are reviewed. Because the traditional deterministic measure cannot provide a quantitative basis for safety, a probabilistic measure of safety, i.e. the structural reliability theory, needs to be used.

2

Fundamentals of Structural Reliability Theory

2.1 The Fundamental Case

One of the principle objectives of structural design is the assurance of structural performance, including safety, within the economical constraint. The reliability of a structure can be defined as its capability to fulfil its design purpose for a specified reference period. The problems of structural reliability may essentially be cast as a problem of supply versus demand. In other words, problems of structural reliability may be formulated as the determination of the capacity of a structural system (supply) to meet certain requirements (demand). In consideration of the safety of a steel bar as shown in Figure 2.1, we are concerned with ensuring that the resistance of the bar (supply) is sufficient to withstand the maximum tensional load (demand).

Traditionally, the structural reliability is achieved through using factors or margins of safety and adopting conservative assumptions in the process of design. That is to ascertain that a 'worst' or minimum supply condition will remain adequate (by some margin) under a 'worst' or maximum demand requirement. What constitutes minimum supply and maximum demand conditions, however, is often based on subjective judgments. Moreover, the adequacy or inadequacy of the applied margins may also be evaluated or calibrated only in terms of the past experiences with similar systems. As described in the chapter entitled Measures of Structural Safety, the traditional approach is difficult to quantify and lacks the logical basis to address the effects of uncertainties. Consequently, the level of safety or reliability cannot be assessed quantitatively. Moreover, the problem of assuring performance would obviously be difficult for new systems when there is no prior basis for calibration.

In fact, it is of course not a simple problem to determine the available supply as well as maximum demand. Estimation and prediction are invariably necessary and uncertainties are unavoidable for the simple reason that engineering information is invariably incomplete. Under the uncertainties, the available supply and actual demand cannot be determined precisely. Instead, the available supply and required demand may be modelled as random variables. In other words, structural reliability may be more realistically measured in terms of probability.

Consider a simple case of the steel bar above, where the reliability of a structure or a structural element is determined by only two statistically independent random variables, i.e. a load effect variable S and a resistance variable R as shown in Figure 2.1. In this case, the

Structural Reliability: Approaches from Perspectives of Statistical Moments, First Edition.
Yan-Gang Zhao and Zhao-Hui Lu.
© 2021 John Wiley & Sons Ltd. Published 2021 by John Wiley & Sons Ltd.

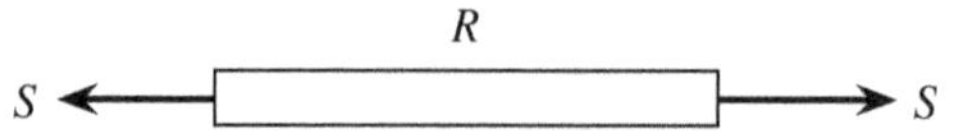

$$R$$

$$S \longleftarrow \boxed{} \longrightarrow S$$

Figure 2.1 A steel bar with tensional load.

objective of reliability analysis is to ensure the event $(R > S)$ throughout the service life, or some specified life of the engineering system. The assurance is expressed in terms of probability $P(R > S)$. For discrete random variables R and S, the probability of failure P_F may be formulated as follows:

$$P_F = P(R \leq S) = \sum_{\text{all } s} P(R \leq S | S = s) P(S = s) = \sum_{\text{all } s} P(R \leq s) P(S = s), \qquad (2.1)$$

and for continuous R and S, if R and S are statistically independent, then the joint probability density function (PDF) of R and S can be expressed as $f_{R, S}(r, s) = f_R(r) f_S(s)$, then the probability of failure is

$$P_F = P(R \leq S) = \int_{r \leq s} f_R(r) f_S(s) dr ds = \int_{-\infty}^{+\infty} \left[\int_{-\infty}^{s} f_R(r) dr \right] f_S(s) ds,$$

i.e.

$$P_F = \int_{-\infty}^{+\infty} F_R(s) f_S(s) ds \qquad (2.2a)$$

Equation (2.2a) is the convolution with respect to s and may be explained with reference to Figure 2.2 as follows: for $S = s$, the conditional probability of failure would be $F_R(s)$, but since $S = s$ is associated with probability $f_S(s) ds$, the integration over all values of S yields Eq. (2.2a). Alternatively, the probability of failure may also be formulated by convolution with respect to r, yielding

$$P_F = \int_{-\infty}^{+\infty} [1 - F_S(r)] f_R(r) dr \qquad (2.2b)$$

As portrayed graphically in Figure 2.2, the overlapping of the curves $f_R(r)$ and $f_S(s)$ represents a qualitative measure of the failure probability P_F, where we can observe that the following (Ang and Tang 1984); the measure of safety or reliability should be a function of the

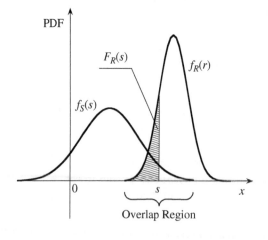

Figure 2.2 PDFs of R and S.

relative positions of $f_R(r)$ and $f_S(s)$ as well as of their degree of dispersions. The relative position between $f_R(r)$ and $f_S(s)$ may be measured by the ratio μ_R/μ_S, which is generally called the *central safety factor* or the difference $(\mu_R-\mu_S)$ which is the mean *safety margin*.

Theoretically, the failure probability P_F will depend on the form of $f_R(r)$ and $f_S(s)$, in which the position of PDFs and degree of dispersions are of course contained.

Example 2.1 To Show that Area of the Overlap Region is not the Probability of Failure

It should be noted that the probability of failure P_F is not equal to the area of the overlap region, which is sometimes confused in concept. As shown in Figure 2.3, assume that $f_S(s)$ and $f_R(r)$ cross at x^*, and the overlap region is divided by x^* into two parts, ω_1 and ω_2; then the area of the overlap region should be $\omega_1 + \omega_2$.

$$\omega_1 = \int_{-\infty}^{x^*} f_R(x)dx = F_R(x^*)$$

$$\omega_2 = \int_{x^*}^{\infty} f_S(x)dx = 1 - F_S(x^*)$$

Using Eq. (2.2a), the probability of failure is expressed as

$$P_F = \int_{-\infty}^{x^*} F_R(x)f_S(x)dx + \int_{x^*}^{+\infty} F_R(x)f_S(x)dx$$

note,

$$\int_{x^*}^{+\infty} F_R(x)f_S(x)dx = [F_R(x)F_S(x)]\big|_{x^*}^{+\infty} - \int_{x^*}^{+\infty} F_S(x)f_R(x)dx$$

$$= 1 - F_R(x^*)F_S(x^*) - \int_{x^*}^{+\infty} F_S(x)f_R(x)dx$$

Figure 2.3 Relationship between P_F and the area of overlap region.

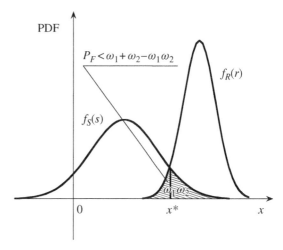

PDF

$P_F < \omega_1 + \omega_2 - \omega_1\omega_2$

$f_R(r)$

$f_S(s)$

$\omega_1 - \omega_2$

0

x^*

x

then,

$$P_F = \int_{-\infty}^{x^*} F_R(x)f_S(x)dx + 1 - F_R(x^*)F_S(x^*) - \int_{x^*}^{+\infty} F_S(x)f_R(x)dx$$

Since, $F_S(x) > F_S(x^*)$ for $x > x^*$ and $F_R(x) < F_R(x^*)$ for $x < x^*$, we have

$$\int_{x^*}^{+\infty} F_S(x)f_R(x)dx > \int_{x^*}^{+\infty} F_S(x^*)f_R(x)dx = F_S(x^*)[1 - F_R(x^*)]$$

and

$$\int_{-\infty}^{x^*} F_R(x)f_S(x)dx < \int_{-\infty}^{x^*} F_R(x^*)f_S(x)dx = F_R(x^*)F_S(x^*)$$

Therefore,

$$P_F < 1 - F_S(x^*)[1 - F_R(x^*)] = 1 - (1 - \omega_2)(1 - \omega_1)$$

That is to say,

$$P_F < \omega_1 + \omega_2 - \omega_1\omega_2$$

The probability of failure P_F is obviously not equal to the area of the overlap region.

Margin of Safety: The above supply-demand problem may also be formulated in terms of the safety margin, $M = R - S$. As R and S are random variables, M is also a random variable with corresponding PDF $f_M(m)$. In this case, failure is clearly the event ($M \leq 0$), and the probability of failure, therefore, is

$$P_F = \int_{-\infty}^{0} f_M(m)dm = F_M(0) \tag{2.3}$$

Graphically, this is represented by the area under $f_M(m)$ below 0 as shown in Figure 2.4.

The Factor of Safety: Another commonly used term in engineering is the factor of safety, which is defined as $\Theta = R/S$. Obviously Θ is also a positive random variable with corresponding PDF $f_\Theta(\theta)$ if R and S are positive random variables. In this case, failure is clearly the event ($\Theta \leq 1$), and thus the corresponding probability of failure is

$$P_F = \int_{0}^{1} f_\Theta(\theta)d\theta = F_\Theta(1.0) \tag{2.4}$$

Figure 2.4 PDF of safety margin M.

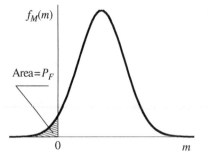

Figure 2.5   PDF of safety factor Θ.

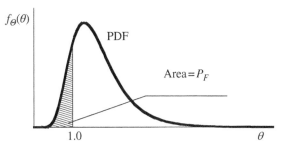

Graphically, this is represented by the area under $f_\Theta(\theta)$ between 0 and 1 as shown in Figure 2.5.

Example 2.2   Reliability Analysis Using Safety Margin/Safety Factor

Consider the basic problem of resistance (R) versus load effect (S). First assume that R and S are statistically independent normal random variables, $N(\mu_R, \sigma_R^2)$, $N(\mu_S, \sigma_S^2)$. The probability distribution of the safety margin $M = R - S$ is thus also a normal distribution $N(\mu_M, \sigma_M^2)$, in which

$$\mu_M = \mu_R - \mu_S, \sigma_M^2 = \sigma_R^2 + \sigma_S^2$$

and the standardised variable

$$M_s = \frac{M - \mu_M}{\sigma_M}$$

is a standard normal variable $N(0, 1)$.
   Since

$$M = \sigma_M M_s + \mu_M$$

We have

$$P_F = P(M \leq 0) = P\left(M_s \leq -\frac{\mu_M}{\sigma_M}\right)$$

That is

$$P_F = \Phi\left(-\frac{\mu_M}{\sigma_M}\right) = 1 - \Phi\left(\frac{\mu_M}{\sigma_M}\right)$$

and

$$P_S = 1 - P_F = \Phi\left(\frac{\mu_M}{\sigma_M}\right)$$

where $\Phi(\cdot)$ is the cumulative distribution function (CDF) of the standard normal variable $N(0, 1)$.

It can be observed that the reliability is a function of the ratio μ_M/σ_M, and the reliability increases with the increase of μ_M/σ_M. Therefore, μ_M/σ_M can be considered for the measure of reliability, is often referred to as the *reliability index* or *safety index*, and is generally denoted by β, where

$$\beta = \frac{\mu_M}{\sigma_M} = \frac{\mu_R - \mu_S}{\sqrt{\sigma_R^2 + \sigma_S^2}}$$

This illustrates the fact that the level of reliability is a function of both the relative position of $f_R(r)$ and $f_S(s)$ as measured by the mean $\mu_M = \mu_R - \mu_S$, and their degree of dispersions as measured in terms of the standard deviation $\sigma_M = \sqrt{\sigma_R^2 + \sigma_S^2}$. Clearly, in this case a quantity that accounts for the effect of both these factors is the reliability index β.

Next assume that R and S are statistically independent lognormal random variables with parameters λ_R, ζ_R and λ_S, ζ_S. The probability distribution of the factor of safety $\Theta = R/S$ is also lognormal with parameters

$$\lambda_\Theta = \lambda_R - \lambda_S, \zeta_\Theta^2 = \zeta_R^2 + \zeta_S^2$$

where λ_Θ and ζ_Θ are the mean and standard deviation of $\ln\Theta$, respectively.

Therefore, according to Eq. (2.4), the probability of failure can be calculated as

$$P_F = F_\Theta(1.0) = \Phi\left(\frac{\ln 1.0 - \lambda_\Theta}{\zeta_\Theta}\right) = \Phi\left(-\frac{\lambda_\Theta}{\zeta_\Theta}\right)$$

In this case, the ratio $\lambda_\Theta/\zeta_\Theta$ is also the reliability index β

$$\beta = \frac{\lambda_\Theta}{\zeta_\Theta}$$

2.2 Performance Function and Probability of Failure

2.2.1 Performance Function

Consider the steel bar shown in Figure 2.1 again, in which the reliability of a structure or a structural element is determined by only two random variables, i.e. the load effect variable S and the resistance variable R. There are three states of the structural member according to the relationship between R and S:

- $R > S$; safe
- $R = S$; limit state
- $R < S$; failure.

In structural reliability analysis, the three states above are generally described by a function

Figure 2.6 Safe and failure states for a 2-dimensional problem.

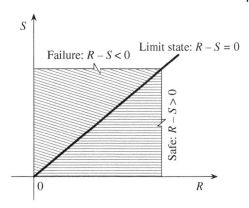

$$Z = G(R,S) = R - S \begin{cases} > 0, & \text{Safe state} \\ = 0, & \text{Limit state} \\ < 0, & \text{Failure state} \end{cases} \qquad (2.5)$$

Equation (2.5) is geometrically shown in Figure 2.6. The function $Z = G(R, S)$ is named *limit state function* or *performance function*.

2.2.2 Probability of Failure

The objective of structural design is to ensure the event $Z = G(R, S) > 0$ throughout the service life or some specified time period, of the structure. This assurance is delivered only in terms of the probability $P[Z = G(R, S) > 0]$. This probability therefore represents a realistic measure of structural reliability. Conversely, the probability of the complimentary event $Z = G(R, S) \leq 0$ is the corresponding measure of unreliability, or failure. These can be expressed as,

$$\text{Probability of safety}: P_S = P[Z = G(R,S) > 0] \qquad (2.6a)$$
$$\text{Probability of failure}: P_F = P[Z = G(R,S) \leq 0] \qquad (2.6b)$$
$$P_F + P_S = 1 \qquad (2.6c)$$

If the joint PDF of the basic variables R and S is $f_{R,S}(r, s)$, the probability of the safe state is

$$P_S = P[Z = G(R,S) > 0] = \int_{r-s>0} f_{R,S}(r,s)drds \qquad (2.7a)$$

Equation (2.7a) is the integral over the safe region. Conversely the probability of the failure state or failure probability, would be corresponding to the integral over the failure region.

$$P_F = P[Z = G(R,S) \leq 0] = \int_{r-s\leq0} f_{R,S}(r,s)drds \qquad (2.7b)$$

Geometrically, Equations (2.7a) and (2.7b) are illustrated in Figure 2.7.

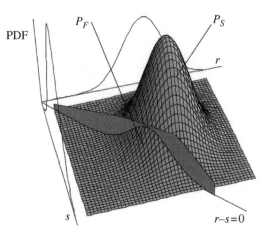

Figure 2.7 Probability of failure and safety.

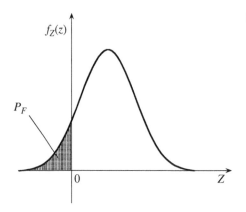

Figure 2.8 Probability of failure.

Since $Z = G(R, S)$ is also a random variable, and its PDF of can be expressed as $f_Z(z)$, the failure probability may then be expressed as

$$P_F = P[Z = G(R, S) \leq 0] = \int_{-\infty}^{0} f_Z(z)dz = F_Z(0) \tag{2.8}$$

Geometrically, this is represented by the area under $f_Z(z)$ below 0, as shown in Figure 2.8.

Generally, structural reliability may involve multiple variables and may be defined as the probability of performing its intended function or mission, where we can define a *performance function* or a *limit state function*

$$G(\mathbf{X}) = G(X_1, X_2, \cdots, X_n) \tag{2.9}$$

where $\mathbf{X} = (X_1, X_2, ..., X_n)$ is a vector of basic variables of the system and the function $G(\mathbf{X})$ denotes the performance function or limit state of the system. Accordingly, the limiting performance requirement may be defined as $G(\mathbf{X}) = 0$, which is the *limit-state equation* of the system. Therefore, Eq. (2.5) in such case is rewritten as

$$Z = G(\mathbf{X}) = G(X_1, X_2, \cdots, X_n) \begin{cases} > 0, & \text{Safe state} \\ = 0, & \text{Limit state} \\ < 0, & \text{Failure state} \end{cases} \qquad (2.10)$$

Geometrically, the limit-state equation, $Z = G(\mathbf{X}) = 0$, is an n-dimensional surface that may be called the *limit state surface*. One side of the limit state surface is the safe state, $Z = G(\mathbf{X}) > 0$, whereas the other side of the surface corresponds to the failure state, $Z = G(\mathbf{X}) < 0$. Hence, if the joint PDF of the basic random vector $\mathbf{X}$ is $f_{\mathbf{X}}(\mathbf{x})$, the probability of the safety is

$$P_S = P[Z = G(\mathbf{X}) > 0] = \int_{G(\mathbf{x}) > 0} f_{\mathbf{X}}(\mathbf{x}) d\mathbf{x} \qquad (2.11a)$$

Equation (2.11a) is simply the volume integral over the safe region. Correspondingly, the probability of the failure state or failure probability is the volume integral over the failure region.

$$P_F = P[Z = G(\mathbf{X}) \le 0] = \int_{G(\mathbf{x}) \le 0} f_{\mathbf{X}}(\mathbf{x}) d\mathbf{x} \qquad (2.11b)$$

Evaluation of the probability through Eq. (2.11a) or Eq. (2.11b), however, is generally computationally formidable.

If the PDF of Z is obtained as $f_Z(z)$, then, the failure probability may then be expressed as

$$P_F = P[Z = G(\mathbf{X}) \le 0] = \int_{-\infty}^{0} f_Z(z) dz = F_Z(0) \qquad (2.12)$$

2.2.3 Reliability Index

For Eq. (2.12), assume that the performance function Z is a normal random variable with mean value μ_Z and standard deviation σ_Z, then the standardised variable

$$Z_s = \frac{Z - \mu_Z}{\sigma_Z}$$

is a standard normal variable $N(0, 1)$.

Since

$$Z = \sigma_Z Z_s + \mu_Z$$

We have

$$P_F = P(Z \le 0) = P(\sigma_Z Z_s + \mu_Z \le 0) = P\left(Z_s \le -\frac{\mu_Z}{\sigma_Z}\right)$$

That is,

$$P_F = \Phi\left(-\frac{\mu_Z}{\sigma_Z}\right) = 1 - \Phi\left(\frac{\mu_Z}{\sigma_Z}\right) \qquad (2.13a)$$

and the reliability is

$$P_S = 1 - P_F = \Phi\left(\frac{\mu_Z}{\sigma_Z}\right) \tag{2.13b}$$

As we discussed in Example 2.2, if the performance function Z is a normal random variable, the probability of failure or the reliability is a function of the ratio μ_Z/σ_Z. The reliability index increases with the increase of μ_Z/σ_Z. Therefore, μ_Z/σ_Z can be considered as a measure of reliability and is called the *reliability index* or *safety index*, and is generally denoted by β as,

$$\beta = \frac{\mu_Z}{\sigma_Z} \tag{2.14}$$

In general, the reliability index can be further defined as

$$\beta = \Phi^{-1}(1 - P_F) = -\Phi^{-1}(P_F) \tag{2.15a}$$

The probability of safety, therefore, becomes

$$P_S = \Phi(\beta) \tag{2.15b}$$

and the corresponding probability of failure is

$$P_F = 1 - \Phi(\beta) = \Phi(-\beta) \tag{2.16}$$

Figure 2.9 shows the graphical relationship between the reliability index and the probability of failure.

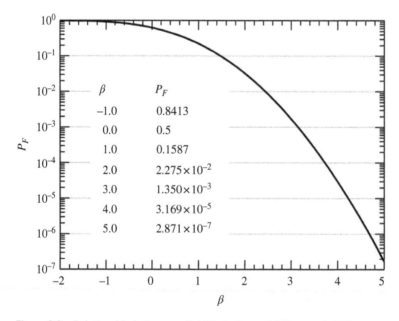

β	P_F
−1.0	0.8413
0.0	0.5
1.0	0.1587
2.0	2.275×10^{-2}
3.0	1.350×10^{-3}
4.0	3.169×10^{-5}
5.0	2.871×10^{-7}

Figure 2.9 Relationship between reliability index and failure probability.

Example 2.3 Consider the following performance function,

$$Z = G(\mathbf{X}) = \ln(R/S) = \ln R - \ln S$$

Suppose that R and S are statistical independent lognormal random variables with mean values of μ_R and μ_S, and standard deviations of σ_R and σ_S. The corresponding parameters of the lognormal distributions would be

$$\zeta_R^2 = \ln\left(1 + \frac{\sigma_R^2}{\mu_R^2}\right), \zeta_S^2 = \ln\left(1 + \frac{\sigma_S^2}{\mu_S^2}\right), \lambda_R = \ln\mu_R - \frac{1}{2}\zeta_R^2, \lambda_S = \ln\mu_S - \frac{1}{2}\zeta_S^2$$

The relationship between lognormal and normal variables gives

$$\ln R = \zeta_R U_R + \lambda_R, \ln S = \zeta_S U_S + \lambda_S,$$

where U_R and U_S are standard normal variable corresponding to R and S, respectively. Therefore,

$$Z = \zeta_R U_R - \zeta_S U_S + \lambda_R - \lambda_S$$

is a normal variable, and its mean and standard deviation are given as

$$\mu_Z = \lambda_R - \lambda_S, \sigma_Z = \sqrt{\zeta_R^2 + \zeta_S^2}$$

Since Z is normal, according to Eq. (2.14), the reliability index is derived as

$$\beta = \frac{\mu_Z}{\sigma_Z} = \frac{\lambda_R - \lambda_S}{\sqrt{\zeta_R^2 + \zeta_S^2}}$$

Example 2.4 Consider the following performance function

$$Z = G(\mathbf{X}) = a_0 + \sum_{i=1}^{n} a_i X_i, \tag{2.17}$$

where X_i, $i = 1, 2,..., n$ are statistical independent normal random variables with mean values and standard deviations of μ_i and σ_i.

Since Z is a linear function of normal random variables, Z is also a normal variable. The mean and standard deviation of Z are given as

$$\mu_Z = a_0 + \sum_{i=1}^{n} a_i \mu_i, \sigma_Z = \sqrt{\sum_{i=1}^{n} a_i^2 \sigma_i^2}$$

Since Z is a normal random variable, according to Eq. (2.14), the reliability index is given as

$$\beta = \frac{\mu_Z}{\sigma_Z} = \frac{a_0 + \sum_{i=1}^{n} a_i \mu_i}{\sqrt{\sum_{i=1}^{n} a_i^2 \sigma_i^2}} \tag{2.18}$$

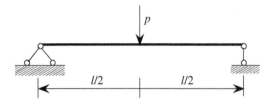

Figure 2.10 A simple beam.

Example 2.5 Consider the beam in Figure 2.10 subjected to a concentrated load p at midspan.

The maximum deflection is

$$\Delta_{\max} = \frac{pl^3}{48EI}$$

where l, E, and I are the length, the Young's modulus, and the moment of inertia of the beam, respectively. Suppose that p, l, E, and I are statistically independent lognormal random variables with mean values of $\mu_p = 4$ kN, $\mu_l = 4$ m, $\mu_E = 2 \times 10^7$ kN/m², $\mu_I = 4 \times 10^{-5}$ m⁴, and standard deviations of $\sigma_p = 1.6$ kN, $\sigma_l = 0.4$ m, $\sigma_E = 1 \times 10^6$ kN/m², $\sigma_I = 8 \times 10^{-6}$ m⁴, then the corresponding parameters of the lognormal distributions are

$$\zeta_p^2 = \ln\left(1 + \frac{\sigma_p^2}{\mu_p^2}\right) = \ln\left(1 + 0.4^2\right) = 0.1484, \quad \zeta_l^2 = \ln\left(1 + \frac{\sigma_l^2}{\mu_l^2}\right) = \ln\left(1 + 0.1^2\right) = 0.01,$$

$$\zeta_E^2 = \ln\left(1 + \frac{\sigma_E^2}{\mu_E^2}\right) = \ln\left(1 + 0.05^2\right) = 0.0025, \quad \zeta_I^2 = \ln\left(1 + \frac{\sigma_I^2}{\mu_I^2}\right) = \ln\left(1 + 0.2^2\right) = 0.039,$$

$$\lambda_p = \ln\mu_p - \frac{1}{2}\zeta_p^2 = 1.312, \lambda_l = \ln\mu_l - \frac{1}{2}\zeta_l^2 = 1.381,$$

$$\lambda_E = \ln\mu_E - \frac{1}{2}\zeta_E^2 = 16.811, \lambda_I = \ln\mu_I - \frac{1}{2}\zeta_I^2 = -10.127.$$

Suppose that the safe state is defined as

$$\frac{\Delta_{\max}}{l} = \frac{pl^2}{48EI} < \frac{1}{100}$$

Then the performance function may be expressed as

$$Z = G(\mathbf{X}) = \ln\frac{1}{100} - \left(\ln p + \ln l^2 - \ln E - \ln I - \ln 48\right) = \ln E + \ln I - \ln p - 2\ln l - 0.734$$

Since $\ln E$, $\ln I$, $\ln p$, and $\ln l$ become normal random variables, Z is also a normal variable with mean and standard deviation calculated as

$$\mu_Z = \lambda_E + \lambda_I - \lambda_p - 2\lambda_l - 0.734 = 1.876, \sigma_Z = \sqrt{\zeta_E^2 + \zeta_I^2 + \zeta_p^2 + 4\zeta_l^2} = 0.4581$$

Then the reliability index and the corresponding failure probability are given as

$$\beta = \frac{\mu_Z}{\sigma_Z} = 4.095, \quad P_F = 1 - \Phi(\beta) = 2.112 \times 10^{-5}$$

Example 2.6 Consider a simple performance function $Z = G(\mathbf{X}) = X - Y$, where X and Y are statistically dependent normal random variables, $N(\mu_X, \sigma_X^2)$, $N(\mu_Y, \sigma_Y^2)$, with correlation coefficient of ρ. To obtain the reliability index, the distribution properties of function Z should be first studied. Using Eq. (A.107) and Eq. (A.115a), the CDF of Z is

$$F_Z(z) = \frac{1}{2\pi\sigma_X\sigma_Y\sqrt{1-\rho^2}} \cdot \iint\limits_{x-y\leq z} \exp\left\{\frac{-1}{2(1-\rho^2)}\left[\left(\frac{x-\mu_X}{\sigma_X}\right)^2 - 2\rho\left(\frac{x-\mu_X}{\sigma_X}\right)\left(\frac{y-\mu_Y}{\sigma_Y}\right) + \left(\frac{y-\mu_Y}{\sigma_Y}\right)^2\right]\right\}dxdy$$

$$= \frac{1}{2\pi\sigma_X\sigma_Y\sqrt{1-\rho^2}} \cdot \int_{-\infty}^{\infty}\int_{-\infty}^{z} \exp\left\{\frac{-1}{2(1-\rho^2)}\left[\left(\frac{z+y-\mu_X}{\sigma_X}\right)^2 - 2\rho\left(\frac{z+y-\mu_X}{\sigma_X}\right)\left(\frac{y-\mu_Y}{\sigma_Y}\right) + \left(\frac{y-\mu_Y}{\sigma_Y}\right)^2\right]\right\}dzdy$$

thus the PDF of Z is

$$f_Z(z) = \frac{1}{2\pi\sigma_X\sigma_Y\sqrt{1-\rho^2}} \exp\left\{\frac{-1}{2(1-\rho^2)}\left[\frac{(z-\mu_X)^2}{\sigma_X^2} + 2\rho\frac{(z-\mu_X)\mu_Y}{\sigma_X\sigma_Y} + \frac{\mu_X^2}{\sigma_Y^2}\right]\right\}$$

$$\int_{-\infty}^{\infty} \exp\left\{\frac{-1}{2(1-\rho^2)}\left[\frac{(\sigma_X^2 - 2\rho\sigma_X\sigma_Y + \sigma_Y^2)y^2 - 2\mu_Y y(\sigma_X^2 - \rho\sigma_X\sigma_Y) - 2(z-\mu_X)y(\sigma_Y^2 + \rho\sigma_X\sigma_Y)}{\sigma_X^2\sigma_Y^2}\right]\right\}dy$$

$$= \frac{1}{2\pi\sigma_X\sigma_Y\sqrt{1-\rho^2}} \exp\left\{\frac{-1}{2(1-\rho^2)}\left[\frac{(z-\mu_X)^2}{\sigma_X^2} + 2\rho\frac{(z-\mu_X)\mu_Y}{\sigma_X\sigma_Y} + \frac{\mu_X^2}{\sigma_Y^2}\right]\right\}\int_{-\infty}^{\infty} \exp\left[-\frac{1}{2}(uy^2 - 2vy)\right]dy$$

where

$$u = \frac{(\sigma_X^2 - 2\rho\sigma_X\sigma_Y + \sigma_Y^2)}{\sigma_X^2\sigma_Y^2(1-\rho^2)}, v = \frac{\mu_Y(\sigma_X^2 - \rho\sigma_X\sigma_Y) + (z-\mu_X)(\sigma_Y^2 + \rho\sigma_X\sigma_Y)}{\sigma_X^2\sigma_Y^2(1-\rho^2)}$$

Let $w = y - v/u$, the last integral above becomes

$$\int_{-\infty}^{\infty} \exp\left[-\frac{1}{2}(uy^2 - 2vy)\right]dy = \exp\left(\frac{v^2}{2u}\right)\int_{-\infty}^{\infty}\exp\left(-\frac{1}{2}uw^2\right)dw = \sqrt{\frac{2\pi}{u}}\exp\left(\frac{v^2}{2u}\right)$$

After some algebraic manipulation, the PDF of Z can be calculated as

$$f_Z(z) = \frac{1}{\sqrt{2\pi}\sqrt{\sigma_X^2 - 2\rho\sigma_X\sigma_Y + \sigma_Y^2}} \exp\left[-\frac{1}{2}\left(\frac{z - (\mu_X - \mu_Y)}{\sqrt{(\sigma_X^2 - 2\rho\sigma_X\sigma_Y + \sigma_Y^2)}}\right)^2\right]$$

which shows that Z is also a normal variable with mean and variance as

$$\mu_Z = \mu_X - \mu_Y, \sigma_Z^2 = \sigma_X^2 - 2\rho\sigma_X\sigma_Y + \sigma_Y^2$$

According to Eq. (2.14), the reliability index is given as

$$\beta = \frac{\mu_X - \mu_Y}{\sqrt{\sigma_X^2 - 2\rho\sigma_X\sigma_Y + \sigma_Y^2}} \tag{2.19}$$

For $\mu_X = 200$, $\sigma_X = 40$, $\mu_Y = 100$, and $\sigma_Y = 30$, the variation of the reliability index with respect to the correlation coefficient ρ is shown in Figure 2.11, from which it can be observed that reliability index increases with the increase of the correlation coefficient.

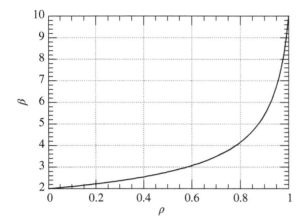

Figure 2.11 Reliability index changed with correlation coefficient.

In general, it can be shown inductively that

$$Z = G(\mathbf{X}) = \sum_{i=1}^{n} a_i X_i \qquad (2.20)$$

where a_i are constants, and X_i are statistically dependent normal variables $N(\mu_i, \sigma_i)$ with correlated coefficient of ρ_{ij}. The performance function Z is also Gaussian with mean and variance

$$\mu_Z = \sum_{i=1}^{n} a_i \mu_i, \sigma_Z^2 = \sum_{i=1}^{n} a_i^2 \sigma_i^2 + \sum_{i=1}^{n} \sum_{i \neq j} \rho_{ij} a_i a_j \sigma_i \sigma_j$$

The reliability index is given as

$$\beta = \frac{\sum\limits_{i=1}^{n} a_i \mu_i}{\sqrt{\sum\limits_{i=1}^{n} a_i^2 \sigma_i^2 + \sum\limits_{i=1}^{n} \sum\limits_{i \neq j} \rho_{ij} a_i a_j \sigma_i \sigma_j}} \qquad (2.21)$$

2.3 Monte Carlo Simulation

2.3.1 Formulation of the Probability of Failure

The Monte Carlo simulation (MCS) is a process of 'numerical or computational experiments' to replicate the real world based on a set of assumptions and conceived models of reality. In the case of structural reliability analysis, it implies to simulate the limit state using a particular set of values of the random variables generated in accordance with the corresponding probability distributions. By repeating the process, samples of solutions, each corresponding to a different set of values of the random variables, are obtained. A sample from a MCS is similar to a sample of experimental observations, which may be treated statistically. Such results may be presented in the form of histograms, and methods of statistical estimation and inference are applicable. In more common practice of structural reliability analysis, MCS is simply used to estimate the probability of failure. For such a case, the limit state function is checked in the repeating process of MCS, and the probability of failure is estimated by counting the number of trials for which the limit state is exceeded. For a performance function $G(\mathbf{X})$, randomly give a sample vector $\mathbf{x}_k$ for $\mathbf{X}$, $G(\mathbf{x}_k) \leq 0$ is then checked. If the performance function is violated, the structure or structural element will be considered to fail. If the experiment is repeated for N times, the probability of failure can be approximated by

$$P_F \approx \frac{n[G(\mathbf{x}_k) \leq 0]}{N} \tag{2.22}$$

where $n[G(\mathbf{x}_k) \leq 0]$ is the number of trials for which $G(\mathbf{x}_k) \leq 0$. Obviously, the number N of trials required is related to the desired accuracy for P_F.

2.3.2 Generation of Random Numbers

A key task in the MCS is the generation of appropriate samples of the random variables in accordance with the respectively prescribed probability distributions. Special devices may be used for simple random variables; for example, tossing a fair coin for random variables with two equally likely values, or rolling a 6-faced die for a random variable with six equally likely possible values. With digital computers, the generation of random numbers is generally accomplished by *the inverse transformation method*, i.e. first generating a uniformly distributed random number between 0 and 1.0 and then obtaining the corresponding random number with the specified probability distribution through appropriate transformations.

Suppose a random variable X with CDF of $F_X(x)$, which must by definition, lie in the range $(0, 1)$. The inverse transformation method is to generate a uniformly distributed random number r_k $(0 \leq r_k \leq 1)$ and equate r_k to $F_X(x)$ as shown in Figure 2.12, and a random number with specified distribution

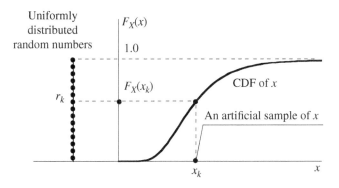

Figure 2.12 Generation of random numbers.

$$F_X(x_k) = r_k \text{ or } x_k = F_X^{-1}(r_k) \tag{2.23}$$

The most common practical approach to generate uniformly distributed random numbers is to employ a suitable pseudo random number generator, available on virtually all computer systems. They are pseudo since they use a formula to generate a sequence of numbers. This sequence is reproducible and repeats after a long cycle interval. For most purposes, a sequence of numbers generated by a suitable modern random number generator is indistinguishable from a sequence of strictly true random numbers (Rubinstein 1981).

For the cases of which analytical expressions exist for the inverse CDF, the inverse transform method is an efficient technique.

The random numbers of a rectangular distribution with the CDF

$$F_X(x) = \frac{x-a}{b-a}, \quad a \le x \le b$$

are generated using the following equation.

$$x_i = (b-a)r_i + a$$

For the exponential distribution with the CDF

$$F_X(x) = 1 - \exp(-\lambda x), \quad x \ge 0$$

the random numbers are generated as

$$x_i = F_X^{-1}(r_i) = -\frac{1}{\lambda} \ln(1 - r_i)$$

For Gumbel distribution (the extreme value distribution) with CDF

$$F_X(x) = \exp\left[-\exp\left(\frac{\zeta - x}{\theta}\right)\right]$$

the random numbers are generated using the following equation

$$x_i = F_X^{-1}(r_i) = \zeta - \theta \ln\left[\ln\left(\frac{1}{r_i}\right)\right]$$

Example 2.7 Methods for generating uniformly distributed random numbers are generally based on recursive calculation of the residues of modulus m from a linear transformation. An example of a recursive relation for this purpose is

$$x_{i+1} = ax_i + c - m \operatorname{int}\left(\frac{ax_i + c}{m}\right)$$

and

$$r_{i+1} = \frac{x_{i+1}}{m}$$

where a, c, and m are nonnegative integers and r_{i+1} is the random number uniformly distributed between 0 and 1.

The procedure is demonstrated for $a = 3$, $c = 2$, $m = 10$, and $x_0 = 1.0$ as follows:

$$k_0 = \text{Int}[(3 \times 1 + 2)/10] = \text{Int}(0.5) = 0, x_1 = 3 \times 1 + 2 - 10 \times 0 = 5, r_1 = 5/10 = 0.5$$

$$k_1 = \text{Int}[(3 \times 5 + 2)/10] = \text{Int}(1.7) = 1, x_2 = 3 \times 5 + 2 - 10 \times 1 = 7, r_2 = 7/10 = 0.7$$

The procedure above may be repeated and the results would be as follows: $i = 1$, $r_i = 0.5$; $i = 2$, $r_i = 0.7$; $i = 3$, $r_i = 0.3$; $i = 4$, $r_i = 0.1$; $i = 5$, $r_i = 0.5$.

Note that the generated pseudo random numbers are cyclic, which implies they are repeated with a given period (four in the case above). The period of the cycle is less than m (Knuth 1969), therefore, the period should be as long as possible to ensure randomness. A large value of m should be assigned in the generation of r_i in practical applications.

The procedure is used in some computer systems as a uniform random number generator with $a = 16\,807$, $c = 0$, and $m = 2^{31} - 1$ (Ang and Tang 1984).

Example 2.8 Suppose $Y = \max(X_1, X_2, ..., X_n)$ where X_i, $i = 1, 2, ..., n$ are independent and identically distributed random variables with CDF $F_X(x)$, the CDF of Y is

$$F_Y(y) = [F_X(y)]^n$$

Then the random numbers of Y are generated as

$$y_i = F_X^{-1}\left(r_i^{1/n}\right)$$

The inverse transformation method is effective if the inverse of the CDF can be analytically expressed. However, there are many probability distributions such as normal and lognormal, in which the CDF cannot be analytically inverted. For such cases, other methods such as *function of random variables method* may be more efficient or effective.

In the *function of random variables method*, suppose a random variable X can be expressed as a function of other random variables Y_1, Y_2, ..., Y_m, that is,

$$X = g(Y_1, Y_2, ..., Y_m) \tag{2.24}$$

and methods for generating values of Y_1 through Y_m are available. Then, a value of X may be determined as

$$x = g(y_1, y_2, ..., y_m) \tag{2.25}$$

where $(y_1, y_2, ..., y_m)$, are random values that have been generated for Y_1, Y_2, ..., Y_m.

A typical application of the method above is the generation of random values for standard normal variables, which is proposed by Box and Muller (1958).

$$U_1 = \sqrt{-2 \ln R_1} \cos 2\pi R_2 \tag{2.26a}$$

$$U_2 = \sqrt{-2 \ln R_1} \sin 2\pi R_2 \tag{2.26b}$$

where R_1 and R_2 are uniformly distributed independent random variables in the interval $[0, 1]$.

The procedure can be verified as follows (e.g. see Ang and Tang 1984).

Let $X = -\ln R_1$ and $Y = R_2$, then,

$$f_X(x) = f_{R_1}(e^{-x}) \frac{dR_1}{dx} = -1 \cdot e^{-x} = -e^{-x}, f_Y(y) = 1$$

as R_1 and R_2 are statistically independent, the joint PDF of X and Y is

$$f_{X,Y}(x,y) = f_X(x) f_Y(y) = -e^{-x}$$

Also, in terms of X and Y,

$$U_1 = (2X)^{1/2} \cos 2\pi Y, U_2 = (2X)^{1/2} \sin 2\pi Y$$

Inversion then yields

$$X = \frac{1}{2} (U_1^2 + U_2^2), Y = \frac{1}{2\pi} \tan^{-1} \left(\frac{U_2}{U_1} \right)$$

The Jacobian of the transformation is

$$J = \begin{vmatrix} \dfrac{\partial x}{\partial u_1} & \dfrac{\partial x}{\partial u_2} \\ \dfrac{\partial y}{\partial u_1} & \dfrac{\partial y}{\partial u_2} \end{vmatrix} = \begin{vmatrix} u_1 & u_2 \\ -\dfrac{u_2}{2\pi(u_1^2 + u_2^2)} & \dfrac{u_1}{2\pi(u_1^2 + u_2^2)} \end{vmatrix} = \frac{1}{2\pi}$$

Therefore, the joint PDF of U_1 and U_2 is

$$f_{U_1,U_2}(u_1, u_2) = f_{X,Y} \left[\frac{1}{2} (u_1^2 + u_2^2), \frac{1}{2\pi} \tan^{-1} \left(\frac{u_2}{u_1} \right) \right] \cdot J = \frac{1}{2\pi} \exp \left(-\frac{u_1^2 + u_2^2}{2} \right)$$

which is the joint PDF of two independent standard normal random variables.

2.3.3 Direct Sampling for Structural Reliability Evaluation

The basis for the Monte Carlo technique sketched in Section 2.3.1 is as follows. The probability of failure may be expressed as

$$P_F = J = \int I[G(\mathbf{x}) \le 0] f_{\mathbf{X}}(\mathbf{x}) d\mathbf{x} \tag{2.27}$$

where $I[\cdot]$ is an indicator function which equals 1 if $[\cdot]$ is 'true' and 0 if $[\cdot]$ is 'false.' Here the indicator function has taken on the role of identifying the integration domain.

It can be observed that the integral represents the expected value of $I[\cdot]$, i.e.

$$P_F = J = E[I(G \le 0)] \tag{2.28}$$

If now $\mathbf{x}_j$ represents the jth vector of random observations from $f_{\mathbf{X}}(\mathbf{x})$, and let $G_j = G(\mathbf{x}_j)$, then it follows directly from sample statistics that

$$P_F \approx J_1 = \frac{1}{N} \sum_{j=1}^{N} I(G_j \le 0) \tag{2.29}$$

is an unbiased estimator of J and hence of P_F. Hence Eq. (2.29) provides a direct estimate of P_F. This is what was done in Section 2.2.

An estimate of the number of simulations may be made as follows for a given confidence level. Since $G(\mathbf{X})$ is a random variable and a function of $\mathbf{X}$, the indicator $I(G \leq 0)$ is also a random variable. With only two possible outcomes, it follows from the central limit theorem that the distribution of J_1 given by the sum of independent sample functions Eq. (2.29) approaches a normal distribution as $N \to \infty$. The mean of this distribution is (Melchers 1987)

$$\mu_{J_1} = E(J_1) = \sum_{j=1}^{N} \frac{1}{N} E[I(G \leq 0)] = E[I(G \leq 0)] \tag{2.30a}$$

which is equal to J, while the variance is given by

$$\sigma_{J_1} = \sqrt{\sum_{j=1}^{N} \frac{1}{N^2} Var[I(G \leq 0)]} = \frac{1}{N^{1/2}} \sigma[I(G \leq 0)] \tag{2.30b}$$

This shows that the standard deviation of Monte Carlo estimate J_1 expressed in Eq. (2.29) varies directly with the standard deviation of $I(\cdot)$ and inversely with $N^{1/2}$. These observations are important for determining the number of simulations required for a particular level of confidence. On the basis that the central limit theorem applies, the following confidence statement can be made for the number N_F of trails in which 'failure' occurs:

$$P\left(-k\sigma_{N_F} < N_F - \mu_{N_F} < k\sigma_{N_F}\right) = C \tag{2.31}$$

where μ_{N_F} and σ_{N_F} are the mean value and standard deviation, respectively, of N_F, expressed in Eqs. (2.30).

$$k = -\Phi^{-1}\left[\frac{1}{2}(1-C)\right] \tag{2.32}$$

For confidence interval $C = 95\%$, k is obtained as 1.96, using the CDF of the standard normal random variable. The following confidence statement can be given for the Monte Carlo estimate J_1 (Shooman 1968; Melchers 1987):

$$P\left(-k\sigma_{J_1} < J_1 - \mu_{J_1} < k\sigma_{J_1}\right) = C \tag{2.33}$$

where

$$\mu_{J_1} = P_F, \quad \sigma_{J_1} = \left[\frac{P_F(1-P_F)}{N}\right]^{1/2} \tag{2.34}$$

If the error between the actual value of P_F and the observed J_1 is denoted by

$$\varepsilon = \frac{J_1 - P_F}{P_F} \tag{2.35}$$

Then the following confidence statement can be given for the error ε:

$$P(-kV_{J_1} < \varepsilon < kV_{J_1}) = C \tag{2.36}$$

where

$$V_{J_1} = \left[\frac{(1-P_F)}{NP_F}\right]^{1/2} \tag{2.37}$$

is the coefficient of variation (COV) of Monte Carlo estimate J_1. For small values of P_F, we have $1 - P_F \approx 1$, then V_{J_1} is given as

$$V_{J_1} = (NP_F)^{-1/2} \tag{2.38}$$

Generally, for a given number N of trials of MCS, the error in Monte Carlo estimate J_1 will be less than the value given in the following equation with a confidence C,

$$\varepsilon < k(NP_F)^{-1/2} = -\Phi^{-1}\left[\frac{1}{2}(1-C)\right](NP_F)^{-1/2} = -\Phi^{-1}\left[\frac{1}{2}(1-C)\right]V_{J_1} \tag{2.39}$$

In particular, with 95% confidence, the error in Monte Carlo estimate J_1 with a given number N of trails of MCS is approximated as (Shooman 1968)

$$\varepsilon < 2(NP_F)^{-1/2} = 2V_{J_1} \tag{2.40}$$

If $N = 10\,000$ samples, and $P_F = 10^{-2}$, then the COV in P_F is estimated as 0.1, and the error in J_1, and hence P_F, will be less than 20% with 95% confidence (and then $k = 1.96$). Thus it is 95% likely that the actual probability of failure will be within 0.01 ± 0.002. The COV and error in Monte Carlo estimate J_1 changed with NP_F is shown in Figure 2.13 and some numerical values of COV and error corresponding to NP_F are listed in Table 2.1.

On the other hand, from Eq. (2.37), in order to limit the error in Monte Carlo estimate J_1 within ε with confidence C, the number of trials should satisfy with

$$NP_F > \left(\frac{k}{\varepsilon}\right)^2 = \left[\frac{1}{\varepsilon}\Phi^{-1}\left(\frac{1}{2}C\right)\right]^2 \tag{2.41}$$

In particular, with 95% confidence, in order to limit the error in Monte Carlo estimate J_1 within ε, the number of trials should satisfy

$$NP_F > \left(\frac{2}{\varepsilon}\right)^2 \tag{2.42}$$

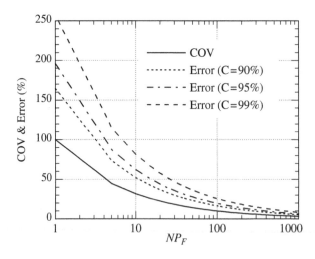

Figure 2.13 COV and error of Monte Carlo estimator changed with NP_F.

Table 2.1 COV and error of Monte Carlo estimator.

NP_F	COV	error		
		90%	95%	99%
10 000	0.01	1.64%	1.95%	2.57%
5000	0.014	2.33%	2.77%	3.64%
1000	0.032	5.20%	6.20%	8.15%
500	0.045	7.36%	8.77%	11.5%
100	0.10	16.4%	19.6%	25.8%
50	0.14	23.3%	27.7%	36.4%
10	0.32	52.0%	62.0%	81.5%
1	1.0	164%	200.0%	258%

If an accuracy of 20% with 95% confidence is desired, NP_F should be larger than 100, and the required number N of simulations should be larger than 10 000 for $P_F = 0.01$.

Example 2.9 Consider the performance function $G(\mathbf{X}) = R - S$, where R and S are statistically independent lognormal random variables with $\mu_R = 500$, $\sigma_R = 100$, $\mu_S = 200$, and $\sigma_S = 80$, then we have

$$\zeta_R^2 = \ln\left(1 + V_R^2\right) = \ln\left(1 + 0.2^2\right) = 0.0392, \quad \lambda_R = \ln\mu_R - \frac{1}{2}\zeta_R^2 = \ln 500 - 0.5 \times 0.0392 = 6.195$$

$$\zeta_S^2 = \ln\left(1 + V_S^2\right) = \ln\left(1 + 0.4^2\right) = 0.1484, \quad \lambda_S = \ln\mu_S - \frac{1}{2}\zeta_S^2 = \ln 200 - 0.5 \times 0.1484 = 5.224$$

In this case, the performance function is equivalent to $G(\mathbf{X}) = R/S$. According to Example 2.2, the reliability index is given as

$$\beta = \frac{\lambda_R - \lambda_S}{\sqrt{\zeta_R^2 + \zeta_S^2}} = \frac{6.195 - 5.224}{\sqrt{0.0392 + 0.1484}} = \frac{0.971}{0.433} = 2.242$$

The corresponding probability of failure is

$$P_F = 1 - \Phi(-\beta) = \Phi(-2.242) = 0.012481$$

Then we will use MCS to obtain the probability of failure. First, using 20 samples, the histogram of the performance function obtained by these samples is shown in Figure 2.14. From these results it can be observed that only one trial violates the limit state. Thus the probability of failure is estimated as $1/20 = 0.05$ with COV of 1.0. The number of trials is too small to properly estimate the failure probability for this problem.

If we will increase the number of trails to 1000, the histogram of the values of the performance function is shown in Figure 2.15. The probability of failure is estimated as $14/1000 = 0.014$ with reliability index of $\beta = 2.197$. The COV of the failure probability is 0.267 and the error is less than 53.4% with confidence level of 95%. One can see that although the results improve very much when compared with those with 20 trials, the number of trails is still too small to estimate the failure probability for this problem.

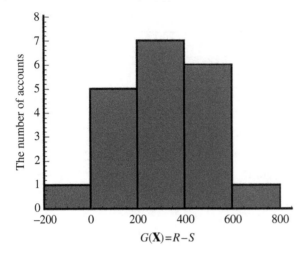

Figure 2.14 Histogram of values of G(**X**) with 20 samples.

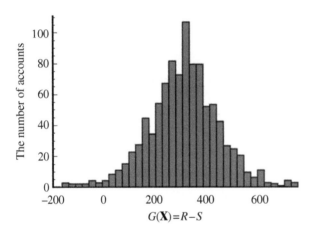

Figure 2.15 Histogram of values of G(**X**) with 1000 samples.

If we further increase the number of trails to 100 000, the histogram is shown in Figure 2.16 for the values of the performance function. The probability of failure is estimated as $1263/100\,000 = 0.01263$ with reliability index of $\beta = 2.237$. The COV of the failure probability is 0.0281 and the error is less than 5.63% with confidence level of 95%. One can see that this result is quite close to the exact result. Using 1 000 000 trials, the probability of failure is estimated as $12\,478/1\,000\,000 = 0.01248$ with reliability index of $\beta = 2.242$. The COV of the failure probability is 0.009 and the error is less than 1.79% with confidence level of 95%. One can see that the result is almost equal to the exact result. The histogram of the values of the performance function is shown in Figure 2.17, where it can be observed that the histogram is quite smooth to approach the PDF of the performance function.

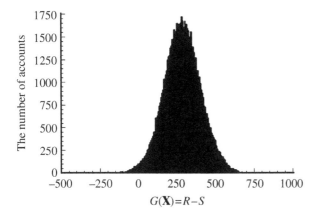

Figure 2.16 Histogram of values of $G(\mathbf{X})$ with 100 000 samples.

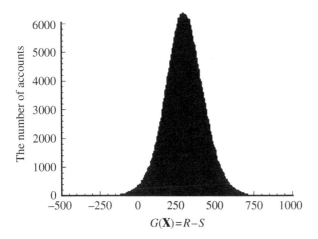

Figure 2.17 Histogram of values of $G(\mathbf{X})$ with 1 000 000 samples.

From Table 2.1 and Example 2.9, one can see that a large number of samples are required to reduce the deviation of the probability of failure P_F, especially for the small value of P_F. Many techniques such as importance sampling (Ayyub and Haldar 1984; Bucher 1988; Mori and Ellingwood 1993a), subset simulation (Au and Beck 2001a), and others have been developed to reduce the deviation and the sample size.

2.4 A Brief Review on Structural Reliability Theory

It is not always so lucky to have the simple formulations we have described in engineering practice. Since $G(\mathbf{X})$ is generally high dimensional and a complicated function of non-normal random variables, computation of probability in Eq. (2.11a) or Eq.(2.11b) is almost impossible. The difficulty has led to the development of various approximation methods, among which the first-order reliability method (FORM) (Hasofer and Lind 1974; Rackwitz and Fiessler 1978; Shinozuka 1983; Zhao and Ono 1999a) is considered as the most reliable computational method. Over the past four decades, contributions from numerous studies have brought FORM to fruition as a basic method for structural reliability which is now used worldwide in many engineering codes (ISO 2394 2015). Therefore, almost all the text books of structural reliability (e.g. Ang and Tang 1984; Madsen et al. 1986; Ditlevsen and Madsen 1996; Nowak and Collins 2000; Melchers and Beck 2018) are focused on reliability methods based on FORM. In the last three decades, researchers have also noted the shortcomings of FORM, such as the accuracy and the difficulties involved in the search of the design point by iteration using the derivatives of the performance function, and the exclusion of random variables with unknown distributions. Several methods have been developed based on FORM to eliminate these shortcomings. Second-order reliability method (SORM) (Fiessler et al. 1979; Breitung 1984; Koyluoglu and Nielson 1994; Der Kiureghian et al. 1987; Zhao and Ono 1999b; Zhao and Ono 1999c), Importance Sampling Monte Carlo simulation (Nie and Ellingwood 2005; Hurtado 2007; Lu et al. 2008; Papaioannou et al. 2016), First-order third-moment (FOTM) (Tichy 1994; Zhao and Ang 2012) have been proposed to improve the accuracy of FORM for nonlinear performance functions. To avoid using the derivatives of the performance function, response surface approach (RSA) (Bucher and Bourgund 1990; Faravelli 1989; Rajashekhar and Ellingwood 1993) and genetic algorithm (Zhao and Jiang 1995) were developed. However, the shortcoming associated with the design points or multiple design points (Kuschel et al. 1998; Der Kiureghian and Dakessian 1998), an inherent drawback in all iteration procedures, appears difficult to overcome. Furthermore, when the PDFs of the basic random variables are unknown (Der Kiureghian and Liu 1986; Zhao and Lu 2007a; Lu et al. 2017a, 2020), neither FORM nor SORM are applicable.

In this book another perspective for structural reliability, i.e. methods of moment, are introduced. As will be illustrated in the following chapters, the methods of moment, being very straightforward, do not involve the design point, thus requiring neither iteration nor the computation of derivatives. Furthermore, the methods of moments are demonstrated to account for random variables with unknown probability distributions. Thus, the methods of moment present efficient and effective approaches for structural reliability analysis.

The framework of the methods of moment-based theory is illustrated in Figure 2.18 for structural reliability presented in this book.

As shown in Figure 2.18, the methods are divided into two categories based on whether the information of the distributions is available or not. In each category, the direct methods of moment directly utilise the first few moments of the original performance function, while those methods of moment based on first/second-order approximations utilise the first few moments of first/second-order approximations at the so called design points of the performance functions in standard normal space or pseudo normal space. In the case that the information of the probability distributions of all the random variables are available, if we use the methods of moment based on the first order approximation of the performance function, the performance function in the standardised normal space will become a normal random variable after first order approximation. The third and fourth moments are zero and three, respectively, that is to say, the first order third and fourth moment methods are not necessary in this case. The method is essentially exactly the same as the so called FORM.

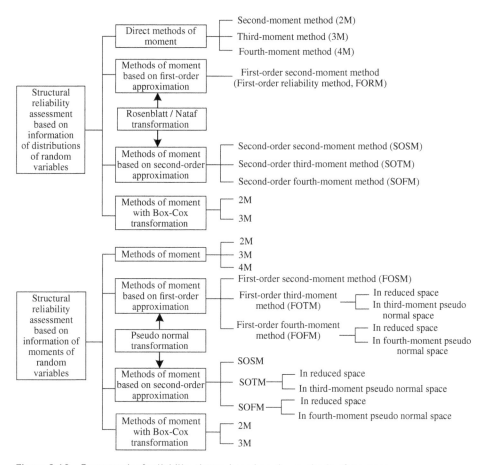

Figure 2.18 Framework of reliability theory based on the methods of moment.

When one use methods of moment with Box-Cox transformation, since the distribution of the new performance function after Box-Cox transformation is quite approach to normal, the third-moment method is generally accurate enough, and the fourth-moment method is unnecessary.

2.5 Summary

In this chapter, the basic concepts of structural reliability including some terms such as performance function, failure probability, and reliability index are introduced through a fundamental performance function of only two random variables, i.e. a load effect variable S and a resistance variable R. Since the accurate integration for failure probability is almost impossible for practical engineering problems, the Monte Carlo simulation is introduced. A brief review on structural reliability theory is also conducted, from which the framework of the methods of moment-based theory of this book is illustrated.

3

Moment Evaluation for Performance Functions

3.1 Introduction

According to the discussion in Section 2.2 in the chapter on Fundamentals of Structural Reliability Theory, if the probability distribution of a performance function $Z = G(\mathbf{X})$ can be obtained, its failure probability is defined as the value of the cumulative distribution function (CDF) of Z at 0. According to Appendix A.9, the probability distribution of the performance function $Z = G(\mathbf{X})$, which is a function of random variables, can theoretically be derived from the probability distributions of its basic random variables. However, such derivations are generally difficult, especially for nonlinear performance functions. In such circumstances, the moments of the performance functions, particularly the first few central moments, may be the only information that is practically obtainable. As main descriptors of a random variable (here specifically, the performance function), these moments are generally closely related to the location and shapes of the distribution of performance function. Since the failure probability is defined as the probability of the performance function being less than or equal to zero, it is natural to understand that the failure probability will be closely related to the location and shapes of the distribution, and thus related to the first few moments of the performance function.

The effects of the first few moments on the probability of failure are illustrated in Figure 3.1, from which the following can be observed.

1) Since the increase of the mean of performance function makes the area under the probability density function (PDF) less than zero decrease as shown in Figure 3.1a, the larger the mean of performance function, the smaller the failure probability.
2) Since the increase of the standard deviation of performance function makes the random variable more widely spread around the central value as shown in Figure 3.1b, the larger the standard deviation, the larger the failure probability.
3) For negative skewness as shown in Figure 3.1c, since the increase of the skewness (absolute value) makes the increase of asymmetry, the values of performance function distribute more in negative direction of the z-axis, which generally leads to higher possibility for the performance function to have negative values. Thus, in general, the larger the skewness (absolute value) of the performance function, the larger the failure probability.

Structural Reliability: Approaches from Perspectives of Statistical Moments, First Edition.
Yan-Gang Zhao and Zhao-Hui Lu.
© 2021 John Wiley & Sons Ltd. Published 2021 by John Wiley & Sons Ltd.

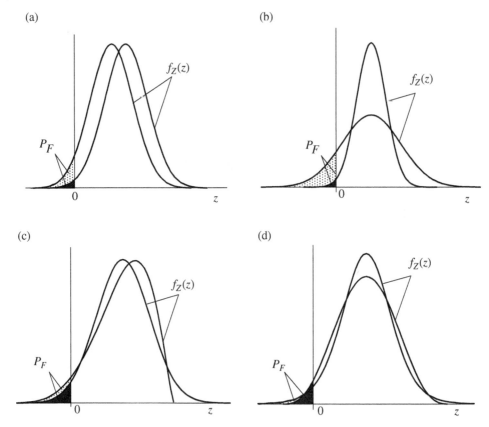

Figure 3.1 Effect of the first four moments on the probability of failure. (a) Effect of mean. (b) Effect of deviation. (c) Effect of skewness. (d) Effect of kurtosis.

4) The kurtosis has significant influence on the left tail of the distribution of the performance function as shown in Figure 3.1d, although weare not sure the clear rule of this influence.

The failure probability can thus be estimated if the relationship can be established between the failure probability and the central moments. In many instances, as will be discussed in the chapter entitled Direct Methods of Moment, this may be sufficiently accurate for most practical applications, even if the actual probability distribution remains undetermined. In this chapter, we will discuss the method and procedure for obtaining the first four moments of a performance function.

In the final analysis, the first few statistical moments of the performance function $Z = G(\mathbf{X})$ are theoretically expressed as the following integrals:

$$\mu_G = E[G(\mathbf{X})] = \int G(\mathbf{x}) f_{\mathbf{X}}(\mathbf{x}) d\mathbf{x} \tag{3.1a}$$

$$\sigma_G^2 = E\{[G(\mathbf{X}) - \mu_G]^2\} = \int [G(\mathbf{X}) - \mu_G]^2 f_{\mathbf{X}}(\mathbf{x}) d\mathbf{x} \tag{3.1b}$$

$$\sigma_G^k \alpha_{kG} = E\left\{[G(\mathbf{X}) - \mu_G]^k\right\} = \int [G(\mathbf{X}) - \mu_G]^k f_{\mathbf{X}}(\mathbf{x}) d\mathbf{x} \quad \text{for} \quad k > 2 \tag{3.1c}$$

where $G(\mathbf{X})$ is the performance function, a function of either a single random variable or multiple basic random variables $\mathbf{X}$; μ_G and σ_G are the mean and standard deviation of $Z = G(\mathbf{X})$, respectively; α_{kG} is the kth order dimensionless central moment (or referred to as the kth order moment ratio as defined in Appendix A.4) of $Z = G(\mathbf{X})$; and $f_{\mathbf{X}}(\mathbf{x})$ is the joint PDF of basic random variables $\mathbf{X}$.

For some simple functions such as linear sum or product of independent random variables, the first few moments can be readily obtained from the definitions in Eqs. (3.1a)–(3.1c). However, in practice, $G(\mathbf{X})$ is generally a complicated and implicit function and the direct integration of Eqs. (3.1a)–(3.1c) is difficult if not impossible. The commonly used method of approximation (Ibrahim 1987; Singh and Lee 1993; Impollonia et al. 1998) involves obtaining the Taylor expansion of the function about the expectations of the random variables (Benjamin and Cornell 1970). This often imposes excessive restrictions on the function (existence and continuity of the first few derivatives) and requires the computation of derivatives (Rosenblueth 1975). Accurate derivatives of $G(\mathbf{X})$ however are generally difficult to obtain (Zhao and Sun 1991) because $G(\mathbf{X})$ is usually implicit and the computation of $G(\mathbf{X})$ involves complicated procedures such as eigenvalue analysis. Furthermore, the Taylor expansions may converge slowly or not at all (Gorman 1980).

In this chapter, we will first discuss the moment computation for simple performance functions. And then point estimates method will be discussed for moment computation for complicated performance functions.

3.2 Moment Computation for some Simple Functions

As shown in Eqs. (3.1a)–(3.1c), the first few moments are generally functionally related to the distribution of the individual basic random variables. In some circumstances, e.g. for some simple performance functions, the moments may be derived as functions of the moments of the basic variables. In this section, we will discuss the first few moments of such simple functions.

3.2.1 Moment Computation for Linear Sum of Random Variables

Consider the moments of linear functions. First of all, suppose that

$$Y = aX + b \tag{3.2}$$

where a and b are constants. According to Eqs. (3.1a)–(3.1c), the mean of Y is given as

$$\mu_Y = E(Y) = E(aX + b) = aE(X) + b$$

That leads to

$$\mu_Y = a\mu_X + b \tag{3.3a}$$

whereas the variance of Y is

$$Var(Y) = E\left[(Y - \mu_Y)^2\right] = E\left[(aX + b - a\mu_X - b)^2\right] = a^2 E\left[(X - \mu_X)^2\right] = a^2 Var(X)$$

Then the standard deviation of Y is given as

$$\sigma_Y = |a|\sigma_X \tag{3.3b}$$

The third central moments of Y is

$$M_{3Y} = \sigma_Y^3 \alpha_{3Y} = E\left[(Y - \mu_Y)^3\right] = E\left[(aX + b - a\mu_X - b)^3\right] = a^3 E\left[(X - \mu_X)^3\right] = a^3 \sigma_X^3 \alpha_{3X}$$

Then the skewness of Y is given as

$$\alpha_{3Y} = \text{sign}(a)\alpha_{3X} \tag{3.3c}$$

Similarly, the fourth central moments of Y is

$$M_{4Y} = \sigma_Y^4 \alpha_{4Y} = a^4 M_{4X} = a^4 \sigma_X^4 \alpha_{4X}$$

and the kurtosis of Y is given as,

$$\alpha_{4Y} = \alpha_{4X} \tag{3.3d}$$

Furthermore, consider a performance function with two independent random variables,

$$G(\mathbf{X}) = a_1 X_1 + a_2 X_2$$

where a_1 and a_2 are constants, and X_1 and X_2 are independent random variables. According to Eqs. (3.1a)–(3.1c), the mean of $G(\mathbf{X})$ is given as

$$\mu_G = E(G) = E(a_1 X_1 + a_2 X_2) = a_1 E(X_1) + a_2 E(X_2) = a_1 \mu_1 + a_2 \mu_2$$

whereas the variance of $G(\mathbf{X})$ is

$$\begin{aligned} Var(G) &= E\left[(G - \mu_G)^2\right] = E\left[(a_1 X_1 + a_2 X_2 - a_1 \mu_1 - a_2 \mu_2)^2\right] = E\left\{[a_1(X_1 - \mu_1) + a_2(X_2 - \mu_2)]^2\right\} \\ &= E\left\{\left[a_1^2(X_1 - \mu_1)^2 + 2a_1 a_2(X_1 - \mu_1)(X_2 - \mu_2) + a_2^2(X_2 - \mu_2)^2\right]\right\} \\ &= a_1^2 Var(X_1) + 2a_1 a_2 E(X_1 - \mu_1)E(X_2 - \mu_2) + a_2^2 Var(X_2) \end{aligned}$$

that is

$$\sigma_G^2 = a_1^2 \sigma_1^2 + a_2^2 \sigma_2^2$$

The third central moments of $G(\mathbf{X})$ is

$$\begin{aligned} M_{3G} &= \sigma_G^3 \alpha_{3G} = E\left[(G - \mu_G)^3\right] = E\left\{[a_1(X_1 - \mu_1) + a_2(X_2 - \mu_2)]^3\right\} \\ &= E\left[a_1^3(X_1 - \mu_1)^3 + 3a_1^2 a_2(X_1 - \mu_1)^2(X_2 - \mu_2)\right. \\ &\quad \left. + 3a_1 a_2^2(X_1 - \mu_1)(X_2 - \mu_2)^2 + a_2^3(X_2 - \mu_2)^3\right] \\ &= a_1^3 E\left[(X_1 - \mu_1)^3\right] + 3a_1^2 a_2 E\left[(X_1 - \mu_1)^2\right]E(X_2 - \mu_2) \\ &\quad + 3a_1 a_2^2 E(X_1 - \mu_1)E\left[(X_2 - \mu_2)^2\right] + a_2^3 E\left[(X_2 - \mu_2)^3\right] \end{aligned}$$

Then the skewness of $G(\mathbf{X})$ is given as,

$$\alpha_{3G}\sigma_G^3 = \alpha_{31} a_1^3 \sigma_1^3 + \alpha_{32} a_2^3 \sigma_2^3$$

Similarly, the fourth central moment of $G(\mathbf{X})$ is

$$\alpha_{4G}\sigma_G^4 = \alpha_{41}a_1^4\sigma_1^4 + \alpha_{42}a_2^4\sigma_2^4 + 6a_1^2a_2^2\sigma_1^2\sigma_2^2$$

Generally, commonly encountered function in structural reliability is expressed as a linear sum of independent random variables expressed as

$$G(\mathbf{X}) = \sum_{i=1}^{n} a_i X_i \tag{3.4}$$

where X_i ($i = 1, ..., n$) are mutually independent random variables; and a_i are coefficients.

According to the definitions in Eqs. (3.1a)–(3.1c), the first four moments of the performance function Eq. (3.4) can be easily obtained as

$$\mu_G = \sum_{i=1}^{n} a_i \mu_i \tag{3.5a}$$

$$\sigma_G^2 = \sum_{i=1}^{n} a_i^2 \sigma_i^2 \tag{3.5b}$$

$$\alpha_{3G}\sigma_G^3 = \sum_{i=1}^{n} \alpha_{3i} a_i^3 \sigma_i^3 \tag{3.5c}$$

$$\alpha_{4G}\sigma_G^4 = \sum_{i=1}^{n} \alpha_{4i} a_i^4 \sigma_i^4 + 6\sum_{i=1}^{n-1}\sum_{j>i}^{n} a_i^2 a_j^2 \sigma_i^2 \sigma_j^2 \tag{3.5d}$$

where μ_i and σ_i are the mean and standard deviation of X_i, respectively; α_{3i} and α_{4i} are the third and fourth dimensionless central moments, i.e. the skewness and kurtosis of X_i; μ_G and σ_G are the mean and standard deviation of $G(\mathbf{X})$; and α_{3G} and α_{4G} are the skewness and kurtosis of $G(\mathbf{X})$, respectively.

Example 3.1 Consider the following simple performance function, which is an elementary reliability model used in many situations.

$$G(\mathbf{X}) = R - S$$

where R is a resistance with mean of $\mu_R = 100$ and standard deviation of $\sigma_R = 20$, and S is a load with mean of $\mu_S = 50$ and standard deviation of $\sigma_S = 20$. Both R and S are assumed to be statistically independent and lognormal random variables. Using Eqs. (A.66a) and (A.66b), the skewness and kurtosis of R and S can be easily obtained using the moment formula of lognormal random variables as $\alpha_{3R} = 0.608$, $\alpha_{4R} = 3.664$, $\alpha_{3S} = 1.264$, and $\alpha_{4S} = 5.969$. Using Eqs. (3.5a)–(3.5d), the first four moments of the performance function can be easily calculated as

$$\mu_G = \mu_R - \mu_S = 100 - 50 = 50$$

$$\sigma_G^2 = \sigma_R^2 + \sigma_S^2 = 20^2 + 20^2 = 800, \sigma_G = 28.284$$

$$\alpha_{3G} = \left(\alpha_{3R}\sigma_R^3 - \alpha_{3S}\sigma_S^3\right)/\sigma_G^3 = \left(0.608 \times 20^3 - 1.264 \times 20^3\right)/28.284^3 = -0.232$$

$$\alpha_{4G} = \left(\alpha_{4R}\sigma_R^4 + 6\sigma_R^2\sigma_S^2 + \alpha_{4S}\sigma_S^4\right)/\sigma_G^4$$

$$= \left(3.664 \times 20^4 + 6 \times 20^2 \times 20^2 + 5.969 \times 20^4\right)/28.284^4 = 3.908$$

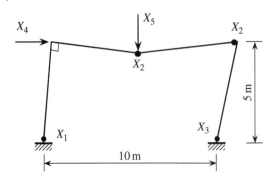

Example 3.2 Consider the performance function for a plastic collapse mechanism of a one-bay frame as shown in Figure 3.2.

$$G(\mathbf{X}) = X_1 + 4X_2 + X_3 - 5X_4 - 5X_5$$

The variables X_i are statistically independent and have lognormal distribution with mean of $\mu_1 = \mu_2 = \mu_3 = 120$, $\mu_4 = 50$, $\mu_5 = 40$; and standard deviations of $\sigma_1 = \sigma_2 = \sigma_3 = 12$, $\sigma_4 = 15$, $\sigma_5 = 12$. Using Eqs. (A.66a) and (A.66b), the corresponding skewness and kurtosis can be easily obtained as $\alpha_{31} = \alpha_{32} = \alpha_{33} = 0.301$, $\alpha_{34} = \alpha_{35} = 0.927$, $\alpha_{41} = \alpha_{42} = \alpha_{43} = 3.162$, and $\alpha_{44} = \alpha_{45} = 4.566$.

Using Eqs. (3.5), the first four moments of $G(\mathbf{X})$ can be calculated as

$$\mu_G = \mu_1 + 4\mu_2 + \mu_3 - 5\mu_4 - 5\mu_5 = 120 + 4 \times 120 + 120 - 5 \times 50 - 5 \times 40 = 270$$

$$\sigma_G^2 = \sigma_1^2 + 4^2 \times \sigma_2^2 + \sigma_3^2 + (-5)^2 \times \sigma_4^2 + (-5)^2 \times \sigma_5^2 = 12^2 + 16 \times 12^2$$

$$+ 12^2 + 25 \times 15^2 + 25 \times 12^2 = 11817$$

$$\sigma_G = 108.706$$

$$\alpha_{3G} = \left(\alpha_{31}\sigma_1^3 + 4^3 \times \alpha_{32}\sigma_2^3 + \alpha_{33}\sigma_3^3 - 5^3 \times \alpha_{34}\sigma_4^3 - 5^3 \times \alpha_{35}\sigma_5^3\right)/\sigma_G^3$$

$$= \left(0.301 \times 12^3 + 64 \times 0.301 \times 12^3 + 0.301 \times 12^3 - 125 \times 0.927 \times 15^3\right.$$

$$\left. - 125 \times 0.927 \times 12^3\right)/108.706^3$$

$$= -0.4336$$

$$\alpha_{4G} = \left[\alpha_{41}\sigma_1^4 + 4^4 \times \alpha_{42}\sigma_2^4 + \alpha_{43}\sigma_3^4 + 5^4 \times \alpha_{44}\sigma_4^4 + 5^4 \times \alpha_{45}\sigma_5^4\right.$$

$$+ 6 \times \left(4^2 \times \sigma_1^2\sigma_2^2 + \sigma_1^2\sigma_3^2 + 5^2 \times \sigma_1^2\sigma_4^2 + 5^2 \times \sigma_1^2\sigma_5^2\right) + 6 \times \left(4^2 \times \sigma_2^2\sigma_3^2 + 4^2 \times 5^2 \times \sigma_2^2\sigma_4^2\right.$$

$$+ 4^2 \times 5^2 \times \sigma_2^2\sigma_5^2\right) + 6 \times \left(5^2 \times \sigma_3^2\sigma_4^2 + 5^2 \times \sigma_3^2\sigma_5^2\right) + 6 \times 5^2 \times 5^2 \times \sigma_4^2\sigma_5^2\right]/\sigma_G^4$$

$$= \left[3.162 \times 12^4 + 4^4 \times 3.162 \times 12^4 + 3.162 \times 12^4 + 5^4 \times 4.566 \times 15^4 + 5^4 \times 4.566 \times 12^4\right.$$

$$+ 6 \times \left(4^2 \times 12^2 \times 12^2 + 12^2 \times 12^2 + 5^2 \times 12^2 \times 15^2 + 5^2 \times 12^2 \times 12^2\right)$$

$$+ 6 \times \left(4^2 \times 12^2 \times 12^2 + 4^2 \times 5^2 \times 12^2 \times 15^2 + 4^2 \times 5^2 \times 12^2 \times 12^2\right)$$

$$+ 6 \times \left(5^2 \times 12^2 \times 15^2 + 5^2 \times 12^2 \times 12^2\right) + 6 \times 5^2 \times 5^2 \times 15^2 \times 12^2\right]/108.706^4 = 3.5064$$

3.2.2 Moment Computation for Products of Random Variables

Another commonly used function in structural reliability is the product of random variables. For the case of only two random variables, the function is expressed as

$$G(\mathbf{X}) = X_1 X_2$$

where X_1 and X_2 are mutually independent random variables.

According to Eqs.(3.1a)–(3.1c), the mean of $G(\mathbf{X})$ is given as

$$\mu_G = E(G) = E(X_1 X_2) = E(X_1)E(X_2) = \mu_1 \mu_2$$

whereas the variance of $G(\mathbf{X})$ is

$$
\begin{aligned}
Var(G) &= E\big[(G-\mu_G)^2\big] = E\big[(X_1 X_2 - \mu_1\mu_2)^2\big] = E\big(X_1^2 X_2^2 - 2X_1 X_2 \mu_1\mu_2 + \mu_1^2\mu_2^2\big) \\
&= E\big(X_1^2\big)E\big(X_2^2\big) - \mu_1^2\mu_2^2 = \big(\sigma_1^2 + \mu_1^2\big)\big(\sigma_2^2 + \mu_2^2\big) - \mu_1^2\mu_2^2 = \mu_1^2\mu_2^2\big[(V_1^2 + 1)(V_2^2 + 1) - 1\big]
\end{aligned}
$$

that is,

$$\sigma_G^2 = \mu_G^2\big[(V_1^2 + 1)(V_2^2 + 1) - 1\big]$$

where V_1 and V_2 are the coefficients of variation (COV) of X_1 and X_2, respectively.

The third central moment of $G(\mathbf{X})$ is

$$
\begin{aligned}
M_{3G} &= \sigma_G^3 \alpha_{3G} = E\big[(G-\mu_G)^3\big] = E\big[(X_1 X_2 - \mu_1\mu_2)^3\big] \\
&= E\big(X_1^3 X_2^3 - 3X_1^2 X_2^2 \mu_1\mu_2 + 3\mu_1^2\mu_2^2 X_1 X_2 - \mu_1^3\mu_2^3\big) \\
&= E\big(X_1^3\big)E\big(X_2^3\big) - 3E\big(X_1^2\big)E\big(X_2^2\big)\mu_1\mu_2 + 2\mu_1^3\mu_2^3
\end{aligned}
$$

Using Eq. (A.31), the equation above becomes

$$
\begin{aligned}
M_{3G} &= \big(\alpha_{31}\sigma_1^3 + 3\sigma_1^2\mu_1 + \mu_1^3\big)\big(\alpha_{32}\sigma_2^3 + 3\sigma_2^2\mu_2 + \mu_2^3\big) - 3\big(\sigma_1^2 + \mu_1^2\big)\big(\sigma_2^2 + \mu_2^2\big)\mu_1\mu_2 + 2\mu_1^3\mu_2^3 \\
&= \mu_1^3\mu_2^3\big[(\alpha_{31}V_1^3 + 3V_1^2 + 1)(\alpha_{32}V_2^3 + 3V_2^2 + 1) - 3(V_1^2 + 1)(V_2^2 + 1) + 2\big]
\end{aligned}
$$

that is

$$\alpha_{3G} = \frac{1}{V_G^3}\big[(\alpha_{31}V_1^3 + 3V_1^2 + 1)(\alpha_{32}V_2^3 + 3V_2^2 + 1) - 3(V_1^2 + 1)(V_2^2 + 1) + 2\big]$$

Similarly, the fourth central moment of $G(\mathbf{X})$ is

$$
\begin{aligned}
\alpha_{4G} = \frac{1}{V_G^4}\big[&(\alpha_{41}V_1^4 + 4\alpha_{31}V_1^3 + 6V_1^2 + 1)(\alpha_{42}V_2^4 + 4\alpha_{32}V_2^3 + 6V_2^2 + 1) \\
&- 4(\alpha_{31}V_1^3 + 3V_1^2 + 1)(\alpha_{32}V_2^3 + 3V_2^2 + 1) + 6(V_1^2 + 1)(V_2^2 + 1) - 3\big]
\end{aligned}
$$

Generally, a commonly encountered function in structural reliability that is the product of independent random variables can be expressed as

$$G(\mathbf{X}) = \Pi_{i=1}^n X_i \tag{3.6}$$

where X_i, ($i = 1, ..., n$) are mutually independent random variables.

According to the definitions in Eqs.(3.1a)–(3.1c), the first four moments of Eq. (3.6) can be also easily obtained as

$$\mu_G = \Pi_{i=1}^n \mu_i \tag{3.7a}$$

$$\sigma_G^2 = \mu_G^2\big[\Pi_{i=1}^n (1 + V_i^2) - 1\big] \tag{3.7b}$$

$$a_{3G} = \frac{1}{V_G^3} \left[\Pi_{i=1}^n \left(\alpha_{3i} V_i^3 + 3V_i^2 + 1 \right) - 3\Pi_{i=1}^n \left(1 + V_i^2 \right) + 2 \right] \tag{3.7c}$$

$$a_{4G} = \frac{1}{V_G^4} \left[\Pi_{i=1}^n \left(\alpha_{4i} V_i^4 + 4\alpha_{3i} V_i^3 + 6V_i^2 + 1 \right) \right.$$

$$\left. - 4\Pi_{i=1}^n \left(\alpha_{3i} V_i^3 + 3V_i^2 + 1 \right) + 6\Pi_{i=1}^n \left(1 + V_i^2 \right) - 3 \right] \tag{3.7d}$$

where V_i and V_G are the COV of X_i and $G(\mathbf{X})$, respectively.

Using Eqs. (3.5a)–(3.5d) and (3.7a)–(3.7d), the first four moments of some commonly encountered performance functions in many structural design codes can be readily obtained. Take the performance function $G(\mathbf{X}) = X_1 + 3X_2X_3 - X_4X_5X_6/4$ for example, one may assume $Y_1 = X_2X_3$ and $Y_2 = X_4X_5X_6$, then $G(\mathbf{X}) = X_1 + 3Y_1 - Y_2/4$. After obtaining the first four moments of Y_1 and Y_2 with the aid of Eqs. (3.7a)–(3.7d), one can obtain those of $G(\mathbf{X})$ using Eqs. (3.5a)–(3.5d).

Example 3.3 Consider the following simple performance function used as a basic reliability model in many situations.

$$G(\mathbf{X}) = DR - S \tag{3.8}$$

where

R = a resistance lognormally distributed with $\mu_R = 100$ and $\sigma_R = 20$;
S = a load lognormally distributed with $\mu_S = 50$ and $\sigma_S = 20$; and
D = a modification of R normally distributed with $\mu_D = 1$ and $\sigma_D = 0.1$.

According to Appendix A.5, the corresponding skewness and kurtosis of D, R, and S are $\alpha_{3D} = 0$, $\alpha_{3R} = 0.608$, $\alpha_{3S} = 1.264$, $\alpha_{4D} = 3$, $\alpha_{4R} = 3.664$, and $\alpha_{4S} = 5.969$.

Equation (3.8) is a combination of Eqs. (3.4) and (3.6). Let $Y = DR$, then, $G(\mathbf{X}) = Y - S$. We can first obtain the moments of Y using Eqs. (3.7a)–(3.7d) and then obtain the moments of $G(\mathbf{X})$ using Eqs. (3.5a)–(3.5d).

Using Eqs. (3.7a)–(3.7d), the first four moments of Y can be calculated as

$$\mu_Y = \mu_D \mu_R = 100 \times 1 = 100$$

$$\sigma_Y^2 = \mu_G^2 \left[\left(1 + V_D^2 \right) \left(1 + V_R^2 \right) - 1 \right] = 100^2 \left[\left(1 + 0.1^2 \right) \left(1 + 0.2^2 \right) - 1 \right] = 504$$

$$\sigma_Y = 22.45, V_Y = 0.225$$

$$\alpha_{3Y} = \left[\left(\alpha_{3d} V_D^3 + 3V_D^2 + 1 \right) \left(\alpha_{3R} V_R^3 + 3V_R^2 + 1 \right) - 3\left(1 + V_D^2 \right) \left(1 + V_R^2 \right) + 2 \right] / V_Y^3$$

$$= \left[\left(0 \times 0.1^3 + 3 \times 0.1^2 + 1 \right) \left(0.608 \times 0.2^3 + 3 \times 0.2^2 + 1 \right) \right.$$

$$\left. - 3\left(1 + 0.1^2 \right) \left(1 + 0.2^2 \right) + 2 \right] / 0.2245^3 = 0.655$$

$$\alpha_{4Y} = \left[\left(\alpha_{4D} V_D^4 + 4\alpha_{3D} V_D^3 + 6V_D^2 + 1 \right) \left(\alpha_{4R} V_R^4 + 4\alpha_{3R} V_R^3 + 6V_R^2 + 1 \right) \right.$$

$$- 4\left(\alpha_{3D} V_D^3 + 3V_D^2 + 1 \right) \left(\alpha_{3R} V_R^3 + 3V_R^2 + 1 \right) + 6\left(1 + V_D^2 \right) \left(1 + V_R^2 \right) - 3 \right] / V_Y^4$$

$$= \left[\left(3 \times 0.1^4 + 4 \times 0 \times 0.1^3 + 6 \times 0.1^2 + 1 \right) \left(3.664 \times 0.2^4 + 4 \times 0.608 \times 0.2^3 + 6 \times 0.2^2 + 1 \right) \right.$$

$$- 4\left(0 \times 0.1^3 + 3 \times 0.1^2 + 1 \right) \left(0.608 \times 0.2^3 + 3 \times 0.2^2 + 1 \right) + 6\left(1 + 0.1^2 \right) \left(1 + 0.2^2 \right) - 3 \right] / 0.225^4$$

$$= 3.737$$

Utilising the first four moments of Y and Eqs. (3.5a)–(3.5d), the first four moments of $G(\mathbf{X})$ can be obtained as $\mu_G = 50$, $\sigma_G = 30.067$, $\alpha_{3G} = -0.0994$, and $\alpha_{4G} = 3.821$.

3.2.3 Moment Computation for Power of a Lognormally Distributed Random Variable

Following Eqs.(3.5a)–(3.5d) and (3.7a)–(3.7d), the first four moments can be readily obtained for some commonly encountered performance functions which are expressed as the sums or products of the random variables. However, the formulations cannot be used when the performance function includes polynomials of a single random variable, such as $G(\mathbf{X}) = 2X_1^3 - X_2X_3/X_4$. There are many such performance functions in structural engineering, including the sum or product of simple function of one random variable (e.g. X_1^3, $1/X_4$). Therefore, it would be convenient if the first four moments can be calculated for such functions of a single random variable. As a commonly used case, the power of a lognormally distributed random variable is first considered as following

$$Y = X^a \tag{3.9}$$

where a is arbitrary non-zero real number.

The first four moments of Eq. (3.9) can be obtained as

$$\mu_Y = \mu^a \left(1 + V^2\right)^{\frac{a(a-1)}{2}} \tag{3.10a}$$

$$V_Y^2 = \left(1 + V^2\right)^{a^2} - 1 \tag{3.10b}$$

$$\alpha_{3Y} = 3V_Y + V_Y^3 \tag{3.10c}$$

$$\alpha_{4Y} = 3 + 16V_Y^2 + 15V_Y^4 + 6V_Y^6 + V_Y^8 \tag{3.10d}$$

where V and V_Y are the COVs of X and Y, respectively.

In particular, when $a = 2$, Eq. (3.9) becomes the square of a lognormally distributed random variable as

$$Y = X^2 \tag{3.11a}$$

And its first four moments are given as

$$\mu_Y = \mu^2 \left(1 + V^2\right) \tag{3.11b}$$

$$V_Y^2 = \left(1 + V^2\right)^4 - 1 \tag{3.11c}$$

$$\alpha_{3Y} = 3V_Y + V_Y^3 \tag{3.11d}$$

$$\alpha_{4Y} = 3 + 16V_Y^2 + 15V_Y^4 + 6V_Y^6 + V_Y^8 \tag{3.11e}$$

When $a = -1$, Eq. (3.9) becomes the reciprocal of a lognormally distributed random variable as

$$Y = \frac{1}{X} \tag{3.12a}$$

And its first four moments are given as

$$\mu_Y = \frac{1}{\mu}(1 + V^2) \tag{3.12b}$$

$$V_Y = V \tag{3.12c}$$

$$\alpha_{3Y} = \alpha_3 \tag{3.12d}$$

$$\alpha_{4Y} = \alpha_4 \tag{3.12e}$$

Example 3.4   The Derivation of Eqs. (3.10a)–(3.10d)

Assume that X is a lognormal random variable with parameter $\lambda = \mu_{\ln X}$ and $\zeta = \sigma_{\ln X}$, applying Eq. (A.58) leads to

$$U = \frac{\ln X - \lambda}{\zeta} \quad \text{and} \quad \mu = \exp\left[\frac{1}{2}\zeta^2 + \lambda\right], \quad V^2 = \exp(\zeta^2) - 1$$

Rewrite Eq. (3.9) as $X = Y^{\frac{1}{a}}$, and one obtains

$$U = \frac{\frac{1}{a}\ln Y - \lambda}{\zeta} = \frac{\ln Y - a\lambda}{a\zeta}$$

This implies that Y is also a lognormal random variable with parameters $\mu_{\ln Y} = a\lambda$ and $\sigma_{\ln Y} = a\zeta$, and therefore,

$$\mu_Y = \exp\left[\frac{1}{2}(a\zeta)^2 + a\lambda\right] = \left[\exp(\zeta^2)\right]^{\frac{1}{2}a^2}\left[\exp(\lambda)\right]^a = (1+V^2)^{\frac{1}{2}a^2}\left[\mu(1+V^2)^{-\frac{1}{2}}\right]^a$$

$$= \mu^a(1+V^2)^{\frac{a(a-1)}{2}}$$

$$V_Y^2 = \exp(a\zeta)^2 - 1 = (1+V^2)^{a^2} - 1$$

Since Y is also a lognormal random variable, according to Eqs. (A.66a) and (A.66b), α_{3Y} and α_{4Y} can be given in terms of V_Y as shown in Eqs. (3.10c) and (3.10d), respectively.

Example 3.5   Consider the following performance function.

$$G(\mathbf{X}) = 567fr - 0.5H^2 \tag{3.13}$$

where

f = a random variable of normal distribution with $\mu_f = 0.6$ and $V_f = 0.131$;
r = a random variable of normal distribution with $\mu_r = 2.18$ and $V_r = 0.03$; and
H = a random variable of lognormal distribution with $\mu_H = 32.8$ and $V_H = 0.03$.

Using the mean and COV for each variable, we have
$\sigma_f = 0.0786$, $\sigma_r = 0.0654$, $\sigma_H = 0.984$
According to Appendix A.5, the corresponding skewness and kurtosis of f, r, and H can be easily obtained as $\alpha_{3f} = 0$, $\alpha_{3r} = 0$, $\alpha_{3H} = 0.09$, $\alpha_{4f} = 3$, $\alpha_{4r} = 3$, and $\alpha_{4H} = 3.0144$, respectively.
Equation (3.13) can be considered as a combination of Eqs. (3.4), (3.6), and (3.9). Let $g_1 = fr$, $g_2 = H^2$, $G(\mathbf{X})$ can then be expressed as $G(\mathbf{X}) = 567g_1 - 0.5g_2$. We can obtain the

first four moments of g_1 using Eqs. (3.7a)–(3.7d) and then the moments of g_2 using Eqs. (3.11b)–(3.11e). Finally, the first four moments of $G(\mathbf{X})$ can be obtained using Eqs. (3.5a)–(3.5d).

First using Eqs. (3.7a)–(3.7d), the first four moments of g_1 can be calculated as

$$\mu_{g_1} = \mu_f \mu_r = 0.6 \times 2.18 = 1.308$$

$$\sigma_{g_1}^2 = \mu_{g_1}^2 \left[\left(1 + V_f^2\right)\left(1 + V_r^2\right) - 1 \right] = 1.308^2 \left[\left(1 + 0.131^2\right)\left(1 + 0.03^2\right) - 1 \right] = 0.03093$$

$$\sigma_{g_1} = 0.1759, V_{g_1} = \frac{\sigma_{g_1}}{\mu_{g_1}} = \frac{0.1759}{1.308} = 0.1344$$

$$\alpha_{3g_1} = \frac{1}{V_{g_1}^3}\left[\left(\alpha_{3f}V_f^3 + 3V_f^2 + 1\right)\left(\alpha_{3r}V_r^3 + 3V_r^2 + 1\right) - 3\left(1 + V_f^2\right)\left(1 + V_r^2\right) + 2\right]$$

$$= \frac{1}{0.1344^3}\left[\left(3 \times 0.131^2 + 1\right)\left(3 \times 0.03^2 + 1\right) - 3 \times \left(1 + 0.131^2\right)\left(1 + 0.03^2\right) + 2\right]$$

$$= 0.03813$$

$$\alpha_{4g_1} = \frac{1}{V_{g_1}^4}\left[\left(\alpha_{4f}V_f^4 + 4\alpha_{3f}V_f^3 + 6V_f^2 + 1\right)\left(\alpha_{4v}V_v^4 + 4\alpha_{3v}V_v^3 + 6V_v^2 + 1\right)\right.$$

$$\left. - 4\left(\alpha_{3f}V_f^3 + 3V_f^2 + 1\right)\left(\alpha_{3v}V_v^3 + 3V_v^2 + 1\right) + 6\left(1 + V_f^2\right)\left(1 + V_v^2\right) - 3\right]$$

$$= \frac{1}{0.1344^4}\left[\left(3 \times 0.131^4 + 6 \times 0.131^2 + 1\right)\left(3 \times 0.03^4 + 6 \times 0.03^2 + 1\right)\right.$$

$$\left. - 4\left(3 \times 0.131^2 + 1\right)\left(3 \times 0.03^2 + 1\right) + 6\left(1 + 0.131^2\right)\left(1 + 0.03^2\right) - 3\right] = 3.0102$$

Then using Eqs. (3.11b)–(3.11e), the first four moments of g_2 can be calculated as

$$\mu_{g_2} = \mu_H^2\left(1 + V_H^2\right) = 32.8^2\left(1 + 0.03^2\right) = 1076.8083$$

$$V_{g_2}^2 = \left(1 + V_H^2\right)^4 - 1 = \left(1 + 0.03^2\right)^4 - 1 = 3.604863 \times 10^{-3}$$

$$V_{g_2} = 0.06004, \sigma_{g_2} = \mu_{g_2} \cdot V_{g_2} = 1076.8083 \times 0.06004 = 64.6521$$

$$\alpha_{3g_2} = 3V_{g_2} + V_{g_2}^3 = 3 \times 0.06004 + 0.06004^3 = 0.1803$$

$$\alpha_{4g_2} = 3 + 16V_{g_2}^2 + 15V_{g_2}^4 + 6V_{g_2}^6 + V_{g_2}^8$$

$$= 3 + 16 \times 0.06004^2 + 15 \times 0.06004^4 + 6 \times 0.06004^6 + 0.06004^8 = 3.0579$$

Finally, using Eqs. (3.5a)–(3.5d) the first four moments of $G(\mathbf{X})$ can be calculated as

$$\mu_G = 567\mu_{g_1} - 0.5\mu_{g_2} = 0567 \times 1.308 - 0.5 \times 1076.8083 = 741.636 - 538.4041 = 203.2319$$

$$\sigma_G^2 = 567^2 \times \sigma_{g_1}^2 + 0.5^2 \times \sigma_{g_2}^2 = 567^2 \times 0.03093 + 0.5^2 \times 64.6521^2 = 10987.4515$$

$$\sigma_G = 104.821$$

$$\alpha_{3G} = \frac{\alpha_{3g_1} \times 567^3 \times \sigma_{g_1}^3 - \alpha_{3g_2} \times 0.5^3 \times \sigma_{g_2}^3}{\sigma_G^3}$$

$$= \frac{0.03813 \times 567^3 \times 0.1759^3 - 0.1803 \times 0.5^3 \times 64.6521^3}{104.821^3} = 0.02753$$

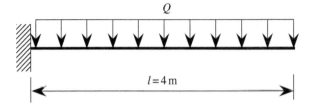

Figure 3.3 A cantilevered beam subjected to a uniformly distributed load of Example 3.6.

$$\alpha_{4G} = \frac{\alpha_{4g_1} \times 567^4 \times \sigma_{g_1}^4 + \alpha_{3g_2} \times 0.5^4 \times \sigma_{g_2}^4 + 6 \times 567^2 \times 0.1759^2 \times \sigma_{g_1}^2 \sigma_{g_2}^2}{\sigma_G^4}$$

$$= \frac{3.0102 \times 567^4 \times 0.1759^4 + 3.0579 \times 0.5^4 \times 64.6521^4 + 6 \times 567^2 \times 0.1759^2 \times 0.5^2 \times 64.6521^2}{104.821^4}$$

$$= 3.0089$$

Example 3.6 Consider a cantilevered beam subjected to a uniformly distributed load q with the beam length $l = 4.0$ m as shown in Figure 3.3.

The allowable deflection at the free end is 1/50 of the beam length. So the performance function can be expressed as

$$G(\mathbf{X}) = \frac{l}{50} - \frac{ql^3}{8EI} = 0.08 - 8\frac{q}{EI}$$

where

q = a uniformly distributed load of lognormal distribution with $\mu_q = 5000$ N/m and $V_q = 0.2$,
E = Young's modulus of the beam materials of lognormal distribution with $\mu_E = 2 \times 10^{10}$ N/m^2 and $V_E = 0.05$, and
I = inertia moment of the beam section of lognormal distribution with $\mu_I = 3.9025 \times 10^{-5}$m^4 and $V_I = 0.1$.

Using the mean and coefficient of variation for each variable, we have

$$\sigma_q = 1000\text{N/m},\ \sigma_E = 10^9\text{N/m}^2,\ \sigma_I = 3.9025 \times 10^{-6}\text{m}^4$$

According to Appendix A.5, the corresponding skewness and kurtosis of q, E, and I can be easily obtained as $\alpha_{3q} = 0.608$, $\alpha_{3E} = 0.1501$, $\alpha_{3I} = 0.301$, $\alpha_{4q} = 3.6644$, $\alpha_{4E} = 3.0401$, $\alpha_{4I} = 3.1615$.

The performance function in this example is a combination of Eqs. (3.4), (3.6), and (3.9). Let $g_1 = 1/E$, and $g_2 = 1/I$, then $G(\mathbf{X})$ can be expressed as $G(\mathbf{X}) = 0.08 - 8qg_1g_2$. We can obtain the first four moments of g_1 and g_2 using Eqs. (3.12b)–(3.12e), then obtain the moments of $G(\mathbf{X})$ using Eqs. (3.7a)–(3.7d) and (3.5a)–(3.5d).

Using Eqs. (3.12b)–(3.12e), the first four moments of g_1 can be calculated as

$$\mu_{g_1} = \frac{1}{\mu_E}(1 + V_E^2) = \frac{1}{2 \times 10^{10}}(1 + 0.05^2) = 5.0125 \times 10^{-11}$$

$$V_{g_1} = V_E = 0.05, \sigma_{g_1} = V_{g_1}\mu_{g_1} = 0.05 \times 5.0125 \times 10^{-11} = 2.5062 \times 10^{-12}$$

$$\alpha_{3g_1} = \alpha_{3E} = 0.1501, \alpha_{4g_1} = \alpha_{4E} = 3.0401$$

Similarly, the first four moments of g_2 can be calculated as

$$\mu_{g_2} = \frac{1}{\mu_I}\left(1 + V_I^2\right) = \frac{1}{3.9025 \times 10^{-5}}\left(1 + 0.1^2\right) = 25880.8456$$

$$V_{g_2} = V_I = 0.1, \sigma_{g_2} = V_{g_2}\mu_{g_2} = 0.1 \times 25880.8456 = 2588.0846$$

$$\alpha_{3g_2} = \alpha_{3I} = 0.301, \alpha_{4g_2} = \alpha_{4I} = 3.1615$$

Let $Y = qg_1g_2$, using Eqs. (3.7a)–(3.7d), the first four moments of Y can be calculated as

$$\mu_Y = 0.00649, \sigma_Y = 0.001494, V_Y = 0.2303, \alpha_{3Y} = 0.703, \alpha_{4Y} = 3.8915$$

Finally, the first four moments of $G(\mathbf{X})$ using Eqs. (3.5a)–(3.5d) can be calculated as

$$\mu_G = 0.08 - 8\mu_Y = 0.08 - 8 \times 0.00649 = 0.0281$$

$$\sigma_G^2 = 8^2 \times \sigma_Y^2 = 64 \times 0.001494^2 = 1.428 \times 10^{-4}, \sigma_G = 0.01195$$

$$\alpha_{3G} = -\alpha_{3Y} = -0.703, \alpha_{4G} = \alpha_{4Y} = 3.8915$$

Example 3.7 Consider a plate containing a crack of length $2A$ and subjected to membrane stress S as shown in Figure 3.4. The performance function may be expressed as (Ang and Tang 1984)

$$G(\mathbf{X}) = K_C - S\sqrt{\pi A}$$

where A, K_C, and S are independent random variables; A is a lognormal variable with $\mu_A = 0.06$ and $V_A = 0.25$; K_C is a normal variable, $\mu_{KC} = 120$ and $V_{KC} = 0.1$; and S is a normal variable with $\mu_S = 100$ and $V_S = 0.52$.

Let $W = \sqrt{A}$, the first four moments of W can be readily obtained as follows with the aid of Eqs. (3.10a)–(3.10d).

$$\mu_W = 0.243, \sigma_W = 0.030, \alpha_{3W} = 0.373, \text{ and } \alpha_{4W} = 3.2479$$

Let $Y = SW$, the first four moments of Y can be readily obtained as follows using Eqs. (3.7a)–(3.7d).

$$\mu_Y = 24.31, \sigma_Y = 13.087, \alpha_{3Y} = 0.167, \text{ and } \alpha_{4Y} = 3.206$$

Using Eqs. (3.5a)–(3.5d), the first four moments of $G(\mathbf{X})$ can be obtained as

$$\mu_G = 76.912, \sigma_G = 26.116 \ \alpha_{3G} = -0.117 \text{ and } \alpha_{4G} = 3.128$$

Figure 3.4 Plate under membrane stress.

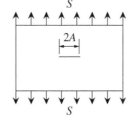

3.2.4 Moment Computation for Power of an Arbitrarily Distributed Random Variable

The power of an arbitrary distributed random variable is

$$Y = X^n, n \geq 1 \tag{3.14a}$$

Since

$$E(Y^k) = E(X^{kn}) = \mu_{kn} \tag{3.14b}$$

For an arbitrary distributed random variable, the first four moments of Y can be generally given as

$$\mu_Y = E(X^n) = \mu_n \tag{3.14c}$$

$$\sigma_Y^2 = \mu_{2n} - \mu_Y^2 \tag{3.14d}$$

$$\alpha_{3Y}\sigma_Y^3 = \mu_{3n} - 3\mu_{2n}\mu_Y + 2\mu_Y^3 \tag{3.14e}$$

$$\alpha_{4Y}\sigma_Y^4 = \mu_{4n} - 4\mu_{3n}\mu_Y + 6\mu_{2n}\mu_Y^2 - 3\mu_Y^4 \tag{3.14f}$$

Particularly when X is a standard normal random variable, according to Eq. (A.49) in Appendix A.5, the first few moments of Y are given as

For $n = 2, \mu_Y = 1, \sigma_Y = \sqrt{2} = 1.414, \alpha_{3Y} = 2\sqrt{2} = 2.828, \alpha_{4Y} = 15,$

For $n = 3, \mu_Y = 0, \sigma_Y = \sqrt{15} = 3.872, \alpha_{3Y} = 0, \alpha_{4Y} = 46.2,$ and

For $n = 4, \mu_Y = 3, \sigma_Y = 4\sqrt{6} = 9.798, \alpha_{3Y} = 33\sqrt{6}/8 = 10.104, \alpha_{4Y} = 207.$

For arbitrary distributed random variables, because the higher (> 4th) order moments of X are required to calculate the first few moments of Y, the equations above are rarely used practically. Here, we only discuss the calculation of the first three moments of Eq. (3.14a) for random variables with small COV.

When the coefficient of variation is less than 0.2 for a random variable, empirical formulas for the first three moments of $Y = X^2$, X^3, and X^4 are available using only the first four moments of X (Zhao and Lu 2007b).

For the square of a random variable expressed as the following function,

$$Y = X^2 \tag{3.15a}$$

With the aid of Eqs. (3.14c) and (3.14d), the mean and standard deviation of Y can be obtained as

$$\mu_Y = \mu_X^2(1 + V_X^2) \tag{3.15b}$$

$$\sigma_Y^2 = 4\mu_X^2\sigma_X^2 + 4\mu_X\alpha_{3X}\sigma_X^3 + (\alpha_{4X} - 1)\sigma_X^4 \tag{3.15c}$$

An empirical formula for estimating the skewness is suggested as (Zhao and Lu 2007b)

$$\alpha_{3Y} = 4\mu_X^2\sigma_X^3[2\mu_X\alpha_{3X} + 3\sigma_X(\alpha_{4X} - 1)]/\sigma_Y^3 \tag{3.15d}$$

For the third power of a random variable expressed as

$$Y = X^3 \tag{3.16a}$$

According to Eq. (3.14c), the mean of Y can be readily obtained as

$$\mu_Y = \mu_X^3 \left(1 + 3V_X^2 + \alpha_{3X} V_X^3\right) \tag{3.16b}$$

Empirical formulas for computing the standard deviation and skewness are given as (Zhao and Lu 2007b)

$$\sigma_Y^2 = 3\mu_X^4 \sigma_X^2 \left[3 + 6V_X \alpha_{3X} + (5\alpha_{4X} - 3)V_X^2\right] \tag{3.16c}$$

$$\alpha_{3Y} = \alpha_{3X} + 3(\alpha_{4X} - 1)V_X \tag{3.16d}$$

For the fourth power of a random variable expressed as

$$Y = X^4 \tag{3.17a}$$

With the aid of Eq. (3.14c), the mean of Y can be obtained as

$$\mu_Y = \mu_X^4 \left(1 + 6V_X^2 + 4\alpha_{3X} V_X^3 + \alpha_{4X} V_X^4\right) \tag{3.17b}$$

Empirical formulas for computing the standard deviation and skewness of Y are suggested as (Zhao and Lu 2007b)

$$\sigma_Y^2 = 4\mu_X^6 \sigma_X^2 \left[4 + 12V_X \alpha_{3X} + (17\alpha_{4X} - 27)V_X^2\right] \tag{3.17c}$$

$$\alpha_{3Y} = \alpha_{3X} + 4.5(\alpha_{4X} - 1)V_X \tag{3.17d}$$

Example 3.8 Consider the simple beam subjected to a uniformly distributed load Q, as shown in Figure 3.5. Assume that the flexural strength of this steel beam is M. Then, at the section with the maximum applied bending moment, the performance function may be defined as

$$G(\mathbf{X}) = M - QL^2/8$$

where M, Q, and L are independent random variables; M is lognormal with $\mu_M = 150$ and $V_M = 0.1$; Q is Gumbel with $\mu_Q = 10$ and $V_Q = 0.3$; and L is normal with $\mu_L = 8$ and $V_L = 0.05$.

The corresponding skewnesses and kurtosis of M, Q, and L are $\alpha_{3M} = 0.301$, $\alpha_{4M} = 3.162$; $\alpha_{3Q} = 1.14$, $\alpha_{4Q} = 5.4$; and $\alpha_{3L} = 0$, $\alpha_{4L} = 3$, respectively.

Let $W = L^2$, then the first three moments of Y can be readily obtained as follows with the aid of Eqs. (3.15b)–(3.15d).

$$\mu_W = \mu_L^2 \left(1 + V_L^2\right) = 8^2 \times \left(1 + 0.05^2\right) = 64.16$$

$$\sigma_W^2 = 4\mu_L^2 \sigma_L^2 + 4\mu_L \alpha_{3L} \sigma_L^3 + (\alpha_{4L} - 1)\sigma_L^4 = 41.01$$

Figure 3.5 A simple beam.

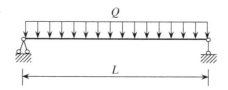

$$\sigma_W = 6.404, V_W = 0.100$$

$$\alpha_{3W} = 4\mu_L^2\sigma_L^3[2\mu_L\alpha_{3L} + 3\sigma_L(\alpha_{4L} - 1)]/\sigma_W^3 = 0.150$$

Let $Y = QW$, using Eqs. (3.7a)–(3.7d) the first three moments of Y can be calculated as

$$\mu_Y = \Pi_{i=1}^2\mu_i = 10 \times 64.16 = 641.6$$

$$\sigma_Y^2 = \mu_Y^2\left[\Pi_{i=1}^2(1 + V_i^2) - 1\right] = 41518.95$$

$$\sigma_Y = 203.76, V_Y = 0.318$$

$$\alpha_{3Y} = \left[\Pi_{i=1}^2(\alpha_{3i}V_i^3 + 3V_i^2 + 1) - 3\Pi_{i=1}^2(1 + V_i^2) + 2\right]/V_Y^3 = 1.164$$

Then, using Eqs.(3.5a)–(3.5d) the first three moments of $G(\mathbf{X}) = M - Y/8$ can be obtained as

$$\mu_G = 69.8, \sigma_G = 29.56, \text{ and } \alpha_{3G} = -0.705$$

3.2.5 Moment Computation for Reciprocal of an Arbitrary Distributed Random Variable

Another commonly used function is the reciprocal of an arbitrary distributed random variable expressed as $Y = 1/X$, which is the same as Eq. (3.12a). Obviously, the general expression of the first few moments is quite complicated. However, if the coefficient of variation of the random variable X is very small, the first four moments can be approximately given as (Zhao and Lu 2007b):

$$\mu_Y = \frac{1}{\mu_X}\left(1 + V_X^2\right) \tag{3.18a}$$

$$\sigma_Y = \frac{V_X}{\mu_X} \tag{3.18b}$$

$$\alpha_{3Y} = 6V_X - \alpha_{3X} + 4.5V_X^2(\alpha_{4X} - 3) \tag{3.18c}$$

$$\alpha_{4Y} = \alpha_{4X} + 4V_X\alpha_{3X} \tag{3.18d}$$

Example 3.9   Derivation of Eqs. (3.18a)–(3.18d)

By Taylor expansion, it can be obtained that

$$E(Y^n) = \frac{1}{\mu^n}\left[1 + \frac{(n+1)!}{(n-1)!2!}\cdot V^2 - \frac{(n+2)!}{(n-1)!3!}\cdot V^3 \cdot \alpha_3 + \frac{(n+3)!}{(n-1)!4!}\cdot V^4 \cdot \alpha_4 - \cdots\right]$$

where $V = \dfrac{\sigma}{\mu}$.

Hence

$$E(Y) = \frac{1}{\mu}\left(1 + V^2 - V^3\alpha_3 + V^4\alpha_4 - \cdots\right), E(Y^2) = \frac{1}{\mu^2}\left(1 + 3V^2 - 4V^3\alpha_3 + 5V^4\alpha_4 - \cdots\right)$$

$$E(Y^3) = \frac{1}{\mu^3}\left(1 + 6V^2 - 10V^3\alpha_3 + 15V^4\alpha_4 - \cdots\right)$$

$$E(Y^4) = \frac{1}{\mu^4}\left(1 + 10V^2 - 20V^3\alpha_3 + 35V^4\alpha_4 - \cdots\right)$$

Then

$$\sigma_Y^2 = E(Y^2) - \mu_Y^2 = \frac{V^2}{\mu^2}\left(1 - 2V\alpha_3 - V^2 + 3V^2\alpha_4 + 2V^3\alpha_3 + \cdots\right)$$

$$\alpha_{3Y}\sigma_Y^3 = E(Y^3) - 3\mu_Y E(Y^2) + 2\mu_Y^3$$

$$= \frac{V^3}{\mu^3}\left(-\alpha_3 - 3V + 3V\alpha_4 + 9V^2\alpha_3 + 2V^3 - 6V^3\alpha_3^2 - 12V^3\alpha_4 + \cdots\right)$$

$$\alpha_{4Y}\sigma_Y^4 = E(Y^4) - 4\mu_Y E(Y^3) + 6\mu_Y^2 E(Y^2) - 3\mu_Y^4$$

$$= \frac{V^4}{\mu^4}\left(\alpha_4 + 4V\alpha_3 + 6V^2 - 4V^2\alpha_3^2 - 12V^2\alpha_4 - 24V^3\alpha_3 + 16V^3\alpha_3\alpha_4 + \cdots\right)$$

Therefore, for small V, Eqs. (3.18a)–(3.18d) may be easily obtained.

3.3 Point Estimates for a Function with One Random Variable

In practice, $G(\mathbf{X})$ is often complicated and implicit, so analytical method described in the previous section may sometimes not be applicable. In this section, the point estimate method is introduced for the first four moments of an arbitrary performance function. As the basis of point estimate for moments of a function of multiple random variables, we will first discuss the computation of the first few moments for a function of single random variable, i.e. $Y = y(X)$,

$$\mu_Y = E[y(X)] = \int y(x)f(x)dx \tag{3.19a}$$

$$\sigma_Y^2 = E\{[y(X) - \mu_Y]^2\} = \int [y(X) - \mu_Y]^2 f(x)dx \tag{3.19b}$$

$$\sigma_Y^k \alpha_{kY} = E\{[y(X) - \mu_Y]^k\} = \int [y(x) - \mu_Y]^k f(x)dx \quad \text{for} \quad k > 2 \tag{3.19c}$$

3.3.1 Rosenblueth's Two-Point Estimate

As is well known, direct integration in Eqs. (3.19a)–(3.19c) is inconvenient. The commonly used method involves obtaining the Taylor expansion of the function about the expectations of the random variable, which imposes excessive restrictions on the function (existence and continuity of the first few derivatives) and requires the computation of derivatives (Rosenblueth 1975). Rosenblueth (1975) has proposed a method to estimate the first few moments, but does not involve derivatives for a function of a single random variable presented in Eqs. (3.19a)–(3.19c). The method uses a weighted sum of the function evaluated at

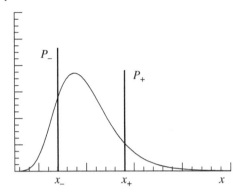

a finite number of points. The weights and the points at which the function is evaluated are chosen to be the weights and the points at which the variable itself must be evaluated in order to give the correct values of its own first few moments.

For a two-point estimate, let the concentrations be x_- and x_+ with corresponding weights of P_- and P_+, respectively. As shown in Figure 3.6, the kth central moment of a function $Y = y(X)$ can be calculated by (Rosenblueth 1975)

$$\mu_Y = P_- y(x_-) + P_+ y(x_+) \tag{3.20a}$$

$$\sigma_Y^2 = P_- [y(x_-) - \mu_Y]^2 + P_+ [y(x_+) - \mu_Y]^2 \tag{3.20b}$$

$$\sigma_Y^k \alpha_{kY} = P_- [y(x_-) - \mu_Y]^k + P_+ [y(x_+) - \mu_Y]^k \tag{3.20c}$$

where x_- and x_+ are the two points for estimation, and P_- and P_+ are their corresponding weights. P_-, x_-, and P_+, x_+ must satisfy the following equations (Rosenblueth 1975):

$$P_+ + P_- = 1 \tag{3.21a}$$

$$P_+ x_+ + P_- x_- = \mu_X \tag{3.21b}$$

$$P_+ (x_+ - \mu_X)^2 + P_- (x_- - \mu_X)^2 = \sigma_X^2 \tag{3.21c}$$

$$P_+ (x_+ - \mu_X)^3 + P_- (x_- - \mu_X)^3 = \alpha_{3X} \sigma_X^3 \tag{3.21d}$$

It can be observed from Eqs. (3.21a)–(3.21d) that there are four equations with four unknown parameters. Algebraic manipulation shows that x_- and x_+ are the solutions of the following equation.

$$(x - \mu_X)^2 - \alpha_{3X} \sigma_X (x - \mu_X) - \sigma_X^2 = 0 \tag{3.21e}$$

Therefore,

$$x_- = \mu_X + \frac{\sigma_X}{2} \left(\alpha_{3X} - \sqrt{\alpha_{3X}^2 + 4} \right), P_- = \frac{1}{2} \left(1 + \frac{\alpha_{3X}}{\sqrt{\alpha_{3X}^2 + 4}} \right) \tag{3.22a}$$

$$x_+ = \mu_X + \frac{\sigma_X}{2} \left(\alpha_{3X} + \sqrt{\alpha_{3X}^2 + 4} \right), P_+ = \frac{1}{2} \left(1 - \frac{\alpha_{3X}}{\sqrt{\alpha_{3X}^2 + 4}} \right) \tag{3.22b}$$

3.3.2 Gorman's Three-Point Estimate

To improve the accuracy of the two-point estimate, Gorman (1980) proposed a three-point estimate. Let the concentrations be x_-, x_0, and x_+ with corresponding weights of P_-, P_0, and P_+, as shown in Figure 3.7, then the kth central moment of $Y = y(X)$ can be calculated by

$$\mu_Y = P_- y(x_-) + P_0 y(x_0) + P_+ y(x_+) \tag{3.23a}$$

$$\sigma_Y^2 = P_- \left[y(x_-) - \mu_y \right]^2 + P_0 \left[y(x_0) - \mu_y \right]^2 + P_+ \left[y(x_+) - \mu_y \right]^2 \tag{3.23b}$$

$$\sigma_Y^k \alpha_{kY} = P_- \left[y(x_-) - \mu_y \right]^k + P_0 \left[y(x_0) - \mu_y \right]^k + P_+ \left[y(x_+) - \mu_y \right]^k \tag{3.23c}$$

where x_-, x_0, and x_+ are the three estimating points; and P_-, P_0, and P_+ are their corresponding weights. P_-, x_-, P_0, x_0, P_+, and x_+ must satisfy the following equations:

$$P_+ + P_0 + P_- = 1 \tag{3.24a}$$

$$P_+ x_+ + P_0 x_0 + P_- x_- = \mu_X \tag{3.24b}$$

$$P_+ (x_+ - \mu_x)^2 + P_0 (x_0 - \mu_x)^2 + P_- (x_- - \mu_x)^2 = \sigma_X^2 \tag{3.24c}$$

$$P_+ (x_+ - \mu_x)^3 + P_0 (x_0 - \mu_x)^3 + P_- (x_- - \mu_x)^3 = \alpha_{3X} \sigma_X^3 \tag{3.24d}$$

$$P_+ (x_+ - \mu_x)^4 + P_0 (x_0 - \mu_x)^4 + P_- (x_- - \mu_x)^4 = \alpha_{4X} \sigma_X^4 \tag{3.24e}$$

In these equations there are a total of five equations with six unknown parameters. Choose x_0 as μ_X, then x_- and x_+ are found to be the solutions of the following equation:

$$(x - \mu_X)^2 - \alpha_{3X} \sigma_X (x - \mu_X) + \sigma_X^2 \alpha_{3X}^2 - \sigma_X^2 \alpha_{4X} = 0 \tag{3.24f}$$

Therefore

$$x_- = \mu_X - \frac{\sigma_X}{2} (\theta - \alpha_{3X}), P_- = \frac{1}{2} \left(\frac{1 + \alpha_{3X}/\theta}{\alpha_{4X} - \alpha_{3X}^2} \right) \tag{3.25a}$$

$$x_0 = \mu_X, P_0 = 1 - \frac{1}{\alpha_{4X} - \alpha_{3X}^2} \tag{3.25b}$$

Figure 3.7 Concentrations of probability density function using three points.

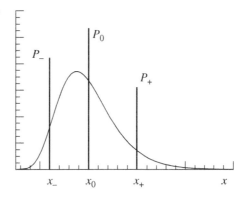

$$x_+ = \mu_X + \frac{\sigma_X}{2}(\theta + \alpha_{3X}), \quad P_+ = \frac{1}{2}\left(\frac{1 - \alpha_{3X}/\theta}{\alpha_{4X} - \alpha_{3X}^2}\right) \tag{3.25c}$$

where

$$\theta = \left(4\alpha_{4X} - 3\alpha_{3X}^2\right)^{1/2} \tag{3.25d}$$

For any distribution, Eq. (A.23) gives

$$\alpha_{4X} - \alpha_{3X}^2 \geq 1 \tag{3.26}$$

One may easily understand that Eqs. (3.25a)–(3.25d) is always operable and

$$\theta \geq \left(\alpha_{3X}^2 + 4\right)^{1/2} \tag{3.27}$$

This is to say, P_-, P_0, and P_+ are all between 0 and 1.
If $P_0 = 0$, i.e.

$$\alpha_{4X} - \alpha_{3X}^2 = 1, \theta = \left(\alpha_{3X}^2 + 4\right)^{1/2} \tag{3.28}$$

Equations (3.25a)–(3.25d) reduce to Eqs. (3.22a)–(3.22b), implying that the two-point estimate can be explained as a special case of the three-point estimate. By writing out the expressions for the moments, it can be seen that the two-point estimate will always give Eq. (3.28), which limits the distribution of x to a line in the $\alpha_3{}^2 - \alpha_4$ plane, i.e. a U-shaped distribution (see Figure A.13). General discussions can be further referred to Harr (1989), Li (1992), and Hong (1998), etc.

Example 3.10 The example considers the following function of a standard normal random variable,

$$Y = \exp\left(X\zeta + \lambda\right) \tag{3.29}$$

where λ and ζ are parameters, and X is a standard normal random variable. Here, $\lambda = 1.5$ and $\zeta = 0.1 \sim 0.7$ are taken in the investigation.

Y is obviously a lognormal variable with parameters λ and ζ, where for $\zeta = 0.4$, the exact first four moments of Y are obtained as $\mu_Y = 4.855$, $\sigma_Y = 2.022$, $\alpha_{3Y} = 1.322$, and $\alpha_{4Y} = 6.260$.

Utilising the information $\mu_X = 0$, $\sigma_X = 1$, and $\alpha_{3X} = 0$, the estimating point and their corresponding weights for the two-point estimate are readily obtained as $x_- = -1$, $P_- = 0.5$, $x_+ = 1$, and $P_+ = .5$; and the first four moments of y are calculated as

$$\mu_Y = P_- y(x_-) + P_+ y(x_+) = P_- \exp(x_-\zeta + \lambda) + P_+ \exp(x_+\zeta + \lambda)$$
$$= 0.5 \times \exp(-1 \times 0.4 + 1.5) + 0.5 \times \exp(1 \times 0.4 + 1.5) = 0.5 \times 3.004 + 0.5 \times 6.686 = 4.845$$
$$\sigma_Y^2 = P_- [y(x_-) - \mu_Y]^2 + P_+ [y(x_+) - \mu_Y]^2 = P_- [\exp(x_-\zeta + \lambda) - \mu_Y]^2 + P_+ [\exp(x_+\zeta + \lambda) - \mu_Y]^2$$
$$= 0.5 \times [\exp(-1 \times 0.4 + 1.5) - 4.845]^2 + 0.5 \times [\exp(1 \times 0.4 + 1.5) - 4.845]^2$$
$$= 0.5 \times (3.004 - 4.845)^2 + 0.5 \times (6.686 - 4.845)^2 = 3.389$$
$$\sigma_Y = 1.841$$

$$\alpha_{3Y}\sigma_Y^3 = P_-[y(x_-) - \mu_Y]^3 + P_+[y(x_+) - \mu_Y]^3 = P_-[\exp(x_-\zeta + \lambda)$$
$$- \mu_Y]^3 + P_+[\exp(x_+\zeta + \lambda) - \mu_Y]^3$$
$$= 0.5 \times [\exp(-1 \times 0.4 + 1.5) - 4.845]^3 + 0.5 \times [\exp(1 \times 0.4 + 1.5) - 4.845]^3$$
$$= 0.5 \times (3.004 - 4.845)^3 + 0.5 \times (6.686 - 4.845)^3 = 0 \quad \alpha_{3Y} = 0/1.841^3 = 0$$

$$\alpha_{4Y}\sigma_Y^4 = P_-[y(x_-) - \mu_Y]^4 + P_+[y(x_+) - \mu_Y]^4 = P_-[\exp(x_-\zeta + \lambda) - \mu_Y]^4$$
$$+ P_+[\exp(x_+\zeta + \lambda) - \mu_Y]^4$$
$$= 0.5 \times [\exp(-1 \times 0.4 + 1.5) - 4.845]^4 + 0.5 \times [\exp(1 \times 0.4 + 1.5) - 4.845]^4$$
$$= 0.5 \times (3.004 - 4.845)^4 + 0.5 \times (6.686 - 4.845)^4 = 11.487 \alpha_{4Y} = 11.487/1.841^4 = 1$$

From these results it can be observed that significant errors exist when using the two-point estimate even for mean and standard deviation. For skewness and kurtosis, this provides quite different values from the exact results.

Utilising the information $\mu_X = 0$, $\sigma_X = 1$, $\alpha_{3X} = 0$, and $\alpha_{4X} = 3$, the estimating point and their corresponding weights for the three-point estimate are readily obtained as $x_- = -1.732$, $P_- = 1/6$, $x_0 = 0$, $P_0 = 2/3$, $x_+ = 1.732$, $P_+ = 1/6$; and the first four moments of y are obtained as

$$\mu_Y = P_- y(x_-) + P_0 y(x_0) + P_+ y(x_+) = P_- \exp(x_-\zeta + \lambda)$$
$$+ P_0 \exp(x_0\zeta + \lambda) + P_+ \exp(x_+\zeta + \lambda)$$
$$= (1/6) \times \exp(-1.732 \times 0.4 + 1.5) + (2/3) \times \exp(0 \times 0.4 + 1.5)$$
$$+ (1/6) \times \exp(1.732 \times 0.4 + 1.5)$$
$$= (1/6) \times 2.242 + (2/3) \times 4.482 + (1/6) \times 8.960 = 4.855$$

$$\sigma_Y^2 = P_-[y(x_-) - \mu_Y]^2 + P_0[y(x_0) - \mu_Y]^2 + P_+[y(x_+) - \mu_Y]^2$$
$$= P_-[\exp(x_-\zeta + \lambda) - \mu_Y]^2 + P_0[\exp(x_0\zeta + \lambda) - \mu_Y]^2 + P_+[\exp(x_+\zeta + \lambda) - \mu_Y]^2$$
$$= (1/6) \times [\exp(-1.732 \times 0.4 + 1.5) - 4.855]^2 + (2/3) \times [\exp(0 \times 0.4 + 1.5) - 4.855]^2$$
$$+ (1/6) \times [\exp(1.732 \times 0.4 + 1.5) - 4.855]^2$$
$$= (1/6) \times (2.242 - 4.855)^2 + (2/3) \times (4.482 - 4.855)^2 + (1/6) \times (8.960 - 4.855)^2$$
$$= 4.093 \quad \sigma_Y = 2.010$$

$$\alpha_{3Y}\sigma_Y^3 = P_-[y(x_-) - \mu_Y]^3 + P_0[y(x_0) - \mu_Y]^3 + P_+[y(x_+) - \mu_Y]^3$$
$$= P_-[\exp(x_-\zeta + \lambda) - \mu_Y]^3 + P_0[\exp(x_0\zeta + \lambda) - \mu_Y]^3 + P_+[\exp(x_+\zeta + \lambda) - \mu_Y]^3$$
$$= (1/6) \times [\exp(-1.732 \times 0.411 + 1.5) - 4.855]^3$$
$$+ (2/3) \times [\exp(0 \times 0.4 + 1.5) - 4.855]^3$$
$$+ (1/6) \times [\exp(1.732 \times 0.4 + 1.5) - 4.855]^3$$
$$= (1/6) \times (2.242 - 4.855)^3 + (2/3) \times (4.482 - 4.855)^3$$
$$+ (1/6) \times (8.960 - 4.855)^3 = 8.521$$

$$\alpha_{3Y} = 8.521/2.01^3 = 1.049$$

$$\begin{aligned}
\alpha_{4Y}\sigma_Y^4 &= P_-[y(x_-) - \mu_Y]^4 + P_0[y(x_0) - \mu_Y]^4 + P_+[y(x_+) - \mu_Y]^4 \\
&= P_-[\exp(x_- \zeta + \lambda) - \mu_Y]^4 + P_0[\exp(x_0\zeta + \lambda) - \mu_Y]^4 \\
&\quad + P_+[\exp(x_+ \zeta + \lambda) - \mu_Y]^4 \\
&= (1/6) \times [\exp(-1.732 \times 0.4 + 1.5) - 4.855]^4 \\
&\quad + (2/3) \times [\exp(0 \times 0.4 + 1.5) - 4.855]^4 \\
&\quad + (1/6) \times [\exp(1.732 \times 0.4 + 1.5) - 4.855]^4 \\
&= (1/6) \times (2.242 - 4.855)^4 + (2/3) \times (4.482 - 4.855)^4 \\
&\quad + (1/6) \times (8.960 - 4.855)^4 = 55.109 \quad \alpha_{4Y} = 55.109/2.01^3 = 3.376
\end{aligned}$$

It can be observed that although the three-point estimate provides much better results than the two-point estimate, there are still significant errors in the results of skewness and kurtosis.

Figure 3.8 presents the variations of the first four estimated moments with respect to ζ compared with corresponding theoretical values. It can be observed that the moments obtained from the point estimates agree very well with the mean and standard deviation for small ζ. But for the third- and fourth-order moments, the results obtained from the point estimates can be hardly used as approximation of exact results due to poor accuracy.

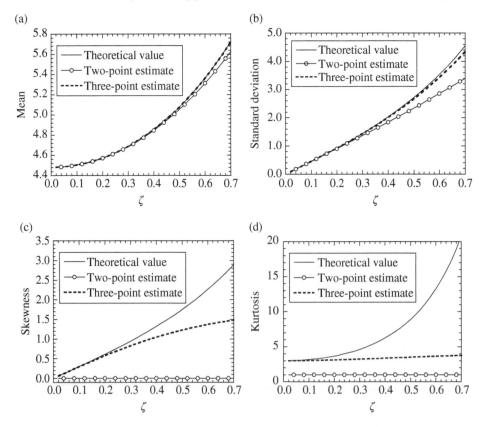

Figure 3.8 Variations of the first four moments for Example 3.10. (a) Mean. (b) Standard deviation. (c) Skewness. (d) Kurtosis.

Example 3.11 Consider the following function of a lognormal random variable

$$Y = \ln(X) \tag{3.30}$$

where X is a lognormal variable with parameters λ and ζ. Here the cases of $\lambda = 1.5$ and $\zeta = 0.1 \sim 0.6$ are considered.

Y is a normal random variable, and the first four moments can be readily calculated exactly as $\mu_Y = \lambda$, $\sigma_Y = \zeta$, $\alpha_{3Y} = 0$, and $\alpha_{4Y} = 3$.

Figure 3.9 presents the comparisons of the first four moments obtained using the two- and three-point estimates with the corresponding theoretical values. It can be observed from Figure 3.9 that for mean and standard deviation, the moments obtained by the point estimation agree very well with the exact results when ζ takes quite a small value. The accuracy decreases with the increase of the orders of the moments to be evaluated by point estimates. For the third- and fourth-order moments, the results obtained by the point estimation can hardly be used as the approximation for the exact results. When ζ is larger than 0.52, the results of point estimates cannot be obtained since the method is not applicable. For example, for $\zeta = 0.6$, the three estimating points are obtained as, $x_- = -1.496$, $x_0 = 5.366$, and $x_+ = 20.21$. Obviously, $x_- = -1.496$ is out of the definition range of X since X is a lognormal random variable. The variations of the estimating points with respect to ζ are shown in

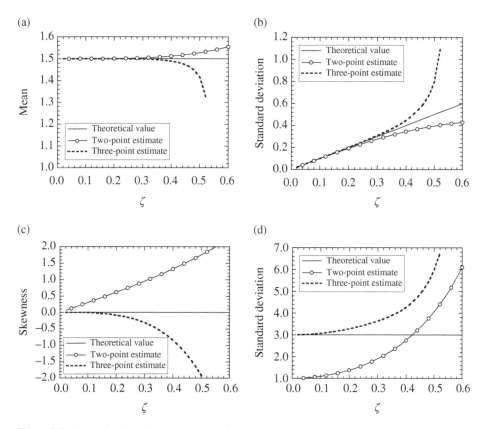

Figure 3.9 Investigation of point estimates by Example 3.11. (a) Mean. (b) Standard deviation. (c) Skewness. (d) Kurtosis.

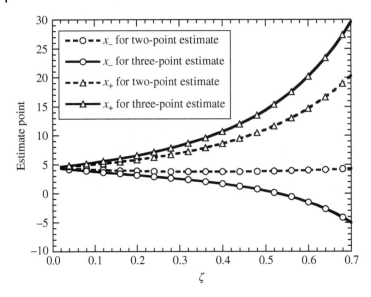

Figure 3.10 Variations of the estimating points with respect to ζ.

Figure 3.10, where it can be observed that x_- becomes negative with the increase of ζ for the three-point estimates. Since the lognormal random variable is defined for positive values of x, the point estimate is not applicable in this case.

The two examples above show that the two- and three-point estimates have the following weaknesses:

1) The accuracy in general cases is quite low (Gorman 1980) especially in the case that the coefficient of variation is large and when these methods are used to evaluate high order moments (Zhao et al. 1999).
2) For some random variables that are variables lognormally or exponentially distributed, if the standard deviation is relatively large, then x_- given by Eq. (3.25a) may be outside of the region in which the random variable is defined, and so the computation becomes impossible.

Because the procedure of point estimates is very simple and does not require the computation of derivatives, if the two weaknesses described above can be improved, the computation of moments for a performance function will become quite straightforward. Therefore, in the next section, we will introduce the new point estimates for probability moments that remove the above two weaknesses.

3.4 Point Estimates in Standardised Normal Space

3.4.1 Formulae of Moment of Functions with Single Random Variable

In order to improve the accuracy of the two- and three-point estimates, a spur-of-the-moment idea is to increase the number of estimating points. For an m point estimate, the kth central moment of $Y = y(X)$ can be calculated by

$$\mu_Y = \sum_{j=1}^{m} P_j y(x_j) \tag{3.31a}$$

$$\sigma_Y^2 = \sum_{j=1}^{m} P_j \left[y(x_j) - \mu_Y \right]^2 \tag{3.31b}$$

$$\sigma_Y^k \alpha_{kY} = \sum_{j=1}^{m} P_j \left[y(x_j) - \mu_Y \right]^k \tag{3.31c}$$

where x_1, x_2, ..., x_m are the estimating points; and P_1, P_2, ..., P_m are their corresponding weights, which are selected using the following equations:

$$\sum_{j=1}^{m} P_j = 1 \tag{3.32a}$$

$$\sum_{j=1}^{m} P_j x_j = \mu_X \tag{3.32b}$$

$$\sum_{j=1}^{m} P_j \left(x_j - \mu_X \right)^2 = \sigma_X^2 \tag{3.32c}$$

$$\sum_{j=1}^{m} P_j \left(x_j - \mu_X \right)^k = \sigma_X^k \alpha_{kX} \tag{3.32d}$$

where μ_X is the mean of X, σ_X is the standard deviation of X, and α_{kX} is the kth dimensionless central moment of X.

For a three-point estimate, assuming that an estimating point is fixed at $x_0 = \mu_X$, a total of five such equations will be needed and the first four moments will be required as shown in Gorman's formula of Eqs. (3.24a)–(3.24e). For a five-point estimate, supposing that an estimating point is fixed at $x_0 = \mu_X$, a total of nine such equations will be needed and the first eight moments will be required. In general, for an m point estimate, supposing that an estimating point is fixed at $x_0 = \mu_X$, $2m - 1$, such equations will be needed and the first $2(m - 1)$ moments of x will be required. However, the high-order moments of an arbitrary random variable are not always easy to obtain, and it is generally hard to accept the use of moments higher than fourth-order in engineering, since only the first four moments (mean, deviation, skewness, and kurtosis) are common in engineering and have clear physical meanings as illustrated in Appendix A.3. Even if the moments higher than fourth-order could be used, the solution of the equations would be very complicated for general α_{kX}, and it would be almost impossible to obtain the general expressions of x_1, x_2, ..., x_m, and P_1, P_2, ..., P_m for an arbitrary random variable. Furthermore, since α_{kX} is dependent on the type of random variable, it is difficult to prevent that the estimating point may fall outside of the region defined for the random variable.

In order to avoid the problems mentioned above, the estimating points obtained in standard normal space (Zhao and Ono 2000a) will be introduced in this section. This is because any random variables can be easily transformed into standard normal random variables through the Rosenblatt transformation (Hohenbichler and Rackwitz 1981) as shown later in the chapter on Transformation of Non-Normal Variables to Independent Normal Variables, in Section 7.3.1:

$$U = T(X) \tag{3.33a}$$

$$X = T^{-1}(U) \tag{3.33b}$$

Using the inverse transformation shown in Eq. (3.33b), Eqs. (3.19a)–(3.19c) can be rewritten as:

$$\mu_Y = \int y[T^{-1}(u)]\phi(u)du \tag{3.34a}$$

$$\sigma_Y^2 = \int \{y[T^{-1}(u)] - \mu_Y\}^2 \phi(u)du \tag{3.34b}$$

$$\sigma_Y^k \alpha_{kY} = \int \{y[T^{-1}(u)] - \mu_Y\}^k \phi(u)du \quad \text{for} \quad k > 2 \tag{3.34c}$$

where $\phi(\cdot)$ is the PDF of a standard normal random variable.

Because the standard normal random variable is one of the simplest and most commonly used random variables, the evaluation of Eqs. (3.34a)–(3.34c) is easier than that of Eqs. (3.19a)–(3.19c). After substituting the probability information of the standard normal random variable into Eqs. (3.32a)–(3.32d), these equations become much easier to solve because of the following characteristics:

1) Because a standard normal distribution is symmetric with respect to the origin point, the number of equations is reduced to half if the estimating points that are symmetrically distributed are selected.
2) All of the central moments of a standard normal random variable can be easily obtained numerically as

$$\mu_u = 0, \sigma_u = 1$$

$$\alpha_{ku} = \begin{cases} 0, & \text{for odd } k \\ (k-1)!!, & \text{for even } k \end{cases}$$

Because of this, the estimating points in standard normal space and their corresponding weights can be obtained numerically even if the general expressions are difficult to obtain. These general expressions are not required since the estimating points in standard normal space are independent to the random variable in original space.
3) No central moments are required for the original random variables when obtaining the estimating points in the standard normal space. Furthermore, because the normal random variable is defined within the whole range of $(-\infty, +\infty)$, the problem that the estimating point may fall outside of the defined region no longer exists.

After obtaining the estimating points $u_1, u_2, \ldots, u_m$ and their corresponding weights $P_1, P_2, \ldots, P_m$, for a function $y = y(x)$, the kth central moment of y can be calculated as:

$$\mu_Y = \sum_{j=1}^m P_j y[T^{-1}(u_j)] \tag{3.35a}$$

$$\sigma_Y^2 = \sum_{j=1}^m P_j \{y[T^{-1}(u_j)] - \mu_Y\}^2 \tag{3.35b}$$

$$\sigma_Y^k \alpha_{kY} = \sum_{j=1}^m P_j \{y[T^{-1}(u_j)] - \mu_Y\}^k \tag{3.35c}$$

Here, $T^{-1}(u_j)$ is the inverse Rosenblatt transformation, which is generally expressed as:

$$x_j = T^{-1}(u_j) = F_X^{-1}\left[\Phi(u_j)\right] \tag{3.36}$$

where $\Phi(\cdot)$ and $F_X(\cdot)$ are the CDFs of u and x, respectively. For specific random variables, Eq. (3.36) can generally be expressed in much simpler form.

It is worth noting that the general expression for the function $y[T^{-1}(U)]$ in Eqs. (3.35a)–(3.35c) is not necessary, and that the Rosenblatt transformation is only required at the estimating points.

3.4.2 Two- and Three-Point Estimates in the Standard Normal Space

For a two-point estimate in the standard normal space, the estimating points u_-, u_+, and their corresponding weights P_-, P_+ are readily obtained by substituting the probability characteristics of the standard normal random variable into Eqs. (3.22a)–(3.22b):

$$u_- = -1, P_- = 0.5 \tag{3.37a}$$
$$u_+ = 1, P_+ = 0.5 \tag{3.37b}$$

For a three-point estimate in the standard normal space, the estimating points u_-, u_0, u_+, and their corresponding weights P_-, P_0, P_+ are readily obtained by substituting the probability characteristics of the standard normal random variable into Eqs. (3.25a)–(3.25d):

$$u_- = -\sqrt{3}, P_- = \frac{1}{6} \tag{3.38a}$$

$$u_0 = 0, P_0 = \frac{2}{3} \tag{3.38b}$$

$$u_+ = \sqrt{3}, \quad P_+ = \frac{1}{6} \tag{3.38c}$$

The concentrations of PDF of standard normal variables using two and three points are shown in Figure 3.11.

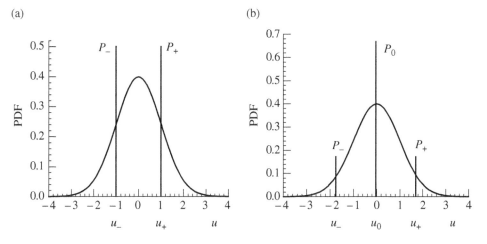

Figure 3.11 Concentrations of PDF of standard normal variable using two and three points. (a) Two points; (b) Three points.

3.4.3 Five-Point Estimate in Standard Normal Space

For a five-point estimate in standard normal space, utilising the symmetry of the standard normal random variable, let $u_0 = 0$, $u_1 = -u_{1-} = u_{1+}$, $u_2 = -u_{2-} = u_{2+}$, $P_1 = P_{1-} = P_{1+}$, and $P_2 = P_{2-} = P_{2+}$; u_1, u_2, P_0, P_1, and P_2 can then be determined from the following five equations:

$$P_0 + 2P_1 + 2P_2 = 1 \tag{3.39a}$$

$$2P_1 u_1^2 + 2P_2 u_2^2 = 1 \tag{3.39b}$$

$$2P_1 u_1^4 + 2P_2 u_2^4 = 3 \tag{3.39c}$$

$$2P_1 u_1^6 + 2P_2 u_2^6 = 15 \tag{3.39d}$$

$$2P_1 u_1^8 + 2P_2 u_2^8 = 105 \tag{3.39e}$$

Let $y_k = u_k^2$, and $k = 1, 2$. After simplification, y_1 and y_2 are found to be the solutions of the following equation:

$$y^2 - 10y + 15 = 0 \tag{3.40}$$

Equation (3.40) is the second-order Hermite polynomial, of which the two solutions are

$$y_1 = 5 - \sqrt{10}, y_2 = 5 + \sqrt{10} \tag{3.41}$$

The five estimating points in standard normal space and their corresponding weights are then obtained as

$$u_{2-} = -\sqrt{5 + \sqrt{10}}, \quad P_{2-} = \frac{7 - 2\sqrt{10}}{60} \tag{3.42a}$$

$$u_{1-} = -\sqrt{5 - \sqrt{10}}, \quad P_{1-} = \frac{7 + 2\sqrt{10}}{60} \tag{3.42b}$$

$$u_0 = 0, \quad P_0 = \frac{8}{15} \tag{3.42c}$$

$$u_{1+} = \sqrt{5 - \sqrt{10}}, \quad P_{1+} = \frac{7 + 2\sqrt{10}}{60} \tag{3.42d}$$

$$u_{2+} = \sqrt{5 + \sqrt{10}}, \quad P_{2+} = \frac{7 - 2\sqrt{10}}{60} \tag{3.42e}$$

The numerical results of the five-point estimate in standard normal space are therefore given as

$$u_{2-} = -2.8569700, \quad P_{2-} = 1.12574 \times 10^{-2}$$
$$u_{1-} = -1.3556262, \quad P_{1-} = 0.2220759$$
$$u_0 = 0, \quad P_0 = 0.5333$$
$$u_{1+} = 1.3556262, \quad P_{1+} = 0.2220759$$
$$u_{2+} = 2.8569700, \quad P_{2+} = 1.12574 \times 10^{-2}$$

The concentrations of PDF of standard normal variable using five points are shown in Figure 3.12a.

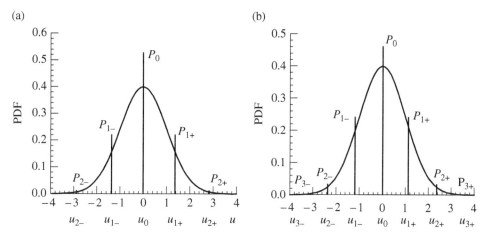

Figure 3.12 Concentrations of PDF of standard normal variable. (a) Five points. (b) Seven points.

3.4.4 Seven-Point Estimate in Standard Normal Space

For a seven-point estimate in standard normal space, utilising the symmetry of standard normal random variable, let $u_0 = 0$, $u_1 = -u_{1-} = u_{1+}$, $u_2 = -u_{2-} = u_{2+}$, $u_3 = -u_{3-} = u_{3+}$, $P_1 = P_{1-} = P_{1+}$, $P_2 = P_{2-} = P_{2+}$, and $P_3 = P_{3-} = P_{3+}$. Let $y_k = u_k^2$, and $k = 1, 2, 3$; then y_1, y_2, y_3 are found to be the solutions of the following equation

$$y^3 - 21y^2 + 105y - 105 = 0 \tag{3.43}$$

Equation (3.43) is the third-order Hermite polynomial, of which the three solutions are

$$y_1 = 7 - \sqrt{14}\cos(\phi/3) - \sqrt{42}\sin(\phi/3) \tag{3.44a}$$

$$y_2 = 7 - \sqrt{14}\cos(\phi/3) + \sqrt{42}\sin(\phi/3) \tag{3.44b}$$

$$y_3 = 7 + 2\sqrt{14}\cos(\phi/3) \tag{3.44c}$$

where

$$\tan\phi = \sqrt{10}/2 \tag{3.44d}$$

Substituting y_1, y_2, and y_3 into Eq. (3.43), and let

$$c = \sqrt{14}\cos(\phi/3), s = \sqrt{42}\sin(\phi/3) \tag{3.45}$$

One obtains

$$P_0 = \frac{16}{35} \tag{3.46a}$$

$$P_1 = \frac{19c(s-c) + 7(7c + 7s + 39)}{70s(3c + s)} \tag{3.46b}$$

$$P_2 = \frac{19c(c+s) - 7(7c - 7s + 39)}{70s(3c - s)} \tag{3.46c}$$

$$P_3 = \frac{19(c^2 - s^2) + 14(39 - 14c)}{70(3c - s)(3c + s)} \tag{3.46d}$$

The numerical results of the seven-point estimate in standard normal space are given as follows:

$$u_{3-} = -3.7504397, \quad P_{3-} = 5.48269 \times 10^{-4}$$
$$u_{2-} = -2.3667594, \quad P_{2-} = 3.07571 \times 10^{-2}$$
$$u_{1-} = -1.1544054, \quad P_{1-} = 0.2401232$$
$$u_0 = 0, \quad P_0 = 0.4571427$$
$$u_{1+} = 1.1544054, \quad P_{1+} = 0.2401232$$
$$u_{2+} = 2.3667594, \quad P_{2+} = 3.07571 \times 10^{-2}$$
$$u_{3+} = 3.7504397, \quad P_{3+} = 5.48269 \times 10^{-4}$$

The concentrations of PDF of the standard normal variable using seven points are shown in Figure 3.12b.

Example 3.12 Consider the following function of an exponential random variable

$$Y = \alpha - \beta \ln(X) \tag{3.47}$$

where X is an exponential random variable with $\mu_X = 1$, $\sigma_X = 1$, $\alpha_{3X} = 2$, $\alpha_{4X} = 9$, and Y is easily understood as a random variable with extreme value distribution (Gumbel distribution). For the case of $\alpha = 4$ and $\beta = 2$, the exact values of the first four moments can be found as (see Appendix A.7.3)

$$\mu_Y = \alpha + 0.57722\beta = 5.15444, \sigma_Y = \sqrt{6\pi\beta}/6 = 0.40825\pi\beta = 2.56511$$

$$\alpha_{3Y} = \sqrt{1.29857} = 1.13955, \alpha_{4Y} = 27/5 = 5.4$$

Substituting the moments of X into Eqs. (3.25d) and (3.25a), the estimating point x_- for Gorman's three-point estimate is obtained as

$$\theta = \left(4\alpha_{4X} - 3\alpha_{3X}^2\right)^{1/2} = \left(4 \times 9 - 3 \times 2^2\right)^{1/2} = 4.899$$
$$x_- = \mu_X - \sigma_X(\theta - \alpha_{3X})/2 = 1 - 1 \times (4.899 - 2)/2 = -0.499$$

It can be observed that x_- gives a value outside of the defined region of X. This means that Gorman's three-point estimate cannot be applied in such an example.

Using the point estimates in standard normal space, the first four moments of y can be easily evaluated without the problem described above. The estimating points for the three-point estimate in standard space are $u_- = -1.732$, $u_0 = 0$, and $u_+ = 1.732$ with corresponding weights of $P_- = 1/6$, $P_0 = 2/3$, and $P_+ = 1/6$. Using inverse Rosenblatt transformation, the estimating points can be easily transformed in original space as $x_- = 0.0425$, $x_0 = 0.693$, and $x_+ = 3.179$. Then, the point estimates of the first four moments of y can be obtained as

$$\mu_Y = P_- y(x_-) + P_0 y(x_0) + P_+ y(x_+) = P_- \exp(4 - 2\ln x_-)$$
$$+ P_0 \exp(4 - 2\ln x_0) + P_+ \exp(4 - 2\ln x_+)$$
$$= (1/6) \times \exp(4 - \ln 0.0425) + (2/3) \times \exp(4 - \ln 0.693)$$
$$+ (1/6) \times \exp(4 - 2\ln 3.179)$$
$$= (1/6) \times 10.316 + (2/3) \times 4.733 + (1/6) \times 1.687 = 5.156$$
$$\sigma_Y^2 = P_- [y(x_-) - \mu_Y]^2 + P_0 [y(x_0) - \mu_Y]^2 + P_+ [y(x_+) - \mu_Y]^2$$
$$= P_- [(4 - 2\ln x_-) - \mu_Y]^2 + P_0 [(4 - 2\ln x_0) - \mu_Y]^2 + P_+ [(4 - 2\ln x_+) - \mu_Y]^2$$
$$= (1/6) \times [(4 - 2\ln 0.0425) - 5.156]^2 + (2/3) \times [(4 - 2\ln 0.693) - 5.516]^2$$
$$+ (1/6) \times [(4 - 2\ln 3.179) - 5.156]^2$$
$$= (1/6) \times (0.316 - 5.156)^2 + (2/3) \times (4.733 - 5.156)^2$$
$$+ (1/6) \times (1.687 - 5.156)^2 = 6.563 \quad \sigma_Y = 2.562$$
$$\alpha_{3Y}\sigma_Y^3 = P_- [y(x_-) - \mu_Y]^3 + P_0 [y(x_0) - \mu_Y]^3 + P_+ [y(x_+) - \mu_Y]^3$$
$$= P_- [(4 - 2\ln x_-) - \mu_Y]^3 + P_0 [(4 - 2\ln x_0) - \mu_Y]^3 + P_+ [(4 - 2\ln x_+) - \mu_Y]^3$$
$$= (1/6) \times [(4 - 2\ln 0.0425) - 5.156]^3 + (2/3) \times [(4 - 2\ln 0.693) - 5.516]^3$$
$$+ (1/6) \times [(4 - 2\ln 3.179) - 5.156]^3$$
$$= (1/6) \times (0.316 - 5.156)^3 + (2/3) \times (4.733 - 5.156)^3$$
$$+ (1/6) \times (1.687 - 5.156)^3 = 15.892$$
$$\alpha_{3Y} = 15.892/2.562^3 = 0.945$$
$$\alpha_{4Y}\sigma_Y^4 = P_- [y(x_-) - \mu_Y]^4 + P_0 [y(x_0) - \mu_Y]^4 + P_+ [y(x_+) - \mu_Y]^4$$
$$= P_- [(4 - 2\ln x_-) - \mu_Y]^4 + P_0 [(4 - 2\ln x_0) - \mu_Y]^4 + P_+ [(4 - 2\ln x_+) - \mu_Y]^4$$
$$= (1/6) \times [(4 - 2\ln 0.0425) - 5.156]^4 + (2/3) \times [(4 - 2\ln 0.693) - 5.516]^4$$
$$+ (1/6) \times [(4 - 2\ln 3.179) - 5.156]^4$$
$$= (1/6) \times (0.316 - 5.156)^4 + (2/3) \times (4.733 - 5.156)^4$$
$$+ (1/6) \times (1.687 - 5.156)^4 = 142.393 \quad \alpha_{4Y} = 142.393/2.562^4 = 3.305$$

The results are listed in Table 3.1 for the three, five, and seven point estimates, along with the exact results for the purpose of comparison. From Table 3.1, it can be observed that the estimates with five and seven estimating points provide good approximations of the exact results.

Table 3.1 Comparison between the results of point estimates and exact results.

Moment	Point-estimate method in standard normal space			
	3 points	5 points	7 points	Exact
μ_Y	5.155 74	5.154 45	5.154 43	5.154 44
σ_Y	2.561 55	2.564 85	2.565 09	2.565 11
α_{3Y}	0.945 19	1.140 61	1.139 66	1.139 55
α_{4Y}	3.304 54	5.407 04	5.399 76	5.4

Example 3.13 Consider the following function of a random variable

$$Y = \sqrt{X} \tag{3.48}$$

where X is an lognormal random variable with $\mu_X = 100$, $\sigma_X = 10$, $\alpha_{3X} = 0.301$, and $\alpha_{4X} = 3.162$.

Applying Eq. (3.10) leads to the exact values of the first four moments of Y as

$$\mu_Y = \sqrt{\mu_X}(1 + V_X^2)^{-\frac{1}{8}} = 9.9876, V_Y^2 = (1 + V_X^2)^{\frac{1}{4}} - 1 = 0.00249, \sigma_Y = \mu_Y V_Y = 0.4984$$

$$\alpha_{3Y} = 3V_Y + V_Y^3 = 0.1498, \quad \alpha_{4Y} = 3 + 16V_Y^2 + 15V_Y^4 + 6V_Y^6 + V_Y^8 = 3.0399$$

Using the point estimates in the standard normal space, the first four moments of Y can be easily evaluated. The results obtained by using the three, five, and seven point estimates are listed in Table 3.2, along with the exact results for comparison. As can be observed from Table 3.2, the estimates with five and seven estimating points give good approximations of the exact results in this example.

3.4.5 General Expression of Estimating Points and Their Corresponding Weights

In order to obtain the point estimate in standard normal space with any number of points, re-express $Y = y(X)$ in terms of u and let

$$z(u) = \{y[T^{-1}(U)] - \mu_Y\}^k \tag{3.49}$$

According to the definition of the moments of $Y = y(X)$ and the concept of point estimate described above, Eqs.(3.31a)–(3.31c) can be formulated as the following equation.

$$\int z(u)\phi(u)du = \Sigma_{j=1}^m P_j z(u_j) \tag{3.50}$$

i.e.

$$\int z(u) \exp\left(-\frac{1}{2}u^2\right) du = \sqrt{2\pi}\Sigma_{j=1}^m P_j z(u_j) \tag{3.51}$$

Table 3.2 Comparison between the results of point estimates and exact results.

Moment	Point-estimate method in standard normal space			Exact
	3 points	5 points	7 points	
μ_Y	9.9876	9.9876	9.9876	9.9876
σ_Y	0.4984	0.4984	0.4984	0.4984
α_{3Y}	0.1493	0.1498	0.1498	0.1498
α_{4Y}	3.0074	3.0399	3.0399	3.0399

The left side of Eq. (3.51) is a Gaussian Type integration (Abramowitz and Stegun 1972). According to the principle of numerical integration, this type of integration can be approximately computed using the abscissas and weights of Hermite integration with weight function $\exp(-x^2/2)$ (Abramowitz and Stegun 1972).

$$\int z(u)\exp\left(-\frac{1}{2}u^2\right)du = \sum_{j=1}^{m} w_j z(u_j) \tag{3.52}$$

where u_j and w_j are the abscissas and weights of Hermite polynomials $H_{em}(u)$ with weight function $\exp(-x^2/2)$.

From the comparison between Eqs. (3.51) and (3.52), one can easily understand that the estimating points u_j is equal to the abscissas of the Hermite polynomials, i.e. u_j are the roots of the equation $H_{em}(u) = 0$.

Let $u = \sqrt{2}t$, Eq. (3.52) then becomes,

$$\int z\left(\sqrt{2}t\right)\exp\left(-t^2\right)dt = \frac{1}{\sqrt{2}}\sum_{j=1}^{m} w_j z\left(\sqrt{2}t_j\right) \tag{3.53}$$

The left side of Eq. (3.53) is a Gaussian integration with weight function of $\exp(-t^2)$, the integration can be computed using the following equation.

$$\int z\left(\sqrt{2}t\right)\exp\left(-t^2\right)dt = \sum_{j=1}^{m} w'_j z\left(\sqrt{2}t_j\right) \tag{3.54}$$

where t_j and w'_j are the abscissas and weights for Hermite polynomials $H_m(t)$ with weight function $\exp(-t^2)$. The weights corresponding to the abscissas with any number m are calculated as,

$$w'_j = \frac{2^{m-1}m!\sqrt{\pi}}{m^2 H_{m-1}^2(t_j)} \tag{3.55}$$

The comparisons among Eqs. (3.51), (3.52), and (3.54) lead to the relationship among P_j, w_j, and w'_j as following

$$P_j = \frac{1}{\sqrt{2\pi}}w_j = \frac{1}{\sqrt{\pi}}w'_j \tag{3.56}$$

Since

$$H_m(x) = \frac{(-1)^m}{\exp\left(-x^2\right)}\frac{d^m\left[\exp\left(-x^2\right)\right]}{dx^m}, H_{em}(u) = \frac{(-1)^m}{\phi(u)}\frac{d^m[\phi(u)]}{du^m} \tag{3.57}$$

The relationship between $H_m(t)$ and $H_{em}(u)$ can be expressed as

$$H_m(t) = 2^{m/2}H_{em}(u) \tag{3.58}$$

Substituting Eq. (3.58) into Eq. (3.55), the formula of P_j corresponding to u_j can be obtained as

$$P_j = \frac{m!}{m^2 H_{em-1}^2(u_j)} \tag{3.59}$$

Particularly, for odd m, since $He_{m-1}(0) = (m-2)!!$, the weight corresponding to $u_j = 0$ is easily obtained as

$$P_0 = \frac{(m-1)!!}{m!!} \tag{3.60}$$

For example, for $m = 3$, $P_0 = 2/3$; for $m = 5$, $P_0 = 8/15$; for $m = 7$, $P_0 = 16/35$; and for $m = 9$, $P_0 = 128/315$. One can see that P_0 decrease as m increases.

Using the method described above, the estimating points and corresponding weights up to $m = 21$ are listed in Table 3.3. The relationship between the weights and the PDF of a standard normal random variable with odd m and even m are shown in Figures 3.13 and 3.14, respectively.

3.4.6 Accuracy of the Point Estimate

Since there are various types of performance functions, it is quite difficult to investigate the accuracy of the point estimate for any type of performance function. Here we will focus on the accuracy of the point estimate according to the order of u in the performance function.

For a point estimate with m estimating points, the error of Eq. (3.54) can be expressed as (Abramowitz and Stegun 1972)

$$R_m = \frac{m!\sqrt{\pi}}{(2m)!} z^{(2m)}\left(\sqrt{2}\xi\right), \qquad -\infty < \xi < +\infty \tag{3.61}$$

where $z^{(2m)}\left(\sqrt{2}\xi\right)$ is the value of the $2m$th derivate of $z\left(\sqrt{2}t\right)$ in ξ, and ξ is an any real number.

According to Eq. (3.61), for a point estimate with m estimating points, since the $2m$th derivative of function $z(u) = u^{2m-1}$ is 0, the error of Eq. (3.54) will become 0. This implies that a point estimate with m estimating points is accurate for the mean of function $z(u) = u^{2m-1}$. Therefore, in order to estimate the mean of a function u^r or the rth moments of u, the number of estimating points should be $(r+1)/2$.

Generally speaking, in order to estimate the kth moment of a function u^r, the minimum number of estimating points will be

$$m = \frac{kr + 1}{2} \tag{3.62}$$

For example, to estimate the fourth moment of U^4, the number of estimating points should be 9; and to estimate the sixth moment of U^6, the number of estimating points should be 19. After the Rosenblatt transformation, it may be difficult to understand the exact order of U in the performance function, but one can understand that the order of U is generally higher with the increase of the nonlinearity of the performance function. Therefore, more effort is necessary to estimate the moment of performance function with very strong nonlinearity.

Table 3.3 Estimating points and weights for point estimates in standard normal space.

u_i	P_i
$n = 2$	
±1.000 000 00	$5.000\,000\,00 \times 10^{-1}$
$n = 3$	
0.000 000 00	$6.666\,666\,67 \times 10^{-1}$
±1.732 050 81	$1.666\,666\,67 \times 10^{-1}$
$n = 4$	
±0.741 963 78	$4.541\,241\,45 \times 10^{-1}$
±2.334 414 22	$4.587\,585\,48 \times 10^{-2}$
$n = 5$	
0.000 000 00	$5.333\,333\,33 \times 10^{-1}$
±1.355 626 18	$2.220\,759\,22 \times 10^{-1}$
±2.856 970 01	$1.125\,741\,13 \times 10^{-2}$
$n = 6$	
±0.616 706 59	$4.088\,284\,70 \times 10^{-1}$
±1.889 175 88	$8.861\,574\,60 \times 10^{-2}$
±3.324 257 43	$2.555\,784\,40 \times 10^{-3}$
$n = 13$	
0.000 000 00	$3.409\,923\,41 \times 10^{-1}$
±0.856 679 49	$2.378\,715\,23 \times 10^{-1}$
±1.725 418 38	$7.916\,895\,59 \times 10^{-2}$
±2.620 689 97	$1.177\,056\,05 \times 10^{-2}$
±3.563 444 38	$6.812\,363\,50 \times 10^{-4}$
±4.591 398 45	$1.152\,659\,65 \times 10^{-5}$
±5.800 167 25	$2.722\,627\,64 \times 10^{-8}$
$n = 14$	
±0.412 590 46	$3.026\,346\,27 \times 10^{-1}$
±1.242 688 96	$1.540\,833\,40 \times 10^{-1}$
±2.088 344 75	$3.865\,010\,88 \times 10^{-2}$
±2.963 036 58	$4.428\,919\,11 \times 10^{-3}$
±3.886 924 58	$2.003\,395\,54 \times 10^{-4}$
±4.896 936 40	$2.660\,991\,34 \times 10^{-6}$
±6.087 409 55	$4.868\,161\,26 \times 10^{-9}$
$n = 15$	
0.000 000 00	$3.182\,595\,18 \times 10^{-1}$
±0.799 129 07	$2.324\,622\,94 \times 10^{-1}$
$n = 19$	
0.000 000 00	$2.837\,731\,92 \times 10^{-1}$
±0.712 085 04	$2.209\,417\,12 \times 10^{-1}$
±1.428 876 68	$1.036\,036\,57 \times 10^{-1}$
±2.155 502 76	$2.866\,669\,10 \times 10^{-2}$
±2.898 051 28	$4.507\,235\,42 \times 10^{-3}$
±3.664 416 55	$3.785\,021\,09 \times 10^{-4}$
±4.465 872 63	$1.535\,114\,60 \times 10^{-5}$
±5.320 536 38	$2.532\,220\,03 \times 10^{-7}$
±6.262 891 16	$1.220\,370\,85 \times 10^{-9}$
±7.382 579 02	$7.482\,830\,05 \times 10^{-13}$
$n = 20$	
±0.346 964 16	$2.607\,930\,63 \times 10^{-1}$
±1.042 945 35	$1.617\,393\,34 \times 10^{-1}$
±1.745 247 32	$6.150\,637\,21 \times 10^{-2}$
±2.458 663 61	$1.399\,783\,74 \times 10^{-2}$
±3.189 014 82	$1.830\,103\,13 \times 10^{-3}$
±3.943 967 35	$1.288\,262\,80 \times 10^{-4}$
±4.734 581 33	$4.402\,121\,09 \times 10^{-6}$
±5.578 738 81	$6.127\,490\,26 \times 10^{-8}$

(Continued)

Table 3.3 (Continued)

u_i	P_i
$n = 7$	
0.000 000 00	$4.571\,428\,57 \times 10^{-1}$
±1.154 405 39	$2.401\,231\,79 \times 10^{-1}$
±2.366 759 41	$3.075\,712\,40 \times 10^{-2}$
±3.750 439 72	$5.482\,688\,56 \times 10^{-4}$
$n = 8$	
±0.539 079 81	$3.730\,122\,58 \times 10^{-1}$
±1.636 519 04	$1.172\,399\,08 \times 10^{-1}$
±2.802 485 86	$9.635\,220\,12 \times 10^{-3}$
±4.144 547 19	$1.126\,145\,38 \times 10^{-4}$
$n = 9$	
0.000 000 00	$4.063\,492\,06 \times 10^{-1}$
±1.023 255 66	$2.440\,975\,03 \times 10^{-1}$
±2.076 847 98	$4.991\,640\,68 \times 10^{-2}$
±3.205 429 00	$2.789\,141\,32 \times 10^{-3}$
±4.512 745 86	$2.234\,584\,40 \times 10^{-5}$
$n = 10$	
±0.484 935 71	$3.446\,423\,35 \times 10^{-1}$
±1.465 989 09	$1.354\,837\,03 \times 10^{-1}$
±2.484 325 84	$1.911\,158\,05 \times 10^{-2}$

u_i	P_i
$n = 7$ (continued)	
±1.606 710 07	$8.941\,779\,54 \times 10^{-2}$
±2.432 436 83	$1.736\,577\,45 \times 10^{-2}$
±3.289 082 42	$1.567\,357\,50 \times 10^{-3}$
±4.196 207 71	$5.642\,146\,41 \times 10^{-5}$
±5.190 093 59	$5.975\,419\,60 \times 10^{-7}$
±6.363 947 89	$8.589\,649\,90 \times 10^{-10}$
±6.510 590 16	$2.482\,062\,36 \times 10^{-10}$
±7.619 048 54	$1.257\,800\,67 \times 10^{-13}$
$n = 16$	
±0.386 760 60	$2.865\,685\,21 \times 10^{-1}$
±1.163 829 10	$1.583\,383\,73 \times 10^{-1}$
±1.951 980 35	$4.728\,475\,24 \times 10^{-2}$
±2.760 245 05	$7.266\,937\,60 \times 10^{-3}$
±3.600 873 62	$5.259\,849\,27 \times 10^{-4}$
±4.492 955 30	$1.530\,003\,22 \times 10^{-5}$
±5.472 225 71	$1.309\,473\,22 \times 10^{-7}$
±6.630 878 20	$1.497\,814\,72 \times 10^{-10}$
$n = 17$	
0.000 000 00	$2.995\,383\,70 \times 10^{-1}$
±0.751 842 60	$2.267\,063\,08 \times 10^{-1}$
±1.509 883 31	$9.740\,637\,12 \times 10^{-2}$
±2.281 019 44	$2.308\,665\,70 \times 10^{-2}$
±3.073 797 18	$2.858\,946\,06 \times 10^{-3}$

u_i	P_i
$n = 21$	
0.000 000 00	$2.702\,601\,83 \times 10^{-1}$
±0.678 045 69	$2.153\,337\,16 \times 10^{-1}$
±1.359 765 82	$1.083\,922\,86 \times 10^{-1}$
±2.049 102 47	$3.395\,272\,98 \times 10^{-2}$
±2.750 592 98	$6.439\,697\,05 \times 10^{-3}$
±3.469 846 69	$7.080\,477\,95 \times 10^{-4}$
±4.214 343 98	$4.219\,234\,74 \times 10^{-5}$
±4.994 963 94	$1.225\,354\,84 \times 10^{-6}$
±5.829 382 01	$1.450\,661\,28 \times 10^{-8}$
±6.751 444 72	$4.975\,368\,60 \times 10^{-11}$
±7.849 382 90	$2.098\,991\,22 \times 10^{-14}$
$n = 22$	
±0.331 179 32	$2.502\,435\,97 \times 10^{-1}$
±0.995 162 42	$1.619\,062\,93 \times 10^{-1}$
±1.664 248 4	$6.719\,631\,14 \times 10^{-2}$
±2.341 760 00	$1.756\,907\,29 \times 10^{-2}$
±3.032 404 23	$2.808\,761\,05 \times 10^{-3}$
±3.741 496 35	$2.622\,833\,03 \times 10^{-4}$

Table 3.3 (Continued)

u_i	P_i	u_i	P_i	u_i	P_i
±3.581 823 48	7.580 709 34 × 10⁻⁴	±3.900 065 72	1.684 914 32 × 10⁻⁴	±4.476 361 98	1.334 597 71 × 10⁻⁵
±4.859 462 83	4.310 652 63 × 10⁻⁶	±4.778 531 59	4.012 679 45 × 10⁻⁶	±5.247 724 43	3.319 853 75 × 10⁻⁷
$n = 11$		±5.744 460 08	2.808 016 12 × 10⁻⁸	±6.073 074 95	3.366 514 16 × 10⁻⁹
0.000 000 00	3.694 083 69 × 10⁻¹	±6.889 122 44	2.584 314 92 × 10⁻¹¹	±6.985 980 42	9.841 378 98 × 10⁻¹²
±0.928 869 00	2.422 403 00 × 10⁻¹	$n = 18$		±8.074 029 98	3.479 460 65 × 10⁻¹⁵
±1.876 035 02	6.613 874 61 × 10⁻²	±0.365 245 76	2.727 832 35 × 10⁻¹	$n = 23$	
±2.865 123 16	6.720 285 24 × 10⁻³	±1.098 395 52	1.606 853 04 × 10⁻¹	0.000 000 00	2.585 097 41 × 10⁻¹
±3.936 166 61	1.956 719 30 × 10⁻⁴	±1.839 779 92	5.489 663 25 × 10⁻²	±0.648 471 15	2.099 596 70 × 10⁻¹
±5.188 001 22	8.121 849 79 × 10⁻⁷	±2.595 833 69	1.051 651 78 × 10⁻²	±1.299 876 47	1.120 733 83 × 10⁻¹
$n = 12$		±3.374 736 54	1.065 484 80 × 10⁻³	±1.957 327 55	3.886 718 37 × 10⁻²
±0.444 403 00	3.216 643 62 × 10⁻¹	±4.188 020 23	5.179 896 14 × 10⁻⁵	±2.624 232 63	8.579 678 39 × 10⁻³
±1.340 375 20	1.469 670 48 × 10⁻¹	±5.054 072 69	1.021 552 40 × 10⁻⁶	±3.305 040 02	1.167 628 64 × 10⁻³
±2.259 464 45	2.911 668 79 × 10⁻²	±6.007 745 91	5.905 488 48 × 10⁻⁹	±4.004 775 32	9.340 818 61 × 10⁻⁵
±3.223 709 83	2.203 380 69 × 10⁻³	±7.139 464 85	4.416 588 77 × 10⁻¹²	±4.730 724 20	4.089 977 24 × 10⁻⁶
±4.271 825 85	4.837 184 92 × 10⁻⁵			±5.493 473 99	8.775 062 48 × 10⁻⁸
±5.500 901 70	1.499 927 17 × 10⁻⁷			±6.310 349 85	7.670 888 86 × 10⁻¹⁰
				±7.214 659 44	1.922 935 31 × 10⁻¹²
				±8.293 386 03	5.732 383 17 × 10⁻¹⁶

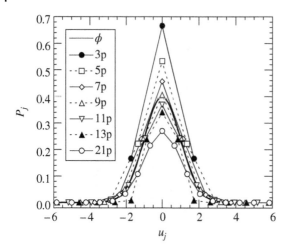

Figure 3.13 Estimating points and weights of odd m.

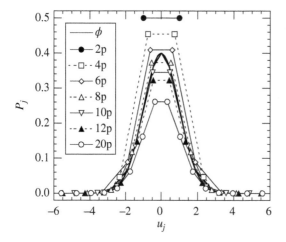

Figure 3.14 Estimating points and weights of even m.

Example 3.14 Consider the following function of a lognormal random variable

$$Y = X^2 \tag{3.63}$$

where X is a lognormal random variable with mean $\mu_X = 1$.

When the coefficient of variance of X, V_X is 0.2, the mean and standard deviation of $\ln(X)$ are $\lambda = -0.0196$ and $\zeta = 0.198$, respectively. The inverse Rosenblatt transformation is simply expressed as,

$$X = \exp\left(\zeta U + \lambda\right) = \exp\left(0.198U - 0.0196\right) \tag{3.64}$$

From Table 3.3, the estimating points in standard normal space for $m = 5$ are $u_1 = -2.856\,97$, $u_2 = -1.355\,63$, $u_3 = 0$, $u_4 = 1.355\,63$, and $u_5 = 2.856\,97$ with corresponding weights of $P_1 = 0.011\,257$, $P_2 = 0.222\,08$, $P_3 = 0.5333$, $P_4 = 0.222\,08$, and $P_5 = 0.011\,257$. Using Eq. (3.64), the estimating points in original space are obtained as $x_1 = 0.5569$, $x_2 = 0.7497$, $x_3 = 0.9806$, $x_4 = 1.2826$, and $x_5 = 1.7267$; and the corresponding function values are obtained as 0.3101, 0.5621, 0.9616, 1.6451, and 2.9815. Substituting these function values and the corresponding weights into Eqs. (3.35a)–(3.35c), the first six moments of y are obtained as $\mu = 1.04$, $\sigma = 0.4286$, $\alpha_3 = 1.3037$, $\alpha_4 = 6.0577$, $\alpha_5 = 22.159$, and $\alpha_6 = 99.652$.

Using V_X as a parameter, the variations of the point estimated first six moments are shown in Figures 3.15a–f. From these figures, it can be observed that:

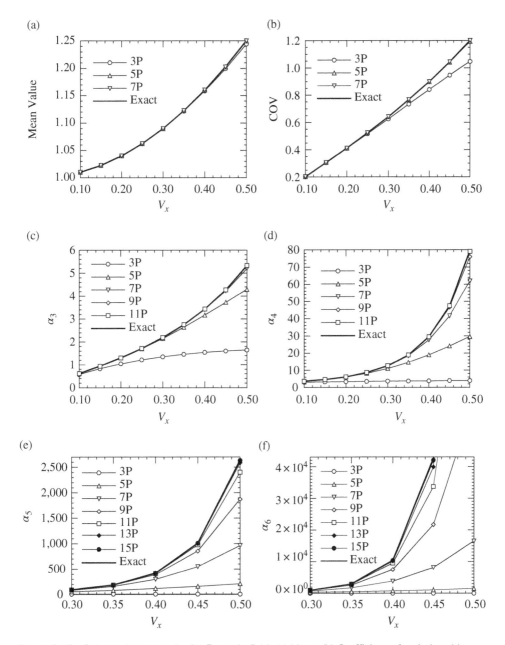

Figure 3.15 Point estimate results for Example 3.14. (a) Mean; (b) Coefficient of variation; (c) Skewness; (d) Kurtosis; (e) The fifth moment; (f) The sixth moment.

1) Since the nonlinearity of the function increases with the increase of V_X, the accuracy of point estimate decreases with the increase of V_X. The increase of the number of estimating points provides good improvement of the estimation.

2) The necessary number of estimate points increases with the increase of the order of the moment. For this example, five- and seven-point estimates are adequate for estimating the mean and standard deviation. For estimating the third- and fourth-order moments, 11 estimating points will be required, and for fifth and sixth moments, 15 estimating points will be required.

3.5 Point Estimates for a Function of Multiple Variables

3.5.1 General Expression of Point Estimates for a Function of *n* Variables

The procedure for a single random variable described above can be generalised for a function of many variables. For a function $G = g(X,Y)$, using a two-point estimate for each variable, let the concentrations be $x_-, x_+, y_-,$ and y_+, as shown in Figures 3.16 and 3.17a, the kth central moment of a function $G = g(X, Y)$, can be calculated by

$$\mu_G = P_- P_- g(x_-, y_-) + P_- P_+ g(x_-, y_+) + P_+ P_- g(x_+, y_-) + P_+ P_+ g(x_+, y_+)$$
(3.65a)

$$\sigma_G^2 = P_- P_- [g(x_-, y_-) - \mu_G]^2 + P_- P_+ [g(x_-, y_+) - \mu_G]^2$$
$$+ P_+ P_- [g(x_+, y_-) - \mu_G]^2 + P_+ P_+ [g(x_+, y_+) - \mu_G]^2$$
(3.65b)

$$\sigma_G^k \alpha_{kG} = P_- P_- [g(x_-, y_-) - \mu_G]^k + P_- P_+ [g(x_-, y_+) - \mu_G]^k$$
$$+ P_+ P_- [g(x_+, y_-) - \mu_G]^k + P_+ P_+ [g(x_+, y_+) - \mu_G]^k$$
(3.65c)

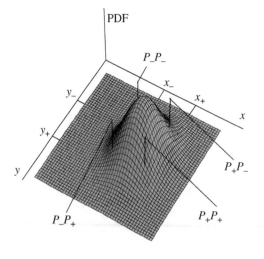

Figure 3.16 Concentrations of two dimensional PDF using two points.

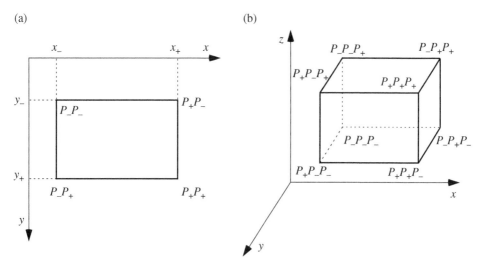

Figure 3.17 Concentrations of two and three dimensional PDFs using two points. (a) Two-dimensional PDF. (b) Three-dimensional PDF.

For a function $G = g(X, Y, Z)$, using two-point estimate for each variable, let the concentrations be $x_-, x_+, y_-, y_+, z_-,$ and z_+; and as shown in Figure 3.17b, the kth central moment of a function $G = g(X, Y, Z)$, can be calculated by formulas similar to Eqs. (3.65a)–(3.65c). Generally, for $Z = G(\mathbf{X})$, where $\mathbf{X} = (x_1, x_2, ..., x_n)$, the joint probability density is assumed to be concentrated at points in the m^n hyperquadrants of the space defined by the n random variables, where m is the number of estimating points used in the point estimates for functions of single random variables. Using the standard point estimate, the first k moments are estimated as

$$\mu_G = \sum \Pi_{i=1}^n P_{ci} \left\{ G\left[T^{-1}(u_{c1}, u_{c2}, ..., u_{cn})\right] \right\} \tag{3.66a}$$

$$\sigma_G^2 = \sum \Pi_{i=1}^n P_{ci} \left\{ G\left[T^{-1}(u_{c1}, u_{c2}, ..., u_{cn})\right] - \mu_G \right\}^2 \tag{3.66b}$$

$$\alpha_{kG}\sigma_G^k = \sum \Pi_{i=1}^n P_{ci} \left\{ G\left[T^{-1}(u_{c1}, u_{c2}, ..., u_{cn})\right] - \mu_G \right\}^k \tag{3.66c}$$

where n is the number of random variables, c is a combination of n items from a group $[1, 2, ..., m]$, m is the number of estimating points, ci is the ith iterm of c, u_{ci} is the cith estimating point, and P_{ci} is the weight corresponding to u_{ci}.

The combination of n items from a group $[1, 2, ..., m]$ can be done by using nested Fortran loops (Gentleman 1975). Since there are m^n combinations of n items from a group of m items, m^n points will be required for Eqs. (3.66a)–(3.66c). For $m = 5$, $n = 2$; 25 points have to be used, as shown in Figure 3.18. For $m = 5$, $n = 6$; more than 10^4 points have to be used. One can see that the computation becomes excessive when n is large. Therefore, Eqs. (3.66a)–(3.66c) are often applied to problems with quite a small number of random variables.

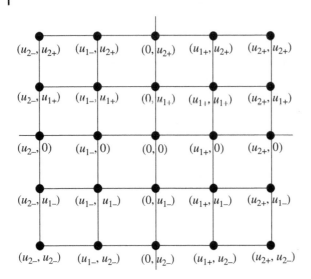

Figure 3.18 Estimating points in standard normal space for $m = 5$, $n = 2$.

3.5.2 Approximate Point Estimates for a Function of n Variables

Some approximations were developed to avoid excessive computation when n takes large values. The first route is to approximate the performance function $G(\mathbf{X})$. When the random variables are mutually independent, Rosenblueth (1975) approximates $G(\mathbf{X})$ by the following function

$$Z = G'(\mathbf{X}) = G^* \Pi_{i=1}^n \left(\frac{Z_i}{G^*} \right)$$

where G^* is the function evaluated at the mean of all the variables, and Z_i are the functions computed as though x_i was the only random variable, with the other variables set equal to their mean. It has been shown that there are significant errors in this approximation (Zhao and Ono 2000a).

In this section, the function $G(\mathbf{X})$ is approximated by the following function (Zhao and Ono 2000a):

$$Z = G'(\mathbf{X}) = \sum_{i=1}^n (G_i - G_\mu) + G_\mu \tag{3.67a}$$

where

$$G_\mu = G(\boldsymbol{\mu}) \tag{3.67b}$$

$$G_i = G[T^{-1}(U_i)] \tag{3.67c}$$

where $\boldsymbol{\mu} = [\mu_1, \mu_2, ..., \mu_n]^T$ represents the vector in which all the random variables take their mean; and $\mathbf{U}_i = [u_{\mu 1}, u_{\mu 2}, ..., u_{\mu i-1}, u_i, u_{\mu i+1}, ..., u_{\mu n}]^T$, where $u_{\mu k}$, $k = 1, ..., n$ except i, is the kth value of u_μ, which is the vector in u-space corresponding to $\boldsymbol{\mu}$. G_μ is a constant and G_i is a function of only u_i. $T^{-1}(\cdot)$ is the inverse Rosenblatt transformation.

Note that u_i, $i = 1, ..., n$ are independent and G_i is a function of only u_i; therefore, G_i, $i = 1, ..., n$, are also independent of each other. Hence, the first four moments of $G'(\mathbf{X})$ of Eq. (3.67a) can be expressed as

$$\mu_G = \sum_{i=1}^{n} (\mu_i - G_\mu) + G_\mu \tag{3.68a}$$

$$\sigma_G^2 = \sum_{i=1}^{n} \sigma_i^2 \tag{3.68b}$$

$$\alpha_{3G}\sigma_G^3 = \sum_{i=1}^{n} \alpha_{3i}\sigma_i^3 \tag{3.68c}$$

$$\alpha_{4G}\sigma_G^4 = \sum_{i=1}^{n} \alpha_{4i}\sigma_i^4 + 6\sum_{i=1}^{n}\sum_{j>i}^{n} \sigma_i^2\sigma_j^2 \tag{3.68d}$$

where μ_i, σ_i, α_{3i}, and α_{4i} are the mean, standard deviation, skewness, and kurtosis of G_i, respectively.

Since G_i is a function of single standard normal random variable u_i, the first four moments μ_i, σ_i, α_{3i}, and α_{4i} can be point-estimated from Eqs. (3.35a)–(3.35c). For a performance function $G(\mathbf{X})$ with n variables, if the probability moments of G_i are estimated using m-point estimate, only $mn + 1$ function calls of $G(\mathbf{X})$ are required for estimating the first few moments of $G(\mathbf{X})$. The distribution of estimating points for $m = 5$, $n = 2$ is demonstrated in Figure 3.19a. Note that since $\mathbf{U} = T(\mathbf{X})$ are mutually independent, the assumption that $\mathbf{X}$ should be mutually independent is not required in Eq. (3.67c).

In particular, for independent random variables $\mathbf{X}$, G_i can simply be expressed as (Ono and Idota 1986)

$$G_i = G(\mu_1, \mu_2, \cdots, \mu_{i-1}, x_i, \mu_{i+1}, \cdots, \mu_n) \tag{3.69}$$

(a) (b)

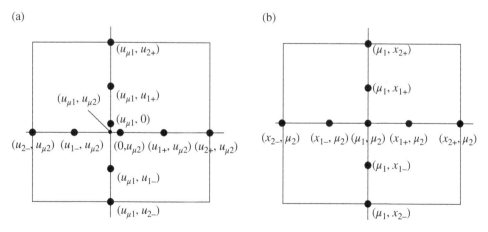

Figure 3.19 Estimating points for $m = 5$, $n = 2$ using the mean of random variables. (a) In standard normal space. (b) In original space.

Then G_i is a function of single random variable x_i, its first four moments μ_i, σ_i, α_{3i}, and α_{4i} can be obtained using the point estimates presented in the previous section. Since $\boldsymbol{\mu} = [\mu_1, \mu_2,..., \mu_n]^T$ is a common point to estimate the moments of all G_i, $i = 1, 2, ..., n$, only $(m - 1)$ $n + 1$ points are required when using an m-point point estimate for the moments of G_i. For $m = 3$, $2n + 1$ points are required. The distribution of estimating points for $m = 5$, $n = 2$ is demonstrated in Figure 3.19b.

The approximation in Eq. (3.67a) may be considered as a generalisation of the following: If $G(\mathbf{X})$ is of the form

$$G(\mathbf{X}) = \sum_{i=1}^{n} a_i x_i \text{ or } G(\mathbf{X}) = \sum_{i=1}^{n} y_i(x_i)$$

where a_i is a constant and y_i is an arbitrary function of x_i, Eq. (3.67a) will become exact, i.e. $G'(\mathbf{X}) = G(\mathbf{X})$.

Example 3.15 Derivation of Eq. (3.67a)

Equation (3.67a) can be derived as follows (Lu et al. 2014):

Ignoring the cross term of derivatives, the Taylor series of $G(\mathbf{X})$ at the mean is expressed as

$$G(\mathbf{X}) = G(\boldsymbol{\mu}) + \sum_{i=1}^{n} \sum_{j=1}^{\infty} \frac{1}{j!} \frac{\partial^j G(\mathbf{X})}{\partial x_i^j} \Big|_{\mathbf{X}=\boldsymbol{\mu}} (x_i - \mu_i)^j \tag{3.70}$$

For the ith random variable x_i, by introducing a function G_i expressed as Eq. (3.69), in which all random variables take their mean expect for x_i, then the Taylor expansion of G_i can be rewritten as

$$G_i = G_i(\mu_i) + \sum_{j=1}^{\infty} \frac{1}{j!} \frac{d^j G_i}{dx_i^j} \Big|_{x_i=\mu_i} (x_i - \mu_i)^j \tag{3.71}$$

That is

$$\sum_{j=1}^{\infty} \frac{1}{j!} \frac{d^j G_i}{dx_i^j} \Big|_{x_i=\mu_i} (x_i - \mu_i)^j = G_i - G_i(\mu_i) \tag{3.72}$$

Since

$$G_i(\mu_i) = G(\boldsymbol{\mu}), \frac{d^j G_i}{dx_i^j} \Big|_{x_i=\mu_i} = \frac{\partial^j G(\mathbf{X})}{\partial x_i^j} \Big|_{\mathbf{X}=\boldsymbol{\mu}}$$

Equation (3.72) can also be expressed as

$$\sum_{j=1}^{\infty} \frac{1}{j!} \frac{\partial^j G(\mathbf{X})}{\partial x_i^j} \Big|_{\mathbf{X}=\boldsymbol{\mu}} (x_i - \mu_i)^j = G_i - G(\boldsymbol{\mu}) \tag{3.73}$$

Substituting Eq. (3.73) into Eq. (3.70) leads to

$$G(\mathbf{X}) = G(\boldsymbol{\mu}) + \sum_{i=1}^{n} \sum_{j=1}^{\infty} \frac{1}{j!} \frac{\partial^j G(\mathbf{X})}{\partial x_i^j} \Big|_{\mathbf{X}=\boldsymbol{\mu}} (x_i - \mu_i)^j = G(\boldsymbol{\mu}) + \sum_{i=1}^{n} [G_i - G(\boldsymbol{\mu})] \tag{3.74}$$

Therefore, Eq. (3.67a) is the first order approximation with ignoring the cross terms of derivatives.

Example 3.16 Consider the simple performance function in Example 3.3 again. G_i and G_μ in Eq. (3.67a) are readily derived as,

$$G_1 = D\mu_R - \mu_S = 100D - 50, G_2 = \mu_D R - \mu_S = R - 50, G_3 = \mu_D\mu_R - S = 100 - S, G_\mu$$

$$= \mu_D\mu_R - \mu_S = 1 \times 100 - 50 = 50$$

Using Gorman's three-point estimate for the first four moments of each single variable function G_i, the points and weights are given as follows.

For G_1;

$$\theta = \left(4\alpha_{4D} - 3\alpha_{3D}^2\right)^{1/2} = 2\sqrt{3} = 3.464$$

$$D_- = \mu_D - \frac{\sigma_D}{2}(\theta - \alpha_{3D}) = 1 - (0.1/2) \times 3.464 = 0.8268,$$

$$P_- = \frac{1}{2}\left(\frac{1 + \alpha_{3D}/\theta}{\alpha_{4D} - \alpha_{3D}^2}\right) = 1/6 = 0.167$$

$$D_+ = \mu_D + \frac{\sigma_D}{2}(\theta + \alpha_{3D}) = 1 + (0.1/2) \times 3.464 = 1.1732,$$

$$P_+ = \frac{1}{2}\left(\frac{1 + \alpha_{3D}/\theta}{\alpha_{4D} - \alpha_{3D}^2}\right) = 1/6 = 0.167$$

$$D_0 = \mu_D = 1, P_0 = 1 - \frac{1}{\alpha_{4D} - \alpha_{3D}^2} = 2/3 = 0.667$$

$$\mu_1 = P_- G_1(D_-) + P_0 G_1(D_0) + P_+ G_1(D_+) = 50$$

$$\sigma_1^2 = P_- [G_1(D_-) - \mu_1]^2 + P_0[G_1(D_0) - \mu_1]^2 + P_+ [G_1(D_+) - \mu_1]^2 = 100, \sigma_1 = 10$$

$$\alpha_{31} = \frac{1}{\sigma_1^3}\left\{P_- [G_1(D_-) - \mu_1]^3 + P_0[G_1(D_0) - \mu_1]^3 + P_+ [G_1(D_+) - \mu_1]^3\right\} = 0$$

$$\alpha_{41} = \frac{1}{\sigma_1^4}\left\{P_- [G_1(D_-) - \mu_1]^4 + P_0[G_1(D_0) - \mu_1]^4 + P_+ [G_1(D_+) - \mu_1]^4\right\} = 3$$

Similarly for G_2;
$\theta = 3.681$

$$R_- = 69.29, \quad P_- = 0.177, R_0 = 100, \quad P_0 = 0.696, R_+ = 142.9, P_+ = 0.127$$

$\mu_2 = 50, \sigma_2 = 20, \alpha_{32} = 0.608, \alpha_{42} = 3.664$
Similarly for G_3;
$\theta = 4.369$

$$S_- = 18.95, P_- = 0.147, S_0 = 50, P_0 = 0.771, S_+ = 106.3, P_+ = 0.081$$

$\mu_3 = 50, \sigma_3 = 20, \alpha_{33} = -1.264, \alpha_{43} = 5.969$
Using Eqs. (3.68a)–(3.68d), the first four moments of the performance function are obtained as

$$\mu_G = \sum_{i=1}^{n}(\mu_i - G_\mu) + G_\mu = (50-50) + (50-50) + (50-50) + 50 = 50$$

$$\sigma_G^2 = \sum_{i}^{n}\sigma_i^2 = 10^2 + 20^2 + 20^2 = 900, \sigma_G = 30$$

$$\alpha_{3G} = \frac{1}{\sigma_G^3}\sum_{i=1}^{n}\sigma_i^3\alpha_{3i} = \frac{1}{30^3}\left[10^3 \times 0 + 20^3 \times 0.608 + 20^3 \times (-1.264)\right] = -0.194$$

$$\alpha_{4G} = \frac{\sum_{i=1}^{n}\sigma_i^4\alpha_{4i} + 6\sum_{i=1}^{n-1}\sum_{j>i}^{n}\sigma_i^2\sigma_j^2}{\sigma_G^4}$$

$$= \frac{1}{30^4}\left[10^4 \times 3 + 20^4 \times 3.664 + 20^4 \times 5.969 + 6(10^2 \times 20^2 + 10^2 \times 20^2 + 20^2 \times 20^2)\right]$$

$$= 3.718$$

It can be observed that the estimated mean and standard deviation are very close to the exact results ($\mu_G = 50$ and $\sigma_G = 30.067$) from Example 3.3, while the estimated skewness and kurtosis are slightly different from the exact results ($\alpha_{3G} = -0.09927$ and $\alpha_{4G} = 3.821$). This can be attributed to: (i) only three points are used above, and more points are needed to improve the accuracy; (ii) the approximation of Eq. (3.67a) may produce unneglectable error when applied to the performance function with cross-term variables.

Example 3.17   Consider the following simple performance function:

$$G(\mathbf{X}) = X_1^2 + \ln(X_1) - \sqrt{X_2}$$

where X_1 is a lognormal random variable with $\mu_1 = 20$ and $\sigma_1 = 2$; and X_2 is a Weibull random variable with $\mu_2 = 40\,000$ and $\sigma_2 = 16\,000$. G_μ and G_i in Eq. (3.67a) can be written as

$$G(\boldsymbol{\mu}) = \mu_1^2 + \ln(\mu_1) - \sqrt{\mu_2} = 203$$

$$G_1 = X_1^2 + \ln(X_1) - \sqrt{\mu_2} = X_1^2 + \ln(X_1) - 200$$

$$G_2 = \mu_1^2 + \ln(\mu_1) - \sqrt{X_2} = 403 - \sqrt{X_2}$$

Using seven-point estimate in standardised normal space leads to the following estimates for the first four moments of each single variable function G_i.

For G_1;
$\mu_1 = 206.99$, $\sigma_1 = 81.51$, $\alpha_{31} = 0.612$, $\alpha_{41} = 3.673$
and for G_2;
$\mu_2 = 207.4$, $\sigma_2 = 41.8$, $\alpha_{32} = 0.305$, $\alpha_{42} = 2.941$
Using Eqs. (3.68a)–(3.68d), the first four moments of the performance function are obtained as $\mu_G = 211.411$, $\sigma_G = 91.606$, $\alpha_{3G} = 0.460$, and $\alpha_{4G} = 3.419$.

3.5.3 Dimension Reduction Integration

An alternative method for evaluating the first few moments of a performance function is the generalised multivariate dimension-reduction method proposed by Xu and Rahman (2004), in which the n-dimensional performance function is approximated by the summation of a series of, at most, D-dimensional functions ($D < n$).

Here, the bivariate- and one-dimension reduction ($D = 2$ and 1) are introduced. For a performance function $Z = G(\mathbf{X})$, using the inverse Rosenblatt transformation, the kth moments about zero of Z can be defined as (Zhao and Ono 2000a)

$$\mu_{kG} = E\left\{[G(\mathbf{X})]^k\right\} = \int_{-\infty}^{\infty} \cdots \int_{-\infty}^{\infty} \{G[\mathbf{x}]\}^k f_X(\mathbf{x}) d\mathbf{x} = \int_{-\infty}^{\infty} \cdots \int_{-\infty}^{\infty} \left\{G[T^{-1}(\mathbf{u})]\right\}^k \phi(\mathbf{u}) d\mathbf{u} \tag{3.75}$$

Let $L(\mathbf{u}) = G[T^{-1}(\mathbf{u})]$ in Eq. (3.75). By the bivariate dimension reduction method (Xu and Rahman 2004)

$$L(\mathbf{u}) \cong L_2 - (n-2)L_1 + \frac{(n-1)(n-2)}{2}L_0 \tag{3.76}$$

where

$$L_0 = L(0, ..., 0, ..., 0) \tag{3.77a}$$

$$L_1 = \sum_{i=1}^{n} L_i \tag{3.77b}$$

$$L_2 = \sum_{i<j} L_{ij} \tag{3.77c}$$

$$L_i = L(0, ..., U_i, ..., 0) = L_i(U_i) \tag{3.77d}$$

$$L_{ij} = L(0, ..., U_i, ..., U_j, ..., 0) = L_{ij}(U_i, U_j) \tag{3.77e}$$

where $i, j = 1, 2, ..., n$ and $i < j$. It is noted that L_1 is a summation of n one-dimensional functions L_i and L_2 is a summation of $[n(n-1)]/2$ two-dimensional functions L_{ij}.

Substituting Eq. (3.76) into Eq. (3.75) helps reduce the n-dimensional integral of Eq. (3.75) into a summation of, at most, two-dimensional integrals

$$\mu_{kG} = E\left(\left\{G[T^{-1}(\mathbf{U})]\right\}^k\right) \cong E\left([L(\mathbf{U})]^k\right) = \sum_{i<j} \mu_{k-L_{ij}} - (n-2)\sum_{i} \mu_{k-L_i}$$

$$+ \frac{(n-1)(n-2)}{2} L_0^k \tag{3.78}$$

where

$$\mu_{k-L_{ij}} = \int_{-\infty}^{\infty} \int_{-\infty}^{\infty} \left[L_{ij}(u_i, u_j)\right]^k \phi(u_i) \phi(u_j) du_i du_j \tag{3.79a}$$

$$\mu_{k-L_i} = \int_{-\infty}^{\infty} [L_i(u_i)]^k \phi(u_i) du_i \tag{3.79b}$$

$$L_0^k = [L(0, ..., 0, ..., 0)]^k \tag{3.79c}$$

Using the Gauss-Hermite integration, the one-dimensional integral in Eq. (3.79b) can be approximated by the following equation.

$$\mu_{k-L_i} = \sum_{r=1}^{m} P_r [L_i(u_{ir})]^k \tag{3.80}$$

The estimating points u_{ir} and the corresponding weights P_r can be readily obtained from Table 3.3.

Similarly, the two-dimensional integral in Eq. (3.79a) can be approximated by

$$\mu_{k-L_{ij}} = \sum_{r_1=1}^{m} \sum_{r_2=1}^{m} P_{r_1} P_{r_2} \left[L_{ij}(u_{ir1}, u_{ir2}) \right]^k \tag{3.81}$$

Since the performance function is generally expressed in original space, it is more convenient and efficient to use the following equations instead of Eqs. (3.77a) through (3.81).

$$L_0 = G(\mu_1, ..., \mu_i, ..., \mu_n) \tag{3.82a}$$

$$L_i = G(m_1, ..., X_i, ..., m_n) = G_i(X_i) \tag{3.82b}$$

$$L_{ij} = G(m_1, ..., x_i, ..., x_j, ..., m_n) = G_{ij}(X_i, X_j) \tag{3.82c}$$

$$L_0^k = \left[G_\mu(\mu_1, ..., \mu_i, ..., \mu_n) \right]^k \tag{3.83}$$

$$\mu_{k-L_i} = \sum_{r=1}^{m} P_r \left\{ G_i \left[T^{-1}(u_{ir}) \right] \right\}^k \tag{3.84a}$$

$$\mu_{k-L_{ij}} = \sum_{r_1=1}^{m} \sum_{r_2=1}^{m} P_{r_1} P_{r_2} \left\{ G_{ij} \left[T^{-1}(u_{ir1}), T^{-1}(u_{jr2}) \right] \right\}^k \tag{3.84b}$$

Finally, the mean, standard deviation, skewness, and kurtosis of a performance function $G(\mathbf{X})$ with n random variables can be obtained as

$$\mu_G = \mu_{1G} \tag{3.85a}$$

$$\sigma_G = \sqrt{\mu_{2G} - \mu_{1G}^2} \tag{3.85b}$$

$$\alpha_{3G} = \left(\mu_{3G} - 3\mu_{2G}\mu_{1G} + 2\mu_{1G}^3 \right) / \sigma_G^3 \tag{3.85c}$$

$$\alpha_{4G} = \left(\mu_{4G} - 4\mu_{3G}\mu_{1G} + 6\mu_{2G}\mu_{1G}^2 - 3\mu_{1G}^4 \right) / \sigma_G^4 \tag{3.85d}$$

Using the one-dimension reduction method (Rahman and Xu 2004), Eq. (3.76) reduces to

$$L(\mathbf{u}) \cong L_1 - (n-1)L_0 \tag{3.86}$$

which is essentially the same as Eq. (3.67a).

Example 3.18 Consider again the simple performance function in Example 3.3. L_0, L_i, and L_{ij} composed in Eqs. (3.82a)–(3.82c) are easily understood as,

$$L_0 = \mu_D \mu_R - \mu_S = 1 \times 100 - 50 = 50$$

$$L_1 = D\mu_R - \mu_S = 100D - 50, L_2 = \mu_D R - \mu_S = R - 50, L_3 = \mu_D \mu_R - S = 100 - S$$

$$L_{1,2} = DR - \mu_S = DR - 50, L_{1,3} = D\mu_R - S = 100D - S, L_{2,3} = \mu_D R - S = R - S$$

Using Gorman's three-point estimate for the first four moments of each single variable function L_i, the points and weights are given as

$$D_- = 0.8268, P_- = 0.167, D_0 = 1, P_0 = 0.667, D_+ = 1.1732, P_+ = 0.167$$

$$R_- = 69.29, P_- = 0.177, R_0 = 100, P_0 = 0.696, R_+ = 142.9, P_+ = 0.127$$

$$S_- = 18.95, P_- = 0.147, S_0 = 50, P_0 = 0.771, S_+ = 106.3, P_+ = 0.081$$

For L_1, according to Eq. (3.80);

$$\mu_{1-L_1} = \sum_{r=1}^{m} P_r \{G_1 [T^{-1}(u_{1r})]\} = P_- G_1(D_-) + P_0 G_1(D_0) + P_+ G_1(D_+)$$

$$= 0.167 \times (100 \times 0.8268 - 50) + 0.667 \times (100 \times 1 - 50)$$

$$+ 0.167 \times (100 \times 1.1732 - 50) = 50.05$$

$$\mu_{2-L_1} = \sum_{r=1}^{m} P_r \{G_1 [T^{-1}(u_{1r})]\}^2 = P_- [G_1(D_-)]^2 + P_0 [G_1(D_0)]^2 + P_+ [G_1(D_+)]^2$$

$$= 0.167 \times (100 \times 0.8268 - 50)^2 + 0.667 \times (100 \times 1 - 50)^2$$

$$+ 0.167 \times (100 \times 1.1732 - 50)^2 = 2602.694$$

$$\mu_{3-L_1} = \sum_{r=1}^{m} P_r \{G_1 [T^{-1}(u_{1r})]\}^3 = 140154.118$$

$$\mu_{4-L_1} = \sum_{r=1}^{m} P_r \{G_1 [T^{-1}(u_{1r})]\}^4 = 7789218.297$$

Similarly for L_2;

$$\mu_{1-L_2} = 50.01, \mu_{2-L_2} = 2901.9245, \mu_{3-L_2} = 190094.6525, \mu_{4-L_2} = 13833972.73$$

Similarly for L_3;

$$\mu_{1-L_3} = 49.954, \quad \mu_{2-L_3} = 2896.373, \mu_{3-L_3} = 174621.3326, \mu_{4-L_3} = 11162384.42$$

For $L_{1,2}$, according to Eq. (3.81);

$$\mu_{1-L_{1,2}} = \sum_{r_1=1}^{m} \sum_{r_2=1}^{m} P_{r_1} P_{r_2} \{G_{1,2} [T^{-1}(u_{1r1}), T^{-1}(u_{2r2})]\}$$

$$= 0.167 \times [0.177 \times (0.8268 \times 69.29 - 50) + 0.696 \times (0.8268 \times 100 - 50)$$

$$+ 0.127 \times (0.8268 \times 142.9 - 50)] + 0.667 \times [0.177 \times (1 \times 69.29 - 50)$$

$$+ 0.696 \times (1 \times 100 - 50) + 0.127 \times (1 \times 142.9 - 50)]$$

$$+ 0.167 \times [0.177 \times (1.1732 \times 69.29 - 50)$$

$$+ 0.696 \times (1.1732 \times 100 - 50) + 0.127 \times (1.1732 \times 142.9 - 50)] = 50.063$$

$$\mu_{2-L_{1,2}} = \sum_{r_1=1}^{m} \sum_{r_2=1}^{m} P_{r_1} P_{r_2} \left\{ G_{1,2} \left[T^{-1}(u_{1r1}), T^{-1}(u_{2r2}) \right] \right\}^2$$

$$= 0.167 \times \left[0.177 \times (0.8268 \times 69.29 - 50)^2 + 0.696 \times (0.8268 \times 100 - 50)^2 \right.$$
$$+ 0.127 \times (0.8268 \times 142.9 - 50)^2 \right] + 0.667 \times \left[0.177 \times (1 \times 69.29 - 50)^2 \right.$$
$$+ 0.696 \times (1 \times 100 - 50)^2 + 0.127 \times (1 \times 142.9 - 50)^2 \right] + 0.167$$

$$\times \left[0.177 \times (1.1732 \times 69.29 - 50)^2 + 0.696 \times (1.1732 \times 100 - 50)^2 \right.$$
$$+ 0.127 \times (1.1732 \times 142.9 - 50)^2 \right] = 3009.06$$

$$\mu_{3-L_{1,2}} = \sum_{r_1=1}^{m} \sum_{r_2=1}^{m} P_{r_1} P_{r_2} \left\{ G_{1,2} \left[T^{-1}_*(u_{1r1}), T^{-1}(u_{2r2}) \right] \right\}^3 = 208480$$

$$\mu_{4-L_{1,2}} = \sum_{r_1=1}^{m} \sum_{r_2=1}^{m} P_{r_1} P_{r_2} \left\{ G_{1,2} \left[T^{-1}(u_{1r1}), T^{-1}(u_{2r2}) \right] \right\}^4 = 16296434.958$$

Similarly for $L_{1,3}$;

$$\mu_{1-L_{1,3}} = 50.004, \mu_{2-L_{1,3}} = 2999.36, \quad \mu_{3-L_{1,3}} = 189811.2604, \mu_{4-L_{1,3}} = 12944770.4912$$

Similarly for $L_{2,3}$;

$$\mu_{1-L_{2,3}} = 49.967, \mu_{2-L_{2,3}} = 3297.896, \quad \mu_{3-L_{2,3}} = 239670.872, \mu_{4-L_{2,3}} = 19700243.97$$

According to Eq. (3.78);

$$\mu_{1G} = \sum_{i<j} \mu_{1-L_{ij}} - (n-2) \sum_{i} \mu_{1-L_i} + \frac{(n-1)(n-2)}{2} L_0$$

$$= 50.063 + 50.004 + 49.967 - (3-2) \times (50.05 + 50.01 + 49.954)$$
$$+ (3-1) \times (3-2) \times 50/2 = 50.02$$

$$\mu_{2G} = \sum_{i<j} \mu_{2-L_{ij}} - (n-2) \sum_{i} \mu_{2-L_i} + \frac{(n-1)(n-2)}{2} L_0^2 = 3405.32$$

$$\mu_{3G} = \sum_{i<j} \mu_{3-L_{ij}} - (n-2) \sum_{i} \mu_{3-L_i} + \frac{(n-1)(n-2)}{2} L_0^3 = 258092.029$$

$$\mu_{4G} = \sum_{i<j} \mu_{4-L_{ij}} - (n-2) \sum_{i} \mu_{4-L_i} + \frac{(n-1)(n-2)}{2} L_0^4 = 22405873.972$$

According to Eqs. (3.85a)–(3.85d);

$$\mu_G = \mu_{1G} = 50.02$$

$$\sigma_G = \sqrt{\mu_{2G} - \mu_{1G}^2} = 30.055$$

$$\alpha_{3G} = \left(\mu_{3G} - 3\mu_{2G}\mu_{1G} + 2\mu_{1G}^3 \right)/\sigma_G^3 = -0.0961$$

$$\alpha_{4G} = \left(\mu_{4G} - 4\mu_{3G}\mu_{1G} + 6\mu_{2G}\mu_{1G}^2 - 3\mu_{1G}^4 \right)/\sigma_G^4 = 3.8084$$

It can be observed that the bivariate dimension reduction method provides better estimated results when compared with the exact results ($\mu_G = 50$, $\sigma_G = 30.067$, $\alpha_{3G} = -0.099$ 27, and $\alpha_{4G} = 3.821$) than those obtained by the univariate dimension reduction method as shown previously in Example 3.16.

Example 3.19 Consider the following function of multi-random variables;

$$y = x_1^2 x_2^2 + 2x_3^4 \tag{3.87}$$

where x_1, x_2, and x_3 are independent lognormal random variables with $\mu_{x1} = \mu_{x2} = \mu_{x3} = 1$ and $V_{x1} = V_{x2} = V_{x3} = V$.

In order to obtain the first few moments accurately, let $z_1 = x_1^2$, $z_2 = x_2^2$, $z_{12} = z_1 z_2$, and $z_3 = x_3^4$; then $y = z_1 z_2 + 2z_3 = z_{12} + 2z_3$. We can obtain the moments of $z_1 = x_1^2$, $z_2 = x_2^2$, and $z_3 = x_3^4$ using Eqs. (3.10a)–(3.10d) and then obtain the moments of $z_{12} = z_1 z_2$ using Eqs. (3.7a)–(3.7d), and finally obtain the moments of $y = z_{12} + 2z_3$ using Eqs. (3.5a)–(3.5d).

The moments can also be estimated using the nm-point estimate, where G_μ and G_i in Eqs. (3.67a)–(3.67c) are easily understood as

$$G_\mu = \mu_1^2 \mu_2^2 + 2\mu_3^4 = 3 \tag{3.88a}$$

$$G_1 = x_1^2 \mu_2^2 + 2\mu_3^4 = x_1^2 + 2 \tag{3.88b}$$

$$G_2 = \mu_1^2 x_2^2 + 2\mu_3^4 = x_2^2 + 2 \tag{3.88c}$$

$$G_3 = \mu_1^2 \mu_2^2 + 2x_3^4 = 1 + x_3^4 \tag{3.88d}$$

Using five-point estimate, the first four moments of G_1, G_2, and G_3 are obtained as

$$\mu_1 = 3.01, \quad \sigma_1 = 0.2035, \quad \alpha_{31} = 0.6127, \quad \alpha_{41} = 3.763,$$
$$\mu_2 = 3.01, \quad \sigma_2 = 0.2035, \quad \alpha_{32} = 0.6127, \quad \alpha_{42} = 3.763,$$
$$\mu_3 = 3.123, \quad \sigma_3 = 0.882, \quad \alpha_{33} = 1.3151, \quad \alpha_{43} = 6.1109.$$

Using Eqs. (3.68a)–(3.68d), the first four moments of G are obtained as $\mu_G = 3.143$, $\sigma_G = 0.9277$, $\alpha_{3G} = 1.1428$, and $\alpha_{4G} = 5.544$.

The estimated results are presented in Table 3.4 for the two methods with 5, 7, 9, and 11 estimating points. The values in bold mean that their errors are less than 2%. It can be observed that both methods provide results quite close to the exact ones. Generally, the bivariate dimension reduction provides better approximations than the univariate dimension reduction method.

From Table 3.4, the results obtained by all the approximations are observed to be close to the exact results for mean μ_y and standard deviation σ_y. For the skewness α_{3y} and kurtosis α_{4y}, results obtained by the five-point estimate are observed to be quite different from the exact results, while those obtained by the nine-point estimate are closer to the exact values. It is worth noting that for the case of $V = 0.3$, more estimating points are needed to obtain a good accuracy.

Table 3.4 Comparison between the results of point estimates and exact results for Example 3.19.

Moment	v	Univariate dimension reduction				Bivariate dimension reduction				Exact
		5p	7p	9p	11p	5p	7p	9p	11p	
μ_y	0.1	3.1430	3.1430	3.1430	3.1430	3.1431	3.1431	3.1431	3.1431	3.1431
	0.2	3.6160	3.6160	3.6160	3.6160	3.6122	3.6122	3.6122	3.6122	3.6122
	0.3	4.5438	4.5342	4.5342	4.5342	4.5419	4.5423	4.5423	4.5423	4.5423
σ_y	0.1	0.9277	0.9277	0.9277	0.9277	0.9295	0.9295	0.9295	0.9295	0.9296
	0.2	2.4368	2.4409	2.4409	2.4409	2.4481	2.4522	2.4522	2.4522	2.4539
	0.3	5.7114	5.8567	5.8646	5.8648	5.7396	5.8843	5.8921	5.8923	5.9008
α_{3y}	0.1	1.1428	1.1452	1.1453	1.1453	1.1468	1.1493	1.1493	1.1493	1.1537
	0.2	3.0121	3.2858	3.3027	3.3033	2.9751	3.2452	3.2619	3.2624	3.2762
	0.3	5.5551	8.6631	9.6763	9.8432	5.4670	8.5381	9.5375	9.7021	9.7069
α_{4y}	0.1	5.5440	5.6479	5.6492	5.6492	5.4849	5.5881	5.5894	5.5894	5.6397
	0.2	18.147	27.284	29.181	29.381	17.600	26.728	28.590	28.787	28.876
	0.3	43.157	160.12	290.95	363.70	42.197	156.97	285.37	356.76	385.28

3.6 Point Estimates for a Function of Correlated Random Variables

In previous sections, the evaluation of the first few moments of performance functions is focused on independent random variables. However, for analysis of engineering systems under uncertainties, it is often necessary to consider the correlations among input variables. The procedure for evaluating the first few moments of performance functions involving correlated random variables is almost the same as that for independent random variables except where the transformation procedure from correlated random variables into independent standard normal ones is required. If the joint PDF of input variables is known, Rosenblatt transformation is available (details can be referred to the chapter on Transformation of Non-Normal Variables to Independent Normal Variables in Section 7.3.1). If the joint PDF of the input variables is unknown except for their marginal PDFs and correlation coefficients, the Nataf transformation can be utilised (details can be referred to the same chapter as previously mentioned in Section 7.3.2). If only the information of the first few moments and correlation matrix is known, the transformation can be achieved by the third- and fourth-order moment pseudo normal transformation as described in the same chapter in Section 7.5. Here, only some numerical examples are presented.

Example 3.20   Consider the following performance function,

$$G(\mathbf{X}) = 18 - 3X_1 - 2X_2 \tag{3.89}$$

where the joint PDF of X_1 and X_2 is

$$f_{12}(\mathbf{x}) = (x_1 + x_2 + x_1 x_2)e^{(-x_1 - x_2 - x_1 x_2)}, \quad x_1, x_2 \geq 0$$

The marginal PDFs of X_1 and X_2 are obtained as

$$f_1(x_1) = \int_0^{+\infty} f_{12}(\mathbf{x})dx_2 = e^{-x_1}, x_1 \geq 0$$

$$f_2(x_2) = \int_0^{+\infty} f_{12}(\mathbf{x})dx_1 = e^{-x_2}, x_2 \geq 0$$

It can be observed that both X_1 and X_2 follow the exponential distribution with parameter 1. The first four moments of each variable and the correlation coefficient can be calculated as

$$\mu_{X_1} = \mu_{X_2} = 1, \sigma_{X_1} = \sigma_{X_2} = 1, \alpha_{3X_1} = \alpha_{3X_2} = 2, \alpha_{4X_1} = \alpha_{4X_2} = 9, \rho_{12} = -0.404$$

1) **Direct integration method:** Since the joint PDF and the performance function are available, the mean of the performance function can be obtained by numerical integral over the original space: $\mu_G = \int G(\mathbf{x})f_{12}(\mathbf{x})d\mathbf{x} = 13.0$
   Similarly, one can obtain $\sigma_G = 2.8599, \alpha_{3G} = -1.5161, \alpha_{4G} = 7.4853$.
2) **Point estimate method based on Rosenblatt transformation:** Since the joint PDF is known, the point estimate method based on Rosenblatt transformation can be used to compute the first four moments of the performance function in Eq. (3.89). When the transformation order of $X_1 \rightarrow X_2$, is adopted in Rosenblatt transformation, one has

$$\Phi(u_1) = F_1(x_1) = \int_{-\infty}^{x_1} f_1(x_1)dx_1 = 1 - \exp(-x_1)$$

$$\Phi(u_2) = F_2(x_2|x_1) = \int_{-\infty}^{x_2} \frac{f_{12}(u, x_1)}{f_1(x_1)} du = [1 - (1 + x_2)] \cdot \exp(-x_2 - x_1x_2)$$

The point of original space can be obtained by putting the point of standard normal space into the formula, such as, putting $u_1 = 0$, $u_2 = 0$ into the formula, as follows;

$$\Phi(0) = 0.5 = 1 - \exp(-x_1)$$
$$\Phi(0) = 0.5 = [1 - (1 + x_2)] \cdot \exp(-x_2 - x_1x_2)$$

Solving the equations above, the original estimates point can be obtained as

$$x_1 = 0.693147, \quad x_2 = 0.734724$$

$$P = P_1P_2 = 0.457142857 \times 0.457142857 = 0.209$$

$$G(\mathbf{X}) = 18 - 3X_1 - 2X_2 = 18 - 3 \times 0.693172 - 2 \times 0.734724 = 14.451$$

Similarly, we can obtain the other estimating points in the original space and the corresponding weights, part of which are listed in Table 3.5.

Then, the first four moments of the performance function in Eq. (3.89) is obtained as

$$\mu_G = \sum P \cdot G(\mathbf{x}) = 13.1151$$

$$\sigma_G^2 = \sum P \cdot [G(\mathbf{x}) - \mu_G]^2 = 8.57346, \sigma_G = 2.8559$$

$$\alpha_{3G}\sigma_G^3 = \sum P \cdot [G(\mathbf{x}) - \mu_G]^3 = -35.6883, \alpha_{4G}\sigma_G^4 = \sum P \cdot [G(\mathbf{x}) - \mu_G]^4 = 525.8997$$

$$\alpha_{3G} = -1.5161; \alpha_{4G} = 7.1547$$

However, if the transformation $(X_2 \rightarrow X_1)$ order is selected, we find

$$\Phi(u_2) = F_2(x_2) = \int_{-\infty}^{x_2} f_2(x_2)dx_2 = 1 - \exp(-x_2)$$

$$\Phi(u_1) = F_1(x_1|x_2) = \int_{-\infty}^{x_1} \frac{f_{12}(x_1, u)}{f_2(x_2)} du = [1 - (1 + x_1)] \cdot \exp(-x_1 - x_1x_2)$$

Similarly, the first four moments of the performance function can be obtained as

$$\mu_G = 13.1727, \sigma_G = 2.8698, \alpha_{3G} = -1.3222, \alpha_{4G} = 6.8020$$

It can be observed that they are different from those obtained in the transformation order of $X_1 \rightarrow X_2$.

3) **Point estimate method based on fourth-order moment pseudo normal transformation:** Substituting the skewness and kurtosis of each variable into Eqs. (7.72a) and (7.72b) in the chapter on Transformation of Non-Normal Variables to Independent Normal Variables, the four undetermined parameters a, b, c, and d can be obtained

$$a_1 = a_2 = -0.313749, b_1 = b_2 = 0.826324, c_1 = c_2 = 0.313749, d_1 = d_2 = 0.0227066$$

Therefore, X_1 and X_2 can be expressed as

Table 3.5 The estimate points in original space and the corresponding weights for Example 3.20.

u_1	P_1	x_1	u_2	P_2	x_2	P	$G(X)$
0	0.457	0.693	0	0.457	0.734	0.208	14.451
			1.154	0.240	1.850	0.109	12.218
			−1.154	0.240	0.172	0.109	15.576
			2.366	0.030	3.697	0.014	8.525
			−2.366	0.030	0.012	0.014	15.894
			3.750	5.482×10^{-4}	6.720	2.506×10^{-4}	2.479
			−3.750	5.482×10^{-4}	1.273×10^{-4}	2.506×10^{-4}	15.920
1.154	0.240	2.086	0	0.457	0.312	0.109	11.116
			1.154	0.240	0.880	0.057	9.980
			−1.154	0.240	0.062	0.057	11.616
			2.366	0.030	1.868	7.385×10^{-3}	8.003
			−2.366	0.030	4.315×10^{-3}	7.385×10^{-3}	11.732
			3.750	5.482×10^{-4}	3.513	1.316×10^{-4}	4.715
			−3.750	5.482×10^{-4}	4.231×10^{-5}	1.316×10^{-4}	11.741
−1.154	0.240	0.132	0	0.457	1.376	0.109	14.849
			1.154	0.240	3.084	0.057	11.433
			−1.154	0.240	0.437	0.057	16.727
			2.366	0.030	5.862	7.385×10^{-3}	5.877
			−2.366	0.030	0.056	7.385×10^{-3}	17.489
			3.750	5.482×10^{-4}	10.390	1.316×10^{-4}	−3.178
			−3.750	5.482×10^{-4}	6.641×10^{-4}	1.316×10^{-4}	17.600

$$X_1 = 1 + \left(-0.313749 + 0.826324 Z_1 + 0.313749 Z_1^2 + 0.0227066 Z_1^3\right)$$

$$X_2 = 1 + \left(-0.313749 + 0.826324 Z_2 + 0.313749 Z_2^2 + 0.0227066 Z_2^3\right)$$

where Z_1 and Z_2 are correlated standard normal variable with the correlation coefficient ρ_{012}, which can be determined by Algorithm 7.1 (Section 7.5.3 in the same chapter just mentioned) as

$$\rho_{012} = -0.589$$

Then, the covariance matrix of Z_1 and Z_2 can be written as

$$\mathbf{C}_0 = \begin{bmatrix} 1 & -0.589 \\ -0.589 & 1 \end{bmatrix}$$

Obtaining the lower triangular matrix $\mathbf{L}_0$ by Cholesky decomposition of the matrix $\mathbf{C}_0$,

$$\mathbf{L}_0 = \begin{bmatrix} 1 & 0 \\ -0.589 & 0.808 \end{bmatrix}$$

Therefore,

$$Z_1 = U_1, Z_2 = -0.589 U_1 + 0.808 U_2$$

Then the performance function can be expressed as the function of independent standard normal variables. The point estimates method can be used to evaluate the first four moments of function as

$$\mu_G = 13.0, \sigma_G = 2.8599, \alpha_{3G} = -1.5909, \alpha_{4G} = 7.8042$$

It can be observed that the results agree well with the exact values.

Example 3.21 Consider the following performance function, which has been investigated by Fan et al. (2016)

$$G(\mathbf{X}) = X_1 - \frac{75000 X_2}{X_3^3} \tag{3.90}$$

where X_1 is a Weibull variable with scale parameter 20 and shape parameter 25, and X_2 and X_3 are normal random variables with $X_2 = N(1000, 250)$ and $X_3 = N(250, 25)$. The correlation coefficient between X_1 and X_2 is assumed as $\rho_{12} = 0.5$.

The exact values of the first four moments can be obtained by numerical integral of the performance function, given by Fan et al. (2016), which are $\mu_G = 14.4569$, $\sigma_G = 2.0814$, $\alpha_{3G} = -1.3292$, and $\alpha_{4G} = 7.2539$. The results obtained by using the point estimate method based on the fourth-order moment pseudo normal transformation are $\mu_G = 14.4569$, $\sigma_G = 2.0693$, $\alpha_{3G} = -1.3284$, and $\alpha_{4G} = 7.0817$. It can be observed that the point estimate method based on the fourth-order moment pseudo normal transformation provides relatively accurate results for this nonlinear performance function involving correlated random variables.

Example 3.22 Consider a one-story one-bay elastoplastic frame as shown in Figure 3.20. The performance functions that correspond to four most likely failure modes from stochastic limit analysis are derived as

$$g_1(\mathbf{X}) = 2M_1 + 2M_3 - 4.5S \tag{3.91a}$$

$$g_2(\mathbf{X}) = 2M_1 + M_2 + M_3 - 4.5S \tag{3.91b}$$

$$g_3(\mathbf{X}) = M_1 + M_2 + 2M_3 - 4.5S \tag{3.91c}$$

$$g_4(\mathbf{X}) = M_1 + 2M_2 + M_3 - 4.5S \tag{3.91d}$$

The performance function of a series system, $G(\mathbf{X})$, can be expressed as the minimum of the performance functions that corresponds to all the potential failure modes (Zhao and Ang 2003)

$$G(\mathbf{X}) = \min\{g_1(\mathbf{X}), g_2(\mathbf{X}), g_3(\mathbf{X}), g_4(\mathbf{X})\} \tag{3.92}$$

where M_i and S are lognormal random variables with mean and standard deviations of $\mu_{Mi} = 200$, $\sigma_{Mi} = 30$, $\mu_S = 50$, $\sigma_S = 20$, respectively. Here, M_i ($i = 1, 2, 3$) and S are assumed to be correlated with ρ, while M_1, M_2, and M_3 are independent random variables.

Figure 3.21 presents the ratio of the first four moments of the performance function considering the correlation coefficient to those with independent variables for the range of the

Figure 3.20 A one-story one-bay elastoplastic frame.

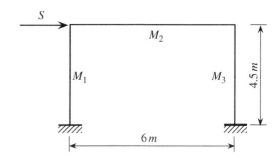

Figure 3.21 The ratio of first four moments.

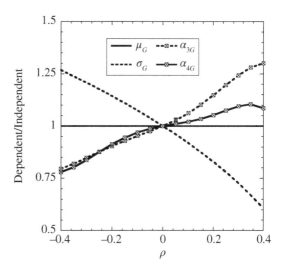

correlation coefficient (from -0.3 to 0.3). As can be observed from Figure 3.21, the correlation coefficient has great influence on the standard deviation, skewness, and kurtosis, while it has little effect on the mean of the performance function.

3.7 Hybrid Dimension-Reduction Based Point Estimate Method

Consider the fact that the sensitivity to the whole performance function $G(\mathbf{X})$ of each variable x_i, which is affected by distribution types, variability, and inherent characters is different. This means more computational quantities should be focused on the random variables with high sensitivity rather than those with relatively insensitivity to the performance function.

One of the most well-known methods of probabilistic sensitivity analysis is a variance-based method, referred to as the analysis of variance, which is used to determine whether there is a statistical association between an output and one or more inputs (Krishnaiah 1981). Based on Sobol's decomposition theory (Sobol 1993), the first order sensitivity index (S_i) of random variable x_i in a performance function $G(\mathbf{X})$ can be expressed as

$$S_i = \frac{\sigma^2_{G_i}}{\displaystyle\sum_{i=1}^{n} \sigma^2_{G_i}} \tag{3.93}$$

where σ_{G_i} is the standard deviation of G_i and G_i is a function of single random variable x_i, in which all the basic random variables except for x_i are deterministic (usually take the mean values).

Using point-estimate method in standard normal space (Zhao and Ono 2000a), $\sigma^2_{G_i}$ can be given by

$$\sigma^2_{G_i} = \sum_{k=1}^{m} P_k \cdot \left\{ G\left[\mu_1, ..., T^{-1}(u_i), ..., \mu_n\right] - \mu_{G_i} \right\}^2 \tag{3.94}$$

where

$$\mu_{G_i} = G(\mu_1, ..., \mu_i, ..., \mu_n) \tag{3.95}$$

Considering the balance between computational efficiency and accuracy to achieve the first four moments of performance functions, here we simply classify the random variables of performance functions as two types; the sensitive variables $\mathbf{X}_S$ and the dull variables $\mathbf{X}_D$ playing insensitive roles in performance function, and the number of these two types are N_S and N_D, respectively ($N_S + N_D = n$). The classification method for the two types is given as follows.

According to Eq. (3.93), the mean sensitivity index of an n-variable function can be obtained as

$$S_m = \frac{1}{n} \times 100\% \tag{3.96}$$

The sensitive variables $\mathbf{X}_S$ (i.e. $S_i > S_m$) and the dull variables $\mathbf{X}_D$ (i.e. $S_i < S_m$) can then be simply separated according to the mean sensitivity index.

According to the univariate-dimension reduction moment method, the performance function $G(\mathbf{X})$ is approximated by the following function

$$G(\mathbf{X}) \cong G_1(\mathbf{X}_S) + G_2(\mathbf{X}_D) - G_\mu \tag{3.97}$$

where

$$G_1(\mathbf{X}_S) = G_1\left(\mathbf{X}_S, \mu_{\mathbf{X}_D}\right) \tag{3.98}$$

$$G_2(\mathbf{X}_D) = G_2\left(\mu_{\mathbf{X}_S}, \mathbf{X}_D\right) \tag{3.99}$$

$$G_\mu = G\left(\mu_{\mathbf{X}_S}, \mu_{\mathbf{X}_D}\right) = G(\mu_1, ..., \mu_i, ..., \mu_n) \tag{3.100}$$

where G_μ represents the value of $G(\mathbf{X})$ when all the random variables take their mean values; $G_1\left(\mathbf{X}_S, \mu_{\mathbf{X}_D}\right)$ is denoted as the performance function of type 1, which means only dull variables $\mathbf{X}_D$ take their mean values; and $G_2\left(\mu_{\mathbf{X}_S}, \mathbf{X}_D\right)$ is denoted as the performance function of type 2, in which only sensitive variables $\mathbf{X}_S$ take their mean values.

For the performance function of type 1, the bivariate-dimension reduction method as described in Section 3.5.3, i.e. Eqs. (3.75)–(3.85d), is used to calculate its first four moments; for the performance function of type 2, the univariate-dimension reduction method as described in Section 3.5.2, i.e. Eqs. (3.67a)–(3.67c), (3.68a)–(3.68c), and (3.35a)–(3.35c), is used to obtain its first four moments; and finally using Eqs. (3.5a)–(3.5d), the first four moments of the performance function of Eq. (3.97) can be obtained.

Example 3.23 Consider the following performance function

$$G(\mathbf{X}) = \min(g_1, g_2, g_3, g_4, g_5) \tag{3.101}$$

where

$g_1 = \max[T_1 - 1.127F_1 + 0.826F_2, \min(T_2 - 4/3F_1, T_3 - F_1 + 3/4F_2, T_4 - 5/4F_2, T_5 - 5/3F_1 + 5/4F_2)]$

$g_2 = \max[T_2 - 0.206F_1 - 0.826F_2, \min(T_1 - 4/3F_1, T_3 - 3/4F_2, T_4 - 5/3F_1 - 5/4F_2, T_5 - 5/3F_1 + 5/4F_2)]$

$g_3 = \max[T_3 - 0.155F_1 + 0.13F_2, \min(T_1 - 4/3F_1 + F_1, T_2 - F_2, T_4 - 5/3F_1)]$

$g_4 = \max[T_4 - 1.409F_1 - 0.217F_2, \min(T_1 - F_2, T_2 - 4/3F_1 - F_2, T_3 - F_1, T_5 - 5/3F_2)]$

$g_5 = \max[T_5 - 0.258F_1 + 0.217F_2, \min(T_1 - 4/3F_1 + F_2, T_2 - F_2, T_4 - 5/3F_1)]$

where fracture strength T_i and load F_i are independent lognormal random variables with means and standard deviations of $\mu_{T_1} = \mu_{T_2} = 40t$, $\mu_{T_3} = 10t$, $\mu_{T_4} = \mu_{T_5} = 20t$, $\mu_{F_1} = 7t$, and $\mu_{F_2} = 2t$; and standard deviations of $\sigma_{T_1} = \sigma_{T_2} = 6t$, $\sigma_{T_3} = 1.5t$, $\sigma_{T_4} = \sigma_{T_5} = 3t$, $\sigma_{F_1} = 2.1t$, and $\sigma_{F_2} = 0.6t$.

The sensitive indices of each variable are determined by using Eq. (3.93) using the seven-point estimate, and are listed in Table 3.6. The mean sensitivity index is obtained as 14.3%; therefore, the random variables of T_4 and load F_1 belong to sensitive variables, and the others are dull random variables.

Then the performance function in Eq. (3.101) can be approximated as

$$G^*(\mathbf{X}) = G_1^*(T_4, F_1) + G_2^*(T_1, T_2, T_3, T_5, F_2) - 9.175 \tag{3.102}$$

Table 3.6 Results of sensitivity analysis for Example 3.23.

Basic variables	Standard deviation, σ_{Gi}	Sensitive index, S_i (%)	Types of basic variables
T_1, T_2, T_5	0	0	Dull
F_2	0.069	1.2	Dull
T_3	0.511	8.8	Dull
T_4	0.133	43.5	Sensitive
F_1	2.687	46.5	Sensitive

Table 3.7 Evaluation of statistical moments with different methods for Example 3.23.

Methods	μ_G	σ_G	α_{3G}	α_{4G}	Number of points or samples
Hybrid-dimension reduction	9.031	3.863	−0.075	3.824	84
Bivariate-dimension reduction	9.143	3.859	−0.051	3.812	799
Monte Carlo simulation	9.182	3.844	−0.065	3.822	10^7

Using the bivariate-dimension reduction based point estimate method, i.e. Eqs. (3.75)–(3.85d), the first four moments of G_1^* can be obtained as $\mu_{G_1^*} = 9.114$, $\sigma_{G_1^*} = 3.828$, $\alpha_{3G_1^*} = -0.076$, and $\alpha_{4G_1^*} = 3.855$. Using Eqs. (3.67a)–(3.67c), (3.68a)–(3.68c), and (3.35a)–(3.35c) the first four moments of G_2^* can be obtained as $\mu_{G_2^*} = 9.091$, $\sigma_{G_2^*} = 0.516$, $\alpha_{3G_2^*} = -0.394$, and $\alpha_{4G_2^*} = 1.921$. Using Eqs. (3.5a)–(3.5d) the first four moments of the performance function $G^*(\mathbf{X})$ are obtained as $\mu_G = 9.031$, $\sigma_G = 3.863$, $\alpha_{3G} = -0.075$, $\alpha_{4G} = 3.824$, which are listed in Table 3.7.

The results of hybrid-dimension reduction, bivariate-dimension reduction, and Monte Carlo simulation (MCS) are also shown in Table 3.7, where the results of MCS are taken as exact values. We can observe from Table 3.7 that the hybrid dimension-reduction based point estimate can achieve a good balance between accuracy and efficiency.

3.8 Summary

The moment computation for performance functions is discussed in this chapter. For some explicit functions, the first few moments can be easily calculated from the definitions, as shown in Section 3.2. However, this could be difficult for many complicated performance functions in practical engineering. The point estimate is a simple and efficient approach for a good behaviour function of only one variable. Rosenblueth's two-point estimate and Gorman's three-point estimate are introduced as shown in Section 3.3. Since two- and three-point estimates may be inadequate for practical use, the point estimate method in standard normal space with arbitrary number of points is further discussed. With this method, the first few moments for a function of a single variable can be quickly obtained with required accuracy. For functions of multiple random variables, hyper quadrants could be used, but might lead to excessive computational efforts when the number of random variables is large. Two approximation methods are introduced in Section 3.5. The moment computation for the performance function of correlated random variables is introduced in Section 3.6 using the second- and third-order polynomial transformations. The point estimate method in standard normal space generally gives good results, but its robustness cannot be ensured. Further studies are required for more efficient and effective methods to compute the first few moments of a general performance function. Many new methods have been developed for moment evaluation of performance functions in recent years, and can be referred to He et al. (2014), Fan et al. (2016), Xu and Lu (2017), Cai et al. (2019), etc.

4

Direct Methods of Moment

4.1 Basic Concept of Methods of Moment

4.1.1 Integral Expression of Probability of Failure

As described in the chapter on Fundamentals of Structural Reliability Theory, the fundamental problem in structural reliability theory is to compute the probability of failure which is generally expressed as the multi-fold probability integral

$$P_F = P[Z = G(\mathbf{X}) \le 0] = \int_{G(\mathbf{X}) \le 0} f_{\mathbf{x}}(\mathbf{x})d\mathbf{x} \tag{4.1}$$

where $\mathbf{X} = [X_1, ..., X_n]^T$, in which the superposed T = transpose, is an n-dimensional vector of random variables representing uncertain quantities such as loads, material properties, geometric dimensions, and boundary conditions; $f_{\mathbf{X}}(\mathbf{x})$ denotes the joint probability density function (PDF) of $\mathbf{X}$; $G(\mathbf{X})$ is the performance function defined such that the domain of integration $G(\mathbf{X}) \le 0$ denotes the failure set; and P_F is the probability of failure.

As also shown in that chapter, the evaluation of the probability of failure P_F through Eq. (4.1) is generally a formidable task. Another route to evaluate P_F is to compute the following one-dimensional integral

$$P_F = P[Z = G(\mathbf{X}) \le 0] = \int_{-\infty}^{0} f_Z(z)dz = F_Z(0) \tag{4.2}$$

where $f_Z(z)$ is the PDF of the performance function $Z = G(\mathbf{X})$, which is also a single random variable.

Integration in Eq. (4.2) seems easier than that in Eq. (4.1) since it is one dimensional. According to Appendix A.9, the probability distribution for a function of random variables can, theoretically, be derived from the probability distribution of the basic random variables $\mathbf{X}$. However, the derivation of the probability distribution is generally difficult for a function (here specifically, the performance function) of random variables, especially when the function is nonlinear. In such circumstances, the central moments, particularly the first few central moments such as the mean, standard deviation, skewness, and kurtosis of the performance function, may be the only information practically available. As shown in Figure 3.1, the first four central moments of the performance function are closely related

Structural Reliability: Approaches from Perspectives of Statistical Moments, First Edition.
Yan-Gang Zhao and Zhao-Hui Lu.
© 2021 John Wiley & Sons Ltd. Published 2021 by John Wiley & Sons Ltd.

to the failure probability, since these moments are closely related to the location and shapes of the distribution of the performance function. Therefore, the failure probability can be evaluated directly by utilising the first few moments of the performance function. If the central moments of the performance function can be obtained, the failure probability, which is defined as the probability when the performance function is less than or equal to zero as shown in Eq. (4.2), can be expressed as a function of the central moments. The failure probability can thus be computed through establishing its relationship with the central moments. In many instances, this may be sufficiently accurate for many practical applications even if the correct probability distributions are undetermined.

4.1.2 The Second-Moment Method

For a performance function $Z = G(\mathbf{X})$, assuming that the first two moments are known, and $Z = G(\mathbf{X})$ follows a normal distribution, the reliability index β_{2M} and the failure probability P_F are expressed as

$$\beta_{2M} = \frac{\mu_G}{\sigma_G} \tag{4.3}$$

$$P_F = \Phi(-\beta_{2M}) \tag{4.4}$$

where μ_G and σ_G are the mean and standard deviation of $Z = G(\mathbf{X})$, respectively; and $\Phi(\cdot)$ is the cumulative distribution function (CDF) of a standard normal random variable.

Generally, the reliability method based on Eqs. (4.3) and (4.4) is referred to as the second-moment method since only the first two moments are used. It is worth noting that the method is accurate only in the case that $Z = G(\mathbf{X})$ is normal. The reliability index expressed in Eq. (4.3) is hardly used as an accurate safety measure corresponding to the probability of failure, since $Z = G(\mathbf{X})$ obeys different types of probability distributions for general cases. For example, suppose $\beta_{2M} = \mu_G/\sigma_G = 2$, different types of distribution of $Z = G(\mathbf{X})$ will lead to different probability of failure as follows:

Weibull $P_F = 4.0 \times 10^{-2}$; Normal $P_F = 2.27 \times 10^{-2}$; lognormal $P_F = 5.0 \times 10^{-3}$;
Gumbel $P_F = 5.0 \times 10^{-3}$; Frechet $P_F = 1.0 \times 10^{-7}$.

It can be observed that the failure probability varies significantly with different kinds of distribution of $Z = G(\mathbf{X})$.

Example 4.1 Consider the performance function $Z = G(\mathbf{X}) = R - S$. If R and S are statistically independent random variables with mean and standard deviations of $\mu_R = 100$, $\sigma_R = 10$ and $\mu_S = 50$, $\sigma_S = 20$, then the first two central moments of $Z = G(\mathbf{X}) = R - S$ are $\mu_G = \mu_R - \mu_S = 50$ and $\sigma_G = \sqrt{\sigma_R^2 + \sigma_S^2} = 22.361$.

Then, the *reliability index* based on the second-moment method, regardless of the probability distributions of R and S, is given as

$$\beta_{2M} = \frac{\mu_G}{\sigma_G} = \frac{50}{22.361} = 2.236$$

The corresponding probability of failure is given as 0.0127 according Eq. (4.4).

Table 4.1 Probability of failure with different distributions of random variables.

R	S	P_F	COV of P_F	β_{MCS}
Normal	Normal	0.0129	0.88%	2.229
Normal	Lognormal	0.0283	0.59%	1.906
Normal	Weibull	0.0159	0.79%	2.148
Normal	Gumbel	0.0273	0.60%	1.922
Lognormal	Normal	0.0119	0.92%	2.261
Lognormal	Lognormal	0.0273	0.60%	1.922
Weibull	Normal	0.0142	0.84%	2.192
Weibull	Lognormal	0.0287	0.59%	1.900
Gumbel	Normal	0.0110	0.95%	2.290
Gumbel	Lognormal	0.0270	0.61%	1.927
Gamma	Normal	0.0124	0.90%	2.245
Gamma	Lognormal	0.0277	0.60%	1.916

The results of the second-moment method above utilised only the information of the first two central moments. In order to investigate the accuracy of the second-moment method, the probability of failure is estimated using Monte Carlo simulation (MCS) with 1000000 samples under various combinations of distributions. The results are listed in Table 4.1.

From Table 4.1, one may observe that only in the case that both R and S are normal, i.e. $Z = G(\mathbf{X})$ is also normal, does the second moment reliability index have good agreement with that obtained by MCS.

Formula variance of the second moment method. Another weakness of the second moment method is its formula variance. This will be illustrated by the Example 4.2.

Example 4.2 For the lognormal random variables R and S with $\mu_R = 200$, $\mu_S = 100$, $V_R = 0.1$, and $V_S = 0.2$, consider performance functions $Z = G(\mathbf{X}) = R - S$, $Z = G(\mathbf{X}) = \ln R - \ln S$, $Z = G(\mathbf{X}) = 1 - S/R$, $Z = G(\mathbf{X}) = R/S - 1$, and $Z = G(\mathbf{X}) = 1/S - 1/R$. Let $\theta = \mu_R/\mu_S$, the second moment reliability indices are obtained as follows.

For $Z = G(\mathbf{X}) = R - S$,

$$\beta_{2M} = \frac{\mu_G}{\sigma_G} = \frac{\mu_R - \mu_S}{\sqrt{\sigma_R^2 + \sigma_S^2}} = \frac{\theta - 1}{\sqrt{\theta^2 V_R^2 + V_S^2}} = 3.536$$

For $Z = G(\mathbf{X}) = \ln R - \ln S$,

$$\beta_{2M} = \frac{\mu_G}{\sigma_G} = \frac{\lambda_R - \lambda_S}{\sqrt{\zeta_R^2 + \zeta_S^2}}$$

Since V_R and V_S are small, we have $\zeta_R \approx V_R$, $\zeta_S \approx V_S$, then

$$\beta_{2M} \approx \frac{\ln\theta - \frac{1}{2}\ln\left(\frac{1+V_R^2}{1+V_S^2}\right)}{\sqrt{V_R^2 + V_S^2}} \approx \frac{\ln\theta}{\sqrt{V_R^2 + V_S^2}} = 3.094$$

For $Z = G(\mathbf{X}) = 1 - S/R$, let $Y = 1/R$, using Eqs. (3.12a)–(3.12c) from the chapter on Moment Evaluation, we have

$$\mu_Y = \frac{1}{\mu_R}\left(1 + V_R^2\right), V_Y = V_R$$

With the aid of Eqs. (3.7a)–(3.7b) from the same chapter, one obtains

$$\beta_{2M} = \frac{\mu_G}{\sigma_G} = \frac{1 - \frac{\mu_S}{\mu_R}\left(1 + V_R^2\right)}{\frac{\mu_S}{\mu_R}\left(1 + V_R^2\right)\sqrt{\left(1 + V_R^2\right)\left(1 + V_S^2\right)} - 1} = 4.365$$

For $Z = G(\mathbf{X}) = R/S - 1$, let $Y = 1/S$, using Eqs. (3.12a)–(3.12c) again, we have

$$\mu_Y = \frac{1}{\mu_S}\left(1 + V_S^2\right), V_Y = V_S$$

With the aid of Eqs. (3.7a)–(3.7b) again, one obtains

$$\beta_{2M} = \frac{\mu_G}{\sigma_G} = \frac{\frac{\mu_R}{\mu_S}\left(1 + V_S^2\right) - 1}{\frac{\mu_R}{\mu_S}\left(1 + V_S^2\right)\sqrt{\left(1 + V_R^2\right)\left(1 + V_S^2\right)} - 1} = 2.313$$

For $Z = G(\mathbf{X}) = 1/S - 1/R$, let $X = 1/S$, $Y = 1/R$ then,

$$\mu_X = \frac{1}{\mu_S}\left(1 + V_S^2\right), \quad \mu_Y = \frac{1}{\mu_R}\left(1 + V_R^2\right), \quad V_X = V_S, \quad V_Y = V_R$$

$$\beta_{2M} = \frac{\mu_G}{\sigma_G} = \frac{\frac{1}{\mu_S}\left(1 + V_S^2\right) - \frac{1}{\mu_R}\left(1 + V_R^2\right)}{\sqrt{\frac{V_S^2}{\mu_S^2}\left(1 + V_S^2\right)^2 + \frac{V_R^2}{\mu_R^2}\left(1 + V_R^2\right)^2}} = \frac{\theta\left(1 + V_S^2\right) - \left(1 + V_R^2\right)}{\sqrt{\theta^2 V_S^2\left(1 + V_S^2\right)^2 + V_R^2\left(1 + V_R^2\right)^2}}$$

$$= 2.4995$$

Since the five performance functions present the same limit state, the failure probability or the reliability indices should be the same. However, the results above show that the second moment reliability index varies with the form of the performance functions though the limit states are the same. This is often called the *formula variance* of the second-moment method.

4.1.3 General Expressions for Methods of Moment

As described above, there are mainly two weaknesses in the second-moment method. First, the failure probability varies with the distribution of the performance function $Z = G(\mathbf{X})$; even the second moment reliability index remains the same. Second, the second-moment reliability index changes with the form of the performance functions even for the same limit states. Both of the two weaknesses may be attributed to the fact that too limited information is utilised in the second-moment method. In other words, the first two moments are generally inadequate. In order to improve the reliability methods based on moments of performance functions, higher order moments need to be used.

Without loss of generality, we standardise the performance function $Z = G(\mathbf{X})$ using its mean and standard deviation with the aid of the following standardised variable

$$Z_S = \frac{Z - \mu_G}{\sigma_G} \tag{4.5}$$

Then the definition of the probability of failure leads to the following

$$P_F = P(Z \le 0) = P(\sigma_G Z_s + \mu_G \le 0) = P(Z_s \le -\mu_G/\sigma_G)$$

that is

$$P_F = P(Z_s \le -\beta_{2M}) \tag{4.6}$$

Suppose the CDF and PDF of Z_s are F_{Z_s} and f_{Z_s}, respectively, then

$$P_F = F_{Z_s}(-\beta_{2M}) = \int_{-\infty}^{-\beta_{2M}} f_{Z_s}(z_s)dz_s \tag{4.7}$$

That is to say, the probability of failure is the value of CDF of Z_s at $-\beta_{2M}$, which is also expressed as the area under the PDF curve less than $-\beta_{2M}$ as illustrated in Figure 4.1.

Based on the theory underlying the normal and inverse normal transformations, the CDF of Z_S should be equal to that of the standard normal random variable (Hohenbichler and Rackwitz 1981), i.e.

$$F_{Z_s}(z_s) = \Phi(u) \tag{4.8}$$

According to the equation above, the relationship between the standardised variable Z_s and the standard normal variable U may be expressed as the following functions of the first few moments of $Z = G(\mathbf{X})$,

$$U = S(Z_s, \mathbf{M}) \tag{4.9a}$$

$$Z_s = S^{-1}(U, \mathbf{M}) \tag{4.9b}$$

where $\mathbf{M}$ is a vector denoting the first few moments of $Z = G(\mathbf{X})$ and S^{-1} is the inverse function of S.

Figure 4.1 Probability of failure.

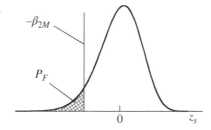

Obviously, the accurate function $S^{-1}(U, \mathbf{M})$ has to be determined from F_{Z_s}, the CDF of Z_s. However, the CDF of Z_s is sometimes unknown and the central moments, particularly the first few central moments such as the mean, standard deviation, skewness, and kurtosis, of Z_s, may be the only practically available information. In such a circumstance, $S^{-1}(U, \mathbf{M})$ is generally determined by making the first few moments of the right side to be equal to those of the left side of Eq. (4.9b). Although Z_s may be a non-normal random variable, after the transformation of Eq. (4.9a) using the information of the central moments of $Z = G(\mathbf{X})$, it becomes a standard normal random variable. Therefore, Eq. (4.9a) is called the *moment standardisation function* of $Z = G(\mathbf{X})$ and Eq. (4.9b) is called the *inverse moment standardisation function*. This will be further explained in Sections 7.4.2 and 7.4.3 in detail.

Substituting Eq. (4.9a) into Eq. (4.8) leads to

$$F_{Z_s}(z_s) = \Phi(u) = \Phi[S(z_s, \mathbf{M})] \tag{4.10}$$

Therefore, Eq. (4.7) can be expressed as

$$P_F = F_{Z_s}(-\beta_{2M}) = \Phi[S(-\beta_{2M}, \mathbf{M})] \tag{4.11}$$

And the reliability index is then expressed as

$$\beta = -\Phi^{-1}(P_F) = -S(-\beta_{2M}, \mathbf{M}) \tag{4.12}$$

Note that here, the CDF of Z_s is only used to deduce Eq. (4.12), no F_{Z_s} is used in the calculation of β. Since the first few moments of $G(\mathbf{X})$ are used in Eq. (4.12), the reliability index expressed in Eq. (4.12) is referred to as a moment reliability index.

In particular, if only the first two moments of $Z = G(\mathbf{X})$ are available, then Eq. (4.9b) becomes $Z_s = U$ and Eq. (4.12) becomes $\beta = -\Phi^{-1}(P_F) = \beta_{2M}$. This indicates that the moment reliability index reduces to the second-moment reliability index described in Section 4.1.2.

4.2 Third-Moment Reliability Method

4.2.1 General Formulation of the Third-Moment Reliability Index

As described in Appendix A.3.3, the third moment, i.e. the skewness, expresses the degree and direction of asymmetry of the PDF of a random variable. The PDFs are shown in Figure 4.2 for some standardised random variables with different skewness, from which it can be observed that the skewness has a great effect on the probability of failure.

In order to account for the effect of the skewness on probability of failure for a performance function, one may assume that the performance function follows a three-parameter distribution, i.e. the distribution can be determined using three independent parameters (mean, standard deviation, and skewness).

When standardising the performance function $Z = G(\mathbf{X})$ to Z_s using Eq. (4.5), if the first three moments are obtained, suppose the relationship between the standardised variable Z_s and the standard normal variable U can be expressed as the following functions of the first three moments of $Z = G(\mathbf{X})$,

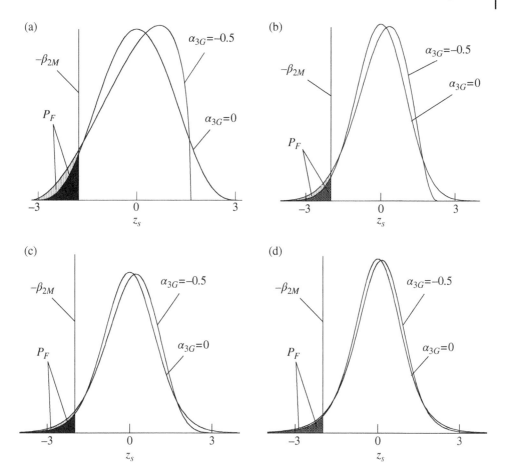

Figure 4.2 Effect of skewness on failure probability in different cases of kurtosis. (a) α_{4G} = 2.5. (b) α_{4G} = 3.0. (c) α_{4G} = 3.5. (d) α_{4G} = 5.0.

$$Z_s = S^{-1}(U, \alpha_{3G}) \tag{4.13a}$$

$$U = S(Z_s, \alpha_{3G}) \tag{4.13b}$$

where α_{3G} is the skewness, i.e. the third dimensionless central moment of $Z = G(\mathbf{X})$
According to Eq. (4.12), the third-moment reliability index, β_{3M}, is expressed as

$$\beta_{3M} = -S(-\beta_{2M}, \alpha_{3G}) \tag{4.14}$$

Two such three-parameter distributions are introduced in Appendix B, including the three-parameter (3P) lognormal distribution (Tichy 1994) and the square normal distribution (Zhao et al. 2001). The next section will discuss the reliability indices derived from the 3P lognormal distribution and the square normal distribution.

4.2.2 Third-Moment Reliability Indices

4.2.2.1 Third-Moment Reliability Index Based on the 3P Lognormal Distribution

The third-moment reliability index can be derived from the 3P lognormal distribution (see in detail in Appendix B.2). The distribution belongs to the Pearson system (see Appendix C.2) and is introduced by Tichy (1994) to obtain the first-order third-moment method.

Assuming that the standardised variable Z_s follows the 3P lognormal distribution (Tichy 1994), the standard normal random variable U can be expressed as the following function (Zhao and Ono 2001):

$$U = \frac{\text{sign}(\alpha_{3G})}{\sqrt{\ln(A)}} \ln\left[\sqrt{A}\left(1 - \frac{Z_s}{u_b}\right)\right] \tag{4.15a}$$

where u_b is the standardised bound of the distribution; and A is a function of u_b, which is given by

$$A = 1 + \frac{1}{u_b^2} \tag{4.15b}$$

$$u_b = (a+b)^{\frac{1}{3}} + (a-b)^{\frac{1}{3}} - \frac{1}{\alpha_{3G}} \tag{4.15c}$$

$$a = -\frac{1}{\alpha_{3G}}\left(\frac{1}{2} + \frac{1}{\alpha_{3G}^2}\right), \quad b = \frac{1}{2\alpha_{3G}^2}\sqrt{\alpha_{3G}^2 + 4} \tag{4.15d}$$

Then according to Eq. (4.14), the reliability index and failure probability based on the 3P lognormal distribution are obtained as (Zhao and Ono 2001)

$$\beta_{3M} = \frac{-\text{sign}(\alpha_{3G})}{\sqrt{\ln(A)}} \ln\left[\sqrt{A}\left(1 + \frac{\beta_{2M}}{u_b}\right)\right] \tag{4.16a}$$

$$P_F = \Phi(-\beta_{3M}) \tag{4.16b}$$

For $-1 < \alpha_{3G} < 1$, the standardised bound u_b is simplified as (Zhao and Ang 2003)

$$u_b = \frac{3}{\alpha_{3G}}$$

Eq. (4.15a) can be simplified as the following approximation (see in detail in Appendix B.2)

$$U = \frac{\alpha_{3G}}{6} + \frac{3}{\alpha_{3G}} \ln\left(1 + \frac{1}{3}\alpha_{3G}Z_s\right) \tag{4.17}$$

and the third-moment reliability index expressed in Eq. (4.16a) becomes

$$\beta_{3M-L} = -\frac{\alpha_{3G}}{6} - \frac{3}{\alpha_{3G}} \ln\left(1 - \frac{1}{3}\alpha_{3G}\beta_{2M}\right) \tag{4.18}$$

Hereafter, the simplified third-moment reliability index defined by Eq. (4.18) is referred to as β_{3M-L}.

In particular, with the aid of a second-order Taylor expansion, it can be observed that as x approaches 0, $\ln(1+x) = x - x^2/2$. Eq. (4.18) becomes

$$\beta_{3M-L} = -\frac{\alpha_{3G}}{6} - \frac{3}{\alpha_{3G}} \ln\left(1 - \frac{1}{3}\alpha_{3G}\beta_{2M}\right) = -\frac{\alpha_{3G}}{6} - \frac{3}{\alpha_{3G}}\left[-\frac{\alpha_{3G}}{3}\beta_{2M} - \frac{1}{2}\left(-\frac{1}{3}\alpha_{3G}\beta_{2M}\right)^2\right]$$

That is

$$\beta_{3M} = \beta_{2M} + \frac{1}{6}\alpha_{3G}\left(\beta_{2M}^2 - 1\right) \tag{4.19}$$

This implies that β_{3M-L} approaches β_{2M} for extremely small values of α_{3G}.

For negative α_{3G}, Eq. (4.18) is valid for any values of β_{2M}. However, Eq. (4.18) is valid only if $\beta_{2M} < 3/\alpha_{3G}$ when α_{3G} is positive. It can also be observed that Eq. (4.18) is always monotonically increasing for $\beta_{2M} > 0$ since $u_b > 0$ for $\alpha_{3G} < 0$ and $u_b < 0$ for $\alpha_{3G} > 0$.

Note that when $\alpha_{3G} = 0$, Eq. (4.18) cannot directly give appropriate results. However, since the limitation of Z_s is U (Appendix B.2), i.e.

$$\lim_{\alpha_{3G} \to 0} Z_s = U$$

One can easily understand that the limitation of β_{3M-L} is β_{2M}, i.e.

$$\lim_{\alpha_{3G} \to 0} \beta_{3M-L} = \beta_{2M}$$

Therefore, in this case, the third-moment method can be directly expressed as

$$\beta_{3M-L} = \beta_{2M} \quad \text{for} \quad \alpha_{3G} = 0$$

4.2.2.2 Third-Moment Reliability Index Based on the Square Normal Distribution

Another rational and practical approach to obtain the third-moment reliability index is to use the square normal distribution (see in detail in Appendix B.3), from which the x-u and u-x transformations are expressed as (Zhao and Ono 2000b),

$$U = \frac{1}{2\lambda}\left(\sqrt{1 + 2\lambda^2 + 4\lambda X_s} - \sqrt{1 - 2\lambda^2}\right) \tag{4.20a}$$

$$X_s = -\lambda + \sqrt{1 - 2\lambda^2}U + \lambda U^2 \tag{4.20b}$$

where

$$\lambda = \text{sign}(\alpha_{3G})\sqrt{2}\cos\left[\frac{\pi + |\theta|}{3}\right], \theta = \tan^{-1}\left(\frac{\sqrt{8 - \alpha_{3G}^2}}{\alpha_{3G}}\right) \tag{4.20c}$$

The signs in Eq. (4.20c) are taken to be the sign of α_{3G}.

Then, according to Eq. (4.14), the third-moment reliability index based on the square normal distribution is obtained as

$$\beta_{3M} = \frac{1}{2\lambda}\left(\sqrt{1 - 2\lambda^2} - \sqrt{1 + 2\lambda^2 - 4\lambda\beta_{2M}}\right) \tag{4.21}$$

For $-1 < \alpha_{3G} < 1$, Eq. (4.20c) can further be simplified as the following equation with an error of less than 2% (Zhao et al. 2001).

$$\lambda = \frac{\alpha_{3G}}{6}$$

The x-u transformation can be simplified as (see details in Appendix B.3)

$$U = \frac{1}{\alpha_{3X}}\left(\sqrt{9 + \alpha_{3X}^2 + 6\alpha_{3X}X_s} - 3\right) \tag{4.22}$$

and the third-moment reliability index given by Eq.(4.21) becomes

$$\beta_{3M-S} = \frac{1}{\alpha_{3G}}\left(3 - \sqrt{9 + \alpha_{3G}^2 - 6\alpha_{3G}\beta_{2M}}\right) \tag{4.23}$$

Hereafter, the simplified third-moment reliability index defined by Eq. (4.23) is referred to as β_{3M-S}.

For very small $|\alpha_{3G}|$, applying a second-order Taylor expansion to the square root term in Eq. (4.23) leads to

$$\beta_{3M-S} = \frac{3}{\alpha_{3G}}\left\{1 - \left[1 + \frac{1}{2} \times \frac{(\alpha_{3G}^2 - 6\alpha_{3G}\beta_{2M})}{9} - \frac{1}{8} \times \frac{(\alpha_{3G}^2 - 6\alpha_{3G}\beta_{2M})^2}{9^2}\right]\right\}$$

$$= \beta_{2M} + \frac{1}{6}\alpha_{3G}(\beta_{2M}^2 - 1)$$

which is observed to be exactly the same as Eq. (4.19). This implies that, for very small $|\alpha_{3G}|$, both the third-moment reliability indices can be expressed as Eq. (4.19).

It can be observed that Eq. (4.23) is valid for any values of β_{2M} when α_{3G} is negative. However, for positive α_{3G}, in order to make β_{3M-S} operable, β_{2M} and α_{3G} should satisfy the following.

$$\beta_{2M} \leq \frac{3}{2\alpha_{3G}} + \frac{1}{6}\alpha_{3G} \tag{4.24}$$

Example 4.3 Consider the following performance function previously discussed in Example 3.5

$$G(\mathbf{X}) = 567fr - 0.5H^2$$

The first four moments of $G(\mathbf{X})$ have been calculated as

$$\mu_G = 203.2319, \sigma_G = 104.821, \alpha_{3G} = 0.02753, \alpha_{4G} = 3.0089$$

Using the first two moments, the second-moment reliability index is given as

$$\beta_{2M} = \frac{\mu_G}{\sigma_G} = \frac{203.23}{104.82} = 1.939$$

When using the reliability index based on the 3P lognormal distribution, i.e. Eq. (4.16a), the related parameters are firstly obtained as

$$a = -\frac{1}{\alpha_{3G}}\left(\frac{1}{2} + \frac{1}{\alpha_{3G}^2}\right) = -\frac{1}{0.02753}\left(\frac{1}{2} + \frac{1}{0.02753^2}\right) = -47945.3$$

$$b = \frac{1}{2\alpha_{3G}^2}\sqrt{\alpha_{3G}^2 + 4} = \frac{1}{2 \times 0.02753^2}\sqrt{0.02753^2 + 4} = 1319.56$$

$$u_b = (a+b)^{\frac{1}{3}} + (a-b)^{\frac{1}{3}} - \frac{1}{\alpha_{3G}} = (-46625.74)^{\frac{1}{3}} + (-49264.86)^{\frac{1}{3}} - \frac{1}{0.02753} = -108.975$$

$$A = 1 + \frac{1}{u_b^2} = 1 + \frac{1}{(-108.975)^2} = 1.00008$$

Then, the third-moment reliability index and failure probability are obtained as

$$\beta_{3M} = \frac{-\operatorname{sign}(\alpha_{3G})}{\sqrt{\ln(A)}}\ln\left[\sqrt{A}\left(1 + \frac{\beta_{2M}}{u_b}\right)\right]$$

$$= \frac{-1}{\sqrt{\ln(1.00008)}}\ln\left[\sqrt{1.00008}\left(1 + \frac{1.9389}{-108.975}\right)\right] = 1.9518$$

$$P_F = \Phi(-1.9518) = 0.02548$$

Using Eq. (4.18), β_{3M-L} can be obtained as

$$\beta_{3M-L} = -\frac{\alpha_{3G}}{6} - \frac{3}{\alpha_{3G}}\ln\left(1 - \frac{1}{3}\alpha_{3G}\beta_{2M}\right) = 1.952$$

While using the reliability index based on the square normal distribution, i.e. Eq. (4.21), the related parameters are firstly obtained as

$$\theta = \tan^{-1}\left(\frac{\sqrt{8 - \alpha_{3G}^2}}{\alpha_{3G}}\right) = \tan^{-1}\left(\frac{\sqrt{8 - 0.02753^2}}{0.02753}\right) = 1.5611$$

$$\lambda = \operatorname{sign}(\alpha_{3G})\sqrt{2}\cos\left(\frac{\pi + |\theta|}{3}\right) = \sqrt{2}\cos\left(\frac{\pi + 1.5611}{3}\right) = 0.0045884$$

Then, the third-moment reliability index and failure probability are obtained as

$$\beta_{3M} = \frac{\sqrt{1 - 2\lambda^2} - \sqrt{1 + 2\lambda^2 - 4\lambda\beta_{2M}}}{2\lambda}$$

$$= \frac{\sqrt{1 - 2 \times 0.0046^2} - \sqrt{1 + 2 \times 0.0046^2 - 4 \times 0.0046 \times 1.9389}}{2 \times 0.0046} = 1.9517$$

$$P_F = \Phi(-1.9517) = 0.02548$$

Using Eq. (4.23), the β_{3M-S} reliability index can be obtained as

$$\beta_{3M-S} = \frac{1}{\alpha_{3G}}\left(3 - \sqrt{9 + \alpha_{3G}^2 - 6\alpha_{3G}\beta_{2M}}\right) = 1.952$$

As can be observed, β_{3M-L} and β_{3M-S} provide almost the same results for this example.

Example 4.4 Consider the performance function in Example 3.6,

$$G(\mathbf{X}) = \frac{l}{50} - \frac{ql^3}{8EI} = 0.08 - 8\frac{q}{EI}$$

The first four moments of $G(\mathbf{X})$ have been calculated as

$$\mu_G = 0.0281, \sigma_G = 0.01195, \alpha_{3G} = -0.703, \alpha_{4G} = 3.8915$$

Using the first two moments, the second-moment reliability index is given as

$$\beta_{2M} = \frac{\mu_G}{\sigma_G} = \frac{0.0281}{0.01195} = 2.352$$

Using Eq. (4.18), β_{3M-L} can be obtained as

$$\beta_{3M-L} = -\frac{\alpha_{3G}}{6} - \frac{3}{\alpha_{3G}} \ln\left(1 - \frac{1}{3}\alpha_{3G}\beta_{2M}\right) = 1.991$$

Using Eq. (4.23), β_{3M-S} can be obtained as

$$\beta_{3M-S} = \frac{1}{\alpha_{3G}}\left(3 - \sqrt{9 + \alpha_{3G}^2 - 6\alpha_{3G}\beta_{2M}}\right) = 2.0$$

For this example, it can be observed that the two third-moment reliability indices are slightly different.

4.2.3 Empirical Applicable Range of Third-Moment Method

Obviously, the third-moment reliability method is an approximation, and thus it is expected to have a range for its applicability. The typical PDFs of 3P distributions are depicted in Figure A.11. It can be observed from Figure A.11 that the left tail of the PDF is long for negative α_{3G}, while the right tail is long for positive α_{3G}. Because the probability of failure is integrated in the left tail according to Eq. (4.7), it is easy to understand that the third-moment method provides better results for negative α_{3G} than positive α_{3G}.

The reliability index of β_{3M-S} derived from Eq. (4.20b) means that the performance function is approximated as a second order polynomial of u. That is to say, the first three moments only determine a second polynomial of u. Since the performance function is approximated using second polynomials of u in the third moment standardisation, the third-moment method may not be applicable to a performance function with more than second power of u.

Because a practical reliability problem should have only one solution, the two third-moment reliability indices are expected to give same results of failure probability. If the relative differences between β_{3M-L} and β_{3M-S} are beyond the allowable value, the third-moment method is considered to be no longer applicable. Similarly, the applicability of the third-moment method can be determined by the rule that the relative differences between β_{3M-L} and β_{3M-S} should be below the allowable value.

From Eqs. (4.18) and (4.23), it can be observed that although β_{3M-L} and β_{3M-S} are based on different probability distributions and expressed in different forms, they are both functions of β_{2M} and α_{3G}. Thus, the applicability of the third-moment reliability index can be determined based on the values of β_{2M} and α_{3G}.

Figures 4.3a and b present the variations of β_{3M-L} and β_{3M-S} with respect to β_{2M} for α_{3G} equal to −0.6, −0.4, −0.2, −0.05, 0, 0.05, 0.2, 0.4, and 0.6. From these figures, it can be observed that the smaller the β_{2M} and $|\alpha_{3G}|$, the smaller the differences between the two third-moment reliability indices, and that both of the two indices become closer to β_{2M} with

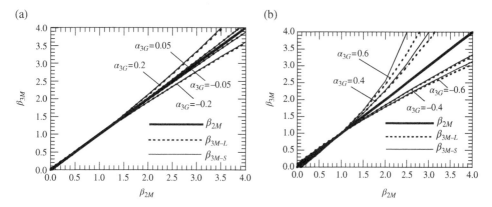

Figure 4.3 Third-moment reliability indices with respect to β_{2M}. (a) $\alpha_{3G} = \pm 0.05$, ± 0.2. (b) $\alpha_{3G} = \pm 0.4$, ± 0.6.

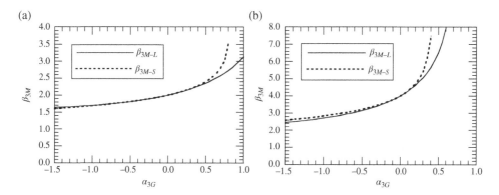

Figure 4.4 Third-moment reliability indices with respect to α_{3G}. (a) $\beta_{2M} = 2.0$. (b) $\beta_{2M} = 4.0$.

the decrease of β_{2M} and $|\alpha_{3G}|$. It can also be observed that the differences between the two indices are much larger for positive α_{3G} than those for negative α_{3G}. This is because the third-moment method is more suitable for negative α_{3G} than positive α_{3G}, as discussed earlier.

Figures 4.4a and b presents the variations of β_{3M-L} and β_{3M-S} with respect to α_{3G} for $\beta_{2M} = 2.0$ and 4.0, respectively. It can be observed that β_{3M-L} and β_{3M-S} are almost the same for negative α_{3G}. For positive values of α_{3G}, β_{3M-L} and β_{3M-S} are almost the same when α_{3G} is small. As α_{3G} increases, the difference between β_{3M-L} and β_{3M-S} also increases and becomes remarkable.

The relative differences between β_{3M-L} and β_{3M-S} is defined as

$$r = \frac{|\beta_{3M-L} - \beta_{3M-S}|}{(\beta_{3M-L} + \beta_{3M-S})/2} = \frac{2|\beta_{3M-L} - \beta_{3M-S}|}{\beta_{3M-L} + \beta_{3M-S}}$$

For practical cases, β_{2M} is generally considered to be not very small, and the discussion in this section is therefore focus on cases of $\beta_{2M} \geq 1$.

Using the means of non-linear fit with a large amount of data of relative differences between β_{3M-L} and β_{3M-S}, α_{3G} satisfying the allowable relative difference of r is approximately given as (Zhao et al. 2006a)

$$\alpha_{3G} \leq 40r/\beta_{2M} \qquad \text{for} \qquad \alpha_{3G} > 0 \tag{4.25a}$$

$$\alpha_{3G} \geq -120r/\beta_{2M} \qquad \text{for} \qquad \alpha_{3G} < 0 \tag{4.25b}$$

Thus the applicable range of the third-moment method for $\beta_{2M} \geq 1$ can be expressed as:

$$-120r/\beta_{2M} \leq \alpha_{3G} \leq 40r/\beta_{2M} \tag{4.26a}$$

Particularly, for $r = 2\%$, Eq. (4.26a) reduces to

$$-2.4/\beta_{2M} \leq \alpha_{3G} \leq 0.8/\beta_{2M} \tag{4.26b}$$

For the case of $\beta_{2M} = 2.0$, the applicable range of the third-moment method is $-1.2 \leq \alpha_{3G} \leq 0.4$, while for the case of $\beta_{2M} = 4.0$, the applicable range of the third-moment method is $-0.6 \leq \alpha_{3G} \leq 0.2$.

It should be noted that r here is considered only for a reference error index for third-moment reliability index, but cannot be used to assess the error in the third-moment reliability index. When Eq. (4.26b) is satisfied, the relative difference between the third-moment reliability indices will be less than 2%. However, the real error may be larger than 2% in each reliability index.

The applicable range of the third-moment method is illustrated in Figure 4.5 for $r = 0.5\%$, 1%, and 2%. One may observe that the third-moment reliability method is only valid when the absolute value of α_{3G} is small.

When $r = 2\%$, the range of the third-moment reliability index is shown in Figure 4.6. One may observe that the applicable third-moment reliability method defines a triangular area in which the relative difference between the third-moment reliability indices is less than the allowable error.

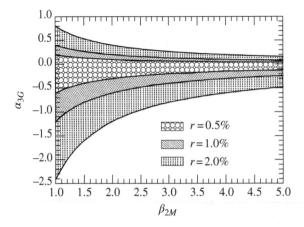

Figure 4.5 Empirical applicable range for third-moment reliability method.

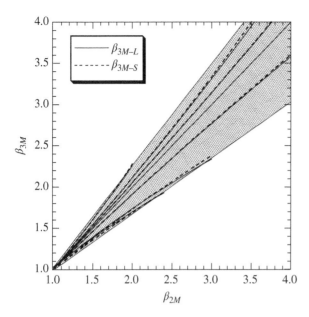

Figure 4.6 Empirical applicable range for third-moment reliability method when $r = 2\%$.

4.2.4 Simplification of Third-Moment Reliability Index

As described above, since the third-moment reliability index is only valid when the absolute value of α_{3G} is quite small, it will be simplified under the assumption of small skewness for the performance functions under investigation.

Within the applicable range of the third-moment method, a simple third-moment reliability index, using a trial and error method, can be derived as

$$\beta_{3M-E} = \beta_{2M} + \frac{\alpha_{3G}}{6}\left(\beta_{2M}^2 - 1\right)\left(1 + \frac{\alpha_{3G}}{6}\beta_{2M}\right) \tag{4.27}$$

where β_{3M-E} is the suggested third-moment reliability index for practical application.

Compared with β_{3M-L} and β_{3M-S}, the simplified formula in Eq. (4.27) does not include either logarithmic term or square root or unknown denominator. Furthermore, there is no other mathematical limitation. In other words, with the calculation error already constrained within an acceptable range, Eq. (4.27) is theoretically applicable to calculate the third-moment reliability index for any cases.

Figure 4.7a presents the variations of β_{3M-L}, β_{3M-S}, and β_{3M-E} with respect to α_{3G} for β_{2M} equal to 1.0, 2.0, 3.0, and 4.0. Similarly, the variations of β_{3M-L}, β_{3M-S}, and β_{3M-E} with respect to β_{2M} are presented in Figure 4.7b for α_{3G} equal to −0.3, −0.6, 0.3, 0.0, and 0.6. As can be observed from Figures 4.7a and b, the differences among the three third-moment reliability indices are smaller for negative values of α_{3G} than those for positive α_{3G}. As the values of α_{3G} and β_{2M} increase, the difference among β_{3M-L}, β_{3M-S}, and β_{3M-E} becomes larger. When α_{3G} and β_{2M} are large enough, β_{3M-L} and β_{3M-S} cannot be used as shown in Eqs. (4.18) and (4.23), in such circumstances, however, Eq. (4.27) can still be used to calculate the reliability index. β_{3M-E} can also be rewritten as

$$\beta_{3M} = \beta_{2M} + \frac{\alpha_{3G}}{6}\left(\beta_{2M}^2 - 1\right) + \frac{\alpha_{3G}^2}{36}\beta_{2M}\left(\beta_{2M}^2 - 1\right) \tag{4.28}$$

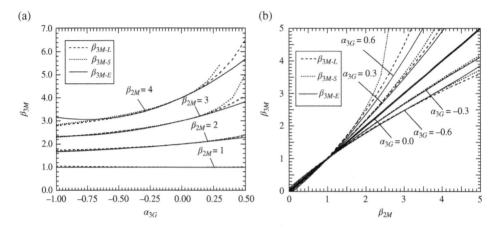

Figure 4.7 Changes of the three third-moment reliability indices. (a) β_{3M} with respect to α_{3G}. (b) β_{3M} with respect to β_{2M}.

When α_{3G} is small enough, β_{3M-E} can be simplified as

$$\beta_{3M-E} = \beta_{2M} + \frac{\alpha_{3G}}{6}\left(\beta_{2M}^2 - 1\right)$$

which is the same with Eq. (4.19). Furthermore, in the case of $\alpha_{3G} \to 0$, Eq. (4.27) reduces to $\beta_{3M} = \beta_{2M}$.

Example 4.5 Consider the following performance function

$$G(X) = R - S \tag{4.29}$$

where R and S are the resistance and the load effect, respectively.

Six cases are investigated under the assumption that R and S follow different probability distributions, and the statistical parameters of R and S are listed in Table 4.2, where μ_R and μ_S, and V_R and V_S are the mean and coefficient of variation of R and S, respectively.

The three third-moment reliability indices, i.e. β_{3M-L}, β_{3M-S}, and β_{3M-E}, changed with the variation of the central factor of safety $=(\mu_R/\mu_S)$ are shown in Figure 4.8, together with those obtained using MCS for 10^6 samples and the second-moment method. For cases 1–6, as

Table 4.2 The probability distribution information of R and S in different cases.

Case	R	S	μ_R	μ_S	V_R	V_S
1	Normal	Lognormal	30 ~ 120	30	0.2	0.4
2	Normal	Weibull	30 ~ 120	30	0.2	0.4
3	Gumbel	Normal	30 ~ 120	30	0.2	0.4
4	Lognormal	Weibull	30 ~ 120	30	0.2	0.4
5	Weibull	Gamma	30 ~ 120	30	0.2	0.4
6	Weibull	Gumbel	30 ~ 120	30	0.2	0.4

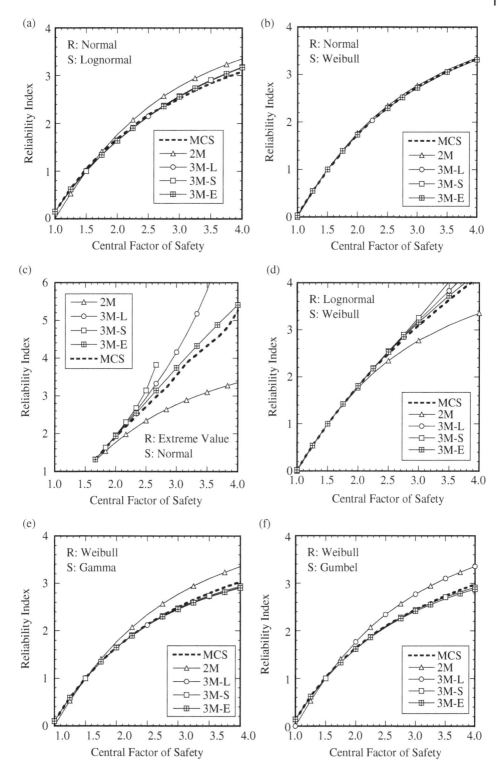

Figure 4.8 Reliability indices changed with the variation of the central factor of safety (= μ_R/μ_S). (a) Case 1. (b) Case 2. (c) Case 3. (d) Case 4. (e) Case 5. (f) Case 6.

shown in Figure 4.8, it can be observed that as the value of μ_R/μ_S increases, the results of the third-moment method move away from the 'exact values' from MCS. It is obvious that the second-moment method is not accurate enough, while the simplified third-moment method (β_{3M-E}) is in close agreement with the MCS for most cases. For cases 1, 2, 4, 5, and 6, it can be seen that the simplified third-moment method has, at least, the same accuracy as β_{3M-L} and β_{3M-S}. In case 3, when μ_R is larger than 76, β_{3M-S} is out of its applicable range, and the result of β_{3M-E} is closer to the MCS results than that of β_{3M-L}.

Example 4.6 Consider the following performance function, for a plastic collapse mechanism of a one-bay frame as shown in Figure 4.9, which has been used by Der Kiureghian et al. (1987).

$$G(\mathbf{X}) = X_1 + 2X_2 + 2X_3 + X_4 - 5X_5 - 5X_6$$

The variables X_i are statistically independent and lognormally distributed with the means $\mu_1 = \mu_2 = \mu_3 = \mu_4 = 120$, $\mu_5 = 50$, $\mu_6 = 40$, and standard deviations $\sigma_1 = \sigma_2 = \sigma_3 = \sigma_4 = 12$, $\sigma_5 = 15$, and $\sigma_6 = 12$.

Using Eqs. (3.5a)–(3.5c) in the chapter on Moment Evaluation, the first three moments of $G(\mathbf{X})$ can be easily obtained as $\mu_G = 270$, $\sigma_G = 103.271$, and $\alpha_{3G} = -0.5284$.

Using Eqs. (4.3) and (4.4), the second-moment reliability index and the corresponding failure probability are readily obtained as

$$\beta_{2M} = 270/103.271 = 2.615 \text{ and } P_F = 0.00447.$$

Using Eq. (4.26b), the application range of β_{3M-L} and β_{3M-S} is given as

$$-2.4/\beta_{2M} = -0.918 < \alpha_{3G} < 0.8/\beta_{2M} = 0.306$$

Obviously, $\alpha_{3G} = -0.528$ falls into the applicable range of the third-moment reliability method. Using Eq. (4.27), the third-moment reliability index β_{3M-E} is readily obtained as

$$\beta_{3M-E} = \beta_{2M} + \frac{\alpha_{3G}}{6}\left(\beta_{2M}^2 - 1\right)\left(1 + \frac{\alpha_{3G}}{6}\beta_{2M}\right) = 2.219$$

The probability of failure corresponding to the third-moment reliability index is equal to 0.01324.

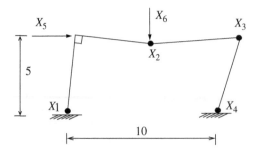

Figure 4.9 A plastic collapse mechanism of a one-bay frame of Example 4.6.

The exact value of the failure probability is $P_f = 0.0121$ (Der Kiureghian et al. 1987) and the corresponding reliability index is equal to 2.254. It can be observed that the results of the third-moment reliability index are in good agreement with the exact ones.

Example 4.7 Consider Example 4.4 again.
Using Eq. (4.27), the third-moment reliability index is readily obtained as $\beta_{3M} = 1.967$ with corresponding failure probability of $P_F = 0.0246$. It can be observed that the result of β_{3M-E} is close to those of β_{3M-L} and β_{3M-S}.

4.2.5 Applicable Range of the Second-Moment Method

It is well known that the second-moment method is only suitable when the performance function $G(\mathbf{X})$ can be approximately expressed by a normal random variable, of which the skewness α_{3G} is quite small. This section will explore the applicable range of the second-moment method.

As shown in the previous sections, when $|\alpha_{3G}|$ is small enough, all the third-moment reliability indices can be expressed as Eq. (4.19), i.e.

$$\beta_{3M} = \beta_{2M} + \frac{1}{6}\alpha_{3G}\left(\beta_{2M}^2 - 1\right)$$

If the relative errors between β_{3M} and β_{2M} are required to be below the allowable value r, as shown in Eq. (4.30), the second-moment method is expected to provide good results.

$$\left|\frac{\beta_{3M} - \beta_{2M}}{\beta_{3M}}\right| \leq r \tag{4.30}$$

Substituting Eq. (4.19) into the equation above, and, for $\beta_{2M} > 1$, it can be obtained as

$$|\alpha_{3G}| \leq \frac{6 \cdot r}{(\beta_{2M} - 1/\beta_{2M})} \tag{4.31a}$$

and then Eq. (4.31a) defines the applicable range of the second-moment method. Particularly, if r is assumed as 2%, then

$$|\alpha_{3G}| \leq \frac{0.12}{(\beta_{2M} - 1/\beta_{2M})} \tag{4.31b}$$

The applicable range of the second-moment method is presented in Figure 4.10 in comparison with that of the simple third-moment reliability index. From Figure 4.10, it can be observed that the applicable range is very small when $\beta_{2M} > 2$, and when β_{2M} is close to 1.0, the range of α_{3G} tends to same as that of the third-moment method.

For Example 4.3, using Eq. (4.31b), the range of skewness for which the second-moment reliability index is applicable is given as

$$|\alpha_{3G}| \leq \frac{0.12}{(\beta_{2M} - 1/\beta_{2M})} = \frac{0.12}{(1.939 - 1/1.939)} = 0.0843$$

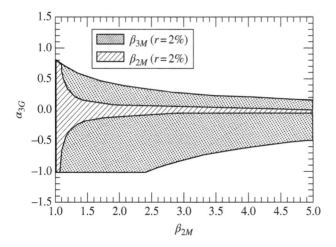

Figure 4.10 Applicable range of β_{3M} and β_{2M}.

Obviously, $\alpha_{3G} = 0.02753$ falls into the applicable range of the second-moment reliability method. Therefore, the second-moment reliability index gives good results.

In Example 4.6 for error $r = 2\%$, Eq. (4.31b) gives the application range of the second-moment reliability method as

$$|\alpha_{3G}| \leq \frac{0.12}{(\beta_{2M} - 1/\beta_{2M})} = \frac{0.12}{(2.615 - 1/2.615)} = 0.0537$$

Obviously, $\alpha_{3G} = -0.528$ falls out of the applicable range of the second-moment reliability method. Therefore, the second-moment reliability index deviates significantly from the exact value.

Example 4.8 Consider the following performance function from Example 3.3 in the chapter on Moment Evaluation

$$G(\mathbf{X}) = DR - S$$

The first three moments of $G(\mathbf{X})$ has been obtained as $\mu_G = 50$, $\sigma_G = 30.067$, and $\alpha_{3G} = -0.0994$.

Using Eq. (4.3), the second-moment reliability index and the corresponding failure probability are readily obtained as $\beta_{2M} = 1.663$, and $P_F = 0.04816$.

For error $r = 2\%$, Using Eq. (4.31b), the application range of the second-moment reliability method is given as

$$|\alpha_{3G}| \leq 0.113$$

Using Eq. (4.26b), the application range of the third-moment reliability method is given as

$$\text{Maximum}\,(-1, -1.443) = -1 < \alpha_{3G} < 0.481$$

Obviously, $\alpha_{3G} = -0.0994$ falls within the applicable range of both the third-moment reliability method the second-moment reliability method. Using Eq. (4.27), the third-moment reliability index is readily obtained as $\beta_{3M} = 1.635$, and the corresponding probability of failure is equal to 0.0551.

Using MCS with 1 000 000 trails, the failure probability is given as $P_f = 0.0477$ with COV of 0.45%, and the corresponding reliability index is equal to 1.668. Since the skewness is in the applicable range of the third-moment and second-moment reliability methods, both reliability indices are in close agreement with those from MCS.

Example 4.9   Consider the performance function $Z = G(\mathbf{X}) = R - S$ in Example 4.1, where R and S are statistical independent random variables with mean and standard deviations of $\mu_R = 100$, $\sigma_R = 10$ and $\mu_S = 50$, $\sigma_S = 20$. The first two central moments of $Z = G(\mathbf{X})$ are obtained as $\mu_G = 50$, $\sigma_G = 22.361$, and the second-moment reliability index is given as $\beta_{2M} = \mu_G/\sigma_G = 2.236$ with the corresponding probability of failure of $P_F = 0.0127$. As shown in Example 4.1, since the results of the second-moment method utilised the only the information of the first two central moments, it cannot match up the probability of failure under various cases of different combination of distributions.

Here, the probability of failure for various cases of different combination of distributions is re-estimated using the third-moment reliability method. The applicable range of the third-moment method is estimated as $-1 < \alpha_{3G} < 0.358$ for $r = 2\%$, and the results are presented in Table 4.3.

Table 4.3   Probability of failure with different distributions of random variables.

R	S	α_{3G}	β_{3M}	P_{F-3M}	P_{F-MCS}	β_{MCS}	Error in β_{3M} (%)
Normal	Normal	0.0	2.236	0.0127	0.0129	2.229	0.31
Normal	Lognormal	−0.904	1.836	0.0331	0.0283	1.906	−3.67
Normal	Weibull	−0.198	2.114	0.0173	0.0159	2.148	−1.58
Normal	Gumbel	−0.815	1.857	0.0316	0.0273	1.922	−3.38
Lognormal	Normal	0.0269	2.254	0.0121	0.0119	2.261	−0.31
Lognormal	Lognormal	−0.878	1.842	0.0327	0.0273	1.922	−4.16
Weibull	Normal	−0.064	2.194	0.0141	0.0142	2.192	0.09
Weibull	Lognormal	−0.968	1.824	0.0341	0.0287	1.900	−4.00
Gumbel	Normal	0.102	2.307	0.0105	0.0110	2.290	0.74
Gumbel	Lognormal	−0.802	1.861	0.0313	0.0270	1.927	−3.43
Gamma	Normal	0.0179	2.248	0.0123	0.0124	2.245	0.13
Gamma	Lognormal	−0.887	1.840	0.0329	0.0277	1.916	−3.97

From Table 4.3, one may observe that the failure probability obtained from the third-moment reliability index generally agree closely with the results of MCS. For some cases of relatively larger skewness, the error in the third-moment reliability index become relatively large, but may still be considered in applicable range for practical engineering use. It should also be noted that the error in the third-moment reliability indices is obtained using the MCS reliability index as reference values. Since the MCS reliability indices have inherent errors, these data are only used for the purpose of comparison.

Example 4.10 In order to investigate the applicable range affected by the probability distribution of random variables, consider the performance function again:

$$G(\mathbf{X}) = R - S$$

where R is resistance and S is load effect.

Because only two basic random variables are involved in the linear function, FORM generally gives good results (For details of FORM, See the chapter entitled Methods of Moment Based on First- and Second-Order Transformation).

In the following three cases, the coefficient of variation is taken to be 0.2 for R and 0.4 for S. It is also assumed that R and S follow different probability distributions.

- Case 1, R is normal with $\alpha_{3R} = 0.0$ and S is Weibull with $\alpha_{3S} = 0.2768$.
- Case 2, R is normal with $\alpha_{3R} = 0.0$ and S is lognormal with $\alpha_{3S} = 1.264$.
- Case 3, R is lognormal with $\alpha_{3R} = 0.608$ and S is Weibull with $\alpha_{3S} = 0.2768$.

Figures 4.11–4.13 present the variations of the reliability indices, the skewness, and the applicable range of the second-moment and third-moment methods with respect to μ_R/μ_S (the means of R and S, respectively) for the cases 1–3, respectively.

For Case 1, in Figure 4.11, it can be observed that the results of both the second-moment and third-moment methods are in close agreement with those of FORM for the entire investigation range since α_{3G} is in both the applicable ranges of the second-moment and third-moment methods.

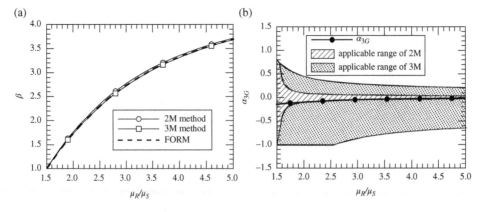

Figure 4.11 Variations of the reliability indices and the skewness for Case 1. (a) Reliability indices. (b) Applicable ranges.

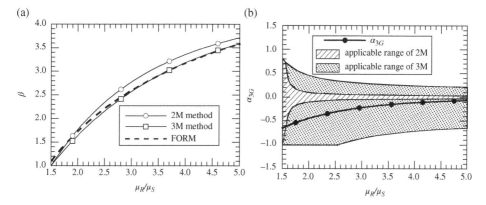

Figure 4.12 Variations of the reliability indices and the skewness for Case 2. (a) Reliability indices. (b) Applicable ranges.

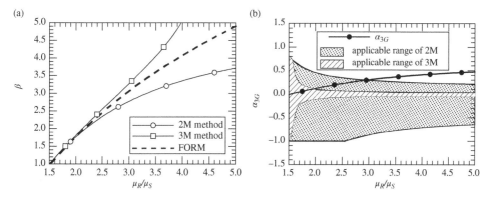

Figure 4.13 Variations of the reliability indices and the skewness for Case 3. (a) Reliability indices. (b) Applicable ranges.

From Figure 4.12, it can be observed that the results of the third-moment method for Case 2 are in close agreement with those of FORM for the entire investigation range because α_{3G} is always in the applicable range. The second-moment method, meanwhile, gives a good approximation for the results of FORM when μ_R/μ_S is small and has moderate errors when μ_R/μ_S is large when the skewness exceeds the applicable range.

For Case 3, it can be observed from Figure 4.13 that the results of both the second-moment and third-moment methods agree very closely with the FORM results when μ_R/μ_S is small. When μ_R/μ_S is large, the third-moment method has significant errors (especially when $\mu_R/\mu_S > 3.0$) due to α_{3G} being out of the applicable range, and the second-moment method also produces significant errors when $\mu_R/\mu_S > 2.0$, since the skewness of the performance function is out of the application range of the second-moment reliability method.

Another observation from the three cases above is that the skewness of the performance function varies significantly with μ_R/μ_S although the skewness of both the random variable remains constant. Also, although the skewness of S, one of the random variables in the performance function is quite large (Case 2), the skewness of the performance function is always in the applicable range in the whole investigation range. Needless to say, the applicable range of the third-moment reliability method depends on the skewness of the performance function not that of one of the random variables.

Example 4.11 Consider a performance function of $G(\mathbf{X}) = 1/S - 1/R$, where R and S are lognormal random variables with $\mu_R = 200$, $\mu_S = 100$, $V_R = 0.1$ and $V_S = 0.2$.
Let $X = 1/S$, $Y = 1/R$ then,

$$\mu_X = \frac{1}{\mu_S}\left(1 + V_S^2\right) = 0.0104, \mu_Y = \frac{1}{\mu_R}\left(1 + V_R^2\right) = 0.00505$$

$$V_X = V_S = 0.2, V_Y = V_R = 0.1, \alpha_{3X} = \alpha_{3S} = 0.608, \alpha_{3Y} = \alpha_{3R} = 0.301$$

With the aid of Eqs. (3.5a)–(3.5c) from the chapter on Moment Evaluation, one obtains

$$\beta_{2M} = 2.4995, \alpha_{3G} = 0.554$$

The third-moment reliability index is calculated as 3.096 with corresponding failure probability of 0.000981.

4.3 Fourth-Moment Reliability Method

4.3.1 General Formulation of the Fourth-Moment Reliability Index

As described in Appendix A.3.4, the fourth moment of a random variable, i.e. kurtosis, expresses the sharpness of its probability distribution. The PDFs of some standardised random variables with different kurtosis are shown in Figure 4.14. It can be observed that the kurtosis has a large effect on the probability of failure. Therefore, in order to achieve more accurate estimation of the probability of failure, the fourth-order moment of performance function is explored, in addition to the first three moments.

In order to account for the effect of the kurtosis on probability of failure, one may estimate the failure probability using Eq. (4.11) or evaluate the reliability index using Eq. (4.12). From Eqs. (4.11) and (4.12), one can easily understand that there are generally three routes to estimate the failure probability or the reliability index using the first four central moments of the performance function, i.e. to use the expression of the CDF or PDF of the performance function, or the expression of x–u or u–x transformation.

As the first way to directly use the expression of the CDF of the performance function, a simple idea for estimating failure probability through Eq. (4.11) is to use the Edgeworth expansion or Gram-Charlier series (Stuart and Ord 1987) where the CDF is directly expressed as explicit polynomials of the standardised performance function, Z_S. The Gram-Charlier series have been investigated by Hong (1996) and it is found that this series provides unsuitable results. Using the Edgeworth expansion, the probability distribution function of the standardised variable in Eq. (4.5) is expressed by the first four moments using the following expansion:

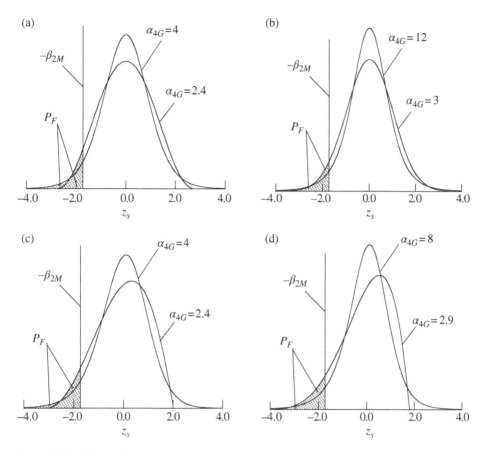

Figure 4.14 Effects of kurtosis on failure probability. (a) $\alpha_{3G} = 0$ ($\alpha_{4G} = 2.4, 4.0$). (b) $\alpha_{3G} = 0$ ($\alpha_{4G} = 3.0$, 12.0). (c) $\alpha_{3G} = -0.3$ ($\alpha_{4G} = 2.4, 4.0$). (d) $\alpha_{3G} = -0.6$ ($\alpha_{4G} = 2.9, 8.0$).

$$F_{Z_s}(z_s) = \Phi(z_s) - \phi(z_s)\left[\frac{1}{6}\alpha_{3G}H_2(z_s) + \frac{1}{24}(\alpha_{4G} - 3)H_3(z_s) + \frac{1}{72}\alpha_{3G}^2 H_5(z_s)\right] \quad (4.32)$$

where

$H_2(x) = x^2 - 1$; $H_3(x) = x^3 - 3x$; and $H_5(x) = x^5 - 10x^3 + 15x$ are the second-, third- and fifth-order Hermite polynomials, respectively.

Using the relationship described in Eq. (4.11), the failure probability based on the fourth-moment method can be expressed as

$$P_F = \Phi(-\beta_{2M}) - \phi(-\beta_{2M})\left[\frac{1}{6}\alpha_{3G}H_2(-\beta_{2M}) + \frac{1}{24}(\alpha_{4G} - 3)H_3(-\beta_{2M}) + \frac{1}{72}\alpha_{3G}^2 H_5(-\beta_{2M})\right]$$

$$(4.33)$$

when $\alpha_{3G} = 0$, $\alpha_{4G} = 3$, Eq. (4.33) reduces to $P_F = \Phi(-\beta_{2M})$.

It has been shown by Zhao and Ono (2001) that Eq. (4.33) is generally monotonically increasing. However, the reliability index has significant errors when the central factor of safety is large and cannot always be obtained due to the fact that the definition of the Edgeworth expansion sometimes does not satisfy the definition of the probability distribution function. Therefore, the fourth-moment reliability method based on Edgeworth expansion or Gram-Charlier series will not be further discussed in this chapter.

As the second way of estimating the failure probability using the expression of the PDF of the performance function, a usual way is utilising the existing systems of frequency curves, such as the Pearson, Johnson, and Burr systems (Stuart and Ord 1987), and Ramberg's Lambda distribution (Ramberg and Schmeiser 1974). The Johnson system has been investigated by Parkinson (1978) and Hong (1996), and the Lambda distribution has been investigated by Grigoriu (1983) and Zhao et al. (2006b). It has been shown that the quality of approximation for the tail area of a distribution is relatively insensitive to the family selected (Pearson et al. 1979) and it is required to solve nonlinear equations in order to determine the parameters of the Johnson and Burr systems (Slifker and Shapiro 1980), and the Lambda distribution (Zhao et al. 2006b). This section will focus on the Pearson system.

The third way is to use the expression of the u–x/x–u transformation. For the standardised performance function Z_s, if the first four moments are obtained, the relationship between the standardised variable Z_s and the standard normal variable U can be expressed as the following functions using the first four moments of $Z = G(\mathbf{X})$

$$U = S(Z_s, \alpha_{3G}, \alpha_{4G}) \tag{4.34a}$$

$$Z_s = S^{-1}(U, \alpha_{3G}, \alpha_{4G}) \tag{4.34b}$$

where α_{3G} and α_{4G} are the third and fourth moments of $Z = G(\mathbf{X})$, respectively; and S^{-1} is the inverse function of S.

According to Eq. (4.12), the fourth-moment reliability index, β_{4M}, can thus be expressed as

$$\beta_{4M} = -S(-\beta_{2M}, \alpha_{3G}, \alpha_{4G}) \tag{4.35}$$

In this Section, we will derive the reliability index using Pearson system of distributions and explicit expressions of the u–x/x–u transformation.

4.3.2 Fourth-Moment Reliability Index on the Basis of the Pearson System

For the standardised performance function Z_S in Eq. (4.5), f, the PDF of Z_S, satisfies the following differential equation in the Pearson system (See Appendix C.2).

$$\frac{1}{f}\frac{df}{dZ_S} = -\frac{aZ_S + b}{c + bZ_S + dZ_S^2} \tag{4.36}$$

where the parameters a, b, c, and d are expressed as

$$a = 10\alpha_{4G} - 12\alpha_{3G}^2 - 18 \tag{4.37a}$$

$$b = \alpha_{3G}(\alpha_{4G} + 3) \tag{4.37b}$$

$$c = 4\alpha_{4G} - 3\alpha_{3G}^2 \tag{4.37c}$$

$$d = 2\alpha_{4G} - 3\alpha_{3G}^2 - 6 \tag{4.37d}$$

Using the relationship in Eq. (4.12), the fourth-moment reliability index based on the Pearson system β_{4M-P} is given as:

$$\beta_{4M-P} = -\Phi^{-1}(P_F) = -\Phi^{-1}\left[\int_{-\infty}^{-\beta_{2M}} f(z_s)dz_s\right] \tag{4.38}$$

Hereafter, the reliability index expressed by Eq. (4.38) is denoted as the reliability index β_{4M-P}.

Solving Eq. (4.36), $f(\cdot)$, the PDF of Z_S, is as follows, depending on the values of the parameters a, b, c, and d (see Appendix C.2).

$$f(z_s) = K(z_s - r_2)^{\frac{-ar_2-b}{\sqrt{\Delta}}}(r_1 - z_s)^{\frac{ar_1+b}{\sqrt{\Delta}}} \qquad \text{for } \Delta > 0, d < 0 \tag{4.39a}$$

$$f(z_s) = K(c + bz_s)^{\frac{ac-b^2}{b^2}}\exp\left[-\frac{az_s}{b}\right] \qquad \text{for } \Delta > 0, d = 0 \tag{4.39b}$$

$$f(z_s) = K|z_s - r_1|^{\frac{ar_1+b}{\sqrt{\Delta}}}|z_s - r_2|^{\frac{-ar_2-b}{\sqrt{\Delta}}} \qquad \text{for } \Delta > 0, d > 0 \tag{4.39c}$$

$$f(z_s) = K|z_s - r_0|^{-\frac{a}{d}}\exp\left[\frac{ar_0+b}{d(z_s-r_0)}\right] \qquad \text{for } \Delta = 0 \tag{4.39d}$$

$$f(z_s) = K(c + bz_s + dz_s^2)^{-\frac{a}{2d}}\exp\left[\frac{ab-2bd}{d\sqrt{-\Delta}}\tan^{-1}\left(\frac{b+2dz_s}{\sqrt{-\Delta}}\right)\right] \qquad \text{for } \Delta < 0 \tag{4.39e}$$

where K is determined from $F(\infty) = 1$, and

$$\Delta = b^2 - 4cd, r_1 = \frac{-b-\sqrt{\Delta}}{2d}, r_2 = \frac{-b+\sqrt{\Delta}}{2d}, r_0 = \frac{-b}{2d} \tag{4.39f}$$

One may note that when $\alpha_{3G} = 0$ and $\alpha_{4G} = 3$, Z_S becomes a standard normal variable; in this case, $\beta_{4M-P} = \beta_{2M}$.

Example 4.12 The example is the most likely failure mode of an elasto-plastic frame structure with six stories and three bays as shown in Figure 4.15, with probabilistic characteristics of the member strength and load listed in Table 4.4. The corresponding performance function is

$$G(X) = 2M_1 + 2M_4 + 2M_7 + 2M_{10} + 2M_{13} + M_{14} + M_{15} - 3.8S_2 - 7.6S_3 - 11.4S_4$$
$$- 15.2S_5 - 19S_6 \tag{4.40}$$

The skewness and kurtosis of the variables for member strength and load are also listed in Table 4.4. Using Eqs. (3.5a)–(3.5d) from the chapter on Moment Evaluation, the mean, standard deviation, skewness, and kurtosis of $G(X)$ are readily obtained as $\mu_G = 619$, $\sigma_G = 154.285$, $\alpha_{3G} = -0.694$, and $\alpha_{4G} = 4.084$. The second-moment reliability index is readily obtained as $\beta_{2M} = 4.012$.

Using Eqs. (4.37a)–(4.37d), the coefficients of the Pearson system can be obtained as

$$a = 10\alpha_{4G} - 12\alpha_{3G}^2 - 18 = 10 \times 4.084 - 12 \times (-0.694)^2 - 18 = 17.604$$

$$b = \alpha_{3G}(\alpha_{4G} + 3) = -0.694 \times (4.084 + 3) = -4.9163$$

$$c = 4\alpha_{4G} - 3\alpha_{3G}^2 = 4 \times 4.084 - 3 \times (-0.694)^2 = 14.8911$$

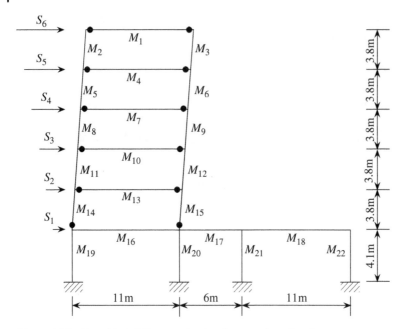

Figure 4.15 Most likely failure mode of a six-story three-bay frame.

Table 4.4 Random variables in Example 4.12.

Variables (Independent and Lognormal)	Mean	Coefficient of variation	Skewness	Kurtosis
M_1, M_4, M_7, M_{17}, M_{18}	90.8 tm	0.1	0.301	3.1615
M_2, M_3, M_5, M_6	145.2 tm	0.1	0.301	3.1615
M_8, M_9, M_{21}, M_{22}	145.2 tm	0.1	0.301	3.1615
M_{10}, M_{13}, M_{16}	103.4 tm	0.1	0.301	3.1615
M_{11}, M_{12}, M_{14}	162.8 tm	0.1	0.301	3.1615
M_{15}, M_{19}, M_{20}	162.8 tm	0.1	0.301	3.1615
S_1	2.5 t	0.4	1.264	5.969
S_2	5.0 t	0.4	1. 264	5.969
S_3	7.5 t	0.4	1.264	5.969
S_4	10.0 t	0.4	1.264	5.969
S_5	12.5 t	0.4	1.264	5.969
S_6	15.0 t	0.4	1.264	5.969

$$d = 2a_{4G} - 3a_{3G}^2 - 6 = 2 \times 4.084 - 3 \times (-0.694)^2 - 6 = 0.7231$$

Then with the aid of Eq. (4.39f), Δ can be calculated as

$$\Delta = b^2 - 4cd = (-4.9163)^2 - 4 \times 14.8911 \times 0.7231 = -18.9$$

Since $\Delta < 0$, the PDF of $G(\mathbf{X})$ is the same as Eq. (4.39e) with

$$r_0 = \frac{-b}{2d} = \frac{4.9163}{2 \times 0.7231} = 3.399$$

$$K = \frac{1}{\int_{-\infty}^{+\infty} (c + bz_s + dz_s^2)^{-\frac{a}{2d}} \exp\left[\frac{ab - 2bd}{d\sqrt{-\Delta}} \tan^{-1}\left(\frac{b + 2dz_s}{\sqrt{-\Delta}}\right)\right] dz_s} = 29496.1$$

According to Eq. (4.38), one obtains that

$$\beta_{4M-P} = -\Phi^{-1}(P_F) = -\Phi^{-1}\left[\int_{-\infty}^{-4.0121} f(z_s) dz_s\right] = 2.950$$

$$P_F = \Phi(-2.950) = 0.00159$$

Using the method of MCS with 500 000 samples, the probability of failure for this performance function is obtained as 0.001598 with corresponding reliability index of $\beta = 2.948$. The FORM (see the chapter on Methods of Moment Based on First- and Second-Order Transformation) reliability index is $\beta_F = 3.100$, which corresponds to a failure probability of $P_f = 0.000\,968$. It can be observed that the probability of failure obtained using the fourth-moment reliability index of β_{4M-P} is closer to the result of MCS than that from FORM.

4.3.3 Fourth-Moment Reliability Index Based on Third-Order Polynomial Transformation

For the standardised performance function Z_S, if the first four moments are obtained, the third-order polynomial transformation is given as (Fleishman 1978; Hong and Lind 1996; Zhao et al. 2002b; Chen and Tung 2003; Zhao and Lu 2007a; Zhao et al. 2018a)

$$Z_s = S^{-1}(U) = a_1 + a_2 U + a_3 U^2 + a_4 U^3 \tag{4.41}$$

where Z_S is the standardised variable described in Eq. (4.5); U is the standard normal random variable; and a_1, a_2, a_3, and a_4 are deterministic coefficients that can be obtained by making the first four central moments of $S^{-1}(U)$ equal to those of Z_S. The detailed procedure for determining the four coefficients is presented in the chapter entitled Transformation of Non-Normal Variables to Independent Normal Variables .

As described in Appendix C, a probability distribution can be defined based on Eq. (4.41) and it is referred to as cubic normal distribution (Zhao and Lu 2008; Zhao et al. 2018b). Therefore, the reliability index based on Eq. (4.41) is denoted as β_{4M-C}.

According to Eq. (4.12), the fourth-moment reliability index β_{4M-C} is obtained as

$$\beta_{4M-C} = -S(-\beta_{2M}, a_{3G}, a_{4G}) \tag{4.42}$$

where $S(\cdot)$ is the inverse function of $S^{-1}(\cdot)$ in Eq. (4.41), and its complete monotonic expression (Zhao et al. 2018a) can be found in Section 7.4.3.3 in the chapter entitled Transformation of Non-Normal Variables to Independent Normal Variables.

Then, the corresponding probability of failure is given by

$$P_F = \Phi(-\beta_{4M-C}) \tag{4.43}$$

For most combinations of skewness and kurtosis, the fourth moment reliability index can be obtained using the expression of $S(\cdot)$ in Eq. (7.75), which is expressed as

$$\beta_{4M-C} = a/3 - \sqrt[3]{A} - \sqrt[3]{B} \tag{4.44a}$$

In general, with the aid of complete monotonic expression of $S(\cdot)$, the fourth-moment reliability index β_{4M-C} can be derived as

$$\beta_{4M-C} = \begin{cases} a/3 + 2r\cos\left[(\theta + \pi)/3\right], & a_4 < 0 \text{ and } -J_{1s}^* < \beta_{2M} < -J_{2s}^* \\ a/3 - 2r\cos(\theta/3), & a_4 > 0 \text{ and } p < 0 \text{ and } \alpha_{3G} \geq 0 \text{ and } -J_{2s}^* < \beta_{2M} < -J_{1s}^* \\ a/3 - \sqrt[3]{A} - \sqrt[3]{B}, & a_4 > 0 \text{ and } p < 0 \text{ and } \alpha_{3G} \geq 0 \text{ and } \beta_{2M} \leq -J_{2s}^* \\ a/3 + 2r\cos\left[(\theta - \pi)/3\right], & a_4 > 0 \text{ and } p < 0 \text{ and } \alpha_{3G} < 0 \text{ and } -J_{2s}^* < \beta_{2M} < -J_{1s}^* \\ a/3 - \sqrt[3]{A} - \sqrt[3]{B}, & a_4 > 0 \text{ and } p < 0 \text{ and } \alpha_{3G} < 0 \text{ and } \beta_{2M} \geq -J_{1s}^* \\ a/3 - \sqrt[3]{A} - \sqrt[3]{B}, & a_4 > 0 \text{ and } p \geq 0 \\ \left[a_2 - \sqrt{a_2^2 + 4a_3(a_3 - \beta_{2M})}\right]/(2a_3), & a_4 = 0 \text{ and } \alpha_{3G} \neq 0 \text{ and } a_2^2 + 4a_3(a_3 - \beta_{2M}) \geq 0 \\ \beta_{2M}, & a_4 = 0 \text{ and } \alpha_{3G} = 0 \end{cases} \tag{4.44b}$$

where

$$A = -\frac{q}{2} + \sqrt{\Delta}, B = -\frac{q}{2} - \sqrt{\Delta}, \theta = \arccos\left(\frac{-q}{2r^3}\right), r = \sqrt{-\frac{p}{3}} \tag{4.45}$$

$$\Delta = \left(\frac{p}{3}\right)^3 + \left(\frac{q}{2}\right)^2, p = c - a^2/3, q = \frac{2}{27}a^3 - \frac{ac}{3} - a + \frac{\beta_{2M}}{a_4} \tag{4.46}$$

$$J_{1s}^* = a_4\left(-2r^3 + \frac{2}{27}a^3 - \frac{ac}{3} - a\right), J_{2s}^* = a_4\left(2r^3 + \frac{2}{27}a^3 - \frac{ac}{3} - a\right) \tag{4.47}$$

$$a = \frac{a_3}{a_4}, c = \frac{a_2}{a_4}, \beta_{2M} = \frac{\mu_G}{\sigma_G} \tag{4.48}$$

Example 4.13 This example is the most likely failure mode of an elasto-plastic frame structure with two stories and two bays as shown in Figure 4.16, with the statistical parameters of the member strength and load listed in Table 4.5. The corresponding performance function is

$$G(X) = 2M_1 + 2M_2 + 2M_3 - 15S_1 - 15S_2$$

The skewness and kurtosis of the variables of member strength and load are also listed in Table 4.5. The mean, standard deviation, skewness and kurtosis of $G(X)$ are obtained as

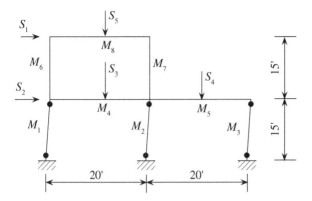

Figure 4.16 Most likely failure mode of a two-story two-bay frame.

Table 4.5 Random variables in Example 4.13.

Variables (Lognormal)	Mean	Coefficient of variation	Skewness	Kurtosis
M_1, M_2, M_3	70 K-ft	0.15	0.4534	3.3677
M_6, M_7	70 K-ft	0.15	0.4534	3.3677
M_4	150 K-ft	0.15	0.4534	3.3677
M_5	120 K-ft	0.15	0.4534	3.3677
M_8	90 K-ft	0.15	0.4534	3.3677
S_1	5 K	0.25	0.7656	4.0601
S_2	10 K	0.25	0.7656	4.0601
S_3	26.5 K	0.15	0.4534	3.3677
S_4	18 K	0.25	0.7656	4.0601
S_5	14 K	0.25	0.7656	4.0601

$\mu_G = 195$, $\sigma_G = 55.505$, $\alpha_{3G} = -0.192$, and $\alpha_{4G} = 3.257$. The second-moment reliability index is readily obtained as $\beta_{2M} = 3.513$.

Using Eqs. (7.72a)–(7.72f) in the chapter on Transformation of Non-Normal Variables to Independent Normal Variables, one obtains that $a_1 = 0.03052$, $a_2 = 0.974647$, $a_3 = -0.03052$, and $a_4 = 0.008075$, then according to Eqs. (4.45)-(4.48) one obtains that

$$a = a_3/a_4 = -0.03052/0.008075 = -3.77957, c = a_2/a_4 = 0.974647/0.008075 = 120.6993$$

$$p = c - a^2/3 = 120.6993 - (-3.77957)^2/3 = 115.9376$$

$$q = 2a^3/27 - ac/3 - a + \beta_{2M}/a_4$$
$$= 2 \times (-3.77957)^3/27 - (-3.77957) \times 120.6993/3 - (-3.77957) + 3.513/0.008075$$
$$= 586.8903$$

$$\Delta = \left(\frac{p}{3}\right)^3 + \left(\frac{q}{2}\right)^2 = \left(\frac{115.9375}{3}\right)^3 + \left(\frac{586.8903}{2}\right)^2 = 143827.8$$

$$A = -\frac{q}{2} + \sqrt{\Delta} = -\frac{586.8903}{2} + \sqrt{143827.8} = 85.8012$$

$$B = -\frac{q}{2} - \sqrt{\Delta} = -\frac{586.8903}{2} - \sqrt{143827.8} = -672.692$$

According to Eq. (4.44a), one obtains

$$\beta_{4M-C} = a/3 - \sqrt[3]{A} - \sqrt[3]{B} = (-3.77957)/3 - \sqrt[3]{85.8012} - \sqrt[3]{-67.692} = 3.092$$

The corresponding probability of failure is equal to 0.0009953.

Using the method of MCS with 500 000 samples, the probability of failure for this performance function is obtained as 0.001002 with corresponding reliability index of $\beta = 3.0896$. The reliability index can also be obtained using the method of FORM with $\beta_F = 3.099$, which corresponds to a failure probability of $P_f = 0.000971$. It can be observed that the probability of failure obtained using the fourth-moment reliability method is closer to the result of MCS for this example than that of FORM.

4.3.4 Applicable Range of Fourth-Moment Method

The fourth-moment reliability method is an approximation method, and is thus expected to have a range of applications. As shown in Figure A.11, the left tail of the PDF is long for negative α_{3G}, and the right tail is long for positive α_{3G}. Because the failure probability is integrated in the left tail according to Eq. (4.7), it is easy to understand that the fourth-moment reliability method is more suitable for negative α_{3G} than positive α_{3G}. This is the same as we described in the applicable range of the third-moment reliability method.

It can be observed clearly from Eq. (4.41) that the first four moments only determine a third polynomial of u. Since it is difficult to approximate a performance function with the third power of u using third polynomials of u, the fourth-moment reliability method may not be applicable to a performance function with more than the third power of u.

Because a practical reliability problem should have only one solution, both the fourth-moment reliability indices are expected to give similar results of failure probability for a specific reliability problem. If the relative differences between β_{4M-P} and β_{4M-C} are beyond the allowable value, it is thought that the fourth-moment method is out of its applicable range. Similarly, the applicable range of the fourth-moment method can be determined by the rule that the relative differences between β_{4M-P} and β_{4M-C} are below the tolerance value.

From Eqs. (4.38) and (4.42), it can be observed that although β_{4M-P} and β_{4M-C} are based on different probability distributions and described in different forms, they are functions of β_{2M}, α_{3G}, and α_{4G}. Thus, the applicable range of fourth-moment reliability index will be determined using β_{2M}, α_{3G}, and α_{4G} as parameters.

Figure 4.17 presents the variation of β_{4M-P} and β_{4M-C} with respect to β_{2M}. It can be observed that for positive α_{3G}, the smaller the β_{2M}, the smaller the differences between the two fourth-moment reliability indices, and the smaller the α_{3G}, the smaller the differences between the two fourth-moment reliability indices. While for negative α_{3G}, β_{4M-P} is in

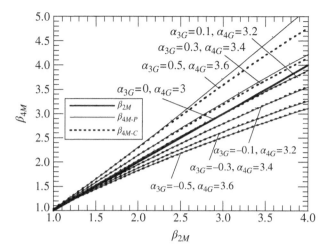

Figure 4.17 Fourth-moment reliability indices changed with respect to β_{2M}.

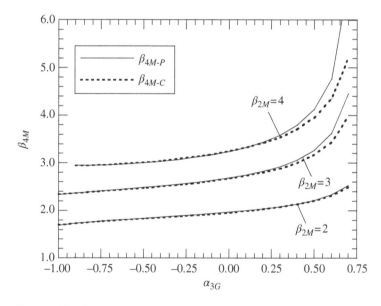

Figure 4.18 Fourth-moment reliability indices with respect to α_{3G} ($\alpha_{4G} = 4$).

close agreement with β_{4M-C} for the entire range and both of them become closer to β_{2M} with the decrease of β_{2M} or $|\alpha_{3G}|$. It can also be clearly observed that the differences between the two indices for positive α_{3G} are much larger than those for negative α_{3G}. This is because the fourth-moment reliability method is more suitable for negative α_{3G} than positive α_{3G}, as discussed earlier.

Figure 4.18 presents the variation of β_{4M-P} and β_{4M-C} with respect to α_{3G}. It can be observed that for negative α_{3G}, β_{4M-P} and β_{4M-C} are almost the same, while for positive

α_{3G}, β_{4M-P} and β_{4M-C} are almost the same when α_{3G} is small; however, as α_{3G} becomes larger the differences between β_{4M-P} and β_{4M-C} become larger.

Figure 4.19 presents the variation of β_{4M-P} and β_{4M-C} with respect to α_{4G}. It can be observed that β_{4M-P} and β_{4M-C} are almost the same when α_{4G} is close to 3.0. However, as α_{4G} goes lower than 3.0, the differences between β_{4M-P} and β_{4M-C} become larger. It can also be observed that the differences between the two indices for positive α_{3G} are much larger than those for negative α_{3G}, especially when $|\alpha_{3G}|$ is relatively large.

Consider the relative difference given as

$$r = 2(|\,\beta_{4M-P} - \beta_{4M-C}\,|)/(\beta_{4M-P} + \beta_{4M-C}) \tag{4.49}$$

For practical cases, β_{2M} is generally considered to be not very small, and therefore the discussion here is concentrated on cases of $\beta_{2M} \geq 1$.

For the case of $\alpha_{3G} > 0$, the range of α_{3G} and α_{4G}, for which the relative differences between β_{4M-P} and β_{4M-C} are less than the allowable relative difference $r = 2\%$, is given as (Zhao and Lu 2007c)

$$2.9 + \alpha_{3G} \leq \alpha_{4G} \leq 5.2 - 2.5\alpha_{3G}, 0 < \alpha_{3G} \leq 1.2/\beta_{2M} \tag{4.50}$$

While for case of $\alpha_{3G} < 0$, the range of α_{3G} and α_{4G}, for which the relative differences between β_{4M-P} and β_{4M-C} are less than the allowable relative difference $r = 2\%$, is given as (Zhao and Lu 2007c)

$$2.9 + \alpha_{3G}^2 \leq \alpha_{4G} \leq 5.2 + \alpha_{3G}^2, \alpha_{3G} \leq 0 \tag{4.51}$$

The applicable range of the fourth-moment reliability method is shown in Figure 4.20 for an allowable value $r = 2\%$. Figure 4.20a indicates the applicable range of the skewness α_{3G},

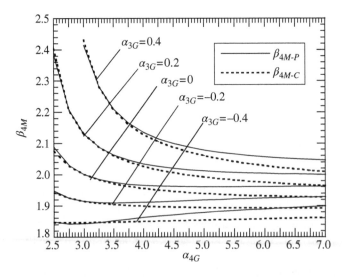

Figure 4.19 Fourth-moment reliability indices with respect to α_{4G}.

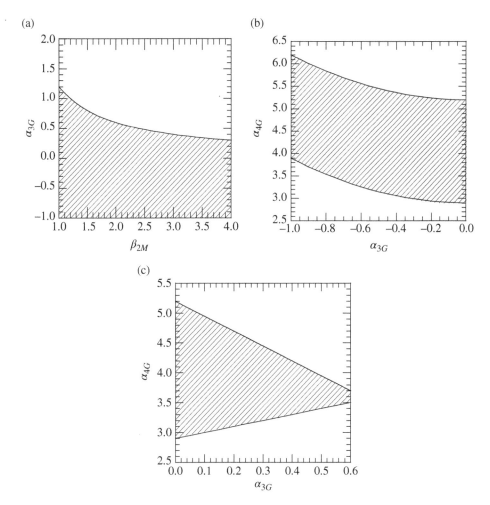

Figure 4.20 The applicable range of the fourth-moment method ($r = 2\%$). (a) α_{3G} v.s. β_{2M}. (b) α_{4G} v.s. $-\alpha_{3G}$. (c) α_{4G} v.s. α_{3G}.

and Figures 4.20b and c indicate the applicable range of the kurtosis α_{4G} for negative α_{3G} and positive α_{3G}, respectively.

Example 4.14 Consider a single rectangular reinforced concrete beam with width $b = 250$ mm, distance from extreme compression fibre to the centroid of tension reinforcement $d = 500$ mm, and area of tension reinforcement $A_s = 1529$ mm^2. The flexure capacity of the beam, M_f, is given by Eq. (4.52a) (MacGregor 1988).

$$M_f = A_s f_y d \left[1 - 0.59 A_s f_y / \left(f'_c b d \right)\right] \tag{4.52a}$$

where f_y is the yield strength of reinforcement and f'_c is the compressive strength of concrete.

The limit state function under dead and live loads, $G(\mathbf{X})$, is given as

$$G(\mathbf{X}) = B_m A_s f_y d \left[1 - 0.59 A_s f_y / \left(f_c' b d \right) \right] - D - L \qquad (4.52b)$$

where B_m is the modelling uncertainty factor for flexure, D is the dead load effect, and L is the maximum live load effect during 50 years. Assumptions of the uncertain variables are shown in Table 4.6. The assumed coefficients of variation for the uncertain parameters are in agreement with the values in the literature (Mirza and MacGregor 1982; Ellingwood et al. 1980; Israel 1986).

Using MCS with 1 000 000 samples, the probability of failure for this system is 2.76×10^{-3} with a corresponding reliability index of $\beta = 2.775$ (Hong and Lind 1996).

Using the seven-point estimate in the chapter on Moment Evaluation for Performance Functions, the first four moments of $G(\mathbf{X})$ are approximately $\mu_G = 99.221$, $\sigma_G = 34.347$, $\alpha_{3G} = 9.740 \times 10^{-3}$, and $\alpha_{4G} = 3.209$. The second-moment reliability index is readily obtained as $\beta_{2M} = 2.889$ with $P_f = 1.93 \times 10^{-3}$. Using Eq. (4.27), the third-moment reliability index is given as $\beta_{3M} = 2.901$ with $P_f = 1.86 \times 10^{-3}$. One may observe that the third-moment reliability index is almost equal to that of the second-moment method since the skewness is very small.

Using Eq. (4.50), the range of α_{4G} for which the fourth-moment reliability method is applicable is given as $[2.9, 5.2]$, it can be observed that the value of α_{4G} of this example is just in the range. The two fourth-moment reliability indices are obtained as

$$\beta_{4M-P} = 2.7924, P_F = 0.00262$$
$$\beta_{4M-C} = 2.7917, P_F = 0.00262$$

It can be observed that both the fourth-moment reliability indices are almost the same and give good improvement of the accuracy of the third-moment reliability index. For comparison, the FORM reliability index is $\beta_F = 2.886$, which corresponds to a failure probability of $P_f = 1.95 \times 10^{-3}$. It can be concluded that the probability of failure obtained using the fourth-moment reliability method is closer to the result of MCS for this example than that of FORM.

Table 4.6 Random variables in Example 4.14.

Variables	Mean	Coefficient of variation	Distribution
B_m	1.01	0.06	Normal
f_y	400 MPa	0.10	Lognormal
f_c'	20 MPa	0.18	Normal
D	95.87 kNm	0.10	Normal
L	67.11 kNm	0.25	Gumbel

Example 4.15 Consider a beam of circular cross section as shown in Figure 4.21 (Xu and Cheng 2003). The beam is simply supported at both ends and subjected to a distributive load q. There exists a spring support at the middle. Assuming that the vertical displacement at the middle of the beam is limited to 1 cm, the performance function can be given as

$$G(\mathbf{X}) = 1 - \frac{5qL^4}{8KL^3 + 6\pi ED^4} \tag{4.53a}$$

or

$$G(\mathbf{X}) = 8K + 6\pi ED^4/L^3 - 5qL \tag{4.53b}$$

where

K is the spring stiffness of lognormal distribution with $\mu_K = 10$ N/cm, $V_K = 0.1$;
E is the Young's modulus of lognormal distribution with $\mu_E = 2.1 \times 10^7$ N/cm^2, $V_E = 0.05$;
D is the diameter of the cross section of lognormal distribution with $\mu_D = 10$ cm, $V_D = 0.05$;
L is the length of the beam with deterministic value of $L = 200$ cm; and
q is the distributive load of lognormal distribution with $\mu_q = 200$ N/cm, $V_q = 0.3$.

Using MCS with 1 000 000 samples, the probability of failure for this system is 4.22×10^{-3} with a corresponding reliability index of $\beta = 2.634$.

Using the formulas in Section 3.2 of the chapter on Moment Evaluation, the first four moments of $G(\mathbf{X})$ are obtained as $\mu_G = 302\,349$, $\sigma_G = 120\,577$, $\alpha_{3G} = 0.301$, and $\alpha_{4G} = 3.516$. The second-moment reliability index is readily obtained as $\beta_{2M} = 2.508$ with $P_f = 0.006$. Using Eq. (4.27), the third-moment reliability index is given as $\beta_{3M} = 2.819$ with $P_f = 0.0024$.

Using Eq. (4.50), the range of α_{4G} for fourth-moment reliability method being applicable is given as [3.1994, 4.4515], it can be observed that the value of α_{4G} of this example is just in the range. The two fourth-moment reliability indices are obtained as

$$\beta_{4M-P} = 2.6513, P_F = 0.004$$
$$\beta_{4M-C} = 2.6373, P_F = 0.0042$$

It can be observed that both the fourth-moment reliability indices are almost the same with that from MCS and give good improvement of the accuracy of the third-moment reliability index.

When $L = 100$ cm, using the formulas in Section 3.2 of the chapter on Moment Evaluation, the first four moments of $G(\mathbf{X})$ are given as $\mu_G = 3\,918\,240$, $\sigma_G = 837\,247$, $\alpha_{3G} = 0.6326$,

Figure 4.21 The beam in Example 4.15.

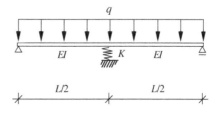

and $\alpha_{4G} = 3.7226$. The second-moment reliability index is readily obtained as $\beta_{2M} = 4.6799$ with $P_f = 1.435 \times 10^{-6}$. The range of α_{3G} for which the third-moment reliability method is applicable is given as $[-0.513, 0.171]$ and the range of α_{4G} for which the fourth-moment reliability method is applicable is given as $[3.532, 3.618]$. It can be observed that the values of α_{3G} and α_{4G} of this example are out of the applicable range of third-moment and fourth-moment reliability method. For reference, the two fourth-moment reliability indices are obtained as

$$\beta_{4M-P} = \text{cannot be given}$$

$$\beta_{4M-C} = 7.3, P_F = 1.436 \times 10^{-13}$$

That is to say, the fourth-moment reliability method cannot be used in this case.

4.3.5 Simplification of Fourth-Moment Reliability Index

For the fourth-moment reliability index based on the Pearson system, although the parameters in the expression can be directly determined by the first four central moments, there are 12 kinds of PDF in close form (see Appendix C) and the integration is necessary. While for the fourth-moment reliability index based on the cubic normal distribution has a single expression, the parameters should be determined by solving nonlinear equations (see the chapter entitled Transformation of Non-Normal Variables to Independent Normal Variables).

For obvious reasons, the fourth-moment reliability index for users or designers in practical engineering should be as simple and accurate as possible. A simple and accurate formula in close form for approximating Eq. (4.41) is given as (Zhao and Lu 2007a)

$$Z_s = -l_1 + k_1 U + l_1 U^2 + k_2 U^3 \tag{4.54a}$$

where

$$l_1 = \frac{\alpha_{3G}}{6(1 + 6l_2)}, l_2 = \frac{1}{36}\left(\sqrt{6\alpha_{4G} - 8\alpha_{3G}^2 - 14} - 2\right) \tag{4.54b}$$

$$k_1 = \frac{1 - 3l_2}{\left(1 + l_1^2 - l_2^2\right)}, k_2 = \frac{l_2}{\left(1 + l_1^2 + 12l_2^2\right)} \tag{4.54c}$$

From Eq. (4.54b), in order to make Eq. (4.54a) operable, the following condition should be satisfied

$$\alpha_{4G} \geq \left(7 + 4\alpha_{3G}^2\right)/3 \tag{4.54d}$$

Thus, the fourth-moment reliability index based on Eq. (4.54a) and the corresponding probability of failure can be expressed as follows (Lu et al. 2017b):

$$\beta_{4M} = \frac{\sqrt[3]{2p}}{\sqrt[3]{-q_0 + \Delta_0}} - \frac{\sqrt[3]{-q_0 + \Delta_0}}{\sqrt{2}} + \frac{l_1}{3k_2} \tag{4.55a}$$

$$P_F = \Phi(-\beta_{4M}) \tag{4.55b}$$

where

$$\Delta_0 = \sqrt{q_0^2 + 4p^3}, \quad p = \frac{3k_1k_2 - l_1^2}{9k_2^2}, \quad q_0 = \frac{2l_1^3 - 9k_1k_2l_1 + 27k_2^2(-l_1 + \beta_{2M})}{27k_2^3} \tag{4.56}$$

Hereafter, the fourth moment reliability index defined by Eq. (4.55a) is referred to as β_{4M}, and β_{4M} is the simple fourth-moment reliability index suggested for practical application in engineering.

Example 4.16 To investigate the insensitivity of the fourth-moment reliability method to the formulations of the limit-states function, consider the reliability problem shown in Table 4.7, of which both R and S are lognormal variables with mean value, standard deviation, skewness, and kurtosis of $\mu_R = 200$, $\sigma_R = 30$, $\alpha_{3R} = 0.453$, $\alpha_{4R} = 3.368$, $\mu_S = 100$, $\sigma_S = 20$, $\alpha_{3S} = 0.608$, and $\alpha_{4S} = 3.664$. Since both R and S are positive, the five performance functions in Table 4.7 are equivalent. The first four central moments of the performance functions obtained using the methods described in Section 3.2 of the chapter on Moment Evaluation are listed in Table 4.7 with the results of the second-moment and the fourth-moment reliability indices.

From Table 4.7, it can be observed that the fourth-moment reliability method is generally insensitive to the different formulations in its applicable range. For Case 4, the result of the fourth-moment reliability method shows moderate error. This is attributed to the fact that the reformulations of the performance function make the skewness beyond the applicable range of the fourth-moment method. Therefore, the insensitivity of the fourth-moment reliability method to the formulation of the limit-states should be limited in the applicable range.

In contrast, the second-moment method is very different for the various formulations, and it gives an exact result only for Case 2 because its performance function $G(X)$ is nearly a normal random variable in this case. As for FORM, it has almost the same results for the different formulations with the value of $\beta_F = 2.83$ and it gives good results for this example.

Table 4.7 Formula insensitivity of the fourth-moment method.

	G(X)	μ_G	σ_G	α_{3G}	α_{4G}	β_{2M}	β_{4M}
Case 1	$R - S$	100	36.06	0.16	3.24	2.77	2.84
Case 2	$\ln R - \ln S$	0.70	0.25	0	3.00	2.83	2.83
Case 3	$1 - S/R$	0.49	0.13	−0.77	4.08	3.80	2.84
Case 4	$R/S - 1$	1.08	0.52	0.77	4.08	2.06	2.77
Case 5	$1/S - 1/R$	5.29×10^{-3}	2.22×10^{-3}	0.48	3.52	2.39	2.82

Example 4.17   In order to investigate the application of fourth-moment reliability method for performance function with multiple design point in FORM, consider the following parabolic performance function by Der Kiureghian and Dakessian (1998).

$$G(\mathbf{X}) = b - X_2 - k(X_1 - e)^2$$

where $b = 5$, $k = 0.5$, and $e = 0.1$; X_1 and X_2 are standard normal random variables.

If FORM is used to solve this problem, there are two design points which are successfully obtained by Der Kiureghian and Dakessian (1998) as: $\mathbf{X}_1^* = [-2.741, 0.965]^T$ with $\beta_1 = 2.906$, and $\mathbf{X}_2^* = [2.916, 1.036]^T$ with $\beta_2 = 3.094$.

Using the formulas in Section 3.4 in the chapter on Moment Evaluation, the first four moments of $G(\mathbf{X})$ are easily obtained exactly as $\mu_G = 4.495$, $\sigma_G = 1.229$, $\alpha_{3G} = -0.555$, and $\alpha_{4G} = 4.368$. With the aid of Eqs. (4.27) and (4.55a), the moment-based reliability indices are

$$\beta_{2M} = 3.6575 \text{ with } P_F = 1.273 \times 10^{-4}$$
$$\beta_{3M} = 2.899 \text{ with } P_F = 0.00186$$
$$\beta_{4M} = 2.764 \text{ with } P_F = 0.0029$$

The reliability index using MCS obtained by Der Kiureghian and Dakessian (1998) is $\beta = 2.751$, and the corresponding failure probability is 2.97×10^{-3}. It can be observed that the result of the fourth-moment reliability method is in good agreement with the MCS result, whereas the second-moment method has significant error. Apparently, the moment method is applicable for the problems of multi-design points.

4.4   Summary

Direct methods of moment are introduced for structural reliability in this chapter. These methods are very straightforward and do not have disadvantages of FORM as we will describe in the chapter entitled Methods of Moment Based on First- and Second-Order Transformation, such as the design points and derivative-based iteration computation. Direct methods of moment thus are expected to be convenient for structural reliability analysis. The applicable ranges are determined for the second-, third-, and fourth-moment reliability methods. Simplified third- and fourth-moment reliability indices are discussed for structural reliability analysis in engineering.

Compared with the FORM based reliability methods that have been developed over a period of more than five decades, the methods of moment for structural reliability has been rarely studied. However, being very simple, the methods of moment are expected to be more convenient and more reliable for many performance functions in practical engineering.

5

Methods of Moment Based on First- and Second-Order Transformation

5.1 Introduction

For the direct methods of moment, the reliability index or the probability of failure is estimated directly using the first few moments of the original performance function without any transformation. As shown in the chapters entitled Moment Evaluation for Performance Functions and Direct Methods of Moment, the first few moments of an arbitrary performance function cannot always be accurately obtained; and the second-, third-, and fourth-moment reliability methods have their range of applicability, respectively. The direct methods of moment generally can give good approximation of the structural reliability, however, when outside the applicable range, the formula will prevent the application of the direct methods of moment. In order to overcome such problems, this Chapter presents the methods of moment based on first- and second-order transformation of the performance function.

5.2 First-Order Reliability Method

5.2.1 The Hasofer-Lind Reliability Index

Consider the following performance function,

$$Z = G(\mathbf{X}) = R - S \tag{5.1a}$$

If R and S are statistically independent normal random variables of $N(\mu_R, \sigma_R^2)$ and $N(\mu_S, \sigma_S^2)$, the limit state line is then illustrated in Figure 5.1a, and the probability of failure and the reliability index are obtained as

$$P_F = \Phi\left(-\frac{\mu_Z}{\sigma_Z}\right) = 1 - \Phi\left(\frac{\mu_Z}{\sigma_Z}\right) \tag{5.1b}$$

and

$$\beta = \frac{\mu_Z}{\sigma_Z} = \frac{\mu_R - \mu_S}{\sqrt{\sigma_R^2 + \sigma_S^2}} \tag{5.1c}$$

Structural Reliability: Approaches from Perspectives of Statistical Moments, First Edition.
Yan-Gang Zhao and Zhao-Hui Lu.
© 2021 John Wiley & Sons Ltd. Published 2021 by John Wiley & Sons Ltd.

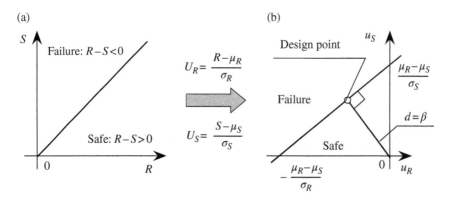

Figure 5.1 Limit state line in original and u-space. (a) Original space. (b) u-space.

In order to investigate the geometrical meaning of the reliability index in Eq. (5.1c), Eq. (5.1a) can be rewritten in terms of standard normal random variables U_R and U_S,

$$U_R = \frac{R - \mu_R}{\sigma_R}, U_S = \frac{S - \mu_S}{\sigma_S} \tag{5.2a}$$

From Eq. (5.2a), we have

$$R = \sigma_R U_R + \mu_R, S = \sigma_S U_S + \mu_S \tag{5.2b}$$

Substituting Eq. (5.2b) into Eq. (5.1a) leads to

$$g(U_R, U_S) = \sigma_R U_R - \sigma_S U_S + \mu_R - \mu_S \tag{5.2c}$$

The limit state line corresponding to Eq. (5.2c) is illustrated in Figure 5.1b.

It is easy to observe that the distance from the origin in the u-space to the limit state line is

$$d = \frac{\mu_R - \mu_S}{\sqrt{\sigma_R^2 + \sigma_S^2}} \tag{5.3}$$

Comparing Eq. (5.3) with Eq. (5.1c) it can be observed that the reliability index is equal to the distance from the origin in the u-space to the limit state line.

It can also be observed that as the limit state line $g(U_R, U_S) = 0$ moves further or closer to the origin, the safe region, $g(U_R, U_S) > 0$, increases or decreases accordingly. Therefore, the position of the failure surface relative to the origin should determine the safety or reliability of the corresponding engineering problem. The position of the failure surface may be represented by the minimum distance from the origin in u-space to the line $g(U_R, U_S) = 0$ (Hasofer and Lind 1974; Ditlevsen 1979a).

The particular point in the limit state line, which has the shortest distance from origin in the u_R-u_S plane to the limit state line, is often referred to as the *checking point* or *design point* (u_R^*, u_S^*).

Obviously, the coordinates of the design point in Figure 5.1b is given by

$$\left(-\frac{\sigma_R(\mu_R - \mu_S)}{\sigma_R^2 + \sigma_S^2}, \frac{\sigma_S(\mu_R - \mu_S)}{\sigma_R^2 + \sigma_S^2} \right)$$

(a)

(b)

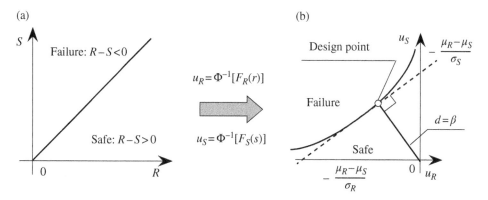

Figure 5.2 Limit state line with non-normal random variables. (a) Original space. (b) Standard normal space.

If R and S are not normal random variables, R and S can be transformed into standard normal random variables of U_R and U_S using the so-called Rosenblatt transformation, which will be demonstrated in Section 7.3.1 in the chapter entitled Transformation of Non-Normal Variables to Independent Normal Variables. The performance function $G(R, S)$ can then be expressed as a function of U_R and U_S. The limit sate line in the original space and in the u_R–u_S plane are illustrated in Figure 5.2a,b, respectively, where the limit state line in u_R–u_S plane is not a straight line, and the reliability index is defined as

$$\beta = min\sqrt{u_R^2 + u_S^2} \tag{5.4}$$

where (u_R, u_S) represents the coordinates of any point on the limit state line.

Similar to the case of normal random variables, the particular point on the limit state line, where Eq. (5.4) is satisfied and the tangential line of the limit state line is perpendicular to β at the *checking point* or *design point* $\mathbf{u}^*$.

It has been shown that the point on the failure line with minimum distance to the origin is the *most probable failure point* (Shinozuka 1983). Thus, in some approximate sense, this minimum distance may be used as a measure of reliability. The reliability index defined in this way is called the Hasofer-Lind reliability index or first order reliability index.

5.2.2 First-Order Reliability Method

For a general performance function $G(\mathbf{X})$, $\mathbf{X}$ can be transformed into standard normal random vector of $\mathbf{U}$ using the Rosenblatt transformation. $G(\mathbf{X})$ can then be rewritten as a function of $\mathbf{U}$, i.e. $g(\mathbf{U})$. In the standard normal space, generally $\mathbf{U}$ space, the required minimum distance may be determined as follows (Shinozuka 1983). The distance from the origin to a point $\mathbf{u} = (u_1, u_2, ..., u_n)$ on the failure surface $g(\mathbf{u}) = 0$ is

$$d = \sqrt{u_1^2 + u_2^2 + ... + u_n^2} = \left(\mathbf{u}^T\mathbf{u}\right)^{1/2}$$

The point on the failure surface, $\mathbf{u}^* = (u_1{}^*, u_2{}^*, ..., u_n{}^*)$ that has the minimum distance to the origin may be determined by minimising the function d, subject to the constraint $g(\mathbf{u}) = 0$, that is,

Minimise $d = (\mathbf{u}^T\mathbf{u})^{1/2}$

Subject to $g(\mathbf{u}) = 0$

For this purpose, the method of Lagrange's multiplier may be applied, where the modified function can be written as

$$L = (\mathbf{u}^T\mathbf{u})^{1/2} + \lambda g(\mathbf{u})$$

where λ is a Lagrangian multiplier.

Taking the differentiation of L leads to the following set of $n + 1$ equations with $n + 1$ unknowns

$$\frac{\partial L}{\partial u_i} = u_i(\mathbf{u}^T\mathbf{u})^{-1/2} + \lambda\frac{\partial g(\mathbf{u})}{\partial u_i} = 0, i = 1, 2, ..., n \qquad (5.5a)$$

and

$$\frac{\partial L}{\partial \lambda} = g(\mathbf{u}) = 0 \qquad (5.5b)$$

which may be written compactly as

$$\frac{\mathbf{u}}{d} + \lambda\nabla g(\mathbf{u}) = 0 \qquad (5.5c)$$

and

$$g(\mathbf{u}) = 0 \qquad (5.5d)$$

Solving of the above set of equations should yield the most probable failure point

$$\mathbf{u}^* = (u_1{}^*, u_2{}^*, ..., u_n{}^*)$$

Equation (5.5c) immediately produces

$$\mathbf{u}^* = -\lambda d\nabla g(\mathbf{u}^*) \qquad (5.6a)$$

where $\nabla g(\mathbf{u}^*)$ are the derivatives $\nabla g(\mathbf{u})$ evaluated at $\mathbf{u}^* = (u_1{}^*, u_2{}^*, ..., u_n{}^*)$.

Therefore

$$d = (\mathbf{u}^{*T}\mathbf{u}^*)^{1/2} = \left\{[-\lambda d\nabla g(\mathbf{u}^*)]^T[-\lambda d\nabla g(\mathbf{u}^*)]\right\}^{1/2} = \lambda d[\nabla^T g(\mathbf{u}^*)\nabla g(\mathbf{u}^*)]^{1/2}$$

and thus

$$\lambda = [\nabla^T g(\mathbf{u}^*)\nabla g(\mathbf{u}^*)]^{-1/2}$$

Using the last result in Eq. (5.6a), yields

$$\mathbf{u}^* = \frac{-d\nabla g(\mathbf{u}^*)}{[\nabla^T g(\mathbf{u}^*)\nabla g(\mathbf{u}^*)]^{1/2}} \qquad (5.6b)$$

Alternatively [pre-multiplying Eq. (5.6b) by $\nabla^T g(\mathbf{u}^*)$],

$$d = \frac{-\nabla^T g(\mathbf{u}^*)\mathbf{u}^*}{\left[\nabla^T g(\mathbf{u}^*)\nabla g(\mathbf{u}^*)\right]^{1/2}} \tag{5.7}$$

Whether the point $\mathbf{u}^*$ represents a minimum or maximum depends on the nature of $g(\mathbf{u})$. If $g(\mathbf{u})$ is linear or regular and convex towards the origin, the point $\mathbf{u}^*$ is clearly a minimum point (Lind 1979). Since many limit state functions depart only slightly from linearity, it is assumed in what follows that the point $\mathbf{u}^*$ does locate the minimum point.

It will now be shown that d is equal to β and that β is therefore the minimum distance from the origin to the limit state $g(\mathbf{u}) = 0$, provided that $g(\mathbf{u})$ is linearised at $\mathbf{u}^*$ (Lind 1979; Shinozuka 1983).

Let the performance function $g(\mathbf{u})$ be linearised at the design point $\mathbf{u}^*$ by means of a Taylor series expansion (i.e. in first-order terms only). This approximation provides a tangent (hyper) plane $G_L(\mathbf{u})$ at $\mathbf{u}^*$ to the function $g(\mathbf{u})$:

$$Z = G_L(\mathbf{u}) = g(\mathbf{u}^*) + \sum_{i=1}^{n}(u_i - u_i^*)\frac{\partial g}{\partial u_i}\bigg|_{\mathbf{u}^*} \tag{5.8a}$$

Since $\mathbf{u}^*$ is on the limit state surface, $g(\mathbf{u}^*) = 0$. Then, Eq. (5.8a) becomes

$$Z = G_L(\mathbf{u}) = \sum_{i=1}^{n}(u_i - u_i^*)\frac{\partial g}{\partial u_i}\bigg|_{\mathbf{u}^*} = (\mathbf{u} - \mathbf{u}^*)^T \nabla g(\mathbf{u}^*) \tag{5.8b}$$

Considering that $\mathbf{U}$ is a vector of normal random variable, it is obvious that $Z = G_L(\mathbf{u})$ is a normal random variable. The probability of failure and the reliability index corresponding to the limit state function $Z = G_L(\mathbf{u})$ can therefore be accurately given by

$$P_F = \Phi\left(-\frac{\mu_{GL}}{\sigma_{GL}}\right), \beta = \frac{\mu_{GL}}{\sigma_{GL}}$$

where μ_{GL} and σ_{GL} are the mean value and standard deviation of $Z = G_L(\mathbf{u})$, respectively, and can be obtained as

$$\mu_{GL} = -\sum_{i=1}^{n}u_i^*\frac{\partial g}{\partial u_i}\bigg|_{\mathbf{u}^*} = -\nabla^T g(\mathbf{u}^*)\mathbf{u}^* \tag{5.9a}$$

and

$$\sigma_{GL}^2 = \sum_{i=1}^{n}\left(\frac{\partial g}{\partial u_i}\bigg|_{\mathbf{u}^*}\right)^2 = \nabla^T g(\mathbf{u}^*)\nabla g(\mathbf{u}^*) \tag{5.9b}$$

Then, the reliability index is given as

$$\beta = \frac{\mu_{GL}}{\sigma_{GL}} = \frac{-\sum_{i=1}^{n}u_i^*\frac{\partial g}{\partial u_i}\big|_{\mathbf{u}^*}}{\sqrt{\sum_{i=1}^{n}\left(\frac{\partial g}{\partial u_i}\big|_{\mathbf{u}^*}\right)^2}} = \frac{-\nabla^T g(\mathbf{u}^*)\mathbf{u}^*}{\sqrt{\nabla^T g(\mathbf{u}^*)\nabla g(\mathbf{u}^*)}} \tag{5.10a}$$

which is identical to Eq. (5.7). This implies that the reliability index β is the minimum distance from the origin to the limit state $g(\mathbf{u}) = 0$. Note, in a more strict sense, the reliability index β in Eq. (5.10a) is not for limit state function $g(\mathbf{u})$ but for its first-order approximation $G_L(\mathbf{u})$ at the design point $\mathbf{u}^*$. Therefore, the reliability index is called first-order reliability index, which is rewritten as

$$\beta_F = \frac{\mu_{GL}}{\sigma_{GL}} = \frac{-\nabla^T g(\mathbf{u}^*)\mathbf{u}^*}{\sqrt{\nabla^T g(\mathbf{u}^*)\nabla g(\mathbf{u}^*)}} \tag{5.10b}$$

Let

$$\boldsymbol{\alpha} = \frac{\nabla g(\mathbf{u}^*)}{\sqrt{\nabla^T g(\mathbf{u}^*)\nabla g(\mathbf{u}^*)}} = \frac{\nabla g(\mathbf{u}^*)}{|\nabla g(\mathbf{u}^*)|} \tag{5.11}$$

Then, we have

$$\mathbf{u}^* = -\boldsymbol{\alpha}\beta_F \tag{5.12a}$$

$$\beta_F = -\boldsymbol{\alpha}^T u^* \tag{5.12b}$$

In the scalar form, the components of $\mathbf{u}^*$, Eq. (5.12a), are

$$u_i^* = -\alpha_i \beta_F, i = 1, 2, ..., n \tag{5.13a}$$

$$\beta_F = -\sum_{i=1}^{n} \alpha_i u_i^* \tag{5.13b}$$

where

$$\alpha_i = \frac{\left.\frac{\partial g}{\partial u_i}\right|_{\mathbf{u}^*}}{\sqrt{\sum_{i=1}^{n}\left(\left.\frac{\partial g}{\partial u_i}\right|_{\mathbf{u}^*}\right)^2}} \tag{5.13c}$$

is the direction cosine along the axes u_i.

Therefore, the first-order approximation of the limit state function can be also expressed as

$$G_L(\mathbf{U}) = \beta_F + \mathbf{U}^T \boldsymbol{\alpha} \tag{5.14}$$

In this section, the reliability analysis method is based on the linearisation or the first-order approximation of the transformed performance function in the standard normal space at its design point, therefore it is referred to the first-order reliability method (FORM).

Example 5.1 Consider a linear limit state function in an n-dimensional u-space

$$G(\mathbf{U}) = a_0 + \sum_{i=1}^{n} a_i U_i \tag{5.15a}$$

where U_i, $i = 1, 2,..., n$, are standard normal random variables.

From Eq. (5.15a), we have

$$\frac{\partial G}{\partial u_i} = a_i$$

and

$$\alpha_i = \frac{\left.\dfrac{\partial G}{\partial u_i}\right|_{\mathbf{u}^*}}{\sqrt{\displaystyle\sum_{i=1}^{n}\left(\left.\dfrac{\partial G}{\partial u_i}\right|_{\mathbf{u}^*}\right)^2}} = \frac{a_i}{\sqrt{a_1^2 + a_2^2 + \dots + a_n^2}}$$

Using Eq. (5.13b) leads to

$$\beta_F = -\sum_{i=1}^{n} \alpha_i u_i^* = -\frac{\displaystyle\sum_{i=1}^{n} a_i u_i^*}{\sqrt{a_1^2 + a_2^2 + \dots + a_n^2}}$$

Since $\mathbf{u}^*$ is on the limit state surface $G(\mathbf{u}) = 0$, we have

$$\sum_{i=1}^{n} a_i u_i^* = -a_0$$

then the first-order reliability index is given as

$$\beta_F = \frac{a_0}{\sqrt{a_1^2 + a_2^2 + \dots + a_n^2}} \tag{5.15b}$$

and the design point is given as

$$u_i^* = -\alpha_i \beta_F = -\frac{a_i a_0}{a_1^2 + a_2^2 + \dots + a_n^2}, i = 1, 2, \dots, n \tag{5.15c}$$

Example 5.2 Consider the following second-order limit state function in u-space

$$G(\mathbf{U}) = U_1^2 - 80U_2 + 240 \tag{5.16a}$$

where U_1 and U_2 are standard normal random variables.

From Eq. (5.16a), the derivatives of G to U_i, $i = 1, 2$ and the direction cosines are obtained as

$$\frac{\partial G}{\partial u_1} = 2u_1, \frac{\partial G}{\partial u_2} = -80, \alpha_1 = \frac{2u_1}{\sqrt{4u_1^2 + 6400}}, \alpha_2 = \frac{-80}{\sqrt{4u_1^2 + 6400}}$$

Using Eqs. (5.13a) and (5.13b) gives

$$\sqrt{u_1^2 + u_2^2} = \frac{2u_1^2 - 80u_2}{\sqrt{4u_1^2 + 6400}} \tag{5.16b}$$

Since $\mathbf{u}^*$ is on the limit state surface $G(\mathbf{u}) = 0$, we have

$$u_1^2 = 80u_2 - 240$$

Substituting the equation above into Eq. (5.16b) leads to

$$\sqrt{u_2^2 + 80u_2 - 240} = \frac{80u_2 - 480}{\sqrt{320u_2 + 5440}}$$

Rearranging the equation above we have

$$u_2^3 + 77u_2^2 + 1360u_2 - 4800 = 0$$

The root of the cubic equation above is $u_2 = 3$, then $u_1 = 0$ can be easily obtained, and therefore the design point is $\mathbf{u}^* = (0, 3)^T$. The first-order reliability index is then given as $\beta_F = 3$.

At the design point $\mathbf{u}^* = (0, 3)^T$, the direction cosines are obtained as

$$\alpha_1 = \frac{2u_1}{\sqrt{4u_1^2 + 6400}} = 0, \qquad \alpha_2 = \frac{-80}{\sqrt{4u_1^2 + 6400}} = -1$$

Then, the first-order approximation of the limit state function can be obtained as

$$G_L(\mathbf{u}) = \beta_F + \mathbf{u}^T \boldsymbol{\alpha} = 3 + u_1 \cdot \alpha_1 + u_2 \cdot \alpha_2 = 3 - u_2$$

The original performance function and its first-order approximation are illustrated in Figure 5.3.

In this example, the design points are analytically derived. However, for most performance functions in practical engineering, especially those with a large number of input variables and complex limit state functions, the analytical solution of the design point is difficult, if not impossible, to obtain. For such cases, numerical solution is necessary when applying FORM, which will be discussed in the following section. From Figure 5.3, it can also be observed that the first-order approximation is not a good

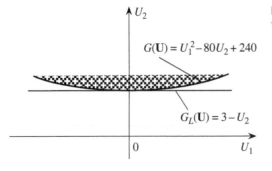

Figure 5.3 The original performance function and its first-order approximation.

approximation of the original performance function. A higher order, e.g. second-order, approximation of the performance function would be necessary, which will be discussed in Section 5.3.

Example 5.3 Consider the following performance function in Example 3.3 in the chapter on Moment Evaluation

$$G(\mathbf{X}) = DR - S$$

where D, R, and S = independent random variables with normal distributions of which the mean and standard deviation are: $\mu_D = 1$ and $\sigma_D = 0.1$, $\mu_R = 100$ and $\sigma_R = 20$, $\mu_S = 50$ and $\sigma_S = 20$.

Using the FORM, the design point is obtained as $(-0.4596, -1.1338, 1.1885)$ in standard normal space and $(0.954, 77.324, 73.77)$ in original space. The first-order approximation of the limit state function in standard normal space are given as

$$G_L(\mathbf{U}) = \beta_F + \mathbf{U}^T \boldsymbol{\alpha} = 1.7057 + 0.2694 U_D + 0.6647 U_R - 0.6968 U_S$$

Therefore, the FORM reliability index is obtained as 1.706.

Recall that as shown in Example 3.3 in the chapter on Moment Evaluation, the first two moments of the $G(\mathbf{X})$ without any linearisation are given as $\mu_G = 50$, $\sigma_G = 30.067$, and the direct second-moment reliability index is given as $\beta_{2M} = 50/30.067 = 1.663$. It can be noticed that the direct second-moment method has a relatively large error.

5.2.3 Numerical Solution for FORM

In this section, we will discuss numerical solution for FORM. In Section 5.2.2, the discussion is limited with the performance function including only normal random variables. For general cases of performance functions with random variables other than standard normal random variables, the transformation is necessary from any distributed random variables to standard normal variables. The transformation can be generally realised by the Rosenblatt transformation (Rosenblatt 1952; Hohenbichler and Rackwitz 1981), which will be discussed in Section 7.3.1 in the chapter entitled Transformation of Non-Normal Variables to Independent Normal. For a general case, the numerical solution for FORM is generally realised using the procedure proposed by Hohenbichler and Rackwitz (1981), which includes the following steps.

1) For a performance function $G(\mathbf{X})$, assume an initial checking point $\mathbf{x}_0$.
2) Using the Rosenblatt transformation (see the section mentioned above), obtain the corresponding checking point in the u-space, that is $\mathbf{u}_0$.
3) Determine the Jacobian Matrix

$$\mathbf{J} = \frac{\partial \mathbf{x}}{\partial \mathbf{u}} = \frac{\partial(x_1, ..., x_n)}{\partial(u_1, ..., u_n)} \tag{5.17}$$

evaluated at $\mathbf{x}_0$.

4) Evaluate the performance function and gradient vector at $\mathbf{u}_0$.

$$g(\mathbf{u}_0) = G(\mathbf{x}_0) \tag{5.18a}$$

$$\nabla g(\mathbf{u}_0) = \mathbf{J}^T \nabla G(\mathbf{x}_0) \tag{5.18b}$$

in which the gradient vector $\nabla G(\mathbf{x}_0)$ can be computed directly by taking the derivative for explicit performance functions. While for implicit performance functions, the numerical differentiation methods may be used to determine the gradient vector $\nabla G(\mathbf{x}_0)$. If the central difference method is used, $\nabla G(\mathbf{x}_0)$ can be expressed as:

$$\nabla G(\mathbf{x}_0) = \begin{pmatrix} \dfrac{G(x_{01} + \Delta, x_{02}, \cdots, x_{0n}) - G(x_{01} - \Delta, x_{02}, \cdots, x_{0n})}{2\Delta} \\ \vdots \\ \dfrac{G(x_{01}, \cdots, x_{0i} + \Delta, \cdots, x_{0n}) - G(x_{01}, \cdots, x_{0i} - \Delta, \cdots, x_{0n})}{2\Delta} \\ \vdots \\ \dfrac{G(x_{01}, x_{02}, \cdots, x_{0n} + \Delta) - G(x_{01}, x_{02}, \cdots, x_{0n} - \Delta)}{2\Delta} \end{pmatrix} \tag{5.19}$$

5) Obtain a new checking point

$$\mathbf{u}^{(k+1)} = \frac{1}{\nabla^T g(\mathbf{u}^{(k)}) \nabla g(\mathbf{u}^{(k)})} \left[\nabla^T g\left(\mathbf{u}^{(k)}\right) \mathbf{u}^{(k)} - g\left(\mathbf{u}^{(k)}\right) \right] \nabla g\left(\mathbf{u}^{(k)}\right) \tag{5.20}$$

and the checking point in the original space is

$$\mathbf{x}^{(k+1)} = \mathbf{x}^{(k)} + \mathbf{J}\left(\mathbf{u}^{(k+1)} - \mathbf{u}^{(k)}\right) \tag{5.21}$$

6) Calculate reliability index

$$\beta_F = \sqrt{\mathbf{u}^{*T}\mathbf{u}^*} \tag{5.22}$$

7) Repeat steps 2 through 6 using the above $\mathbf{x}^*$ as the new checking point until convergence is achieved.

For correlated random variables, it should be noted that $\dfrac{\partial x_i}{\partial u_j}$ is not easily evaluated since there is no explicit expression for Eq. (7.8) in the chapter on Transformation of Non-Normal Variables to Independent Normal Variables. One first needs to obtain the components of $\mathbf{J}^{-1}$, i.e. $\dfrac{\partial u_i}{\partial x_j}$. Through implicit differentiation of Eq. (7.5) in the same chapter, i.e. the Rosenblatt transformation, the partial derivatives of in the $\mathbf{J}^{-1}$ of step 3 are

$$\frac{\partial u_i}{\partial x_j} = \frac{\partial \Phi^{-1}[F(x_i|x_1, \cdots x_{i-1})]}{\partial x_j} = \frac{1}{\phi(u_i)} \frac{\partial F(x_i|x_1, \cdots x_{i-1})}{\partial x_j} \tag{5.23}$$

Since

$$\frac{\partial u_i}{\partial x_j} = 0, \text{ for } i < j \tag{5.24}$$

The $\mathbf{J}^{-1}$ will be a lower triangular matrix, and thus its inverse $\mathbf{J}$ in step 3 is easily obtained from $\mathbf{J}^{-1}$ through back substitution.

Example 5.4 To demonstrate the procedure of numerical solution for FORM, consider a linear limit state function

$$G(\mathbf{X}) = X_2 - X_1$$

where X_1 and X_2 are independent random variables. X_1 is a normal variable with mean of $\mu_1 = 100$ and standard deviation of $\sigma_1 = 20$, and X_2 is a lognormal variable with mean of $\mu_2 = 300$ and standard deviation of $\sigma_2 = 120$.

The two parameters of the lognormal random variable X_2 can be easily given as

$$\zeta = \sqrt{\ln(1 + V^2)} = 0.3853, \lambda = \ln \mu - \frac{1}{2}\zeta^2 = 5.63$$

Step 1; Assume the mean value as an initial checking point

$$\mathbf{x}^{(0)} = \{100, 300\}$$

First Iteration
Step 2; Using the Rosenblatt transformation, obtain the corresponding checking point in the u-space,

$$u_1^{(0)} = \frac{x_1^{(0)} - \mu_1}{\sigma_1} = 0, u_2^{(0)} = \frac{\ln x_2^{(0)} - \lambda_2}{\zeta_2} = 0.1926$$

That is, $\mathbf{u}^{(0)} = \{0, 0.1926\}$
The first order reliability index at the current stage is given as

$$\beta_F^{(0)} = \sqrt{\mathbf{u}^{(0)T}\mathbf{u}^{(0)}} = 0.1926$$

Step 3; Determine the Jacobian Matrix evaluated at the checking point $\mathbf{u}^{(0)}$.
Since X_1 and X_2 are independent,

$$\frac{\partial x_1}{\partial u_2} = \frac{\partial x_2}{\partial u_1} = 0, \frac{\partial x_1}{\partial u_1} = \sigma_1 = 20, \frac{\partial x_2}{\partial u_2} = \zeta x_2 = 115.6$$

That is,

$$\mathbf{J} = \frac{\partial \mathbf{x}}{\partial \mathbf{u}}\Big|_{\mathbf{x}^{(0)}} = \begin{bmatrix} 20 & 0 \\ 0 & 115.6 \end{bmatrix}$$

Step 4; Evaluate the performance function and gradient vector at $\mathbf{u}_0$.

$$g\left(u^{(0)}\right) = G\left(x^{(0)}\right) = 200, \nabla G(\mathbf{x}_0) = \begin{Bmatrix} -1 \\ 1 \end{Bmatrix},$$

$$= \mathbf{J}^T \nabla G(\mathbf{x}_0) = \begin{bmatrix} 20 & 0 \\ 0 & 115.6 \end{bmatrix} \begin{Bmatrix} -1 \\ 1 \end{Bmatrix} = \begin{Bmatrix} -20 \\ 115.6 \end{Bmatrix}$$

Step 5; Obtain a new checking point

$$\mathbf{u}^{(1)} = \frac{1}{\nabla^T g(\mathbf{u}_0) \nabla g(\mathbf{u}_0)} \left[\nabla^T g(\mathbf{u}_0) \mathbf{u}_0 - g(\mathbf{u}_0) \right] \nabla g(\mathbf{u}_0)$$

$$= \frac{1}{[-20 \ \ 115.6] \begin{Bmatrix} -20 \\ 115.6 \end{Bmatrix}} \left[[-20 \ \ 115.6] \begin{Bmatrix} 0 \\ 0.1926 \end{Bmatrix} - 200 \right] \begin{Bmatrix} -20 \\ 115.6 \end{Bmatrix} = \begin{Bmatrix} 0.2584 \\ -1.4931 \end{Bmatrix}$$

and in the space of the original variables, the checking point is

$$\mathbf{x}^{(1)} = \mathbf{x}^{(0)} + \mathbf{J}\left(\mathbf{u}^{(1)} - \mathbf{u}^{(0)} \right) = \begin{Bmatrix} 100 \\ 300 \end{Bmatrix} + \begin{bmatrix} 20 & 0 \\ 0 & 115.6 \end{bmatrix} \left(\begin{Bmatrix} 0.2584 \\ -1.4931 \end{Bmatrix} - \begin{Bmatrix} 0 \\ 0.1926 \end{Bmatrix} \right)$$

$$= \begin{Bmatrix} 105.168 \\ 105.168 \end{Bmatrix}$$

Step 6; Calculate reliability index

$$\beta_F^1 = \sqrt{\mathbf{u}^{(1)T} \mathbf{u}^{(1)}} = \sqrt{[0.2584 \ \ -1.4931] \begin{Bmatrix} 0.2584 \\ -1.4931 \end{Bmatrix}} = 1.5153$$

Step 7; The relative difference between $\beta_F^{(1)}$ and $\beta_F^{(0)}$ is given as

$$\varepsilon = \frac{\left| \beta_F^{(1)} - \beta_F^{(0)} \right|}{\beta_F^{(1)}} = \frac{|1.5153 - 0.1926|}{1.5153} = 0.873$$

Since the difference is too large, repeat steps 2 through 6 using the above $\mathbf{x}^{(1)}$ as the new checking point.

$$\mathbf{x}^{(1)} = \begin{Bmatrix} 105.168 \\ 105.168 \end{Bmatrix}$$

Second Iteration

Step 2; The corresponding checking point in the u-space is

$$\mathbf{u}^{(1)} = \begin{Bmatrix} 0.2584 \\ -2.5283 \end{Bmatrix}$$

Step 3; The Jacobian Matrix evaluated at the checking point $\mathbf{u}^{(1)}$

$$\mathbf{J} = \begin{bmatrix} 20 & 0 \\ 0 & 40.516 \end{bmatrix}$$

Step 4; The performance function and gradient vector at $\mathbf{u}^{(1)}$.

$$g\left(\mathbf{u}^{(1)}\right) = 0, \nabla G\left(\mathbf{x}^{(1)}\right) = \left\{ \begin{matrix} -1 \\ 1 \end{matrix} \right\}, \nabla g\left(\mathbf{u}^{(1)}\right) = \begin{bmatrix} 20 & 0 \\ 0 & 40.516 \end{bmatrix} \left\{ \begin{matrix} -1 \\ 1 \end{matrix} \right\} = \left\{ \begin{matrix} -20 \\ 40.516 \end{matrix} \right\}$$

Step 5; Obtain a new checking point

$$\mathbf{u}^{(2)} = \left\{ \begin{matrix} 1.054 \\ -2.135 \end{matrix} \right\}$$

and in the space of the original variables, the checking point is

$$\mathbf{x}^{(2)} = \left\{ \begin{matrix} 121.08 \\ 121.08 \end{matrix} \right\}$$

Step 6; Calculate reliability index

$$\beta_F^{(2)} = 2.381$$

Step 7; The relative difference between $\beta_F^{(2)}$ and $\beta_F^{(1)}$ is given as

$$\varepsilon = \frac{\left|\beta_F^{(2)} - \beta_F^{(1)}\right|}{\beta_F^{(2)}} = \frac{|2.381 - 1.5153|}{2.381} = 0.364$$

Since the difference is still large, repeat steps 2 through 6 using the above $\mathbf{x}^{(2)}$ as the new checking point.

Third Iteration

$$\mathbf{u}^{(2)} = \left\{ \begin{matrix} 1.054 \\ -2.162 \end{matrix} \right\}, \qquad \mathbf{J} = \begin{bmatrix} 20 & 0 \\ 0 & 46.647 \end{bmatrix}, \qquad g\left(\mathbf{u}^{(2)}\right) = 0,$$

$$\nabla g\left(\mathbf{u}^{(2)}\right) = \left\{ \begin{matrix} -20 \\ 46.647 \end{matrix} \right\}, \qquad \mathbf{u}^{(3)} = \left\{ \begin{matrix} 0.947 \\ -2.208 \end{matrix} \right\}, \qquad \mathbf{x}^{(3)} = \left\{ \begin{matrix} 118.94 \\ 118.94 \end{matrix} \right\}$$

$$\beta_F^{(3)} = 2.403, \qquad\qquad \varepsilon = 0.0089$$

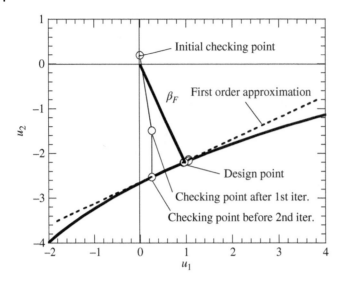

Figure 5.4 FORM process for Example 5.4.

If one does not satisfy with this difference, one may repeat steps 2 through 6 using the above $\mathbf{x}^{(3)}$ as the new checking point, then the results of the fourth iteration are:

$$\mathbf{u}^{(3)} = \left\{ \begin{array}{c} 0.947 \\ -2.209 \end{array} \right\}, \quad J = \left[\begin{array}{cc} 20 & 0 \\ 0 & 45.821 \end{array} \right], \quad g\left(\mathbf{u}^{(3)}\right) = 0,$$

$$\nabla g\left(\mathbf{u}^{(3)}\right) = \left\{ \begin{array}{c} -20 \\ 45.821 \end{array} \right\}, \quad \mathbf{u}^{(4)} = \left\{ \begin{array}{c} 0.961 \\ -2.203 \end{array} \right\}, \quad \mathbf{x}^{(4)} = \left\{ \begin{array}{c} 119.23 \\ 119.23 \end{array} \right\}$$

$$\beta_F^{(3)} = 2.403, \qquad \varepsilon = 0.0001$$

The convergence has been achieved after the fourth iteration with the first-order reliability index of 2.403 for this problem. The corresponding failure probability is $P_F = 0.0081$. The iteration process is shown in Figure 5.4.

Example 5.5 Consider the following limit state function

$$G(\mathbf{X}) = X_1 X_2 - X_3$$

where X_1, X_2, and X_3 are independent random variables. X_1 is a normal variable with mean of $\mu_1 = 1$ and standard deviation of $\sigma_1 = 0.2$; X_2 is a lognormal variable with mean of $\mu_2 = 400$ and standard deviation of $\sigma_2 = 80$; and X_3 is a lognormal variable with mean of $\mu_3 = 100$ and standard deviation of $\sigma_3 = 40$.

The two parameters of the lognormal random variable X_2 and X_3 can be calculated as

$$\zeta_2 = \sqrt{\ln\left(1 + V^2\right)} = 0.198, \lambda_2 = \ln\mu - \frac{1}{2}\zeta^2 = 5.972$$

$$\zeta_3 = \sqrt{\ln\left(1 + V^2\right)} = 0.385, \lambda_3 = \ln\mu - \frac{1}{2}\zeta^2 = 4.531$$

The iterations are summarised as follows:

Iter.No.	$\mathbf{x}^*$	$\mathbf{u}^*$	$\nabla G(\mathbf{x})$	$\mathbf{J}$	$\nabla g(\mathbf{u})$	$\mathbf{u}$	β_F
1	$\begin{Bmatrix} 1 \\ 400 \\ 100 \end{Bmatrix}$	$\begin{Bmatrix} 0 \\ 0.099 \\ 0.193 \end{Bmatrix}$	$\begin{Bmatrix} 400 \\ 1 \\ -1 \end{Bmatrix}$	$\begin{bmatrix} 0.2 & 0 & 0 \\ 0 & 79.22 & 0 \\ 0 & 0 & 38.53 \end{bmatrix}$	$\begin{Bmatrix} 80 \\ 79.22 \\ -38.53 \end{Bmatrix}$	$\begin{Bmatrix} -1.693 \\ -1.676 \\ 0.815 \end{Bmatrix}$	2.518
2	$\begin{Bmatrix} 0.661 \\ 259.4 \\ 124.0 \end{Bmatrix}$	$\begin{Bmatrix} -1.693 \\ -2.088 \\ 0.751 \end{Bmatrix}$	$\begin{Bmatrix} 259.4 \\ 0.661 \\ -1 \end{Bmatrix}$	$\begin{bmatrix} 0.2 & 0 & 0 \\ 0 & 51.37 & 0 \\ 0 & 0 & 47.76 \end{bmatrix}$	$\begin{Bmatrix} 51.88 \\ 33.98 \\ -47.76 \end{Bmatrix}$	$\begin{Bmatrix} -2.051 \\ -1.343 \\ 1.888 \end{Bmatrix}$	3.094
3	$\begin{Bmatrix} 0.590 \\ 297.7 \\ 178.3 \end{Bmatrix}$	$\begin{Bmatrix} -2.051 \\ -1.393 \\ 1.694 \end{Bmatrix}$	$\begin{Bmatrix} 297.7 \\ 0.590 \\ -1 \end{Bmatrix}$	$\begin{bmatrix} 0.2 & 0 & 0 \\ 0 & 58.95 & 0 \\ 0 & 0 & 68.70 \end{bmatrix}$	$\begin{Bmatrix} 59.53 \\ 34.77 \\ -68.70 \end{Bmatrix}$	$\begin{Bmatrix} -1.786 \\ -1.043 \\ 2.061 \end{Bmatrix}$	2.920
4	$\begin{Bmatrix} 0.642 \\ 318.3 \\ 203.5 \end{Bmatrix}$	$\begin{Bmatrix} -1.786 \\ -1.055 \\ 2.037 \end{Bmatrix}$	$\begin{Bmatrix} 318.3 \\ 0.643 \\ -1 \end{Bmatrix}$	$\begin{bmatrix} 0.2 & 0 & 0 \\ 0 & 63.04 & 0 \\ 0 & 0 & 78.41 \end{bmatrix}$	$\begin{Bmatrix} 63.66 \\ 40.52 \\ -78.41 \end{Bmatrix}$	$\begin{Bmatrix} -1.705 \\ -1.086 \\ 2.099 \end{Bmatrix}$	2.915

The first-order reliability index is $\beta_F = 2.914$ with corresponding failure probability of 0.00178.

5.2.4 The Weakness of FORM

It has been shown that the iteration procedure can fail in some circumstances (Fiessler 1979). Here, we will demonstrate some weaknesses of FORM through some examples.

Example 5.6 Consider the minus of the linear limit state function in Example 5.4

$$G(\mathbf{X}) = X_1 - X_2$$

where X_1 and X_2 are independent random variables with the same distributions as those in Example 5.4.

The iterations are summarised as follows:

Iter.No.	$\mathbf{x}^*$	$G(\mathbf{x}^*)$	$\mathbf{u}^*$	$\nabla G(\mathbf{x})$	$\mathbf{J}$	$\nabla g(\mathbf{u})$	$\mathbf{u}$	β_F
1	$\begin{Bmatrix} 100 \\ 300 \end{Bmatrix}$	-200	$\begin{Bmatrix} 0 \\ 0.1926 \end{Bmatrix}$	$\begin{Bmatrix} 1 \\ -1 \end{Bmatrix}$	$\begin{bmatrix} 20 & 0 \\ 0 & 115.6 \end{bmatrix}$	$\begin{Bmatrix} 0.258 \\ -2.528 \end{Bmatrix}$	$\begin{Bmatrix} 0.258 \\ -1.493 \end{Bmatrix}$	1.515
2	$\begin{Bmatrix} 105.17 \\ 105.17 \end{Bmatrix}$	0	$\begin{Bmatrix} 0.258 \\ -2.528 \end{Bmatrix}$	$\begin{Bmatrix} 1 \\ -1 \end{Bmatrix}$	$\begin{bmatrix} 20 & 0 \\ 0 & 40.516 \end{bmatrix}$	$\begin{Bmatrix} 20 \\ -40.52 \end{Bmatrix}$	$\begin{Bmatrix} 1.054 \\ -2.135 \end{Bmatrix}$	2.381
3	$\begin{Bmatrix} 121.08 \\ 121.08 \end{Bmatrix}$	0	$\begin{Bmatrix} 1.054 \\ -2.162 \end{Bmatrix}$	$\begin{Bmatrix} 1 \\ -1 \end{Bmatrix}$	$\begin{bmatrix} 20 & 0 \\ 0 & 46.647 \end{bmatrix}$	$\begin{Bmatrix} 20 \\ -46.65 \end{Bmatrix}$	$\begin{Bmatrix} 0.947 \\ -2.208 \end{Bmatrix}$	2.403
4	$\begin{Bmatrix} 118.94 \\ 118.94 \end{Bmatrix}$	0	$\begin{Bmatrix} 0.947 \\ -2.209 \end{Bmatrix}$	$\begin{Bmatrix} 1 \\ -1 \end{Bmatrix}$	$\begin{bmatrix} 20 & 0 \\ 0 & 45.821 \end{bmatrix}$	$\begin{Bmatrix} 20 \\ -45.82 \end{Bmatrix}$	$\begin{Bmatrix} 0.961 \\ -2.203 \end{Bmatrix}$	2.403

The convergence has been achieved after the fourth iteration. Therefore the first-order reliability index is 2.403 for this problem with corresponding failure probability of $P_F = 0.0081$.

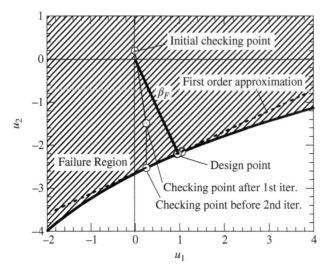

Figure 5.5 FORM process for Example 5.6.

However, since the performance function in this example is the minus of the performance function in Example 5.4, the original point is in the failure region as shown in Figure 5.5. Therefore, the distance 2.403 from the original point to the limit state curve is not the first-order reliability index.

Since the original point is in the failure region, the failure probability can be given as

$$P_F = \Phi(2.403) = 0.991875$$

and the reliability index is −2.403. It is therefore worth noting whether the original point is in the failure region or not when using the FORM procedure.

Example 5.7 Consider the following linear limit state function

$$G(\mathbf{X}) = 60 - X_1 + X_2$$

where X_1 and X_2 are independent random variables. X_1 is a normal variable with mean of $\mu_1 = 50$ and standard deviation of $\sigma_1 = 10$; and X_2 is a lognormal variable with mean of $\mu_2 = 80$ and standard deviation of $\sigma_2 = 40$.

The two parameters of the lognormal random variable X_2 can be easily given as

$$\zeta = \sqrt{\ln(1 + V^2)} = 0.472, \quad \lambda = \ln \mu - \frac{1}{2}\zeta^2 = 4.270$$

In the first iteration, we obtain that

$$\mathbf{x}^{(0)} = \left\{ \begin{array}{c} 50 \\ 80 \end{array} \right\}, \quad \mathbf{u}^{(0)} = \left\{ \begin{array}{c} 0 \\ 0.236 \end{array} \right\}, \quad \mathbf{J} = \left[\begin{array}{cc} 10 & 0 \\ 0 & 37.79 \end{array} \right], \quad g(\mathbf{u}^{(0)}) = 90$$

$$\nabla G(\mathbf{x}^{(0)}) = \left\{ \begin{array}{c} -1 \\ 1 \end{array} \right\}, \quad \nabla g(\mathbf{u}^{(0)}) = \left\{ \begin{array}{c} -10 \\ 37.79 \end{array} \right\}, \quad \mathbf{u}^{(1)} = \left\{ \begin{array}{c} 0.531 \\ -2.005 \end{array} \right\}$$

and in the space of the original variables, the checking point is

$$\mathbf{x}^{(1)} = \mathbf{x}^{(0)} + \mathbf{J}\left(\mathbf{u}^{(1)} - \mathbf{u}^{(0)}\right) = \begin{Bmatrix} 50 \\ 80 \end{Bmatrix} + \begin{bmatrix} 10 & 0 \\ 0 & 37.79 \end{bmatrix} \left(\begin{Bmatrix} 0.531 \\ -2.005 \end{Bmatrix} - \begin{Bmatrix} 0 \\ 0.236 \end{Bmatrix}\right) = \begin{Bmatrix} 59.31 \\ -4.695 \end{Bmatrix}$$

$$(5.25)$$

It can be observed that the coordinate of X_2 in the checking point is negative. Since X_2 is a lognormal random variable, it is obvious that the value of -4.695 falls out the definition region of the lognormal variable; then the iteration cannot be continued. This difficulty can be overcome by using the following equation instead of Eq. (5.21).

$$\mathbf{x}^{(k+1)} = \mathbf{x}^{(k)} + \delta\mathbf{J}\left(\mathbf{u}^{(k+1)} - \mathbf{u}^{(k)}\right)$$

$$(5.26)$$

where $0 < \delta \leq 1$ is a modification factor.

Taking $\delta = 0.8$, Eq. (5.25) becomes

$$\mathbf{x}^{(1)} = \mathbf{x}^{(0)} + \mathbf{J}\left(\mathbf{u}^{(1)} - \mathbf{u}^{(0)}\right) = \begin{Bmatrix} 50 \\ 80 \end{Bmatrix} + 0.8\begin{bmatrix} 10 & 0 \\ 0 & 37.79 \end{bmatrix} \left(\begin{Bmatrix} 0.531 \\ -2.005 \end{Bmatrix} - \begin{Bmatrix} 0 \\ 0.236 \end{Bmatrix}\right) = \begin{Bmatrix} 54.24 \\ 12.24 \end{Bmatrix}$$

It can be observed at this time that the coordinate of X_2 in the checking point is positive in the definition region of the lognormal variable; then the iteration can be continued. The iterations are shown in Figure 5.6 and summarised as follows:

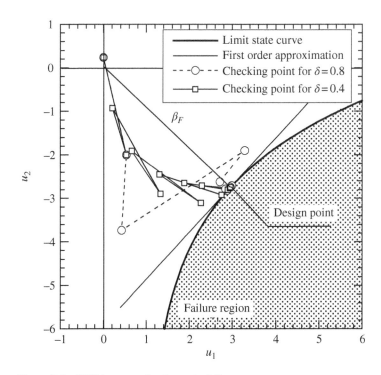

Figure 5.6 FORM process for Example 5.7.

Iter.No.	$\mathbf{x}^*$	$G(\mathbf{x}^*)$	$\mathbf{u}^*$	$\nabla G(\mathbf{x})$	$\mathbf{J}$		$\nabla g(\mathbf{u})$	$\mathbf{u}$	β_F
1	$\begin{Bmatrix} 50 \\ 80 \end{Bmatrix}$	90	$\begin{Bmatrix} 0 \\ 0.236 \end{Bmatrix}$	$\begin{Bmatrix} -1 \\ 1 \end{Bmatrix}$	$\begin{bmatrix} 10 & 0 \\ 0 & 37.79 \end{bmatrix}$		$\begin{Bmatrix} -10 \\ 37.79 \end{Bmatrix}$	$\begin{Bmatrix} 0.531 \\ -2.005 \end{Bmatrix}$	3.212
2	$\begin{Bmatrix} 54.24 \\ 12.24 \end{Bmatrix}$	18	$\begin{Bmatrix} 0.424 \\ -3.737 \end{Bmatrix}$	$\begin{Bmatrix} -1 \\ 1 \end{Bmatrix}$	$\begin{bmatrix} 10 & 0 \\ 0 & 5.784 \end{bmatrix}$		$\begin{Bmatrix} -10 \\ 5.784 \end{Bmatrix}$	$\begin{Bmatrix} 3.287 \\ -1.90 \end{Bmatrix}$	3.797
3	$\begin{Bmatrix} 77.14 \\ 20.74 \end{Bmatrix}$	3.6	$\begin{Bmatrix} 2.714 \\ -2.621 \end{Bmatrix}$	$\begin{Bmatrix} -1 \\ 1 \end{Bmatrix}$	$\begin{bmatrix} 10 & 0 \\ 0 & 9.798 \end{bmatrix}$		$\begin{Bmatrix} -10 \\ 9.798 \end{Bmatrix}$	$\begin{Bmatrix} 2.879 \\ -2.821 \end{Bmatrix}$	4.030
4	$\begin{Bmatrix} 78.46 \\ 19.18 \end{Bmatrix}$	0.72	$\begin{Bmatrix} 2.846 \\ -2.787 \end{Bmatrix}$	$\begin{Bmatrix} -1 \\ 1 \end{Bmatrix}$	$\begin{bmatrix} 10 & 0 \\ 0 & 9.060 \end{bmatrix}$		$\begin{Bmatrix} -10 \\ 9.060 \end{Bmatrix}$	$\begin{Bmatrix} 2.989 \\ -2.708 \end{Bmatrix}$	4.034

Example 5.8 Consider the following performance function defined in the standard normal space,

$$G(\mathbf{U}) = U_2^2 - 10U_1 + 20$$

The limit state curve is depicted in Figure 5.7, where one can observe that the first-order approximation is much different from the original limit state curve.

As shown in Figure 5.7, the first-order reliability index is $\beta_F = 2$, which corresponds to the probability of failure $P_F = 0.02275$. Using the Monte Carlo simulation with 500 000 trials, the probability of failure is obtained as $P_F = 0.01838$ with coefficient of variation of 1.03% and reliability index $\beta = 2.0884$. It can be observed that the first-order reliability results are different from those of the Monte Carlo simulation.

5.3 Second-Order Reliability Method

5.3.1 Necessity of Second-Order Reliability Method

As described in the last section, FORM is based on the linearisation of the transformed performance function in the standard normal space at its design point. The method has been applied successfully to a large number of engineering problems. Despite its conceptual simplicity, practical examples usually require the use of computers due to the complexity of the engineering problem.

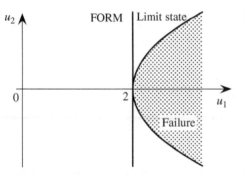

Figure 5.7 Limit state curve corresponding to Example 5.8.

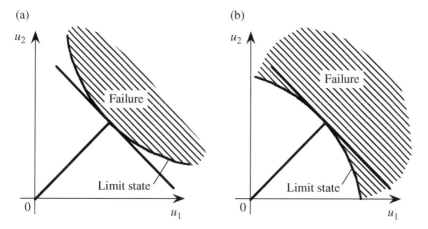

Figure 5.8 First approximation of limit state curves convex and concave to the origin. (a) FORM for convex limit state curve. (b) FORM for concave limit state curve.

Questions have been raised as to the accuracy of FORM since a linear approximation of the true failure surface appeared to be rather crude (Fiessler 1979). As shown in Figure 5.8a and b, the FORM will overestimate the probability of failure for a limit state curve convex to the origin, but underestimate it for that concave to the origin.

5.3.2 Second-Order Approximation of the Performance Function

In order to improve the accuracy of FORM, the performance function is approximated by a second-order function at the design point, which is usually referred to as Second-Order Reliability Method (SORM). For the SORM, the limit state surface is approximated by a second-order surface, the curvature radii of the second-order surfaces are generally equal to those of the true limit state surfaces, as illustrated in Figure 5.9a and b. For a general performance function $G(\mathbf{U})$, let $G(\mathbf{U})$ be approximated at the design point $\mathbf{u}^*$ by means of a second-order Taylor series expansion. This approximation provides a second-order surface $G_S(\mathbf{U})$ at $\mathbf{u}^*$ to the function $G(\mathbf{U})$:

$$G_S(\mathbf{U}) = \beta_F + \mathbf{U}^T \alpha + \frac{1}{2}(\mathbf{U} - \mathbf{u}^*)^T \mathbf{B}(\mathbf{U} - \mathbf{u}^*) \tag{5.27}$$

where

$$\alpha = \frac{\nabla G(\mathbf{u}^*)}{|\nabla G(\mathbf{u}^*)|}, \quad \mathbf{B} = \frac{\nabla^2 G(\mathbf{u}^*)}{|\nabla G(\mathbf{u}^*)|}, \quad \beta_F = -\mathbf{u}^{*T}\alpha$$

where $\mathbf{u}^* = (u_1{}^*, u_2{}^*, ..., u_n{}^*)$ is the design point obtained by FORM. $\nabla G(\mathbf{u}^*)$ and $\nabla^2 G(\mathbf{u}^*)$ are the derivatives $\nabla G(\mathbf{U})$ and $\nabla^2 G(\mathbf{U})$, respectively, evaluated at $\mathbf{u}^* = (u_1{}^*, u_2{}^*, ..., u_n{}^*)$.

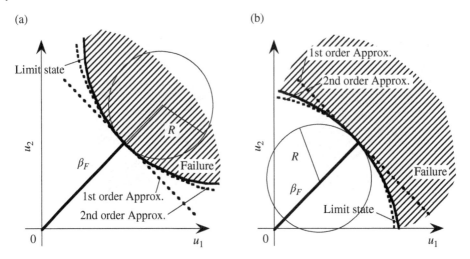

Figure 5.9 Second-order approximation for limit state curve. (a) Convex limit state curve. (b) Concave limit state curve.

The performance function of (5.27) can be expressed as (5.28) without loss of generality.

$$G_S(\mathbf{Y}) = a_0 + \sum_{i=1}^{n} \left(\gamma_i Y_i + \lambda_i Y_i^2 \right) \tag{5.28}$$

where Y_i are independent standard normal random variables with number of n.

The exact integral expression for the probability content of the quadratic set is given by Tvedt (1990). Since the expression for the failure probability that corresponds to a general second-order surface is quite complicated, this failure probability is usually estimated using a parabolic surface. Two approximations have been developed. The first is the general parabolic approximation, which approximates the second-order surface using a parabola defined by the principal curvatures of the limit state function at the design point (Breitung 1984; Tvedt 1983; Der Kiureghian et al. 1987; Hohenbichler and Rackwitz 1988; Der Kiureghian and De Stefano 1991; Koyluoglu and Nielsen 1994; Cai and Elishakoff 1994). The other is the simple parabolic approximation, which was developed to obtain the principal curvature at the design point without the need for rotational transformation and eigenvalue analysis of the Hessian matrix (Zhao and Ono 1999b).

5.3.2.1 General Parabolic Approximation

By rotating of $\mathbf{U}$ into a new set of mutually independent standard normal random variables $\mathbf{X} = \mathbf{HU}$, where the nth row of the rotation matrix $\mathbf{H}$ is α, one obtain

$$G_S(\mathbf{X}) = -(X_n - \beta_F) + \frac{1}{2} \left\{ \begin{matrix} \mathbf{X}' \\ X_n - \beta_F \end{matrix} \right\}^T \mathbf{A} \left\{ \begin{matrix} \mathbf{X}' \\ X_n - \beta_F \end{matrix} \right\} \tag{5.29}$$

where $\mathbf{X}' = (X_1, X_2, ..., X_{n-1})$ and $\mathbf{A} = \mathbf{HBH}^T$.

Approximating Eq. (5.29) using the equation

$$G_S(\mathbf{X}) = -(X_n - \beta_F) + \frac{1}{2} \mathbf{X}'^T \mathbf{A}' \mathbf{X}' \tag{5.30}$$

Without loss of generality, the performance function in Eq. (5.30) can be expressed as

$$G_S(\mathbf{Y}) = -(Y_n - \beta_F) + \frac{1}{2}\sum_{i=1}^{n-1} k_i Y_i^2 \tag{5.31}$$

where k_i, $i = 1, ..., n-1$ are principle curvatures at the design point which are determined as the eigenvalues of $\mathbf{A}'$.

$$|\mathbf{A}' - k\mathbf{I}| = 0 \tag{5.32}$$

where $\mathbf{I}$ is the unit matrix.

$\mathbf{Y}$ is also a set of mutually independent standard normal random variables obtained by another rotation of $\mathbf{X}$, $\mathbf{Y} = \mathbf{RX}$, where $\mathbf{R}$ is a matrix with the eigenvectors of $\mathbf{A}'$ as column vectors.

One may note the difference between the principal curvatures of Eqs. (5.29) and (5.30). In order to obtain the principal curvatures k_i in Eq. (5.31), two matrix rotations are needed, i.e. the rotation from $\mathbf{B}$ to obtain $\mathbf{A}$ in Eq. (5.29) and the eigenvalue analysis in Eq. (5.32).

Example 5.9   To demonstrate the procedure of SORM above, consider the following performance function in standard normal space.

$$G(\mathbf{U}) = 7U_1^3 + 100U_1^2 - 1000U_1 - 1000U_2 - 200U_1U_2 + 3000$$

After five iterations, the first order reliability index is obtained as $\beta_F = 1.9803$, with corresponding a probability of failure of $P_F = 0.02384$. The design point is obtained as $\mathbf{u}^* = (1.251, 1.535)^T$ and the directional cosines is obtained as $\alpha = (-0.6317, -0.7752)^T$. According to Eq. (5.14), the first order approximation is expressed as

$$G(\mathbf{U}) = 1.9803 - 0.6317U_1 - 0.7752U_2$$

The Hessian matrix at the design point can be obtained as

$$\mathbf{B} = \begin{bmatrix} 0.1564 & -0.1238 \\ -0.1238 & 0 \end{bmatrix}$$

Then the second-order approximation of the performance function is obtained as

$$G(\mathbf{U}) = 1.9803 + (-0.6317, \quad -0.7752)\begin{Bmatrix} U_1 \\ U_2 \end{Bmatrix}$$

$$+ \frac{1}{2}(U_1 - 1.251, \quad U_2 - 1.535)\begin{bmatrix} 0.1564 & -0.1238 \\ -0.1238 & 0 \end{bmatrix}\begin{Bmatrix} U_1 - 1.251 \\ U_2 - 1.535 \end{Bmatrix}$$

The second-order approximation is schematically shown in Figure 5.10a.

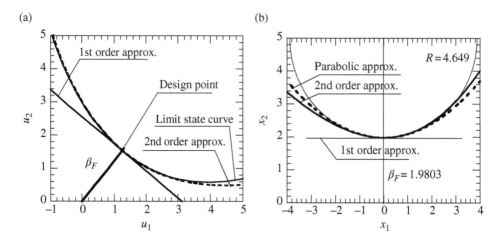

Figure 5.10 The second-order approximation for Example 5.9. (a) In original *u*-space. (b) In rotated *u*-space.

Using the Gram-Schmidt procedure, the rotation matrix and the transformed Hessian matrix are derived as

$$\mathbf{H} = \begin{bmatrix} -0.7752 & 0.6317 \\ -0.6317 & -0.7752 \end{bmatrix}, \mathbf{A} = \mathbf{HBH}^T = \begin{bmatrix} 0.2153 & 0.05157 \\ 0.05157 & -0.05889 \end{bmatrix}$$

and the second-order approximation in the rotated space is expressed as

$$G(\mathbf{U}) = 1.9803 - U_2 + \frac{1}{2}(U_1, \quad U_2 - 1.9803) \begin{bmatrix} 0.2153 & 0.05157 \\ 0.05157 & -0.05889 \end{bmatrix} \begin{Bmatrix} U_1 \\ U_2 - 1.9803 \end{Bmatrix}$$

Deleting the second row and second column of $\mathbf{A}$ leads to $\mathbf{A}' = 0.2153$. Since there is only one element in $\mathbf{A}'$, the eigenvalue value of $\mathbf{A}'$ is itself, i.e. the principle curvature is $k = 0.2153$. Then the parabolic approximation can be derived and expressed as

$$G(\mathbf{U}) = 1.9803 - U_2 + 0.1076U_1^2$$

The second-order approximation in rotated *u*-space is depicted in Figure 5.10b.

Example 5.10 Consider the following limit state function in Example 5.5.

$$G(\mathbf{X}) = X_1 X_2 - X_3$$

As shown in Example 5.5, after six iterations, the first-order reliability index is obtained as $\beta_F = 2.914$, with corresponding probability of failure of $P_F = 0.00178$. The design point, the directional cosines, and the Hessian matrix at the design point are obtained as

$$\mathbf{u}^* = \begin{Bmatrix} -1.656 \\ -1.096 \\ 2.132 \end{Bmatrix}, \alpha = \begin{Bmatrix} 0.568 \\ 0.376 \\ -0.732 \end{Bmatrix}, \mathbf{B} = \begin{bmatrix} 0 & 0.1125 & 0 \\ 0.1125 & 0.0744 & 0 \\ 0 & 0 & -0.281 \end{bmatrix}$$

Then the second-order approximation of the performance function is obtained as

$$g(\mathbf{U}) = 2.914 + \left\{ \begin{array}{c} 0.568 \\ 0.376 \\ -0.732 \end{array} \right\}^T \left\{ \begin{array}{c} U_1 \\ U_2 \\ U_3 \end{array} \right\} + \frac{1}{2} \left\{ \begin{array}{c} U_1 + 1.656 \\ U_2 + 1.096 \\ U_3 - 2.132 \end{array} \right\}^T \left[\begin{array}{ccc} 0 & 0.1125 & 0 \\ 0.1125 & 0.0744 & 0 \\ 0 & 0 & -0.281 \end{array} \right]$$
$$\left\{ \begin{array}{c} U_1 + 1.656 \\ U_2 + 1.096 \\ U_3 - 2.132 \end{array} \right\}$$

Using Gram-Schmidt procedure, the rotation matrix and the transformed Hessian matrix are obtained as

$$\mathbf{H} = \left[\begin{array}{ccc} 0.7898 & 0 & 0.6133 \\ -0.2307 & 0.9265 & 0.2971 \\ 0.5683 & 0.3762 & -0.7318 \end{array} \right], \mathbf{A} = \mathbf{HBH}^T = \left[\begin{array}{ccc} -0.1057 & 0.0311 & 0.1595 \\ 0.0311 & -0.0090 & 0.1365 \\ 0.1595 & 0.1365 & -0.0918 \end{array} \right]$$

and the second-order approximation in the rotated space is expressed as

$$g(\mathbf{U}) = 2.914 - U_3 + \frac{1}{2} \left\{ \begin{array}{c} U_1 \\ U_2 \\ U_3 - 2.914 \end{array} \right\}^T \left[\begin{array}{ccc} -0.1057 & 0.0311 & 0.1595 \\ 0.0311 & -0.0090 & 0.1365 \\ 0.1595 & 0.1365 & -0.0918 \end{array} \right] \left\{ \begin{array}{c} U_1 \\ U_2 \\ U_3 - 2.914 \end{array} \right\}$$

Deleting the third row and third column of $\mathbf{A}$ leads to

$$\mathbf{A} = \left[\begin{array}{cc} -0.1057 & 0.0311 \\ 0.0311 & -0.0090 \end{array} \right]$$

The eigenvalues of $\mathbf{A}'$, i.e. the principle curvatures, are obtained as $k_1 = -0.1148$ and $k_2 = 0.00017$. Then the parabolic approximation can be derived and expressed as

$$g(\mathbf{U}) = 2.914 - U_3 + \frac{1}{2} \left(-0.1148 U_1^2 + 0.00017 U_2^2 \right)$$

5.3.2.2 Simple Parabolic Approximation

From Examples 5.9 and 5.10 two matrix rotations are needed, i.e. the rotation from $\mathbf{B}$ to $\mathbf{A}$ in Eq. (5.29) and the eigenvalue analysis in Eq. (5.32). In order to avoid the two matrix operations, consider the second-order Taylor expansion in standardised normal space for Eq. (5.29). Forming a plane surface in the x_n–x_j plane through the design point $\mathbf{x}^*$ as illustrated in Figure 5.11a–c, the intersection curve between the limit state surface of Eq. (5.29) and the plane can be expressed as (Zhao and Ono 1999b):

$$g(x_j, x_n) = a_{jj}x_j^2 + 2a_{nj}x_j(x_n - \beta_F) + a_{nn}(x_n - \beta_F)^2 - 2(x_n - \beta_F) = 0 \qquad (5.33)$$

Since the design point is $(\mathbf{0}, \beta_F)$, then we have

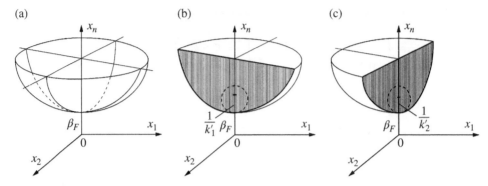

Figure 5.11 Simple parabolic approximation.

$$\frac{\partial g}{\partial x_n}\Big|_{(0,\beta_F)} = \left[2a_{nj}x_j + 2a_{nn}(x_n - \beta_F) - 2\right]\Big|_{(0,\beta_F)} = -2$$

$$\frac{\partial g}{\partial x_j}\Big|_{(0,\beta_F)} = \left[2a_{jj}x_j + 2a_{nj}(x_n - \beta_F)\right]\Big|_{(0,\beta_F)} = 0$$

$$\frac{\partial^2 g}{\partial x_n^2}\Big|_{(0,\beta_F)} = 2a_{nn}, \quad \frac{\partial^2 g}{\partial x_j^2}\Big|_{(0,\beta_F)} = 2a_{jj}, \quad \frac{\partial^2 g}{\partial x_n \partial x_j}\Big|_{(0,\beta_F)} = 2a_{nj}, \quad \frac{dx_n}{dx_j} = -\frac{\partial g}{\partial x_j}\Big/\frac{\partial g}{\partial x_n} = 0$$

$$\frac{d^2 x_n}{dx_j^2} = -\frac{\dfrac{\partial^2 g}{\partial x_j^2}\left(\dfrac{\partial g}{\partial x_n}\right)^2 - 2\dfrac{\partial^2 g}{\partial x_n \partial x_j}\dfrac{\partial g}{\partial x_j}\dfrac{\partial g}{\partial x_n} + \dfrac{\partial^2 g}{\partial x_n^2}\left(\dfrac{\partial g}{\partial x_j}\right)^2}{\left(\dfrac{\partial g}{\partial x_n}\right)^3}$$

$$= -\frac{2a_{jj}(-2)^2 - 2\dfrac{\partial^2 g}{\partial x_n \partial x_j}0\dfrac{\partial g}{\partial x_n} + \dfrac{\partial^2 g}{\partial x_n^2}0^2}{(-2)^3} = a_{jj}$$

The curvature of this curve at the design point is then obtained as

$$k_j' = \frac{\dfrac{d^2 x_n}{dx_j^2}}{\left(1 + \left(\dfrac{\partial g}{\partial x_n}\right)^2\right)^{\frac{3}{2}}} = \frac{a_{jj}}{(1+0)^{\frac{3}{2}}} = a_{jj}$$

That is

$$k_j' = a_{jj} \tag{5.34}$$

in which a_{jj}, $j = 1, ..., n - 1$ are diagonal elements of **A**.

For the limit state surface of Eq. (5.29), the sum of the principal curvatures k_j, $j = 1, ...,$ $n - 1$ at the design point can be expressed as the following equation according to differential geometry (Kobayasi 1977):

$$K_S = \sum_{j=1}^{n-1} k_j = \sum_{j=1}^{n-1} k'_j = \sum_{j=1}^{n} a_{jj} - a_{nn} \tag{5.35}$$

Because **A** in Eq. (5.29) is transformed from **B** in Eq. (5.27) by using orthogonal transformation, the following equation holds true according to linear algorithms.

$$\sum_{j=1}^{n} a_{jj} = \sum_{j=1}^{n} b_{jj}, a_{nn} = \alpha^T \mathbf{B} \alpha \tag{5.36}$$

where $b_{jj}, j = 1, ..., n-1$ are the diagonal elements of **B**.

Substituting Eq. (5.36) into Eq. (5.35), K_s in Eq. (5.35) can be further expressed as:

$$K_S = \sum_{j-1}^{n} b_{jj} - \alpha^T \mathbf{B} \alpha \tag{5.37}$$

Approximating the limit state surface by a rotational parabolic surface of diameter $2R$, where R is the average principal curvature radius expressed as:

$$R = \frac{n-1}{K_S} \tag{5.38}$$

The performance functions in standardised space can then be expressed as Eq. (5.39), which may be considered as a special form of Eq. (5.31) (Zhao and Ono 1999b),

$$G(\mathbf{U}) = -(U_n - \beta_F) + \frac{1}{2R} \sum_{j=1}^{n-1} U_j^2 \tag{5.39}$$

From Eqs. (5.37)–(5.39), it can be observed that, for the simple parabolic approximation, the average principal curvature radius R is directly obtained from the Hessian matrix in Eq. (5.27), and does not require any matrix rotation or eigenvalue analysis. The simple parabolic approximation is therefore much easier to implement than the general parabolic approximation. It is easy to understand that the limit state surface of Eq. (5.39) is convex to the origin when R is positive and is concave to the origin when R is negative.

Using the same procedure above for Eq. (5.30), then the intersection curve between the limit state surface of Eq. (5.30) and the x_n–x_j plane can be expressed as:

$$g(x_j, x_n) = a_{jj}x_j^2 - 2(x_n - \beta_F) = 0 \text{ or } x_n = \beta_F + \frac{1}{2}a_{jj}x_j^2 = 0$$

At the design point, $\mathbf{x}^* = (\mathbf{0}, \beta_F)$, we have

$$\frac{dx_n}{dx_j}\bigg|_{\mathbf{x}^*} = a_{jj}x_j\big|_{\mathbf{x}^*} = 0, \frac{d^2x_n}{dx_j^2}\bigg|_{\mathbf{x}^*} = a_{jj}$$

Then the curvature of this curve at the design point is obtained as

$$k'_j = \frac{d^2x_n}{dx_j^2} = a_{jj}$$

One can see that the result is as same as that obtained from Eq. (5.29). Thus, the simple parabolic approximation for Eqs. (5.29) and (5.30) are the same. Therefore, K_s is equal to the sum of the principle curvatures of the general parabolic approximation.

$$K_S = \sum_{j=1}^{n-1} k_j$$

where $k_j, j = 1, ..., n-1$ are principle curvatures at the design point which are determined as the eigenvalues of $\mathbf{A}'$.

Example 5.11 To demonstrate the procedure of the simple parabolic approximation described above, consider Example 5.8 again. The Hessian matrix at the design point has been obtained as

$$\mathbf{B} = \begin{bmatrix} 0.1564 & -0.1238 \\ -0.1238 & 0 \end{bmatrix}$$

Using the Hessian matrix, one obtains that

$$\sum_{j=1}^{n} b_{jj} = 0.1564 + 0 = 0.1564$$

$$\alpha^T \mathbf{B} \alpha = (-0.6317, \ -0.7752) \begin{bmatrix} 0.1564 & -0.1238 \\ -0.1238 & 0 \end{bmatrix} \left\{ \begin{matrix} -0.6317 \\ -0.7752 \end{matrix} \right\} = -0.05889$$

$$K_S = \sum_{j=1}^{n} b_{jj} - \alpha^T \mathbf{B} \alpha = 0.1564 - (-0.05889) = 0.2153$$

The average principal curvature radius is given by:

$$R = \frac{n-1}{K_S} = \frac{1}{0.2153} = 4.645$$

Then the simple parabolic approximation expressed as

$$G(\mathbf{U}) = 1.9803 - U_2 + 0.1076 U_1^2$$

which is the same as that of general parabolic approximation shown in Example 5.9, although this procedure is much more simplified.

Example 5.12 Consider the Example 5.10 again. The direction of cosines and the Hessian matrix at the design point have been obtained as

$$\alpha = \left\{ \begin{matrix} 0.568 \\ 0.376 \\ -0.732 \end{matrix} \right\}, \mathbf{B} = \begin{bmatrix} 0 & 0.1125 & 0 \\ 0.1125 & 0.0744 & 0 \\ 0 & 0 & -0.281 \end{bmatrix}$$

using the Hessian matrix, one obtains that

$$K_S = \sum_{j=1}^{n} b_{jj} - \alpha^T \mathbf{B} \alpha = -0.2065 - (-0.0918) = -0.1147$$

Note that value of K_s above is equal to the sum of the principle curvatures k_i in the general parabolic approximation in Example 5.10.

The average principal curvature radius is given by:

$$R = \frac{n-1}{K_S} = \frac{2}{-0.1147} = -17.44$$

Then the simple parabolic approximation can be derived and expressed as

$$G(\mathbf{U}) = 2.914 - U_3 - 0.0287\left(U_1^2 + U_2^2\right)$$

The two examples above demonstrate that the procedure of the simple parabolic approximation is much simpler than that of the general parabolic approximation.

5.3.2.3 Point-Fitting Second Order Approximation

As described above, the second-order approximation of a performance function requires the computation of the Hessian matrix, which can be prohibitively costly when the number of random variables is large and the performance function involves complicated numerical algorithms. Difficulty in computing the Hessian matrix has led to the development of another type of SORM approximation, i.e. the point-fitting SORM method (Der Kiureghian et al. 1987; Der Kiureghian and De Stefano 1991), in which an efficient algorithm is developed to determine the principal curvatures without computing the Hessian matrix. In this section, a point-fitting method (Zhao and Ono 1999c) for second-order approximation is presented, where the performance function is iteratively point-fitted using a second-order polynomial of standard normal random variables. Neither computation of the Hessian matrix or the gradients of the performance function are required.

Consider the limit state surface in standard normal space expressed by a performance function $G(\mathbf{U})$ and define the second-order surface approximation in terms of a set of fitting points on the limit state surface in the vicinity of the design point as shown in Figure 5.12. These points, $2n + 1$ in number, are selected along the coordinate axes in standard normal space. Along each axis u_j, $j = 1, ..., n$, two points having the coordinates $(\mathbf{u}'^*, u_j^* - \delta)$ and $(\mathbf{u}'^*, u_j^* + \delta)$ are selected, where $\mathbf{u}'^* = \{u_k^*, k = 1, ..., n \text{ except } j\}$ represents the coordinates of the design point along all the axes except the j-axis, and δ is a factor which represents the distance from the design point to the fitting point. The point-fitted performance function is expressed as a second-order polynomial of standard normal random variables, including $2n + 1$ regression coefficients.

Figure 5.12 The concept of point-fitting second order approximation.

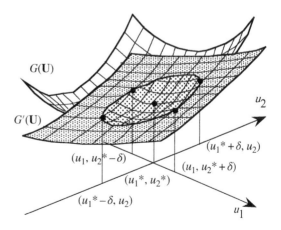

$$G'(\mathbf{U}) = a_0 + \sum_{i=1}^{n}(b_i U_i + c_i U_i^2) \tag{5.40}$$

where a_0, b_j, and c_j are $2n + 1$ regression coefficients.

Fitting the practical performance function $G(\mathbf{U})$ by $G'(\mathbf{U})$ at the fitting points described above, the regression coefficients a_0, b_j, and c_j can be determined from linear equations of a_0, b_j, and c_j obtained at each fitting point.

Computation of the Hessian matrix may be difficult for $G(\mathbf{U})$, but can be computed quite easily for $G'(\mathbf{U})$. Once the point-fitted performance function (5.40) is obtained, the Hessian matrix corresponding to (5.40) can be readily given as

$$\mathbf{B} = \frac{2}{|\nabla G'|}\begin{bmatrix} c_1 & \cdots & 0 \\ \vdots & \ddots & \vdots \\ 0 & \cdots & c_n \end{bmatrix} \tag{5.41}$$

where

$$|\nabla G'| = \sqrt{\sum \left(b_j + 2c_j u_j^*\right)^2} \tag{5.42}$$

The sum of the principal curvatures and the average principal curvature radius of the limit state surface at the design point $\mathbf{u}^*$ can be expressed as (Zhao and Ono 1999c):

$$K_S = \frac{2}{|\nabla G|}\sum c_j\left[1 - \frac{1}{|\nabla G|^2}\left(b_j + 2c_j u_j^*\right)^2\right] \tag{5.43}$$

Although the procedure just described can be used to obtain the second-order reliability index conveniently, without computation of the Hessian matrix corresponding to $G(\mathbf{U})$, and without rotational transformation and eigenvalue analysis of the Hessian matrix corresponding to $G'(\mathbf{U})$, implementing this method requires the design point, which is generally not known beforehand. Computation of the design point for $G(\mathbf{U})$ requires the gradients of the limit state surface, which may also be prohibitively costly to compute when the number of random variables is large and the performance function involves complicated numerical algorithms. To avoid this problem, applying the iterative response surface approach (Bucher and Bourgund 1990; Rajashekhar and Ellingwood 1993) to the second order approximation of performance function in standard normal space, an iterative point-fitting procedure can be described as following to obtain the design point in the iterative procedure.

1) Select an initial central point $\mathbf{u}_c$ in standard normal space (generally, the point corresponding to the mean value in original space is recommended).
2) Select fitting points along the coordinate axes. Along each axis u_j, $j = 1, ..., n$, two points having the coordinates $(\mathbf{u}_c', u_{cj} - \delta)$ and $(\mathbf{u}_c', u_{cj} + \delta)$ are selected, where $\mathbf{u}_c' = \{u_{ck}, k = 1, ..., n$ except $j\}$ represents the coordinates of the design point along all the axes except the j-axis, and δ is a factor which represents the distance from the central point to the fitting point.

3) Transform the fitting points to the original space using the Rosenblatt transformation, and fit the original performance function by the performance function approximation in Eq. (5.40) at these points. The regression coefficients included in Eq. (5.40) can now be obtained.

4) For the point-fitted performance function Eq. (5.40), conduct FORM iteration and obtain the design point $\mathbf{u}^*$ corresponding to Eq. (5.40).

5) Substituting $\mathbf{u}^*$ for $\mathbf{u}_c$ in step (2), repeat steps (2)–(4) until convergence.

6) After obtaining the design point, the failure probability or the second-order reliability index can be obtained using the reliability index that will be described in Section 5.3.3.

Example 5.13   The object of the example is to investigate the convergence of the iterative point-fitting procedure described above. Consider the following performance function that includes only two random variables.

$$G(\mathbf{X}) = 80 - X_1 + X_2 \tag{5.44}$$

where X_1 is a normal random variable with mean value of $\mu = 50$ and standard deviation of $\sigma = 25$; and X_2 is a lognormal random variable with mean value of $\mu = 80$ and standard deviation of $\sigma = 64$. For the purpose of comparison, the problem is first solved using gradient-based optimisation algorithms, and the design point is obtained as $(2.1334, -1.4000)$ in the standard normal space with the first-order reliability index of $\beta_F = 2.5518$. Using differential geometry, the curvature radius is obtained as $R = 3.7072$, and the central point of the tangent circle is obtained as $(5.2232, -3.4346)$ in standard normal space, as depicted by the thin solid line in Figure 5.13. Using the integration method, the failure probability is obtained as 4.020×10^{-3} with the corresponding reliability index of $\beta = 2.6504$.

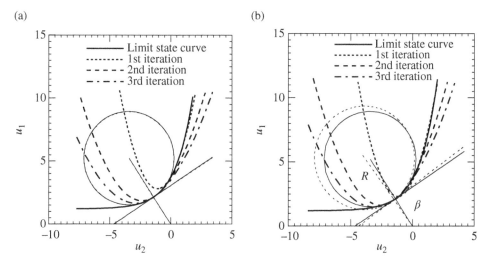

Figure 5.13   Procedure of point-fitting second approximation. (a) $\delta = 0.2$. (b) $\delta = 1.5$.

When the analysis is conducted, the initial central point is taken to be the point transformed from the mean value point, and the distance from the central point to the fitting point is taken as $\delta = 0.2$. Convergence is obtained after four iterations with an error tolerance of 0.0001. The results obtained in each iteration are summarised as follows:

Iter.No.	Central point	$g'(\mathbf{U})$	Design point	β_F	R
1	$\left\{\begin{array}{c} 0 \\ 0.352 \end{array}\right\}$	$g(\mathbf{U}) = 92.598 - 25U_1$ $+ 32.513U_2 + 19.821U_2^2$	$\left\{\begin{array}{c} 2.802 \\ -0.938 \end{array}\right\}$	2.9546	0.7395
2	$\left\{\begin{array}{c} 2.802 \\ -0.938 \end{array}\right\}$	$g(\mathbf{U}) = 90.715 - 25U_1$ $+ 37.805U_2 + 8.004U_2^2$	$\left\{\begin{array}{c} 2.100 \\ -1.450 \end{array}\right\}$	2.5520	2.8039
3	$\left\{\begin{array}{c} 2.100 \\ -1.450 \end{array}\right\}$	$g(\mathbf{U}) = 87.314 - 25U_1$ $+ 32.081U_2 + 5.580U_2^2$	$\left\{\begin{array}{c} 2.150 \\ -1.374 \end{array}\right\}$	2.5513	3.7459
4	$\left\{\begin{array}{c} 2.150 \\ -1.374 \end{array}\right\}$	$g(\mathbf{U}) = 87.924 - 25U_1$ $+ 32.948U_2 + 5.887U_2^2$	$\left\{\begin{array}{c} 2.121 \\ -1.418 \end{array}\right\}$	2.5516	3.6955
Exact values		$g(\mathbf{U}) = 30 - 25U_1$ $+ \exp(U_2 + 4.1347)$	$\left\{\begin{array}{c} 2.133 \\ -1.400 \end{array}\right\}$	2.5518	3.7072

From these results, one can see that although the results obtained in the first iteration are much different from the exact results, when convergence is reached (the fourth iteration), the design point is obtained as $\mathbf{u}^* = (2.121, -1.418)$ with corresponding first-order reliability index $\beta_F = 2.5516$, which is in good agreement with the exact results $\mathbf{u}^* = (2.133, -1.400)$ and $\beta_F = 2.5518$. Applying formulas in Eqs. (5.43) and (5.38) to the point-fitted performance function obtained in the fourth iteration, the average curvature radius is obtained as $R' = 3.6955$, which is very close to the exact results $R = 3.7072$. The failure probability corresponding to the point-fitted performance function in the fourth iteration is obtained as $P_F = 4.013 \times 10^{-3}$ with the corresponding reliability index of $\beta_S = 2.6510$, which are also close to the exact results $P_F = 4.020 \times 10^{-3}$, $\beta_S = 2.6504$. (The failure probability and reliability index corresponding to second-order performance function will be describe in the next section.)

The point-fitted limit state surfaces are depicted in Figure 5.13a for each iteration, except the fourth iteration since it is almost identical to that obtained in the third iteration. Figure 5.13a also shows that the point fitted limit state surface gradually approaches the original limit state surface at design point with the increase of the number of iterations. The tangent circle at the design point of the point-fitted limit state surface obtained in the fourth iteration is depicted by the thin dashed line. This circle is nearly invisible because it almost completely coincides with that of the original limit state surface. The centre point of the circle is obtained as $(5.2195, -3.4323)$, which is almost identical to the exact centre point $(5.2232, -3.4346)$.

In order to investigate the effects of the fitting points, the previous problem is solved using different fitting points ranging from $\delta = 0.1$ to $\delta = 2.0$ with an interval of 0.1. All of the computations converged within six iterations. The first- and second-order reliability indices obtained with different δ are shown in Figure 5.14, and the corresponding curvature radii

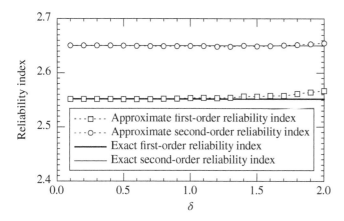

Figure 5.14 Reliability indices affected by δ.

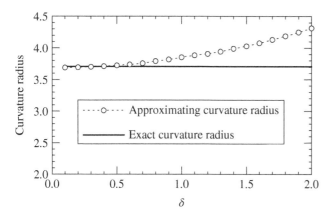

Figure 5.15 Curvature radius affected by δ.

are shown in Figure 5.15. From Figure 5.14, it can be observed that the first- and second-order reliability indices are only slightly affected by the fitting points, except when δ is very large. When δ is larger than 1.0, the first-order reliability index increased slightly with the increase of δ. In contrast, the second-order reliability index remains almost unchanged. This can be attributed to the fact that, in these cases, the curvature radius R also becomes large (see Figure 5.15) and the modification effect of R becomes weak. As an example of large δ, the point-fitted second-order surface obtained with $\delta = 1.5$ is depicted in Figure 5.13b. It can be observed that the tangent circle (depicted by a thin dashed line) of the point-fitted limit state surface does not closely approach that of the original limit state surface (depicted by a thin solid line). The approximating curvature radius is obtained as $R = 4.0305$ which is very different comparing with the exact value $R = 3.7072$. The first- and second-order reliability index are obtained as $\beta_F = 2.5562$ and $\beta_S = 2.6493$ corresponding to the point-fitted limit state surface, respectively, which are still close to the exact values of $\beta_F = 2.5518$ and $\beta_S = 2.6504$.

Figures 5.14 and 5.15 show that the value of δ should be between 0.1 and 0.5 to achieve acceptable accuracy.

5.3.3 Failure Probability for Second-Order Performance Function

Once the second-order approximation of performance function in standard normal space has been obtained, we can then compute its failure probability. Different methods have been developed including the integration method, the Inverse Fast Fourier Transformation (IFFT) method, the approximate second-order reliability indices in close form, and the moment approximation. In this section, the integration method, the IFFT method, the approximate second order reliability indices will be introduced, and moment approximation will be discussed latter in Section 5.3.4.

5.3.3.1 Integration Method

For the simple parabolic approximation in Eq. (5.39), the performance function $Z = G(\mathbf{U})$ is the combination of a standardised normal random variable and a random variable of a central chi-square (χ^2) distribution with $n-1$ degrees of freedom (Fiessler et al. 1979). The probability $P_F = \text{Prob}\{G(\mathbf{U}) < 0\}$ is computed using

$$P_F = \int_0^\infty \Phi\left(\frac{t}{2R} - \beta_F\right) f_{\chi^2_{n-1}}(t)dt \tag{5.45}$$

where

$$f_{\chi^2_{n-1}}(t) = \frac{1}{\Gamma\left(\dfrac{n-1}{2}\right)2^{\frac{n-1}{2}}} t^{\frac{n-3}{2}} \exp\left(-\frac{t}{2}\right) \tag{5.46}$$

For the general second-order approximation in Eq. (5.29), the exact probability density function (PDF) of the quadratic form is generally not available in the closed form. Direct numerical integration methods, like the saddle-point integration method in Tvedt (1990) can be used.

5.3.3.2 Inverse Fast Fourier Transformation (IFFT) Method

The IFFT method can be used to evaluate the failure probability exactly and efficiently. For the IFFT method, the PDF and the characteristic function of a random variable are expressed as a pair of Fourier transformations (Lin 1967; Sakamoto and Mori 1995; Sakamoto and Mori 1997)

$$Q(t) = \int_{-\infty}^\infty f(x) \exp(itx)dx \tag{5.47}$$

$$f(x) = \frac{1}{2\pi} \int_{-\infty}^\infty Q(t) \exp(-itx)dt \tag{5.48}$$

where $f(x)$ and $Q(t)$ are the PDF and the characteristic function of a random variable x, respectively. For the general second order approximation (Eq. 5.29) of point-fitted performance function, the characteristic function can be explicitly obtained as (Tvedt 1990):

$$Q(t) = \exp{(ia_0 t)} \prod_{j=1}^{n} \frac{\exp{\left[t^2 \gamma_j^2 / 2 (1 - 2it\lambda_j)\right]}}{\sqrt{1 - 2it\lambda_j}} \tag{5.49}$$

Particularly for the performance function in the parabolic approximation of Eq. (5.31), the characteristic function is expressed as:

$$Q(t) = \exp{(i\beta_F t)} \exp{\left[t^2 / 2\right]} \prod_{j=1}^{n-1} \frac{1}{\sqrt{1 - itk_j}} \tag{5.50}$$

Using the evenly distributed values $Q(t_s)$, $s = 1, ..., N$ of Eq. (5.49) or Eq. (5.50) in the interval of $[t_1, t_N]$, where N is the number of discrete data and $[t_1, t_N]$ can be selected by evaluating the effective range of $Q(t)$, the discrete values of inverse Fourier coefficients are readily obtained as F_r, $r = 1, ..., N$, using the IFFT. According to the definition of discrete Fourier transformation and PDF, the discrete values of PDF can be obtained as (Zhao and Ono 1999c):

$$f(x_r) = \frac{t_N - t_1}{2\pi\sqrt{N}} F_r \exp{(-itx_r)} \qquad \text{for} \quad x_r \geq 0 \tag{5.51a}$$

where

$$x_r = \frac{2\pi(r-1)}{t_N - t_1} \tag{5.51b}$$

Because discrete values of $f(x)$ are obtained for only positive values of x, the failure probability for $\text{Prob}\{x < 0\}$ can be readily obtained by the numerical integration of the discrete values of $f(x)$.

$$P_F = 1 - \pi \sum_{r=1}^{N-1} \frac{f(x_r) + f(x_{r+1})}{(t_N - t_1)} \tag{5.52}$$

Example 5.14 To investigate the efficiency of the IFFT method, consider the following performance function in standardised space (Der Kiureghian et al. 1987).

$$G(\mathbf{U}) = \beta_F - U_n + \frac{1}{2} \sum_{j=1}^{n-1} ja U_j^2 \tag{5.53}$$

The real and imaginary parts of the characteristic function are easily obtained as Figures 5.16 and 5.17 respectively, for $a = 0.01, 0.05, 0.1, 0.2, 0.3, 0.4, 0.5$. As the by-product of the IFFT method, the corresponding PDFs in Figure 5.18 show that the effective range of the PDF increases with the increase of a.

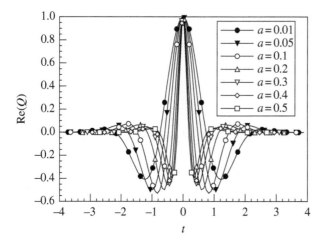

Figure 5.16 Real part of characteristic function.

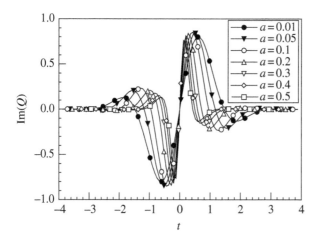

Figure 5.17 Imaginary part of characteristic function.

The sum of the principal curvatures can be readily obtained as $K_S = an(n-1)/2$ for the limit state surface for Eq. (5.53), and the corresponding average curvature radius is obtained as $R = 2/na$. Using R, n, and β_F as parameters, the failure probabilities are obtained using the IFFT method and listed in Table 5.1, in comparison with the results from Monte Carlo Simulation with 500 000 samples. The parameters β_F, n, and R are listed in columns (1), (2), and (3), respectively; the reliability index β and the corresponding failure probability P_F obtained using the IFFT method are listed in columns (4) and (5), respectively; and those obtained using the Monte Carlo simulation are listed in columns (6) and (7), respectively. From Table 5.1, it can be observed that there is very good agreement between the results

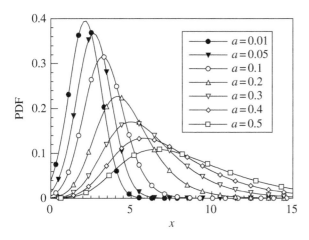

Figure 5.18 PDFs of performance function for Example 5.14.

Table 5.1 Comparison between results obtained by IFFT and MCS for Example 5.14.

Parameters			IFFT		Monte Carlo	
β_F	n	R	β	P_F	β	P_F
(1)	(2)	(3)	(4)	(5)	(6)	(7)
0.0	8	5.0	0.642	2.604×10^{-1}	0.643	2.603×10^{-1}
0.4	8	5.0	1.019	1.541×10^{-1}	1.019	1.541×10^{-1}
1.0	8	5.0	1.588	5.617×10^{-2}	1.589	5.603×10^{-2}
1.8	8	5.0	2.352	9.338×10^{-3}	2.339	9.662×10^{-3}
2.6	8	5.0	3.121	9.008×10^{-4}	3.123	8.941×10^{-4}
2.0	2	5.0	2.083	1.862×10^{-2}	2.083	1.864×10^{-2}
2.0	4	5.0	2.241	1.252×10^{-2}	2.240	1.255×10^{-2}
2.0	6	5.0	2.394	8.331×10^{-3}	2.396	8.302×10^{-3}
2.0	24	5.0	3.650	1.313×10^{-4}	3.648	1.320×10^{-4}
2.0	30	5.0	4.032	2.764×10^{-5}	4.029	2.801×10^{-5}
2.0	8	10.0	2.307	1.051×10^{-2}	2.308	1.051×10^{-2}
2.0	8	5.0	2.544	5.482×10^{-3}	2.541	5.530×10^{-3}
2.0	8	3.3	2.731	3.155×10^{-3}	2.726	3.210×10^{-3}
2.0	8	2.5	2.885	1.958×10^{-3}	2.886	1.950×10^{-3}
2.0	8	2.0	3.014	1.287×10^{-3}	3.026	1.240×10^{-3}

obtained using the IFFT method and those obtained using Monte Carlo simulation with any value of curvature radius R, number of random variables n, and first-order reliability index β_F. In other words, the IFFT method can be used to accurately compute the failure probability corresponding to quadratic performance functions in normal space.

Table 5.2 Formulas of reliability index corresponding to the second-order approximation.

Researchers	Formula
Breitung (1984)	$P_F = \Phi(-\beta_F) \prod_{j=1}^{n-1} \left(1 + \beta_F k_j\right)^{-\frac{1}{2}}$
Koyluoglu and Nielsen (1994)	$P_F = \Phi(-\beta_F) \prod_{j=1}^{n-1} \left(1 + k_j \dfrac{\phi(\beta_F)}{\Phi(\beta_F)}\right)^{-\frac{1}{2}}$ for $k_j > 0, j = 1, 2, ..., n-1$
	$P_F = 1 - \Phi(\beta_F) \prod_{j=1}^{n-1} \left(1 - k_j \dfrac{\phi(\beta_F)}{\Phi(\beta_F)}\right)^{-\frac{1}{2}}$ for $k_j < 0, j = 1, 2, ..., n-1$
Cai and Elishakoff (1994)	$P_F = \Phi(-\beta_F) - \phi(-\beta_F)(D_1 + D_2 + D_3)$
	$D_1 = \sum_{j=1}^{n-1} \lambda_j, D_2 = -\dfrac{1}{2}\beta_F \left(3\sum_{j=1}^{n-1} \lambda_j^2 + \sum_{j\neq l}^{n-1} \lambda_j \lambda_l\right)$
	$D_3 = -\dfrac{1}{6}(\beta_F^2 + 1)\left(15\sum_{j=1}^{n-1} \lambda_j^3 + \sum_{j\neq l}^{n-1} \lambda_j^2 \lambda_l + \sum_{j\neq l\neq m}^{n-1} \lambda_j \lambda_l \lambda_m\right), \lambda_j = \dfrac{1}{2}k_j, \ j = 1, 2, ..., n-1$
Tvedt (1983)	$P_F = \Phi(-\beta_F) \prod_{j=1}^{n-1} \left(1 + \beta_F k_j\right)^{-\frac{1}{2}} + A_2 + A_3$
	$A_2 = [\beta_F \Phi(-\beta_F) - \phi(-\beta_F)]\left\{\prod_{j=1}^{n-1} \left(1 + \beta_F k_j\right)^{-\frac{1}{2}} - \prod_{j=1}^{n-1} \left(1 + (\beta_F + 1)k_j\right)^{-\frac{1}{2}}\right\}$
	$A_3 = (\beta_F + 1)[\beta_F \Phi(-\beta_F) - \phi(-\beta_F)]\left\{\prod_{j=1}^{n-1} \left(1 + \beta_F k_j\right)^{-\frac{1}{2}} - \text{Re}\left[\prod_{j=1}^{n-1} \left(1 + (\beta_F + i)k_j\right)^{-\frac{1}{2}}\right]\right\}$
The empirical formula (Zhao and Ono 1999b)	$\beta_s = \left(1 - \dfrac{K_s}{3\beta_F + 3(n-1)/K_s + 1}\right)\beta_F + \dfrac{1}{2}K_s, K_s \geq 0$
	$\beta_s = \left(1 - \dfrac{K_s^2}{3(n - \beta_F + 3)}\right)\beta_F + \dfrac{1}{2}K_s, K_s < 0$
	When K_s is very small $\beta_S = \beta_F + \dfrac{1}{2}K_s$

5.3.3.3 Approximate Second-Order Reliability Indices

Computation of the failure probability is challenging for the second-order performance function. Various approximate formulas of reliability index have therefore been proposed using the second order approximation, which includes Breitung's (1984), Koyluoglu and Nielsen's (1994), Cai and Elishakoff's (1994), and Tvedt's formula (1983), and the empirical formula (Zhao and Ono 1999b). These formulas are summarised in Table 5.2.

Example 5.15 In order to investigate the accuracy of the approximate second-order reliability indices listed in Table 5.2, consider the following performance function in standardised space, which is the general case of the practical examples used by Cai and Elishakoff (1994) and Koyluoglu and Nielsen (1994).

$$G(\mathbf{U}) = R^2 - \sum_{j=1}^{n} (U_j - \lambda_j)^2 \tag{5.54}$$

The limit state surface in Eq. (5.54) is a hypersphere concave to the origin with radius R and centre at point $(\lambda_j, j = 1, ..., n)$. $y = G(\mathbf{U})$ is a random variable having the non-central chi-squared distribution, and the exact value of probability $P_F = Prob\{G(\mathbf{U}) < 0\}$ is computed directly using this distribution (Sankaran 1959, 1963). The distance from origin to the spherical centre δ is expressed as:

$$\delta^2 = \sum_{i=1}^{n} \lambda_i^2 = (R - \beta_F)^2$$

The design point $\mathbf{u}^*$ and the directional vector α at $\mathbf{u}^*$ are expressed as:

$$\mathbf{u}^* = \left\{ -\frac{\beta_F}{\delta} \lambda_j, j = 1, \cdots, n \right\}, \alpha = \left\{ -\frac{\lambda_j}{\delta}, j = 1, \cdots, n \right\}$$

from which the scaled Hessian matrix is obtained as:

$$\mathbf{B} = \frac{1}{R} \mathbf{I}$$

When using previous SORM formulas, a rotation matrix $\mathbf{H}$ should be established to obtain $\mathbf{A}$ in Eq. (5.28), $\mathbf{A}$ is then simplified to $\mathbf{A}'$ and the principal curvatures are obtained from the eigenvalue analysis of $\mathbf{A}'$ using Eq. (5.29). The results are obtained as

$$k_1 = k_2 = \cdots = k_{n-1} = \frac{1}{R}$$

When using the empirical reliability index, the sum of the principal curvatures can be readily obtained as $K_S = (n - 1)/R$ without matrix rotation and eigenvalue analysis.

The variations of the second-order reliability indices with respect to the number of random variables are shown in Figure 5.19a, in which the first-order reliability index is taken to be 3.0, and curvature radius is taken to be -10. The variations of the second-order reliability indices with respect to the curvature radius are shown in Figure 5.19b, in which the first-order reliability index is taken to be 3.0, and the number of variables is taken to be 8. From Figure 5.19, it can be observed that for small number of random variables and large curvature radius, all of the formulas give good approximations of the exact results and show good improvement of the first order reliability index. However, significant errors can also be observed for large number of random variables and small curvature radius.

5.3.4 Methods of Moment for Second-Order Approximation

The evaluation of the probability of failure P_F corresponding the second-order approximation of $S = G_S(\mathbf{U})$ of performance function $G(\mathbf{U})$ can be expressed as the following one-dimensional integral

$$P_F = P[S = G_S(\mathbf{U}) \leq 0] = \int_{-\infty}^{0} f_S(s)ds = F_S(0) \tag{5.55}$$

where $f_S(s)$ is the density function of the performance function $S = G_S(\mathbf{U})$, which is also a single random variable.

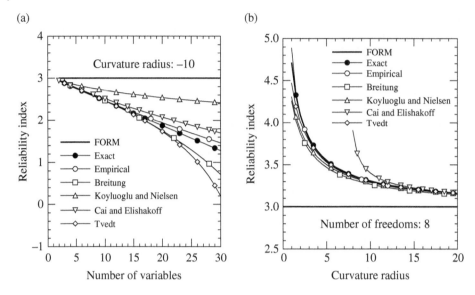

Figure 5.19 Comparison of second-order reliability indices in close form for Example 5.15. (a) Variation with number of variables. (b) Variation with curvature radius.

In this section, the failure probability will be evaluated directly by utilising the statistical moment information of the second-order approximation of performance function at the design point, $S = G_S(\mathbf{U})$. From Eq. (5.27), it is observed that $S = G_S(\mathbf{U})$ is an explicit function of the first-order reliability index and standard normal random variables. Therefore, the first few moments of $S = G_S(\mathbf{U})$ can be easily obtained using the methods described in the chapter on Moment Evaluation for Performance Functions, and thus the direct methods of moment can be applied easily.

5.3.4.1 Second-Order Second-Moment Method

For the second-order performance function $S = G_S(\mathbf{U})$, if the first two moments are obtained, assuming that $Z = G(\mathbf{X})$ obeys normal distribution, the second moment reliability index is given as (Zhao et al. 2002a)

$$\beta_{SOSM} = \frac{\mu_S}{\sigma_S} \tag{5.56}$$

where μ_S and σ_S are the mean value and standard deviation of $S = G_S(\mathbf{U})$, respectively.

Since the first two moments of the second-order approximation of the performance function are used, Eq. (5.56) is hereafter referred to as second-order second-moment (SOSM) reliability index.

For the simple parabolic performance function expressed in Eq. (5.39), the mean and standard deviation of $S = G_S(\mathbf{U})$ are given as:

$$\mu_S = \beta_F + \frac{n-1}{2R} \tag{5.57a}$$

$$\sigma_S^2 = 1 + \frac{n-1}{2R^2} \tag{5.57b}$$

where β_F is the first order reliability index, R is the average curvature radius expressed as Eq. (5.38), and n is the number of variables. Substituting Eqs. (5.57a) and (5.57b) into Eq. (5.56), the SOSM reliability index is given as

$$\beta_{SOSM} = \frac{\beta_F + \dfrac{n-1}{2R}}{\sqrt{1 + \dfrac{n-1}{2R^2}}} \tag{5.58a}$$

Equation (5.58a) can also be derived from asymptotic distribution method (Adhikari 2005). When R is large enough, it can be simplified as

$$\beta_{SOSM} = \beta_F + \frac{n-1}{2R} \tag{5.58b}$$

For the general parabolic approximation function expressed in Eq. (5.31), the mean value and standard deviation of $S = G_S(\mathbf{U})$ are given as:

$$\mu_S = \beta_F + \frac{1}{2}\sum_{i=1}^{n-1} k_i \tag{5.59a}$$

$$\sigma_S^2 = 1 + \frac{1}{2}\sum_{i=1}^{n-1} k_i^2 \tag{5.59b}$$

where β_F is the first order reliability index, where $k_j, j = 1, ..., n-1$ are principle curvatures at the design point and n is the number of variables. Substituting Eqs. (5.59a) and (5.59b) into Eq. (5.56), the SOSM reliability index corresponding to the general parabolic approximation is given as

$$\beta_{SOSM} = \frac{\beta_F + \frac{1}{2}\sum_{i=1}^{n-1} k_i}{\sqrt{1 + \frac{1}{2}\sum_{i=1}^{n-1} k_i^2}} \tag{5.60a}$$

when $k_j, j = 1, ..., n-1$ are small enough, it can be simplified as

$$\beta_{SOSM} = \beta_F + \frac{1}{2}\sum_{i=1}^{n-1} k_i = \beta_F + \frac{1}{2}K_S \tag{5.60b}$$

where K_S is the sum of all the principle curvatures. One may note that Eq. (5.60b) is the same as the Eq. (5.53), which is an empirical formula.

Example 5.16 Consider the limit state function in Example 5.2 again. The parabolic form of the limit state function is

$$G(\mathbf{U}) = 3 - U_2 + \frac{1}{80}U_1^2$$

The mean and standard deviation of the performance function above are given as

$$\mu_S = \beta_F + \frac{n-1}{2R} = 3 + \frac{1}{80} = 3.0125$$

$$\sigma_S = \sqrt{1 + \frac{n-1}{2R^2}} = \sqrt{1 + \frac{1}{3200}} = 1.0$$

The SOSM reliability index is obtained as

$$\beta_{SOSM} = \frac{\mu_S}{\sigma_S} = \frac{3.0125}{1.0} = 3.0125$$

The corresponding probability of failure is $P_F = 1.30 \times 10^{-3}$.

The first-order reliability index has been given as $\beta_F = 3$, the probability of failure $P_F = 1.35 \times 10^{-3}$. Using MCS with 1 000 000 samples, the failure probability is obtained and the corresponding reliability index is 3.0147. It can be observed that the SOSM reliability index gives improvement of FORM reliability index and has better agreement with the result of Monte Carlo simulation.

Example 5.17 For the performance function in Example 5.8 again, the parabolic approximation is easily obtained as

$$G(\mathbf{U}) = 2 - U_2 + \frac{1}{10} U_1^2$$

It can be derived that $\beta_F = 2, R = 5, n = 2$. The mean and standard deviation of the second-order approximate performance function above are given as

$$\mu_S = \beta_F + \frac{n-1}{2R} = 2 + \frac{1}{10} = 2.1, \sigma_S = \sqrt{1 + \frac{n-1}{2R^2}} = \sqrt{1 + \frac{1}{50}} = 1.01$$

The SOSM reliability index is obtained as

$$\beta_{SOSM} = \frac{\mu_S}{\sigma_S} = \frac{2.1}{1.01} = 2.079$$

The corresponding probability of failure is $P_F = 0.0188$. One can see that the SOSM reliability index gives improvement of FORM reliability index and has better agreement with the result of the Monte Carlo simulation in Example 5.8 ($P_F = 0.01838, \beta = 2.0884$).

Example 5.18 Again consider the performance function in Example 5.9; the parabolic approximation has been obtained as

$$G(\mathbf{U}) = 1.9803 - U_2 + 0.1076 \, U_1^2$$

with average curvature radius $R = 4.645$ and first order reliability index $\beta_F = 1.9803$. The SOSM reliability index is calculated as

$$\mu_S = \beta_F + \frac{n-1}{2R} = 1.9803 + \frac{1}{9.29} = 2.091, \sigma_S = \sqrt{1 + \frac{n-1}{2R^2}} = \sqrt{1 + \frac{1}{43.15}} = 1.0115$$

$$\beta_{SOSM} = \frac{\mu_S}{\sigma_S} = \frac{2.091}{1.0115} = 2.067$$

The corresponding probability of failure is $P_F = 0.01937$. Using the Monte Carlo simulation with 500 000 trials, the probability of failure is obtained as $P_F = 0.01918$ with coefficient of variation of 1.01% and reliability index $\beta = 2.0711$. One can see that the SOSM reliability index has better agreement with the result of Monte Carlo simulation then FORM.

Example 5.19 Again consider the performance function in Example 5.10; the parabolic approximation has been obtained as

$$\text{General approximation}: G(\mathbf{U}) = 2.914 - U_3 + \frac{1}{2}\left(-0.1148U_1^2 + 0.00017U_2^2\right)$$

$$\text{Simple approximation}: G(\mathbf{U}) = 2.914 - U_3 - 0.0287\left(U_1^2 + U_2^2\right)$$

For the general second order approximation, the principle curvatures are obtained as $k_1 = -0.1148$, $k_2 = 0.00017$, the first order reliability index is $\beta_F = 2.589$. The SOSM reliability index is obtained as

$$\beta_{SOSM} = \frac{\beta_F + \frac{1}{2}\sum_{i=1}^{n-1} k_i}{\sqrt{1 + \frac{1}{2}\sum_{i=1}^{n-1} k_i^2}} = \frac{2.914 + \frac{1}{2}(-0.1148 + 0.00017)}{\sqrt{1 + \frac{1}{2}(0.1148^2 + 0.00017^2)}} = 2.847$$

For the simple second order approximation, the average curvature radius is $R = -17.44$, the SOSM reliability index is calculated as

$$\beta_{SOSM} = \frac{\beta_F + \frac{n-1}{2R}}{\sqrt{1 + \frac{n-1}{2R^2}}} = \frac{2.914 + \frac{2}{-17.44}}{\sqrt{1 + \frac{2}{608.3}}} = 2.852$$

Both simple and general parabolic approximation give similar results. Using the Monte Carlo simulation with 1 000 000 trials, the probability of failure is obtained as $P_F = 0.00227$ with coefficient of variation of 2.1% and reliability index $\beta = 2.838$.

5.3.4.2 Second-Order Third-Moment Method

In order to derive the second-order third-moment (SOTM) method (Zhao et al. 2002a), the moment properties are first investigated for the simple parabolic performance function. Beside the first two moments expressed in Eqs. (5.57a) and (5.57b), the third, fourth, and fifth central moments for G_S in Eq. (5.39) are obtained as:

$$\alpha_{3S}\sigma_S^3 = \frac{n-1}{R^3} \tag{5.61a}$$

$$\alpha_{4S}\sigma_S^4 = 3 + \frac{3(n-1)}{R^2}\left(1 + \frac{n+3}{4R^2}\right) \tag{5.61b}$$

$$\alpha_{5S}\sigma_S^5 = \frac{10(n-1)}{R^3}\left(1 + \frac{5n+7}{10R^2}\right) \tag{5.61c}$$

where α_{jS} is the jth dimensionless central moment of the second-order performance function $G_S(\mathbf{U})$.

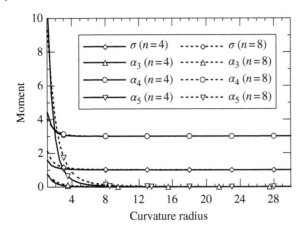

Figure 5.20 Moments of the simple parabolic approximation.

Using Eqs. (5.57b), (5.61a), (5.61b), and (5.61c), the variations of σ_S, α_{3S}, α_{4S}, and α_{5S} are depicted in Figure 5.20 with respect to the average curvature radius R. From Figure 5.20, it can be seen that σ_S, α_{3S}, α_{4S}, and α_{5S} rapidly approach the values 1, 0, 3, 0, respectively, as R increases. Because $\sigma_S = 1$, $\alpha_{3S} = 0$, $\alpha_{4S} = 3$, and $\alpha_{5S} = 0$ are the moment properties of a normal random variable, this suggests that the simple parabolic approximation rapidly converges to a normal random variable as the average curvature radius R increases. In order to confirm this assumption, the rth cumulant K_r of Eq. (5.39) is obtained as

$$K_r = \frac{(r-1)!(n-1)}{2R^r} \quad \text{for } r > 2 \tag{5.62}$$

Equation (5.62) shows that K_r rapidly approaches 0 with an increase in R. In other words, G_S satisfies the prerequisite of the Cornish-Fisher expansion (Fisher and Cornish 1960, see section 7.4.3.4).

For a standardised random variable

$$X_S = \frac{G_S - \mu_S}{\sigma_S}$$

It can be approximately transformed using the following first polynomial of the inverse Cornish-Fisher expansion according to section 7.4.3.4:

$$U = X_S - \frac{1}{6}\alpha_{3S}(X_S^2 - 1)$$

Since

$$P_F = P(G_S \leq 0) = P\left(X_s \leq -\frac{\mu_S}{\sigma_S}\right) = P(X_s \leq -\beta_{SOSM})$$

Since the abnormality is small after the second-order approximation, the reliability index and failure probability corresponding to the simple parabolic approximation can be expressed as

$$\beta_{SOTM1} = \beta_{SOSM} + \frac{1}{6}\alpha_{3S}\left(\beta_{SOSM}^2 - 1\right) \tag{5.63}$$

where

$$\beta_{SOSM} = \frac{\mu_S}{\sigma_S}$$

μ_S and σ_S are the mean value and standard deviation of G_S, respectively.

For the simple parabolic approximation in Eq. (5.39), since the first three moments are used in Eq. (5.63), respectively, they are referred to as SOTM reliability index.

In particular, Eq. (5.63) degenerates to $\beta_{SOTM} = \beta_{SOSM}$ when α_{3S} is equal to 0. When the curvature radius is sufficiently large, σ_S approaches 1, and α_{3S} will approaches 0, and then the SOTM and SOSM reliability indices degenerate to $\beta_{SOTM} = \beta_{SOSM} = \beta_F$.

It will be shown in the next sections that although Eq. (5.63) is very simple and has the same form for positive and negative principal curvatures, it has compatible accuracy with other formulae.

For the general parabolic approximation given in Eq. (5.31), the rth cumulant K_r is given as

$$K_r = \frac{(r-1)!}{2}\sum_{j=1}^{n-1} k_j^r \tag{5.64}$$

Equation (5.64) shows that K_r rapidly approaches 0 with the decrease of k_j, $j = 1, ..., n - 1$. In other words, the general parabolic approximation also satisfies the prerequisite of the Cornish-Fisher expansion. Substituting the first three moments of the general parabolic approximation into Eq. (5.63), the SOTM reliability index can be easily obtained. The third central moment of Eq. (5.31) is given by:

$$\alpha_{3S}\sigma_S^3 = \sum_{i=1}^{n-1} k_i^3 \tag{5.65}$$

A more accurate but somewhat complicated SOTM reliability index could use the third moment reliability indices given by in Section 4.2 in the chapter on Direct Methods of Moment, e.g. Eq. (4.27) in the same chapter, and then the SOTM reliability index is:

$$\beta_{SOTM2} = \beta_{SOSM} + \frac{\alpha_{3S}}{6}\left(\beta_{SOSM}^2 - 1\right)\left(1 + \frac{\alpha_{3S}}{6}\beta_{SOSM}\right) \tag{5.66}$$

As investigated in Section 4.2 just mentioned, the application range of the reliability index is

$$-2.4/\beta_{SOSM} \leq \alpha_{3S} \leq 0.8/\beta_{SOSM}$$

for allowable error of $r = 2\%$. β_{SOTM2} will approach to β_{SOTM1} when α_{3S} is very small.

For the general polynomial expression of Eq. (5.28), the first three central moments of the general polynomial expression above are:

$$\mu_S = a_0 + \sum_{i=1}^{n} \lambda_i \tag{5.67a}$$

$$\sigma_S^2 = \sum_{i=1}^{n} \left(\gamma_i^2 + 2\lambda_i^2 \right) \tag{5.67b}$$

$$\alpha_{3S}\sigma_S^3 = 2\sum_{i=1}^{n} \lambda_i \left(3\gamma_i^2 + 4\lambda_i^2 \right) \tag{5.67c}$$

As shown in the following sections, since the moment properties do not satisfy the prerequisite of the Cornish-Fisher expansion, Eq. (5.63) is not applicable to Eq. (5.30).

Example 5.20   An Investigation into Cai and Elishakoff's Formula
For the general parabolic approximation, Cai and Elishakoff (1994) proposed a series formula as shown in Table 5.2. A three-term approximation is suggested for purpose of practical application.

According to the definition by Cai and Elishakoff (1994), let

$$\lambda_j = \frac{1}{2} k_j, \, j = 1, 2, ..., n-1$$

Then Eq. (5.31) is rewritten as

$$G_S(\mathbf{Y}) - \beta_F = -y_n + \sum_{i=1}^{n-1} \lambda_i y_i^2$$

According to the definition of moment, one obtains that

$$E(G_S - \beta_F) = \sum_{j=1}^{n-1} \lambda_j$$

$$E\left[(G_S - \beta_F)^2 \right] = 3\sum_{j=1}^{n-1} \lambda_j^2 + \sum_{j\neq l}^{n-1} \lambda_j \lambda_l$$

$$E\left[(G_S - \beta_F)^3 \right] = 15\sum_{j=1}^{n-1} \lambda_j^3 + \sum_{j\neq l}^{n-1} \lambda_j^2 \lambda_l + \sum_{j\neq l\neq m}^{n-1} \lambda_j \lambda_l \lambda_m$$

Then, Cai and Elishakoff's formula in Table 5.2 can be rewritten as

$$P_F = \Phi(-\beta_F) - \phi(-\beta_F)\left\{ E(G_S - \beta_F) - \frac{1}{2}\beta_F E\left[(G_S - \beta_F)^2 \right] - \frac{1}{6}(\beta_F^2 + 1)E\left[(G_S - \beta_F)^3 \right] \right\} \tag{5.68}$$

where $E[G_S - \beta_F]$, $E[(G_S - \beta_F)^2]$, and $E[(G_S - \beta_F)^3]$ are the first three moments of G_S about β_F.

In other words, Cai and Elishakoff's formula can be also interpreted as a third-moment formula, although it uses the moments of G_S about β_F rather than the central moments (needless to say, moments about the mean of G_S).

Example 5.21   To investigate the variation and the comparison among the moment reliability index for the second-order approximation, consider the simple parabolic approximation [i.e. Eq. (5.39)] directly. It has been shown that for this performance function, the empirical second-order reliability index provides better approximations than other reliability indices of closed form (Zhao and Ono 1999b).

Variations of the computed second-order reliability indices are shown in Figure 5.21a– c with respect to the number of random variables n, to the first-order reliability index β_F, and to the curvature radius R, respectively, for the exact results, the results using the empirical formula (shown in Table 5.2), and for the results obtained using the SOSM and SOTM reliability indices given in Eqs. (5.58a), (5.61a), and (5.63). In Figure 5.21a–c, the exact results are obtained using direct integration, where the curves above the line of the first-order reliability index show the results of the limit state surface with a positive curvature radius (convex to the origin) and those below the first-order reliability index show the results of the limit state surface with a negative curvature radius (concave to the origin). Figure 5.21a–c reveals the following:

1) In general, both the empirical second-order reliability index and the SOTM reliability index improve the first-order reliability index and give good approximations of the exact results for both cases of positive and negative curvature radii in the investigated ranges of n, β_F, and R. This implies that the SOTM reliability index can be used as alternative second-order reliability index of the empirical reliability index even though it is quite simple and has the same form for positive and negative curvature radii.
2) For cases of relatively small n or large absolute R, even the simple SOSM reliability index gives very good approximations of the exact results, especially for limit state surfaces which are concave to the origin.

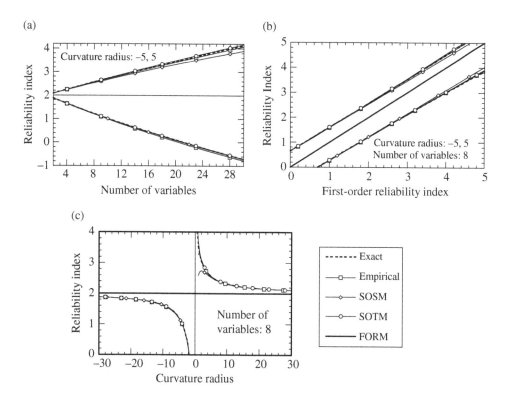

Figure 5.21 Variation of the reliability index for Example 5.21. (a) The number of random variables n. (b) The first-order reliability index β_F. (c) The curvature radius R.

Example 5.22 Considering the performance function in Examples 5.8 and 5.17 again, we have obtained that

$$\beta_F = 2, R = 5, n = 2, \beta_{SOSM} = 2.079$$

Using Eq. (5.61a), the third central moment is obtained as

$$\alpha_{3S} = \frac{n-1}{R^3 \sigma_S^2} = \frac{1}{5^3 \times 1.01^2} = 0.0078$$

Using Eqs. (5.63) and (5.66), the SOTM reliability indices are obtained as

$$\beta_{SOTM1} = \beta_{SOSM} + \frac{1}{6}\alpha_{3S}(\beta_{SOSM}^2 - 1) = 2.079 + \frac{0.0078}{6}(2.079^2 - 1) = 2.083$$

$$\beta_{SOTM2} = \beta_{SOSM} + \frac{\alpha_{3S}}{6}(\beta_{SOSM}^2 - 1)\left(1 + \frac{\alpha_{3S}}{6}\beta_{SOSM}\right) = 2.083$$

One can see that both SOTM reliability indices give same results and both of them corresponding to a probability of failure of $P_F = 0.0186$ and have a good agreement with the result of Monte Carlo simulation in Example 5.8 ($P_F = 0.01838, \beta = 2.0884$).

Example 5.23 Consider the performance function in Examples 5.9 and 5.18; previously we have obtained that

$$\beta_F = 1.9803, R = 4.645, n = 2, \beta_{SOSM} = 2.067$$

Using Eq. (5.61a), the third central moment is

$$\alpha_{3S} = \frac{n-1}{R^3 \sigma_S^2} = \frac{1}{4.645^3 \times 1.0115^2} = 0.00964$$

Using Eqs. (5.63) and (5.66), the SOTM reliability indices are obtained as

$$\beta_{SOTM1} = \beta_{SOSM} + \frac{1}{6}\alpha_{3S}(\beta_{SOSM}^2 - 1) = 1.9803 + \frac{0.0096}{6}(1.9803^2 - 1) = 2.0723$$

$$\beta_{SOTM2} = \beta_{SOSM} + \frac{\alpha_{3S}}{6}(\beta_{SOSM}^2 - 1)\left(1 + \frac{\alpha_{3S}}{6}\beta_{SOSM}\right) = 2.0723$$

It can be observed that the two SOTM reliability indices give the same results and both correspond to a probability of failure of $P_F = 0.01925$. Both the reliability indices and the probability of failure agree well with those of Monte Carlo simulation in Example 5.9 ($P_F = 0.01918, \beta = 2.0711$).

As shown in Example 5.9, the second order approximation of the performance function is obtained as

$$G(\mathbf{U}) = 1.9803 + (-0.6317, \ -0.7752)\begin{Bmatrix} U_1 \\ U_2 \end{Bmatrix}$$
$$+ \frac{1}{2}(U_1 - 1.251, \ U_2 - 1.535)\begin{bmatrix} 0.1564 & -0.1238 \\ -0.1238 & 0 \end{bmatrix}\begin{Bmatrix} U_1 - 1.251 \\ U_2 - 1.535 \end{Bmatrix}$$

The eigenvalues of the Hessian matrix are obtained as (0.2246, −0.0683) with corresponding standardised eigenvector matrix of

$$H = \begin{bmatrix} 0.8757 & -0.4828 \\ 0.4828 & 0.8757 \end{bmatrix}$$

Substituting $Y = HU$ into the second-order approximation above leads to the general second order approximation

$$G(Y) = 2.0756 - 0.460Y_1 - 0.879Y_2 + 0.1123Y_1^2 - 0.0341Y_2^2$$

where Y is a vector of independent standard normal variables. Applying Eq. (5.61), the first three central moments are obtained as

$$\mu_S = 2.154, \sigma_S = 1.0059, \alpha_{3S} = -0.00458$$

$$\beta_{SOSM} = \frac{\mu_S}{\sigma_S} = \frac{2.154}{1.0059} = 2.141$$

Using Eqs. (5.63) and (5.66), the SOTM reliability indices are obtained as $\beta_{SOTM1} = \beta_{SOTM2} = 2.1383$ with corresponding probability of failure of $P_F = 0.01624$. The results are not as good as those obtained from the parabolic approximation above. However, it should be noted that the SOTM should be generally used for parabolic approximation rather than the general second-order approximation. Since the asymptotic characteristics of the parabolic approximated performance function have been clearly derived as shown in Figure 5.20, similar property has not been confirmed for the general second-order approximation.

Example 5.24 Consider the performance function in Example 5.10 again; the principle curvatures are obtained as $k_1 = -0.1148$ and $k_2 = 0.00017$ for the general second-order approximation, and the first-order reliability index is $\beta_F = 2.914$. The SOSM reliability index is obtained as $\beta_{SOSM} = 2.847$. The third dimensionless moment is obtained as $\alpha_{3G} = -0.0015$, using Eqs. (5.63) and (5.66), the SOTM reliability indices are obtained as $\beta_{SOTM1} = \beta_{SOTM2} = 2.845$ with corresponding probability of failure is $P_F = 0.0222$.

For the simple second-order approximation, the average curvature radius is $R = -17.44$, and the SOSM reliability index is obtained as $\beta_{SOSM} = 2.852$. The third dimensionless moment is obtained as $\alpha_{3G} = -0.00038$, using Eqs. (5.63) and (5.66), and the SOTM reliability indices are obtained as $\beta_{SOTM1} = \beta_{SOTM2} = 2.851$ with corresponding probability of failure equal to $P_F = 0.0218$.

Since the third dimensionless moment is too small, the SOSM gives sufficiently accurate result in this example.

5.3.4.3 Second-Order Fourth-Moment Method
Similar to the SOTM method, the second-order fourth-moment method (SOFM) can also be developed in terms of the first four dimensionless central moments of the second-order approximation of the performance function (Lu et al. 2017b). Using Eq. (4.55a) in the chapter on Direct Methods of Moment, a second-order fourth-moment reliability index is given as

$$\beta_{SOFM} = \frac{\sqrt[3]{2}p}{\sqrt[3]{-q_0 + \Delta_0}} - \frac{\sqrt[3]{-q_0 + \Delta_0}}{\sqrt[3]{2}} + \frac{l_1}{3k_2} \tag{5.69}$$

where

$$\Delta_0 = \sqrt{q_0^2 + 4p^3}, p = \frac{3k_1k_2 - l_1^2}{9k_2^2}, q_0 = \frac{2l_1^3 - 9k_1k_2l_1 + 27k_2^2(-l_1 + \beta_{SOSM})}{27k_2^3},$$

$$l_1 = \frac{\alpha_{3S}}{6(1 + 6l_2)}, k_1 = \frac{1 - 3l_2}{(1 + l_1^2 - l_2^2)}, k_2 = \frac{l_2}{(1 + l_1^2 + 12l_2^2)}, l_2 = \frac{1}{36}\left(\sqrt{6\alpha_{4S} - 8\alpha_{3S}^2 - 14} - 2\right)$$

Here the fourth dimensionless central moment is given by Eq. (5.61b) for the simple parabolic approximation, and that for the general parabolic approximation of Eq. (5.31) is given as

$$\alpha_{4S}\sigma_S^4 = 3 + 3\sum_{i=1}^{n-1}\left(\frac{5}{4}k_i^4 + k_i^2\right) + \frac{3}{2}\sum_{i=1}^{n-2}\sum_{j>i}^{n-1}k_i^2k_j^2 \qquad (5.70)$$

For the general polynomial expression of Eq. (5.28), the fourth order moment is given as

$$\alpha_{4S}\sigma_S^4 = 3\sum_{i=1}^{n}(\gamma_i^4 + 20\gamma_i^2\lambda_i^2 + 20\lambda_i^4) + 6\sum_{i=1}^{n-1}\sum_{j>i}^{n}\left(\gamma_i^2\gamma_j^2 + 4\lambda_i^2\lambda_j^2 + 2\gamma_i^2\lambda_j^2 + 2\gamma_j^2\lambda_i^2\right)$$

$$(5.71)$$

Example 5.25 Investigation of General Paraboloid with Unevenly Distributed Curvatures

Considers the following performance function in standard normal space; (for this example, investigated by Zhao and Ono (1999b), none of the currently used SORM formulae, including the empirical second-order reliability index, gives satisfactory results),

$$G(\mathbf{U}) = \beta_F - U_8 + \frac{1}{2}\sum_{j=1}^{7} a^j U_j^2$$

where a is a factor ranging from -0.5 to 0.5. Because a^j changes according to j, the paraboloid defined by the equation above has unevenly distributed curvatures.

Figure 5.22a shows the variation of the sums of the curvature with respect to a, where the sum of the curvature increases rapidly with the increase of $|a|$. Equations (4.31a) in the chapter on Direct Methods of Moment and (5.65) define the shadow region that second-moment reliability index (here, second-moment reliability index is the SOSM reliability index since the performance function is second-order approximated) is accurate under the condition of tolerance error of 2%. That is to say, the SOSM reliability index will produce a significant error when $|a|$ takes large value.

The variations in the second-order reliability indices are shown in Figure 5.22b with respect to a. This figure shows the results of the empirical formula (shown in Table 5.2) and the SOSM and SOTM reliability indices. The first-order reliability index was assumed to be 2.0, and exact results were obtained by the IFFT method. Figure 5.22b shows that for $a > 0$, which implies that the curvatures have the same signs, both the empirical and the SOTM reliability indices provide good approximations of the exact reliability index, while

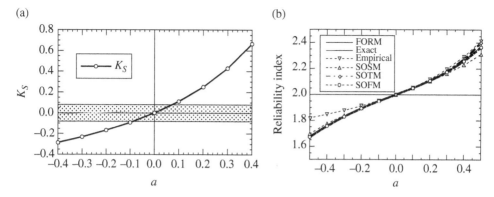

Figure 5.22 Variation of the reliability index with respect to parameter a for Example 5.24. (a) Variation of the sums of the curvature. (b) Variation of the reliability index.

the SOSM reliability index produces significant errors as described above. For $a < 0$, which implies that the curvatures have different signs and that the total principal curvature is negative, both the SOSM and the SOTM reliability indices provide good approximations of the exact reliability index, while the empirical reliability index produces significant errors. This implies that the SOTM reliability index can be applied to problems with unevenly distributed principal curvatures. Using Eq. (5.69), the SOFM reliability index is calculated and presented in Figure 5.22b, which is observed to be that the SOFM reliability index is in agreement with the exact result in the entire range of investigation.

Example 5.26 Investigation of the Variation of SOFM Reliability Index with Respect to the Number of Random Variables
Considers the following performance function in standardised space, which has been used as the fourth example by Der Kiureghian et al. (1987) and as the second example by Zhao and Ono (1999b).

$$G(\mathbf{U}) = \beta_F - U_n + \frac{1}{2}\sum_{j=1}^{n-1}(a \cdot j)U_j^2$$

where the first-order reliability index β_F was assumed to be 3.0 and a is taken to be 0.01.

The variations of the SOSM, SOTM, and SOFM reliability indices are shown in Figure 5.23 with respect to the number of random variables, together with exact results obtained by the IFFT method (Zhao and Ono 1999c). From Figure 5.23, the SOFM method is observed to be in close agreement with the correct results for the entire investigation range. The SOTM method is observed to overestimate the reliability index when the number of random variables is relatively large, while the SOSM method leads to significant underestimation in this case.

5.3.5 Applicable Range of FORM

The FORM is considered to be one of the most reliable computational methods (Ang and Tang 1984; Bjerager 1991) and has been developed as a basic reliability method in engineering. Its accuracy has been examined by many studies through a large number of examples

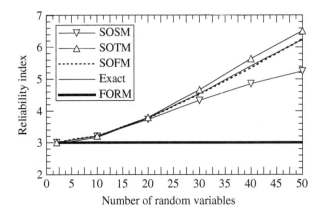

Figure 5.23 Variation of reliability index with number of random variables for Example 5.26.

(Fiessler et al. 1979; Tichy 1994; Der Kiureghian and De Stefano 1991). However, its parameter ranges for sufficient accuracy are rarely reported. Without this information, it is difficult to determine whether or not the results of FORM are sufficiently accurate, or whether SORM or another more accurate method should be used. An empirical investigation was conducted, but under only one level of accuracy (Zhao and Ono 1999c). Using the SOSM and SOTM reliability indices introduced in this section, the applicable range of FORM can be explored analytically.

Since FORM is accurate only in cases where the curvature radius is very large, in the range of R where the accuracy of FORM was investigated, the SOSM reliability index is sufficiently accurate. As shown in Figure 5.20, the standard deviation σ_S and the third dimensionless central moment α_{3S} approach 1 and 0, respectively. As the curvature radius R increases, the second-order reliability index β_S used in the investigation of the applicable range for FORM can be expressed as

$$\beta_S = \beta_F + \frac{1}{2}K_S = \beta_F + \frac{n-1}{2R} \tag{5.72}$$

where K_s and R are the total principal curvature and the average curvature radius, respectively, of the limit state surface at the design point.

The following criterion is used for the first-order reliability index to determine whether FORM is sufficiently accurate,

$$\frac{|\beta_S - \beta_F|}{\beta_S} \leq \gamma \tag{5.73}$$

where γ is the tolerance error of FORM.

Since in the range of R where the accuracy of FORM was investigated, β_F is approximately equal to β_S, the range of the average curvature radius or the sum of the principal curvature that satisfies Eq. (5.73) can be given as (Zhao et al. 2002a)

$$|R| \geq \frac{n-1}{2\gamma\beta_F} \tag{5.74a}$$

or

$$|K_S| \le 2\gamma\beta_F \qquad (5.74b)$$

Using Eqs. (5.74a) and (5.74b), the applicable range for which FORM is sufficiently accurate can be judged quite conveniently. For example, if γ is taken to be 2%, then

$$|K_S| \le 0.04\beta_F$$

is necessary, and furthermore, for $n = 5$,

$$|R| \ge 100/\beta_F$$

is also necessary. Consequently, FORM is more accurate for performance functions that have a small number of random variables and large curvature radius as shown in Eq. (5.74a).

Similarly, for the general parabolic approximation, the second-order reliability index β_S used in the investigation of the applicable range for FORM can be expressed as

$$\beta_S = \beta_F + \frac{1}{2}\sum_{i=1}^{n-1} k_i \qquad (5.75)$$

and the applicable range for FORM can be easily expressed as:

$$|K_S| = \left|\sum_{i=1}^{n-1} k_i\right| \le 2\gamma\beta_F \qquad (5.76)$$

Example 5.27 Consider performance function in Example 5.9.

The sum of the principal curvatures of the limit state surface can be readily obtained as $K_S = an(n-1)/2$. Based on Eq. (5.76), the range of K_S for which FORM is accurate can be readily obtained as $K_S \le 0.08$ for $\beta_F = 2$ and $\gamma = 2\%$. Using the average curvature radius $R = 2/(na)$ as a parameter, the variations of K_S with respect to R are shown in Figure 5.24a for $n = 2, 4, 8$, where the shadow region indicates K_S for which the error of β_F is less than $\gamma = 2\%$. Figure 5.24b presents the variations of the SOSM and SOTM reliability indices compared with the exact results by the IFFT method (Zhao and Ono 1999c). From Figure 5.24b, both are observed to approach β_F value as R increases and the SOTM reliability index provides good approximation of the exact reliability index. It can be also seen that the SOSM reliability index gives good approximation for relatively large values of R.

From Figure 5.24a, it can be observed that K_S decreases as R increases. When R is greater than 12.5, 37.5, and 87.5 for $n = 2, 4$, and 8, respectively, K_S falls into the shadow region for which the error of β_F is less than $\gamma = 2\%$. For the critical values of $R = 12.5, 37.5$, and 87.5 for $n = 2, 4$, and 8, respectively, the corresponding values $\beta_{SOSM} = 2.0368, 2.0387, 2.0394$ and $\beta_{SOTM} = 2.0372, 2.0388, 2.0394$, respectively, are obtained. One can observe that the β_{SOSM} values are almost equal to β_{SOTM}, and the differences between β_F and β_{SOTM} are less than $\gamma = 2\%$. This implies that the assumption used in deriving Eq. (5.76) was appropriate and that Eq. (5.76) can be used to estimate the applicable range of FORM.

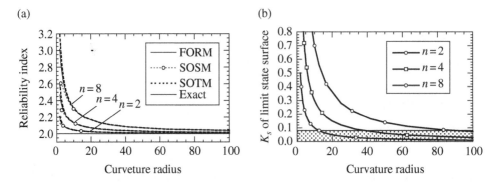

Figure 5.24 Variation of the reliability index and principal curvature for Example 5.25. (a) Variation of the reliability index. (b) Variation of the principal curvature.

Example 5.28 Investigation of the Accuracy of FORM for a Simple Performance Function

Consider the following performance function used in many situations,

$$G(\mathbf{X}) = R - S$$

where R is a resistance of normal random variable with a coefficient of variance of 0.2, and S is a load of a lognormal random variable with a mean value of 100 and coefficient of variance of 0.4. In the following investigations, the exact results of reliability index and the corresponding probability of failure are obtained using MCS with 5 000 000 samplings.

Using the point-fitting method (Zhao and Ono 1999b), the total principal curvature of the limit state surface is derived and shown in Figure 5.25a where the shadow region indicates K_S for which the error of β_F is less than $\gamma = 2\%$. From Figure 5.25a, it can be observed that the total principal curvature falls outside the applicable range of FORM for $\gamma = 2\%$. The variation in the reliability index is shown in Figure 5.25b with respect to the central factor of safety, from which visual errors can be observed for the FORM, although the performance function is very simple. Both the SOSM and SOTM reliability indices have very good agreement with those obtained by MCS.

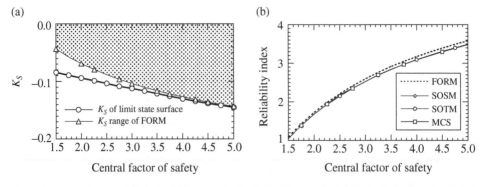

Figure 5.25 Variation of the reliability index and principal curvature for Example 5.28. (a) The total principal curvature. (b) The reliability index.

5.4 Summary

Methods of moment based on the first-order and second-order approximation of the performance function are introduced in this Chapter. It is shown that for linear performance functions, the failure probability can be obtained directly from the first-order reliability index. If the performance function is non-linear, the first-order reliability index can still be defined, but only with respect to a tangent hyperplane to approximate the non-linear limit state function. Under the first-order approximation of performance function in standardised normal space, the first-order second-moment reliability index is equal to the shortest distance from the hyperplane and therefore perpendicular to it in the standardised normal space. The corresponding point on the hyper plane was termed the checking or design point.

After obtained the design point, the performance function in the standardised normal space can also be approximated by second-order polynomials. In this circumstance, the SORM is developed. Many second-order reliability indices have been proposed. It is shown in this Chapter that methods of moment are suitable for obtaining the failure probability corresponding to the second-order approximation of the performance function. Therefore, the SOSM method, second-order third-moment (SOTM) method, and the second-order fourth-moment (SOFM) method, are developed. It should be noted that first-order higher moment methods are unnecessary since the first-order approximation of the performance function is normal. The applicable range of FORM is also discussed.

6

Structural Reliability Assessment Based on the Information of Moments of Random Variables

6.1 Introduction

As shown in the previous chapters, the calculation of the probability of safety, or probability of failure, requires the knowledge of the probability distributions, i.e., the joint probability density function (PDF) or the joint cumulative distribution function (CDF) of basic variables **X**. The integration of the joint PDF in the failure region directly gives probability of failure. In FORM/SORM, the normal transformation (x–u transformation) and its inverse transformation (u–x transformation) are generally achieved by using the Rosenblatt transformation (Hohenbichler and Rackwitz 1981) or Nataf transformation (Liu and Der Kiureghian 1986) based on the CDF/PDF of basic random variables. In the Monte Carlo simulation (MCS), CDF/PDFs are used to generate the random samples of the basic random variables. In reality, however, due to the lack of statistical data, the CDF/PDFs of some basic random variables are often unknown, and the probabilistic characteristics of these variables are often expressed using their statistical moments. Under such conditions, neither the Rossenblatt nor Nataf transformations are applicable, nor does a strict evaluation of the probability of failure need to be applied. An alternative measure of reliability is therefore necessary.

Most earlier studies concerning this subject dealt with second-moment methods, in which only the mean values and standard deviations of the basic random variables are known (Ang and Cornell 1974; Hasofer and Lind 1974; Ditlevsen 1979a; Madsen et al. 1986). In such a case, the variables are commonly transformed into a set of standard variables with zero means and unit standard deviation. A comprehensive framework was proposed by Der Kiureghian and Liu (1986) for structural reliability analysis, in which incomplete probability information on random variables, including moments, bounds, marginal distributions, and partial joint distributions are incorporated in reliability analysis under stipulated requirements of consistency, invariance, operability, and simplicity. The method was found to be consistent with full distribution structural reliability theory and has been used for structural safety measure under imperfect states of knowledge (Der Kiureghian 1989).

When the PDF/CDF of a random variable is unknown, a method has been proposed for estimating complex distributions using B-spline functions (Zong and Lam 1998), in which the estimation of PDF is summarised as a nonlinear programming problem. One may

Structural Reliability: Approaches from Perspectives of Statistical Moments, First Edition.
Yan-Gang Zhao and Zhao-Hui Lu.
© 2021 John Wiley & Sons Ltd. Published 2021 by John Wiley & Sons Ltd.

alternatively use the approach based on Bayesian theory (Der Kiureghian and Liu 1986), of which the distribution is assumed to be a weighted average of all candidate distributions. In this type of modelling, the weights represent the subjective probabilities of each candidate distribution being the true distribution. For a variable x_1 with k candidate distributions $F_{1i}(x_1)$, $i = 1, ..., k$, the distribution is written in the form of

$$F_1(x_1) = \sum_{i=1}^{k} p_{1i} F_{1i}(x_1) \tag{6.1}$$

where p_{1i} are the weights and satisfy

$$\sum_{i=1}^{k} p_{1i} = 1 \tag{6.2}$$

Furthermore, all candidate distributions are assumed to have the same mean and variance because these are assumed to be known quantities.

After obtaining the distribution in Eq. (6.1), the reliability analysis can be conducted in a manner similar to the full distribution structural reliability theories such as FORM/SORM described in the chapter entitled Methods of Moment Based on First- and Second-Order Transformation.

Another way to conduct structural reliability analysis involving random variables with unknown CDF/PDFs relies on the u–x and x–u transformations directly using the first few moments of the random variable, which can be easily obtained from the statistical data. This method can be divided into two routes, one is to use the distribution families, and the other is to use the polynomial normal transformation. Either the Burr, John, or Pearson systems (Stuart and Ord 1987; Hong 1996), or the Lambda distribution (Ramberg and Schmeiser 1974; Grigoriu 1983) can be used as the distribution families. Since the quality of approximating the tail area of a distribution is relatively insensitive to the selected distributions family (Pearson et al. 1979) and the parameters require solution of nonlinear equation for Burr and John systems and Lambda distribution, Pearson system is generally used.

In this chapter, the reliability evaluation based on the information of moments of random variables is discussed, in which the first few moments of the random variable will be used instead of their full probability distributions. In particular, the first part of this chapter will consider the special case of reliability estimation in which each variable is represented only by its first two moments, i.e. by its mean and standard deviation. This is known as the *second-moment* level of representation. Higher order moments, such as skewness and kurtosis, etc., of the distribution, are then considered.

When the CDF/PDFs of the basic random variables are known, we have discussed the direct methods of moment as described in the chapter on Direct Methods of Moment and the methods of moment based on first- and second-order transformation in the chapter on Methods of Moment Based on First- and Second-Order Transformation. In this chapter, we will investigate the two methods under the circumstance that only the first few moments of the basic random variables are known.

6.2 Direct Methods of Moment without Using Probability Distribution

6.2.1 Second-Moment Formulation

Not infrequently, the available information or data may be only sufficient to evaluate the first- and second-moments; namely the mean values and variances of the respective random variables. Practical measures of safety or reliability, therefore, must often be limited to functions of these first two moments. Under this circumstance, the implementation of reliability concepts must necessarily be limited to a formulation based on the first- and second-moments of the random variables, i.e. restricted to the second-moment formulation (Cornell 1969; Ang and Cornell 1974). It should be emphasised that the second-moment approach can be interpreted as if each random variable is represented by the normal distribution as shown in the chapter on Direct Methods of Moment.

Because of their inherent simplicity, the second-moment approach has become very popular, particularly in design code calibration work (Melchers 1987). Early works by Freudenthal (1956) and Rzhanitzyn (1957) contained second-moment concepts. Not until the late 1960s, however, was the time ripe for the ideas to gain a measure of acceptance (Cornell 1969; Melchers 1987).

With the second-moment approach, the reliability may be measured entirely with a function of the first- and second-moments of the design variables, namely, the reliability index, β, when there is no information on the probability distributions.

Recall the safety margin $Z = G(\mathbf{X}) = R - S$, where R and S are independent random variables. In this term, the safe state of a system may be defined as $Z > 0$, whereas the failure state is $Z < 0$. The boundary separating the safe and failure states is the limit-state defined by the equation $Z = 0$.

As shown in the chapter on Moment Evaluation, no matter what kinds of the distribution, the first two moments of Z are expressed as

$$\mu_Z = \mu_R - \mu_S, \sigma_Z^2 = \sigma_R^2 + \sigma_S^2$$

and the second moment reliability index is given as

$$\beta_{2M} = \frac{\mu_Z}{\sigma_Z} = \frac{\mu_R - \mu_S}{\sqrt{\sigma_R^2 + \sigma_S^2}}$$

As has been shown in the chapter on Moment Evaluation, for some simple functions, the first two moments of the performance function can be accurately obtained from the first two moments of the basic variables. In general, the first two moments of a non-linear performance function are difficult to be obtained accurately. In such cases, point estimate described in Section 3.3 of the previously mentioned chapter is sometimes a suitable selection.

If the first two moments of a basic variable are the only known information, the two-point estimate can be used. Let the concentrations be P_-, x_- and P_+, x_+ from Eqs.(3.22a)–(3.22b) in the same chapter, the estimating points x_-, x_+ and the corresponding weights P_-, P_+ are given as

$$x_- = \mu_x - \sigma_x, \quad P_- = \frac{1}{2} \tag{6.3a}$$

$$x_+ = \mu_x + \sigma_x, \quad P_+ = \frac{1}{2} \tag{6.3b}$$

Using Eqs. (3.20a) and (3.20b) from the same chapter (Rosenblueth 1975), the first two central moment of a function $Y = y(x)$ can be calculated.

As a function of multiple basic variables, the performance function $G(\mathbf{X})$, may be approximated by Eq. (3.67a) or Eq. (3.76), and its first two central moments can be computed by Eqs. (3.68a–b) or Eqs. (3.85a–b), all in the same chapter.

6.2.2 Third-Moment Formulation

As has been shown in the chapter on Direct Methods of Moment, if the first three moments of the performance function are known, the third-moment reliability index is given as

$$\beta_{3M} = \beta_{2M} + \frac{\alpha_{3G}}{6}\left(\beta_{2M}^2 - 1\right)\left(1 + \frac{\alpha_{3G}}{6}\beta_{2M}\right) \tag{6.4}$$

For some simple functions, the first three moments of the performance function can be accurately obtained from the first three moments of the basic variables. In general, the first three moments of a non-linear performance function are difficult to be obtained accurately. In such cases, one may first approximate the performance function $G(\mathbf{X})$ using Eq. (3.67a), and then evaluate the first three moments of G_i in Eqs. (3.68a–c) using Eqs. (3.20a–c) in the chapter we have been discussing, where only the first three moments of each random variable are used.

Meanwhile, the random variable with the first three moments known can also be approximated by the second-order pseudo normal transformation [i.e. Eq. (7.44) in the chapter on Transformation of Non-Normal Variables to Independent Normal Variables], and point estimate described in Section 3.3 in the chapter on Moment Evaluation is often required. As a function of multiple basic variables, the performance function $G(\mathbf{X})$ can be transformed into standard normal space using Eq. (7.44) in the chapter on Transformation of Non-Normal Variables to Independent Normal Variables, which may be approximated by Eq. (3.67a) or Eq. (3.76) and the first three moments of the performance function including random variables with unknown probability distributions can be computed by Eqs. (3.68a–c) or Eqs. (3.85a–c), all in the chapter on Moment Evaluation.

6.2.3 Fourth-Moment Formulation

As has been shown in the chapter on Direct Methods of Moment, if the first four moments of the performance function are known, the fourth-moment reliability index is given as

$$\beta_{4M} = \frac{\sqrt[3]{2p}}{\sqrt[3]{-q_0 + \Delta_0}} - \frac{\sqrt[3]{-q_0 + \Delta_0}}{\sqrt[3]{2}} + \frac{l_1}{3k_2} \tag{6.5}$$

where

$$\Delta_0 = \sqrt{q_0^2 + 4p^3}, \quad p = \frac{3k_1k_2 - l_1^2}{9k_2^2}, \quad q_0 = \frac{2l_1^3 - 9k_1k_2l_1 + 27k_2^2(-l_1 + \beta_{2M})}{27k_2^3},$$

$$l_1 = \frac{\alpha_{3G}}{6(1 + 6l_2)}, \quad k_1 = \frac{1 - 3l_2}{(1 + l_1^2 - l_2^2)}, \quad k_2 = \frac{l_2}{(1 + l_1^2 + 12l_2^2)}, \quad l_2 = \frac{1}{36}\left(\sqrt{6\alpha_{4G} - 8\alpha_{3G}^2 - 14} - 2\right)$$

For some simple functions, the first four moments of the performance function can be accurately obtained from the first four moments of the basic variables. In general, the first four moments of a non-linear performance function are difficult to be obtained accurately. In such cases, one may first approximate the performance function $G(X)$ using Eq. (3.67a), and then evaluate the first four moments of G_i in Eqs. (3.68a–d) using Eqs. (3.23a–c) as previously mentioned, where only the first four moments of each random variable are used.

Meanwhile, the random variable with the first four moments known can be approximated by the third-order polynomial of standard normal random variable [i.e. Eq. (7.70) as previously mentioned], and point estimate described in Section 3.3 in the chapter on Moment Evaluation is often required. As a function of multiple basic variables, the performance function $G(X)$ can be transformed into standard normal space using Eq. (7.70), which may be approximated by Eq. (3.67a) or Eq. (3.76) (as previously mentioned) and the first four moments of the performance function including random variables with unknown probability distributions can be computed by Eqs. (3.68a–d) or Eqs. (3.85a–d) as previously mentioned.

6.3 First-Order Second-Moment Method

In this section, we still continue to consider that the only information we know about the random variables is their first two moments, namely the mean and the standard deviation.

As has been shown in the chapter on Moment Evaluation, the first two moments can be accurately obtained only for a few simple functions. In general, the first two moments of a non-linear performance function are difficult to be obtained accurately.

Consider the safety margin again for $Z = G(X) = R - S$, where R and S are independent random variables. Introduce the reduced variables

$$R' = \frac{R - \mu_R}{\sigma_R}, S' = \frac{S - \mu_S}{\sigma_S}$$

In the space of these reduced variables, the safe state and failure state may be represented as shown in Figure 6.1. Also, in terms of the reduced variables, the limit state equation, $Z = 0$, becomes

$$\sigma_R R' - \sigma_S S' + \mu_R - \mu_S = 0$$

which is a straight line as shown in Figure 6.1. The distance from the limit state line to the origin, 0, is in itself a measure of reliability, this distance, d, is given as

$$d = \frac{\mu_R - \mu_S}{\sqrt{\sigma_R^2 + \sigma_S^2}}$$

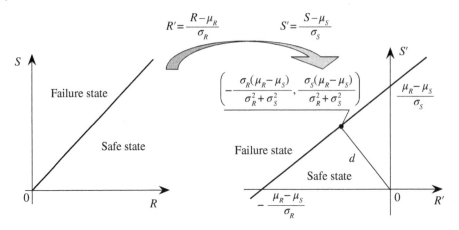

Figure 6.1 Limit state line in the space of reduced variables R' and S'.

Indeed, according to Eq. (5.1c) in the chapter on Methods of Moment Based on First- and Second-Order Transformation , it can be observed that for normal R and S, this distance d is also the reliability index, that is $d = \beta$ and thus the failure probability is

$$P_F = \Phi(-d)$$

For a general performance function $G(\mathbf{X})$, if $X = (X_1, X_2, \cdots, X_n)$ are uncorrelated random variables, introduce the set of uncorrelated reduced variables (Freudenthal 1956)

$$X'_i = \frac{X_i - \mu_{X_i}}{\sigma_{X_i}}, \qquad i = 1, 2, \cdots, n$$

Obviously, the limit state surface may also be portrayed in the space of the above reduced variables, separated by the appropriate performance function. In terms of the reduced variables, the limit state equation would be

$$G\left(\sigma_{X_1}X'_1 + \mu_{X_1}, \cdots, \sigma_{X_n}X'_n + \mu_{X_n}\right) = 0$$

Observe that as the limit state surface $G(\mathbf{X}') = 0$ moves further or closer to the origin, the safe region, $G(\mathbf{X}') > 0$, increases or decreases accordingly. Therefore, the position of the failure surface relative to the origin should determine the safety or reliability of the system. The position of the failure surface may be represented by the minimum distance from the surface $G(\mathbf{X}') = 0$ to the origin in X'-space (Hasofer and Lind 1974; Ditlevsen 1979a). Thus, similar with the case described in Section 5.2.2 in the chapter on Methods of Moment Based on First- and Second-Order Transformation, in some approximate sense, this minimum distance may be used as a measure of reliability. The reliability index defined in such sense is called Hasofer-Lind reliability index or first-order reliability index. The required minimum distance may be determined as follows (Shinozuka 1983). The distance from a point $\mathbf{X}' = (X'_1, X'_2, \cdots, X'_n)$ on the failure surface $G(\mathbf{X}') = 0$ to the origin is, of course.

$$d = \sqrt{X'^2_1 + X'^2_2 + \ldots + X'^2_n} = \left(\mathbf{X}'^T\mathbf{X}'\right)^{1/2}$$

The point on the failure surface, $\mathbf{x}'^* = \left(x_1'^*, x_2'^*, \cdots, x_n'^*\right)$ having the minimum distance to the origin may be determined by minimising the function d, subject to the constraint $G(\mathbf{X}) = 0$, that is,

$$\text{Minimize} : d = \left(\mathbf{X}'^T \mathbf{X}'\right)^{1/2} \tag{6.6a}$$

$$\text{Subject to} : G(\mathbf{X}') = 0 \tag{6.6b}$$

Similar to Section 5.2.2 in the chapter on Methods of Moment Based on First- and Second-Order Transformation, the minimum distance is obtained as

$$d = \frac{-\nabla^T G(\mathbf{x}'^*)\mathbf{x}'^*}{\left[\nabla^T G(\mathbf{x}'^*)\nabla G(\mathbf{x}'^*)\right]^{1/2}} \tag{6.7a}$$

where

$$\nabla G(\mathbf{X}') = \left(\frac{\partial G(\mathbf{X})}{\partial X_1'}, \frac{\partial G(\mathbf{X})}{\partial X_2'}, \cdots, \frac{\partial G(\mathbf{X})}{\partial X_n'}\right), \frac{\partial G(\mathbf{X})}{\partial X_i'} = \frac{\partial G(\mathbf{X})}{\partial X_i}\frac{\partial X_i}{\partial X_i'} = \sigma_{X_i}\frac{\partial G(\mathbf{X})}{\partial X_i} \tag{6.7b}$$

and the point $\mathbf{x}'^* = \left(x_1'^*, x_2'^*, \cdots, x_n'^*\right)$ represents a minimum or maximum depends on the nature of $G(\mathbf{X})$.

Let the performance function $G(\mathbf{X}')$ be linearized at the design point $\mathbf{x}'^*$ by means of a Taylor series expansion (i.e. in first-order terms only). This approximation provides a tangent (hyper) plane $G_L(\mathbf{X}')$ at $\mathbf{x}'^*$ to the function $G(\mathbf{X}')$:

$$Z = G_L(\mathbf{X}') = G(\mathbf{x}'^*) + \sum_{i=1}^{n}(X_i' - x_i'^*)\frac{\partial G}{\partial x_i'}\bigg|_{\mathbf{x}'^*} \tag{6.8a}$$

Since $\mathbf{x}'^*$ is on the limit state surface, $G(\mathbf{x}'^*) = 0$. Eq. (6.8a) becomes

$$Z = G_L(\mathbf{X}') = \sum_{i=1}^{n}(X_i' - x_i'^*)\frac{\partial G}{\partial x_i'}\bigg|_{\mathbf{x}'^*} = (\mathbf{X}' - \mathbf{x}'^*)^T \nabla G(\mathbf{x}'^*) \tag{6.8b}$$

Suppose that $Z = G_L(\mathbf{X}')$ is a normal random variable, the probability of failure and the reliability index corresponding to limit state function $Z = G_L(\mathbf{X}')$ can be accurately derived as

$$P_F = \Phi\left(-\frac{\mu_{GL}}{\sigma_{GL}}\right), \beta = \frac{\mu_{GL}}{\sigma_{GL}}$$

where μ_{GL} and σ_{GL} are the mean value and standard deviation of $Z = G_L(\mathbf{X}')$ respectively. And μ_{GL} and σ_{GL} are easily obtained as

$$\mu_{GL} = -\sum_{i=1}^{n}x_i'^*\frac{\partial G}{\partial x_i'}\bigg|_{\mathbf{x}^*} = -\nabla^T G(\mathbf{x}'^*)\mathbf{x}'^* \tag{6.9a}$$

and

$$\sigma_{GL}^2 = \sum_{i=1}^{n}\left(\frac{\partial G}{\partial x_i'}\bigg|_{\mathbf{x}'^*}\right)^2 = \nabla^T G(\mathbf{x}'^*)\nabla G(\mathbf{x}'^*) \tag{6.9b}$$

Then the reliability index is given as

$$\beta = \frac{\mu_{GL}}{\sigma_{GL}} = \frac{-\sum_{i=1}^{n} x_i'^* \frac{\partial G}{\partial x_i}\Big|_{\mathbf{x}'^*}}{\sqrt{\sum_{i=1}^{n} \left(\frac{\partial G}{\partial x_i}\Big|_{\mathbf{x}'^*}\right)^2}} = \frac{-\nabla^T G(\mathbf{x}'^*)\mathbf{x}'^*}{\sqrt{\nabla^T G(\mathbf{x}'^*)\nabla G(\mathbf{x}'^*)}} \tag{6.10a}$$

which is identical with Eq. (6.7a). That is, the reliability index β is the minimum distance from the origin to the limit state $G(\mathbf{X}') = 0$. Note, in the strict sense, the reliability index β in Eq. (6.10a) is not for limit state function $G(\mathbf{X}')$ but for its first-order approximation $G_L(\mathbf{X}')$ at $\mathbf{x}'^*$, therefore, the reliability index is called *first-order second-moment reliability index*, which is rewritten as

$$\beta_{FOSM} = \frac{\mu_{GL}}{\sigma_{GL}} = \frac{-\nabla^T G(\mathbf{x}'^*)\mathbf{x}'^*}{\sqrt{\nabla^T G(\mathbf{x}'^*)\nabla G(\mathbf{x}'^*)}} \tag{6.10b}$$

Let

$$\boldsymbol{\alpha} = \frac{\nabla G(\mathbf{x}'^*)}{\sqrt{\nabla^T G(\mathbf{x}'^*)\nabla G(\mathbf{x}'^*)}} = \frac{\nabla G(\mathbf{x}'^*)}{|\nabla G(\mathbf{x}'^*)|} \tag{6.11}$$

Then we have

$$\mathbf{x}'^* = -\boldsymbol{\alpha}\beta_{FOSM} \tag{6.12a}$$

$$\beta_{FOSM} = -\boldsymbol{\alpha}^T\mathbf{x}'^* \tag{6.12b}$$

In scalar form, the components of $\mathbf{x}'^*$ in Eq. (6.12a) are

$$x_i'^* = -\alpha_i\beta_{FOSM}, i = 1, 2, ..., n \tag{6.13a}$$

$$\beta_{FOSM} = -\sum_{i=1}^{n} \alpha_i x_i'^* \tag{6.13b}$$

in which

$$\alpha_i = \frac{\frac{\partial G}{\partial x_i'}\Big|_{\mathbf{x}'^*}}{\sqrt{\sum_{i=1}^{n} \left(\frac{\partial G}{\partial x_i'}\Big|_{\mathbf{x}'^*}\right)^2}}, \qquad i = 1, 2, ..., n \tag{6.13c}$$

are the direction cosines along the axes x_i'.

The first-order approximation of the limit state function can be also expressed as

$$G_L(\mathbf{X}') = \beta_{FOSM} + \mathbf{X}'^T\boldsymbol{\alpha} \tag{6.14}$$

Example 6.1 Consider the following simple linear limit state function

$$G(\mathbf{X}) = X_1 - X_2 \tag{6.15}$$

where X_1 and X_2 are independent random variables. X_1 is a normal variable with mean value of $\mu_1 = 300$ and standard deviation of $\sigma_1 = 60$, and X_2 is a lognormal variable with mean value of $\mu_2 = 100$ and standard deviation of $\sigma_2 = 40$.

For the purpose of demonstration, we will solve the problem using only the information of the first two moments, although the probability distributions of the basic random variables are known.

First, the direct second-moment reliability index is obtained as

$$\beta_{2M} = \frac{300 - 100}{\sqrt{60^2 + 40^2}} = 2.744$$

Second, the problem will be solved by FOSM in reduced space. The iteration process is illustrated in Figure 6.2. For this example, the limit state curve is also linear in the reduced space, and the procedure is finished after the first iteration. The performance function in the reduced space is expressed as

$$G_L(\mathbf{X}') = 2.774 + 0.832x_1' - 0.555x_2' \tag{6.16}$$

With the aid of Eq. (6.14), the first order second moment reliability index in reduced space is obtained as $\beta_{FOSM} = 2.774$ with corresponding failure probability of $P_F = 0.00277$. It can be observed that β_{FOSM} is equal to β_{2M} because the performance function of Eq. (6.15) is linear. The first-order reliability index (FORM, see Section 5.2 in the chapter on Methods of Moment Based on First- and Second-Order Transformation) is easily given as $\beta_F = 2.580$ with corresponding failure probability of $P_F = 0.00494$. Using MCS with 1 000 000 samples, the reliability index is obtained as 2.5304 and the corresponding failure probability of

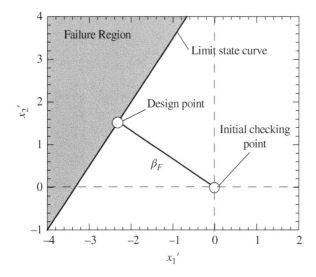

Figure 6.2 Iteration processes of FOSM in reduced space.

$P_F = 0.00569$. Obviously, the accuracy of FOSM is not sufficient since only the first two moments of the basic random variables are used.

In earlier works (e.g. Cornell 1969; Ang and Cornell 1974), the first-order approximations were evaluated at the mean values $\mu_X = (\mu_{X_1}, \mu_{X_2}, \cdots, \mu_{X_n})$, that is, to linearize the performance function $G(\mathbf{X})$ using the first order Taylor series expansion a point μ_X,

$$G(\mathbf{X}) = G(X_1, X_2, \cdots, X_n) = G(\mu_{X_1}, \mu_{X_2}, \ldots, \mu_{X_n}) + \sum_{i=1}^{n} (X_i - \mu_{X_i}) \frac{\partial G}{\partial X_i}\bigg|_{\mu_X} \tag{6.17}$$

where the derivatives are evaluated at $\mu_X = (\mu_{X_1}, \mu_{X_2}, \cdots, \mu_{X_n})$.

From Eq. (6.17), one obtains

$$\mu_G = G(\mu_{X_1}, \mu_{X_2}, \cdots, \mu_{X_n}) \tag{6.18a}$$

$$\sigma_G^2 = E(G - \mu_G)^2 = \sum_{i=1}^{n} \sigma_{X_i}^2 \left(\frac{\partial G}{\partial X_i}\bigg|_{\mu_X}\right)^2 + 2\sum_{i=1}^{n} \sum_{j=i+1}^{n} c_{ij} \frac{\partial G}{\partial X_i} \frac{\partial G}{\partial X_j}\bigg|_{\mu_X} \tag{6.18b}$$

where

$$c_{ij} = E\left[(X_i - \mu_i)(X_j - \mu_j)\right] \tag{6.18c}$$

For uncorrelated random variables, the variance of $G(\mathbf{X})$ is given as

$$\sigma_G^2 = \sum_{i=1}^{n} \sigma_{X_i}^2 \left(\frac{\partial G}{\partial X_i}\bigg|_{\mu_X}\right)^2 \tag{6.18d}$$

Then the second-moment reliability index is given as

$$\beta_{2M} = \frac{\mu_G}{\sigma_G} \tag{6.19}$$

It will be shown that the equation above could lead to significant errors.

Example 6.2 Consider the following performance function from Example 3.3 in the chapter on Moment Evaluation

$$G(\mathbf{X}) = DR - S \tag{6.20}$$

where R, S, D = independent random variables with unknown probability distribution, the only information known about them are their first two moments.

$$\mu_R = 100 \text{ and } \sigma_R = 20, \mu_S = 50 \text{ and } \sigma_S = 20, \mu_D = 1 \text{ and } \sigma_D = 0.1$$

The derivatives evaluated at point $\mu = (\mu_R, \mu_D, \mu_S)$ are given as

$$\frac{\partial G}{\partial D}\bigg|_{\mu} = \mu_R, \quad \frac{\partial G}{\partial R}\bigg|_{\mu} = \mu_D, \quad \frac{\partial G}{\partial S}\bigg|_{\mu} = -1 \tag{6.21}$$

The first-order approximation of at point μ is given as

$$G(\mathbf{X}) = \mu_D \mu_R - \mu_S + (R - \mu_R)\mu_D + (D - \mu_D)\mu_R - (S - \mu_S) \tag{6.22}$$

The first two moments of $G(\mathbf{X})$ above are given as

$$\mu_G = \mu_D\mu_R - \mu_S = 1 \times 100 - 50 = 50 \tag{6.23a}$$

$$\sigma_G^2 = \sigma_R^2\mu_D^2 + \sigma_D^2\mu_R^2 + \sigma_S^2(-1)^2 = 20^2 \times 1^2 + 0.1^2 \times 100^2 + 20^2 = 900 \tag{6.23b}$$

The second moment reliability index is given as

$$\beta_{2M}|_\mu = \frac{50}{30} = 1.667 \tag{6.24}$$

Using the first-order second-moment method (FOSM), the design point in the reduced space is obtained as $(-0.4596, -1.1338, 1.1885)$ corresponding to that in original space of $(0.954, 77.324, 73.77)$, and the first two moments of the first-order Taylor series expansion $G(\mathbf{X})$ at the design point are obtained as 48.968 and 28.703. The corresponding FOSM reliability index is 1.706, and the first order approximation of the limit state function in the original and the reduced space are given as

$$G_L(\mathbf{X}) = -73.7704 + 77.3238D + 0.95405R - S \tag{6.25a}$$

$$G_L(\mathbf{X}') = 1.706 + 0.2694D' + 0.6647R' - 0.6968S' \tag{6.25b}$$

It can be observed that the second-moment reliability index obtained at the expansion point of mean ($=1.667$) is much different from that obtained at the design point in the reduced space ($=1.706$). The FOSM results are the same as those of FORM as shown in Example 5.3 in the chapter on Methods of Moment Based on First- and Second-Order Transformation, since all the random variables are normal in that example.

6.4 First-Order Third-Moment Method

Using the information of the first three moments, we have two choices of spaces to conduct the first order approximation of performance functions. Thus we will have two first-order third-moment methods (FOTMs); in reduced space and in third-moment pseudo space.

6.4.1 First-Order Third-Moment Method in Reduced Space

Consider the first-order approximation of the limit state function Eq. (6.14), the first three moments are given as

$$\mu_{GL} = \beta_{FOSM} \tag{6.26a}$$

$$\sigma_{GL} = 1 \tag{6.26b}$$

$$\alpha_{3GL} = \sum_{i=1}^{n} \alpha_{3i}\alpha_i^3 \tag{6.26c}$$

With the first three moments above, a FOTM is applied as

$$\beta_{FOTM} = \beta_{FOSM} + \frac{\alpha_{3GL}}{6}\left(\beta_{FOSM}^2 - 1\right)\left(1 + \frac{\alpha_{3GL}}{6}\beta_{FOSM}\right) \tag{6.27}$$

Example 6.3   Consider the linear limit state function in Example 6.1 again.

The first-order approximation performance function has been obtained in the reduced space as expressed in Eq. (6.16). Beside the first two moments of the random variables used in FOSM, here the information of the third moment will be included. Using Eq. (6.26c), the skewness of the performance function of Eq. (6.16) is given as $\alpha_{3GL} = -0.216$. With the aid of Eq. (6.27), the first-order third-moment reliability index in reduced space is obtained as 2.572 with corresponding failure probability of $P_F = 0.00506$. It can observed that the FOTM in reduced space agrees well with the MCS and FORM results as shown in Example 6.1.

6.4.2   First-Order Third-Moment Method in Third-Moment Pseudo Standard Normal Space

If the first three moments of the independent basic random variables are known, the third-moment pseudo normal transformation and its inverse transformation (Zhao and Ono 2000b) can be accomplished by Eqs. (7.48) and (7.47) in the chapter on Transformation of Non-Normal Variables to Independent Normal Variables, respectively. While for correlated random variables, if their first three moments and the correlation matrix are known, the third-moment pseudo normal transformation and its inverse transformation (Lu et al. 2017a) can be realised by Eqs. (7.117) and (7.116) in the same chapter, respectively.

The basic idea of FOTM in the third-moment pseudo standard normal space can be described as follows:

Using the first three moments and the correlation matrix of arbitrary random variables **X** (continuous or discontinuous and correlated for general case) with unknown PDFs/CDFs, independent standard normal random variables **U** can be obtained using Eq. (7.117) and the random variables **X′** corresponding to **U** can be obtained from Eq. (7.116), in the same chapter as mentioned in the last paragraph. Since **U** are independent continuous random variables, **X′** will be correlated continuous random variables. Although **X** and **X′** are different random vectors, they correspond to the same independent standard normal random vector and have the same first three central moments, correlation matrix, and the same statistical information source. Therefore, $f_{\mathbf{X}'}(\mathbf{x}')$ can be considered to be an anticipated joint PDF of **X**. Using this joint PDF, the x–u and u–x transformations will become operable and the general procedure of FORM can be conducted. Since $f_{\mathbf{X}'}(\mathbf{x}')$ is equivalent to $f_{\mathbf{X}}(\mathbf{x})$ under the condition of the equivalence between their first three moments and correlation matrixes, and since the transformation holds true for the first three moments, the FORM procedure using this transformation is essentially a FOTM in the third-moment pseudo normal space. Because the u–x and x–u transformations are realised directly by using Eqs. (7.117) and (7.116) as previously mentioned, the specific form of $f_{\mathbf{X}'}(\mathbf{x}')$ is not required in FOTM.

The computation procedure for FOTM is shown as follows:

1) Obtain the first three moments of each random variable and original correlation matrix $\mathbf{C_X}$ by the probability information (e.g. the joint PDF or the marginal PDFs and correlation matrix, or statistical moments and correlation matrix).
2) Determine the polynomial coefficients using Eq. (7.45) and the equivalent correlation coefficients ρ_{0ij} using Table 7.12, both in the same chapter we have been discussing. Then, determine the lower triangular matrix $\mathbf{L}_0$ and its inverse matrix $\mathbf{L}_0^{-1}$.

3) Assume an initial checking point $\mathbf{x}_0$ (generally is the mean vector of $\mathbf{X}$).
4) Obtain the corresponding checking point in the third-moment pseudo standard normal space, $\mathbf{u}_0$, again using Eq. (7.117), and determine the initial reliability index β_0.
5) Determine the Jacobian matrix $\mathbf{J} = \partial \mathbf{X}/\partial \mathbf{U}$ evaluated at $\mathbf{u}_0$, where the element of Jacobian matrix derived from Eq. (7.116) is again given by

$$\frac{\partial X_i}{\partial U_j} = \sigma_{X_i} l_{ij} \left[b_i + 2c_i \sum_{k=1}^{i} l_{ik} U_k \right], \quad (i,j = 1, 2, ..., n) \tag{6.28}$$

6) Evaluate the performance function and gradient vector at $\mathbf{u}_0$:

$$g(\mathbf{u}_0) = G(\mathbf{x}_0), \nabla g(\mathbf{u}_0) = \mathbf{J}^T \cdot \nabla G(\mathbf{x}_0) \tag{6.29}$$

where $\nabla g(\mathbf{u}_0)$ is the gradient vector of performance function $g(\mathbf{U})$ at $\mathbf{u}_0$; and $\nabla G(\mathbf{x}_0)$ is the gradient vector of performance function $G(\mathbf{X})$ at $\mathbf{x}_0$. For implicit performance functions, the numerical differentiation methods as shown in Eq. (5.19) in the chapter on Methods of Moment Based on First- and Second-Order Transformation can be used to determine $\nabla G(\mathbf{x}_0)$.
7) Obtain new checking point $\mathbf{u}_1$ in the third-moment pseudo standard normal space:

$$\mathbf{u}_1 = \frac{1}{\nabla^T g(\mathbf{u}_0) \cdot \nabla g(\mathbf{u}_0)} \left[\nabla^T g(\mathbf{u}_0) \cdot \mathbf{u}_0 - g(\mathbf{u}_0) \right] \nabla g(\mathbf{u}_0) \tag{6.30}$$

Then, the corresponding reliability index can be determined as $\beta = \left(\mathbf{u}_1^T \cdot \mathbf{u}_1 \right)^{1/2}$.
8) Calculate the absolute difference between β and β_0, i.e. $\varepsilon_r = \text{abs}\,(\beta - \beta_0)$. If $\varepsilon_r > \varepsilon$, where ε is the permissible error (generally $\varepsilon = 10^{-6}$), determine the new checking point in original space using the following equation:

$$\mathbf{x}_1 = \mathbf{x}_0 + \mathbf{J}(\mathbf{u}_1 - \mathbf{u}_0) \tag{6.31}$$

Repeat the steps 4 through step 8 using $\mathbf{x}_1$ as the new point until convergence is achieved.

Particularly, if the performance function $G(\mathbf{X})$ is explicit, again substituting Eq. (7.116) into the performance function, it can be formulated as:

$$G(\mathbf{X}) = G(X_1, X_2, \cdots, X_n) = g(U_1, U_2, \cdots, U_n) = g(\mathbf{U}) \tag{6.32}$$

where $g(\mathbf{U})$ is an explicit function of only independent standard normal random variables. The reliability analysis can thus be readily conducted using the general FORM.

Example 6.4   Consider the following simple linear limit state function in Example 6.1 again. The first-order approximations in standard normal space is obtained as

$$G_L(\mathbf{U}) = 2.595 + 0.604u_1 - 0.797u_2$$

The first-order reliability index is readily calculated as $\beta_F = 2.595$ with the corresponding failure probability of $P_F = 0.00473$, while that in the pseudo normal space is obtained as

$$G_L(\mathbf{U}) = 2.580 + 0.637u_1 - 0.771u_2$$

Table 6.1 Comparison of FORM and FOTM procedures for Example 6.4.

	Using PDF/CDF				Using first three moments			
	Checking point		Jacobian		Checking point		Jacobian	
Iteration (1)	{x} (2)	{u} (3)	{dx/du} (4)	β (5)	{x} (6)	{u} (7)	{dx/du} (8)	β (9)
1	300	0	60	0.1926	300	0	60	0.2178
	100	0.1926	38.53		100	0.2178	41.85	
2	153.13	−1.8939	60	2.656	159.33	−1.8192	60	2.611
	153.13	1.8621	58.99		159.33	1.873	61.77	
3	186.37	−1.6687	60	2.602	190.85	−1.6787	60	2.581
	186.37	1.9968	71.80		190.85	1.9611	70.09	
4	199.88	−1.5953	60	2.595	199.28	−1.6495	60	2.580
	199.88	2.0474	−7.00		199.28	1.9837	72.16	

The first-order reliability index is calculated as $\beta_F = 2.580$ with corresponding failure probability of $P_F = 0.00494$. It can be observed that the two approximations lead to results that differ only by 4.25%.

The detailed results obtained while determining the design point using the PDF/CDF of X_1 and the first three moments of X_1 are listed in Table 6.1. Table 6.1 shows that the checking point (in original and standard normal spaces), the Jacobians, and the first-order reliability index obtained in each iteration using the first three moments of X_1 (Columns (6)–(9)) are generally close to those obtained in each iteration using the PDF/CDF of X_1 (Columns (2)–(5)).

The comparison of iteration processes is presented in Figure 6.3a–d between FORM and FOTM in pseudo normal space and in original space, from which it can be observed that checking points and the first-order approximation in the FOTM process in both pseudo normal and original spaces are almost same to those obtained in the FORM process.

The first-order third-moment reliability index in reduced space has been obtained as 2.572 with corresponding failure probability of $P_F = 0.00506$ in this Example. It can be concluded that the results obtained by FOTM result in reduced space is consistent with that of the FOTM in pseudo normal space. Both are in close to the FORM and MCS results as shown in Example 6.1.

Example 6.5 Consider the following performance function, which is an elementary reliability model that has several applications

$$G(\mathbf{X}) = dR - S \tag{6.33}$$

where

R = a resistance having $\mu_R = 500$ and $\sigma_R = 100$;
S = a load with a coefficient of variation of 0.4;
d = a modification of R having normal distribution, $\mu_d = 1$ and $\sigma_d = 0.1$.

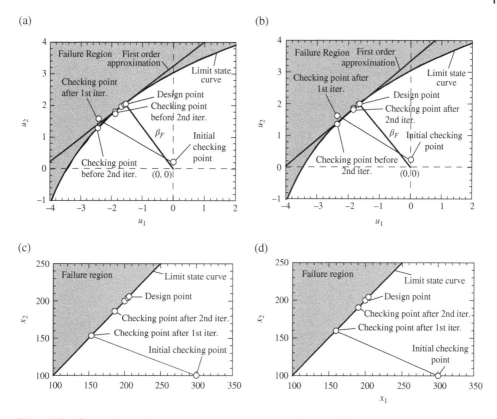

Figure 6.3 Comparison of iteration processes between FORM and FOTM for Example 6.4.

The following four cases are investigated under the assumption that R and S follow different probability distributions.

Case-1: R is lognormal and S is Weibull (Type III – smallest).
Case-2: R is Gamma and S is lognormal.
Case-3: R is Weibull and S is lognormal.
Case-4: R is Frechet (Type II – largest) and S is Exponential.

Because all of the random variables in the performance function have a known PDF/CDF, the first-order reliability index for the four cases just described can be readily obtained using FORM where the full distributions are utilised. In order to investigate the efficiency of the FOTM reliability method including random variables with unknown PDF/CDFs, the PDF/CDF of random variable R in the four cases is assumed unknown, and only its first three moments are known. Considering the first three moments, the u–x and x–u transformations in FORM can be performed easily using the FOTM in third-moment pseudo normal space, and then the first-order reliability index including random variables that have an unknown PDF/CDF can also be readily obtained.

The skewnesses of R are easily obtained as 0.608, 0.4, −0.352, and 2.353 for Cases 1–4, respectively. The first-order reliability indices obtained using the PDF/CDF of R and those using only the first three moments of R are presented in Figure 6.4 for mean values of S in the range of 100–500. Figure 6.4 shows that for all four cases, the results of the first-order reliability index obtained using only the first three moments of R are very close to those obtained using the PDF/CDF of R. This implies that the FOTM in third-moment pseudo normal space is accurate enough to include random variables with an unknown PDF/CDF.

For Case 3, Table 6.2 presents the detailed results obtained while determining the design point using the PDF/CDF of R and the first three moments of R. It can be observed from Table 6.2 that the checking point (in original and standard normal space), the Jacobians, and the first-order reliability index obtained in each iteration using the first three moments

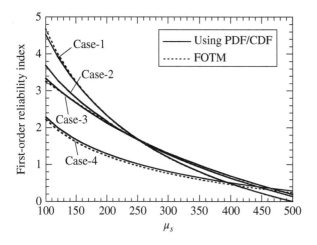

Figure 6.4 Result comparison for Example 6.5.

Table 6.2 Comparison of FORM procedure with known and unknown PDF/CDF for Example 6.5.

	Using PDF/CDF				Using first three moments			
	Checking point		Jacobian		Checking point		Jacobian	
Iteration (1)	{x} (2)	{u} (3)	{dx/du} (4)	β (5)	{x} (6)	{u} (7)	{dx/du} (8)	β (9)
	1	0	0.1		1	0	0.1	
1	500	−0.6283	101.71	0.2044	500	−0.0588	100.35	0.2014
	150	0.1926	57.788		150	0.1926	57.788	
	0.8863	−1.1371	0.1	2.6291	0.8846	−1.1536	0.1	
2	271.67	−2.0865	115.89		273.59	−2.0766	124.07	2.6340
	214.82	1.1249	82.759		215.91	1.1381	83.182	
	0.9508	−0.4922	0.1	2.6655	0.9529	−0.4713	0.1	
5	293.42	−1.9002	117.35		284.75	−1.9863	123.01	2.6766
	278.98	1.8033	107.49		271.33	1.7311	104.53	

of R [Columns (6)–(9)] are generally close to those obtained in each iteration using the PDF/CDF of R [Columns (2)–(5)].

Example 6.6 Consider the following performance function corresponding to the horizontal displacement of point A in the frame shown in Figure 6.5.

$$G(\mathbf{X}) = \Delta_A - \frac{(D + L_S + S)HL^3}{12EI} \tag{6.34}$$

where Δ_A is the allowable displacement of point A; D, L, and S are the dead, live, and snow loads, respectively; and E, I, H, and L are Young's modules, section inertial moment, height of the story, and the length of the span, respectively. The properties of the random variables are listed in Table 6.3.

Figure 6.5 A frame considered in Example 6.6.

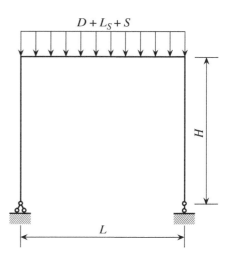

Table 6.3 Properties of random variables in Example 6.6.

Random variable	Mean value	Standard deviation	Distribution
Δ_A	2.5 cm	0.25 cm	Normal
E	2×10^{11} N/m^2	2×10^{10} N/m^2	Lognormal
I	6.06×10^{-4} m^4	3.03×10^{-5} m^4	Lognormal
H	3 m	0.15 m	Lognormal
L	6 m	0.3 m	Lognormal
D	2×10^4 N/m	4×10^3 N/m	Normal
L_S	2×10^3 N/m	6×10^2 N/m	Lognormal
S	3×10^3 N/m	1.5×10^3 N/m	Weibul

The first-order approximations of the performance function is obtained in standard normal space as

$$G_L(\mathbf{U}) = 3.029 + 0.436U_\Delta + 0.377U_E + 0.189U_I - 0.189U_H - 0.567U_L$$
$$- 0.477U_D - 0.0715U_{Ls} - 0.205U_S \tag{6.35}$$

The first-order reliability index results 3.029 with corresponding failure probability of $P_F = 0.001227$, while the first-order approximation of the performance function in pseudo normal space is obtained as

$$G_L(\mathbf{U}) = 3.072 + 0.405U_\Delta + 0.317U_E + 0.163U_I - 0.189U_H - 0.662U_L - 0.456U_D$$
$$- 0.0706U_{Ls} - 0.1473U_S \tag{6.36}$$

The first-order reliability index in pseudo normal space leads to 3.072 with corresponding failure probability of $P_F = 0.00106$.

Finally, the first-order approximation of the performance function is obtained in the reduced space as

$$G_L(\mathbf{X}^*) = 3.035 + 0.440X_\Delta^* + 0.440X_E^* + 0.1965X_I^* - 0.1854X_H^* - 0.529X_L^*$$
$$- 0.4483X_D^* - 0.0725X_{Ls}^* - 0.145X_S^* \tag{6.37}$$

Using Eq. (6.26c), the skewness of performance function Eq. (6.37) is given as $\alpha_{3GL} = 0.0023$. Using Eq. (6.27), the first order third moment reliability index in the reduced space is obtained as 3.038 with corresponding failure probability of $P_F = 0.00119$. It can be observed that the results obtained from FORM, FOTM in pseudo normal space, and FOTM in the reduced space are in comparatively good agreement.

Example 6.7 Consider the following performance function as shown in Example 3.20 again,

$$G(\mathbf{X}) = 18 - 3X_1 - 2X_2 \tag{6.38}$$

where the joint PDF of X_1 and X_2 is

$$f_{X_1X_2}(\mathbf{x}) = (x_1 + x_2 + x_1x_2) \cdot \exp(-x_1 - x_2 - x_1x_2), x_1 \geq 0 \text{ and } x_2 \geq 0 \tag{6.39}$$

According to the joint PDF of X_1 and X_2, their first three moments and correlation matrix can be readily obtained as:

$$\mu_{X_1} = \mu_{X_2} = 1.0, \sigma_{X_1} = \sigma_{X_2} = 1.0, \alpha_{3X_1} = \alpha_{3X_2} = 2.0 \tag{6.40a}$$

$$\mathbf{C_X} = \begin{bmatrix} 1.0 & -0.404 \\ -0.404 & 1.0 \end{bmatrix} \tag{6.40b}$$

The performance function in the third-moment pseudo normal space is obtained as:

$$G(\mathbf{X}) = 18 - 3X_1 - 2X_2 = 14.83 - 1.26U_1 - 1.53U_1^2 + 0.72U_1U_2 - 1.10U_2 - 0.30U_2^2 = g(\mathbf{U}) \tag{6.40c}$$

The reliability index of function Eq. (6.40c) can be easily obtained using the general FORM. The calculations for this example using FOTM in pseudo normal space are summarised in Table 6.4. The initial reliability index and starting design point are selected to be $\beta_0 = 0$ and $\mathbf{U}_0 = (0, 0)^T$, respectively, and convergence is achieved in 6 steps with $\beta = 2.719$ at the design point in pseudo standard normal space $\mathbf{u}^* = (2.7037, -0.2744)^T$ corresponding to the design point in original space $\mathbf{x}^* = (5.6229, 0.5657)^T$.

Since the joint PDF and the performance function are available, the probability of failure, P_f, can be obtained directly by numerical integral:

$$P_f = \int_{G(\mathbf{X}) \le 0} f_{X_1 X_2}(\mathbf{x})d\mathbf{x} = 1 - \int_{G(\mathbf{X}) > 0} f_{X_1 X_2}(\mathbf{x})d\mathbf{x}$$

$$= 1 - \int_0^6 \int_0^{(18-3x_1)/2} [(x_1 + x_2 + x_1 x_2) \cdot \exp(-x_1 - x_2 - x_1 x_2)]dx_2 dx_1 = 2.94486 \times 10^{-3}$$

$$(6.41)$$

and the corresponding reliability index is 2.754.

With the aid of the joint PDF of X_1 and X_2, the reliability index of performance function Eq. (6.38) can also be obtained by FORM based on Rosenblatt transformation, which are listed in Table 6.5. Since the marginal PDFs and correlation matrix can be obtained from

Table 6.4 Summary of the iteration computation for Example 6.7.

| | Design point | | Reliability index |
Iteration	$\mathbf{u}^*$	$\mathbf{x}^*$	β
1	$(6.6865, 5.8550)^T$	$(22.720, 0.1478)^T$	8.88766
2	$(3.7367, -0.0391)^T$	$(8.9417, 1.2169)^T$	3.73686
3	$(2.8098, -0.3578)^T$	$(5.9277, 0.6738)^T$	2.83244
4	$(2.7037, -0.3058)^T$	$(5.6229, 0.5766)^T$	2.72093
5	$(2.7034, -0.2871)^T$	$(5.6221, 0.5669)^T$	2.71863
6	$(2.7037, -0.2844)^T$	$(5.6229, 0.5657)^T$	2.71861

Table 6.5 Comparison of results by different methods for Example 6.7.

Method	Transformation order	Design Point $(x_1^*, x_2^*)^T$	Reliability index β
FORM based on Rosenblatt transformation	$(X_1 \to X_2)$	$(5.9149, 0.1276)^T$	2.784
	$(X_2 \to X_1)$	$(5.7805, 0.2685)^T$	2.613
FORM based on Nataf transformation	$(X_1 \to X_2)$	$(2.7957, 0.0661)^T$	2.797
	$(X_2 \to X_1)$	$(2.7957, 0.0661)^T$	2.797
FOTM in 3M pseudo normal space	$(X_1 \to X_2)$	$(5.6229, 0.5657)^T$	2.719
	$(X_2 \to X_1)$	$(5.6229, 0.5657)^T$	2.719
Numerical integral			2.754

the joint PDF, the reliability index can also be obtained by FORM based on Nataf transformation, which are also listed in Table 6.5, together with the results obtained by direct integration and FOTM in pseudo normal space.

From Table 6.5, it can be observed that the results of FORM based on Rosenblatt transformation are different with the variation of transformation orders of input random variables ($X_1 \rightarrow X_2$, $\beta = 2.784$; $X_2 \rightarrow X_1$, $\beta = 2.613$). It can also be found that FOTM in pseudo normal space provides more accurate results when compared with the numerical integral method than that of FORM based on Nataf transformation, and does not vary with the transformation order of random variables.

Example 6.8 Consider the following performance function (Sorensen 2004; Piric 2015),

$$G(\mathbf{X}) = X_1 - X_2 X_3^2 \tag{6.42}$$

where X_1, X_2, and X_3 are random variables, and their probability distributions and correlation matrix are listed in Table 6.6.

Because all random variables follow the normal distribution and the marginal PDFs and correlation matrix are available, the reliability index of Eq. (6.42) can be obtained using FOTM in pseudo normal space and FORM based on the Rosenblatt transformation or the Nataf transformation. The results are listed in Table 6.7, where it can be observed that when all random variables are normal distribution, FOTM in pseudo normal space gives the same results with those obtained by FORM based on the Rosenblatt transformation or Nataf transformation.

Table 6.6 Probability distributions and correlation matrix of random variables for Example 6.8.

Variables	Distribution	Mean	Standard deviation	Correlation matrix
X_1	Normal	25	6.25	$\begin{pmatrix} 1.0 & 0.5 & 0.2 \\ 0.5 & 1.0 & 0.4 \\ 0.2 & 0.4 & 1.0 \end{pmatrix}$
X_2	Normal	0.4	0.08	
X_3	Normal	2	0.2	

Table 6.7 Comparison of results by different methods for Example 6.8.

Method	Design point $(x_1{}^*, x_2{}^*, x_3{}^*)^T$	Reliability index β	Failure probability P_f
FORM based on Rosenblatt transformation	$(0.9182, 0.2596, 1.8805)^T$	3.86	5.67×10^{-5}
FORM based on Nataf transformation	$(0.9182, 0.2596, 1.8805)^T$	3.86	5.67×10^{-5}
FOTM in 3M pseudo normal space	$(0.9182, 0.2596, 1.8805)^T$	3.86	5.67×10^{-5}

Example 6.9 Consider the following performance function, which has been investigated by Der Kiuregian and Liu (1986),

$$G(\mathbf{X}) = X_1 - X_2 - X_3 \tag{6.43}$$

where X_1 is lognormally distributed with parameters $\lambda = 1.590$ and $\zeta = 0.198$, and X_2 and X_3 are bivariate exponentially distributed

$$f_{X_2X_3}(x_2,x_3) = (x_2 + x_3 + x_2x_3)\exp\left[-(x_2 + x_3 + x_2x_3)\right]; x_2, x_3 \geq 0 \tag{6.44}$$

and the correlation coefficients are $\rho_{12} = 0.3$ and $\rho_{13} = -0.2$.

According to the probability information of the random variables X_1, X_2, and X_3, their first three moments can be readily obtained as:

$$\mu_{X_1} = 5, \sigma_{X_1} = 1, \alpha_{3X_1} = 0.608$$

$$\mu_{X_2} = \mu_{X_3} = 1, \sigma_{X_2} = \sigma_{X_3} = 1, \alpha_{3X_2} = \alpha_{3X_3} = 2$$

and correlation matrix $\mathbf{C_X}$ is

$$\mathbf{C_X} = \begin{bmatrix} 1.0 & 0.3 & -0.2 \\ 0.3 & 1.0 & -0.4 \\ -0.2 & -0.4 & 1.0 \end{bmatrix}$$

The performance function in the third-moment pseudo normal space is obtained as:

$$G(\mathbf{X}) = 3.63 + 0.90U_1 + 0.04U_1^2 - 0.36U_1U_2 + 0.12U_1U_3 - 0.19U_2 \\ - 0.51U_2^2 + 0.34U_2U_3 - 0.56U_3 - 0.16U_3^2 = g(\mathbf{U}) \tag{6.45}$$

For the function $g(\mathbf{U})$ in Eq. (6.45), the reliability index can be calculated using the general FORM. The starting point is selected at $\mathbf{U} = (0,0,0)^T$ and convergence is achieved with $\beta_1 = 2.06$. A second local minimum-distance point is obtained if one started the iteration at $\mathbf{U} = (1,1,1)^T$ and the result is $\beta_2 = 2.46$. Using bounds method (Ditlevsen 1979b), the generalised reliability index is $\beta_g = 1.94$.

Following the method presented by Der Kiuregian and Liu (1986), the reliability indices are obtained as $\beta_{1g} = 1.89$ and $\beta_{2g} = 1.93$ dependent on the ordering of X_2 and X_3. It can be observed that when the available information of the basic variables included, in addition to the marginal PDFs, the joint PDF for a subset of these variables, FOTM in pseudo normal space can also provide sufficiently accurate results and does not vary with the transformation order of random variables.

Example 6.10 Consider the following performance function (Fan et al. 2016):

$$G(\mathbf{X}) = X_1 - \frac{75000X_2}{X_3^3} \tag{6.46}$$

where X_1 is a Weibull distributed variable with scale parameter 20 and shape parameter 25, and X_2 and X_3 are independent normal distributed variables with the mean and standard deviation of $\mu_2 = 1000$, $\sigma_2 = 250$, and $\mu_3 = 250$, $\sigma_3 = 25$, respectively. Assuming that the correlation coefficient between X_1 and X_2 is $\rho_{1,2}$, and X_1 and X_3 is independent.

Because the joint PDF of random variables cannot be available, the reliability analysis based on Rosenblatt transformation can no longer be used. Since the marginal PDFs and correlation matrix are known, reliability analysis can be conducted using Nataf transformation and the reliability indices vary with respect to the correlation coefficient $\rho_{1,2}$ are shown in Figure 6.6, together with the results obtained using the FOTM in pseudo normal space.

From Figure 6.6, it can be observed that if the marginal PDFs and correlation matrix are known, FOTM in pseudo normal space can give almost the same results with those obtained by FORM based on Nataf transformation.

Example 6.11 Consider the following performance function:

$$G(\mathbf{X}) = \frac{1}{50} \cdot \sum_{i=1}^{n} h_i - \sum_{i=1}^{n} h_i \cdot X_i \tag{6.47}$$

where $h_i = $ constant $= 3.6$ m $(i = 1, 2, ..., n)$; X_i $(i = 1, 2, ..., n)$ are lognormal distributed variables with the mean value and standard deviation of $\mu_{X_i} = 0.015$ and $\sigma_{X_i} = 0.003$, respectively; and the correlation coefficient between X_i and X_j is $\rho_{ij} = 0.1$.

Since the marginal PDFs and correlation matrix are available, reliability analysis can be conducted using FORM based on Nataf transformation and the first-order reliability indices changed with respect to the number of random variables n are shown in Figure 6.7.

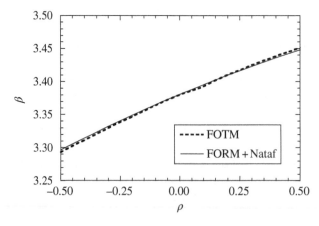

Figure 6.6 Comparison of the first-order reliability indices for Example 6.10.

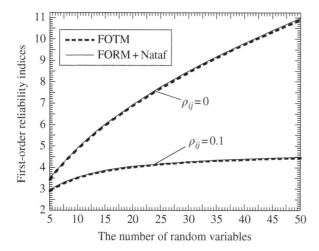

Figure 6.7 Comparison of the first-order reliability indices for Example 6.11.

It can be seen from Figure 6.7 that the results obtained by FOTM in pseudo normal space are in good agreement with those obtained by FORM based on Nataf transformation for the entire investigation range.

For the purpose of comparison, the analysis is also conducted for the case of the correlation coefficient between X_i and X_j is $\rho_{ij} = 0$, and the results of reliability analysis are also presented in Figure 6.7, from which it can be observed that the reliability indices with correlated random variables are smaller than those with independent random variables for this example.

Example 6.12 Consider the following strongly nonlinear performance function:

$$G(\mathbf{X}) = X_1^4 + X_2^2 - 120 \tag{6.48}$$

where X_1 is a lognormal variable with mean value of 5 and standard deviation of 1; X_2 is a Gumbel variable with mean value of 10 and standard deviation of 10; and the correlation coefficient between X_1 and X_2 is 0.25.

The results obtained by FOTM in pseudo normal space and FORM based on the Nataf transformation are listed in Table 6.8, together with the results from MCS. As can be observed from Table 6.8, the first-order reliability indices differ greatly from those obtained by MCS due to the strong nonlinearity of the performance function in Eq. (6.48).

Table 6.8 Comparison of results by different methods for Example 6.12.

	Transformation methods					
	Third-moment transformation			Nataf transformation		
Reliability methods	FOTM	SOTM	MCS	FORM	SORM	MCS
Reliability indices	2.076	2.277	2.283	2.035	2.279	2.279

Using the second-order reliability method (SORM) (Breitung 1984), the second-order reliability indices based on the third-moment transformation (Zhao and Ono 2000b) and the Nataf transformation are also listed in Table 6.8. It can be seen from Table 6.8 that the second-order reliability index obtained using the third-moment transformation is almost the same with that obtained using the Nataf transformation and are in close agreement with MCS results.

Example 6.13 This example considers a slope shown in Figure 6.8, and the performance function for safety of the slope is expressed as (Baecher and Christian 2003; Low 2007):

$$G(\mathbf{X}) = \frac{cA + N' \tan \phi}{W \left(\sin \psi_p + \alpha \cos \psi_p \right) + V \cos \psi_p - T \sin \theta} - 1 \tag{6.49}$$

in which

$$A = (H - z)/ \sin \psi_p \tag{6.50a}$$

$$z = H \left(1 - \sqrt{\cot \psi_f \tan \psi_p} \right) \tag{6.50b}$$

$$N' = W \left(\cos \psi_p - \alpha \sin \psi_p \right) - U - V \sin \psi_p + T \cos \theta \tag{6.50c}$$

$$W = 0.5\gamma H^2 \left\{ \left[1 - \left(\frac{z}{H} \right)^2 \right] \cot \psi_p - \cot \psi_f \right\} \tag{6.50d}$$

$$U = 0.5\gamma_w \cdot r \cdot z \cdot A \tag{6.50e}$$

$$r = z_w / z \tag{6.50f}$$

$$V = 0.5\gamma_w r^2 z^2 \tag{6.50g}$$

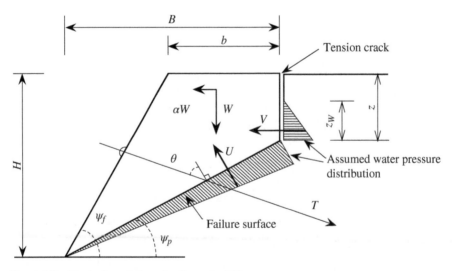

Figure 6.8 Illustration of slope for Example 6.13.

Table 6.9 Probability distributions of random variables and their first three moments for Example 6.13.

Variables	Distribution	Mean	Standard deviation	Skewness
c	Lognormal	140 kPa	28 kPa	0.6080
ϕ	Lognormal	35°	5°	0.4315
r	Truncated exponential (0, 1)	0.3435	0.2626	0.6800
z	Lognormal	14 m^2	3 m^2	0.6527
α	Truncated exponential (0, 0.16)	0.05496	0.04202	0.6800

where $\gamma = 2.6 \times 10^4$ N/m^3, $\gamma_W = 1.0 \times 10^4$ N/m^3, $\psi_f = 50°$, $\psi_p = 35°$, $T = 0$, $\theta = 0$, $H = 60$ m, and c, ϕ, r, z, and α are random variables. Their probability distributions and first three moments are listed in Table 6.9.

If c, ϕ, r, z, and α are mutually independent, using the method of MCS with 1 000 000 samples, the probability of failure P_f is obtained as 8.877×10^{-3} (the coefficients of variation of P_f is 1.06%) with corresponding reliability index of $\beta_{MCS} = 2.37$. Using the FOTM in pseudo normal space, the reliability index can be obtained as 2.40. Good agreement can be observed between reliability indices obtained by FOTM in pseudo normal space and the MCS method.

In engineering practice, c and ϕ, and r and z are often considered correlated. In this example, assume that c and ϕ, and r and z are correlated with correlation coefficient $\rho_{c,\phi} = -0.5$ and $\rho_{r,z} = -0.5$, respectively. Since the joint PDF is unknown, FORM based on Rosenblatt transformation cannot be applied. Although FORM based on Nataf transformation is theoretically available because the marginal PDFs and correlation matrix are known. However, this method requires evaluations of the equivalent correlation coefficients in standard normal space, which involves solving two-dimensional nonlinear integral equations. Using FOTM in pseudo normal space instead, the reliability index can be easily obtained as 3.04. The probability of failure P_f obtained by MCS based on the third-moment transformation is 1.132×10^{-3} (the coefficient of variation of P_f is 2.10%) with corresponding reliability index of $\beta_{MCS} = 3.05$.

Example 6.14 This example considers a two-story two-bay frame structure as shown in Figure 6.9a, in which M_i and S_i denote the member strength and load, respectively, and the probabilistic information of random variables of M_i and S_i are listed in Table 6.10. The most likely failure mode is shown Figure 6.9b and the corresponding performance function is (Zhao and Ono 1998)

$$G(\mathbf{X}) = 2M_1 + 2M_2 + 2M_3 - 4.5S_1 - 4.5S_2 \tag{6.51}$$

It is assumed that all M_i ($i = 1, 2, 3$) are independent of all S_i ($i = 1, 2$), and M_i are correlated with correlation coefficient $\rho_M = -0.1$ and S_i are correlated with correlation coefficient $\rho_S = 0.5$.

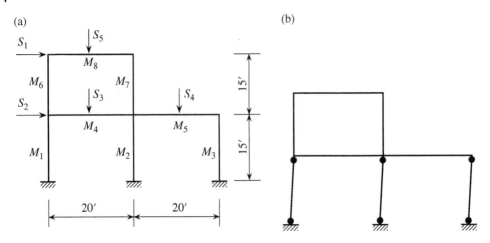

Figure 6.9 Frame structure considered in Example 6.14.

Table 6.10 The first three moments of the random variables for Example 6.14.

Variables	Mean	Standard deviation	Skewness
M_1, M_2, M_3	700 kN·m	105 kN·m	0.453
M_6, M_7	700 kN·m	105 kN·m	0.453
M_4	1500 kN·m	225 kN·m	0.453
M_5	1200 kN·m	180 kN·m	0.453
M_8	90 kN·m	13.5 kN·m	0.453
S_1	100 kN	25 kN	0.765
S_2	200 kN	50 kN	0.765
S_3	265 kN	66.3 kN	0.765
S_4	180 kN	45 kN	0.765
S_5	140 kN	35 kN	0.765

For this example, only the first three moments and correlation matrix are available, reliability analysis based on Rosenblatt transformation and the Nataf transformation cannot be used. However, using the third-moment transformation, the performance function Eq. (6.51) can be expressed as a function $g(\mathbf{U})$ of independent standard normal variables:

$$G(\mathbf{X}) = 2845.76 + 166.50U_1 + 16.26U_1^2 - 2.85U_1U_2 - 3.19U_1U_3 + 184.31U_2$$
$$+ 15.97U_2^2 - 3.53U_2U_3 + 206.39U_3 + 15.57U_3^2 - 223.06U_4$$
$$- 22.02U_4^2 - 25.42U_4U_5 - 190.51U_5 - 21.53U_5^2 = g(\mathbf{U})$$

$$(6.52)$$

where U_i ($i = 1, 2, ..., 5$) are independent third-moment pseudo standard normal random variables.

For the performance function $g(\mathbf{U})$ in Eq. (6.52), the reliability index can be easily calculated using FOTM as 5.27. The probability of failure P_f of Eq. (6.52) obtained by MCS with $100\,000\,000$ samples is 6.00×10^{-8} with corresponding reliability index of $\beta_{MCS} = 5.29$. It can be observed that the reliability analysis with unknown joint PDF and marginal PDFs can be realised by using FOTM in pseudo normal space, which complements existing methods for reliability assessment including correlated variables.

6.5 First-Order Fourth-Moment Method

6.5.1 First-Order Fourth-Moment Method in Reduced Space

Consider the first-order approximation of the limit state function Eq. (6.14), the first four moments are given as

$$\mu_{GL} = \beta_{FOSM} \tag{6.53a}$$

$$\sigma_{GL} = 1 \tag{6.53b}$$

$$\alpha_{3GL} = \sum_{i=1}^{n} \alpha_{3i} \alpha_i^3 \tag{6.53c}$$

$$\alpha_{4GL} = \sum_{i=1}^{n} \alpha_{4i} \alpha_i^4 + 6 \sum_{i=1}^{n-1} \sum_{j>i}^{n} \alpha_i^2 \alpha_j^2 \tag{6.53d}$$

With these first four moments, a first-order fourth-moment method is given as

$$\beta_{4M} = \frac{\sqrt[3]{\sqrt{2}p}}{\sqrt[3]{-q_0 + \Delta_0}} - \frac{\sqrt[3]{-q_0 + \Delta_0}}{\sqrt[3]{2}} + \frac{l_1}{3k_2} \tag{6.54}$$

where

$$\Delta_0 = \sqrt{q_0^2 + 4p^3}, \quad p = \frac{3k_1 k_2 - l_1^2}{9k_2^2}, \quad q_0 = \frac{2l_1^3 - 9k_1 k_2 l_1 + 27k_2^2(-l_1 + \beta_{FOSM})}{27k_2^3},$$

$$l_1 = \frac{\alpha_{3GL}}{6(1 + 6l_2)}, \quad k_1 = \frac{1 - 3l_2}{(1 + l_1^2 - l_2^2)}, \quad k_2 = \frac{l_2}{(1 + l_1^2 + 12l_2^2)},$$

$$l_2 = \frac{1}{36}\left(\sqrt{6\alpha_{4GL} - 8\alpha_{3GL}^2 - 14} - 2\right)$$

Example 6.15 Consider again the linear limit state function as shown in Example 6.1.
The first-order approximation performance function in the reduced space has been obtained as expressed in Eq. (6.16).

Beside the first two moments of the random variables used in FOSM, here the information of the first four moments will be included. Using Eqs. (6.53c) and (6.53d), the skewness and kurtosis of performance function Eq. (6.16) are given as $\alpha_{3GL} = -0.216$ and $\alpha_{4GL} = 3.216$. With the aid of Eq. (6.54), the first-order fourth-moment (FOFM) reliability index in reduced space is obtained as 2.572 with corresponding failure probability of $P_F = 0.00506$.

One can see the FOFM result in reduce space is consistent with the MCS and FORM results as shown in Example 6.1.

The first-order reliability index (FORM, see Section 5.2.2 in the chapter on Methods of Moment Based on First- and Second-Order Transformation) is easily given as $\beta_F = 2.580$ with corresponding failure probability of $P_F = 0.00494$. Using MCS with 1 000 000 samples, the reliability index is obtained as 2.5304 and the corresponding failure probability of $P_F = 0.00569$. Obviously, the accuracy of FOSM is not sufficient since only the first two moments of the basic random variables are used.

6.5.2 First-Order Fourth-Moment Method in Fourth-Moment Pseudo Standard Normal Space

If the first four moments of the independent basic random variables are known, the fourth-moment pseudo normal transformation and its inverse transformation (Zhao et al. 2018a) can be applied by Eq. (7.75) or Table (7.9) and Eq. (7.70), respectively, in the chapter on Transformation of Non-Normal Variables to Independent Normal Variables. While for correlated random variables, if their first four moments and correlation matrix are obtained, the fourth-moment pseudo normal transformation and its inverse transformation (Lu et al. 2020) can be achieved by Eqs. (7.130) and (7.129), respectively, in the same chapter.

Using the first four moments of arbitrary random variables X (continuous or discontinuous and correlated for general case) with unknown PDFs/CDFs, independent standard normal random variables U can be obtained using Eq. (7.130) and the random variables X' corresponding to U can be obtained from Eq. (7.129). Since U are independent continuous random variables, X' will be correlated continuous random variables. Although X and X' are different random vectors, they correspond to the same independent standard normal random vector and have the same fourth central moments, correlation matrix, and the same statistical information source. Therefore, $f_{X'}(x')$ can be considered to be an anticipated joint PDF of X. Using this joint PDF, the x–u and u–x transformations will become operable and the general procedure of FORM can be conducted. Since $f_{X'}(x')$ is equivalent to $f_X(x)$ under the condition of the equivalence between their first four moments and correlation matrix, and since the transformation holds true for the first four moments, the FORM procedure using this transformation is essentially a FOFM in fourth-moment pseudo normal space. Because the u–x and x–u transformations are conducted directly using Eqs. (7.130) and (7.129) from above, the specific form of $f_{X'}(x')$ is not required in FOFM.

The computation procedure for FOFM is similar to that of FOTM as described in Section 6.4.2, where the third-moment pseudo transformations are replaced by the fourth-moment pseudo transformations.

Example 6.16 Consider the following performance function, which is an elementary reliability model with several applications:

$$G(\mathbf{X}) = dR - S \tag{6.55}$$

where R = resistance having $\mu_R = 500$ and $\sigma_R = 100$; S = load with coefficient of variation of 0.4; and d = modification of R having normal distribution, $\mu_d = 1$ and $\sigma_d = 0.1$.

The following six cases are investigated under the assumption that R and S follow different probability distributions:

Case 1 : R is Gumbel (Type I – largest) and S is Weibull (Type III – smallest)
Case 2 : R is Gamma and S is normal
Case 3 : R is lognormal and S is Gamma
Case 4 : R is lognormal and S is Gumbel
Case 5 : R is Weibull and S is lognormal
Case 6 : R is Frechet (Type II – largest) and S is exponential

Because all of the random variables in the performance function have known CDF/PDFs, the first-order reliability index can be readily obtained using FORM for the six cases described here. In order to explore the efficiency of the FOFM reliability method, including random variables with unknown CDF/PDFs, the CDF/PDF of random variables R in the six cases is assumed to be unknown, and only its first four moments are known. With the first four moments, the x–u and u–x transformations in FORM can be performed easily using the fourth-moment pseudo normal transformation and its inverse transformation, and then the first order reliability index, including random variables that have an unknown CDF/PDF, can also be readily obtained.

The skewness and kurtosis of R are easily obtained as 1.14 and 5.4, 0.4 and 3.24, 0.608 and 3.664, 0.608 and 3.664, −0.352 and 3.004, and 2.353 and 16.43 for Cases 1–6, respectively. The first-order reliability index obtained using the CDF/PDF of R and using only first four moments of R are presented in Figure 6.10 for mean values of S in the range of 100–500. Figure 6.10 shows that, for all six cases, the results of the first-order reliability index obtained using only the first four moments of R are in agreement with those obtained using the CDF/PDF of R. This implies that FOFM method is accurate enough to include random variables with unknown CDF/PDFs.

For Case 4, the detailed results obtained while determining the design point using the CDF/PDF of R and using the first four moments of R are listed in Table 6.11, where we

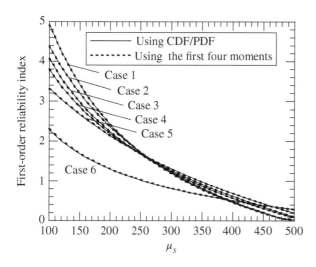

Figure 6.10 Comparisons of first-order reliability index with known and unknown CDF/PDFs.

Table 6.11 Comparisons of FORM Procedure with known and unknown CDF/PDFs for Example 6.16.

	Using CDF/PDF				Using the first four moments			
	Checking point				Checking point			
Iteration (1)	{x} (2)	{u} (3)	Jacobian {dx/du} (4)	β (5)	{x} (6)	{u} (7)	Jacobian {dx/du} (8)	β (9)
1	1	0	0.1	2.2543	1	0	0.1	2.2541
	500	0.09902	99.0211		500	0.09935	99.0174	
	200	0.17734	76.4924		200	0.17734	76.4924	
2	0.91635	−0.83652	0.1	2.2193	0.91635	−0.83646	0.1	2.2242
	326.151	−2.05833	64.5917		326.142	−2.05495	65.4841	
	284.326	1.10206	107.917		284.319	1.10201	107.914	
5	94.722	−0.52783	0.1	2.1897	0.94720	−0.52805	0.1	2.1901
	403.013	−0.98983	79.8136		403.132	−0.98918	79.752	
	381.741	1.88058	143.638		381.845	1.8813	143.674	

see that the checking point (in original and standard normal space), the Jacobians, and the first-order reliability index obtained in each iteration using the first four moments of R [Columns (6)–(9)] are generally close to those obtained in each iteration using the CDF/PDF of R [Columns (2)–(5)].

Example 6.17 As an application of Example 6.16, this example considers the following performance function of an H-shape steel column:

$$G(\mathbf{X}) = AY - C \tag{6.56}$$

where A is the section area, Y is the yield stress, and C is the compressive stress. The CDFs of A and Y are unknown, the only information about them are their first four moments (Ono et al. 1986), i.e. $\mu_A = 71.656 \text{ cm}^2$, $\sigma_A = 3.691 \text{ cm}^2$, $\alpha_{3A} = 0.709$, $\alpha_{4A} = 3.692$, $\mu_Y = 3.055 \text{ t/cm}^2$, $\sigma_Y = 0.364$, $\alpha_{3Y} = 0.512$, and $\alpha_{4Y} = 3.957$. C is assumed as a lognormal variable with mean value $\mu_C = 100 \text{ t}$ and standard deviation $\sigma_C = 40 \text{ t}$. The skewness and kurtosis of C can be readily obtained as $\alpha_{3C} = 1.264$ and $\alpha_{4C} = 5.969$.

Although the CDFs of A and Y are unknown, since the first four moments are known, the x–u and u–x transformations can be easily realised using the fourth-moment pseudo normal transformation and its inverse transformation instead of Rosenblatt transformation and FORM can be readily conducted with results of $\beta_F = 2.079$ and $P_F = 0.0188$. Furthermore, using Eq. (7.70) in the chapter on Transformation of Non-Normal Variables to Independent Normal Variables, the random sampling of A and Y can be easily generated without using their CDFs, and MCS can be thus easily conducted. The probability of failure of this performance function is obtained as $P_F = 0.0188$ and the corresponding reliability index is equal to 2.079 when the number of samplings is taken to be 10 000. It can be observed that the same results are obtained by the two methods for this example.

Example 6.18 Consider the performance function in Example 6.7 again

$$G(\mathbf{X}) = 18 - 3X_1 - 2X_2 \tag{6.57}$$

where the joint PDF of X_1 and X_2 is

$$f_{X_1 X_2}(\mathbf{x}) = (x_1 + x_2 + x_1 x_2) \cdot \exp(-x_1 - x_2 - x_1 x_2); x_1, x_2 \geq 0 \tag{6.58}$$

As given in Example 6.7, the exact probability of failure and reliability index are 2.94486×10^{-3} and 2.7539, respectively. And it has been solved using FORM based on both the Rosenblatt transformation and Nataf transformation. Here, the problem is solved again using the pseudo normal transformation described above. According to the joint PDF of X_1 and X_2, their first four central moments and correlation matrix can be readily obtained as:

$$\mu_{X_1} = \mu_{X_2} = 1, \sigma_{X_1} = \sigma_{X_2} = 1, \alpha_{3X_1} = \alpha_{3X_2} = 2, \alpha_{4X_1} = \alpha_{4X_2} = 9$$

$$\mathbf{C_X} = \begin{bmatrix} 1.0 & -0.404 \\ -0.404 & 1.0 \end{bmatrix}$$

Table 6.12 Summary of the iteration procedure for Example 6.18.

Step	$U = \{U_1, U_2\}^T$	$X = \{X_1, X_1\}^T$	β
1	$(6.6865, 5.8550)^T$	$(22.7198, 0.1478)^T$	8.88766
2	$(3.7367, -0.03913)^T$	$(8.9417, 1.2169)^T$	3.73686
3	$(2.8098, -0.3578)^T$	$(5.9277, 0.6738)^T$	2.83244
4	$(2.7037, -0.3058)^T$	$(5.6229, 0.5766)^T$	2.72093
5	$(2.7034, -0.2871)^T$	$(5.6221, 0.5669)^T$	2.71863
6	$(2.7037, -0.2844)^T$	$(5.6229, 0.5657)^T$	2.71861

The performance function in the third-moment pseudo normal space is:

$$g_{3M-1}(\mathbf{U}) = 14.83 - 1.26U_1 - 1.53U_1^2 + 0.72U_1U_2 - 1.10U_2 - 0.30U_2^2 \tag{6.59}$$

Comparing Eq. (6.57) with Eq. (6.59), it can be observed that the performance function of correlated random variables can be expressed as a function of independent standard normal variables by the third-moment pseudo standard normal transformation. The reliability index of function Eq. (6.59) can be easily obtained using the general FORM. The iteration procedure for this example is summarised in Table 6.12. The initial reliability index and starting design point are selected to be $\beta_0 = 0$ and $\mathbf{U}_0 = (0, 0)^T$, respectively, and convergence is achieved in six steps with $\beta = 2.72$ at $\mathbf{u}^* = (2.7037, -0.2744)^T$ corresponding to $\mathbf{x}^* = (5.6229, 0.5657)^T$.

Using the method of MCS with 1 000 000 samples, the probability of failure P_f is obtained as 0.00411 (the coefficients of variation of P_f is 1.10%) with corresponding reliability index of $\beta_{MCS} = 2.643$. Based on the point estimate method, the first four central moments of Eq. (6.59) are obtained as $\mu_Z = 13$, $\sigma_Z = 2.8559$, $\alpha_{3Z} = -1.9379$, and $\alpha_{4Z} = 9.7288$. Therefore, the third moment reliability index and fourth moment reliability index for Eq. (6.59) are obtained as $\beta_{3M} = 2.745$ and $\beta_{4M} = 2.656$, respectively. It can be observed that the results of FORM, MCS, and direct moment method are almost the same for this example.

In order to investigate the influence of transformation order on the result of the third-moment pseudo normal transformation, the problem is resolved by using different order of transformation (X_2 to X_1). Obviously, the correlation matrix is the same since there are only two random variables, and the performance function in the third-moment pseudo normal space is expressed as:

$$g_{3M-2}(\mathbf{U}) = 14.83 + 0.26U_1 - 1.34U_1^2 - 1.65U_2 + 1.08U_1U_2 - 0.45U_2^2 \tag{6.60}$$

The reliability index corresponding to Eq. (6.60) is also listed in Table 6.13. It can be observed that the results obtained by using the third-moment pseudo normal transformations provide the same results even though the transformation order is different.

Similarly, using the method of MCS with 1 000 000 samples, the probability of failure P_f is obtained as 0.00412 (the coefficients of variation of P_f is 1.55%) with corresponding reliability index of $\beta_{MCS} = 2.643$. Based on the point estimate method, the first four central moments of Eq. (6.60) are obtained as $\mu_Z = 13$, $\sigma_Z = 2.8559$, $\alpha_{3Z} = -1.9379$, and

Table 6.13 Comparison of results by different methods for Example 6.18.

Transformations in FORM	Order of X	Design point $(u_1{}^*, u_2{}^*)$	Design point $(x_1{}^*, x_2{}^*)$	β
Rosenblatt transformation	$(X_1 \rightarrow X_2)$	(2.7840, 0.0262)	(5.9149, 0.1276)	2.7842
	$(X_2 \rightarrow X_1)$	(2.4820, −0.8167)	(5.7805, 0.2685)	2.6129
Nataf transformation	$(X_1 \rightarrow X_2)$	(2.7957, 0.0661)	(5.9564, 0.0654)	2.7965
	$(X_2 \rightarrow X_1)$	(−1.5277, 2.3423)	(5.9564, 0.0654)	2.7965
Third-moment pseudo normal transformation	$(X_1 \rightarrow X_2)$	(2.7037, −0.2844)	(5.6229, 0.5657)	2.7186
	$(X_2 \rightarrow X_1)$	(−2.2548, 1.5189)	(5.6229, 0.5657)	2.7186
Fourth-moment pseudo normal transformation	$(X_1 \rightarrow X_2)$	(2.7973, −0.01343)	(5.9499, 0.0752)	2.7973
	$(X_2 \rightarrow X_1)$	(−1.6590, 2.2523)	(5.9499, 0.0752)	2.7973
Numerical integral				2.7539

$\alpha_{4Z} = 9.7288$. Therefore, the third moment and fourth moment reliability indices are obtained for Eq. (6.60) as $\beta_{3M} = 2.745$ and $\beta_{4M} = 2.656$, respectively. It can be observed that the results of FORM, MCS and direct moment method are almost exactly same even though the transformation order is different.

The limit state curves expressed by Eqs. (6.59) and (6.60) are illustrated in Figure 6.11. From Figure 6.11, it can be observed that the two curves are different although the corresponding reliability indices are the same. From Table 6.13, it can be seen that the design points for the corresponding two curves are different in pseudo standard normal space but those in original space are the same.

Figure 6.11 Limit state curves in third-moment pseudo normal space.

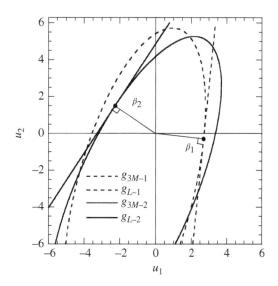

The problem is then solved by using the fourth-moment pseudo standard normal transformation. According to the first four central moments of X_1 and X_2, the polynomial coefficients of the correlated pseudo standard normal variables Z_1 and Z_2, are determined as,

$$a_1 = -0.3137, b_1 = 0.8263, c_1 = 0.3137, d_1 = 0.0227$$
$$a_2 = -0.3137, b_2 = 0.8263, c_2 = 0.3137, d_2 = 0.0227$$

According to the correlation coefficient between X_1 and X_2, ρ_{12} the correlation coefficient between the correlated pseudo standard normal variables Z_1 and Z_2, ρ_{012}, is determined as:

$$\rho_{012} = -0.5892$$

The correlation matrix of correlated pseudo standard normal variables can then be written as,

$$\mathbf{C}_Z = \begin{bmatrix} 1 & -0.5892 \\ -0.5892 & 1 \end{bmatrix}$$

Utilising Cholesky decomposition, the lower triangular matrix $\mathbf{L}_0$ can be obtained as,

$$\mathbf{L}_0 = \begin{bmatrix} 1 & 0 \\ -0.5892 & 0.8080 \end{bmatrix}$$

Following Eq. (7.129) from the chapter on Transformation of Non-Normal Variables to Independent Normal Variables, X_1 and X_2 can be expressed as

$$X_1 = 1 + 1 \cdot \left(-0.3137 + 0.8263 \cdot U_1 + 0.3137 \cdot U_1^2 + 0.0227 \cdot U_1^3 \right) \tag{6.61a}$$

$$X_2 = 1 + 1 \cdot [-0.3137 + 0.8263 \cdot (-0.5892 U_1 + 0.8080 U_2)$$
$$+ 0.3137 \cdot (-0.5892 U_1 + 0.8080 U_2)^2 + 0.0227 \cdot (-0.5892 U_1 + 0.8080 U_2)^3] \tag{6.61b}$$

Substituting Eqs. (6.61a) and (6.61b) into the performance function of Eq. (6.57), leads to

$$g_{4M-1}(\mathbf{U}) = 14.57 - 1.51 U_1 - 1.12 U_1^2 - 0.06 U_1^3 - 1.34 U_2 + 0.60 U_1 U_2$$
$$- 0.04 U_1^2 U_2 - 0.41 U_2^2 + 0.05 U_1 U_1^2 - 0.02 U_2^3 \tag{6.62}$$

The reliability index corresponding to Eq. (6.62) is listed in Table 6.13.

Using MCS with 1 000 000 samples, the probability of failure P_f is obtained as 0.003027 (the coefficients of variation of P_f is 1.81%) with corresponding reliability index of $\beta_{MCS} = 2.745$. Based on the point estimate method, the first four central moments of Eq. (6.62) are obtained as $\mu_Z = 13$, $\sigma_Z = 2.8559$, $\alpha_{3Z} = -1.5909$, and $\alpha_{4Z} = 7.8043$. Therefore, the third-moment and fourth-moment reliability indices for Eq. (6.62) are obtained as $\beta_{3M} = 2.859$ and $\beta_{4M} = 2.745$, respectively.

In order to investigate the influence of transformation order on the result of the fourth-moment pseudo normal transformation, the problem is resolved by using different order of transformation (X_2 to X_1). Obviously, the correlation matrix is the same since there are only two random variables, and the performance function in the fourth-moment pseudo normal space is expressed as

$$g_{4M-2}(\mathbf{U}) = 14.57 - 0.19U_1 - 0.95U_1^2 - 0.03U_1^3 - 2.0U_2 + 0.90U_1U_2$$
$$- 0.06U_1^2U_2 - 0.61U_2^2 + 0.08U_1U_1^2 - 0.04U_2^3 \qquad (6.63)$$

The reliability index corresponding to Eq. (6.63) is also listed in Table 6.13. It can be observed that the results obtained by using the fourth-moment pseudo normal transformations provide exactly same results even though the transformation order is different.

Using MCS with 1 000 000 samples, the probability of failure P_f is obtained as 0.003003 (the coefficients of variation of P_f is 1.82%) with corresponding reliability index of $\beta_{MCS} = 2.745$. Based on the point estimate method, the first four central moments of Eq. (6.63) are obtained as $\mu_Z = 13$, $\sigma_Z = 2.8559$, $\alpha_{3Z} = -1.5909$, and $\alpha_{4Z} = 7.8043$. Therefore, the third moment and fourth moment reliability indices are obtained for Eq. (6.63) as $\beta_{3M} = 2.859$ and $\beta_{4M} = 2.745$, respectively.

The limit state curves are illustrated in Figure 6.12 for Eqs. (6.62) and (6.63). It can be observed that the two curves are different despite same reliability indices. From Table 6.13, it can be observed that the design points are different for the corresponding two curves in pseudo standard normal space but same in original space.

The results of FORM based on Rosenblatt transformation and Nataf transformation are also listed in Table 6.13, together with the results obtained by direct integration, and the third-moment and the fourth-moment pseudo standard normal transformations. For the purpose of comparison, the results with different transformation order of random variables are also listed in Table 6.13.

From Table 6.13, it can be observed that the results of FORM based on Rosenblatt transformation are highly dependent on the transformation order. While those using pseudo standard normal transformations and Nataf transformation provide same results even the transformation order is different. It can also be observed that the pseudo standard normal transformations provide almost the same results with the numerical integral one although only the information of the first three or four moments of random variables is used.

Figure 6.12 Limit state curves in fourth-moment pseudo normal space.

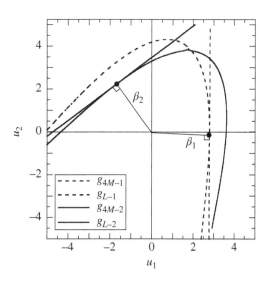

Table 6.14 Comparison of results using third- and fourth-moment pseudo normal transformations.

			Direct moment method			
Reliability method		FORM	Third-moment	Fourth-moment	MCS	Numerical integral
Reliability index	Third-moment pseudo normal transformation	2.719	2.745	2.656	2.643	2.754
	Fourth-moment pseudo normal transformation	2.797	2.859	2.745	2.745	

The results obtained using FORM, direct moment method, and MCS based on the third- and fourth-moment pseudo normal transformations are summarised in Table 6.14, together with the exact one obtained from numerical integration. As can be observed from Table 6.14, the results are generally almost same with the exact one. It can therefore be concluded that the results based on the fourth-moment pseudo normal transformations are closer to the exact one than those based on the third-moment pseudo normal transformations.

Example 6.19 Consider the following performance function previously investigated in Example 6.9

$$G(\mathbf{X}) = X_1 - X_2 - X_3 \tag{6.64}$$

where X_1 is lognormally distributed with parameters of $\lambda = 1.590$ and $\zeta = 0.198$, X_2 and X_3 are bivariate exponentially distributed

$$f_{X_2X_3}(x_2, x_3) = (x_2 + x_3 + x_2x_3) \exp\left[-(x_2 + x_3 + x_2x_3)\right]; x_2, x_3 \geq 0 \tag{6.65}$$

and the correlation coefficients are $\rho_{12} = 0.3$ and $\rho_{13} = -0.2$.

This problem has been solved using FORM based on Nataf transformation. Here, the problem is solved again using the pseudo normal transformation described above.

According to the probability information of the random variables X_1, X_2, and X_3, their first four moments can be readily obtained as:

$$\mu_{X_1} = 5, \sigma_{X_1} = 1, \alpha_{3X_1} = 0.608, \alpha_{4X_1} = 3.664$$

$$\mu_{X_2} = \mu_{X_3} = 1, \sigma_{X_2} = \sigma_{X_3} = 1, \alpha_{3X_2} = \alpha_{3X_3} = 2, \alpha_{4X_2} = \alpha_{4X_3} = 9$$

Using Eq. (6.65), the correlation coefficient between X_2 and X_3 is readily obtained as $\rho_{13} = -0.4$, and the correlation matrix is

$$\mathbf{C_X} = \begin{bmatrix} 1.0 & 0.3 & -0.2 \\ 0.3 & 1.0 & -0.4 \\ -0.2 & -0.4 & 1.0 \end{bmatrix}$$

Using the third-moment pseudo transformation, the correlation matrix of correlated third-moment pseudo standard normal variables, $\mathbf{C}_{\mathbf{Z}\text{-3M}}$, can then be written as,

$$\mathbf{C}_{\mathbf{Z}-3\mathrm{M}} = \begin{bmatrix} 1.0 & 0.344 & -0.241 \\ 0.344 & 1.0 & -0.766 \\ -0.241 & -0.766 & 1.0 \end{bmatrix}$$

Then, the performance function Eq. (6.64) can be expressed as a function $g(\mathbf{U})$ of independent standard normal variables:

$$g_{3M}(\mathbf{U}) = 3.63 + 0.90U_1 + 0.04U_1^2 - 0.36U_1U_2 + 0.12U_1U_3 - 0.19U_2$$
$$- 0.51U_2^2 + 0.34U_2U_3 - 0.56U_3 - 0.16U_3^2 \tag{6.66}$$

where U_1, U_2, and U_3 are independent standard normal variables.

For the function $g(\mathbf{U})$ in Eq. (6.66), the reliability index can be calculated using the general FORM. The starting point is selected at $\mathbf{U} = (0, 0, 0)^T$ and convergence is achieved as $\beta_1 = 2.06$ at the design point $\mathbf{u}^* = (-1.0374, -1.4431, 1.0428)^T$ with direction cosine of $\alpha_1 = (0.5035, 0.7003, -0.5061)$. A second local minimum-distance point is obtained if one started the iteration at $\mathbf{U} = (1,1,1)^T$ and the result is $\beta_2 = 2.46$ at the design point $\mathbf{u}^* = (0.02140, 2.4407, -0.3330)^T$ with direction cosine of $\alpha_2 = (-0.008687, -0.9908, 0.1352)$. In order to obtain the general reliability index, using Eq. (5.14) in the chapter on Methods of Moment Based on First- and Second-Order Transformation, the first-order approximation of the performance function Eq. (6.66) at the two design points are given as

$$g_{L-1}(\mathbf{U}) = \beta_1 + \mathbf{U}^T\alpha_1 = 2.06 + 0.5035U_1 + 0.7003U_2 - 0.5061U_3 \tag{6.67a}$$

$$g_{L-2}(\mathbf{U}) = \beta_2 + \mathbf{U}^T\alpha_2 = 2.46 - 0.008687U_1 - 0.9908U_2 + 0.1352U_3 \tag{6.67b}$$

Using Eq. (8.24) in the chapter on System Reliability Assessment, the correlation coefficient between g_{L-1} and g_{L-2} is given as

$$\rho_{g_{L-1},g_{L-2}} = \frac{\alpha_1^T\alpha_2}{|\alpha_1||\alpha_2|} = -0.7666$$

Since β_1 is close to β_2, the joint failure probability can be approximated by Eq. (8.30) in the same chapter just mentioned,

$$P_{f1,2} = \left[\Phi(-\beta_1)\Phi\left(-\frac{\beta_2 - \rho_{1,2}\beta_1}{\sqrt{1-\rho_{1,2}^2}}\right) + \Phi(-\beta_2)\Phi\left(-\frac{\beta_1 - \rho_{1,2}\beta_2}{\sqrt{1-\rho_{1,2}^2}}\right)\right]\left[1 - \arccos\left(\rho_{1,2}\right)/\pi\right]$$

$$= 1.2519 \times 10^{-12}$$

According to Figure 8.14a in the same chapter, the failure probability of system is obtained as

$$P_{f,sys} = \Phi(-\beta_1) + \Phi(-\beta_2) - P_{f1,2} = 0.02655$$

The generalised reliability index is then obtained as $\beta_g = 1.934$.

The influence of different transformation order of X_1, X_2, and X_3 on reliability analysis has also been investigated, and the results are listed in Table 6.15. It is shown that the results are independent of the transformation order even for such a multiple design point problem.

Table 6.15 Results of different transformation orders using the third-moment pseudo normal transformation.

Order of X	Design point $(u_1{}^*, u_2{}^*, u_3{}^*)$	Design point $(x_1{}^*, x_2{}^*, x_3{}^*)$	β	β_g
$X_1 \rightarrow X_2 \rightarrow X_3$	(−1.0372, −1.4444, 1.0412) (0.02193, 2.4403, −0.3361)	(3.9822, 0.2424, 3.7398) (4.9206, 4.5355, 0.3851)	2.0606 2.4634	1.9341
$X_1 \rightarrow X_3 \rightarrow X_2$	(−1.0372, 1.7720, −0.1746) (0.02193, −2.0524, 1.3621)	(3.9822, 0.2424, 3.7398) (4.9206, 4.5355, 0.3851)	2.0606 2.4634	1.9341
$X_2 \rightarrow X_1 \rightarrow X_3$	(−1.7130, −0.4772, 1.0412) (2.2990, −0.8186, −0.3362)	(3.9822, 0.2424, 3.7398) (4.9206, 4.5355, 0.3851)	2.0606 2.4634	1.9341
$X_2 \rightarrow X_3 \rightarrow X_1$	(−1.7130, 1.0229, −0.5152) (2.2990, −0.3661, −0.8057)	(3.9822, 0.2424, 3.7398) (4.9206, 4.5355, 0.3851)	2.0606 2.4634	1.9341
$X_3 \rightarrow X_1 \rightarrow X_2$	(1.9699, −0.5789, −0.1746) (−1.9971, −0.4741, 1.3621)	(3.9822, 0.2424, 3.7398) (4.9206, 4.5355, 0.3851)	2.0606 2.4634	1.9341
$X_3 \rightarrow X_2 \rightarrow X_1$	(1.9699, −0.3164, −0.5152) (−1.9971, 1.1962, −0.8057)	(3.9822, 0.2424, 3.7398) (4.9206, 4.5355, 0.3851)	2.0606 2.4634	1.9341

Using the method of MCS with 1 000 000 samples, the probability of failure P_f is obtained as 0.02924 (the coefficients of variation of P_f is 0.58%) with corresponding reliability index of $\beta_{MCS} = 1.892$. Based on the point estimate method, the first four central moments of Eq. (6.66) are obtained as $\mu_Z = 3$, $\sigma_Z = 1.4116$, $\alpha_{3Z} = -0.7507$, and $\alpha_{4Z} = 5.1317$. Therefore, the third-moment and fourth-moment reliability indices are obtained for Eq. (6.66) as $\beta_{3M} = 1.868$ and $\beta_{4M} = 1.876$, respectively.

Using the fourth-moment pseudo transformation, the correlation matrix of correlated fourth-moment pseudo standard normal variables, $\mathbf{C}_{Z-4M}$, can then be written as,

$$\mathbf{C}_{Z-4M} = \begin{bmatrix} 1.0 & 0.331 & -0.229 \\ 0.331 & 1.0 & -0.583 \\ -0.229 & -0.583 & 1.0 \end{bmatrix}$$

And the performance function Eq. (6.64) can be expressed as a function g(U) of independent standard normal variables:

$$\begin{aligned} g_{4M}(\mathbf{U}) = {} & 3.53 + 0.89U_1 + 0.05U_1^2 + 0.01U_1^3 - 0.34U_2 - 0.34U_2^2 - 0.02U_2^3 - 0.67U_3 - 0.21U_3^2 \\ & - 0.01U_3^3 - 0.27U_1U_2 + 0.12U_1U_3 + 0.27U_2U_3 - 0.01U_1^2U_2 - 0.02U_1U_{23}^2 \\ & - 0.003U_1^2U_3 + 0.016U_2^2U_3 + 0.01U_1U_3^2 + 0.024U_2U_3^2 - 0.014U_1U_2U_3 \end{aligned}$$

(6.68)

For the function g(U) in Eq. (6.68), the reliability index can be calculated using the general FORM. The starting point is selected at $\mathbf{U} = (0, 0, 0)^T$ and convergence is achieved with $\beta_1 = 2.13$. A second local minimum-distance point is obtained if one starts the iteration at $\mathbf{U} = (1,1,1)^T$ and the result is $\beta_2 = 2.55$. Using the similar method in the procedure of the fourth-moment pseudo transformation, the generalised reliability index is $\beta_g = 2.017$. Similar investigation has shown that the results are independent of the transformation order when using the fourth-moment pseudo transformation.

Table 6.16 Comparison of results using third- and fourth-moment pseudo normal transformations.

| | | | Direct moment method | | |
	Reliability method	FORM	Third-moment	Fourth-moment	MCS
Reliability index	Third-moment pseudo normal transformation	1.934	1.868	1.876	1.892
	Fourth-moment pseudo normal transformation	2.017	1.932	1.927	1.931

Using the method of MCS with 1 000 000 samples, the probability of failure P_f is obtained as 0.02677 (the coefficients of variation of P_f is 0.60%) with corresponding reliability index of $\beta_{MCS} = 1.931$. Based on the point estimate method, the first four central moments of Eq. (6.68) are obtained as $\mu_Z = 3$, $\sigma_Z = 1.4143$, $\alpha_{3Z} = -0.4397$, and $\alpha_{4Z} = 3.7502$. Therefore, the third-moment reliability index and fourth-moment reliability index for Eq. (6.68) are obtained as $\beta_{3M} = 1.932$ and $\beta_{4M} = 1.927$, respectively.

Obviously, both moment methods and MCS for predicting the reliability indices of Eqs. (6.66) and (6.68) do not have issues when applying to problem with multiple design points.

The results obtained using FORM, direct moment method, and MCS based on the third- and fourth-moment pseudo normal transformations are summarised in Table 6.16. As can be observed from Table 6.16, generally, the results are almost the same for these methods.

Example 6.20 Consider a slope shown in Figure 6.8, and the performance function for safety of the slope is expressed as (Baecher and Christian 2003; Low 2007):

$$G(\mathbf{X}) = \frac{cA + N' \tan \phi}{W\left(\sin \psi_p + \alpha \cos \psi_p\right) + V \cos \psi_p - T \sin \theta} - 1 \tag{6.69}$$

in which

$$A = (H - z)/\sin \psi_p \tag{6.70a}$$

$$z = H\left(1 - \sqrt{\cot \psi_f \tan \psi_p}\right) \tag{6.70b}$$

$$N' = W\left(\cos \psi_p - \alpha \sin \psi_p\right) - U - V \sin \psi_p + T \cos \theta \tag{6.70c}$$

$$W = 0.5\gamma H^2 \left\{\left[1 - (z/H)^2\right] \cot \psi_p - \cot \psi_f\right\} \tag{6.70d}$$

$$U = 0.5\gamma_w \cdot r \cdot z \cdot A \tag{6.70e}$$

$$r = z_w/z \tag{6.70f}$$

$$V = 0.5\gamma_w r^2 z^2 \tag{6.70g}$$

Table 6.17 Probability distributions of random variables and their first three moments for Example 6.20.

Variables	Distribution	Mean	Standard deviation	Skewness	Kurtosis
c	Lognormal	140 kPa	28 kPa	0.6080	3.66439
ϕ	Lognormal	35°	5°	0.4315	3.33283
r	Truncated exponential (0, 1)	0.3435	0.2626	0.6800	2.45052
z	Lognormal	14 m²	3 m²	0.6527	3.76691
α	Truncated exponential (0, 0.16)	0.05496	0.04202	0.6800	2.45052

where $\gamma = 2.6 \times 10^4$ N/m³, $\gamma_W = 1.0 \times 10^4$ N/m³, $\psi_f = 50°$, $\psi_p = 35°$, $T = 0$, $\theta = 0$, $H = 60$ m, and c, ϕ, r, z, and α are random variables. Their probability distributions and first three moments are listed in Table 6.17.

If c, ϕ, r, z, and α are mutually independent, using the method of MCS with 1 000 000 samples, the probability of failure P_f is obtained as 8.877×10^{-3} (the coefficients of variation of P_f is 1.06%) with corresponding reliability index of $\beta_{MCS} = 2.37$. Using FOFM in pseudo normal space, the reliability index can be obtained as 2.40. Good agreement can be observed between reliability indices obtained by the FOFM in pseudo normal space and by the MCS method.

In engineering practice, c and ϕ, and r and z are often considered correlated. In this example, assume that c and ϕ, and r and z are correlated with correlation coefficient $\rho_{c,\phi} = -0.5$ and $\rho_{r,z} = -0.5$, respectively. Since the joint PDF is unknown, FORM based on Rosenblatt transformation cannot be applied; although FORM based on Nataf transformation is theoretically available because the marginal PDFs and correlation matrix are known. However, this method requires evaluations of the equivalent correlation coefficients in standard normal space, which involves solving two-dimensional nonlinear integral equations. Based on the fourth-moment pseudo normal transformations, the reliability indices obtained by FOFM in pseudo normal space and MCS are 2.951, and 3.054, respectively. It can be concluded that the FOFM in pseudo normal space is very simple and thus is convenient to be applied to structural reliability analysis.

Example 6.21 This example considers a two-story two-bay frame structure as shown in Figure 6.9a, in which M_i and S_i denote the member strength and load, respectively, and the probabilistic information of random variables of M_i and S_i are listed in Table 6.18. The most likely failure mode is shown Figure 6.9b and the corresponding performance function is (Zhao and Ono 1998)

$$G(\mathbf{X}) = 2M_1 + 2M_2 + 2M_3 - 4.5S_1 - 4.5S_2 \tag{6.71}$$

It is assumed that all M_i ($i = 1, 2, 3$) are independent of all S_i ($i = 1, 2$), and M_i are correlated with correlation coefficient $\rho_M = -0.1$ and S_i are correlated with correlation coefficient $\rho_S = 0.5$.

Table 6.18 The first four moments of the random variables for Example 6.21.

Variables	Mean	Standard deviation	Skewness	Kurtosis
M_1, M_2, M_3	700 kN·m	105 kN·m	0.453	3.36766
M_6, M_7	700 kN·m	105 kN·m	0.453	3.36766
M_4	1500 kN·m	225 kN·m	0.453	3.36766
M_5	1200 kN·m	180 kN·m	0.453	3.36766
M_8	90 kN·m	13.5 kN·m	0.453	3.36766
S_1	100 kN	25 kN	0.765	4.06007
S_2	200 kN	50 kN	0.765	4.06007
S_3	265 kN	66.3 kN	0.765	4.06007
S_4	180 kN	45 kN	0.765	4.06007
S_5	140 kN	35 kN	0.765	4.06007

For this example, only the first four moments and correlation matrix are available, so reliability analysis based on Rosenblatt transformation and the Nataf transformation cannot be used. Based on the fourth-moment pseudo normal transformations, the reliability indices obtained by FOFM in pseudo normal space and MCS are 5.016, and 5.026, respectively. It can be observed that the reliability analysis with unknown marginal PDFs can be conducted by using the fourth-moment pseudo normal transformations, which complements existing methods for reliability assessment including correlated variables with unknown probability distributions.

6.6 Monte Carlo Simulation Using Statistical Moments of Random Variables

When using MCS in structural reliability analysis, the basic random variables which represent uncertain quantities, such as loads, environmental factors, material properties, and structural dimensions are generally assumed to have known CDFs or PDFs, and the generation of random samples of these random variables are generally conducted based on the CDFs of the random variables. In practice, however, the PDF/CDF of some basic random variables is often unknown and the probabilistic characteristics of these variables are often expressed using only statistical moments. Under such circumstances, the MCS cannot be applied and strict evaluation of the probability of failure is not possible (Zhao et al. 2002b). Thus, an alternative measure is required.

For the cases that the first four statistical moments and correlation matrix of the basic random variables $\mathbf{X}$ are known, using the third- and fourth-moment pseudo transformations, i.e. Eq. (7.116) and Eq. (7.129) as previously mentioned, the performance function with correlated variables, $G(\mathbf{X})$, can be rewritten as

$$G(\mathbf{X}) = G(X_1, X_2, \cdots, X_n) = g(U_1, U_2, \cdots, U_n) = g(\mathbf{U}) \tag{6.72}$$

where $g(\mathbf{U})$ is a function of only independent standard normal random variables.

Therefore, MCS method for the evaluation of the probability of failure using statistical moments of random variables can be conducted, and the computation procedure is summarised as follows.

1) Obtain the first three or four moments of each variable and original correlation matrix by the given probability information (e.g. the joint PDF or the marginal PDFs and correlation matrix or statistical data).
2) Determine polynomial coefficients using Eq. (7.45) in the chapter on Transformation of Non-Normal Variables to Independent Normal Variables for third-moment pseudo transformations and Eq. (7.72) in the same chapter for fourth-moment pseudo transformations. Obtain the equivalent correlation matrix $\mathbf{C_z}$ using Eq. (7.105) with Table 7.12 in the same chapter for third-moment pseudo transformations and with Algorithm 7.1, also in the same chapter, for fourth-moment pseudo transformations. Then compute lower triangular matrix $\mathbf{L_0}$ using Cholesky decomposition using Eq. (7.106), in the same chapter.
3) Set $j = 1$;
4) Generate N groups of random samples from independent standard normal space, $\{u_{im}\}$, where $i = 1, 2, ..., n$ and $m = 1, 2, ..., N$; obtain the random samples of $\mathbf{X}$, $\{x_{im}\}$, using Eq. (7.116) for third-moment pseudo transformations and Eq. (7.129) in the above-mentioned chapter for fourth-moment pseudo transformations. Calculate the corresponding values of the performance function of Eq. (6.72). Note that the general expression of $g(\mathbf{U})$ is not necessary;
5) Evaluate the failure probability P_f by $P_f = N_f/(j \times N)$, in which N_f is the total number of performance function values lesser than zero; and then obtain the coefficient of variation (COV) of P_f, V_P, using $V_P = \sqrt{(1 - P_f)/N_f}$; and
6) Output P_f and V_P if $V_P \leq 5.0\%$. Otherwise, replace j by $j + 1$ and return to Step 4.

For a single random variable and the performance functions including only independent random variables, Eq. (7.44) for third-moment pseudo transformations and Eq. (7.70) for fourth-moment pseudo transformations in the same chapter can be used for generating random samples, and therefore MCS can be realised even when the probability distributions of the basic random variables are unknown.

Example 6.22   Random Samples Generation for some Commonly Used Random Variables

To investigate the efficiency of the method for the case with only the first three known moments, the random numbers of the six commonly used random variables (i.e. Gamma, Lognormal, Weibull, Gumbel, and Rayleigh distributions) are generated using only their first three moments. The probabilistic information and the first three moments of the six cases are listed in Table 6.19, and (with equations from the chapter on Transformation of Non-Normal Variables to Independent Normal Variables) using the first three moments and, Eq. (7.45), the three parameters a_1, a_2, and a_3 in Eq. (7.44) are obtained in Table 6.19. For the number of sampling of 40 000, the histograms obtained using Eq. (7.44) from the same chapter with the aid of the random samples of u are shown in Figure 6.13a–f for the six cases, in which the thick solid lines indicate the PDF exactly obtained from the

Table 6.19 Corresponding distributions and parameters of Figure 6.13.

		Moments			Parameters		
	Case distribution	Mean	COV	α_3	a_1	a_2	a_3
(a)	Gamma	100	0.3	0.6	−0.1007	0.9898	0.1007
(b)	Lognormal	100	0.3	0.927	−0.1571	0.9750	0.1571
(c)	Weibull	100	0.2	−0.3519	0.0588	0.9965	−0.0588
(d)	Weibull	100	0.6	0.8496	−0.1436	0.9792	0.1436
(e)	Gumbel	100	0.4	1.1396	−0.1949	0.9613	0.1949
(f)	Rayleigh	100	0.523	0.6311	−0.1060	0.9887	0.1060

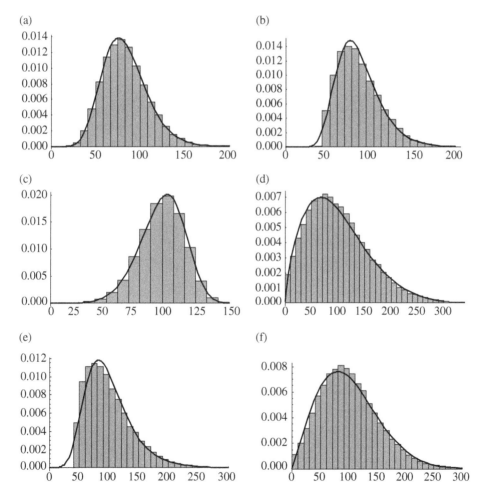

Figure 6.13 Random number generation for some commonly used distributions using the first three moments.

six distributions. In Figure 6.13, the histograms have been scaled using the following equation in order to compare with the exact PDF.

$$h_i = H_i/\Delta x \tag{6.73}$$

where H_i is the relative frequency, Δx is the category interval of the histogram, and h_i is the scaled relative frequency.

From Figure 6.13, it can be observed that the histograms obtained using the same Eq. (7.44) as above generally fit the exact PDF very well for all the six distributions. Considering the fact that only the information of first three moments are used in Eq. (7.44), it can be concluded that Eq. (7.44) can be used to generate the random numbers of a random variable in the case that only the first three moments are known.

The percentile points of 1, 5, 10, 25, 50, 75, 90, 95, and 99% evaluated using Eq. (7.44) for the six cases are listed in Table 6.20 along with the exact results obtained directly using the CDFs of the six distributions. Table 6.20 reveals that although only the first three moments are used in Eq. (7.44), it provides good approximations of the exact percentile points for all the six cases at all the percent levels that are larger than 5%. For cases b, d, e, and f at percent level 1%, the percentile points obtained from Eq. (7.44) have significant errors since only the first three moments are used. Therefore, Eq. (7.44) is only recommended for the case that the information of only the first three moments are known.

Furthermore, to investigate the efficiency of generating random samples using the first four moments, six commonly used random variables that have known CDFs are selected. The first four moments of the six cases are listed in Table 6.21, and with equations from the same chapter mentioned previously using the first four moments and Eq. Eqs.(7.72a–f), the four parameters a_1, a_2, a_3, and a_4 in Eq. (7.70) are obtained in Table 6.21. After the parameters a_1, a_2, a_3, and a_4 are obtained, the random numbers of the six distributions can be easily generated using Eq. (7.70) with the aid of the random numbers of u. The histograms obtained using Eq. (7.70) are shown in Figure 6.14a–f for the six cases, in which the thick solid lines indicate the PDF exactly obtained from the six distributions. In Figure 6.14, the number of sampling is taken to be 40 000 for all the cases, and the histograms have been scaled using Eq. (7.70).

From Figure 6.14, it can be observed that the histograms obtained using Eq. (7.70) fit the exact PDF very well for all the six cases. This implies that random number generation method using Eq. (7.70) is suitable and accurate.

According to Eq. (7.70), the percentile point of a distribution with first four moments corresponding to a percentage value α can be explicitly evaluated as

$$x_\alpha = \sigma\left(a_1 + a_2 u_\alpha + a_3 u_\alpha^2 + a_4 u_\alpha^3\right) + \mu \tag{6.74}$$

where u_α is the percentile point of a standard normal random variable.

The percentile points of 1, 5, 10, 25, 50, 75, 90, 95, and 99% evaluated using Eq. (6.74) are listed in Table 6.21 for the six distributions along with the exact results obtained directly using the CDFs of the six distributions. Table 6.22 reveals that although only the first four moments are used in Eq. (7.70), it provides good approximations of the exact percentile points for all the percent levels and all the six cases.

Table 6.20 Percentile point evaluation using first three moments.*

Case %	Gamma (a) Eq. (7.44)	Gamma (a) exact	Lognormal (b) Eq. (7.44)	Lognormal (b) exact	Weibull (c) Eq. (7.44)	Weibull (c) exact	Weibull (d) Eq. (7.44)	Weibull (d) exact	Gumbel (e) Eq. (7.44)	Gumbel (e) exact	Rayleigh (f) Eq. (7.44)	Rayleigh (f) exact
1	44.25	43.59	52.74	48.38	48.45	48.84	1.332	7.697	44.94	34.37	4.211	11.31
5	56.31	56.27	59.92	59.10	65.21	64.70	18.06	19.89	50.05	47.78	24.44	25.56
10	63.89	63.99	65.54	65.75	73.70	73.25	30.24	30.24	55.73	55.99	37.33	36.63
25	78.33	78.48	77.70	78.58	87.20	87.11	55.68	54.29	69.82	71.81	62.12	60.52
50	96.98	97.02	95.29	95.78	101.2	101.4	91.39	90.59	92.21	93.43	94.46	93.94
75	118.4	118.3	117.2	116.8	114.1	114.3	134.9	135.7	121.7	120.9	131.9	132.9
90	140.0	139.9	140.5	139.5	124.8	124.7	180.8	182.3	154.3	152.2	169.8	171.2
95	154.0	153.9	156.2	155.2	130.8	130.5	211.3	212.5	176.5	174.6	194.5	195.3
99	182.4	182.7	188.8	189.6	141.2	140.5	274.7	272.9	223.8	225.5	244.7	242.1

*Using the equation from the chapter on Transformation of Non-Normal Variables to Independent Normal Variables.

Table 6.21 Corresponding distributions and parameters of Figure 6.14.

Case distribution	Moments				Parameters			
	Mean	COV	α_3	α_4	a_1	a_2	a_3	a_4
(a) Gamma	100	0.3	0.6	3.54	−0.0992	0.9827	0.0992	0.00245
(b) Lognormal	100	0.3	0.927	4.5659	−0.1426	0.9308	0.1426	0.01594
(c) Weibull	100	0.2	−0.3519	3.0039	0.0614	1.0178	−0.0614	−0.00726
(d) Weibull	100	0.6	0.8496	3.7320	−0.1543	1.0105	0.1543	−0.01167
(e) Gumbel	100	0.4	1.1396	5.4	−0.1683	0.8969	0.1683	0.02418
(f) Rayleigh	100	0.523	0.6311	3.2451	−0.1156	1.0282	0.1156	−0.01407

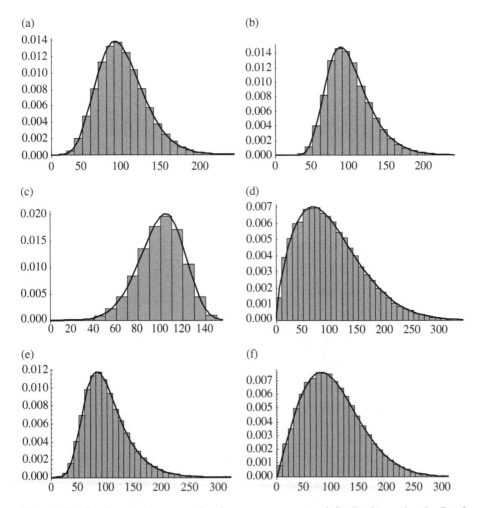

Figure 6.14 Random number generation for some common used distributions using the first four moments.

Table 6.22 Percentile point evaluation using first four moments.[*]

Case %	Gamma (a)		Lognormal (b)		Weibull (c)		Weibull (d)		Gumbel (e)		Rayleigh (f)	
	Eq. (7.70)	exact	Eq. (7.70)	exact	Eq. (7.70)	exact	Eq. (7.70)	exact	Eq. (7.70)	exact	Eq. (7.70)	exact
1	43.62	43.59	47.90	48.38	49.05	48.84	8.630	7.697	34.05	34.37	10.88	11.31
5	56.25	56.27	59.24	59.10	65.07	64.70	19.18	19.89	48.16	47.78	25.17	25.56
10	63.97	63.99	65.95	65.75	73.43	73.25	29.72	30.24	56.31	55.99	36.55	36.63
25	78.47	78.48	78.69	78.58	86.98	87.11	54.27	54.29	71.84	71.81	60.68	60.52
50	97.02	97.02	95.72	95.78	101.2	101.4	90.74	90.59	93.27	93.43	93.96	93.94
75	118.3	118.3	116.6	116.8	114.4	114.3	135.6	135.7	120.8	120.9	132.7	132.9
90	139.8	139.9	139.5	139.5	125.0	124.7	182.2	182.3	152.3	152.2	171.2	171.2
95	153.9	153.9	155.4	155.2	130.7	130.5	212.4	212.5	174.8	174.6	195.4	195.3
99	182.6	182.7	189.9	189.6	140.1	140.5	273.1	272.9	225.3	225.5	242.4	242.1

[*]Using the equation from the chapter on Transformation of Non-Normal Variables to Independent Normal Variables.

Example 6.23   Statistical Properties of the Random Samples Generated by the Fourth-Moment Pseudo Transformations for Correlated Random Variables

This example discusses the statistical correspondence between the random samples generated by the fourth-moment pseudo transformations and the assigned random variables, four cases including normal, lognormal, Gumbel, Weibull, and Gamma distributions are considered and listed in Table 6.23, together with the first four central moments of these distributions. Assume that the correlation coefficient between X_1 and X_2 is $\rho_{12} = 0.5$. Based on the first four central moments and Eq. (7.72) in the chapter on Transformation of Non-Normal Variables to Independent Normal Variables, the polynomial coefficients in Eq. (7.129) can be determined and are listed in Table 6.24. According to Algorithm 7.1 in the chapter mentioned, the equivalent correlation coefficients between Z_1 and Z_2, i.e. ρ_{012}, for Case 1 to 4 are determined to be 0.5049, 0.5170, 0.5348, and 0.5041, respectively. The lower triangular matrix $\mathbf{L}_0$ can be determined from the Cholesky decomposition of the correlation matrix of $\mathbf{Z}$. From the preceding computation, the polynomial coefficients in Eq. (7.129) are determined. The random samples of X_1 and X_2 can be generated using those of independent standard normal random variables. The scatters and histograms of X_1 and X_2 for all cases

Table 6.23   Probability distributions and first four central moments of random variables considered in Example 6.23.

	Variable X_1				Variable X_2				
Case	distribution	mean	Standard deviation	skewness	kurtosis	distribution	mean	Standard deviation	skewness	kurtosis
1	normal	100	20	0	3	lognormal	50	10	0.608	3.664
2	lognormal	100	40	1.264	5.969	Gumbel	50	25	1.14	5.4
3	Weibull	100	10	−0.715	3.780	Gumbel	50	15	1.14	5.4
4	Gamma	100	30	0.6	3.540	Weibull	50	20	0.277	2.788

Table 6.24   Probability distributions and first four central moments of random variables considered in Example 6.23.

	Variable X_1				Variable X_2			
Case	a_1	b_1	c_1	d_1	a_2	b_2	c_2	d_2
1	0.0	1.0	0.0	0.0	−0.0979	0.9696	0.0979	0.0069
2	−0.1818	0.8753	0.1818	0.0294	−0.1683	0.8969	0.1683	0.0242
3	0.1175	0.9741	−0.1175	0.004	−0.1683	0.8969	0.1683	0.0242
4	−0.0992	0.9827	0.0992	0.0025	−0.0507	1.0420	0.0507	−0.0151

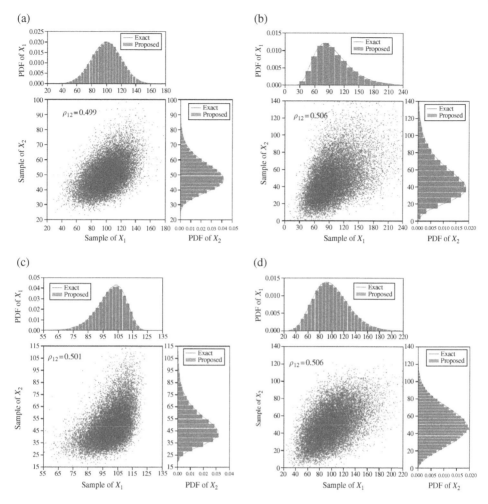

Figure 6.15 Scatters and histograms for Example 6.23.

are depicted in Figure 6.15a–d, together with the given marginal PDFs for comparison. The correlation coefficients obtained from the random samples are also depicted in Figure 6.15. The sample size is 20 000 for all cases. From Figure 6.15, it can be observed that, for all cases, the histograms obtained from the random samples generated using the fourth-moment pseudo transformations are in close agreement with the given marginal PDFs, and the correlation coefficients of the random samples are almost identical with the assigned correlation coefficients. Therefore, the fourth-moment pseudo transformations can be applied to generate the random samples of correlated random variables.

Another question that needs to be considered is whether the statistical properties of the random samples generated by the fourth-moment pseudo transformation are always identical with those of assigned random variables for different correlation coefficients. Here, different correlation coefficients are considered; both negative and positive. The comparison between the assigned correlation coefficient and that obtained from the random samples is

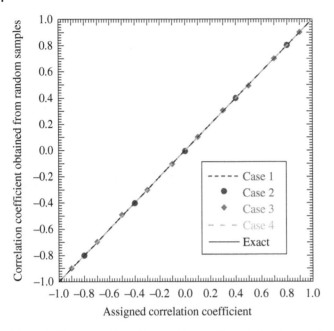

Figure 6.16 Comparison of correlation coefficients for Example 6.23.

depicted in Figure 6.16, where it can be observed that the correlation coefficients obtained from the random samples are approximately equal to those assigned over the entire investigation range. Then, the Kolmogorov-Smirnov (KS) test is used to examine the distributions of random samples. With 1000 simulation runs and 10 000 random samples generated in each run, the average significance probabilities of the KS test are determined and listed in Table 6.25 for the correlation coefficients of X_1 and X_2 being equal to −0.5, 0.0, and 0.5, respectively. From Table 6.25, it can be observed that the average significant probabilities of KS test (the minimum value is 47.45%) are much greater than the generally used rejection levels of 1.0–5.0%; and there are no significant differences for the different correlation coefficients. Therefore, the fourth-moment pseudo transformation is always efficient for different correlation coefficients.

Table 6.25 Significance probability of KS test for Example 6.23.

		Significance probability			
Correlation coefficient	Variable	Case 1	Case 2	Case 3	Case 4
0.5	X_1	51.19%	46.69%	50.30%	50.14%
	X_2	50.18%	50.53%	51.23%	49.22%
0.0	X_1	51.17%	48.68%	48.40%	51.48%
	X_2	49.78%	50.07%	50.90%	50.82%
−0.5	X_1	50.29%	48.71%	49.31%	49.20%
	X_2	51.61%	50.79%	49.98%	47.45%

Example 6.24   Reliability Analysis of a Column

This example considers the following performance function, a simple compressive limit state of a structural column.

$$G(\mathbf{X}) = AX_1X_2 - X_3 \tag{6.75}$$

where A is the nominal section area, X_1 is a random variable presenting the uncertainty included in A, X_2 is yield stress, and X_3 is a compressive load. The column is made of H-shaped structural steel with a section area of $A = 72.38 \text{ cm}^2$. The CDFs of X_1 and X_2 are unknown, the only information about them are their first four moments (Ono et al. 1986), i.e. $\mu_1 = 0.990$, $\sigma_1 = 0.051$, $\alpha_{31} = 0.709$, $\alpha_{41} = 3.692$, $\mu_2 = 3.055 \text{ t/cm}^2$, $\sigma_2 = 0.364$ t/cm^2, $\alpha_{32} = 0.512$, $\alpha_{42} = 3.957$. X_3 is assumed as a lognormal variable with mean $\mu_3 = 100 \text{ t}$ and standard deviation $\sigma_4 = 40 \text{ t}$.

Although the CDFs of X_1 and X_2 are unknown, using Eq. (7.70) as previously, the random sampling of X_1 and X_2 can be easily generated without using their CDFs, and the MCS can be thus easily conducted. The probability of failure is obtained as $P_f = 0.0188$ $(\beta = 2.079)$ when the number of samplings is taken to be 10 000.

Example 6.25   This example considers the following performance function as discussed in Example 6.7:

$$G(\mathbf{X}) = 18 - 3X_1 - 2X_2 \tag{6.76}$$

where the joint PDF of X_1 and X_2 is

$$f_{X_1X_2}(\mathbf{x}) = [(x_1 + x_2 + x_1x_2) - 0.5] \cdot \exp(-x_1 - x_2 - x_1x_2), \text{for } x_1 \geq 0; x_2 \geq 0 \tag{6.77}$$

As given in Example 6.7, the exact probability of failure and reliability index are 2.94486×10^{-3} and 2.7539, respectively. According to the joint PDF of X_1 and X_2, the marginal PDFs of X_1 and X_2 can be obtained as:

$$f_{X_1}(x_1) = \exp(-x_1), \quad \text{for} \quad x_1 > 0; f_{X_2}(x_2) = \exp(-x_2), \quad \text{for} \quad x_2 > 0 \tag{6.78a}$$

and the correlation coefficient is $\rho_{12} = -0.404$. The first four moments of X_1 and X_2 can be obtained as:

$$\mu_{X_1} = \mu_{X_2} = 1.0; \quad \sigma_{X_1} = \sigma_{X_2} = 1.0; \alpha_{3X_1} = \alpha_{3X_2} = 2.0; \quad \alpha_{4X_1} = \alpha_{4X_2} = 9.0 \tag{6.78b}$$

Again with the aid of Eq. (7.129), the random numbers of the correlated random variables X_1 and X_2 can be generated by the random numbers of independent standard normal random variables. Figure 6.17 shows the histograms of random samples of X_1 and X_2 obtained using Eq. (7.129) with the aid of the random samples of independent standard random variables with 1 000 000 samples, in which the thick solid lines indicate the marginal PDFs expressed in Eq. (6.78a). The correlation coefficient of these random numbers of X_1 and X_2 is obtained as -0.4028, which is in good agreement with the correlation coefficient obtained using the joint PDF. Substituting these random numbers into performance

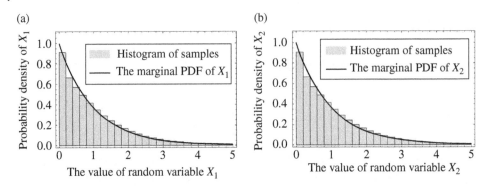

Figure 6.17 The histograms of random numbers of X_1 and X_2.

function, the probability of failure is obtained as 2.595×10^{-3}, and the corresponding reliability index is $\beta_{MCS} = 2.795$, which is almost the same with the exact value.

Figure 6.17a,b shows that the histogram obtained from Eq. (7.129) are in good agreement with the marginal PDFs obtained from the joint PDFs. Eq. (7.129) is shown to enable the generation of the random samples of the correlated random variables with the joint PDF known.

Example 6.26 This example considers an elasto-plastic truss structure in Figure 6.18, where T_i and F_i denote the member strength and load, respectively. The statistical information of random variables T_i and F_i is listed in Table 6.26. The performance functions corresponding to the eight most likely failure modes are given in Eqs. (6.79a)–(6.79d) (Zhao and Ang 2003).

$$g_1 = 0.7071T_4 + 0.7071T_5 - 2.2F_1; \quad g_2 = T_6 + 0.7071T_{10} - 1.2F_1 - F_2 \tag{6.79a}$$

$$g_3 = T_3 + 0.7071T_5 + 0.7071T_{10} - 2.2F_1; \quad g_4 = T_8 + 0.7071T_{10} - 1.2F_1 \tag{6.79b}$$

$$g_5 = T_6 + T_7 - 1.2F_1; \quad g_6 = T_3 + 0.7071T_5 - 1.2F_1 - F_2 \tag{6.79c}$$

$$g_7 = 0.7071T_9 + 0.7071T_{10} - 1.2F_1; \quad g_8 = T_1 + 0.7071T_5 - 3.4F_1 - F_2 \tag{6.79d}$$

The performance function of the system can be defined as the minimum of the above (Zhao and Ang 2003):

$$G(\mathbf{X}) = \min\{g_1, g_2, \cdots, g_8\} \tag{6.80}$$

The correlation matrix of random variables is given as:

$$\mathbf{C_X} = \begin{array}{c} F_1 \\ F_2 \\ T_1 \\ T_2 \\ \vdots \\ T_{10} \end{array} \begin{pmatrix} 1 & & & & & \\ 0.1 & 1 & & & & \\ 0 & 0 & 1 & & & \\ 0 & 0 & 0.5 & 1 & & \\ \vdots & \vdots & \vdots & \ddots & \ddots & \\ 0 & 0 & 0.5 & 0.5 & \cdots & 1 \end{pmatrix} \tag{6.81}$$

Figure 6.18 Two-story one-bay truss structure in Example 6.26.

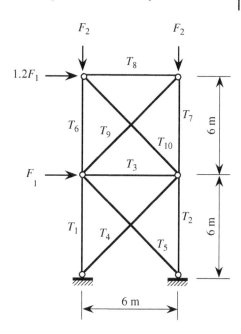

Table 6.26 Probabilistic information of the basic random variables for Example 6.26.

Variable	Mean	Standard deviation	Skewness	Kurtosis
T_1, T_2	90 kN	9.0 kN	0.301	3.1615
T_3	9 kN	0.9 kN	0.301	3.1615
T_4, T_5	48 kN	4.8 kN	0.301	3.1615
T_6, T_7	21 kN	2.1 kN	0.301	3.1615
T_8	15 kN	1.5 kN	0.301	3.1615
T_9, T_{10}	30 kN	3.0 kN	0.301	3.1615
F_1	11 kN	2.2 kN	1.1396	5.40
F_2	3.6 kN	0.72 kN	1.1396	5.40

For this example, only the first four moments and the correlation matrix are available, so reliability analysis based on Rosenblatt transformation and Nataf transformation cannot be used. However, the reliability index of the performance function $G(\mathbf{X})$ in Eq. (6.80) can be easily obtained using the MCS method based on the fourth-moment pseudo transformation. The resulted failure probability P_f with 10 000 000 samples is 4.50×10^{-5} (the COV of the MCS result is 4.71%) with the corresponding reliability index of 3.92. It can be observed that the reliability analysis can be achieved by using the fourth-moment pseudo transformation for the performance function that includes correlated random variables with unknown distributions.

6.7 Subset Simulation Using Statistical Moments of Random Variables

The Subset Simulation (SS) method is a useful tool for improving the computational efficiency of MCS, especially in evaluating the small failure probabilities (Au and Beck 2001a; Au and Beck 2003; Du et al. 2019). The basic idea of SS is to express the small failure probability as a product of a sequence of large conditional probabilities. Based on SS, the failure probability is expressed as (Au and Beck 2001a):

$$P_f = P(F_1) \prod_{j=1}^{M-1} P(F_{j+1} \mid F_j) \tag{6.82}$$

where $F_j = [G(\mathbf{X}) \leq b_j], j = 1, 2, ..., M$, is the intermediate failure event, and $b_1 > b_2 > ... > b_M = 0$ is a decreasing sequence of the threshold values of failure events. Based on Eq. (6.72), the intermediate failure event F_j can be rewritten as $F_j = [g(\mathbf{U}) \leq b_j]$. It should note that the general expression of $g(\mathbf{U})$ is not necessary because the relationship between $\mathbf{X}$ and $\mathbf{U}$ is explicitly expressed in third- and fourth-moment pseudo transformations, i.e. the previously used Eq. (7.116) and Eq. (7.129). The probability $P(F_1)$ can be determined by direct MCS given in Section 6.6. To evaluate the conditional probability $P(F_{j+1} \mid F_j)$, one must generate the random samples from the conditional PDF $q(\mathbf{u} \mid F_j)$, which is expressed as:

$$q(\mathbf{u} \mid F_j) = \frac{\phi_n(\mathbf{u}) I_{F_j}(\mathbf{u})}{P(F_j)} \tag{6.83}$$

where $\phi_n(\mathbf{u}) = \prod_{i=1}^{n} \phi(u_i)$, $\phi(\cdot)$ is the joint PDF of independent standard normal random variables; $I_{F_j}(\mathbf{u})$ is the indicator function of F_j, $I_{F_j}(\mathbf{u}) = 1$ for $\mathbf{u} \in F_j$, and $I_{F_j}(\mathbf{u}) = 0$ for $\mathbf{u} \notin F_j$. The generation of the random samples from $q(\mathbf{u} \mid F_j)$ given in Eq. (6.83) can be performed using the modified Metropolis algorithm (Au and Beck 2001a) combining with Eq. (6.83). In general, the probability of each level, i.e. $P(F_1)$ and $P(F_{j+1} \mid F_j)$, is pre-established as p_0 (e.g. $p_0 = 0.1$).

Similar to the third- and fourth-moment pseudo transformations based MCS, the SS method for the evaluation of the probability of failure using statistical moments of random variables can also be conducted, and the computation procedure is summarised as follows.

1) Evaluate the first four central moments and the correlation matrix $\mathbf{C_X}$ of correlated random variables through the statistical information (e.g. the joint PDF or the marginal PDFs and correlation matrix, or statistical data);
2) Determine polynomial coefficients using (all from the chapter on Transformation of Non-Normal Variables to Independent Normal Variables) Eq. (7.45) for third-moment pseudo transformations and Eq. (7.72) for fourth-moment pseudo transformations. Obtain the equivalent correlation matrix $\mathbf{C_z}$ using Eq. (7.105) with Table 7.12 for third-moment pseudo transformations, and with Algorithm 7.1 for fourth-moment pseudo transformations. Then compute lower triangular matrix $\mathbf{L_0}$ using Cholesky decomposition using Eq. (7.106).

3) Generate N groups of random samples from n-dimensional independent standard normal space, $u_m^{(0)}$, where $m = 1, 2, ..., N$; obtain the random samples of $\mathbf{X}$, $x_m^{(0)}$, again using Eq. (7.116) for third-moment pseudo transformations and Eq. (7.129) for fourth-moment pseudo transformations; and calculate the corresponding values of the performance function $\{g_m^{(0)}\}$;

4) Sort $\{g_m^{(0)}\}$ in ascending order to give the list $\{\tilde{g}_m^{(0)}\}$, in which $\tilde{g}_1^{(0)} > \tilde{g}_2^{(0)} > ... > \tilde{g}_{Nc}^{(0)}$. Set $j = 1$; $b_j = \tilde{g}_{Nc}^{(j-1)}$, in which $N_c = p_0 N$;

5) Select seeds $u_m^{(j)}$ according to $\tilde{g}_m^{(j-1)}$, where $m = 1, 2, ..., N_c$;

6) For each seed $u_m^{(j)}$ ($m = 1, 2, ..., N_c$), generate $N_s = 1/p_0 - 1$ groups of random samples from the conditional PDF $q(\mathbf{u} \mid F_j)$ given in Eq. (6.83) using the modified Metropolis algorithm (Au and Beck 2001a) combining with Eq. (6.72), and calculate the corresponding values of performance function $\{g_m^{(j)}\}$. Note that the total number of random samples is also equal to N groups because of $N_c(N_s + 1) = N$;

7) Sort $\{g_m^{(j)}\}$ in ascending order to give the list $\{\tilde{g}_m^{(j)}\}$, in which $\tilde{g}_1^{(0)} > \tilde{g}_2^{(0)} > ... > \tilde{g}_{Nc}^{(0)}$ and obtain $b_{j+1} = \tilde{g}_{Nc}^{(j)}$; and

8) If $b_{j+1} > 0$, set $j = j + 1$ and repeat Steps 5 through 7. If $b_{j+1} \leq 0$, calculate the failure probability P_f by $P_f = p_0^{j+1} N_f / N$, in which N_f is the number of $\tilde{g}_m^{(j)} \leq 0$. The final j is $M-1$ in Eq.(6.82).

Example 6.27   Roof Truss Reliability Assessment with Available Joint PDF

To investigate the applicability of the developed MCS and SS for structural reliability assessment involving correlated random variables with the joint PDF available, this example considers a typical roof truss structure (Song et al. 2009), as shown in Figure 6.19. The truss top and compression bars are made of concrete, and the truss bottom and tension bars are made of steel. A uniformly distributed load q is applied to the truss, which can be transformed into

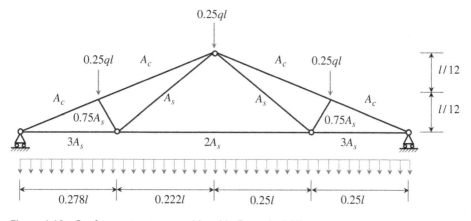

Figure 6.19 Roof truss structure considered in Example 6.27.

nodal loads $P = 0.25ql$. Assume that the failure is reached when the vertical displacement of the peak of structure node exceeds 3 cm. The performance function of the roof truss is explicitly expressed as (Song et al. 2009):

$$G(\mathbf{X}) = 0.03 - \frac{ql^2}{2}\left(\frac{3.81}{A_cE_c} + \frac{1.13}{A_sE_s}\right) \tag{6.84}$$

where q is the uniformly distributed load; l is the roof span; A_s and A_c are the cross-sectional areas of steel and concrete members, respectively; and E_s and E_c are the elastic modulus of steel and concrete members, respectively. The statistical distribution and the first four central moments of the basic random variables are listed in Table 6.27. Except for q and A_c, which are correlated with correlation coefficient $\rho = 0.25$, the other random variables are mutually independent.

Because all random variables are normally distributed, their joint PDF can be uniquely determined from the marginal PDFs and correlation matrix. Generally, based on the Metropolis-Hasting (MH) algorithm (Metropolis et al. 1953; Hastings 1970) and joint PDF, their random samples can be generated. For multivariate correlated normal random variables, Rosenblatt transformation (Hohenbichler and Rackwitz 1981) is a more convenient alternative method to generate such random samples. Here, Rosenblatt transformation is used, and 1.0×10^6 random samples are generated. The failure probability is obtained as $P_f = 6.310 \times 10^{-3}$ (the COV of P_f is 1.25%), and the corresponding reliability index is $\beta = 2.494$, which are considered as the exact values for comparison. The failure probability obtained by the fourth-moment pseudo transformations based MCS with 1.0×10^6 samples is $P_{f-\text{MCS}} = 6.291 \times 10^{-3}$ (the COV of P_f is 1.24%), and the corresponding reliability index is $\beta_{\text{MCS}} = 2.495$. Using the fourth-moment pseudo transformations based SS with $p_0 = 0.1$ and $N = 10000$ (the total number of random samples is 28 000), the failure probability is obtained as $P_{f-\text{SS}} = 6.157 \times 10^{-3}$, and the corresponding reliability index is $\beta_{\text{SS}} = 2.503$. It can be observed that for correlated multivariate normal distribution the results obtained from the fourth-moment pseudo transformations based MCS and SS are almost the same with those determined by MCS based on Rosenblatt transformation.

Assume that all random variables for this example are lognormally distributed, and the corresponding first four central moments are given in Table 6.28. The joint PDF of these random variables can also be uniquely determined from the marginal PDFs and

Table 6.27 Statistical distribution and first four moments of random variables for Example 6.27.

Variable	Distribution	Mean	Standard deviation	Skewness	Kurtosis
q (N/m)	Normal	20 000	5000	0.0	3.0
l (m)	Normal	12	0.6	0.0	3.0
A_s (m²)	Normal	9.82×10^{-4}	0.589×10^{-4}	0.0	3.0
A_c (m²)	Normal	400×10^{-4}	48×10^{-4}	0.0	3.0
E_s (N/m²)	Normal	1.0×10^{11}	5.0×10^9	0.0	3.0
E_c (N/m²)	Normal	2.0×10^{10}	2.0×10^9	0.0	3.0

Table 6.28 First four moments of random variables (lognormally distributed) for Example 6.27.

Variable	Distribution	Mean	Standard deviation	Skewness	Kurtosis
q (N/m)	Lognormal	5000	2500	1.625	8.035
l (m)	Lognormal	12	0.6	0.15	3.04
A_s (m^2)	Lognormal	9.82×10^{-4}	0.589×10^{-4}	0.18	3.058
A_c (m^2)	Lognormal	400×10^{-4}	48×10^{-4}	0.362	3.233
E_s (N/m^2)	Lognormal	1.0×10^{11}	5.0×10^{9}	0.15	3.04
E_c (N/m^2)	Lognormal	2.0×10^{10}	2.0×10^{9}	0.301	3.162

correlation matrix. The failure probability and reliability index obtained from MCS based on Rosenblatt transformation with 1.0×10^8 are $P_f = 8.5 \times 10^{-6}$ (the COV of P_f is 3.43%) and $\beta = 4.301$, respectively. The results obtained from the fourth-moment pseudo transformations based MCS with 1.0×10^8 samples and SS with $p_0 = 0.1$ and $N = 10000$ (the total number of samples is 55 000) are $P_{f-\text{MCS}} = 6.87 \times 10^{-6}$ (the COV of P_f is 3.82%), $\beta_{\text{MCS}} = 4.348$, and $P_{f-\text{SS}} = 7.837 \times 10^{-6}$, $\beta_{\text{SS}} = 4.319$, respectively. It can be observed that for correlated multivariate lognormal distribution the results obtained from the fourth-moment pseudo transformations based MCS and SS are almost the same with those determined by MCS based on Rosenblatt transformation.

This example reveals that when the joint PDF of correlated random variables is uniquely determined by their marginal PDFs and correlation matrix, the fourth-moment pseudo transformations based MCS and SS provide nearly identical results compared with MCS based on Rosenblatt transformation; and SS is much more efficient than MCS.

Example 6.28 Nonlinear RC Frame Structure Reliability Analysis with Known Marginal PDFs and Correlation Matrix

To investigate the applicability of the fourth-moment pseudo transformations based MCS and SS for structural reliability assessment involving correlated random variables with only the marginal PDFs and correlation matrix available, this example considers a three-bay seven-story nonlinear RC frame structure, as shown in Figure 6.20a. The total height of the frame structure is $H = 21.5$ m, wherein the height of the first story is 3.5 m and the individual heights of the second story to seventh story are 3.0 m. The span of each bay is 6.0 m. The sections and arrangement of the reinforcements of column and beam members are shown in Figures 6.20b–d. Here, the serviceability of this frame structure under lateral loads is considered. The failure is reached when the horizontal displacement of the upper floor exceeds the allowable limit Δ_{lim}. The performance function is implicitly expressed as:

$$G(\mathbf{X}) = \Delta_{\text{lim}} - \Delta(\mathbf{X}) \tag{6.85}$$

where $\Delta_{\text{lim}} = H/350$; and $\Delta(\mathbf{X})$ is the horizontal displacement of the upper floor shown in Figure 6.20a, which can be determined from the nonlinear finite element analysis of the structure.

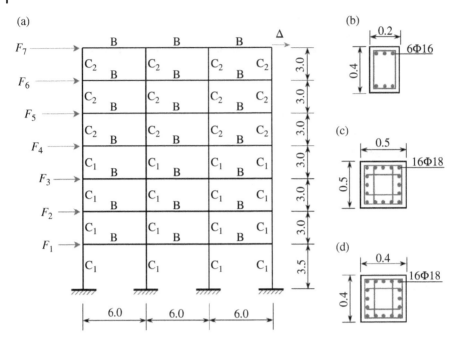

Figure 6.20 RC frame structure considered in Example 6.28 (unit: m).

The finite element model of this RC frame structure is built with OpenSees software. The inelastic fibre elements (Spacone et al. 1996) composed of the steel bars and concrete, as shown Figure 6.21a, are used to simulate the sections of the column and beam members. The Mander (Mander et al. 1988) stress-strain curve, as shown in Figure 6.21b, is used to simulate the unconfined and confined concrete; and the Giuffre-Menegotto-Pinto (Filippou et al. 1983) stress-strain curve, as shown in Figure 6.21c, is applied to simulate the longitudinal steel bar.

The horizontal displacement $\Delta(\mathbf{X})$ involves 17 random variables reflecting the properties of the lateral loads, concrete, and steel. Their probability distributions and first four central moments are given in Table 6.29. Herein, only all loads, F_i $(i = 1, 2, ..., n)$, are correlated with a correlation coefficient of 0.5.

In this example, because the joint PDF of basic random variables cannot be determined from the available probability information, MCS based on MH algorithm or Rosenblatt transformation can no longer be used. The exact failure probability cannot be evaluated because of the incomplete probability information. To enable the use of MCS, the Nataf distribution model can be applied to assess the joint PDF. Reference (Chang et al. 1994) gives the detailed computational procedure for MCS based on Nataf distribution model. The failure probability and corresponding reliability index obtain from MCS based on Nataf distribution are listed in Table 6.30. Based on the fourth-moment pseudo transformations based MCS and SS, the results can be determined and also listed in Table 6.30.

From Table 6.30, it can be observed that if only the marginal PDFs and correlation matrix available, the results obtained from the fourth-moment pseudo transformations based MCS and SS are almost the same with those obtained from MCS based on Nataf distribution model, and SS is much more efficient than MCS.

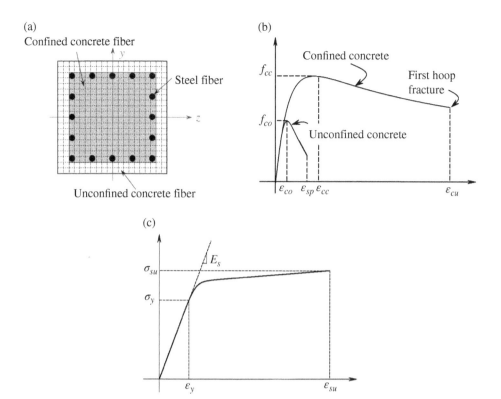

(a)

Confined concrete fiber

Steel fiber

Unconfined concrete fiber

(b)

f_{cc}

Confined concrete

First hoop fracture

f_{co}

Unconfined concrete

ε_{co} ε_{sp} ε_{cc}

ε_{cu}

(c)

σ_{su}

E_s

σ_y

ε_y

ε_{su}

Figure 6.21 Element and stress-strain curves used in Example 6.28.

Table 6.29 Statistical distribution of random variables in Example 6.28.

Variable	Distribution	Mean	Standard deviation	Skewness	Kurtosis
F_1 (kN)	Gumbel	30.0	7.5	1.14	5.4
F_2 (kN)	Gumbel	30.0	7.5	1.14	5.4
F_3 (kN)	Gumbel	45.0	11.25	1.14	5.4
F_4 (kN)	Gumbel	65.0	16.25	1.14	5.4
F_5 (kN)	Gumbel	83.0	20.75	1.14	5.4
F_6 (kN)	Gumbel	90.0	22.50	1.14	5.4
F_7 (kN)	Gumbel	95.0	23.75	1.14	5.4
f_{cc} (MPa)	Lognormal	35	5.25	0.453	3.368
ε_{cc}	Lognormal	4.4×10^{-3}	4.4×10^{-4}	0.453	3.368
ε_{cu}	Lognormal	1.82×10^{-2}	2.73×10^{-3}	0.453	3.368
f_c (MPa)	Lognormal	30.0	4.5	0.453	3.368
ε_c	Lognormal	2.0×10^{-3}	3.0×10^{-4}	0.453	3.368
ε_{sp}	Lognormal	3.3×10^{-3}	4.95×10^{-4}	0.453	3.368
E_c (MPa)	Lognormal	3.0×10^4	4.5×10^3	0.453	3.368
F_y (MPa)	Lognormal	400.0	20.0	0.15	3.04
E_S (GPa)	Lognormal	200.0	10.0	0.15	3.04
b	Lognormal	7.0×10^{-3}	3.5×10^{-4}	0.15	3.04

Table 6.30 Statistical distribution of random variables in Example 6.28.

Method	Sample number	Failure probability	COV of P_f	Reliability index
Nataf distribution based MCS	1.5×10^5	5.09×10^{-3}	3.61%	2.5699
Fourth-moment pseudo transformations based MCS	1.5×10^5	5.13×10^{-3}	3.61%	2.5672
Fourth-moment pseudo transformations based SS	2.8×10^4	5.78×10^{-3}		2.5252

6.8 Summary

In this chapter, the reliability evaluation with inclusion of random variable with unknown probability distribution is discussed, in which the first few moments of the random variable will be used instead of probability distribution. The first part of this chapter considers a special case of reliability estimation in which each variable is represented only by its first two moments, i.e. by its mean and standard deviation. This is known as the *second-moment* level of representation. Higher order moments, which describe skewness and kurtosis of the distribution, are considered in the second part of the chapter. The second- and third-order polynomial normal transformation techniques are then investigated using the first three and four central moments, respectively. Explicit third- and fourth-moment standardisation functions are introduced. Using these methods, the normal transformation for random variables (independent or correlated) with unknown probability distributions can be achieved without using the Rosenblatt or Nataf transformations. Through the numerical examples presented, the methods are demonstrated to be sufficiently accurate to include the random variables with unknown CDF/PDFs not only in the direct methods of moment as described in the chapter with the same title but also in methods of moment based on first- and second-order transformation as described in the chapter with that title, and also in the MCS with little extra computational efforts.

7

Transformation of Non-Normal Variables to Independent Normal Variables

7.1 Introduction

As shown in chapters on Moment Evaluation, Methods of Moment Based on First- and Second-Order Transformation, and Structural Reliability Assessment, there are many cases in structural reliability analysis where it is necessary to transform non-normal random variables into independent standard normal random variables, and vice versa. When the information of the full distribution can be utilised, the transformation can be achieved by normal tail transformation for a single random variable (or independent random variables), and for correlated random variables, the transformation can be realised by Rosenblatt transformation or Nataf transformation. When only the information of the first few moments is known, the transformation can be achieved by the so-called pseudo normal transformation methods. These transformation techniques will be introduced in detail in this chapter.

7.2 The Normal Transformation for a Single Random Variable

For a single non-normal random variable, the normal transformation and its inverse transformation can be realised by the normal tail transformation (Ditlevsen 1981). Assume a random variable X with cumulative distribution function (CDF) $F_X(x)$, a standard normal variable U, can be obtained from the following equation:

$$\Phi(u) = F_X(x) \tag{7.1}$$

Inverting Eq. (7.1) leads to the desired normal variable U

$$u = \Phi^{-1}[F_X(x)] \tag{7.2}$$

The inverse transformation of Eq. (7.2) can be obtained by

$$x = F_X^{-1}[\Phi(u)] \tag{7.3}$$

The transformation is schematically shown in Figures 7.1a–b.

Structural Reliability: Approaches from Perspectives of Statistical Moments, First Edition.
Yan-Gang Zhao and Zhao-Hui Lu.
© 2021 John Wiley & Sons Ltd. Published 2021 by John Wiley & Sons Ltd.

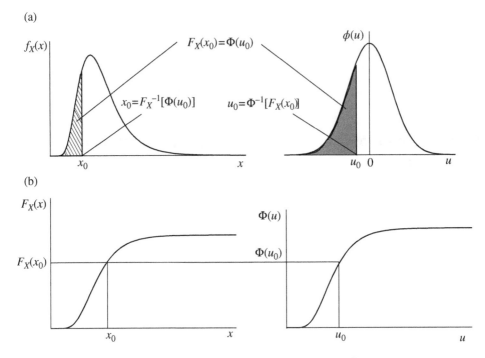

Figure 7.1 Normal transformation for a single variable. (a) PDF. (b) CDF.

In particular, for a normal variable X of $N(\mu_X, \sigma_X)$, the transformation becomes

$$U = \frac{X - \mu_X}{\sigma_X}$$

$$X = \sigma_X U + \mu_X$$

and for a lognormal variable with parameters λ and ζ, the transformation becomes

$$U = \frac{\ln X - \lambda}{\zeta}$$

$$X = \exp(\zeta U + \lambda)$$

7.3 The Normal Transformation for Correlated Random Variables

7.3.1 Rosenblatt Transformation

7.3.1.1 Normal Transformation

When the joint PDF of the correlated non-normal random variables is known, the normal transformation and its inverse transformation can be realised by the Rosenblatt transformation (Rosenblatt 1952; Hohenbichler and Rackwitz 1981). Suppose a set of n random variables $\mathbf{X} = (X_1, X_2, ..., X_n)$ with a joint CDF $F_{\mathbf{X}}(\mathbf{x})$. A set of statistically independent standard normal variables $\mathbf{U} = (U_1, U_2, ..., U_n)$ can be obtained from the following equations:

$$\Phi(u_1) = F(x_1)$$
$$\Phi(u_2) = F(x_2|x_1)$$
$$\vdots \tag{7.4}$$
$$\Phi(u_n) = F(x_n|x_1, x_2, ..., x_{n-1})$$

Inverting these equations successively leads to the desired normal variables $\mathbf{U}$

$$u_1 = \Phi^{-1}[F(x_1)]$$
$$u_2 = \Phi^{-1}[F(x_2|x_1)]$$
$$\vdots \tag{7.5}$$
$$u_n = \Phi^{-1}[F(x_n|x_1, x_2, ..., x_{n-1})]$$

Equation (7.5) constitutes the Rosenblatt transformation.

The conditional CDFs in Eq. (7.4) may be obtained from the joint PDFs as follows. Since

$$f(x_i|x_1, x_2, ..., x_{i-1}) = \frac{f(x_1, x_2, ..., x_i)}{f(x_1, x_2, ..., x_{i-1})} \tag{7.6}$$

the required CDF may be obtained as

$$F(x_i|x_1, x_2, ..., x_{i-1}) = \frac{\int_{-\infty}^{x_i} f(x_1, x_2, ..., x_{i-1}, s_i) ds_i}{f(x_1, x_2, ..., x_{i-1})} \tag{7.7}$$

7.3.1.2 The Inverse Transformation

The inverse transformation of Eq. (7.5) can be obtained by sequentially inverting the one-dimensional relations, that is

$$x_1 = F^{-1}[\Phi(u_1)]$$
$$x_2 = F^{-1}[\Phi(u_2|x_1)]$$
$$\vdots \tag{7.8}$$
$$x_n = F^{-1}[\Phi(u_n|x_1, x_2, ..., x_{n-1})]$$

In general, these inverse relations may only be obtained numerically.

7.3.2 Nataf Transformation

7.3.2.1 Normal Transformation

When the joint PDF of basic random variables is unknown while the marginal PDFs and correlation matrix are known, the normal transformation and its inverse transformation can be achieved by the Nataf transformation (Nataf 1962; Liu and Der Kiureghian 1986).

If the joint distribution of $\mathbf{X} = (X_1, X_2, ..., X_n)$ is unknown except for the marginal distributions and the covariance matrix $\mathbf{C_X}$, it is reasonable to let $\mathbf{Z} = (Z_1, Z_2, ..., Z_n)$ have an n-dimensional normal distribution with a correlation matrix $\mathbf{C_Z}$. In this way, a unique

n-dimensional probability density function $f_X(x_1, x_2, ..., x_n)$ is induced in the x-space. The PDF is

$$f_X(x_1, x_2, ..., x_n) = \frac{\partial Z}{\partial X} \phi_n(Z, C_Z) \tag{7.9}$$

where

$$\frac{\partial Z}{\partial X} = \frac{\partial(z_1, z_2, ..., z_n)}{\partial(x_1, x_2, ..., x_n)} \tag{7.10}$$

is the Jacobin, while

$$\phi_n(Z, C_Z) = \frac{1}{\sqrt{(2\pi)^n \det C_Z}} \exp\left[-\frac{1}{2} z^T C_Z z\right] \tag{7.11}$$

is the n-dimensional normal PDF corresponding to the mean values 0, the variance 1, and the correlation matrix C_Z.

According to Liu and Der Kiureghian (1986), the correlated variables vector X with correlation matrix C_X can be first transformed into correlated standard normal variables vector $Z = (Z_1, Z_2, ..., Z_n)$ with correlation matrix C_Z,

$$\Phi(Z_i) = F_{X_i}(X_i), Z_i = \Phi^{-1}[F_{X_i}(X_i)](i = 1, 2, ..., n) \tag{7.12}$$

where $F_{X_i}(X_i)$ is the marginal CDF of X_i, and $\Phi(\cdot)$ and $\Phi^{-1}(\cdot)$ are the CDF and inverse CDF of a standard normal variable, respectively.

Based on the Nataf transformation theory, the joint PDF of random variables vector X can be expressed as:

$$f_X(x) = \prod_{i=1}^{n} f_{x_i}(x_i) \frac{\phi_n(Z, C_Z)}{\prod\limits_{i=1}^{n} \phi(z_i)} \tag{7.13}$$

where $f_{x_i}(x_i)$ is the marginal CDF of X_i, and $\phi(\cdot)$ is the joint PDF of correlate standard normal variables.

Assume that the correlation coefficient between X_i and X_j is ρ_{ij}, and the correlation coefficient between Z_i and Z_j is ρ_{0ij}, and C_X and C_Z can be written as

$$C_X = \begin{bmatrix} 1 & \rho_{12} & \cdots & \rho_{1n} \\ \rho_{12} & 1 & \cdots & \rho_{2n} \\ \vdots & \vdots & \ddots & \vdots \\ \rho_{n1} & \rho_{n2} & \cdots & 1 \end{bmatrix}, \quad C_Z = \begin{bmatrix} 1 & \rho_{012} & \cdots & \rho_{01n} \\ \rho_{012} & 1 & \cdots & \rho_{02n} \\ \vdots & \vdots & \ddots & \vdots \\ \rho_{0n1} & \rho_{0n2} & \cdots & 1 \end{bmatrix} \tag{7.14}$$

According to the definition of correlation coefficient and Eqs. (7.13) and (7.11), one can obtain,

$$\begin{aligned} \rho_{ij} &= \int_{-\infty}^{+\infty}\int_{-\infty}^{+\infty} \frac{x_i - \mu_{x_i}}{\sigma_{x_i}} \cdot \frac{x_j - \mu_{x_j}}{\sigma_{x_j}} f_{x_i x_j}(x_i, x_j)\, dx_i dx_j \\ &= \int_{-\infty}^{+\infty}\int_{-\infty}^{+\infty} \frac{F^{-1}[\Phi(z_i)] - \mu_{x_i}}{\sigma_{x_i}} \cdot \frac{F^{-1}[\Phi(z_j)] - \mu_{x_j}}{\sigma_{x_j}} \phi_2\left(z_i, z_j, \rho_{0ij}\right) dz_i dz_j \end{aligned} \tag{7.15}$$

The unknown ρ_{0ij} in Eq. (7.15) is required to solve the two-dimensional nonlinear integral equation. Liu and Der Kiureghian (1986) proposed the following empirical formula:

$$\rho_{0ij} = F \cdot \rho_{ij} \tag{7.16}$$

where F is a function of the correlation coefficient ρ_{ij}, which is dependent on the type of marginal PDFs of random variables X_i and X_j and can be determined from Tables 7.1–7.6.

Using Cholesky decomposition, the correlation matrix ρ_0 can be rewritten as,

$$\mathbf{C_Z} = \mathbf{L_0} \mathbf{L_0^T} \tag{7.17}$$

where $\mathbf{L_0}$ is the lower triangular matrix obtained from Cholesky decomposition and $\mathbf{L_0^T}$ is the transpose matrix of $\mathbf{L_0}$.

Using the matrix $\mathbf{L_0}$, the independent standard normal vector $\mathbf{U} = (U_1, U_2, ..., U_n)$ can then be given as:

$$\mathbf{U} = \mathbf{L_0^{-1}} \mathbf{Z} \tag{7.18}$$

where $\mathbf{L_0^{-1}}$ is the inverse matrix of $\mathbf{L_0}$.

Table 7.1 Selected two-parameter marginal distribution.

Group 1 distributions		Group 2 distributions	
Name	Symbol	Name	Symbol
Uniform	UN	Lognormal	LN
Shifted exponential	SE	Gamma	GM
Shifted Rayleigh	SR	Type II – largest value	T2L
Type I – largest value	T1L	Type III – smallest value	T3S
Type I – smallest value	T1S		

Table 7.2 Category 1 Formulas: F = constant, for Normal X_i and X_j belonging to Group 1 distributions.

X_j	F = constant	M.E. (%)
Uniform	1.023	0.0
Shifted exponential	1.107	0.0
Shifted Rayleigh	1.014	0.0
Type I – largest value	1.031	0.0
Type I – smallest value	1.031	0.0

Note: M.E. = Maximum Error.

Table 7.3 Category 2 Formulas: $F = F(V_j)$, for Normal X_i and X_j belonging to Group 2 distributions.

X_j	$F = F(V_j)$	M.E. (%)
Lognormal	$\dfrac{V_j}{\sqrt{\ln\left(1 + V_j^2\right)}}$	Exact
Gamma	$1.001 - 0.007V_j + 0.1186V_j^2$	0.0
Type II – largest value	$1.030 + 0.238V_j + 0.364V_j^2$	0.1
Type III – smallest value	$1.031 - 0.195V_j + 0.328V_j^2$	0.1

Note: Range of coefficient of variation is $V_j = 0.1$–0.5.

7.3.2.2 The Inverse Nataf Transformation

In order to obtain the inverse Nataf transformation, the independent standard normal vector **U** is firstly transformed into correlated standard normal vector **Z** by

$$\mathbf{Z} = \mathbf{L}_0 \mathbf{U} \tag{7.19}$$

Using Eq. (7.12), the correlated random vector of **X** can be obtained as:

$$X_i = F_{X_i}^{-1}[\Phi(Z_i)], (i = 1, 2, ..., n) \tag{7.20}$$

Example 7.1 Investigation with Known Joint PDF of Correlated Random Variables

Consider the following performance function as discussed in Example 6.7 again.

$$G(\mathbf{X}) = 18 - 3X_1 - 2X_2 \tag{7.21}$$

where the joint PDF of X_1 and X_2 is

$$f_{X_1 X_2}(\mathbf{x}) = (x_1 + x_2 + x_1 x_2) \cdot \exp\left(-x_1 - x_2 - x_1 x_2\right); x_1, x_2 \geq 0 \tag{7.22}$$

As the exact result of this example, the probability of failure, P_f, can be obtained directly by numerical integration as 0.002945, and the corresponding reliability index is calculated 2.754.

The problem is first solved using Rosenblatt transformation. Since the joint PDF is known, when the transformation order of $X_1 \rightarrow X_2$ is adopted in Rosenblatt transformation, one has

$$\Phi(u_1) = F_1(x_1) = \int_{-\infty}^{x_1} f_1(x_1) dx_1 = 1 - \exp\left(-x_1\right)$$

$$\Phi(u_2) = F_2(x_2|x_1) = \int_{-\infty}^{x_2} \frac{f_{12}(x_2, x_1)}{f_1(x_1)} dx_2 = [1 - (1 + x_2)] \cdot \exp\left(-x_2 - x_1 x_2\right)$$

Table 7.4 Category 3 Formulas: $F = F(\rho_{ij})$, for both X_i and X_j belonging to Group 1 distributions.

X_i \ X_j	UN	SE	SR	T1L	T1S
UN (M.E.)	$1.047 - 0.047\rho_{ij}^2$ (0.0%)				
SE (M.E.)	$1.133 + 0.029\rho_{ij}^2$ (0.0%)	$1.229 - 0.367\rho_{ij} + 0.153\rho_{ij}^2$ (1.5%)			
SR (M.E.)	$1.038 - 0.008\rho_{ij}^2$ (0.0%)	$1.123 - 0.100\rho_{ij} + 0.021\rho_{ij}^2$ (0.1%)	$1.028 - 0.029\rho_{ij}$ (0.0%)		
T1L (M.E.)	$1.055 + 0.015\rho_{ij}^2$ (0.0%)	$1.142 - 0.154\rho_{ij} + 0.031\rho_{ij}^2$ (0.2%)	$1.046 - 0.045\rho_{ij} + 0.006\rho_{ij}^2$ (0.0%)	$1.064 - 0.069\rho_{ij} + 0.005\rho_{ij}^2$ (0.0%)	
T1S (M.E.)	$1.055 + 0.015\rho_{ij}^2$ (0.0%)	$1.142 + 0.154\rho_{ij} + 0.031\rho_{ij}^2$ (0.2%)	$1.046 + 0.045\rho_{ij} + 0.006\rho_{ij}^2$ (0.0%)	$1.064 + 0.069\rho_{ij} + 0.005\rho_{ij}^2$ (0.0%)	$1.064 - 0.069\rho_{ij} + 0.005\rho_{ij}^2$ (0.0%)

Table 7.5 Category 4 Formulas: $F = F(\rho_{ij}, V_j)$, for X_i belonging to Group 1 distributions and X_j belonging to Group 2 distributions.

X_i \ X_j	UN	SE	SR	T1L	T1S
LN	$1.019 + 0.014V_j$ $+ 0.010\rho_{ij}^2 + 0.249V_j^2$	$1.098 + 0.003\rho_{ij} - 0.019V_j$ $+ 0.025\rho_{ij}^2 + 0.303V_j^2 - 0.437\rho_{ij}V_j$	$1.011 + 0.001\rho_{ij} + 0.014V_j$ $+ 0.004\rho_{ij}^2 + 0.231V_j^2 - 0.130\rho_{ij}V_j$	$1.029 + 0.001\rho_{ij} + 0.014V_j$ $+ 0.004\rho_{ij}^2 + 0.233V_j^2 - 0.197\rho_{ij}V_j$	$1.029 - 0.001\rho_{ij} + 0.014V_j$ $+ 0.004\rho_{ij}^2 + 0.233V_j^2 + 0.197\rho_{ij}V_j$
(M.E.)	(0.7%)	(1.6%)	(0.4%)	(0.3%)	(0.3%)
GM	$1.023 - 0.007V_j$ $+ 0.002\rho_{ij}^2 + 0.127V_j^2$	$1.104 + 0.003\rho_{ij} - 0.008V_j$ $+ 0.014\rho_{ij}^2 + 0.173V_j^2 - 0.296\rho_{ij}V_j$	$1.014 + 0.001\rho_{ij} - 0.007V_j$ $+ 0.002\rho_{ij}^2 + 0.126V_j^2 - 0.090\rho_{ij}V_j$	$1.031 + 0.001\rho_{ij} - 0.132V_j$ $+ 0.003\rho_{ij}^2 + 0.131V_j^2 - 0.132\rho_{ij}V_j$	$1.031 - 0.001\rho_{ij} - 0.007V_j$ $+ 0.003\rho_{ij}^2 + 0.131V_j^2 + 0.132\rho_{ij}V_j$
(M.E.)	(0.1%)	(0.9%)	(0.9%)	(0.3%)	(0.3%)
T2L	$1.033 + 0.305V_j$ $+ 0.074\rho_{ij}^2 + 0.405V_j^2$	$1.109 - 0.152\rho_{ij} + 0.361V_j$ $+ 0.130\rho_{ij}^2 + 0.455V_j^2 - 0.728\rho_{ij}V_j$	$1.036 - 0.038\rho_{ij} + 0.266V_j$ $+ 0.028\rho_{ij}^2 + 0.383V_j^2 - 0.229\rho_{ij}V_j$	$1.056 - 0.060\rho_{ij} + 0.263V_j$ $+ 0.020\rho_{ij}^2 + 0.383V_j^2 - 0.332\rho_{ij}V_j$	$1.056 + 0.060\rho_{ij} + 263V_j$ $+ 0.020\rho_{ij}^2 + 0.383V_j^2 + 0.332\rho_{ij}V_j$
(M.E.)	(2.1%)	(4.5%)	(1.2%)	(1.0%)	(1.0%)
T3S	$1.061 - 0.237V_j$ $- 0.005\rho_{ij}^2 + 0.379V_j^2$	$1.147 + 0.145\rho_{ij} - 0.271V_j$ $+ 0.010\rho_{ij}^2 + 0.495V_j^2 - 0.467\rho_{ij}V_j$	$1.047 + 0.042\rho_{ij} - 0.212V_j$ $+ 0.353\rho_{ij}^2 - 0.136\rho_{ij}V_j$	$1.064 + 0.065\rho_{ij} - 0.210V_j$ $+ 0.003\rho_{ij}^2 + 0.356V_j^2 - 0.211\rho_{ij}V_j$	$1.064 - 0.065\rho_{ij} - 0.210V_j$ $+ 0.003\rho_{ij}^2 + 0.356V_j^2 + 0.211\rho_{ij}V_j$
(M.E.)	(0.5%)	(0.4%)	(0.2%)	(0.2%)	(0.2%)

Note: Range of coefficient of variation is $V_j = 0.1$–0.5.

Table 7.6 Category 5 Formulas: $F = F(\rho_{ij}, V_i, V_j)$, for both X_i and X_j belonging to Group 2 distributions.

X_i / X_j	LN	GM	T2L	T3S
LN	$\dfrac{\ln\left(1+\rho_{ij}V_iV_j\right)}{\rho_{ij}\sqrt{\ln\left(1+V_i^2\right)\ln\left(1+V_j^2\right)}}$			
(M.E.)	(0.0%)			
GM	$1.001 + 0.33\rho_{ij} + 0.004V_i - 0.016V_j$ $+ 0.002\rho_{ij}^2 + 0.223V_i^2 + 0.130V_j^2$ $- 0.104\rho_{ij}V_i + 0.029V_iV_j - 0.119\rho_{ij}V_j$	$1.002 + 0.022\rho_{ij} - 0.012(V_i + V_j)$ $+ 0.001\rho_{ij}^2 + 0.125\left(V_i^2 + V_j^2\right)$ $- 0.077\rho_{ij}(V_i + V_j) + 0.015V_iV$		
(M.E.)	(4.0%)	(4.0%)		
T2L	$1.026 + 0.082\rho_{ij} - 0.019V_i + 0.222V_j$ $+ 0.018\rho_{ij}^2 + 0.288V_i^2 + 0.379V_j^2$ $- 0.441\rho_{ij}V_i + 0.126V_iV_j - 0.277\rho_{ij}V_j$	$1.029 + 0.056\rho_{ij} - 0.030V_i + 0.225V_j$ $+ 0.012\rho_{ij}^2 + 0.174V_i^2 + 0.379V_j^2$ $- 0.313\rho_{ij}V_i + 0.075V_iV_j - 0.182\rho_{ij}V_j$	$1.086 + 0.054\rho_{ij} + 0.104(V_i + V_j) - 0.055\rho_{ij}^2$ $+ 0.662\left(V_i^2 + V_j^2\right) - 0.570\rho_{ij}(V_i + V_j) + 0.203V_iV_j$ $- 0.020\rho_{ij}^3 - 0.218\left(V_i^3 + V_j^3\right) - 0.371\rho_{ij}\left(V_i^2 + V_j^2\right)$ $+ 0.257\rho_{ij}^2(V_i + V_j) + 0.141V_iV_j(V_i + V_j)$	
(M.E.)	(4.3%)	(4.2%)	(4.3%)	
T3S	$1.031 + 0.052\rho_{ij} + 0.011V_i - 0.210V_j$ $+ 0.002\rho_{ij}^2 + 0.220V_i^2 + 0.350V_j^2$ $+ 0.005\rho_{ij}V_i + 0.009V_iV_j - 0.147\rho_{ij}V_j$	$1.032 + 0.034\rho_{ij} - 0.007V_i - 0.202V_j$ $+ 0.121V_i^2 + 0.339V_j^2$ $- 0.006\rho_{ij}V_i + 0.003V_iV_j - 0.111\rho_{ij}V_j$	$1.065 + 0.146\rho_{ij} + 0.241V_i - 0.259V_j$ $+ 0.013\rho_{ij}^2 + 0.372V_i^2 + 0.435V_j^2$ $+ 0.005\rho_{ij}V_i + 0.034V_iV_j - 0.481\rho_{ij}V_j$	$1.063 - 0.004\rho_{ij} - 0.200(V_i + V_j)$ $- 0.001\rho_{ij}^2 + 0.337\left(V_i^2 + V_j^2\right)$ $+ 0.007\rho_{ij}(V_i + V_j) - 0.007V_iV_j$
(M.E.)	(2.4%)	(4.0%)	(3.8%)	(2.6%)

Based on the above transformation, FORM can be applied. The design points are obtained as $\mathbf{u}^* = (2.7840, 0.0262)$ and $\mathbf{x}^* = (5.9149, 0.1276)$ for standard normal space and the original space, respectively, and the first-order reliability index is obtained as 2.7842.

When the transformation order of $X_2 \rightarrow X_1$ is adopted in Rosenblatt transformation, one has

$$\Phi(u_1) = F_2(x_2) = \int_{-\infty}^{x_2} f_2(x_2)dx_2 = 1 - \exp(-x_2)$$

$$\Phi(u_2) = F_2(x_1|x_2) = \int_{-\infty}^{x_1} \frac{f_{12}(x_1, x_2)}{f_2(x_2)} dx_1 = [1 - (1 + x_1)] \cdot \exp(-x_1 - x_1 x_2)$$

Applying the FORM leads to the design points of $\mathbf{u}^* = (2.4820, -0.8167)$ and $\mathbf{x}^* = (5.7805, 0.2685)$ for the standard normal space and the original space, respectively, and the first-order reliability index is obtained as 2.6129.

It can be observed that the transformation order has significant influence on the results of FORM when Rosenblatt transformation is applied.

The problem is then solved using Nataf transformation. When the transformation order of $X_1 \rightarrow X_2$ is adopted in Nataf transformation, the correlated random variables X_1 and X_2 can be transformed into two dependent standard normal variables Z_1 and Z_2

$$\Phi(z_1) = F_1(x_1) = \int_{-\infty}^{x_1} f_1(x_1)dx_1 = 1 - \exp(-x_1), Z_1 = \Phi^{-1}[1 - \exp(-X_1)]$$

$$\Phi(z_2) = F_2(x_2) = \int_{-\infty}^{x_2} f_2(x_2)dx_2 = 1 - \exp(-x_2), \quad Z_2 = \Phi^{-1}[1 - \exp(-X_2)]$$

From Eq. (7.22), the correlation coefficient of X_1 and X_2, ρ_{12}, is readily obtained as -0.4037. According to Table 7.4, the equivalent correlation coefficient $\rho_{0,12}$, between Z_1 and Z_2 (Liu and Der Kiureghian 1986) is

$$\rho_{0,12} = \rho_{12}(1.229 - 0.367\rho_{12} + 0.153\rho_{12}^2) = -0.5660$$

Utilising Cholesky decomposition, the lower triangular matrix $\mathbf{L}_0$ of the equivalent correlation coefficient matrix can be obtained as

$$\mathbf{L}_0 = \begin{bmatrix} 1 & 0 \\ -0.5660 & 0.8244 \end{bmatrix}$$

This then leads to

$$Z_1 = U_1, \quad Z_2 = -0.566U_1 + 0.8244U_2$$

Then

$$X_1 = -\ln[1 - \Phi(U_1)], \quad X_2 = -\ln[1 - \Phi(-0.566U_1 + 0.8244U_2)]$$

Therefore, the performance function Eq. (7.21) can be transformed as

$$G(\mathbf{U}) = 18 + 3\ln[1 - \Phi(U_1)] + 2\ln[1 - \Phi(-0.566U_1 + 0.8244U_2)] \qquad (7.23)$$

where U_1 and U_2 are independent normal random variables.

For the performance function in Eq. (7.23), using FORM, the design points are obtained as $\mathbf{u}^* = (2.7957, 0.0661)$ and $\mathbf{x}^* = (5.9564, 0.0654)$ for the standard normal space and the original space, respectively, and the first-order reliability index is obtained as 2.7965.

When the transformation order of $X_2 \rightarrow X_1$ is adopted, the correlated random variables X_1 and X_2 can be transformed into two dependent standard normal variables Z_1 and Z_2

$$\Phi(z_1) = F_2(x_2) = \int_{-\infty}^{x_2} f_2(x_2)dx_2 = 1 - \exp(-x_2), Z_1 = \Phi^{-1}[1 - \exp(-x_2)]$$

$$\Phi(z_2) = F_1(x_1) = \int_{-\infty}^{x_1} f_1(x_1)dx_1 = 1 - \exp(-x_1), Z_2 = \Phi^{-1}[1 - \exp(-x_1)]$$

According to Table 7.4, the equivalent correlation coefficient $\rho_{0,12}$, between Z_1 and Z_2 (Liu and Der Kiureghian 1986) is

$$\rho_{0,12} = \rho_{12}\left(1.229 - 0.367\rho_{12} + 0.153\rho_{12}^2\right) = -0.5660$$

The lower triangular matrix $\mathbf{L}_0$ of the equivalent correlation coefficient matrix can be obtained as,

$$\mathbf{L}_0 = \begin{bmatrix} 1 & 0 \\ -0.5660 & 0.8244 \end{bmatrix}$$

Then, one obtains

$$Z_1 = U_1, \ Z_2 = -0.566U_1 + 0.8244U_2$$

Then

$$X_2 = -\ln[1 - \Phi(U_1)], \quad X_1 = -\ln[1 - \Phi(-0.566U_1 + 0.8244U_2)]$$

Therefore, the performance function Eq. (7.21) can be transformed as

$$G(\mathbf{U}) = 18 + 3\ln[1 - \Phi(-0.566U_1 + 0.8244U_2)] + 2\ln[1 - \Phi(U_1)] \tag{7.24}$$

where U_1 and U_2 are independent normal random variables.

For the performance function in Eq. (7.24), FORM gives the design points of $\mathbf{u}^* = (0.0661, 2.7957)$ and $\mathbf{x}^* = (5.9564, 0.0654)$ for the standard normal space and the original space, respectively, and the first-order reliability index is obtained as 2.7965, which is as same as the previous case when the transformation order of $X_1 \rightarrow X_2$ is adopted. This implies that different order of Nataf transformation leads to the same results.

Example 7.2   Reliability Analysis with Known Marginal and Partial Joint PDFs

Consider the performance function in Example 6.19 again (in the chapter on Structural Reliability Assessment),

$$G(\mathbf{X}) = X_1 - X_2 - X_3 \tag{7.25}$$

where X_1 is lognormally distributed with parameters of $\lambda = 1.590$ and $\zeta = 0.198$, X_2 and X_3 are bivariate exponentially distributed

$$f_{X_2 X_3}(x_2, x_3) = (x_2 + x_3 + x_2 x_3)\exp[-(x_2 + x_3 + x_2 x_3)]; x_2, x_3 \geq 0 \tag{7.26}$$

and the correlation coefficients are $\rho_{12} = 0.3$ and $\rho_{13} = -0.2$. From Eq. (7.26), the correlation coefficient of X_2 and X_3, ρ_{23}, is readily obtained as -0.4037.

Since the joint PDF of X_1, X_2, and X_3 are unknown, the Rosenblatt transformation cannot be applied. This problem can be solved by using Nataf transformation since it only requires the marginal PDFs of basic random variables.

When the transformation order of $X_1 \rightarrow X_2 \rightarrow X_3$ is adopted in Nataf transformation, the correlated random variables X_1, X_2, and X_3 can be transformed into three dependent standard normal variables Z_1, Z_2, and Z_3

$$\Phi(z_1) = F_1(x_1) = \int_{-\infty}^{x_1} f_1(x_1)dx_1, Z_1 = \frac{\ln X_1 - \lambda}{\zeta}$$

$$\Phi(z_2) = F_2(x_2) = \int_{-\infty}^{x_2} f_2(x_2)dx_2 = 1 - \exp(-x_2), Z_2 = \Phi^{-1}[1 - \exp(-X_2)]$$

$$\Phi(z_3) = F_3(x_3) = \int_{-\infty}^{x_3} f_3(x_3)dx_3 = 1 - \exp(-x_3), Z_3 = \Phi^{-1}[1 - \exp(-X_3)]$$

According to Tables 7.4 and 7.5, the equivalent correlation coefficients $\rho_{0,12}$, $\rho_{0,13}$, and $\rho_{0,23}$, among Z_1, Z_2, and Z_3 (Liu and Der Kiureghian 1986)

$$\rho_{0,12} = \rho_{12}(1.098 - 0.003\rho_{12} + 0.019V_1 + 0.025\rho_{12}^2 + 0.303V_1^2 - 0.437\rho_{12}V_1) = 0.3267$$

$$\rho_{0,13} = \rho_{13}(1.098 - 0.003\rho_{13} + 0.019V_1 + 0.025\rho_{13}^2 + 0.303V_1^2 - 0.437\rho_{13}V_1) = -0.2266$$

$$\rho_{0,23} = \rho_{23}(1.229 - 0.367\rho_{23} + 0.153\rho_{23}^2) = -0.5660$$

Utilising Cholesky decomposition, the lower triangular matrix $\mathbf{L}_0$ of the equivalent correlation coefficient matrix can be obtained as

$$\mathbf{L}_0 = \begin{bmatrix} 1 & 0 & 0 \\ 0.3267 & 0.9451 & 0 \\ -0.2266 & -0.5205 & 0.8233 \end{bmatrix}$$

Then, one obtains

$$Z_1 = U_1, \quad Z_2 = 0.3267U_1 + 0.9451U_2, \quad Z_3 = -0.2266U_1 - 0.5205U_2 + 0.8233U_3,$$

Then

$$X_1 = \text{Exp}[\lambda + \zeta U_1], X_2 = -\ln[1 - \Phi(0.3267U_1 + 0.9451U_2)]$$
$$X_3 = -\ln[1 - \Phi(-0.2266U_1 - 0.5205U_2 + 0.8233U_3)]$$

Therefore, the performance function Eq. (7.25) can be transformed as

$$G(\mathbf{U}) = \text{Exp}[\lambda + \zeta U_1] + \ln[1 - \Phi(0.3267U_1 + 0.9451U_2)]$$
$$+ \ln[1 - \Phi(-0.2266U_1 - 0.5205U_2 + 0.8233U_3)] \qquad (7.27)$$

where U_1, U_2, and U_3 are independent normal random variables.

For the performance function in Eq. (7.27), the reliability index can be calculated using the general FORM. The starting point is selected at $\mathbf{U} = (0, 0, 0)^T$ and convergence is achieved as $\beta_1 = 2.1216$ at the design point $\mathbf{u}^* = (-1.0429, -0.8511, 1.6399)^T$.

Table 7.7 Comparison of results by different transformation orders.

Order of X	Design point $(u_1{}^*, u_2{}^*, u_3{}^*)$	Design point $(x_1{}^*, x_2{}^*, x_3{}^*)$	β
$X_1 \rightarrow X_2 \rightarrow X_3$	$(-1.0429, -0.8511, 1.6399)$	$(3.988, 0.1348, 3.8532)$	2.1216
$X_1 \rightarrow X_3 \rightarrow X_2$	$(-1.0429, 1.8409, 0.1569)$	$(3.988, 0.1348, 3.8532)$	2.1216
$X_2 \rightarrow X_1 \rightarrow X_3$	$(-1.1451, -0.7076, 1.6399)$	$(3.988, 0.1348, 3.8532)$	2.1216
$X_2 \rightarrow X_3 \rightarrow X_1$	$(-1.1451, 1.6754, -0.6188)$	$(3.988, 0.1348, 3.8532)$	2.1216
$X_3 \rightarrow X_1 \rightarrow X_2$	$(2.0294, -0.5986, 0.1569)$	$(3.988, 0.1348, 3.8532)$	2.1216
$X_3 \rightarrow X_2 \rightarrow X_1$	$(2.2094, 0.004098, -0.6188)$	$(3.988, 0.1348, 3.8532)$	2.1216

Table 7.7 lists the design points in standard normal space and the original space along with the corresponding reliability indices when different transformation orders of X_1, X_2, and X_3 are considered. As can be observed from Table 7.7, the design points are the same in the original space although they are different in standard normal space. This demonstrates again that the reliability indices are the same even though the transformation order is different when using Nataf transformation. However, it has been shown in Eq. (6.19) (in the chapter on Structural Reliability Assessment) that the design point is dependent on the initial checking point for this problem, and there are two design points for each transformation order. The general reliability index should be determined by using system reliability method, and the computation procedure has been discussed in Example 6.19.

7.4 Pseudo Normal Transformations for a Single Random Variable

As shown in Section 7.2, for a single random variable, the normal transformation (x–u transformation) and its inverse transformation (u–x transformation) are realised by using the normal tail transformation based on its PDF/CDF. As described in the chapter on Structural Reliability Assessment, sometimes in reality we have to conduct the transformation under the condition that the CDF/PDFs of the basic random variables are unknown due to the lack of statistical data. In this section, the transformation is presented in details using the statistical moments of random variables.

7.4.1 Concept of Pseudo Normal Transformation

As described in Section 7.2, the normal tail transformation for a single random variable is based on the principle of equivalent CDF between a standard normal random variable and the arbitrary random variable, i.e.

$$\Phi(u) = F_X(x) \tag{7.28}$$

where $F_X(x)$ is the CDF of a random variable X, and $\Phi(u)$ is the CDF of a standard normal variable U.

Inverting Eq. (7.28), we have obtained the desired normal variable U

$$u = \Phi^{-1}[F_X(x)] \tag{7.29a}$$

The inverse transformation in Eq. (7.29a) can be obtained by

$$x = F_X^{-1}[\Phi(u)] \tag{7.29b}$$

Rewrite Eqs. (7.29a) and (7.29b) as

$$U = N(X) \tag{7.30a}$$

$$X = N^{-1}(U) \tag{7.30b}$$

Since both the transformations are based on Eqs. (7.29a) and (7.29b) using the CDF/PDF of a random variable X, and U is exactly a standard normal variable, the function N in Eq. (7.30a) is an accurate standard normal transformation function, or simply named as the normal transformation (x–u transformation) function. Similarly, the function N^{-1} in Eq. (7.30b) is an accurate inverse standard normal transformation function, or simply named as the inverse normal transformation (u–x transformation) function. In such cases, needless to say, the moments in any order for the both sides in the both equations are the same, i.e.

$$M_{rU} = M_{rN} \tag{7.31a}$$

$$M_{rX} = M_{rN^{-1}} \tag{7.31b}$$

Obviously, Eq. (7.28) is equivalent to Eqs. (7.31a) and (7.31b) holding true for any rth order. Assume that a pair of transformations based on that Eqs. (7.31a) and (7.31b), instead of the CDF/PDF of X, are given as

$$U' = S(X) \tag{7.32a}$$

$$X = S^{-1}(U') \tag{7.32b}$$

If Eqs. (7.32a) and (7.32b) hold true for any rth order, i.e. $r = 1, 2,..., \infty$, all the moments of $S(X)$ are equal to those of the standard normal variable U. Then $S(X)$ is exactly a standard normal transformation.

When Eqs. (7.32a) and (7.32b) hold true only for the first k order, only the first k moments of S are equal to those of U. Although $S(X)$ is not exactly a standard normal variable, since the first k moments of $U' = S(X)$ are equal to those of U (the standard normal variable), we may call $S(X)$ a pseudo standard normal variable under the condition that its first k moments are equal to those of a standard normal variable, and the function $S(X)$ is called pseudo standard normal transformation function under the condition that Eqs. (7.32a) and (7.32b) hold true for the first k order. Or simply, U' and $S(X)$ in Eqs. (7.32a) and (7.32b) can be referred to as the kth moment pseudo normal variable and the kth moment pseudo normal transformation, respectively. The space defined by U' is then called kth order pseudo normal space.

In particular, when Eqs. (7.32a) and (7.32b) hold true only for the first two moments, i.e. the mean value and the standard deviation, Eqs. (7.32a) and (7.32b) become

$$U' = \frac{X - \mu_X}{\sigma_X} \tag{7.33a}$$

$$X = \sigma_X U' + \mu_X \tag{7.33b}$$

where U' may be called the second-moment pseudo normal variable. However, since only the first two moments of U' are equal to those of U, the characteristics of U' are much different from those of U, U' is generally referred to as a reduced variable of X or standardised variable of X, and Eqs. (7.33a) and (7.33b) are generally expressed as

$$X_s = \frac{X - \mu_X}{\sigma_X} \tag{7.34a}$$

$$X = \sigma_X X_s + \mu_X \tag{7.34b}$$

The space defined by X_s is generally called reduced space or standardised space.

For convenience, Eqs. (7.32a) and (7.32b) are often expressed in terms of the reduced random variable in Eq. (7.34a).

$$U' = S(X_s) \tag{7.35a}$$

$$X_s = S^{-1}(U') \tag{7.35b}$$

When Eqs. (7.35a) and (7.35b) hold true only for the first three moments, i.e. the mean value, the standard deviation, and the skewness, the transformation becomes the third-moment pseudo normal transformation (TMNT), and the space defined by the third-moment pseudo random variable is called third-order pseudo normal space.

Similarly, when Eqs. (7.35a) and (7.35b) hold true only for the first four moments, i.e. the mean value, the standard deviation, the skewness, and the kurtosis, the transformation becomes the fourth-moment pseudo normal transformation (FMNT), and the space defined by the fourth-moment pseudo random variable is called fourth-order pseudo normal space.

In the following text, we will not distinguish U' and U in writing, however, one need be aware that normal transformation $S(X_s)$ based on the first few moments is essentially pseudo normal transformation and vice versa.

7.4.2 Third-Moment Pseudo Normal Transformation

A simple third-moment pseudo normal transformation function was developed by Ono and Idota (1986), which is expressed as

$$U = S(X_s) = a_1 + a_2 X_s + a_3 X_s^2 \tag{7.36}$$

for

$$|a_{3X}| < \ < 3(a_{4X} - 1) \tag{7.37}$$

Making the first three moments of right side of Eq. (7.36) be equal to those of the left side (i.e. U), the third-moment pseudo normal transformation function can be approximated as (Zhao and Ono 2001)

$$U = \frac{a_{3X} + 3(a_{4X} - 1)X_s - a_{3X}X_s^2}{\sqrt{\left(9a_{4X} - 5a_{3X}^2 - 9\right)(1 - a_{4X})}} \tag{7.38}$$

Using Eq. (7.36), the third-moment inverse pseudo normal transformation function is given as

$$X_s = \frac{1}{2\alpha_{3X}} \left\{ 3(\alpha_{4X} - 1) - \sqrt{9(\alpha_{4X} - 1)^2 + 4\alpha_{3X} \left[\alpha_{3X} - U\sqrt{(9\alpha_{4X} - 5\alpha_{3X}^2 - 9)(1 - \alpha_{4X})} \right]} \right\}$$

(7.39)

It should be noted that the first four moments have to be used in Eqs. (7.38) and (7.39) although they are third-moment pseudo normal transformations.

Example 7.3 The Derivation for Eq. (7.38)

The third-moment standardisation function for the standardised random variable X_s in Eq. (7.36) is assumed to be (Ono and Idota 1986)

$$Y = X_s + cX_s^2 \tag{7.40a}$$
$$U = (Y - \mu_Y)/\sigma_Y \tag{7.40b}$$

in which c is determined by setting the skewness of Y equal to that of the normal random variable, i.e. c can be determined using the following equation:

$$\mu_Y = c \tag{7.41a}$$

$$\sigma_Y^2 = (\alpha_{4X} - 1)c^2 + 2\alpha_{3X}c + 1 \tag{7.41b}$$

$$\alpha_{3X}\sigma_Y^3 = (\alpha_{6X} - 3\alpha_{4X} + 2)c^3 + 3(\alpha_{5X} - 2\alpha_{3X})c^2 + 3(\alpha_{4X} - 1)c + \alpha_{3X} = 0 \tag{7.41c}$$

Since Eqs. (7.41a-c) are quite complicated and the use of the fifth and sixth moments is uncommon in engineering, assuming $|c| < <1$ according to the investigation by Ono and Idota (1986) in the cases of mild non-normality, then

$$3(\alpha_{4X} - 1)c + \alpha_{3X} = 0 \tag{7.42a}$$

$$c = \frac{-\alpha_{3X}}{3(\alpha_{4X} - 1)} \tag{7.42b}$$

Therefore, Eq. (7.38) will be easily derived.

It can be observed from Eq. (7.36) that the $k(k - 1)$th central moment of X must be determined to obtain the kth moment standardisation function. Even for the third-moment standardisation, the first six moments of X must be determined. As such the standardisation becomes complicated, it becomes more and more difficult to obtain the accurate standardisation function.

Since the x–u and u–x transformations form a pair, one transformation can be obtained from the other. In this section, the transformations are built from a u–x transformation that is assumed to be in the following form (Zhao and Ono 2000b).

$$X_s = S^{-1}(U) = \sum_{j=1}^{k} a_j U^{j-1} \tag{7.43}$$

where a_j, $j = 1, ..., k$, are deterministic coefficients obtained by making the first k central moment of $S^{-1}(U)$ to be equal to that of X_s.

Using Eq. (7.43), to obtain the kth moment standardisation function, only the first k central moments of X_s are needed. For the third-moment standardisation, the u–x transformation is expressed as

$$X_s = S^{-1}(U) = a_1 + a_2 U + a_3 U^2 \tag{7.44}$$

By making the first three central moments of $S^{-1}(U)$ to be equal to those of X_s, the coefficients can be obtained as (Zhao and Ono 2000b)

$$a_1 = -\lambda, a_2 = \sqrt{1 - \lambda^2}, a_3 = \lambda \tag{7.45}$$

where

$$\lambda = \text{sign}(\alpha_{3X}) \sqrt{2} \cos\left[\frac{\pi + |\theta|}{3}\right] \tag{7.46a}$$

$$\theta = \tan^{-1}\left(\frac{\sqrt{8 - \alpha_{3X}^2}}{\alpha_{3X}}\right) \tag{7.46b}$$

Then, u–x transformation is expressed as

$$X_s = \sqrt{1 - 2\lambda^2} U + \lambda(U^2 - 1) \tag{7.47}$$

From Eq. (7.47), the x–u transformation is readily obtained as

$$U = S(X_s) = \frac{1}{2\lambda}\left(-\sqrt{1 - 2\lambda^2} + \sqrt{2\lambda^2 + 4\lambda X_s + 1}\right) \tag{7.48}$$

Since the x–u transformation expressed in Eq. (7.48) hold true for the first three moments, the transformation is referred to as the third-moment pseudo normal transformation (TMNT). Since the second-order polynomial is used in Eqs. (7.44) and (7.48), Eq. (7.48) is called the second-order polynomial normal transformation (SPNT) and Eq. (7.44) is the inverse SPNT.

From Eq. (7.46b), in order to make Eq. (7.46b) operable, α_{3X} should satisfy the following

$$-2\sqrt{2} \le \alpha_{3X} \le 2\sqrt{2} \tag{7.49}$$

The skewness of some commonly used random variables are listed in Table 7.8. These values show that almost all the skewness satisfies Eq. (7.49), i.e. Eq. (7.49) is generally operable in most engineering practice.

Eq. (7.46b) is equivalent to the following equation.

$$\theta = \cot^{-1}\left(\frac{\alpha_{3X}}{\sqrt{8 - \alpha_{3X}^2}}\right) \tag{7.50a}$$

For small α_{3X}, using the first two terms of Maclaurin expansion for Eq. (7.50a), the absolute value of θ can be approximated as

$$|\theta| = \frac{\pi}{2} - \frac{1}{4}\sqrt{2}\alpha_{3X} \tag{7.50b}$$

Table 7.8 Skewness and kurtosis of some commonly used random variables.

Distributions	Coefficient of Variation	Skewness	Kurtosis
Normal	–	0	3.0
Lognormal	0.1	0.301	3.162
	0.2	0.608	3.664
	0.4	1.264	5.969
	0.6	2.016	11.001
	0.7	2.443	15.205
Exponential	1.0	2.0	9.0
Gamma	0.1	0.2	3.06
	0.2	0.4	3.24
	0.4	0.8	3.96
	0.6	1.2	5.16
	0.7	1.4	5.94
Gumbel	–	1.14	5.4
Frechet	0.1	1.662	8.730
	0.2	2.353	16.431
Weibull	0.1	−0.715	3.780
	0.2	−0.352	3.004
	0.3	−0.026	2.723
	0.4	0.277	2.788
	0.5	0.566	3.131
	0.7	1.131	4.593

Substituting Eq. (7.50b) into Eq. (7.46a) and utilising the relationship $\cos(\pi/2 - x) = \text{sign}(x)\sin(x)$, Eq. (7.46a) can be readily simplified as

$$\lambda = \sqrt{2}\sin\left(\frac{1}{12}\sqrt{2}\alpha_{3X}\right) \tag{7.51}$$

Since $\sin(x) = x$ for small values of x, Eq. (7.51) can be further approximated as

$$\lambda = \frac{1}{6}\alpha_{3X} \tag{7.52}$$

In order to investigate this approximation, the variations of parameter λ with respect to skewness α_{3X} obtained using both Eqs. (7.46a) and (7.52) are presented in Figure 7.2. It can be observed that when the absolute value of skewness is small (e.g. $|\alpha_{3X}| \leq 1$), λ is approximately proportional to α_{3X}, and Eq. (7.52) approximates Eq. (7.46a) very well in the range of $|\alpha_{3X}| \leq 1$.

Substitute Eq. (7.52) into the Eqs. (7.47) and (7.48), a pair of simple transformations is given as

$$X_s = -\frac{1}{6}\alpha_{3X} + \sqrt{1 - \frac{1}{18}\alpha_{3X}^2}\,U + \frac{1}{6}\alpha_{3X}U^2 \tag{7.53}$$

$$U = \frac{1}{\alpha_{3X}}\left(-\sqrt{9 - \frac{1}{2}\alpha_{3X}^2} + \sqrt{9 + \frac{1}{2}\alpha_{3X}^2 + 6\alpha_{3X}X_s}\right) \tag{7.54}$$

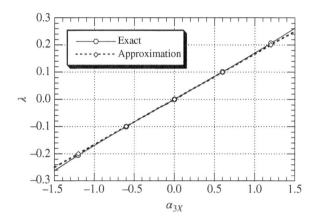

Figure 7.2 The relationship between λ and α_{3X}.

Equations (7.53) and (7.54) can be further simplified as

$$X_s = U + \frac{1}{6}\alpha_{3X}\left(U^2 - 1\right) \tag{7.55}$$

$$U = \frac{1}{\alpha_{3X}}\left(\sqrt{9 + \alpha_{3X}^2 + 6\alpha_{3X}X_s} - 3\right) \tag{7.56}$$

It can be observed that Eq. (7.55) is the same as the Cornish-Fisher expansions using only the first second-order polynomial, which is expressed in Eq. (7.67).

Example 7.4   Derivation for Eq. (7.46a)

For the third-moment standardisation function in Eq. (7.44), the following equations involving a_1, a_2, and a_3 can be derived by making the first three moments of $S^{-1}(U)$ equal to those of X_S.

$$\mu_{su} = a_1 + a_3 = 0 \tag{7.57a}$$

$$\sigma_{su}^2 = a_2^2 + 2a_3^2 = 1 \tag{7.57b}$$

$$\alpha_{3su}\sigma_{su}^3 = 6a_2^2 a_3 + 8a_3^3 = \alpha_{3X} \tag{7.57c}$$

After simplification, the following reduced cubic equation of a_3 is obtained

$$a_3^3 - \frac{3}{2}a_3 + \frac{1}{4}\alpha_{3X} = 0 \tag{7.58}$$

The discriminant is given as

$$D = -\frac{1}{16}\left(\alpha_{3X}^2 - 8\right) \tag{7.59}$$

Thus Eq. (7.59) has three real roots for $\alpha_{3X}^2 \geq 8$ and one real root for $\alpha_{3X}^2 < 8$. From Eq. (7.57b), a_3 should satisfy the following equation in order for a_2 to be real.

$$a_2^2 = 1 - 2a_3^2 \geq 0 \tag{7.60}$$

That is

$$-\frac{\sqrt{2}}{2} \leq a_3 \leq \frac{\sqrt{2}}{2} \tag{7.61}$$

the only root of Eq. (7.58) that satisfies Eq. (7.61) is obtained as

$$a_3 = \text{sign}(\alpha_{3X})\sqrt{2}\cos\left[\frac{\text{sign}(\alpha_{3X})\theta - \pi}{3}\right] \tag{7.62a}$$

where

$$\theta = \arctan\left(\frac{\sqrt{8 - \alpha_{3X}^2}}{-\alpha_{3X}}\right) \tag{7.62b}$$

Let $\lambda = a_3$ and using the relationships $\tan(-\phi) = -\tan(\phi)$ and $\cos(-\phi) = \cos(\phi)$, Eqs. (7.46a) and (7.46b) can be readily obtained.

Example 7.5   Investigation on the Efficiency of the Third-Moment Pseudo Normal Transformations

Consider a lognormal variable X with $\mu = 20$ and $V = 0.2$, the exact normal and inverse normal transformation are given as

$$U = \frac{\ln X - \lambda}{\zeta} \tag{7.63a}$$

$$X = \exp\left(\zeta U + \lambda\right) \tag{7.63b}$$

The normal and inverse normal transformations are illustrated in Figure 7.3a and b, respectively. From these figures, one may understand that the third-moment pseudo normal transformation function is in good agreement with the exact transformation function for relative large U or X_s.

In order to investigate the normality of the pseudo normal variable transformed from Eq. (7.48), the PDF of U obtained from Eq. (7.48) is depicted in Figure 7.4a and b using the first three moments of X. It can be observed that the PDF of the third-moment pseudo normal variable approaches that of the normal variable quite well. This indicates that the pseudo normal variable obtained from Eq. (7.48) can be approximately considered as a normal variable.

Similarly, the PDF of X obtained from Eq. (7.47) is presented in Figure 7.5a, b using the first three moments of X. It can also be observed that the PDF of the inverse third-moment pseudo normal variable approaches that of the lognormal variable quit well. This implies that the inverse pseudo normal variable obtained from Eq. (7.47) can be approximately considered as a lognormal variable if the first three moments of the lognormal variable are used.

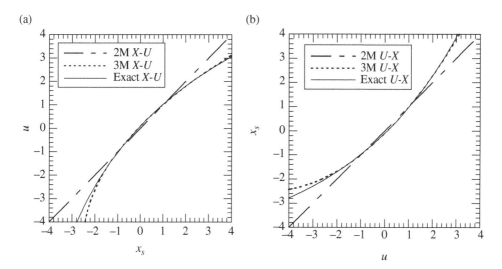

Figure 7.3 Pseudo normal transformation for lognormal variable. (a) X–U. (b) U–X.

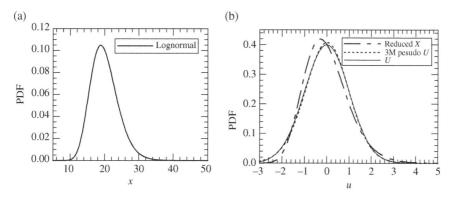

Figure 7.4 PDF after pseudo normal transformation. (a) Original PDF. (b) PDF after transformation.

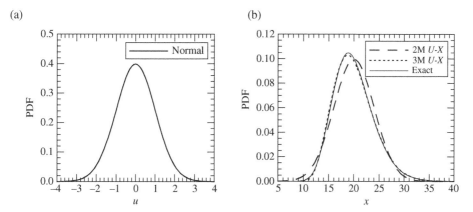

Figure 7.5 PDF after inverse pseudo normal transformation. (a) Original PDF. (b) PDF after transformation.

Example 7.6   The Derivation of the Expressions of PDFs in Figures 7.4 and 7.5

According to Eq. (A.88) in Appendix A, the original PDF is expressed as

$$f_{X_s}(x_s) = f_X[\sigma_X x_s + \mu_X]\sigma_X = \frac{\sigma_X}{\sqrt{2\pi}\zeta(\sigma_X x_s + \mu_X)} \exp\left[-\frac{1}{2}\left(\frac{\ln(\sigma_X x_s + \mu_X) - \lambda}{\zeta}\right)^2\right]$$

(7.64)

Based on Eqs. (7.28) and (7.44), the PDF after the third-moment pseudo normal transformation is

$$f_U(u) = f_{X_s}(x_s)\frac{dx_s}{du} = f_{X_s}[S^{-1}(u)]\frac{dS^{-1}(u)}{du}$$

$$= \frac{\sigma_X\left(\sqrt{1-2\lambda^2} + 2\lambda u\right)}{\sqrt{2\pi}\zeta\left\{\sigma_X\left[\sqrt{1-2\lambda^2}u + \lambda(u^2-1)\right] + \mu_X\right\}}$$

(7.65)

$$\exp\left[-\frac{1}{2}\left(\frac{\ln\left\{\sigma_X\left[\sqrt{1-2\lambda^2}u + \lambda(u^2-1)\right] + \mu_X\right\} - \lambda}{\zeta}\right)^2\right]$$

The PDF in Figure 7.5 can be given similarly.

Example 7.7   Investigation on the Shortcomings of the Third-Moment Pseudo Transformation

In order to investigate the shortcomings of the third-moment pseudo normal transformation method, consider a Weibull random variable. Two cases are investigated including the coefficient of variation taken as $V = 0.1$ and $V = 0.7$. As listed in Table 7.8, the skewness α_{3X} is equal to -0.715 for $V = 0.1$ and 1.131 for $V = 0.7$.

The variations of the u–x transformation function with respect to u are shown in Figure 7.6 for the results obtained from the exact transformation and the third-moment pseudo transformation. Figure 7.6 shows that the third-moment pseudo transformation provides good approximations for the exact result when the absolute value of u is not very large. For $V = 0.1$, which implies that the skewness is negative, the third-moment pseudo transformation leads to significant error when u is larger than 2. For $V = 0.7$, which implies that the skewness is positive, the accurate third-moment transformation produces significant errors when u is less than -3.0. These imply that the specific range of u or x_s for good approximations of the method depends on the value of α_{3X}.

7.4.3   Fourth-Moment Pseudo Normal Transformation

7.4.3.1   The Third-Order Polynomial Transformation

When only the first four moments of a random variable X are available, one may use the Edgeworth and Cornish-Fisher expansions to realise the fourth-moment pseudo normal transformation. For the standardised random variable X_s defined by Eq. (7.34a), the expansions are expressed as follows

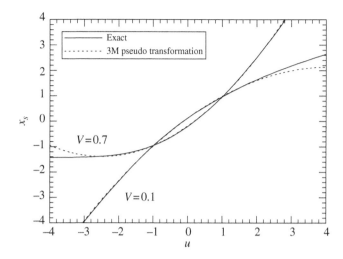

Figure 7.6 *u–x* transformation for Weibull distribution.

$$F(x_s) = \Phi(x_s) - \phi(x_s)\left[\frac{1}{6}\alpha_{3X}h_2(x_s) + \frac{1}{24}(\alpha_{4X} - 3)h_3(x_s) + \frac{1}{72}\alpha_{3X}^2 h_5(x_s)\right] \qquad (7.66)$$

$$U = X_s - \frac{1}{6}\alpha_{3X}h_2(X_s) - \frac{1}{24}(\alpha_{4X} - 3)h_3(X_s) - \frac{1}{36}\alpha_{3X}^2\left(4X_s^3 - 7X_s\right) \qquad (7.67)$$

$$X_s = U + \frac{1}{6}\alpha_{3X}h_2(U) + \frac{1}{24}(\alpha_{4X} - 3)h_3(U) + \frac{1}{36}\alpha_{3X}^2\left(2U^3 - 5U\right) \qquad (7.68)$$

where $h_2(\cdot)$, $h_3(\cdot)$, and $h_5(\cdot)$ are the second-, third-, and fifth-order Hermite polynomials, respectively, and can be expressed as

$$h_2(x) = x^2 - 1, h_3(x) = x^3 - 3x, h_5(x) = x^5 - 10x^3 + 15x \qquad (7.69)$$

The general form of the three polynomials is referred to as an inverse Cornish-Fisher expansion (Stuart and Ord 1987). It has been shown (Hong 1996) that they only provide suitable results when the non-normality of the random variable is very small.

Fleishman (1978) suggested the following third-order polynomial transformation

$$X_s = a_1 + a_2 U + a_3 U^2 + a_4 U^3 \qquad (7.70)$$

where X_s is the standardised random variable; U is the standard normal random variable; and a_1, a_2, a_3, and a_4 are the polynomial coefficients that can be determined by moment-matching method (Fleishman 1978), i.e. making the first four moments of the left side of Eq. (7.70) equal to those of the right side, i.e.

$$a_1 + a_3 = 0 \qquad (7.71a)$$

$$a_2^2 + 2a_3^2 + 6a_2a_4 + 15a_4^2 = 1 \qquad (7.71b)$$

$$6a_2^2a_3 + 8a_3^3 + 72a_2a_3a_4 + 270a_3a_4^2 = \alpha_{3X} \qquad (7.71c)$$

$$3\left(a_2^4 + 20a_2^3a_4 + 210a_2^2a_4^2 + 1260a_2a_4^3 + 3465a_4^4\right) + 12a_3^2\left(5a_2^2 + 5a_3^2 + 78a_2a_4 + 375a_4^2\right) = \alpha_{4X} \qquad (7.71d)$$

Simplification of Eqs. (7.71a)–(7.71d) leads to the following equations of parameters a_2 and a_4

$$2A_1A_2^2 = \alpha_{3X}^2 \tag{7.72a}$$

$$3A_1A_3 + 3A_4 = \alpha_{4X} \tag{7.72b}$$

where

$$A_1 = 1 - a_2^2 - 6a_2a_4 - 15a_4^2 \tag{7.72c}$$

$$A_2 = 2 + a_2^2 + 24a_2a_4 + 105a_4^2 \tag{7.72d}$$

$$A_3 = 5 + 5a_2^2 + 126a_2a_4 + 675a_4^2 \tag{7.72e}$$

$$A_4 = a_2^4 + 20a_2^3a_4 + 210a_2^2a_4^2 + 1260a_2a_4^3 + 3465a_4^4 \tag{7.72f}$$

Since the values α_{3X} and α_{4X} are known, the parameters a_2 and a_4 can be obtained from Eqs. (7.72a)–(7.72f). After the parameters a_2 and a_4 have been obtained, the parameters a_1 and a_3 can be readily solved as

$$a_3 = -a_1 = \frac{\alpha_{3X}}{2A_2} \tag{7.73}$$

Equation (7.70) can be readily used for structural reliability analysis when the coefficients a_1, a_2, a_3, and a_4 are known. However, the determination of the four coefficients is not easy, since the solution of nonlinear equations has to be found when using the moment-matching method (Fleishman 1978). Some other methods such as the least-square method to determine the polynomial coefficients have been reported by Hong and Lind (1996) and Chen and Tung (2003). One may use the common subroutines of nonlinear equations such as the 'FindRoot' function in 'Mathematica' software (Wolfram 2003).

For convenience, the information to approximate the four parameters of a_1, a_2, a_3, and a_4 is given in Table C.2 (see Appendix C) for selected values of α_3 and α_4. In particular, for $\alpha_3 = 0$ and $\alpha_4 = 3$, the parameters are obtained as $a_1 = a_3 = a_4 = 0$, $a_2 = 1$, and Eq. (7.70) reduces to $X_s = U$.

7.4.3.2 Operable Area of the Third-Order Polynomial Transformation

For a specified value of α_3, when the values of α_4 are below a limit value, Eqs. (7.72a) and (7.72b) becomes inoperable. Using the limit value of α_4 for which Eqs. (7.72a) and (7.72b) is inoperable corresponding to the selected α_3, a lower boundary line in the $\alpha_3{}^2 - \alpha_4$ plane can be depicted as shown in Figure 7.7, where the operable area of the third-order polynomial transformation are indicated as the shaded region. The lower boundary line for which Eq. (7.70) is operable is found to be nearly a straight line approximately expressed by (Zhao and Lu 2008)

$$\alpha_4 = 1.88 + 1.55\alpha_3^2 \tag{7.74}$$

In Figure 7.7, the limit for all distributions expressed as $\alpha_4 = \alpha_3{}^2 + 1$ (Johnson and Kotz 1970a) is also depicted along with the $\alpha_3{}^2 - \alpha_4$ relationship for some commonly used distributions, i.e. the normal, Laplace, Gumbel, Rayleigh, and the exponential distribution, which are represented by a single point, the lognormal, the Gamma, and the Weibull

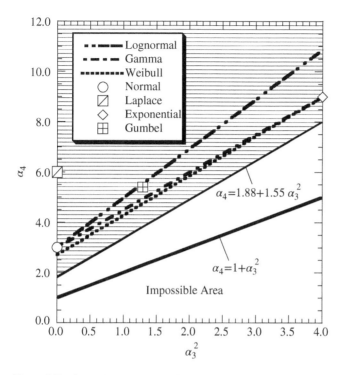

Figure 7.7 Operable area of the third-order polynomial transformation.

distributions, which are represented by a line. It can be observed that the operable area of the third-order polynomial transformation covers a large area in the $\alpha_3{}^2$–α_4 plane, and the $\alpha_3{}^2$–α_4 relationship for most commonly used distributions are in the operable area of this distribution. This implies that the third-order polynomial transformation is generally operable for common engineering applications.

7.4.3.3 Complete Monotonic Expression of the Third-Order Polynomial Normal Transformation

After the four coefficients are obtained, the fourth-moment pseudo normal transformation (FMNT) corresponding to Eq. (7.70) can be readily derived as

$$U = \sqrt[3]{A} + \sqrt[3]{B} - \frac{a}{3} \tag{7.75}$$

where A, B, and a are parameters expressed as follows

$$A = -\frac{q}{2} + \sqrt{\Delta}, B = -\frac{q}{2} - \sqrt{\Delta} \tag{7.76a}$$

$$\Delta = \left(\frac{p}{3}\right)^3 + \left(\frac{q}{2}\right)^2, q = \frac{2}{27}a^3 - \frac{ac}{3} - a - \frac{X_s}{a_4}, a = \frac{a_3}{a_4}, c = \frac{a_2}{a_4} \tag{7.76b}$$

Table 7.9 Complete expression of the fourth-moment pseudo normal transformation $S(x_s)$.

Parameter			Range of x	Normal transformation u	Types
$a_4 < 0$			$J_2^* < x < J_1^*$	$-2r\cos[(\theta + \pi)/3] - a/3$	I
$a_4 > 0$	$p < 0$	$a_{3X} \geq 0$	$J_1^* < x < J_2^*$	$2r\cos(\theta/3) - a/3$	II
			$x \geq J_2^*$	$\sqrt[3]{A} + \sqrt[3]{B} - a/3$	
		$a_{3X} < 0$	$J_1^* < x < J_2^*$	$-2r\cos[(\theta - \pi)/3] - a/3$	III
			$x \leq J_1^*$	$\sqrt[3]{A} + \sqrt[3]{B} - a/3$	
	$p \geq 0$			$\sqrt[3]{A} + \sqrt[3]{B} - a/3$	IV
$a_4 = 0$		$a_{3X} \neq 0$	$a_2^2 + 4a_3(a_3 + x_s) \geq 0$	$\left[-a_2 + \sqrt{a_2^2 + 4a_3(a_3 + x_s)}\right]/2a_3$	V
		$a_{3X} = 0$		x_s	VI

The applicable range of this solution is

$$p = \frac{3a_2a_4 - a_3^2}{3a_4^2} \geq 0 \tag{7.77}$$

In most cases, the solution is applicable. For the pairs of a_3 and a_4 beyond this range, one may select the accurate expression from Table 7.9, which summarises the complete monotonic expressions of the FMNT (Zhao et al. 2018a). The derivation of Table 7.9 is presented in Example 7.8.

In Table 7.9, p, a, A, and B are expressed in Eqs. (7.77), (7.76b), and (7.76a), respectively; a_2, a_3, a_4 are polynomial coefficients given in Eq. (7.70); and the parameters θ, r, J_1^*, and J_2^* are given by

$$\theta = \arccos\left(\frac{-q}{2r^3}\right), r = \sqrt{-\frac{p}{3}} \tag{7.78a}$$

$$J_1^* = \sigma_X a_4\left(-2r^3 + \frac{2}{27}a^3 - \frac{ac}{3} - a\right) + \mu_X, J_2^* = \sigma_X a_4\left(2r^3 + \frac{2}{27}a^3 - \frac{ac}{3} - a\right) + \mu_X \tag{7.78b}$$

Since the x–u transformation expressed in Eq. (7.75) or Table 7.9 hold true for the first four moments, the transformation is referred to as FMNT. Because the FMNT is realised by using the third-order polynomial, it is also called the third-order polynomial normal transformation (TPNT).

The applicable regions of different types of the FMNT with pairs of skewness and kurtosis are depicted in the a_{3X}–a_{4X} plane in Figure 7.8. It can be observed from Figure 7.8 that:

1) The applicable region of type IV of the FMNT covers the largest area in the a_{3X}–a_{4X} plane, which is the appropriate x–u transformation for most combinations of skewness and kurtosis.

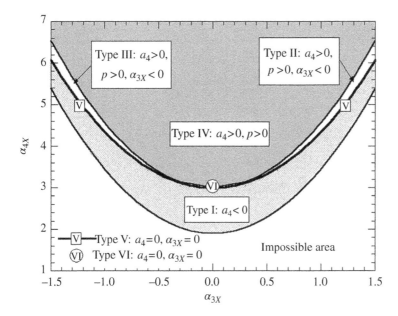

Figure 7.8 The applicable regions of different types of the third order polynomial transformation in the $\alpha_{3x}-\alpha_{4x}$ plane.

2) The applicable region of Type II and Type III are symmetrical and correspond to the cases of $\alpha_{3X} > 0$ and $\alpha_{3X} < 0$, respectively. Type II and Type III are less common for the FMNT since their applicable region is small in the $\alpha_{3X}-\alpha_{4X}$ plane.
3) The applicable region of Type I is much larger than those of Type II and Type III. When the value of α_{4X} is relatively small, Type I of the FMNT is an appropriate $x-u$ transformation.
4) The applicable regions of Type V and Type VI are represented by the dotted line and a single point, respectively. In fact, they are reduced forms of the FMNT, and will be discussed in Example 7.8.

According to Table 7.9, the $x-u$ transformation has six types, i.e. Type I, II, III, IV, V, and VI, and each type has an applicable range of x. Those constitute the complete expression of the FMNT, based on which the $x-u$ transformation can be easily realised. The variation of u with respect to x are presented in Figure 7.9a–f for these six types, respectively, where the suitable values of u are illustrated in thick solid lines and the unsuitable values are shown in dotted lines. As can be observed from Figure 7.9a–f, by eliminating the unsuitable values of u, the variation of the specific values of u with respect to x, i.e. the $x-u$ transformation, is monotonic for each type of FMNT.

7.4.3.4 An Explicit Fourth-Moment Pseudo Normal Transformation
As just described, the Fleishman expression is implicit since the nonlinear equations have to be solved. Thus, the second-order Fisher-Cornish expansion (Fisher and Cornish 1960) is sometimes used, which is expressed as

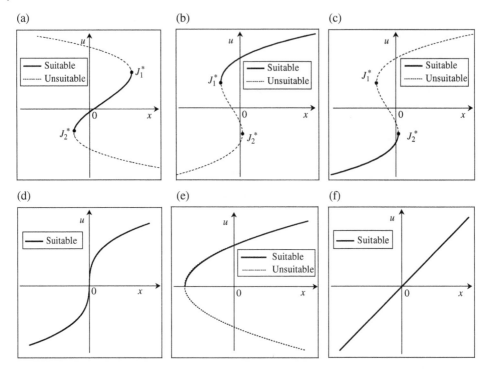

Figure 7.9 Suitable and unsuitable roots for the six types of the FMNT. (a) Type I. (b) Type II. (c) Type III. (d) Type IV. (e) Type V. (f) Type VI.

$$X_s = U + \frac{\alpha_{3X}}{6}\left(U^2 - 1\right) + \frac{1}{24}(\alpha_{4X} - 3)\left(U^3 - 3U\right) \tag{7.79a}$$

and its inverse transformation is

$$U = X_s - \frac{\alpha_{3X}}{6}\left(X_s^2 - 1\right) - \frac{1}{24}(\alpha_{4X} - 3)\left(X_s^3 - 3X_s\right) \tag{7.79b}$$

It can be observed that Eqs. (7.79a,b) are in closed form and are quite easily used. Since the Hermite series are used to express the term in Eqs. (7.79a,b), they are referred to as Hermite model in many references, and Eq. (7.79a) is often written in the following term.

$$X_s = -h_3 + (1 - 3h_4)U + h_3 U^2 + h_4 U^3 \tag{7.80a}$$

where

$$h_3 = \frac{\alpha_{3X}}{6}, h_4 = \frac{\alpha_{4X} - 3}{24} \tag{7.80b}$$

Eq. (7.80a) holds true when $\lim x_s \to u$, and when this cannot be satisfied, the transformation results in relatively large errors since the first four moments of the right side of Eq. (7.80a) are not equal to those of the left side.

Winterstein (1988) developed an expansion expressed as

$$X_s = -\widetilde{k}\widetilde{h}_3 + \widetilde{k}\left(1 - 3\widetilde{h}_4\right)U + \widetilde{k}\widetilde{h}_3 U^2 + \widetilde{k}\widetilde{h}_4 U^3 \tag{7.81}$$

in which

$$\widetilde{h}_3 = \frac{\alpha_{3X}}{4 + 2\sqrt{1 + 1.5(\alpha_{4X} - 3)}}, \widetilde{h}_4 = \frac{\sqrt{1 + 1.5(\alpha_{4X} - 3)} - 1}{18} \tag{7.82a}$$

$$\widetilde{k} = \frac{1}{\sqrt{1 + 2\widetilde{h}_3 + 6\widetilde{h}_4}} \tag{7.82b}$$

Apparently, the Winterstein formula requires $\alpha_{4X} > 7/3$ because of Eq. (7.82a).

It has been demonstrated that the Winterstein formula gives much improvement upon Fisher-Cornish expansion while retaining its simplicity and explicitness. However, as will be discussed later, the transformation is still not convincing since the differences of the first four moments between the two sides of Eq. (7.81) are still large. In general, the transformation in practical engineering should be as simple and accurate as possible.

An explicit fourth-moment standardisation function is suggested as following by trial and error (Zhao and Lu 2007a)

$$X_s = S^{-1}(U) = -l_1 + k_1 U + l_1 U^2 + k_2 U^3 \tag{7.83}$$

where $S^{-1}(U)$ denotes the third polynomial of u; and the coefficients l_1, k_1, and k_2 are given as:

$$l_1 = \frac{\alpha_{3X}}{6(1 + 6l_2)}, \quad k_1 = \frac{1 - 3l_2}{(1 + l_1^2 - l_2^2)}, \quad k_2 = \frac{l_2}{(1 + l_1^2 + 12l_2^2)} \tag{7.84a}$$

$$l_2 = \frac{1}{36}\left(\sqrt{6\alpha_{4X} - 8\alpha_{3X}^2 - 14} - 2\right) \tag{7.84b}$$

From Eq. (7.84b), the following should be satisfied:

$$\alpha_{4X} \geq \left(7 + 4\alpha_{3X}^2\right)/3 \tag{7.84c}$$

Particularly, if $\alpha_{3X} = 0$ and $\alpha_{4X} = 3$, then l_1, l_2, k_1, and k_2 will be obtained as $l_1 = l_2 = k_2 = 0$ and $k_1 = 1$, and the u–x transformation function reduces to $X_s = U$.

Figure 7.10a–i presents the four polynomial coefficients a_1, a_2, a_3, and a_4 determined by Eq. (7.83), when compared with those obtained using Fisher-Cornish expansion, Winterstein formula, and the accurate coefficients obtained from the moment-matching method (Fleishman 1978). The coefficients are expressed as function of α_{4X} for $\alpha_{3X} = 0.0, 0.4$, and 0.8. It can be observed from Figure 7.10 that

1) The coefficients of the Fisher-Cornish expansion have the greatest differences from the accurate coefficients except that the random variable X is nearly a normal random variable.
2) The Winterstein formula improves the Fisher-Cornish expansion very much and gives good results when α_{3X} is small and α_{4X} are within a particular range. However, as α_{3X} becomes larger, especially when α_{3X} is larger than 0.4, the coefficients obtained by the Winterstein formula will have significant differences from the accurate ones.

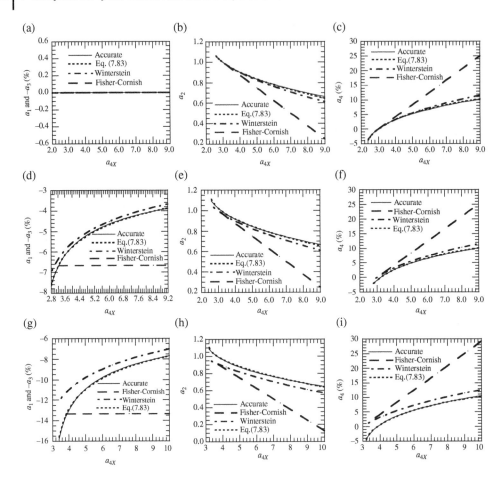

Figure 7.10 Comparisons of the determination of polynomial coefficients using different methods. (a) $a_1, -a_3$ ($\alpha_{3X} = 0.0$). (b) a_2 ($\alpha_{3X} = 0.0$). (c) a_4 ($\alpha_{3X} = 0.0$). (d) $a_1, -a_3$ ($\alpha_{3X} = 0.4$). (e) a_2 ($\alpha_{3X} = 0.4$). (f) a_4 ($\alpha_{3X} = 0.4$). (g) $a_1, -a_3$ ($\alpha_{3X} = 0.8$). (h) a_2 ($\alpha_{3X} = 0.8$). (i) a_4 ($\alpha_{3X} = 0.8$)

3) The coefficients obtained using the Eq. (7.83) formula are in close agreement with the accurate ones in the whole investigation range.

Thus, Eq. (7.83) is the simple and accurate fourth-moment standardisation function. For a random variable, if the first four moments can be obtained, the u–x and x–u transformation can be respectively realised with Eqs. (7.83) and (7.75) or Table 7.9, where a_1, a_2, a_3, and a_4 should be replaced by $l_1, k_1, -l_1$, and k_2, respectively.

Example 7.8 The Derivation of Complete Monotonic Expression of the Fourth-Moment Pseudo Normal Transformation

The FMNT, i.e. the inverse third-order polynomial transformation (x–u transformation), should be conducted by finding the solution to Eq. (7.70). With different combinations of skewness and kurtosis, which result in different combinations of the parameters in

Eq. (7.70), there may be more than one possible value of u corresponding to each value of x. Without clear definition of a complete expression for the inverse transformation and for the corresponding monotonic regions of x or u, the inverse transformation will be inappropriate, even unreliable, to be used in structural reliability.

This example will derive the complete expressions of the inverse transformation for different combinations of skewness and kurtosis, and then investigate the monotonicity of each expression.

Equation (7.70) could be a linear, quadratic, or cubic function with different combinations of skewness and kurtosis of x. Since linear and quadratic functions are a reduced form of the cubic function, we focus on the latter, with the assumption that $a_4 \neq 0$. Noting that $a_1 = -a_3$, Eq. (7.70) can be rewritten as:

$$x' = S_u^*(u) = u^3 + a_3'u^2 + a_2'u - a_3' \tag{7.85}$$

where $S_u^*(u)$ is a cubic function of u; a_2' and a_3' are polynomial coefficients; and x' is the transformed random variable of x_s. So a_2', a_3', and x' are given by

$$a_3' = \frac{a_3}{a_4}, a_2' = \frac{a_2}{a_4}, x' = \frac{x_s}{a_4} \tag{7.86}$$

To identify the proper x–u transformation, it is necessary to find the solution to Eq. (7.85). The numbers in the solution of Eq. (7.85) depend on the monotonicity of $S_u^*(u)$. The shape of monotonic and non-monotonic $S_u^*(u)$ are depicted in Figure 7.11a and b, respectively. The proper x–u transformation can be expressed as the u-value of the intersections between the curve $S_u^*(u)$ and assumed lines $x' = x_i'$, which are introduced to illustrate the problem clearly. In Figure 7.11a, there exist two particular values of x_i', i.e. x_1' and x_2', which are in the two distinguished sections of $S_u^*(u)$ divided by its stationary point J_0. In Figure 7.11b,

(a)

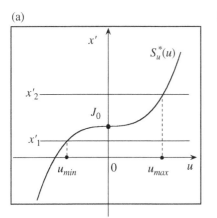

(b)

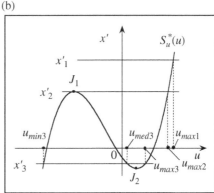

Figure 7.11 Relative position between $S_u^*(u)$ and the line $x = x_i'$ ($i = 1, 2, 3$). (a) One stationary point. (b) Two stationary points.

there exist three particular values of x_i', i.e. x_1', x_2', and x_3', which respectively correspond to the situation when there are one, two, and three intersections between $S^*_u(u)$ and the line $x' = x_i'$.

When $S^*_u(u)$ is monotonic, as shown in Figure 7.11a, there is one stationary point, J_0. The range of u is divided into two monotonic regions: $(-\infty, J_0)$ and $(J_0, +\infty)$. For the convenience of the analysis, the u-values of the intersections between the curve $S^*_u(u)$ and the line $x' = x_i'$ ($i = 1$ and 2), i.e. the solution of Eq. (7.85), referred to in these two sections as u_{min} and u_{max} ($u_{min} < u_{max}$), respectively.

When $S^*_u(u)$ is nonmonotonic, as shown in Figure 7.11b, there are two stationary points, J_1 and J_2. The range of u is divided into three monotonic regions: $(J_2, +\infty)$, $[J_1, J_2]$ and $(-\infty, J_1)$. The u-values of the intersections between the curve $S^*_u(u)$ and the line $x' = x_i'$ ($i = 1, 2,$ and 3), i.e. the solution of Eq. (7.85), in these three sections are referred as u_{max}, u_{med}, and u_{min} ($u_{max} > u_{med} > u_{min}$), respectively.

When there are two stationary points, one may need to select a specific value of u. Since the properties of the derivative of $S^*_u(u)$ at u_{max}, u_{med}, and u_{min} are distinct from each other, the specific value of u can be determined by the characteristic of the derivative of $S^*_u(u)$. The procedure to determine the specific value of u is shown in Figure 7.12 corresponding to each value of x.

According to Figure 7.12, three properties of $S^*_u(u)$ should be discussed to determine the specific value of u corresponding to each x. The relationship between these properties and the first four moments of x are presented in the following.

***Property* 1:** The number of the stationary points of $S^*_u(u)$ is determined by the sign of a parameter, p, which is a function of the skewness and kurtosis of x.

The stationary points of $S^*_u(u)$ are defined as the value of u at which $dS^*_u(u)/du = 0$. According to Eq. (7.85), $dS^*_u(u)/du$ is a quadratic polynomial and expressed as

$$\frac{dS^*_u(u)}{du} = 3u^2 + 2a_3'u + a_2' \tag{7.87a}$$

The number of the stationary points of $S^*_u(u)$, i.e. the number of the real roots solved by making Eq. (7.87a) equal to zero, is determined by the discriminant p of Eq. (7.87b), formulated as

$$p = \frac{3a_2a_4 - a_3^2}{3a_4^2} \tag{7.87b}$$

For $p \geq 0$, there is only one stationary point of $S^*_u(u)$; and for $p < 0$, there are two different stationary points of $S^*_u(u)$. Since p is a function of a_2, a_3, and a_4 which are functions of the first four moments of x_s, the number of the stationary points of $S^*_u(u)$ is then determined by the skewness and kurtosis of x_s.

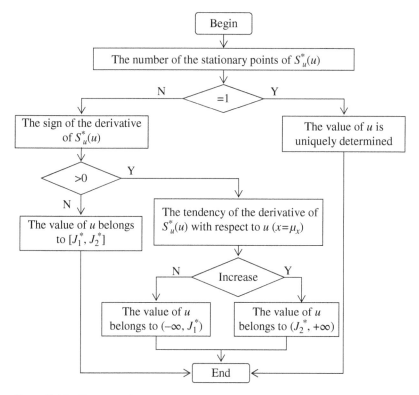

Figure 7.12 The procedure to determine the specific value of u corresponding to one x.

Property 2: The sign of $dS^*_u(u)/du$ is the same as that of a_4.

The derivative of $S^*_u(u)$ can be expressed as follows:

$$\frac{dS^*_u(u)}{du} = \frac{dx'}{du} = \frac{1}{a_4\sigma_x}\frac{dx}{du} \tag{7.88}$$

Assuming that

$$F(x) = \Phi(u) \tag{7.89}$$

where $F(\cdot)$ and $\Phi(\cdot)$ are the CDFs of x and u, respectively.

By differentiating both sides of Eq. (7.89) with respect to u, it can be derived that:

$$\frac{dx}{du} = \frac{\phi(u)}{f(x)} \tag{7.90}$$

Substituting Eq. (7.90) into Eq. (7.88) gives:

$$\frac{dS^*_u(u)}{du} = \frac{1}{a_4\sigma_x}\frac{\phi(u)}{f(x)} \tag{7.91}$$

It can be observed from Eq. (7.91) that, since the value of $f(x)$, $\phi(u)$, and σ_x are positive, the sign of $dS^*_u(u)/du$ is the same as that of a_4.

Property 3: For $a_4 > 0$, the tendency of $dS^*_u(u)/du$ at $u = 0$ decreases with negative skewness (α_{3x}) and increases with positive skewness (α_{3x}).

According to Eq. (7.91), $d^2S^*_u(u)/du^2$ is formulated as:

$$\frac{d^2S^*_u(u)}{du^2} = \frac{1}{a_4\sigma_x} \frac{\phi'(u)f(x) - \phi^2(u)f'(x)/f(x)}{f^2(x)} \tag{7.92}$$

where $f'(x)$ and $\phi'(u)$ represent the derivatives of $f(x)$ with respect to x and $\phi(u)$ with respect to u, respectively.

When $u = 0$, the value of $d^2S^*_u(u)/du^2$ is

$$\frac{d^2S^*_u(u)}{du^2}\bigg|_{x=\mu_x, u=0} = \frac{1}{a_4\sigma_x} \frac{\phi'(0)f(\mu_x) - \phi^2(0)f'(\mu_x)/f(\mu_x)}{f^2(\mu_x)} \tag{7.93}$$

Since $\phi(u)$ reaches its extreme value at $u = 0$, Eq. (7.93) can be simplified as follows

$$\frac{d^2S^*_u(u)}{du^2}\bigg|_{x=\mu_x, u=0} = Kf'(\mu_x) \tag{7.94a}$$

where $f'(\mu_x)$ denotes the derivative of $f(x)$ at $x = \mu_x$, and K is given by

$$K = -\frac{\phi^2(0)}{a_4\sigma_x f^3(\mu_x)} \tag{7.94b}$$

Obviously, when $a_4 > 0$, K is a negative constant. In order to clearly show the relationship between the sign of $f'(\mu_x)$ and α_{3x}, Figure 7.13a–b present representative PDFs with positive and negative α_{3x}, respectively. As can be observed from Figure 7.13, when $\alpha_{3x} > 0$, $f'(\mu_x) < 0$; and when $\alpha_{3x} < 0$, $f'(\mu_x) > 0$.

For $p \geq 0$, $S^*_u(u)$ has only one solution and hence there is only one value of u corresponding to each x. Therefore the FMNT is inherently monotonic and no further investigation is needed. According to the Cardano formula, the explicit analytical expressions of u can be determined from Eq. (7.75). Generally, Eq. (7.75) is an appropriate x–u transformation for most combinations of skewness and kurtosis shown in Figure 7.8.

For $p < 0$, there are two stationary points of $S^*_u(u)$, and there might exist more than one value of u for each value of x. The number of values of u is determined by the value of x as follows

$$\begin{cases} x > \max\left(J^*_1, J^*_2\right) \text{ or } x < \min\left(J^*_1, J^*_2\right), & 1 \text{ root} \\ x = J^*_1 \text{ or } x = J^*_2, & 2 \text{ roots} \\ \min\left(J^*_1, J^*_2\right) < x < \max\left(J^*_1, J^*_2\right), & 3 \text{ roots} \end{cases} \tag{7.95a}$$

(a)   (b)

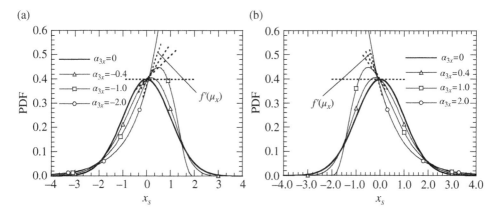

Figure 7.13 The relationship between the sign of $f'(\mu_x)$ and α_{3x}. (a)$\alpha_{3x} \le 0$. (b) $\alpha_{3x} \ge 0$.

$$J_1^* = \sigma_x a_4 \left(-2r^3 + \frac{2}{27}a^3 - \frac{ac}{3} - a \right) + \mu_x, \quad J_2^* = \sigma_x a_4 \left(2r^3 + \frac{2}{27}a^3 - \frac{ac}{3} - a \right) + \mu_x,$$

$$r = \sqrt{-\frac{p}{3}}$$

$$(7.95b)$$

where J_1^* and J_2^* are the values of x when $S^*_u(u)$ reaches its minimum and maximum values for $a_4 > 0$, respectively.

For $x > \max(J_1^*, J_2^*)$, or $x < \min(J_1^*, J_2^*)$, there is only one value of u corresponding to each x as given in Eq. (7.75).

For $x = J_2^*$, or $x = J_1^*$, there are two values of u corresponding to one x, one of which is an extreme point of $S^*_u(u)$. According to Eq. (7.91), the derivative of $S^*_u(u)$ cannot be equal to 0, which makes it impossible for the extreme points of $S^*_u(u)$ to be the specific value of u. Then, the value of u corresponding to each x is uniquely determined in Eq. (7.75).

For $\min(J_1^*, J_2^*) < x < \max(J_1^*, J_2^*)$, there are three values of u corresponding to each x, in which case the theoretical properties of $S^*_u(u)$ will be considered to find the specific value of u. According to Property 2, $dS^*_u(u)/du$ is negative for $a_4 < 0$. Then u_{med} is the specific value given as follows

$$u_{med} = -2r[\cos(\theta + \pi)/3] - \frac{a}{3} \tag{7.96a}$$

$$\theta = \arccos\left(\frac{-q}{2r^3}\right) \tag{7.96b}$$

For $a_4 > 0$, $dS^*_u(u)/du$ is positive, both u_{min} and u_{max} are possible specific values and the tendency of $dS^*_u(u)/du$ needs to be considered. According to Property 3, $dS^*_u(u)/du$ has an increasing tendency for $\alpha_{3x} > 0$. Then u_{max} is the specific value expressed as

$$u_{\max} = 2r\cos\left(\theta/3\right) - a/3 \tag{7.97}$$

For $a_{3x} < 0$, $dS^*_u(u)/du$ has a decreasing tendency. Then $u_{\min}$ is the specific value given as

$$u_{\min} = -2r[\cos\left(\theta - \pi\right)/3] - a/3 \tag{7.98}$$

For $a_4 = 0$, $a_3 \neq 0$, Eq. (7.70) reduces to a quadratic equation

$$x = -a_3 + a_2 u + a_3 u^2 \tag{7.99}$$

The x–u transformation corresponding to Eq. (7.99) has been proposed by Zhao and Ono (2000b) as follows

$$u_1 = \frac{-a_2 + \sqrt{a_2^2 + 4a_3(a_3 + x_s)}}{2a_3} \tag{7.100a}$$

For $a_4 = a_3 = 0$, Eq. (7.70) reduces to a linear equation. For this scenario, $a_2 = 1$, and the x–u transformation becomes

$$u = x_s \tag{7.100b}$$

In summary, the complete expression of the FMNT (i.e. x–u transformation) can be determined, and is summarised in Table 7.9.

Example 7.9   Consider a Gumbel variable X with mean and coefficient of variation of 20 and 0.5, respectively, the exact normal and inverse normal transformations are given as

$$U = X + \frac{\ln\left[-\ln\left(\xi\right)\right]}{\alpha} \tag{7.101a}$$

$$X = U - \frac{\ln\left[-\ln\left(\xi\right)\right]}{\alpha} \tag{7.101b}$$

$$\alpha = \frac{1}{\sqrt{6}}\left(\frac{\pi}{\sigma_x}\right), \quad \xi = \mu_x - \frac{0.5772}{\alpha} \tag{7.101c}$$

The exact normal transformation obtained by Eq. (7.101a) is illustrated in Figure 7.14a, together with that obtained by using the fourth-moment pseudo x–u transformation (Eq. 7.75), the third-moment pseudo x–u transformation (Eq. 7.48), and the second-moment pseudo (reduced) x–u transformation (Eq. 7.33a). Similarly, the inverse normal transformations obtained by Eq. (7.101b) and the second-moment u–x transformation (Eq. 7.33b), the third-moment u–x transformation (Eq. 7.47), and the fourth-moment u–x transformation (Eq. 7.70) are presented in Figure 7.14b. As can be observed from that the fourth-moment pseudo normal transformation function provides better results than the second- and third-moment pseudo normal transformation functions and is in good agreement with the exact transformation function.

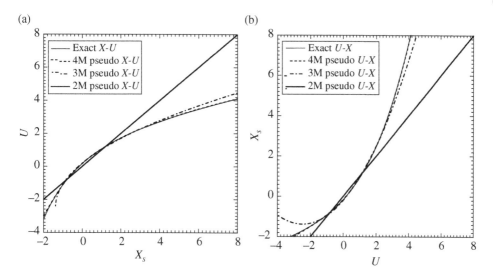

Figure 7.14 Comparisons of x–u and u–x transformations using different transformation methods for a Gumbel variable. (a) x–u transformation. (b) u–x transformation.

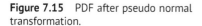

Figure 7.15 PDF after pseudo normal transformation.

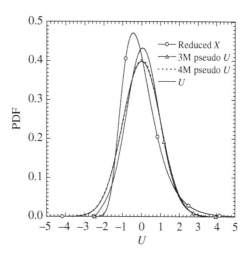

In order to investigate the normality of the pseudo normal variable from Eq. (7.75), the PDF of U' obtained from Eq. (7.75) is depicted in Figure 7.15 using the first four moments of X. From Figure 7.15, it can be observed that the PDF of the fourth-moment pseudo normal variable is better than those of the second- and third-moment pseudo normal variables and approaches that of the normal variable quite well.

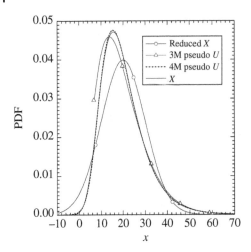

Figure 7.16 PDF after inverse pseudo normal transformation.

Similarly, the PDF of X obtained from Eq. (7.70) is depicted in Figure 7.16 using the first four moments of X. From Figure 7.15, it can be observed that the PDF of the inverse fourth-moment pseudo normal variable approaches that of the Gumbel variable quite well. This is to say, the inverse pseudo normal variable obtained from Eq. (7.70) can be approximately considered as a Gumbel variable if one uses the first four moments of information of the Gumbel variable.

Example 7.10 This example considers four non-normal variables, following Beta, Gamma, Weibull, and Gumbel distributions. The distribution parameters of these four random variables are listed in Table 7.10. Since the PDFs are known, the first four moments of these four random variables can be readily obtained, which are also listed in Table 7.10. Using the first four moments, the parameters and types of the fourth-moment pseudo normal transformation, as well as the coefficients of the Winterstein transformations are listed in Table 7.11.

The variations of the x–u transformation function with respect to x_s are shown in Figure 7.17a–d for the four non-normal random variables, obtained from different methods, i.e. the Rosenblatt transformation (the exact), the FMNT method, and the Winterstein formula.

Table 7.11 and Figure 7.17 reveal the following:

1) The transformation function obtained using the Winterstein formula provides good results when the absolute value of x_s is small. However, when absolute value of x_s is large, the results obtained from the Winterstein formula differ greatly from those obtained from the Rosenblatt transformation, except for the case of Weibull distribution with $\sigma_x = 1$ (Type III of the FMNT).

Table 7.10 Probability distributions and their statistical parameters for Example 7.10.

Distributions	PDFs	Parameters	The first four moments μ_x	σ_x	α_{3x}	α_{4x}
Beta	$\dfrac{(1-x)^{\beta_1-1}x^{\alpha_1-1}}{\text{Beta}[\alpha_1,\beta_1]}, 0<x<1$	$\alpha_1 = 999.1, \beta_1 = 111.01$	0.9	0.009	−0.160	3.033
Gamma	$\dfrac{\alpha_2(\alpha_2 x)^{\beta_2-1}e^{-\alpha_2 x}}{\Gamma(\beta_2)}, x \geq 0$	$\alpha_2 = 0.4, \beta_2 = 25$	10	2	0.4	3.24
Weibull	$\alpha_3\beta_3^{-\alpha_3}x^{\alpha_3-1}\exp\left[-\left(\dfrac{x}{\beta_3}\right)^{\alpha_3}\right], x>0$	$\alpha_3 = 12.153, \beta_3 = 10.43$	10	1	−0.72	3.78
Gumbel	$\dfrac{1}{\alpha_4}\exp\left(\dfrac{\beta_4-x}{\alpha_4}\right)\exp\left[-\exp\left(\dfrac{\beta_4-x}{\alpha_4}\right)\right]$	$\alpha_4 = 2.339, \beta_4 = 8.650$	10	3	1.14	5.40

Table 7.11 Parameters of FMNT and Winterstein formula for Example 7.10.

	Parameters for FMNT						Parameters for Winterstein formula					
a_2	a_3	a_4	p	J_1^*	J_2^*	Type	h_3	h_4	k	a_h	b_h	c_h
0.999	−0.03	-4.9×10^{-5}	-1.19×10^5	0.98	−12.95	I	−0.03	0.001	0.999	−6.5	246.1	8.34×10^6
0.99	0.07	0.001	−311	1.103	10.405	II	0.063	0.009	0.996	2.281	36.10	26 737.05
0.97	−0.12	0.004	−46.39	11.623	12.590	III	−0.103	0.026	0.988	−1.306	12.68	991.508
0.90	0.17	0.024	20.95	None	None	IV	0.137	0.064	0.970	0.720	5.241	51.569

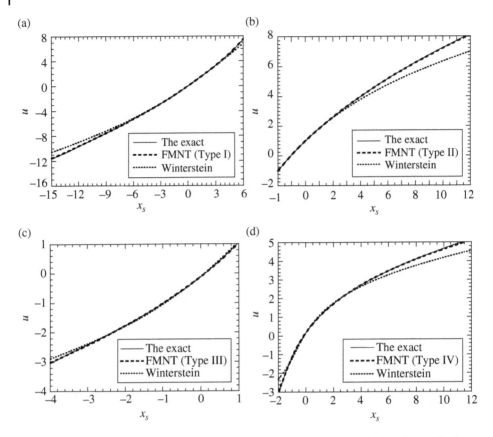

Figure 7.17 Comparison of x–u transformation with different normal transformation methods. (a) Beta. (b) Gamma. (c) Weibull. (d) Gumbel.

2) The x–u transformation function with x_s for the four non-normal random variables, i.e. Beta, Gamma, Weibull, and Gumbel distributions are corresponding to Type I, Type II, Type III, and Type IV of the FMNT, respectively. The FMNT method performs better than the Winterstein formula, and the results of the FMNT are in close agreement with the exact ones in the whole investigation range for all the four cases considered. Take the Gamma random variable, shown in Figure 7.17b, as an example; when $x_s = 12$, the value of u obtained by the Rosenblatt transformation is 7.73, while those obtained by the FMNT and Winterstein formula are 7.70 and 6.70, respectively. Another example considers the Gumbel random variable, shown in Figure 7.17d; when $x_s = -1.75$, the value of u obtained by the Rosenblatt transformation is -2.58, while those obtained by the FMNT and Winterstein formula are -2.53 and -2.08, respectively.

7.5 Pseudo Normal Transformations of Correlated Random Variables

As shown in Section 7.3, Rosenblatt transformation is available to realise the normal transformation when the joint PDF of basic random variables is available and Nataf transformation can be used when the marginal PDFs and correlation coefficients are known. However, the joint PDF and marginal PDFs of some random variables are often unknown in practice, and the probabilistic characteristics of these variables are easier to be expressed using the statistical moments and correlation matrix. In this Section, the transformation for correlated non-normal random variables is presented in details using the statistical moments and correlation matrix of random variables.

7.5.1 Introduction

The pseudo normal transformation has been introduced in the previous sections for independent random variables with unknown distributions. However, in practical engineering, the basic random variables are often correlative. The pseudo normal transformation will be discussed for correlated random variables in this section.

Consider correlated random variables X_i ($i = 1, ..., n$) with correlation matrix $\mathbf{C_X}$

$$
\mathbf{C_X} = \begin{bmatrix} 1 & \rho_{12} & \cdots & \rho_{1n} \\ \rho_{12} & 1 & \cdots & \rho_{2n} \\ \vdots & \vdots & \ddots & \vdots \\ \rho_{n1} & \rho_{n2} & \cdots & 1 \end{bmatrix}
\tag{7.102}
$$

where ρ_{ij} is the correlation coefficient between X_i and X_j.

If the first rth moments of $\mathbf{X}$ are known, the pair of rth moment of pseudo normal transformation for X_i can be given as:

$$
X_{is} = S^{-1}(Z_i, \mathbf{M})
\tag{7.103a}
$$

$$
Z_i = S(X_{is}, \mathbf{M})
\tag{7.103b}
$$

where Z_i is the rth moment of the pseudo standard normal variable.

Substituting Eq. (7.103a) into the definition of ρ_{ij}, one obtains

$$
\rho_{ij} = E(X_{is} \cdot X_{js}) = E\left[S^{-1}(Z_i, \mathbf{M}) \cdot S^{-1}(Z_j, \mathbf{M})\right]
\tag{7.104}
$$

Since $\mathbf{X}$ are correlated random variables, $\mathbf{Z}$ are obviously also correlated random variables. Assuming the correlation coefficient between Z_i and Z_j is ρ_{0ij}, the relationship between ρ_{0ij} and ρ_{ij} can be defined by Eq. (7.104). Then, ρ_{0ij} can be determined from ρ_{ij} and $\mathbf{M}$, and the correlation matrix of rth moment pseudo standard normal variables, $\mathbf{C_Z}$, can be written as:

$$
\mathbf{C_Z} = \begin{bmatrix} 1 & \rho_{012} & \cdots & \rho_{01n} \\ \rho_{012} & 1 & \cdots & \rho_{02n} \\ \vdots & \vdots & \ddots & \vdots \\ \rho_{0n1} & \rho_{0n2} & \cdots & 1 \end{bmatrix}
\tag{7.105}
$$

Using Cholesky decomposition, the correlation matrix $\mathbf{C_Z}$ can be rewritten as,

$$\mathbf{C_Z} = \mathbf{L_0}\mathbf{L_0^T} \tag{7.106}$$

where $\mathbf{L_0}$ is the lower triangular matrix from Cholesky decomposition and $\mathbf{L_0}^T$ is the transpose matrix of $\mathbf{L_0}$.

With the correlated rth moment pseudo standard normal vector $\mathbf{Z}$ obtained from Eq. (7.103b), the independent standard normal vector $\mathbf{U} = (U_1, U_2, ..., U_n)$ can then be given as:

$$\mathbf{U} = \mathbf{L_0^{-1}}\mathbf{Z} \tag{7.107a}$$

In order to obtain the inverse transformation, the independent standard normal vector $\mathbf{U}$ is firstly transformed into correlated rth moment pseudo standard normal vector $\mathbf{Z}$ by

$$\mathbf{Z} = \mathbf{L_0}\mathbf{U} \tag{7.107b}$$

where $\mathbf{L_0}^{-1}$ is the inverse matrix of $\mathbf{L_0}$, and $\mathbf{L_0}$ is expressed as:

$$\mathbf{L_0} = \begin{bmatrix} l_{11} & 0 & \cdots & 0 \\ l_{21} & l_{22} & \cdots & 0 \\ \vdots & \vdots & \ddots & \vdots \\ l_{n1} & l_{n2} & \cdots & l_{nn} \end{bmatrix} \tag{7.108}$$

Using Eq. (7.103a), the reduced random vector of $\mathbf{X}$ can be obtained, and then the u-x_s transformation can be accomplished.

In particular, when both Eqs. (7.103a) and (7.103b) hold true only for the first two moments, i.e. the mean value and the standard deviation, Eqs. (7.103a) and (7.103b) become

$$X_{is} = Z_i \tag{7.109a}$$
$$Z_i = X_{is} \tag{7.109b}$$

In this case, $\rho_{0ij} = \rho_{ij}$.

Then, using the Cholesky decomposition, the correlated pseudo standard variables can be converted to independent standard (reduced) space:

$$\mathbf{U} = \mathbf{L_0^{-1}}\mathbf{X_S} \tag{7.110a}$$
$$\mathbf{X_S} = \mathbf{L_0}\mathbf{U} \tag{7.110b}$$

where $\mathbf{L_0}$ is the same as the matrix shown in Eq. (7.108), and $\mathbf{L_0}^{-1}$ is the inverse matrix of $\mathbf{L_0}$ that is expressed as

$$\mathbf{L_0^{-1}} = \begin{bmatrix} h_{11} & \cdots & 0 & \cdots & 0 \\ \vdots & \ddots & \vdots & \ddots & \vdots \\ h_{i1} & \cdots & h_{ii} & \cdots & 0 \\ \vdots & \ddots & \vdots & \ddots & \vdots \\ h_{n1} & \cdots & h_{ni} & \cdots & h_{nn} \end{bmatrix} \tag{7.111}$$

7.5.2 Third-Moment Pseudo Normal Transformation for Correlated Random Variables

Assume two correlated random variables X_i and X_j with correlative coefficient of ρ_{ij}: using the third-moment transformation, the standardised variable X_{is} and X_{js} of X_i and X_j can be expressed as (Lu et al. 2017a)

$$X_{is} = S^{-1}(Z_i) = a_i + b_i Z_i + c_i Z_i^2 = (a_i, b_i, c_i) \cdot \left(1, Z_i, Z_i^2\right)^T \tag{7.112a}$$

$$X_{js} = S^{-1}(Z_j) = a_j + b_j Z_j + c_j Z_j^2 = (a_j, b_j, c_j) \cdot \left(1, Z_j, Z_j^2\right)^T \tag{7.112b}$$

where the superscript T represents transpose; and Z_i and Z_j are two correlated standard normal variables. The coefficients of a_i (a_j), b_i (b_j), and c_i (c_j) can be determined as shown in Section 7.4.2.

Assuming that the correlation coefficient between Z_i and Z_j is ρ_{0ij}, and according to the definition of correlation coefficient, the following can be derived,

$$\rho_{ij} = E(X_{is} \cdot X_{js}) = E\left[(a_i, b_i, c_i)(1, Z_i, Z_i^2)^T \cdot \left(1, Z_j, Z_j^2\right)(a_j, b_j, c_j)^T\right]$$
$$= (a_i, b_i, c_i)E\left[(1, Z_i, Z_i^2)^T \cdot \left(1, Z_j, Z_j^2\right)\right](a_j, b_j, c_j)^T = (a_i, b_i, c_i)\mathbf{R}(a_j, b_j, c_j)^T \tag{7.113a}$$

$$\mathbf{R} = E\left[(1, Z_i, Z_i^2)^T \cdot \left(1, Z_j, Z_j^2\right)\right] = \begin{bmatrix} 1 & 0 & 1 \\ 0 & \rho_{0ij} & 0 \\ 1 & 0 & 2\rho_{0ij}^2 + 1 \end{bmatrix} \tag{7.113b}$$

Substituting Eq. (7.113b) into Eq. (7.113a) leads to:

$$\rho_{ij} = b_i b_j \rho_{0ij} + 2c_i c_j \rho_{0ij}^2 \tag{7.114a}$$

and ρ_{0ij} can be determined from solving Eq. (7.114a). It is worth noting that the valid solution of ρ_{0ij} should be restricted by the following conditions to satisfy the definition of the correlation coefficient:

$$-1 \le \rho_{0ij} \le 1, \quad \rho_{ij} \cdot \rho_{0ij} \ge 0, \quad \text{and} \quad |\rho_{0ij}| \ge |\rho_{ij}| \tag{7.114b}$$

With Eqs. (7.114a) and (7.114b), the expressions of the equivalent correlation coefficient ρ_{0ij} and the upper and lower bounds of original correlation coefficient ρ_{ij} to ensure the transformation executable are summarised in Table 7.12. The derivation of Table 7.12 is illustrated in Example 7.11.

The preceding procedure can be easily extended to n variables with known statistical moments and correlation matrix. The polynomial coefficients of each variable can be obtained by Eq. (7.45), and for any two correlated variables, the corresponding equivalent correlation coefficients of standard normal variables can be determined from Table 7.12. The equivalent correlation matrix of standard normal variables, $\mathbf{C_Z}$, and the lower triangular matrix $\mathbf{L_0}$ obtained from Cholesky decomposition of $\mathbf{C_Z}$ can then be obtained. Their formulations are similar to those in Eqs. (7.105) and (7.108).

Theoretically, the equivalent correlation matrix ($\mathbf{C_Z}$) is positive semi-definite after the fully correlated variables are excluded. However, small negative eigenvalues of $\mathbf{C_Z}$ might appear because of computational errors during the transformation from correlated non-

Table 7.12 Equivalent correlation coefficient and bounds for third-moment transformation.

Conditions			Application range of ρ_{ij}	ρ_{0ij}
$c_i c_j = 0$			$[-b_i b_j, \, b_i b_j]$	$\dfrac{\rho_{ij}}{b_i b_j}$
$c_i c_j > 0$	$8c_i c_j - b_i^2 b_j^2 > 0$	$4c_i c_j - b_i b_j > 0$	$\left[-b_i^2 b_j^2/(8c_i c_j), \, b_i b_j + 2c_i c_j\right]$	$\dfrac{-b_i b_j + \sqrt{b_i^2 b_j^2 + 8c_i c_j \rho_{ij}}}{4c_i c_j}$
$c_i c_j < 0$	$8c_i c_j + b_i^2 b_j^2 < 0$	$4c_i c_j + b_i b_j < 0$	$\left[2c_i c_j - b_i b_j, \, -b_i^2 b_j^2/(8c_i c_j)\right]$	$\dfrac{-b_i b_j + \sqrt{b_i^2 b_j^2 + 8c_i c_j \rho_{ij}}}{4c_i c_j}$
Otherwise			$[2c_i c_j - b_i b_j, \, b_i b_j + 2c_i c_j]$	$\dfrac{-b_i b_j + \sqrt{b_i^2 b_j^2 + 8c_i c_j \rho_{ij}}}{4c_i c_j}$

normal random vector to correlated normal ones, especially in the cases of highly non-normal random variables. To solve the problem, a method introduced by Ji et al. (2018) is adopted. Under such circumstances, $\mathbf{C_Z}$ may be rewritten as:

$$\mathbf{C_Z} = \mathbf{V \Lambda V}^T \tag{7.115}$$

where $\mathbf{V}$ and $\mathbf{\Lambda}$ are the eigenvector and diagonal eigenvalue matrices of $\mathbf{C_Z}$, respectively. The small negative eigenvalues in $\mathbf{\Lambda}$ are substituted by small positive values, e.g., 0.001, to make Cholesky decomposition ready.

From Eqs. (7.112), (7.107b), (7.108), and Table 7.12, the u–x transformations can be expressed as

$$X_i = \mu_{X_i} + \sigma_{X_i}\left[a_i + b_i \sum_{k=1}^{i} l_{ik} U_k + c_i \left(\sum_{k=1}^{i} l_{ik} U_k\right)^2\right], \quad (i = 1, 2, \cdots, n) \tag{7.116}$$

where l_{ik} is the tth row kth column element of matrix $\mathbf{L_0}$.

From Eqs. (7.107a), (7.111), (7.48), and Table 7.12, the x–u transformations can be expressed as

$$U_i = \sum_{k=1}^{i} h_{ik} S(X_{ks}) = \sum_{k=1}^{i} h_{ik} S\left[(X_k - \mu_{X_k})/\sigma_{X_k}\right], \quad (i = 1, 2, \cdots, n) \tag{7.117}$$

where h_{ik} is the tth row kth column element of matrix $\mathbf{L_0}^{-1}$; and $S(X_s)$ is given by Eq. (7.48).

Example 7.11   The Derivation of Table 7.12

When $c_i c_j$ is not zero, Eq. (7.114a) is a quadratic equation. For brevity, the right side of Eq. (7.114a) is expressed as $h(\rho_{0ij})$, i.e.

$$\rho_{ij} = h\left(\rho_{0ij}\right) = b_i b_j \cdot \rho_{0ij} + 2c_i c_j \cdot \rho_{0ij}^2 \tag{7.118}$$

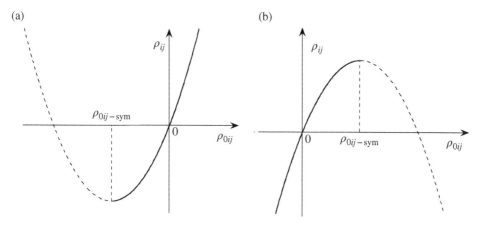

Figure 7.18 The shape of $h(\rho_{0ij})$ for $c_i c_j \neq 0$. (a) $c_i c_j > 0$. (b) $c_i c_j < 0$.

The axis of symmetry of $h(\rho_{0ij})$ is at:

$$\rho_{0ij} = \rho_{0ij-sym} = -b_i b_j / (4 c_i c_j) \qquad (7.119)$$

According to the third-moment pseudo normal transformation, its region of application is given as:

$$-2\sqrt{2} \leq \alpha_{3X_i} \leq 2\sqrt{2} \quad \text{and} \quad -2\sqrt{2} \leq \alpha_{3X_j} \leq 2\sqrt{2} \qquad (7.120)$$

From Eq. (7.45), it can be observed that $b_i b_j$ is non-negative. Therefore, $\rho_{0ij-sym}$ is positive for $c_i c_j < 0$ and negative for $c_i c_j > 0$. The shape of $h(\rho_{0ij})$ for $c_i c_j \neq 0$ is presented in Figure 7.18a-b, in which the solid lines denote the region satisfying the condition that $\rho_{0ij} \cdot \rho_{ij} \geq 0$ and ρ_{0ij} is an increasing function of ρ_{ij}.

From Figure 7.18a, the equivalent correlation coefficient ρ_{0ij} for $c_i c_j > 0$ can be given as:

$$\rho_{0ij} = \frac{-b_i b_j + \sqrt{b_i^2 b_j^2 + 8 c_i c_j \rho_{ij}}}{4 c_i c_j} \qquad (7.121)$$

In order to satisfy $-1 \leq \rho_{0ij} \leq 1$, the ρ_{ij} should have application bounds, i.e. $\rho_{ij} \in [\rho_{ij-min}, \rho_{ij-max}]$, in which ρ_{ij-min} and ρ_{ij-max} are the lower and upper bounds, respectively. According to Figure 7.18a and $|\rho_{0ij}| \geq |\rho_{ij}|$, the upper bound, ρ_{ij-max}, can be readily given as:

$$\rho_{ij-max} = h(1) = b_i b_j + 2 c_i c_j \qquad (7.122a)$$

And the lower bound can be determined as:

$$\rho_{ij-min} = \begin{cases} h\left(\rho_{0ij-sym}\right), & h\left(\rho_{0ij-sym}\right) > -1 \text{ and } \rho_{0ij-sym} > -1 \\ h(-1), & \text{otherwise} \end{cases} \qquad (7.122b)$$

where $h(-1) = 2c_ic_j - b_ib_j$; $h\left(\rho_{0ij-sym}\right) = -b_i^2 b_j^2 / (8c_ic_j)$. Thus, Eq. (7.122b) can be rewritten as:

$$\rho_{ij-min} = \begin{cases} -b_i^2 b_j^2 / (8c_ic_j), & 8c_ic_j - b_i^2 b_j^2 > 0 \text{ and } 4c_ic_j - b_ib_j > 0 \\ 2c_ic_j - b_ib_j, & \text{otherwise} \end{cases} \tag{7.122c}$$

Similarly, from Figure 7.18b, ρ_{0ij} for $c_ic_j < 0$ is also given by Eq. (7.121), and the application bound of ρ_{ij} is expressed as:

$$\rho_{ij-max} = \begin{cases} -b_i^2 b_j^2 / (8c_ic_j), & 8c_ic_j + b_i^2 b_j^2 < 0 \text{ and } 4c_ic_j + b_ib_j < 0 \\ b_ib_j + 2c_ic_j, & \text{otherwise} \end{cases} \tag{7.123a}$$

$$\rho_{ij-min} = 2c_ic_j - b_ib_j \tag{7.123b}$$

When $c_ic_j = 0$, according to Eq. (7.114a), ρ_{0ij} can be readily determined as:

$$\rho_{0ij} = \rho_{ij} / b_ib_j \tag{7.124}$$

and the bound for $c_ic_j = 0$ can be determined as:

$$\rho_{ij-max} = b_ib_j \tag{7.125a}$$

$$\rho_{ij-max} = -b_ib_j \tag{7.125b}$$

Finally, the expressions are summarised in Table 7.12 for the equivalent correlation coefficient ρ_{0ij} and the upper and lower bounds of original correlation coefficient ρ_{ij} to ensure the transformation executable.

7.5.3 Fourth-Moment Pseudo Normal Transformation for Correlated Random Variables

Assume two correlated random variables X_i and X_j with correlative coefficient of ρ_{ij}: using the fourth-moment transformation, the standardised variable X_{is} and X_{js} of X_i and X_j can be expressed as (Lu et al. 2020)

$$X_{is} = S^{-1}(Z_i) = a_i + b_iZ_i + c_iZ_i^2 + d_iZ_i^3 = (a_i, b_i, c_i, d_i)(1, Z_i, Z_i^2, Z_i^3) \tag{7.126a}$$

$$X_{js} = S^{-1}(Z_j) = a_j + b_jZ_j + c_jZ_j^2 + d_jZ_j^3 = (a_j, b_j, c_j, d_j)\left(1, Z_j, Z_j^2, Z_j^3\right) \tag{7.126b}$$

where Z_i and Z_j are two correlated standard normal variables. The coefficients of a_i (a_j), b_i (b_j), c_i (c_j), and d_i (d_j) can be determined as shown in Section 7.4.3.

Assuming that the correlation coefficient between Z_i and Z_j is ρ_{0ij}, and according to the definition of correlation coefficient, the following can be derived,

$$\rho_{ij} = E(X_{is} \cdot X_{js}) = (a_i, b_i, c_i, d_i)\mathbf{R}(a_j, b_j, c_j, d_j)^T \tag{7.127a}$$

where

$$\mathbf{R} = E\left[\left(1, Z_i, Z_i^2, Z_i^3\right)^T \cdot \left(1, Z_j, Z_j^2, Z_j^3\right)\right] = \begin{bmatrix} 1 & 0 & 1 & 0 \\ 0 & \rho_{0ij} & 0 & 3\rho_{0ij} \\ 1 & 0 & 2\rho_{0ij}^2 + 1 & 0 \\ 0 & 3\rho_{0ij} & 0 & 6\rho_{0ij}^3 + 9\rho_{0ij} \end{bmatrix}$$

(7.127b)

Substituting Eq. (7.127b) into Eq. (7.127a) leads to:

$$\rho_{ij} = \left(b_i b_j + 3d_i b_j + 3b_i d_j + 9d_i d_j\right)\rho_{0ij} + 2c_i c_j \rho_{0ij}^2 + 6d_i d_j \rho_{0ij}^3$$

(7.128a)

It can be observed that ρ_{0ij} can be determined from solving Eq. (7.128a). It is worth noting that the valid solution of ρ_{0ij} should be restricted by the following conditions to satisfy the definition of the correlation coefficient:

$$-1 \le \rho_{0ij} \le 1 \rho_{ij} \cdot \rho_{0ij} \ge 0 \quad \text{and} \quad |\rho_{0ij}| \ge |\rho_{ij}|$$

(7.128b)

Following Eqs. (7.128a) and (7.128b), the expressions for the equivalent correlation coefficient ρ_{0ij} are summarised in Algorithm 7.1, together with the applicable bound of original correlation coefficient ρ_{ij}. The derivation of Algorithm 7.1 is illustrated in Example 7.12.

Similarly, the preceding procedure can be easily extended to n variables with known statistical moments and correlation matrix. The polynomial coefficients of each variable can be obtained by Eqs. (7.72a)–(7.72e) and (7.73), and for any two correlated variables, the equivalent correlation coefficients can be determined from Algorithm 7.1. The equivalent correlation matrix of standard normal variables, $\mathbf{C_Z}$, and the lower triangular matrix $\mathbf{L_0}$ obtained from Cholesky decomposition of $\mathbf{C_Z}$ can then be obtained. Their forms are the same as Eq. (7.105) and Eq. (7.108), respectively. For the cases of original correlation matrix with very small eigenvalues, the equivalent correlation matrix might become a non-positive semidefinite matrix. The method for solving the problem (Ji et al. 2018) mentioned in Section 7.5.2 can be adopted to make Cholesky decomposition ready.

From Eqs. (7.107b), (7.108), and (7.126), and Algorithm 7.1, the u–x transformations can be expressed as

$$X_i = \mu_{X_i} + \sigma_{X_i}\left[a_i + b_i \sum_{k=1}^{i} l_{ik} U_k + c_i \left(\sum_{k=1}^{i} l_{ik} U_k\right)^2 + d_i \left(\sum_{k=1}^{i} l_{ik} U_k\right)^3\right], \quad (i = 1, 2, \cdots, n)$$

(7.129)

where l_{ik} is the tth row kth column element of matrix $\mathbf{L_0}$.

From Eq. (7.107a), (7.103b) and (7.111), Table 7.9, and Algorithm 7.1, the x–u transformations can be expressed as

$$U_i = \sum_{k=1}^{i} h_{ik} S(X_{ks}) = \sum_{k=1}^{i} h_{ik} S\left[(X_k - \mu_{X_k})/\sigma_{X_k}\right], \quad (i = 1, 2, \cdots, n)$$

(7.130)

where h_{ik} is the tth row kth column element of matrix $\mathbf{L_0}^{-1}$; and $S(X_s)$ is given by Table 7.9.

Algorithm 7.1 Equivalent Correlation Coefficient and Bounds for Fourth-Moment Transformation

---

As input the algorithm requires:
a_i, b_i, c_i, d_i and a_j, b_j, c_j, d_j: the polynomial coefficients of X_i and X_j

- ρ_{ij}: correlation coefficient between X_i and X_j

The algorithm evaluates the equivalent correlation coefficient ρ_{0ij} and the application range of ρ_{ij}

---

(1) **if** $d_i d_j < 0$:

    $\rho_{0ij} = 2r \cdot \cos[(\theta+\pi)/3] - t_2/3$

    if $(\rho_{0ij-2} < 1 \text{ and } 6d_i d_j \cdot h(\rho_{0ij-2}) < 1)$:

        $\rho_{ij-max} = 6d_i d_j \cdot h(\rho_{0ij-2})$

    else:

        $\rho_{ij-max} = 6d_i d_j \cdot h(1)$

    if $(\rho_{0ij-1} > -1 \text{ and } 6d_i d_j \cdot h(\rho_{0ij-1}) > -1)$:

        $\rho_{ij-min} = 6d_i d_j \cdot h(\rho_{0ij-1})$

    else:

        $\rho_{ij-min} = 6d_i d_j \cdot h(-1)$

(2) **else if** $d_i d_j = 0$:

    if $c_i c_j = 0$:

        $\rho_{0ij} = \rho_{ij} / (b_i b_j + 3b_i d_j + 3d_i b_j)$

        $\rho_{0ij-max} = b_i b_j + 3b_i d_j + 3d_i b_j$

        $\rho_{0ij-min} = -(b_i b_j + 3b_i d_j + 3d_i b_j)$

    else if $c_i c_j > 0$:

        ρ_{0ij} evaluated by Eq. (7.145)

        $\rho_{0ij-max} = h_2(1)$

        if $(\rho_{0ij-sym} > -1 \text{ and } h_2(\rho_{0ij-sym}) > -1)$:

            $\rho_{0ij-min} = h_2(\rho_{0ij-sym})$

        else:

            $\rho_{0ij-min} = h_2(-1)$

    else: $c_i c_j < 0$

        ρ_{0ij} evaluated by Eq. (7.145)

        if $(\rho_{0ij-sym} < 1 \text{ and } h_2(\rho_{0ij-sym}) < -1)$:

            $\rho_{0ij-max} = h_2(\rho_{0ij-sym})$

        else:

            $\rho_{0ij-max} = h_2(1)$

        $\rho_{0ij-min} = h_2(-1)$

(3) **else**: $d_i d_j > 0$

    if $t_2^2 - 3t_1 \le 0$:

        $\rho_{0ij} = A^{1/3} + B^{1/3} - t_2/3$

        $\rho_{ij-max} = 6d_i d_j \cdot h(1)$

        $\rho_{ij-min} = 6d_i d_j \cdot h(-1)$

    else: $t_2^2 - 3t_1 > 0$

        if $t_2 > 0$:

            if $\Delta \ge 0$:

                $\rho_{0ij} = A^{1/3} + B^{1/3} - t_2/3$

            else: $\Delta < 0$

                $\rho_{0ij} = 2r \cdot \cos(\theta/3) - t_2/3$

            $\rho_{ij-max} = 6d_i d_j \cdot h(1)$

            if $(\rho_{0ij-2} > -1 \text{ and } 6d_i d_j \cdot h(\rho_{0ij-2}) > -1)$:

                $\rho_{ij-min} = 6d_i d_j \cdot h(\rho_{0ij-2})$

            else:

                $\rho_{ij-min} = 6d_i d_j \cdot h(-1)$

        else: $t_2 < 0$

            if $\Delta \ge 0$:

                $\rho_{0ij} = A^{1/3} + B^{1/3} - t_2/3$

            else: $\Delta < 0$

                $\rho_{0ij} = -2r \cdot \cos[(\theta-\pi)/3] - t_2/3$

            if $(\rho_{0ij-1} < 1 \text{ and } 6d_i d_j \cdot h(\rho_{0ij-1}) < 1)$:

                $\rho_{ij-max} = 6d_i d_j \cdot h(\rho_{0ij-1})$

            else:

                $\rho_{ij-max} = 6d_i d_j \cdot h(1)$

                $\rho_{ij-min} = 6d_i d_j \cdot h(-1)$

Example 7.12   The Derivation of Algorithm 7.1

If $d_i d_j \ne 0$, Eq. (7.128a) is a cubic equation about ρ_{0ij} and can be equivalently expressed as:

$$\frac{\rho_{ij}}{6d_i d_j} = \frac{b_i b_j + 3b_i d_j + 3d_i b_j + 9d_i d_j}{6d_i d_j}\rho_{0ij} + \frac{c_i c_j}{3d_i d_j}\rho_{0ij}^2 + \rho_{0ij}^3 \tag{7.131}$$

For convenience of exposition, the right side of Eq. (7.131) is expressed as $h(\rho_{0ij})$, i.e.

$$\frac{\rho_{ij}}{6d_i d_j} = h\left(\rho_{0ij}\right) = \rho_{0ij}^3 + t_2\rho_{0ij}^2 + t_1\rho_{0ij} \tag{7.132a}$$

$$t_1 = \frac{b_i b_j + 3b_i d_j + 3d_i b_j + 9d_i d_j}{6d_i d_j}, t_2 = \frac{c_i c_j}{3d_i d_j} \tag{7.132b}$$

The first- and second-order derivatives of $h(\rho_{0ij})$ with respect to ρ_{0ij} are formulated as:

$$h'\left(\rho_{0ij}\right) = dh\left(\rho_{0ij}\right)/d\rho_{0ij} = 3\rho_{0ij}^2 + 2t_2\rho_{0ij} + t_1 \tag{7.133a}$$

$$h''\left(\rho_{0ij}\right) = d^2h\left(\rho_{0ij}\right)/d\rho_{0ij}^2 = 6\rho_{0ij} + 2t_2 \tag{7.133b}$$

According to the compatibility and limitations of the pairs of skewness and kurtosis in the fourth-moment transformation technique for independent non-normal random variables given by Zhao et al. (2018a), the minimum of $b_i b_j + 3b_i d_j + 3d_i b_j + 9d_i d_j$ (i.e. the numerator of t_1) is determined as a positive value. Accordingly, t_1 is positive for $d_i d_j > 0$ and negative for $d_i d_j < 0$.

(1) When $d_i d_j < 0$, there exist two real roots $\rho_{0ij-1} = \left(-t_2 - \sqrt{t_2^2 - 3t_1}\right)/3$ and $\rho_{0ij-2} = \left(-t_2 + \sqrt{t_2^2 - 3t_1}\right)/3$ for the quadratic equation of $h'(\rho_{0ij}) = 0$ due to the fact that its discriminant $D = (2t_2)^2 - 4 \cdot 3 \cdot t_1 = 4(t_2^2 - 3t_1)$ is positive. It should be noted that ρ_{0ij-1} and ρ_{0ij-2} are not correlation coefficients, but just the roots of Eq. (7.133a) equal to zero. Moreover, since $t_1 < 0$ for $d_i d_j < 0$, it can be obtained that $h\left(\rho_{0ij-1}\right) \cdot h\left(\rho_{0ij-2}\right) = t_1^2\left(4t_1 - t_2^2\right)/27 < 0$ and $\rho_{0ij-1} \cdot \rho_{0ij-2} = t_1/3 < 0$. According to the property of cubic function, the shape of $h(\rho_{0ij})$ for $d_i d_j < 0$ is depicted in Figure 7.19a–d, in which the solid line denotes the region satisfying the condition that $\rho_{ij} \cdot \rho_{0ij} \geq 0$ and ρ_{0ij} is an increasing function

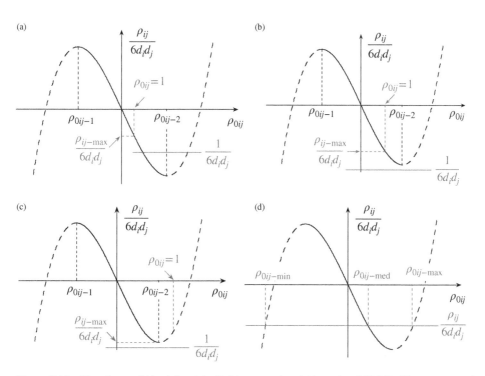

Figure 7.19 The shape of $h(\rho_{0ij})$ for $d_i d_j < 0$. (a) $\rho_{0ij-2} \geq 1$ and $h(\rho_{0ij-2}) \leq 1/(6d_i d_j)$. (b) $\rho_{0ij-2} \geq 1$ and $h(\rho_{0ij-2}) > 1/(66d_i d_j)$. (c) $\rho_{0ij-2} < 1$. (d) Intersections of $h(\rho_{0ij})$ and $\rho_{ij}/(6d_i d_j)$.

of ρ_{ij}. Note that the correlation coefficient ρ_{ij} is a negative value when the vertical axis, $\rho_{ij}/(6d_id_j)$, takes a positive value, as $d_id_j < 0$.

To ensure ρ_{0ij} satisfies the definition of correlation coefficient, i.e. $-1 \le \rho_{0ij} \le 1$, ρ_{ij} should be within a value interval, i.e. $\rho_{ij} \in [\rho_{ij\text{-min}}, \rho_{ij\text{-max}}]$, in which $\rho_{ij\text{-min}}$ and $\rho_{ij\text{-max}}$ are the lower and upper bounds, respectively. The upper bound, $\rho_{ij\text{-max}}$, is first discussed as follows: when $\rho_{0ij-2} \ge 1$, $h(1)$ (i.e. $\rho_{0ij} = 1$) must be larger than $1/(6d_id_j)$ (i.e. $\rho_{ij} \le 1$) because $|\rho_{ij}| \le |\rho_{0ij}|$ is valid for any type of random variables [Liu and Der Kiurghian 1986, Lancaster 1957], as shown in Figure 7.19a,b. From Figure 7.19a,b, the upper bound $\rho_{ij\text{-max}}$ for $\rho_{0ij-2} \ge 1$ can be determined as $6d_id_j \cdot h(1)$ to ensure $\rho_{0ij} \le 1$; when $\rho_{0ij-2} < 1$, as shown in Figure 7.19c, the upper bound can be determined as $\rho_{ij\text{-max}} = 6d_id_j \cdot h(\rho_{0ij-2})$ to ensure ρ_{0ij} increasing with ρ_{ij}. Therefore, when $d_id_j < 0$, the suitable maximum original correlation coefficient, i.e. the upper bound of ρ_{ij}, can be summarised as:

$$\rho_{ij-\max} = \begin{cases} 6d_id_j \cdot h\left(\rho_{0ij-2}\right), & \rho_{0ij-2} < 1 \\ 6d_id_j \cdot h(1), & \text{otherwise} \end{cases} \tag{7.134a}$$

Similarly, the lower bound of ρ_{ij} can be determined as:

$$\rho_{ij-\min} = \begin{cases} 6d_id_j \cdot h\left(\rho_{0ij-1}\right), & \rho_{0ij-1} > -1 \\ 6d_id_j \cdot h(-1), & \text{otherwise} \end{cases} \tag{7.134b}$$

The intersections of $h(\rho_{0ij})$ and $\rho_{ij}/(6d_id_j)$ are depicted in Figure 7.19d. From Figure 7.19d, it can be observed that there are three intersections for $\rho_{ij} \in [\rho_{ij\text{-min}}, \rho_{ij\text{-max}}]$, i.e. $h(\rho_{0ij}) = \rho_{ij}/(6d_id_j)$ have three real roots. The equivalent correlation coefficient ρ_{0ij} is the medial horizontal coordinate of these intersections. According to the solution of cubic equation and its property, ρ_{0ij} is expressed as:

$$\rho_{0ij} = -2r\cos\left(\frac{\theta + \pi}{3}\right) - \frac{t_2}{3} \tag{7.135}$$

(2) When $d_id_j > 0$ and $D = (2t_2)^2 - 4 \cdot 3 \cdot t_1 = 4(t_2^2 - 3t_1) \le 0$, there always exists $h'(\rho_{0ij}) \ge 0$ for arbitrary ρ_{0ij}. Therefore, the tendency of $h(\rho_{0ij})$ is increasing for $\rho_{0ij} \in (-\infty, +\infty)$. The shape of $h(\rho_{0ij})$ for $d_id_j > 0$ and $D \le 0$ is depicted in Figure 7.20.

Because $|\rho_{ij}| \le |\rho_{0ij}|$ is valid for any type of random variables (Liu and Der Kiureghian 1986; Lancaster 1957), the horizontal coordinate of the intersection of $h(\rho_{0ij})$ and $1/6d_id_j$ (i.e. $\rho_{ij} = 1$) is larger than 1, and that of $h(\rho_{0ij})$ and $-1/6d_id_j$ (i.e. $\rho_{ij} = -1$) is lesser than -1, as indicated in Figure 7.20. To ensure $-1 \le \rho_{0ij} \le 1$, the original correlation coefficient ρ_{ij} should be limited in:

$$\rho_{ij-\max} = 6d_id_j \cdot h(1) \tag{7.136a}$$

$$\rho_{ij-\min} = 6d_id_j \cdot h(-1) \tag{7.136b}$$

Because $h(\rho_{0ij})$ is a monotonically increasing function for $\rho_{ij} \in [\rho_{ij\text{-min}}, \rho_{ij\text{-max}}]$, there is only one intersection of $h(\rho_{0ij})$ and $\rho_{ij}/(6d_id_j)$. Therefore, equation $h(\rho_{0ij}) = \rho_{ij}/(6d_id_j)$ has only one real root. According to the solution of cubic equation (Zwillinger 2018), the equivalent correlation coefficient ρ_{0ij} can be expressed as:

$$\rho_{0ij} = \sqrt[3]{A} + \sqrt[3]{B} - \frac{t_2}{3} \tag{7.137}$$

Figure 7.20 The shape of $h(\rho_{0ij})$ for $d_id_j > 0$ and $D \le 0$.

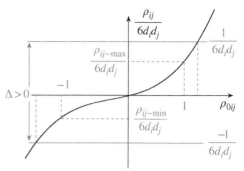

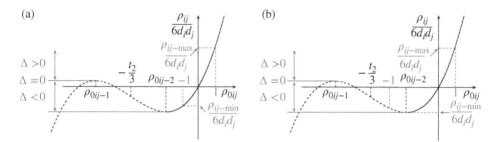

Figure 7.21 The shape of $h(\rho_{0ij})$ for $d_id_j > 0$, $D > 0$, and $t_2 > 0$. (a) $\rho_{0ij-2} \le -1$. (b) $\rho_{0ij-2} > -1$.

(3) When $d_id_j > 0$ and $D = (2t_2)^2 - 4 \cdot 3 \cdot t_1 = 4(t_2^2 - 3t_1) > 0$, equation $h'(\rho_{0ij}) = 0$ has two real roots ρ_{0ij-1} and ρ_{0ij-2}. Therefore, the tendency of $h(\rho_{0ij})$ is increasing for $\rho_{0ij} \in (-\infty, \rho_{0ij-1}) \cup (\rho_{0ij-2}, +\infty)$ and decreasing for $\rho_{0ij} \in [\rho_{0ij-1}, \rho_{0ij-2}]$. According to $h''(\rho_{0ij}) = 0$, the inflexion point of $h(\rho_{0ij})$ is computed as $\rho_{0ij} = -t_2/3$. Moreover, $\rho_{0ij-1} \cdot \rho_{0ij-2} = t_1/3$ is positive because $t_1 > 0$. The shape of $h(\rho_{0ij})$ for this case is depicted in Figure 7.21a,b for $t_2 > 0$ and Figure 7.22a,b for $t_2 < 0$, in which the solid lines denote the region satisfying the condition that $\rho_{ij} \cdot \rho_{0ij} \ge 0$ and ρ_{0ij} is an increasing function of ρ_{ij}.

The derivation of the bounds of ρ_{ij} for $d_id_j > 0$, $D > 0$, and $t_2 > 0$ is similar to that of the preceding cases. As shown in Figure 7.21a,b, the bounds of ρ_{ij} can be expressed as:

$$\rho_{ij-max} = 6d_id_j \cdot h(1) \tag{7.138a}$$

$$\rho_{ij-min} = \begin{cases} 6d_id_j \cdot h\left(\rho_{0ij-2}\right), & \rho_{0ij-2} > -1 \\ 6d_id_j \cdot h(-1), & \text{otherwise} \end{cases} \tag{7.138b}$$

According to the properties of cubic equation, the discriminant Δ expressed in Eq. (7.76b) can be used to determine the number of real roots. The values of Δ for $d_id_j > 0$, $D > 0$, and $t_2 > 0$ are also depicted in Figure 7.21a,b. It can be observed from Figure 7.21a,b that there is only one real root for $\Delta > 0$, two roots for $\Delta = 0$, and three roots for $\Delta < 0$. According to Figure 7.21a,b and the solution of cubic equation (Zwillinger 2018), the equivalent correlation coefficient ρ_{0ij} can be determined: (i) when $\Delta > 0$, there is only one intersection of $h(\rho_{0ij})$ and $\rho_{ij}/(6d_id_j)$, and the solution of ρ_{0ij} is the horizontal coordinate of this intersection, which is identical with Eq. (7.137); (ii) when $\Delta = 0$, there are two intersections, and the solution of ρ_{0ij} is the larger one of the horizontal coordinate of these intersections, which is identical with Eq. (7.137); and (iii) when $\Delta < 0$, there exist three intersections, and the solution of ρ_{0ij} is the largest one among the horizontal coordinate of these intersections, which is expressed as:

$$\rho_{0ij} = 2r \cos\left(\frac{\theta}{3}\right) - \frac{t_2}{3} \tag{7.139}$$

Similarly, as shown in Figure 7.22a,b, the application range of ρ_{ij} for $d_i d_j > 0$, $D > 0$, and $t_2 < 0$ can be determined as:

$$\rho_{ij-\max} = \begin{cases} 6d_i d_j \cdot h\left(\rho_{0ij-1}\right), & \rho_{0ij-1} > 1 \\ 6d_i d_j \cdot h(1), & \text{otherwise} \end{cases} \tag{7.140a}$$

$$\rho_{ij-\min} = 6d_i d_j \cdot h(-1) \tag{7.140b}$$

Based on Figure 7.22 and the solution of cubic equation (Zwillinger 2018), the equivalent correlation coefficient ρ_{0ij} can be determined: (i) when $\Delta > 0$, there is only one intersection of $h(\rho_{0ij})$ and $\rho_{ij}/(6d_i d_j)$, and the solution of ρ_{0ij} is the horizontal coordinate of this intersection, which is identical with Eq. (7.137); (ii) when $\Delta = 0$, there exist two intersections, and the solution of ρ_{0ij} is the smaller one of the horizontal coordinate of these intersections, which is identical with Eq. (7.137); and (iii) when $\Delta < 0$, there exist three intersections, and the solution of ρ_{0ij} is the smallest one among the horizontal coordinate of these intersections, which is expressed as:

$$\rho_{0ij} = -2r \cos\left(\frac{\theta - \pi}{3}\right) - \frac{t_2}{3} \tag{7.141}$$

In summary, the solutions of equivalent correlation coefficient ρ_{0ij} and the applicable bound of original correlation coefficient ρ_{ij} for $d_i d_j \neq 0$ (i.e. Eq. (7.128a) is a cubic equation about ρ_{0ij}) are summarised as indicated in Algorithm 7.1.

(4) If $d_i d_j = 0$ and $c_i c_j \neq 0$, Eq. (7.128a) reduces to a quadratic equation. For brevity, the right side of Eq. (7.128a) is expressed as $h_2(\rho_{0ij})$, i.e.

$$\rho_{ij} = \left(b_i b_j + 3d_i b_j + 3b_i d_j\right) \cdot \rho_{0ij} + 2c_i c_j \rho_{0ij}^2 = h_2\left(\rho_{0ij}\right) \tag{7.142}$$

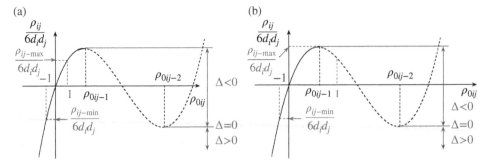

Figure 7.22 The shape of $h(\rho_{0ij})$ for $d_i d_j > 0$, $D > 0$, and $t_2 < 0$. (a) $\rho_{0ij-1} \leq 1$. (b) $\rho_{0ij-1} > 1$.

(a)

(b)

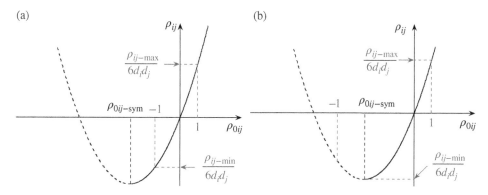

Figure 7.23 The shape of $h_2(\rho_{0ij})$ for $c_i c_j > 0$. (a) $\rho_{0ij-sym} > -1$. (b) $\rho_{0ij-sym} < -1$.

The axis of symmetry of $h_2(\rho_{0ij})$ is at:

$$\rho_{0ij} = \rho_{0ij-sym} = -\frac{b_i b_j + 3b_i d_j + 3d_i d_j}{4c_i c_j} \tag{7.143}$$

Because $b_i b_j + 3b_i d_j + 3d_i b_j$ is larger than zero, $\rho_{0ij-sym}$ is positive for $c_i c_j < 0$ and negative for $c_i c_j > 0$. The shape of $h_2(\rho_{0ij})$ is depicted in Figure 7.23a,b for $c_i c_j > 0$, in which the solid lines denote the region satisfying the condition that $\rho_{0ij} \cdot \rho_{ij} \geq 0$ and ρ_{0ij} is an increasing function of ρ_{ij}.

From Figure 7.23a,b, the application bound of ρ_{ij} for $c_i c_j > 0$ is expressed as:

$$\rho_{ij-max} = h_2(1) \tag{7.144a}$$

$$\rho_{ij-min} = \begin{cases} h_2\left(\rho_{0ij-sym}\right), & \rho_{0ij-sym} > -1 \\ h_2(-1), & \text{otherwise} \end{cases} \tag{7.144b}$$

And the equivalent correlation coefficient ρ_{0ij} is obtained as:

$$\rho_{0ij} = \frac{1}{4c_i c_j}\left[-(b_i b_j + 3b_i d_j + 3d_i b_j) + \sqrt{(b_i b_j + 3b_i d_j + 3d_i b_j)^2 + 8c_i c_j \rho_{ij}}\right] \tag{7.145}$$

Similarly, the solution of ρ_{0ij} for $d_i d_j = 0$ and $c_i c_j < 0$ can be determined, which is identical with Eq. (7.145), and the application bounds of ρ_{ij} are expressed as:

$$\rho_{ij-max} = \begin{cases} h_2\left(\rho_{0ij-sym}\right), & \rho_{0ij-sym} < 1 \\ h_2(1), & \text{otherwise} \end{cases} \tag{7.146a}$$

$$\rho_{ij-min} = h_2(-1) \tag{7.146b}$$

(5) If $d_i d_j = 0$ and $c_i c_j = 0$, Eq. (7.128a) reduces to a linear equation. The equivalent correlation coefficient ρ_{0ij} can then be determined as:

$$\rho_{0ij} = \frac{\rho_{ij}}{b_i b_j + 3b_i d_j + 3d_i b_j} \tag{7.147}$$

And the application bound of ρ_{ij} is expressed as:

$$\rho_{ij-\max} = b_i b_j + 3 b_i d_j + 3 d_i b_j \tag{7.148a}$$

$$\rho_{ij-\min} = -\left(b_i b_j + 3 b_i d_j + 3 d_i b_j\right) \tag{7.148b}$$

Finally, the expressions are summarised in Algorithm 7.1 for the equivalent correlation coefficient ρ_{0ij} and the upper and lower bounds of original correlation coefficient ρ_{ij} to ensure the transformation executable.

7.6 Summary

In this chapter, the transformation of non-normal variables to independent normal variables and its inverse transformation are discussed. When the information of the full distribution can be utilised, the transformation can be achieved by normal tail transformation for a single random variable (or independent random variables), and for correlated random variables with known joint PDF, the transformation can be realised by Rosenblatt transformation. When the marginal PDFs and the correlation matrix of correlated random variables are known, Nataf transformation can be ultilized to achieve the normal transformation and its inverse transformation. When only the information of the first few moments and correlation matrix of the basic random variables are known, the normal transformation and its inverse transformation can be achieved by the so-called pseudo third- and fourth-moment normal transformation methods.

8

System Reliability Assessment by the Methods of Moment

8.1 Introduction

The reliability of a multi-component system is essentially a problem involving multiple modes of failure; that is, the failures of different components, or different sets of components, constitute distinct and different failure modes of the system. The consideration of multiple modes of failure, therefore, is fundamental to the problem of system reliability.

It is obvious that the reliability of a structural system will be a function of the reliability of its individual members, and the reliability assessment of structural systems therefore needs to account for multiple, perhaps correlated, limit states. Methods to deal with such problems will be the subject of this chapter. Discussion in this chapter will be limited to one-dimensional systems such as trusses and rigid frames. Two-dimensional systems such as plates, slabs, and shells will not be considered, nor will three-dimensional continua such as earth embankments and dams. However, for these more complex structural systems, the principles given here are also valid and can be used to develop appropriate calculation techniques.

8.2 Basic Concepts of System Reliability

8.2.1 Multiple Failure Modes

Systems that are composed of multiple components can be classified as series-connected or parallel-connected systems, or combinations thereof. More generally, the failure events, e.g. in the case of multiple failure modes, may also be represented as events in series (union) or in parallel (intersection).

The identification of each individual failure mode and the evaluation of their respective failure probabilities may be problematic in themselves. In the meantime, the reliability of a system will be a function of the redundancy of the system. Indeed, the analysis of reliability depends on whether the system is redundant or non-redundant. The failure of a component in a non-redundant system is tantamount to the failure of the entire system. Generally, the reliability of a system will be improved through the use of redundant components. Redundancy in a system may be either the active type or the standby type. In the case of active

Structural Reliability: Approaches from Perspectives of Statistical Moments, First Edition.
Yan-Gang Zhao and Zhao-Hui Lu.
© 2021 John Wiley & Sons Ltd. Published 2021 by John Wiley & Sons Ltd.

redundancies, all the components of a system are participating, e.g. carrying or sharing loads, whereas for systems with standby redundancies, some of the redundant components are inactive and become activated only when some of the active components have failed.

In the evaluation of system reliability, there are also fundamental differences between an active redundant system and a standby redundant system. For systems with active redundancies, failures of the components will occur sequentially, unless the capacities of the components are perfectly correlated and identically distributed. Therefore, for an active redundant system, all subsequent component failure will involve conditional probabilities. In contrast, for systems with standby redundancies, the component failure events may be statistically independent. Standby redundancy will invariably increase the reliability of a system unless the redundant components are perfectly correlated; in contrast, active redundancy may or may not be very effective in improving the reliability of a system.

8.2.2 Series Systems

Systems that are composed of components connected in series (series systems) are such that the failure of any one or more of these components constitutes the failure of the system. Such systems, therefore, have no redundancy and are also known as 'weakest link' systems. In other words, the reliability or safety of the system requires that none of the components fails. A series system may be represented schematically as in Figure 8.1.

Consider a structural system with k possible failure modes and assume that the performance function for failure mode i is given by

$$g_i(\mathbf{X}) = g_i(X_1, X_2, \cdots, X_n), i = 1, 2, \cdots, k \tag{8.1}$$

where $X_1, X_2, ..., X_n$ are the basic random variables; and $g_i(\cdot)$ is the ith performance function.
Define the failure event for failure mode i as

$$E_i = [g_i(\mathbf{X}) \leq 0] \tag{8.2a}$$

Then, the compliment of E_i is the safe event; that is

$$\overline{E}_i = [g_i(\mathbf{X}) > 0] \tag{8.2b}$$

Since the occurrence of any failure event E_i causes the failure of the structure, the failure event E of the structure is the union of all the possible failure modes, which can be expressed as

$$E = E_1 \cup E_2 \cup \cdots \cup E_k \tag{8.3a}$$

Conversely, the safety of a system is the event in which none of the k potential failure modes occurs; this means

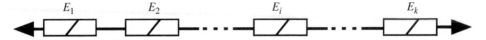

Figure 8.1 Representations of series systems.

Figure 8.2 Basic problem of system reliability in two dimensions.

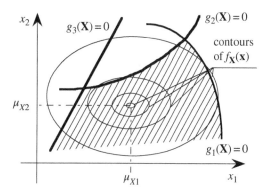

$$\overline{E} = \overline{E}_1 \cap \overline{E}_2 \cap \cdots \cap \overline{E}_k \tag{8.3b}$$

In the case of two variables, the above events are portrayed as in Figure 8.2, which shows three failure modes for the limit state equations $g_i(\mathbf{X}) = 0$, $i = 1, 2, 3$.

In structural reliability theory, the failure probability P_F of a series structural system due to the occurrence of the event E in Eq. (8.3a) involves the following integration

$$P_F = \int \cdots \int_{(E_1 \cup E_2 \cup \cdots \cup E_k)} f_{X_1, X_2, \cdots, X_n}(x_1, x_2, \cdots, x_n) dx_1 dx_2 \cdots dx_n \tag{8.4a}$$

where $f(\cdot)$ is the pertinent joint probability density function.

Whereas the probability of the safety P_S of the system would be

$$P_S = \int \cdots \int_{(\overline{E}_1 \cap \overline{E}_2 \cap \cdots \cap \overline{E}_k)} f_{X_1, X_2, \cdots, X_n}(x_1, x_2, \cdots, x_n) dx_1 dx_2 \cdots dx_n \tag{8.4b}$$

Example 8.1 Consider the idealised series system shown in Figure 8.3. Since the structure is non-redundant, the failure of either member R_i ($i = 1, 2, 3, 4$) will cause the system failure. Suppose the failure of each member is presented by E_i, then

$$E_i = [R_i - S \le 0] \tag{8.5a}$$

The failure of the system would be the event

$$E_F = E_1 \cup E_2 \cup E_3 \cup E_4$$
$$= (R_1 - S \le 0) \cup (R_2 - S \le 0) \cup (R_3 - S \le 0) \cup (R_4 - S \le 0) \tag{8.5b}$$

On the other hand, since the failure of the member with minimum strength will cause the failure of the system, Eq. (8.5b) can be also expressed as

$$E_F = [\min(R_1, R_2, R_3, R_4) - S \le 0] \tag{8.5c}$$

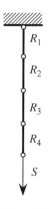

Figure 8.3 An idealised series system for Example 8.1.

In other words, the strength of the series system is the minimum of the strength of all the members.

Example 8.2   Consider the static deterministic truss shown in Figure 8.4a. Since the structure is non-redundant, either the failure of member AB or AC will cause the failure of the structure as shown in Figures 8.4b and c. So the truss is a series system.

Suppose that the strength of the members AB and AC are R_1 and R_2, respectively. The failure of members AB and AC will be represented as

$$E_1 = [g_1 = 3R_1 - 5P \leq 0] \tag{8.6a}$$
$$E_2 = [g_2 = 3R_2 - 4P \leq 0] \tag{8.6b}$$

The failure of the system would be the event

$$E_F = E_1 \cup E_2 = [g_1 \leq 0 \cup g_2 \leq 0] \tag{8.6c}$$

Example 8.3   Consider a one-story one-bay elastoplastic frame shown in Figure 8.5a (Ono et al. 1990). The six most likely failure modes obtained from stochastic limit analysis are shown in Figure 8.5b. Since any of the six failure modes will cause the failure of the structure, the frame is a series system. The six failure modes will be represented as

$$E_1 = [g_1 = 2M_1 + 2M_2 - 15S_1 \leq 0] \tag{8.7a}$$
$$E_2 = [g_2 = M_1 + 3M_2 + 2M_3 - 15S_1 - 10S_2 \leq 0] \tag{8.7b}$$
$$E_3 = [g_3 = 2M_1 + M_2 + M_3 - 15S_1 \leq 0] \tag{8.7c}$$
$$E_4 = [g_4 = M_1 + 2M_2 + M_3 - 15S_1 \leq 0] \tag{8.7d}$$
$$E_5 = [g_5 = M_1 + M_2 + 2M_3 - 15S_1 \leq 0] \tag{8.7e}$$
$$E_6 = [g_6 = M_1 + M_2 + 4M_3 - 15S_1 - 10S_2 \leq 0] \tag{8.7f}$$

The failure of the system would be the event

$$E_F = E_1 \cup E_2 \cup E_3 \cup E_4 \cup E_5 \cup E_6$$
$$= [g_1 \leq 0 \cup g_2 \leq 0 \cup g_3 \leq 0 \cup g_4 \leq 0 \cup g_5 \leq 0 \cup g_6 \leq 0] \tag{8.7g}$$

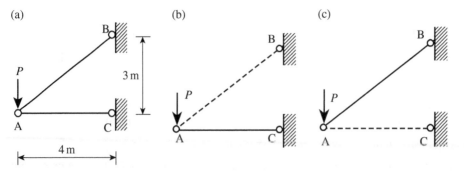

Figure 8.4   A static deterministic truss for Example 8.2. (a) A static deterministic truss. (b) Failure of member of AB. (c) Failure of member of AC.

(a)

(b)

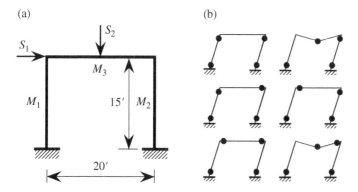

Figure 8.5 One-story one-bay frame for Example 8.3. (a) A one-story one-bay frame. (b) Six failure modes.

8.2.3 Parallel Systems

Systems that are composed of components connected in parallel (parallel systems) are such that the failure of the system requires failures of all the components. In other words, if any one of the components survives, the system remains safe. A parallel system is clearly a redundant system and may be represented schematically as shown in Figure 8.6.

If E_i denotes the failure of component i, the failure of a k-component parallel system, therefore, is

$$E_F = E_1 \cap E_2 \cap \cdots \cap E_k \qquad (8.8a)$$

and the safety of the system would be

$$E_S = \overline{E_F} = \overline{E}_1 \cup \overline{E}_2 \cup \cdots \cup \overline{E}_k \qquad (8.8b)$$

Note that the failure of a series system is the union of the component failures, whereas the safety of parallel system is the union of the survival events of the components. Also, the safety of a series system is the intersection of the safety of the components, whereas the failure of a parallel system is the intersection of the component failures.

Since a parallel system is composed of redundant components, its reliability will depend on whether the redundancies are active or standby. Redundancy in a structural system is invariably of the active type. That is, all the components in a structural system carry loads or are subjected to the effects of the applied load.

Figure 8.6 Representations of parallel systems.

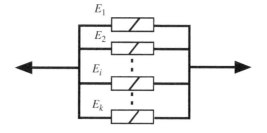

Example 8.4 Standby Redundancy (Adopted from Ang and Tang 1984)

Emergency power provided by diesel generators may be necessary to operate the automatic safety of a plant during earthquake. In order to increase the reliability of the required emergency power generating system, several generators may be installed, each of which has sufficient capacity to provide the emergency power requirement. Suppose four such generating units are installed as shown in Figure 8.7. Since any single unit can supply the emergency power requirement, the failure of the system requires failures of all the four units during an earthquake. Thus the system is a parallel system, with standby redundancy.

If E_i denotes the failure of unit i, then the failure of the system is

$$E_F = E_1 \cap E_2 \cap E_3 \cap E_4$$

Example 8.5 Active Redundancy

Consider the idealised parallel system shown in Figure 8.8a. If the elements are brittle, with different fracture strength R_1 and R_2, then the failure of elements 1 and 2 will be represented as Eqs. (8.9a) and (8.9b), respectively.

$$E_1 = \left[g_1 = R_1 - \frac{1}{2}S \le 0 \right] \tag{8.9a}$$

$$E_2 = \left[g_2 = R_2 - \frac{1}{2}S \le 0 \right] \tag{8.9b}$$

The failure of the system would be the event

$$E_F = E_1 \cap E_2 = \left[\left(R_1 - \frac{1}{2}S \right) \cap \left(R_2 - \frac{1}{2}S \right) \le 0 \right] \tag{8.9c}$$

However, since both elements carry the load S, as element 1 fails, element 2 has to support the entire load S by itself, so the failure of element 2 becomes

$$E_2 = [g_2 = R_2 - S \le 0] \tag{8.9d}$$

The failure of the system in this failure sequence would be the event

$$E_F = \left[\left(R_1 - \frac{1}{2}S \le 0 \right) \cap (R_2 - S \le 0) \right] \tag{8.9e}$$

Similarly, when element 2 fails first, element 1 has to support the whole load S by itself, then the failure of the system in this failure sequence would be the event

$$E_F = \left[(R_1 - S \le 0) \cap \left(R_2 - \frac{1}{2}S \le 0 \right) \right] \tag{8.9f}$$

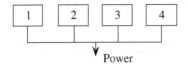

Figure 8.7 Emergency power system for Example 8.4.

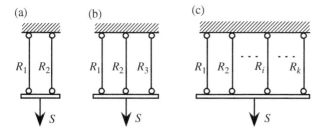

Figure 8.8 Parallel systems for Example 8.5. (a) With two elements. (b) With three elements. (c) With k elements.

As there is no assurance as to which of the elements will fail first, both the two failure sequences should be considered. The failure of the system therefore would be the event

$$E_F = \left[\left(R_1 - \frac{1}{2}S \le 0 \right) \cap (R_2 - S \le 0) \right] \cup \left[\left(R_2 - \frac{1}{2}S \le 0 \right) \cap (R_1 - S \le 0) \right] \quad (8.9\text{g})$$

Then consider the idealised parallel system shown in Figure 8.8b with different brittle fracture strength R_1, R_2, and R_3. According to the analysis above, there are six failure sequences, i.e. 1–2–3, 1–3–2, 2–1–3, 2–3–1, 3–2–1, and 3–1–2. Since there is no assurance as to which of the elements will fail first, all of the six failure sequences should be considered. The failure of the system would be the event

$$\begin{aligned} E_F = &\left[\left(R_1 - \frac{1}{3}S \le 0 \right) \cap \left(R_2 - \frac{1}{2}S \le 0 \right) \cap (R_3 - S \le 0) \right] \\ &\cup \left[\left(R_1 - \frac{1}{3}S \le 0 \right) \cap \left(R_3 - \frac{1}{2}S \le 0 \right) \cap (R_2 - S \le 0) \right] \\ &\cup \left[\left(R_2 - \frac{1}{3}S \le 0 \right) \cap \left(R_3 - \frac{1}{2}S \le 0 \right) \cap (R_1 - S \le 0) \right] \\ &\cup \left[\left(R_2 - \frac{1}{3}S \le 0 \right) \cap \left(R_1 - \frac{1}{2}S \le 0 \right) \cap (R_3 - S \le 0) \right] \\ &\cup \left[\left(R_3 - \frac{1}{3}S \le 0 \right) \cap \left(R_1 - \frac{1}{2}S \le 0 \right) \cap (R_2 - S \le 0) \right] \\ &\cup \left[\left(R_3 - \frac{1}{3}S \le 0 \right) \cap \left(R_2 - \frac{1}{2}S \le 0 \right) \cap (R_1 - S \le 0) \right] \end{aligned} \quad (8.9\text{h})$$

It can be observed that Eq. (8.9h) is quite cumbersome. In order to deal this problem more briefly, we introduce a new set of random variables $R^{(1)}$, $R^{(2)}$, and $R^{(3)}$; where $R^{(1)}$ is the smallest of (R_1, R_2, R_3), $R^{(2)}$ is the second smallest of (R_1, R_2, R_3), and $R^{(3)}$ is the third smallest of (R_1, R_2, R_3), i.e. the largest of (R_1, R_2, R_3); then the failure sequence of the system will be $R^{(1)} - R^{(2)} - R^{(3)}$. Therefore the failure of the system would be the event

$$E_F = \left[\left(R^{(1)} - \frac{1}{3}S \le 0 \right) \cap \left(R^{(2)} - \frac{1}{2}S \le 0 \right) \cap \left(R^{(3)} - \frac{1}{3}S \le 0 \right) \right] \quad (8.9\text{i})$$

Generally, for the parallel system shown in Figure 8.8c with k elements, we introduce a new set of random variables $R^{(i)}$, $i = 1, 2, ..., n$, where $R^{(i)}$ is the ith order and smallest of $(R_i, i = 1, 2, ..., k)$. The failure of the system would be the event

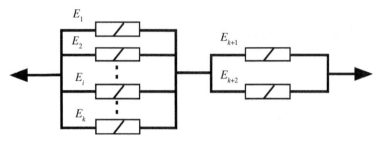

Figure 8.9 Combined systems.

$$E_F = \left[\left(R^{(1)} - \frac{1}{k} S \leq 0 \right) \cap \left(R^{(2)} - \frac{1}{k-1} S \leq 0 \right) \cap \cdots \cap \right.$$
$$\left. \left(R^{(i)} - \frac{1}{k-i+1} S \leq 0 \right) \cap \cdots \cap \left(R^{(k)} - S \leq 0 \right) \right] \qquad (8.9j)$$

8.2.3.1 Combined Series – Parallel Systems

Structural systems may be composed of a combination of series- and parallel-connected components. However, it should be noted that not all systems can be decomposed into such series and parallel components. A general system may, nevertheless, be represented as a combination of failure or safe events in series and/or in parallel. The pertinent events may not simply be the failures or survivals of the individual components. Combined systems may be schematically represented as shown in Figure 8.9.

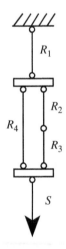

Figure 8.10 A combined system for Example 8.6.

Example 8.6 Consider a simple parallel-chain system shown in Figure 8.10. Assume that the individual components of the system will fail through tensile fracture or compressive buckling, and thus may be assumed to be brittle; i.e. once failure occurs, the strength of a component is reduced to zero. In this case, the system is a non-series system. Suppose that the brittle elements have different fracture strength R_1, R_2, R_3, and R_4. The system is series-composed of R_1 and a sub-system R_{234} that is composed of (R_2, R_3, R_4). Suppose the failure of R_1 and the sub-system R_{234} are presented by event E_1 and E_{234}, then the failure of the system would be the event

$$E_F = [E_1 \cup E_{234}] \qquad (8.10a)$$

For element R_1, the failure is readily expressed as

$$E_1 = (R_1 - S \leq 0) \qquad (8.10b)$$

Note the sub-system R_{234} is parallel-composed of R_4 and another sub-system R_{23} that is composed of (R_2, R_3). So the failure of sub-system R_{234} would be the event

$$E_{234} = \left[\min{(R_4, R_{23})} - \frac{1}{2}S \le 0 \right] \cap [\max{(R_4, R_{23})} - S \le 0] \tag{8.10c}$$

where R_{23} is the strength of the sub-system R_{23}. Note again the sub-system R_{23} is series-composed of R_2 and R_3. So the strength of the sub-system R_{23} would be the minimum of R_2 and R_3, i.e.

$$R_{23} = \min{(R_2, R_3)} \tag{8.10d}$$

Finally, the failure of the system would be the event

$$E_F = [E_1 \cup E_{234}] = (R_1 - S \le 0) \cup \left[\min{(R_4, R_{23})} - \frac{1}{2}S \le 0 \right] \cap [\max{(R_4, R_{23})} - S \le 0]$$

$$= (R_1 - S \le 0) \cup \left\{ \min{[R_4, \min{(R_2, R_3)}]} - \frac{1}{2}S \le 0 \right\} \cap \{ \max{[R_4, \min{(R_2, R_3)}]} - S \le 0 \}$$

$$= (R_1 - S \le 0) \cup \left[\min{(R_2, R_3, R_4)} - \frac{1}{2}S \le 0 \right] \cap \{ \max{[R_4, \min{(R_2, R_3)}]} - S \le 0 \}$$

$$\tag{8.10e}$$

Example 8.7 Consider a statically indeterminate truss with redundancy shown in Figure 8.11a. Assume that the individual components of this truss system will fail through fracture, and thus the truss is a brittle system. The internal forces for all members have been obtained and shown in Figure 8.11b, from which it can be easily derived that the failure of each member can be presented as

$$E_1 = [T_1 - 1.127F_1 + 0.826F_2 \le 0], \quad E_2 = [T_2 - 0.206F_1 - 0.826F_2 \le 0],$$
$$E_3 = [T_3 - 0.155F_1 + 0.13F_2 \le 0], \quad E_4 = [T_4 - 1.409F_1 - 0.217F_2 \le 0], \tag{8.11a}$$
$$E_5 = [T_5 - 0.258F_1 + 0.217F_2 \le 0]$$

Since the truss has only one redundancy, failure of any two members will cause the failure of the system. So there are 10 failure modes as shown in Figure 8.12. However, since the

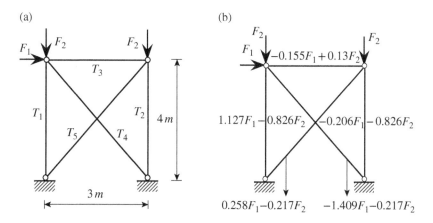

Figure 8.11 A brittle truss for Example 8.7. (a) Load bearing condition and strength of members. (b) Internal forces of all members.

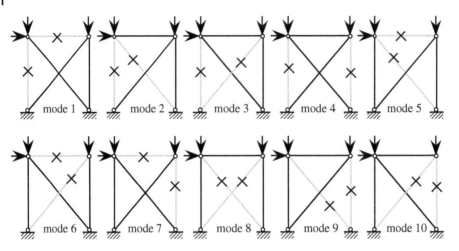

Figure 8.12 The failure modes of the brittle truss.

system is active redundant, the internal forces will change after failure of one member. One should not simply define each mode directly in terms of the intersection of each pair of events listed in Eq. (8.11a). For example, since the internal force in members 1 and 3 will change after one of the two members failed, failure mode 1 cannot be simple defined as

$$E_1 \cap E_3 = [T_1 - 1.127F_1 + 0.826F_2 \leq 0] \cap [T_3 - 0.155F_1 + 0.13F_2 \leq 0]$$

The internal force of each member under the condition that the first member failed is shown in Figure 8.13a, the failure of each member under the condition that the first member failed can be presented as

$$E_2|E_1 = \left[T_2 - \frac{4}{3}F_1 \leq 0\right], \quad E_3|E_1 = \left[T_3 - F_1 + \frac{3}{4}F_2 \leq 0\right],$$

$$E_4|E_1 = \left[T_4 - \frac{5}{4}F_2 \leq 0\right], \quad E_5|E_1 = \left[T_5 - \frac{5}{3}F_1 + \frac{5}{4}F_2 \leq 0\right]$$

$$(8.11b)$$

Under the condition that the first member fails, the failure of any of the four members left will cause the system failure. The failure of the system under the condition that the first member fails would be the event

$$E_F|E_1 = E_1 \cap [(E_2|E_1) \cup (E_3|E_1) \cup (E_4|E_1) \cup (E_5|E_1)]$$
$$= (T_1 - 1.127F_1 + 0.826F_2 \leq 0) \cap$$
$$\left[\left(T_2 - \frac{4}{3}F_1 \leq 0\right) \cup \left(T_3 - F_1 + \frac{3}{4}F_2 \leq 0\right) \cup \left(T_4 - \frac{5}{4}F_2 \leq 0\right) \cup \left(T_5 - \frac{5}{3}F_1 + \frac{5}{4}F_2 \leq 0\right)\right]$$

$$(8.11c)$$

Similarly, the internal force of each member under the condition that the second, third, fourth, and fifth members fail are shown in Figure 8.13b–e respectively, the failure of the system under the condition that the second, third, fourth, and fifth members fail can be presented as

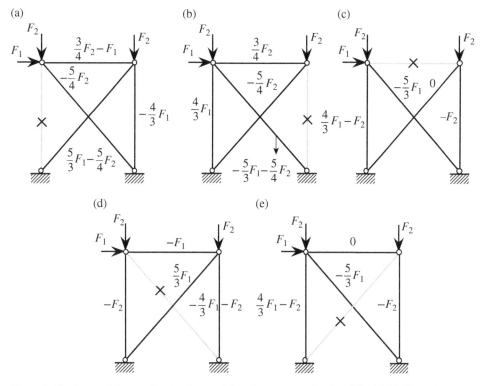

Figure 8.13 Internal forces of under the condition that one member has failed. (a) Failure of member of T_1. (b) Failure of member of T_2. (c) Failure of member of T_3. (d) Failure of member of T_4. (e) Failure of member of T_5.

$$E_F|E_2 = E_2 \cap [(E_1|E_2) \cup (E_3|E_2) \cup (E_4|E_2) \cup (E_5|E_2)]$$
$$= (T_2 - 0.206F_1 - 0.826F_2 \le 0) \cap$$
$$\left[\left(T_1 - \frac{4}{3}F_1 \le 0 \right) \cup \left(T_3 - \frac{3}{4}F_2 \le 0 \right) \cup \left(T_4 - \frac{5}{3}F_1 - \frac{5}{4}F_2 \le 0 \right) \cup \left(T_5 - \frac{5}{4}F_2 \le 0 \right) \right]$$

$$(8.11d)$$

$$E_F|E_3 = E_3 \cap [(E_1|E_3) \cup (E_2|E_3) \cup (E_4|E_3) \cup (E_5|E_3)]$$
$$= (T_3 - 0.155F_1 + 0.13F_2 \le 0) \cap \left[\left(T_1 - \frac{4}{3}F_1 + F_2 \le 0 \right) \right.$$
$$\left. \cup (T_2 - F_2 \le 0) \cup \left(T_4 - \frac{5}{3}F_1 \le 0 \right) \right]$$

$$(8.11e)$$

$$E_F|E_4 = E_4 \cap [(E_1|E_4) \cup (E_2|E_4) \cup (E_3|E_4) \cup (E_5|E_4)]$$
$$= (T_4 - 1.409F_1 - 0.217F_2 \le 0) \cap$$
$$\left[(T_1 - F_2 \le 0) \cup \left(T_2 - \frac{4}{3}F_1 - F_2 \le 0 \right) \cup (T_3 - F_1 \le 0) \cup \left(T_5 - \frac{5}{3}F_1 \le 0 \right) \right]$$

$$(8.11f)$$

$$E_F|E_5 = E_5 \cap [(E_1|E_5) \cup (E_2|E_5) \cup (E_3|E_5) \cup (E_4|E_5)]$$

$$= (T_5 - 0.258F_1 + 0.217F_2 \leq 0) \cap \left[\left(T_1 - \frac{4}{3}F_1 + F_2 \leq 0 \right) \right.$$

$$\left. \cup (T_2 - F_2 \leq 0) \cup \left(T_4 - \frac{5}{3}F_1 \leq 0 \right) \right] \tag{8.11g}$$

Finally, the failure of the system would be the event

$$E_F = [(E_F|E_1) \cup (E_F|E_2) \cup (E_F|E_3) \cup (E_F|E_4) \cup (E_F|E_5)] \tag{8.11h}$$

8.3 System Reliability Bounds

The calculation of the failure probability of a structural system is generally difficult through integration such as Eqs. (8.4a–b), and therefore approximations are always necessary. One of the approximations is to develop upper and lower bounds on the probability of failure of a system.

8.3.1 Uni-Modal Bounds

Suppose the failure of a series system with k failure modes is presented by an event E, and the ith failure mode is presented by the event E_i, then the failure and survival events are presented by

$$E = E_1 \cup E_2 \cup ... \cup E_k$$
$$\bar{E} = \bar{E}_1 \cap \bar{E}_2 \cap ... \cap \bar{E}_k$$

The probability of survival and failure are given by

$$P_S = P(\bar{E}_1 \cap \bar{E}_2 \cap ... \cap \bar{E}_k), \text{ and } P_F = 1 - P_S$$

First, consider events E_i and E_j with failure mode i and j. The relationship between E_i and E_j are shown in Figure 8.14a–c, one can easily understand that

$$P(E_j|E_i) \geq P(E_j)$$
$$P(\bar{E}_j|\bar{E}_i) \geq P(\bar{E}_j)$$

(a)

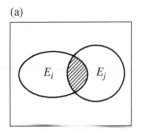

(b)

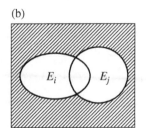

(c)

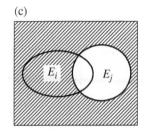

Figure 8.14 Relationship between events E_i and E_j. (a) $E_i \cap E_j$; (b) $\bar{E}_i \cap \bar{E}_j$; (c) $\bar{E}_j$.

Since

$$P(\overline{E}_i \cap \overline{E}_j) \geq P(\overline{E}_j | \overline{E}_i) P(\overline{E}_i)$$

We have

$$P(\overline{E}_i \cap \overline{E}_j) \geq P(\overline{E}_j) P(\overline{E}_i)$$

For k events, this can be generalised to yield

$$P_s = P(\overline{E}_1 \cap \overline{E}_2 \cap \cdots \cap \overline{E}_k) \geq \prod_{i=1}^{k} P(\overline{E}_i) \qquad (8.12a)$$

Conversely, since

$$\overline{E}_1 \cap \overline{E}_2 \cap \ldots \cap \overline{E}_n \subset \overline{E}_i \text{ for any } i$$

We have

$$\overline{E}_1 \cap \overline{E}_2 \cap \ldots \cap \overline{E}_k \subset \min_{1 \leq i \leq k}(\overline{E}_i)$$

Therefore

$$P_S \leq \min_{1 \leq i \leq k}\left[P(\overline{E}_i)\right] \qquad (8.12b)$$

If we denote the reliability against the ith failure mode as P_{Si}, and the probability of failure corresponding to the ith failure mode as P_{fi}, then we have

$$P_{Si} = P(\overline{E}_i), P_{fi} = P(E_i) = 1 - P_{Si}$$

Then the reliability of the system P_S is bounded according to Eqs. (8.12a) and (8.12b) as follows.

$$\prod_{i=1}^{k} P_{Si} \leq P_S \leq \min_{1 \leq i \leq k}[P_{Si}] \qquad (8.13a)$$

Conversely, the corresponding bounds for the failure probability P_F would be (e.g. Cornell 1967, Ang and Amin 1968)

$$\max_{1 \leq i \leq k}(P_{fi}) \leq P_F \leq 1 - \prod_{i=1}^{k}(1 - P_{fi}) \qquad (8.13b)$$

where P_{fi} is the failure probability of the ith failure mode.

The bounds in Eq. (8.13a–b), are often referred to as the 'first-order' or 'uni-modal' bounds on P_F and P_S, in the sense that the lower and upper probability bounds involve single mode probabilities.

Since only the failure probability of a single failure mode is considered, and the correlation of the failure modes is neglected. The above wide-bound estimation method is simple to evaluate. However, for many practical structural systems, especially for complex systems, the first-order bound given by Eq. (8.13b) is too wide to be meaningful (Grimmelt and Schueller 1982). Better bounds have been developed, but at the expense of more computation.

8.3.2 Bi-Modal Bounds

The first-order bounds described can be improved by taking into account the correlation between pairs of potential failure modes. The resulting improved bounds will necessarily require the probabilities of joint events such as

$$E_i \cap E_j \text{ or } \overline{E}_i \cap \overline{E}_j,$$

and thus may be called 'bi-modal' or 'second-order' bounds.

Suppose the failure of a system with k failure modes is represented by the event E, and the ith failure mode is presented by the event E_i, then the failure event is given by

$$
\begin{aligned}
E &= E_1 \cup E_2 \cup \cdots \cup E_k \\
&= E_1 + \left(E_2 \cap \overline{E}_1\right) + \left[E_3 \cap \left(\overline{E}_1 \cap \overline{E}_2\right)\right] + \cdots + \left[E_k \cap \left(\overline{E}_1 \cap \overline{E}_2 \cap \cdots \cap \overline{E}_{k-1}\right)\right]
\end{aligned}
$$

$$(8.14)$$

where $E_2 \cap \overline{E}_1$ and $E_3 \cap \left(\overline{E}_1 \cap \overline{E}_2\right)$ are illustrated in Figure 8.15.

According to the DeMorgan's rule

$$\overline{E}_1 \cap \overline{E}_2 \cap \cdots \cap \overline{E}_{i-1} = \overline{E_1 \cup E_2 \cup \cdots \cup E_{i-1}}$$

for $i = 2, 3, \ldots, k$, we have

$$E_i \cap \left(\overline{E}_1 \cap \overline{E}_2 \cap \cdots \cap \overline{E}_{i-1}\right) = E_i \cap \overline{E_1 \cup E_2 \cup \cdots \cup E_{i-1}}$$

Note that

$$\overline{E_1 \cup E_2 \cup \ldots \cup E_{i-1}} + \left(E_1 \cup E_2 \cup \ldots \cup E_{i-1}\right) = S$$

$$\overline{E_1 \cup E_2 \cup \cdots \cup E_{i-1}} = S - \left(E_1 \cup E_2 \cup \cdots \cup E_{i-1}\right)$$

where S is a certain event.

Then we have

$$
\begin{aligned}
E_i \cap \left(\overline{E}_1 \cap \overline{E}_2 \cap \ldots \cap \overline{E}_{i-1}\right) &= E_i \cap \left[S - \left(E_1 \cup E_2 \cup \ldots \cup E_{i-1}\right)\right] \\
&= E_i \cap S - E_i \cap \left(E_1 \cup E_2 \cup \ldots \cup E_{i-1}\right) = E_i - \left(E_i \cap E_1\right) \cup \left(E_i \cap E_2\right) \cup \cdots \cup \left(E_i \cap E_{i-1}\right)
\end{aligned}
$$

$E_3 \cap \ \left(\overline{E}_1 \cap \overline{E}_2\right)$ $\left(E_2 \cap \overline{E}_1\right)$ **Figure 8.15** Decomposition of E.

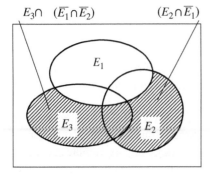

That is

$$P\left[E_i \cap \left(\bar{E}_1 \cap \bar{E}_2 \cap \ldots \cap \bar{E}_{i-1}\right)\right] = P(E_i) - P[(E_i \cap E_1) \cup (E_i \cap E_2) \cup \cdots \cup (E_i \cap E_{i-1})]$$

Note

$$P[(E_i \cap E_1) \cup (E_i \cap E_2) \cup \cdots \cup (E_i \cap E_{i-1})] \leq P(E_i \cap E_1) + P(E_i \cap E_2) + \cdots + P(E_i \cap E_{i-1})$$

Therefore

$$P\left[E_i \cap \left(\bar{E}_1 \cap \bar{E}_2 \cap \ldots \cap \bar{E}_{i-1}\right)\right] \geq P(E_i) - \sum_{j=1}^{i-1} P\left(E_i \cap E_j\right) \tag{8.15}$$

Substituting Eq. (8.15) into Eq. (8.14) leads to

$$P(E) \geq P(E_1) + \sum_{i=2}^{k} \max\left\{\left[P(E_i) - \sum_{j=1}^{i-1} P\left(E_i \cap E_j\right)\right], 0\right\} \tag{8.16}$$

On the other hand, since

$$\bar{E}_1 \cap \bar{E}_2 \cap \ldots \cap \bar{E}_{i-1} \subset \min_{1 \leq j \leq i-1}\left(\bar{E}_j\right)$$

We have

$$E_i \cap \left(\bar{E}_1 \cap \bar{E}_2 \cap \ldots \cap \bar{E}_{i-1}\right) \subset E_i \cap \left[\min_{1 \leq j \leq i-1}\left(\bar{E}_j\right)\right]$$

Note

$$\min_{1 \leq j \leq i-1}\left(\bar{E}_j\right) + \max_{1 \leq j \leq i-1}\left(E_j\right) = S$$
$$\min_{1 \leq j \leq i-1}\left(\bar{E}_j\right) = S - \max_{1 \leq j \leq i-1}\left(E_j\right)$$

we have

$$E_i \cap \left(\bar{E}_1 \cap \bar{E}_2 \cap \ldots \cap \bar{E}_{i-1}\right) \subset E_i \cap \left[S - \max_{1 \leq j \leq i-1}\left(E_j\right)\right]$$
$$E_i \cap \left(\bar{E}_1 \cap \bar{E}_2 \cap \ldots \cap \bar{E}_{i-1}\right) \subset E_i - E_i \cap \max_{1 \leq j \leq i-1}\left(E_j\right)$$

Hence

$$P\left[E_i \cap \left(\bar{E}_1 \cap \bar{E}_2 \cap \ldots \cap \bar{E}_{i-1}\right)\right] \leq P(E_i) - P\left[E_i \cap \max_{1 \leq j \leq i-1}\left(E_j\right)\right]$$

That is

$$P\left[E_i \cap \left(\bar{E}_1 \cap \bar{E}_2 \cap \ldots \cap \bar{E}_{i-1}\right)\right] \leq P(E_i) - \max_{1 \leq j \leq i-1} P\left(E_i \cap E_j\right) \tag{8.17}$$

Substituting Eq. (8.17) into Eq. (8.14), one obtains that

$$P[E] \leq \sum_{i=1}^{k} [P(E_i)] - \sum_{i=2}^{k} \left[\max_{1 \leq j \leq i-1} P\left(E_i \cap E_j\right)\right] \tag{8.18}$$

If we denote the probability of the intersection $E_i \cap E_j$ as P_{fij}, i.e.

$$P_{fij} = P(E_i \cap E_j)$$

Then the probability of failure of the system P_F is bounded according to Eqs. (8.16) and (8.18) as follows (Ditlevsen 1979b).

$$P_{f1} + \sum_{i=2}^{k} \max\left(P_{fi} - \sum_{j=1}^{i-1} P_{fij}, 0\right) \leq P_F \leq \sum_{i=1}^{k} P_{fi} - \sum_{i=2}^{k} \max_{j \leq i}\left(P_{fij}\right) \tag{8.19}$$

where P_{fij} is referred to the joint probability of the simultaneous occurrences of the ith and jth failure modes. The left- and right-hand sides of Eq. (8.19) are the lower bound and upper bound of the failure probability of a series structural system with k potential failure modes, respectively. It can be observed that because the joint probability of simultaneous failures of every pair of failure modes must be evaluated, the resulting bounds of Eq. (8.19) are narrower than those from Eq. (8.13b).

8.3.3 Correlation Between a Pair of Failure Modes

In order to obtain the second order bounds in Eq. (8.19), P_{fij}, the joint probability of the simultaneous occurrences of the ith and jth failure modes, have to be evaluated. Suppose that the failure probability corresponding to failure modes E_i and E_j with correlation coefficient ρ_{ij} are obtained as P_{fi} and P_{fj}, and the corresponding reliability indices are

$$\beta_i = -\Phi^{-1}(P_{fi}), \quad \beta_j = -\Phi^{-1}(P_{fj})$$

Without loss of generality, we introduce two standard normal random variables U_i and U_j with correlation coefficient ρ_{ij}, then the failure modes E_i and E_j can be represented by the following events

$$E_i = [U_i \leq -\beta_i], \quad E_j = \left[U_j \leq -\beta_j\right]$$

Therefore,

$$P_{fij} = P(E_i \cap E_j) = P\left[(U_i \leq -\beta_i) \cap \left(U_j \leq -\beta_j\right)\right]$$

P_{fij} can thus be expressed by

$$P_{fij} = \Phi_2\left(-\beta_i, -\beta_j, \rho_{ij}\right) = \int_{-\infty}^{-\beta_i} \int_{-\infty}^{-\beta_j} \phi_2\left(x_i, x_j, \rho_{ij}\right) dx_i dx_j \tag{8.20}$$

where

$$\phi_2\left(x_i, x_j, \rho_{ij}\right) = \frac{1}{2\pi\sqrt{1-\rho_{ij}^2}} \exp\left(-\frac{1}{2} \cdot \frac{x_i^2 + x_j^2 - 2\rho_{ij}x_i x_j}{1-\rho_{ij}^2}\right) \tag{8.21}$$

where β_i and β_j are the reliability indices corresponding to the ith and jth failure modes, respectively, and ρ_{ij} is the correlation coefficient between the ith and jth failure modes.

$\phi_2(\cdot)$ and $\Phi_2(\cdot)$ are the probability density function and cumulative distribution function, respectively, of the two-dimensional standard normal distribution.

In order to calculate the joint probability of the simultaneous occurrences of the ith and jth failure modes, one needs to calculate the correlation coefficient ρ_{ij}. If the performance functions of ith and jth failure modes are g_i and g_j, respectively, ρ_{ij} is expressed, in definition, as

$$\rho_{ij} = \frac{Cov\left(g_i, g_j\right)}{\sigma_{gi}\sigma_{gj}} \tag{8.22}$$

Example 8. 8   In general, the performance functions of each failure mode of ductile frame structural systems are linear functions. Consider the following performance functions for the ith and jth failure modes

$$g_i = a_0 + a_1 X_1 + a_2 X_2, \ \ g_j = b_0 + b_1 X_1 + b_2 X_2$$

where X_1 and X_2 are uncorrelated with the mean and standard deviations of $\mu_{X1}, \mu_{X2}, \sigma_{X1},$ and σ_{X2}.

The mean values and standard deviations of g_i and g_j are easily obtained as

$$\mu_{gi} = a_0 + a_1\mu_{X1} + a_2\mu_{X2}, \ \ \mu_{gj} = b_0 + b_1\mu_{X1} + b_2\mu_{X2}$$

$$\sigma_{gi} = \sqrt{a_1^2\sigma_{X1}^2 + a_2^2\sigma_{X2}^2}, \ \ \sigma_{gj} = \sqrt{b_1^2\sigma_{X1}^2 + b_2^2\sigma_{X2}^2}$$

The covariance is given as

$$Cov\left(g_i, g_j\right) = E\left[\left(g_i - \mu_{gi}\right)\left(g_j - \mu_{gj}\right)\right]$$
$$= E\{[a_0 + a_1 X_1 + a_2 X_2 - (a_0 + a_1\mu_{X1} + a_2\mu_{X2})][b_0 + b_1 X_1 + b_2 X_2 - (b_0 + b_1\mu_{X1} + b_2\mu_{X2})]\}$$
$$= E\left[a_1 b_1(X_1 - \mu_{X1})^2 + a_2 b_2(X_2 - \mu_{X2})^2 + (a_1 b_2 + a_2 b_1)(X_1 - \mu_{X1})(X_2 - \mu_{X2})\right]$$
$$= a_1 b_1\sigma_{X1}^2 + a_2 b_2\sigma_{X2}^2$$

Then

$$\rho_{ij} = \frac{Cov\left(g_i, g_j\right)}{\sigma_{gi}\sigma_{gj}} = \frac{a_1 b_1\sigma_{X1}^2 + a_2 b_2\sigma_{X2}^2}{\sqrt{a_1^2\sigma_{X1}^2 + a_2^2\sigma_{X2}^2}\sqrt{b_1^2\sigma_{X1}^2 + b_2^2\sigma_{X2}^2}} \tag{8.23}$$

Generally, for the performance functions of the ith and jth failure modes

$$g_i = a_0 + \sum_{i=1}^{n} a_i X_i, \ \ g_j = b_0 + \sum_{i=1}^{n} b_i X_i$$

where X_i, $i = 1, 2, ..., n$, are independent with the mean values and standard deviations of μ_{Xi} and σ_{Xi}, $i = 1, 2, ..., n$.

The correlation coefficient between the ith and jth failure modes, ρ_{ij}, can be expressed as

$$\rho_{ij} = \frac{\sum\limits_{i=1}^{n} a_i b_i \sigma_{Xi}^2}{\sqrt{\sum\limits_{i=1}^{n} a_i^2 \sigma_{Xi}^2}\sqrt{\sum\limits_{i=1}^{n} b_i^2 \sigma_{Xi}^2}} \tag{8.24}$$

8.3.4 Bound Estimation of the Joint Failure Probability of a Pair of Failure Modes

Equation (8.20) is an accurate expression for P_{fij}; numerical integrations would be needed to obtain the results. To avoid such numerical integrations, further approximations are often adopted (involving further bounds for P_{fij}). Specific formulas for evaluating the lower and upper bounds of the joint failure probability P_{fij} were proposed as (Ditlevsen 1979b)

$$\left.\begin{array}{ll} \max\left[P(A), P(B)\right] \le P_{fij} \le P(A) + P(B), & \rho_{ij} \ge 0 \\ 0 \le P_{fij} \le \min\left[P(A), P(B)\right], & \rho_{ij} < 0 \end{array}\right\} \tag{8.25}$$

where

$$P(A) = \Phi(-\beta_i)\Phi\left(-\frac{\beta_j - \rho_{ij}\beta_i}{\sqrt{1-\rho_{ij}^2}}\right), \quad P(B) = \Phi\left(-\beta_j\right)\Phi\left(-\frac{\beta_i - \rho_{ij}\beta_j}{\sqrt{1-\rho_{ij}^2}}\right) \tag{8.26}$$

In which, β_i and β_j are the reliability indices corresponding to the ith and jth failure modes, respectively, and ρ_{ij} is the correlation coefficient between the ith and jth failure modes.

Example 8.9 Derivation of Eq. (8.25)

Consider the following two performance functions in the standard u-space.

$$g_i = a_0 + a_1 U_1 + a_2 U_2, \quad g_j = b_0 + b_1 U_1 + b_2 U_2$$

Using Eq. (8.23), the correlation coefficient can be readily derived as

$$\rho_{ij} = \frac{a_1 b_1 + a_2 b_2}{\sqrt{a_1^2 + a_2^2}\sqrt{b_1^2 + b_2^2}} \tag{8.27}$$

The relationship between the two limit state lines is shown in Figure 8.16a and b, where the joint failure event $E_i \cap E_j$ is the shaded region. From the direction cosines associated with $g_i = 0$ and $g_j = 0$, we have

$$\cos\theta_i = \frac{a_2}{\sqrt{a_1^2 + a_2^2}},$$

$$\cos\theta = \cos\left(\theta_i - \theta_j\right) = \cos\theta_i\cos\theta_j + \sin\theta_i\sin\theta_j = \frac{a_1 b_1 + a_2 b_2}{\sqrt{a_1^2 + a_2^2}\sqrt{b_1^2 + b_2^2}} \tag{8.28}$$

(a)   (b)

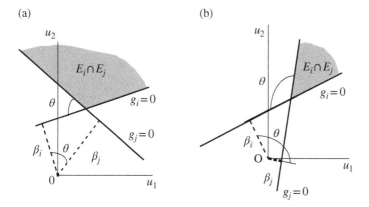

Figure 8.16 Correlation coefficient and limit state lines. (a) $\rho > 0$; (b) $\rho < 0$.

Comparing Eq. (8.27) with Eq. (8.28), one can easily understand that

$$\rho_{ij} = \cos\theta \tag{8.29}$$

where θ is the angle between the two limit state lines.

For positive ρ, $\theta < \pi/2$, and for negative ρ, $\theta > \pi/2$, the relationship between the two limit state lines is shown in Figure 8.16a and b, respectively.

For positive ρ, constructing a perpendicular line to $g_i = 0$ through crossing point between $g_i = 0$ and $g_j = 0$ as shown in Figure 8.17, define the line as a limit state line $g_A = 0$ corresponding to a failure event E_A, then the probability of the event E_A would be

$$P(E_A) = \Phi(-a)$$

where a is the distance from the original O to the limit state line $g_A = 0$.

Denote

$$A = E_A \cap E_i$$

Figure 8.17 Relationship between the two modes ($\rho > 0$).

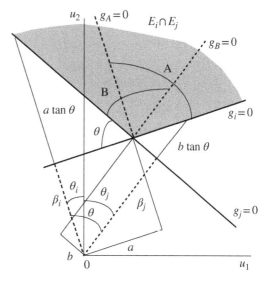

Then A corresponds the area between the lines $g_i = 0$ and $g_A = 0$ as shown in Figure 8.17. Since $g_i = 0$ and $g_A = 0$ are perpendicular, E_A, and E_i, are independent, then one obtains that

$$P(A) = P(E_A \cap E_i) = P(E_i)P(E_A) = \Phi(-\beta_i)\Phi(-a)$$

From the geometrical relationship in Figure 8.17, we have

$$a \tan \theta + \beta_i = \frac{\beta_j}{\cos \theta}$$

That is

$$a \sin \theta + \beta_i \cos \theta = \beta_j$$

Therefore

$$a = \frac{\beta_j - \rho\beta_i}{\sqrt{1-\rho^2}}$$

That is

$$P(A) = \Phi(-\beta_i)\Phi\left(-\frac{\beta_j - \rho\beta_i}{\sqrt{1-\rho^2}}\right)$$

Similarly, constructing a perpendicular line to $g_j = 0$ through the crossing point between $g_i = 0$ and $g_j = 0$ as also shown in Figure 8.17, defines the line as a limit state line $g_B = 0$ corresponding to a failure event E_B.

Denote

$$B = E_B \cap E_j$$

Then we have

$$P(B) = \Phi(-\beta_j)\Phi\left(-\frac{\beta_i - \rho\beta_j}{\sqrt{1-\rho^2}}\right)$$

From the relationship shown in Figure 8.17, one can clearly observe that

$$E_i \cap E_j \supset A$$
$$E_i \cap E_j \supset B$$

where A and B are defined in Figure 8.17.

Therefore, from Figure 8.17, we again obtain the following bounds of the joint failure probability of the two failure modes.

$$\max[P(A), P(B)] \le P_{fij} \le P(A) + P(B)$$

Similar procedures can be conducted for the case of negative ρ, $\theta > \pi/2$, the relationship between the two limit state lines is shown in Figure 8.18 from which one can observe that bounds of the joint failure probability of the two failure modes is given as.

$$0 \le P_{fij} \le \min[P(A), P(B)]$$

Then, Eq. (8.25) can be obtained.

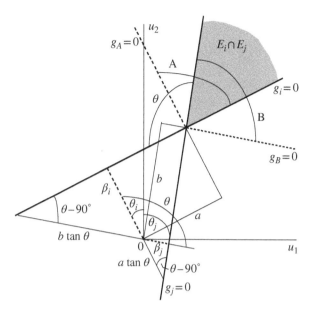

Figure 8.18 relationship between the two modes ($\rho < 0$).

8.3.5 Point Estimation of the Joint Failure Probability of a Pair of Failure Modes

Since Eq. (8.25) is a bound rather than a specific value, it is not convenient to use in Eq. (8.19).

Feng (1989) gave a point estimate for the joint failure probability P_{fij} as

$$P_{fij} = [P(A) + P(B)]\left[1 - \arccos\rho_{ij}/\pi\right] \tag{8.30}$$

where the definitions of $P(A)$ and $P(B)$ are the same as those in Eq. (8.25) and can be also calculated by Eq. (8.26). Since Eq. (8.30) is a specific value rather than a bound, it is considered to have high accuracy when used in Eq. (8.19) for the narrow bounds of the system reliability (Wu and Burnside 1990; Song 1992; Penmetsa and Grandhi 2002; Adduri et al. 2004). As described by Feng (1989), when the correlation coefficient $\rho_{ij} = 0$ or 1, Eq. (8.30) gives accurate solutions; whereas when $0 < \rho_{ij} < 1$, the calculation accuracy is reasonably high, especially when $\rho_{ij} \leq 0.6$. However, as will be shown later, the lower bound obtained with Eq. (8.30) can sometimes be lower than that given by Eq. (8.25). This section will introduce a point estimate of P_{fij} without such restriction of the range of ρ_{ij}.

To express the formulas more conveniently, β_1 and β_2 are used to represent β_i and β_j, respectively, and ρ is used to represent ρ_{ij}. Without loss of generality, assume $0 < \beta_1 \leq \beta_2$. The estimation of the joint failure probability can be written as follows (Zhao et al. 2007).

For $\beta_1/\beta_2 \geq \rho$,

$$P_{f12} = \begin{cases} P_2 + P_0\left(1 - \dfrac{2\theta}{\pi}\right), & \theta_1 \geq \dfrac{\pi}{4} \\[4mm] P_1 + P_2 - P_0\dfrac{2\theta}{\pi}, & \theta_1 < \dfrac{\pi}{4} \end{cases} \tag{8.31a}$$

And for $\beta_1/\beta_2 < \rho$,

$$P_{f12} = \begin{cases} P_{f2} - P_0\dfrac{2\theta}{\pi} & \theta_1 \geq \dfrac{\pi}{4}, \theta_2 \geq \dfrac{\pi}{4} \\ P_{f2} - P_2 + P_0\left(1 - \dfrac{2\theta}{\pi}\right) & \theta_1 \geq \dfrac{\pi}{4}, \theta_2 < \dfrac{\pi}{4} \\ P_{f2} + P_1 - P_2 - P_0\dfrac{2\theta}{\pi} & \theta_1 < \dfrac{\pi}{4} \end{cases} \qquad (8.31b)$$

where

$$P_1 = \Phi(-\beta_1)\Phi\left(-\sqrt{\beta_0^2 - \beta_1^2}\right) \qquad (8.32a)$$

$$P_2 = \Phi(-\beta_2)\Phi\left(-\sqrt{\beta_0^2 - \beta_2^2}\right) \qquad (8.32b)$$

$$\theta = \arccos(\rho), \theta_1 = \arccos(\beta_1/\beta_0), \theta_2 = \arccos(\beta_2/\beta_0) \qquad (8.32c)$$

$$\beta_0 = \sqrt{\frac{\beta_1^2 - 2\rho\beta_1\beta_2 + \beta_2^2}{1 - \rho^2}} \qquad (8.32d)$$

$$P_0 = \Phi^2\left(-\beta_0/\sqrt{2}\right) \qquad (8.32e)$$

$$P_{f2} = \Phi(-\beta_2) \qquad (8.32f)$$

Eqs. (8.32a–b) are almost as the same as $P(A)$ and $P(B)$ in Eq. (8.26) if one uses β_1, β_2, and ρ to represent β_i, β_j, and ρ_{ij}, respectively, in Ditlevsen's formula.

In particular, when $\rho = 0$, it can be seen that $\theta = \arccos(0) = \pi/2$ and $\beta_0 = \sqrt{\beta_1^2 + \beta_2^2}$, then

$$P_1 = P_2 = \Phi(-\beta_1)\Phi(-\beta_2)$$

Since $\theta_1 + \theta_2 = \theta = \pi/2$ and $\theta_1 \geq \theta_2$, one can see that $\theta_1 \geq \pi/4$. Then P_{f12} is given by

$$P_{f12} = \Phi(-\beta_1)\Phi(-\beta_2), \rho = 0 \qquad (8.33)$$

Another special case is when $\rho = 1$. Obviously, P_{f12} cannot be directly given by the equations given above since β_0 is not defined when $\rho = 1$. In this case, when $\rho \to 1$, we have

$$\lim_{\rho \to 1} P_{f12} = P_{f2} \qquad (8.34)$$

Example 8.10   Derivation of Eqs. (8.31a) and (8.31b)

Let Z_1 and Z_2 be the limit state functions in standard normal space corresponding to β_1 and β_2. The geometrical relationship between $Z_1 = 0$ and $Z_2 = 0$ can then be depicted in Figure 8.19a when $\beta_1/\beta_2 \geq \rho$ and in Figure 8.19b when $\beta_1/\beta_2 < \rho$.

Let the angle between OB (β_1) and OC (β_2) be θ, then according to Eq. (8.25), we have

$$\theta = \arccos(\rho)$$

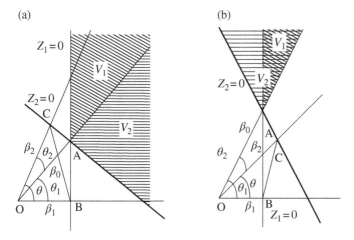

Figure 8.19 Probability calculation of P_{f12} in standard normal space. (a) $\beta_1/\beta_2 \geq \rho$; (b) $\beta_1/\beta_2 < \rho$.

In Figure 8.19a, the crossing point of $Z_1 = 0$ and $Z_2 = 0$ is point A. Define the length of the line segment OA as crossing index β_0, and denote the angle between OA and OB as θ_1, and the angle between OA and OC as θ_2; then θ_1 and θ_2 can be expressed as

$$\theta_1 = \arccos(\beta_1/\beta_0), \theta_2 = \arccos(\beta_2/\beta_0) \tag{8.35a}$$

With the aid of the geometrical relationships of β_0, β_1, and β_2, β_0 can be given as

$$\beta_0 = \sqrt{\frac{\beta_1^2 - 2\rho\beta_1\beta_2 + \beta_2^2}{1 - \rho^2}} \tag{8.35b}$$

Let V_1 denote the area of the failure zone between ray OA and $Z_1 = 0$, and V_2 denote the area of the failure zone between OA and $Z_2 = 0$, shown as the respective shaded zones in Figure 8.19a. The angle $\angle OAB$ is equal to $\pi/2 - \theta_1$ and the angle $\angle OAC$ is equal to $\pi/2 - \theta_2$. Since the joint failure probability $P_{f\,12}$ is the area of the failure zone between $Z_1 = 0$ and $Z_2 = 0$, P_{f12} can be given as the following equation according to the geometrical relations in Figure 8.19a.

$$P_{f12} = \begin{cases} V_1 + V_2, & \beta_1/\beta_2 \geq \rho \\ P_{f2} + V_1 - V_2, & \beta_1/\beta_2 < \rho \end{cases} \tag{8.36}$$

In particular, when $\beta_1/\beta_2 < \rho$, $\beta_0 = \beta_2$. Then one can observe that $V_2 = P_{f2}/2$ from both Figure 8.19a and b, which means that both the formulas in Eq. (8.36) give the same results for $\beta_1/\beta_2 = \rho$.

For $\theta_m \geq \pi/4$ (where $m = 1, 2$), V_m of Eq. (8.36) can be obtained by constructing two perpendicular lines DD′ and EE′ through point A, the crossing point of $Z_1 = 0$ and $Z_2 = 0$. Let the angle $\angle DAO = \angle EAO = \pi/4$, as shown in Figure 8.20.

Obviously, if EE′ and DD′ are considered to be limit state lines, both of their corresponding reliability indices would be $\beta_0/\sqrt{2}$. Since EE′ and DD′ are perpendicular, the probability associated with the area enclosed by $\angle E′AD′$ (shaded area in Figure 8.20) can be obtained as

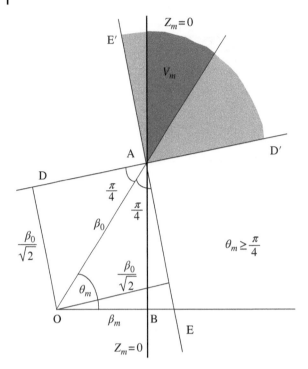

Figure 8.20 Geometrical relations for $v_m \geq \pi/4$.

$\Phi^2(-\beta_0/\sqrt{2})$. Since the angle between the lines of OA and $Z_m = 0$ corresponding to V_m (the darker shaded zone in Figure 8.20) is equal to $\pi/2 - \theta_m$, we have

$$\frac{\pi/2}{\Phi^2(-\beta_0/\sqrt{2})} \approx \frac{\pi/2 - \theta_m}{V_m}$$

Hence, V_m can be given as

$$V_m \approx \Phi^2\left(-\beta_0/\sqrt{2}\right)\left(1 - \frac{2\theta_m}{\pi}\right) \qquad \theta_m \geq \pi/4, \text{where } m = 1, 2 \qquad (8.37a)$$

For $\theta_m < \pi/4$ (where $m = 1, 2$), with the horizontal line FF′ through point A, the reliability index corresponding to limit state line FF′ is equal to $\sqrt{\beta_0^2 - \beta_m^2}$. Since the lines $Z_m = 0$ and FF′ are perpendicular, the probability corresponding to the angle between $Z_m = 0$ and FF′, shown as the shaded area in Figure 8.21a, can be given as $\Phi(-\beta_m)\Phi\left(-\sqrt{\beta_0^2 - \beta_m^2}\right)$. Denote the probability corresponding to the angle OA and FF′ as V' (the darker shaded zone in Figure 8.21a), then V_m can be given by the following equation according to the geometrical relations in Figure 8.21a.

$$V_m \approx \Phi(-\beta_m)\Phi\left(-\sqrt{\beta_0^2 - \beta_m^2}\right) - V'$$

(a)   (b)

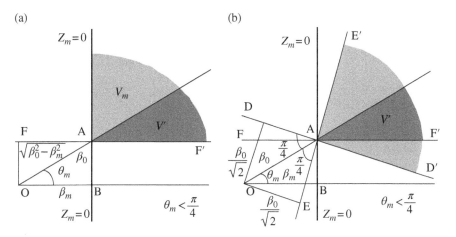

Figure 8.21 Geometrical relations for $\theta_m < \pi/4$. (a) V_m; (b) V'

In order to obtain V', draw two perpendicular lines DD' and EE' through point A. Let the angle $\angle$DAO $= \angle$EAO $= \pi/4$, as shown in Figure 8.21b. Obviously, both reliability indices corresponding to the limit state lines EE' and DD' would be $-\beta_0/\sqrt{2}$ and the probability corresponding to the area defined by the angle $\angle$E'AD' (shaded area in Figure 8.21b) can be obtained as $\Phi^2(-\beta_0/\sqrt{2})$ because EE' and DD' are perpendicular.

Since the angle between OA and FF' (the darker shaded zone in Figure 8.21b) that corresponds to V' is equal to θ_m, one obtains

$$\frac{\frac{\pi}{2}}{\Phi^2\left(-\beta_0/\sqrt{2}\right)} \approx \frac{\theta_m}{V'}$$

and then

$$V' \approx \Phi^2\left(-\beta_0/\sqrt{2}\right)\frac{2\theta_m}{\pi}$$

Therefore

$$V_m \approx \Phi(-\beta_m)\Phi\left(-\sqrt{\beta_0^2-\beta_m^2}\right) - \Phi^2\left(-\beta_0/\sqrt{2}\right)\frac{2\theta_m}{\pi}, \theta_m < \pi/4, \text{where } m = 1,2$$

$$(8.37b)$$

Write

$$P_m = \Phi(-\beta_m)\Phi\left(-\sqrt{\beta_0^2-\beta_m^2}\right), P_0 = \Phi^2\left(-\beta_0/\sqrt{2}\right)$$

$$(8.38)$$

Then Eqs. (8.37a) and (8.37b) can be written as

$$V_m = \begin{cases} P_0\left(1-\dfrac{2\theta_m}{\pi}\right), & \theta_m \geq \dfrac{\pi}{4} \\ P_m - P_0\dfrac{2\theta_m}{\pi}, & \theta_m < \dfrac{\pi}{4} \end{cases} \quad \text{where } m = 1,2$$

$$(8.39)$$

Eqs. (8.36) and (8.39) are the formulas for estimating the joint failure probability P_{f12}. For $\theta_m = \pi/4$, according to Eq. (8.36), $\beta_0 = \sqrt{2}\beta_m$, then $P_m = P_0 = \Phi^2(-\beta_m)$. In this case, the two formulas in Eq. (8.39) give the same results and reduce to Eq. (8.31a).

Example 8.11   The Derivation of Eq. (8.35b)

According to Figure 8.19a and b, $|OA| = \beta_0$, $|OB| = \beta_1$, $|OC| = \beta_2$
  In the triangle $\triangle OBC$, according to the Cosine Law,

$$|BC|^2 = |OB|^2 - 2(\cos \angle BOC)|OB||OC| + |OC|^2 = \beta_1^2 - 2\cos v \cdot \beta_1\beta_2 + \beta_2^2 = \beta_1^2 - 2\rho\beta_1\beta_2 + \beta_2^2$$

$$\because \angle OCA = \angle OBA = \frac{\pi}{2}$$

$\therefore$ Points O, A, B, C lie on the same circle with the center of the circle being at the middle point of line segment OA.
  $\because$ In the same circle, $\angle OAB = \angle OCB$
  In the triangle $\triangle OAB$, according to the Sine Law,

$$\frac{\beta_0}{\sin \dfrac{\pi}{2}} = \frac{\beta_1}{\sin \angle OAB}$$

In the triangle $\triangle OCB$, according to the Sine Law,

$$\frac{\beta_1}{\sin \angle OCB} = \frac{|BC|}{\sin v}$$
$$\because \angle OAB = \angle OCB$$

$$\therefore \frac{\beta_0}{\sin \dfrac{\pi}{2}} = \frac{|BC|}{\sin v}$$

$$\therefore |BC|^2 = \beta_0^2 \sin^2 v = \beta_0^2(1 - \cos^2 v) = \beta_0^2(1 - \rho^2)$$
$$\therefore \beta_1^2 - 2\rho\beta_1\beta_2 + \beta_2^2 = \beta_0^2(1 - \rho^2)$$

$$\therefore \beta_0 = \sqrt{\frac{\beta_1^2 - 2\rho\beta_1\beta_2 + \beta_2^2}{1 - \rho^2}}$$

Example 8.12   Consider the statically deterministic truss shown in Example 8.2, the performance functions of the two modes are as follows

$$g_1 = 3R_1 - 5P, g_2 = 3R_2 - 4P$$

Suppose that R_1, R_2, and P, are independent lognormal random variables with mean values and standard deviation of $\mu_{R1} = 500$, $\mu_{R2} = 450$, $\mu_P = 100$, $\sigma_{R1} = 100$, $\sigma_{R2} = 90$, and $\sigma_P = 40$. The first-order reliability index is obtained as $\beta_1 = 2.663$ for g_1 with corresponding probability of failure of $P_{f1} = 0.00388$, and $\beta_2 = 2.935$ for g_2, with corresponding probability of failure of $P_{f1} = 0.00167$.

Using Eq. (8.24), the correlation coefficient is given as,

$$\rho_{ij} = \frac{\sum\limits_{i=1}^{n} a_i b_i \sigma_{Xi}^2}{\sqrt{\sum\limits_{i=1}^{n} a_i^2 \sigma_{Xi}^2}\sqrt{\sum\limits_{i=1}^{n} b_i^2 \sigma_{Xi}^2}} = \frac{(-5)(-4)40^2}{\sqrt{3^2 100^2 + (-5)^2 40^2}\sqrt{3^2 90^2 + (-4)^2 40^2}} = 0.283$$

According to Eq. (8.26), we have

$$P(A) = \Phi(-\beta_i)\Phi\left(-\frac{\beta_j - \rho_{ij}\beta_i}{\sqrt{1-\rho_{ij}^2}}\right) = \Phi(-2.633)\Phi\left(-\frac{2.935 - 0.283 \times 2.663}{\sqrt{1-0.283^2}}\right)$$

$$= 4.443 \times 10^{-5}$$

$$P(B) = \Phi(-\beta_j)\Phi\left(-\frac{\beta_i - \rho_{ij}\beta_j}{\sqrt{1-\rho_{ij}^2}}\right) = \Phi(-2.935)\Phi\left(-\frac{2.663 - 0.283 \times 2.935}{\sqrt{1-0.283^2}}\right)$$

$$= 4.675 \times 10^{-5}$$

According to Eq. (8.25), the bounds of P_{f12} is obtained as

$$4.675 \times 10^{-5} \le P_{f12} \le 9.118 \times 10^{-5}$$

Then we will obtain the point estimate of P_{f12}.

$$\beta_0 = \sqrt{\frac{\beta_1^2 - 2\rho\beta_1\beta_2 + \beta_2^2}{1-\rho^2}} = \sqrt{\frac{2.663^2 - 2 \times 0.283 \times 2.663 \times 2.935 + 2.935^2}{1 - 0.283^2}} = 3.502$$

$$P_0 = \Phi^2\left(-\beta_0/\sqrt{2}\right) = \Phi^2(-2.477) = 4.406 \times 10^{-5}, \theta = \arccos[\rho] = 1.284$$

$$P_1 = P(A) = 4.443 \times 10^{-5}, P_2 = P(B) = 4.675 \times 10^{-5}$$

Since

$$\beta_1/\beta_2 = 2.633/2.935 = 0.897 > \rho, \theta_1 = \arccos[\beta_1/\beta_2] = 0.707 < \frac{\pi}{4}$$

Then we have

$$P_{f12} = P_1 + P_2 - P_0\frac{2\theta}{\pi} = 4.443 \times 10^{-5} + 4.675 \times 10^{-5} - 4.406 \times 10^{-5} \times \frac{2 \times 1.284}{\pi}$$

$$= 5.517 \times 10^{-5}$$

Then the probability of failure of the system is obtained as

$$P_F = P_{f1} + P_{f2} - P_{f12} = 0.00388 + 0.00167 - 5.517 \times 10^{-5} = 0.004995$$

Since the correlation coefficient ρ is quite small, the joint probability P_{f12} does not have much effect on the probability of failure P_F of the system for this example. For comparison, the integration for P_{f12} is obtained as 5.84×10^{-5}. It can be observed that the point estimate is in good agreement with the integration result.

Example 8.13 Consider a series structural system with only two failure modes. Several cases are examined to compare the results by different methods as follows.

When the reliability indices of the two failure modes are the same, the results for P_{f12} are presented in Figure 8.22a–c with $\beta_1 = \beta_2 = 2$; $\beta_1 = \beta_2 = 3$; and $\beta_1 = \beta_2 = 4$, respectively, and for various correlation coefficient ρ. Figures 8.22a–c show the solutions obtained by

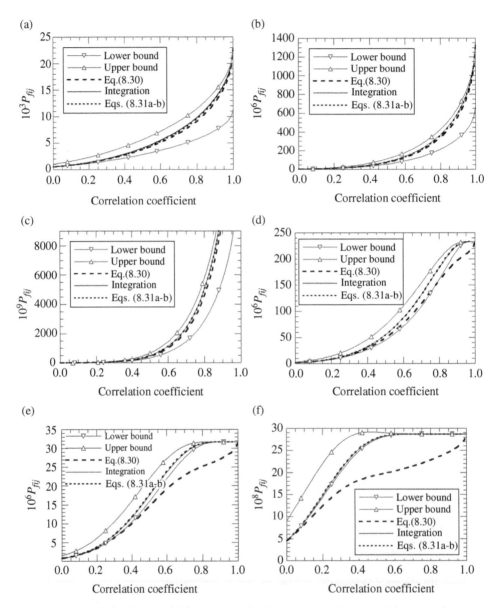

Figure 8.22 Variations of the joint failure probability, P_{f12}, with respect to correlation coefficient. (a) $\beta_1 = \beta_2 = 2$; (b) $\beta_1 = \beta_2 = 3$; (c) $\beta_1 = \beta_2 = 4$; (d) $\beta_1 = 2.5, \beta_2 = 3.5$; (e) $\beta_1 = 2, \beta_2 = 4$; (f) $\beta_1 = 1, \beta_2 = 5$.

integration, the Ditlevsen's bounds (Ditlevsen 1979b), Feng's point estimation (Feng 1989), and the point estimation presented in Eqs. (8.31a–b) (Zhao et al. 2007). From the figures, it can be observed that:

1) Ditlevsen's bounds correctly bound the integration results;
2) The results by Feng's method are quite close to the integration results; and
3) The results given by Eqs. (8.31a–b) have a better agreement with the integration method than those by Feng's method.

When the reliability indices are different for the two failure modes, the variations of the joint failure probability P_{f12} with respect to the correlation coefficient ρ are depicted in Figure 8.22d–f, respectively, for $\beta_1 = 2.5$, $\beta_2 = 3.5$; $\beta_1 = 2$, $\beta_2 = 4$; and $\beta_1 = 1$, $\beta_2 = 5$. From these figures, it can be seen that the integration results are always located between the narrow bound solutions. Both the results by Feng's method and Eqs. (8.31a–b) have good agreements with the integration method when $\rho = 0$ and $\rho = 1$. However, whereas the results by Eqs. (8.31a–b) have a good agreement with those by integration for all the three cases, the results given by Feng's method tend to be lower than the lower bound solution especially for large ρ, and this discrepancy becomes very large leading to significant differences in the two reliability indices (as can be observed in Figure 8.22f).

The variations of the joint failure probability P_{f12} with respect to the difference between the reliability indices for two failure modes are depicted in Figure 8.23a and b, respectively, for $\beta_1 = 2$, $\rho = 0.5$; and for $\beta_1 = 2$, $\rho = 0.8$. From these figures, it can be observed that the larger the difference between the two reliability indices, the narrower the reliability bound width, Feng's method gives good results for small differences between the two reliability indices, and Eqs. (8.31a–b) give gives good approximation of the integration results for the entire investigation range.

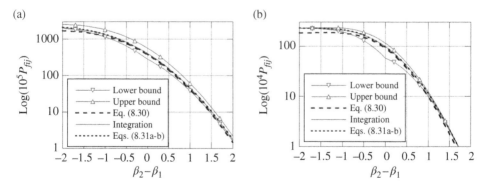

Figure 8.23 Variations of the joint failure probability, P_{f12}, with the difference between the two reliability indices. (a) $\rho = 0.5$, $\beta_1 = 2$; (b) $\rho = 0.8$, $\beta_1 = 2$.

Example 8.14 Next consider a series-connect structural system with four failure modes, in which the first order reliability indices for the four individual failure modes have been obtained as $\beta_1 = 2.5$, $\beta_2 = 2.5$, $\beta_3 = 3.0$, $\beta_4 = 3.5$, and the correlation coefficient between every pair of failure modes is assumed to be $\rho = 0.86$. The joint failure probability, P_{fij}, are calculated by different methods and listed in Table 8.1. The corresponding system failure probability, P_f, are listed in Table 8.2. From Tables 8.1 and 8.2, it can be seen that the results by Eqs. (8.31a–b) are between the lower and upper bounds and are in good agreements with those obtained by numerical integration. Also, the bounds obtained with Eqs. (8.31a–b) are the narrowest among all the methods.

Example 8.15 Consider the one-story one-bay elastoplastic frame in Example 8.3. The loads S_i and member strengths M_i are independent lognormal random variables with mean of $\mu_{M1} = \mu_{M2} = 500$ ft kip, $\mu_{M3} = 667$ ft kip, $\mu_{S1} = 50$ kip, $\mu_{S2} = 100$ kip and standard deviations of $\sigma_{M1} = \sigma_{M2} = 75$ ft kip, $\sigma_{M3} = 100$ ft kip, $\sigma_{S1} = 15$ kip, $\sigma_{S2} = 10$ kip. The performance functions that correspond to the six most likely failure modes obtained from stochastic limit analysis are listed below, with the FORM reliability index for each mode given in parentheses to show the relative dominance of the different modes

$$g_1 = 2M_1 + 2M_2 - 15S_1 (\beta_F = 3.247) \tag{8.40a}$$

$$g_2 = M_1 + 3M_2 + 2M_3 - 15S_1 - 10S_2 (\beta_F = 3.551) \tag{8.40b}$$

$$g_3 = 2M_1 + M_2 + M_3 - 15S_1 (\beta_F = 3.562) \tag{8.40c}$$

$$g_4 = M_1 + 2M_2 + M_3 - 15S_1 (\beta_F = 3.562) \tag{8.40d}$$

$$g_5 = M_1 + M_2 + 2M_3 - 15S_1 (\beta_F = 3.784) \tag{8.40e}$$

$$g_6 = M_1 + M_2 + 4M_3 - 15S_1 - 10S_2 (\beta_F = 3.848) \tag{8.40f}$$

Table 8.1 Calculation of joint failure probability $P_{f\,ij}$ (10^{-4}).

Method	Eqs. (8.31a−b)	Feng's method	Ditlevsen's method		Numerical integration
			Lower bound	Upper bound	
$P_{f\,21}$	25.961	25.385	15.301	30.601	27.340
$P_{f\,31}, P_{f\,32}$	9.5369	8.7634	7.5905	10.5640	9.6690
$P_{f\,41}, P_{f\,42}$	2.1401	1.8334	1.9596	2.2101	2.1150
$P_{f\,43}$	1.5130	1.3802	1.1813	1.6638	1.5270

Table 8.2 The calculation results of system failure probability P_f (10^{-3}).

Method	Eqs. (8.31a-b)	Feng's method	Ditlevsen's method	Numerical integration
Lower bound	9.8234	9.8809	9.3593	$P_f = 9.8912$
Upper bound	10.2386	10.4040	11.5170	
Bound width	0.4152	0.5231	2.1577	

Using these performance functions, the correlation matrix is as follows

$$[C] = \begin{bmatrix} 1 & 0.810 & 0.942 & 0.875 & 0.753 & 0.499 \\ 0.810 & 1 & 0.932 & 0.837 & 0.895 & 0.855 \\ 0.942 & 0.932 & 1 & 0.937 & 0.920 & 0.749 \\ 0.875 & 0.837 & 0.937 & 1 & 0.920 & 0.749 \\ 0.753 & 0.895 & 0.920 & 0.920 & 1 & 0.923 \\ 0.499 & 0.855 & 0.749 & 0.749 & 0.923 & 1 \end{bmatrix}$$

and the joint failure probability for each pair of failure modes are given in the following matrix

$$[P_{fij}] = 10^{-6} \begin{bmatrix} 582.8 & 72.81 & 147.1 & 101.4 & 28.14 & 4.778 \\ 72.81 & 191.8 & 88.45 & 47.40 & 40.39 & 26.43 \\ 147.1 & 88.45 & 183.9 & 89.31 & 46.88 & 13.27 \\ 101.4 & 47.40 & 89.31 & 183.9 & 46.88 & 13.27 \\ 28.14 & 40.39 & 46.88 & 46.88 & 77.14 & 27.83 \\ 4.778 & 26.43 & 13.27 & 13.27 & 27.83 & 59.50 \end{bmatrix}$$

from which the lower and upper bounds of the system failure probability are obtained, respectively, as 7.017×10^{-4} and 9.331×10^{-4}. The corresponding MCS solution using a 10 million sample size is 6.147×10^{-4} with a COV of 1.275%. It can be observed that the MCS result is outside the calculated bounds; this is because the FORM reliability indices used in calculating the above bounds are not accurate for each performance function. Using the fourth-moment approach described in the chapter on Methods of Moment Based on First- and Second-Order Transformation, the reliability indices are more accurately obtained as 3.293, 3.623, 3.629, 3.629, 3.871, and 3.957 corresponding to the six respective performance functions. With these latter reliability indices, the joint failure probability for each pair of failure modes are then obtained as follows

$$[P_{fij}] = 10^{-6} \begin{bmatrix} 495.1 & 56.0 & 115.4 & 79.51 & 20.28 & 3.080 \\ 56.0 & 145.2 & 66.66 & 35.14 & 28.49 & 17.28 \\ 115.4 & 66.66 & 142.3 & 68.13 & 33.46 & 8.628 \\ 79.51 & 35.15 & 68.13 & 142.3 & 33.46 & 8.628 \\ 20.28 & 28.49 & 33.46 & 33.46 & 54.23 & 18.06 \\ 3.080 & 17.28 & 8.628 & 8.628 & 18.06 & 37.86 \end{bmatrix}$$

from which the bounds of the system failure probability become 5.844×10^{-4} and 7.147×10^{-4}. Then, we can observe that the MCS solution for the system failure probability of 6.147×10^{-4} is clearly bounded by the narrow bounds.

Example 8.16 Finally, consider the simple elastoplastic beam-cable system shown in Figure 8.24 (after Ang and Tang 1984). The performance functions of the potential failure modes are listed below with the respective FORM reliability indices in parentheses

$$g_1 = 6M - wL^2/2 (\beta_F = 3.322) \tag{8.41a}$$

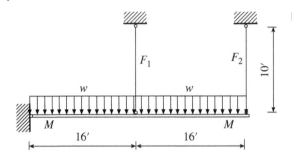

Figure 8.24 A beam-cable system.

$$g_2 = F_1L + 2F_2L - 2wL^2 (\beta_F = 3.647) \tag{8.41b}$$

$$g_3 = M + F_2L - wL^2/2 (\beta_F = 4.515) \tag{8.41c}$$

$$g_4 = 2M + F_1L - wL^2 (\beta_F = 4.515) \tag{8.41d}$$

where M, F_1, F_2, and w are normally distributed with mean values of $\mu_w = 2$ kip/ft, $\mu_{F1} = 60$ kip, $\mu_{F2} = 30$ kip and $\mu_M = 100$ ft kip, and COV's of $V_w = 0.2$ and $V_F = V_M = 0.1$.

Using the performance functions listed in Eqs. (8.41a–d) the correlation matrix is obtained as

$$[C] = \begin{bmatrix} 1 & 0.412 & 0.534 & 0.534 \\ 0.412 & 1 & 0.856 & 0.856 \\ 0.534 & 0.856 & 1 & 0.553 \\ 0.534 & 0.856 & 0.553 & 1 \end{bmatrix}$$

and the joint failure probability of each pair of failure modes are given in the following matrix

$$[P_{fij}] = 10^{-6} \begin{bmatrix} 446.6 & 3.853 & 0.542 & 0.542 \\ 3.853 & 132.6 & 1.324 & 1.324 \\ 0.542 & 1.324 & 3.163 & 0.035 \\ 0.542 & 1.324 & 0.035 & 3.163 \end{bmatrix}$$

from which the bounds of the system failure probability are 5.779×10^{-4} and 5.790×10^{-4}. The result obtained by numerical integration is 5.780×10^{-4}. It can be observed that the bound width is quite narrow and both the lower and upper bounds are close to the solution obtained through numerical integration.

8.4 Moment Approach for System Reliability

From the previous section, as a function of the failure probabilities of the individual modes, the failure probability of a system may be estimated using uni-modal bound techniques. However, the bounds would be wide for a complex system. Bi-modal bound technique (Ditlevsen 1979b; Feng 1989; Zhao et al. 2007) has been developed to improve the bound width of uni-modal bound technique, but mutual correlations and the joint failure probability matrices among the failure modes have to be determined. Since the number of potential failure modes for most practical structures is very large, the determination of the mutual correlations and the joint failure probability matrices is quite cumbersome and difficult. The failure probability of a system may also be estimated approximately using the probabilistic network evaluation technique (PNET) developed by Ang and Ma (1981), where the mutual correlations matrix among the failure modes have also to be computed. Other methods have been proposed or discussed (Moses 1982; Thoft-Christensen and Murotsu 1986; Bennett and Ang 1986; Melchers 1994; Miao and Ghosn 2011; Song and Der Kiureghian 2003; Chang and Mori 2013). In this section, a computationally more effective method using moment approximations is introduced and examined for system reliability of both series and non-series systems.

8.4.1 Performance Function for a System

Consider a structural system with multiple modes of potential failure, e.g. $E_1, E_2, ..., E_k$. For a series structural system, each of the failure modes, E_i, can be defined by a performance function $g_i = g_i(\mathbf{X})$ such that $E_i = (g_i < 0)$ and the failure probability of the system is then:

$$P_F = P[g_1 \leq 0 \cup g_2 \leq 0 \cup \cdots \cup g_k \leq 0]$$

Conversely, the safety of a system is the event in which none of the k potential failure modes occurs; again in the case of a series system, this means

$$P_S = P[g_1 > 0 \cap g_2 > 0 \cap \cdots \cap g_k > 0] = P[\min(g_1, g_2, \cdots g_k) > 0]$$

Thus the performance function of a series system, G, can be expressed as the minimum of the performance functions corresponding to all the potential failure modes; that is,

$$G(\mathbf{X}) = \min[g_1, g_2, \cdots, g_k] \tag{8.42}$$

where $g_i = g_i(\mathbf{X})$ is the performance function of the ith failure mode.

Eq. (8.42) can also be derived using probability integration method and the procedure has been developed by Li et al. (2007), where it is referred to the equivalent extreme-value events.

Similarly, for a parallel structural system, each of the failure modes, E_i, can be defined by a performance function $g_i = g_i(\mathbf{X})$ such that $E_i = (g_i < 0)$ and the failure probability of the system is then:

$$P_F = P[g_1 \leq 0 \cap g_2 \leq 0 \cap \cdots \cap g_k \leq 0] = P[\max(g_1, g_2, \cdots, g_k) \leq 0]$$

Thus the performance function of a parallel system, G, can be expressed as the maximum of the performance functions corresponding to all the potential failure modes; that is,

$$G(\mathbf{X}) = \max[g_1, g_2, \cdots, g_k] \tag{8.43}$$

For a combined series–parallel system, the performance function of the system will generally involve combinations of the maximum and minimum component performance functions. For example, suppose each of the failure modes, E_i, can be defined by a performance function $g_i = g_i(\mathbf{X})$ such that $E_i = (g_i < 0)$, the performance function corresponding to the combined system shown in Figure 8.25, will be given as

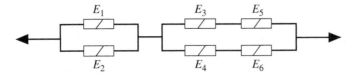

Figure 8.25 Combined systems.

$$G(\mathbf{X}) = \min\{\max(g_1, g_2), \max[\min(g_3, g_5), \min(g_4, g_6)]\} \qquad (8.44)$$

In summary, the performance function of a series system and a parallel system can be expressed as the minimum and the maximum, respectively, of the performance functions corresponding to all the potential failure modes. And the performance functions of a combined series–parallel system generally involve combinations of the maximum and minimum component performance functions. Since the system performance function $G(\mathbf{X})$ will not be smooth even though the performance function of a component is smooth, one may doubt whether the PDF of the system performance function $Z = G(\mathbf{X})$ is smooth or not. We will investigate this problem through several examples.

Example 8.17   A Simple Series System

Consider a series system with only two elements as shown in Figure 8.26, the performance function is defined as

$$G(\mathbf{X}) = \min(R_1 - P, R_2 - P) \qquad (8.45)$$

where R_1, R_2 are independent random variables with mean and standard deviation of $\mu_{R1} = 200$, $\mu_{R2} = 300$, $\sigma_{R1} = 20$, $\sigma_{R2} = 60$, and P is a load with deterministic value of 100. The following four cases are investigated:

Case 1: R_1-normal, R_2-Lognormal;
Case 2: R_1-Lognormal, R_2-Lognormal;
Case 3: R_1-Normal, R_2-Weibull;
Case 4: R_1-Gumbel, R_2-Gamma.

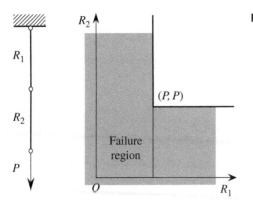

Figure 8.26   A series system of two members.

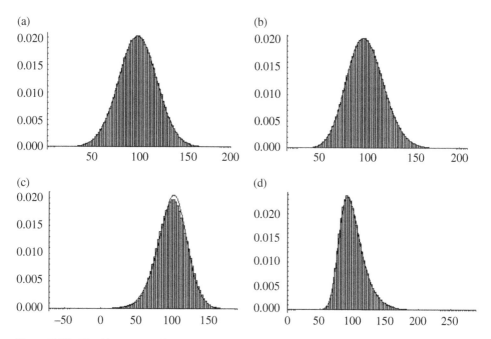

Figure 8.27 The histograms of the performance functions in Example 8.16. (a) Case 1. (b) Case 2. (c) Case 3. (d) Case 4.

Figure 8.27 presents the histograms of the performance function with 1 000 000 samples for the four cases. It can be observed that the histograms have good behaviours and the PDF should be smooth.

Example 8.18 Consider the simple parallel-chain system shown in Example 8.6. The performance function for this system is given as

$$G(\mathbf{X}) = \min \left\{ R_1 - S, \max \left\{ \min (R_2, R_3, R_4) - \frac{1}{2}S, \max [R_4, \min (R_2, R_3)] - S \right\} \right\}$$

(8.46)

where the fracture strengths R_i, and load S are independent lognormal random variables with means of $\mu_{R1} = 2200$ kg, $\mu_{R2} = 2100$ kg, $\mu_{R3} = 2300$ kg, $\mu_{R4} = 2000$ kg, $\mu_S = 1200$ kg, and standard deviations of $\sigma_{R1} = 220$ kg, $\sigma_{R2} = 210$ kg, $\sigma_{R3} = 230$ kg, $\sigma_{R4} = 20$ kg, $\sigma_S = 240$ kg.

It can be observed that the performance function involves combinations of maximum and minimum of the component performance functions. The histograms of the performance function with 1 000 000 is shown in Figure 8.28. It can be observed that the histogram has good behaviours and the PDF should be smooth.

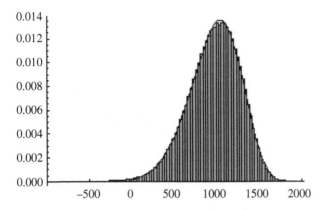

Figure 8.28 The histogram of the performance function in Example 8.17.

8.4.2 Methods of Moment for System Reliability

In this section, direct methods of moment described in the chapter on Direct Methods of Moment will be applied to system reliability evaluation. If the central moments of the system performance function as described in Eq. (8.42) for a series system, Eq. (8.43) for a parallel system, and for a series–parallel system like Eq. (8.44) can be obtained, the failure probability of a system, which is defined as $\text{Prob}[G(\mathbf{X}) < 0]$ can be expressed as a function of the central moments. Because the first two moments are generally inadequate, high-order moments will invariably be necessary. Since the methods including the point-estimate for evaluating the statistical moments of performance functions and the moment based reliability indices for estimating the probability of failure are as same as those described in the chapters on Moment Evaluation for Performance Functions and Direct Methods of Moment, here we will investigate the method through some examples.

Example 8.19 A Simple Series System

Consider the series system with only two elements as shown in Example 8.17 again. The first four moments of the performance functions are readily obtained using the point estimate method and are listed in Table 8.3, from which the PDF defined by the first four moments

Table 8.3 The parameters and moments results of the four cases for Example 8.19.

	Distributions		Moment by MCS			
Case	R_1	R_2	μ_G	σ_G	α_{3G}	α_{4G}
Case1	Normal	Lognormal	99.31	19.77	−0.0161	3.009
Case2	Lognormal	Lognormal	99.30	19.65	0.265	3.136
Case3	Normal	Weibull	97.93	21.39	−0.400	4.252
Case4	Gumbel	Gamma	98.99	19.39	0.948	4.794

can be obtained using the Cubic normal distribution (see Appendix C.3). The PDFs of the performance functions are also illustrated in Figures 8.27. From Figures 8.7a–d, it can be observed that the histograms of the system performance function can be generally approached by the PDF of the Cubic normal distribution, that is, the system reliability can be approximate by the method of moment.

Example 8.20   A Simple Problem of System Reliability

Consider the performance function defined as the minimum value of the following eight linear performance functions as shown in Figure 8.29.

$$G(\mathbf{X}) = \min(g_1, g_2, g_3, g_4, g_5, g_6, g_7, g_8) \tag{8.47}$$

where

$$g_1 = 2X_1 - 2X_2 + 8, (\beta_F = 2.828)$$
$$g_2 = 2.6X_1 - 2X_2 + 9.3, (\beta_F = 2.835)$$
$$g_3 = 1.4X_1 - 2X_2 + 7.2, (\beta_F = 2.949)$$
$$g_4 = 4X_1 - 2X_2 + 14, (\beta_F = 3.131)$$
$$g_5 = 0.7X_1 - 2X_2 + 6.8, (\beta_F = 3.209)$$
$$g_6 = -0.5X_1 - 2X_2 + 8, (\beta_F = 3.881)$$
$$g_7 = -2X_1 - 2X_2 + 11, (\beta_F = 3.889)$$
$$g_8 = -1.5X_1 - 2X_2 + 10, (\beta_F = 4.000)$$

where X_1 and X_2 are independent standard normal random variables.

The example is a series system problem to illustrate the numerical details of the procedure for the assessment of system reliability using the moment method. Since the gradients of the performance function are not convenient to obtain, it is not convenient to solve the example

Figure 8.29 The performance function for Example 8.20.

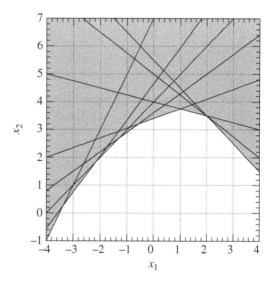

directly using the conventional procedure of FORM described in the chapter on Methods of Moment Based on First- and Second-Order Transformation. The FORM results would be obtained by first locating all the design points and then applying one of the methods to compute the probability on unions. If Ditlevsen's bounds (Ditlevsen 1979b) are used, the correlation matrix and the joint failure probability are obtained as

$$[C] = \begin{pmatrix}
1. & 0.9916 & 0.984784 & 0.948683 & 0.901002 & 0.514496 & 0. & 0.141421 \\
0.9916 & 1. & 0.954035 & 0.981615 & 0.837324 & 0.399267 & -0.129339 & 0.0121942 \\
0.984784 & 0.954035 & 1. & 0.879292 & 0.962682 & 0.655687 & 0.173785 & 0.311308 \\
0.948683 & 0.981615 & 0.879292 & 1. & 0.717581 & 0.21693 & -0.316228 & -0.178885 \\
0.901002 & 0.837324 & 0.962682 & 0.717581 & 1. & 0.835555 & 0.433816 & 0.556876 \\
0.514496 & 0.399267 & 0.655687 & 0.21693 & 0.835555 & 1. & 0.857493 & 0.921635 \\
0. & -0.129339 & 0.173785 & -0.316228 & 0.433816 & 0.857493 & 1. & 0.989949 \\
0.141421 & 0.0121942 & 0.311308 & -0.178885 & 0.556876 & 0.921635 & 0.989949 & 1.
\end{pmatrix}$$

$$[P_{fij}] = 10^6 \begin{pmatrix}
2341.99 & 1943.37 & 1441.39 & 740.392 & 486.883 & 10.6655 & 0.11787 & 0.385213 \\
1943.37 & 2291.28 & 1170.3 & 847.979 & 364.894 & 5.07798 & 0.0172665 & 0.085113 \\
1441.39 & 1170.3 & 1594.02 & 458.581 & 583.31 & 19.272 & 0.590553 & 1.36889 \\
740.392 & 847.979 & 458.581 & 871.061 & 125.313 & 0.571774 & 0.0000783872 & 0.00124889 \\
486.883 & 364.894 & 583.31 & 125.313 & 665.987 & 33.8198 & 2.93761 & 4.87381 \\
10.6655 & 5.07798 & 19.272 & 0.571774 & 33.8198 & 52.0139 & 13.0526 & 15.9204 \\
0.11787 & 0.0172665 & 0.590553 & 0.0000783872 & 2.93761 & 13.0526 & 50.3291 & 28.8034 \\
0.385213 & 0.085113 & 1.36889 & 0.00124889 & 4.87381 & 15.9204 & 28.8034 & 31.6712
\end{pmatrix}$$

the bounds of the failure probability is given as [0.00272, 0.00301], which means that the system reliability index is in the range of [2.747, 2.779].

Using the methods of moment, G_μ and G_i in Eqs. (3.67a–c) in the chapter on Moment Evaluation are readily obtained as

$$G_\mu = 6.8$$

$$G_1 = \min\left(1.4X_1 + 7.2,\ 2X_1 + 8, 2.6X_1 + 9.3,\ -1.5X_1 + 10, 4X_1 + 14, 0.7X_1 \right.$$
$$\left. + 6.8,\ -0.5X_1 + 8,\ -2X_1 + 11\right)$$

$$G_2 = -2X_2 + 6.8$$

Using the seven-point estimates described in Section 3.4.4 in the chapter on Moment Evaluation for Performance Functions, the first four moments of G_1 and G_2 are easily obtained as $\mu_1 = 6.5288$, $\sigma_1 = 0.9087$, $\alpha_{31} = -1.9349$, $\alpha_{41} = 9.0180$, $\mu_2 = 6.8$, $\sigma_2 = 2.0$, $\alpha_{32} = 0$, $\alpha_{42} = 3$. Using Eqs. (3.68a–d), in the chapter just mentioned the first four moments of G are approximately obtained as $\mu_G = 6.5288$, $\sigma_G = 2.19674$, $\alpha_{3G} = -0.1369$, $\alpha_{4G} = 3.1762$. The second-moment reliability index is readily obtained as $\beta_{2M} = 2.972$. With the first three moments of the performance function, Eq. (4.27) in the chapter on Direct Methods of Moment gives the third-moment reliability index as $\beta_{3M} = 2.805$ with corresponding failure probability of $P_F = 2.516 \times 10^{-3}$.

Using Eqs. (4.54b–c), (4.56), and (4.55a–b) from the same chapter, the fourth-moment reliability index is computed as follows.

$$l_2 = \frac{1}{36} \left(\sqrt{6\alpha_{4G} - 8\alpha_{3G}^2 - 14} - 2 \right) = \frac{1}{36} \left(\sqrt{6 \times 3.1762 - 8 \times 0.1369^2 - 14} - 2 \right) = 0.00598$$

$$l_1 = \frac{\alpha_{3G}}{6(1 + 6l_2)} = \frac{-0.1369}{6(1 + 6 \times 0.00598)} = -0.022$$

$$k_1 = \frac{1 - 3l_2}{\left(1 + l_1^2 - l_2^2\right)} = \frac{1 - 3 \times 0.00598}{\left(1 + 0.022^2 - 0.00598^2\right)} = 0.9816$$

$$k_2 = \frac{l_2}{\left(1 + l_1^2 + 12l_2^2\right)} = \frac{0.00598}{\left(1 + 0.022^2 + 12 \times 0.00598^2\right)} = 0.00597$$

$$p = \frac{3k_1 k_2 - l_1^2}{9k_2^2} = \frac{3 \times 0.9816 \times 0.00597 - 0.022^2}{9 \times 0.00597^2} = 53.268$$

$$q_0 = \frac{2l_1^3 - 9k_1 k_2 l_1 + 27k_2^2(-l_1 + \beta_{2M})}{27k_2^3}$$

$$= \frac{2 \times (-0.022)^3 - 9 \times 0.9816 \times 0.00597 \times (-0.022) + 27 \times 0.00597^2(0.022 + 2.972)}{27 \times 0.00597^3}$$

$$= 699.532$$

$$\Delta_0 = \sqrt{q_0^2 + 4p^3} = \sqrt{699.532^2 + 4 \times 53.268^3} = 1045.9$$

$$\beta_{4M} = \frac{\sqrt{2}p}{\sqrt[3]{-q_0 + \Delta_0}} - \frac{\sqrt[3]{-q_0 + \Delta_0}}{\sqrt[3]{2}} + \frac{l_1}{3k_2}$$

$$= \frac{\sqrt{2} \times 53.268}{\sqrt[3]{-699.532 + 1045.9}} - \frac{\sqrt[3]{-699.532 + 1045.9}}{\sqrt[3]{2}} + \frac{-0.022}{3 \times 0.00597} = 2.7531$$

The corresponding probability of failure is equal to 0.00295.

Using the method of MCS with 1 000 000 samples, the probability of failure for this performance function is obtained as 0.00307 with corresponding reliability index of $\beta = 2.740$, and the coefficients of variation of P_F is 1.803%. It is found that the results of the third- and fourth-moment approximations are in close agreement with the MCS results, whereas the second-moment approximation overestimated the reliability index by 8.5%.

It can also be observed that for this example the probability of failure obtained using the fourth-moment method is closer to the result of MCS than that of FORM. Furthermore, the methods of moment can be easily conducted without shortcomings associated with the design points, do not require iteration or computation of derivatives, do not conduct reliability analysis for each mode, and we need not compute the correlation matrix and joint probability.

Example 8.21 The example considers a one-story one-bay elasto-plastic frame shown in Figure 8.30. The FORM reliability indices for the respective failure modes are given in parentheses to indicate the relative dominance of the four different modes.

$$g_1(\mathbf{X}) = 2M_1 + 2M_3 - 4.5S, \ (\beta_F = 3.334) \tag{8.48a}$$

$$g_2(\mathbf{X}) = 2M_1 + M_2 + M_3 - 4.5S, \ (\beta_F = 3.364) \tag{8.48b}$$

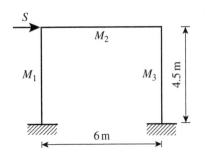

S

M_2

M_1 M_3

4.5 m

6 m

Figure 8.30 One-story one-bay frame structure of Example 8.21.

$$g_3(\mathbf{X}) = M_1 + M_2 + 2M_3 - 4.5S, (\beta_F = 3.364) \tag{8.48c}$$

$$g_4(\mathbf{X}) = M_1 + 2M_2 + M_3 - 4.5S, (\beta_F = 3.364) \tag{8.48d}$$

where M_i, S are independent lognormal random variables with mean and standard deviation of $\mu_M = 200tm$, $\mu_S = 50\,t$, $\sigma_M = 30tm$, $\sigma_S = 20t$.

As this is a series system, the performance function of the system can be defined as the minimum of Eqs. (8.48a) to (8.48d), i.e.

$$G(\mathbf{X}) = \min\{g_1(\mathbf{X}), g_2(\mathbf{X}), g_3(\mathbf{X}), g_4(\mathbf{X})\} \tag{8.48e}$$

Using the mean values of all the variables in Eq. (8.48e), G_μ as defined in Eq. (3.67b) in the chapter on Moment Evaluation for Performance Functions can be obtained as

$$\begin{aligned} G_\mu &= \min\{2 \times 200 + 2 \times 200 - 4.5 \times 50, 2 \times 200 + 200 + 200 - 4.5 \times 50, 200 \\ &\quad + 200 + 2 \times 200 - 4.5 \times 40, 200 + 2 \times 200 + 200 - 4.5 \times 40\} \\ &= \min\{575, 575, 575, 575\} = 575 \end{aligned}$$

Substituting the mean values of all the variables, except M_1, into Eq. (8.48e), G_1 as defined in Eq. (3.67c) of that same chapter becomes

$$\begin{aligned} G_1 &= \min\{2M_1 + 2 \times 200 - 4.5 \times 50, 2M_1 + 200 + 200 - 4.5 \times 50, M_1 + 200 \\ &\quad + 2 \times 200 - 4.5 \times 40, M_1 + 2 \times 200 + 200 - 4.5 \times 40\} \\ &= \min\{2M_1 + 175, 2M_1 + 175, M_1 + 375, M_1 + 375\} = \min\{2M_1 + 175, M_1 + 375\} \end{aligned}$$

Similarly, G_i, $i = 2, 3, 4$, are, respectively

$$G_2 = \min\{575, 2M_2 + 175, M_2 + 375\}, G_3 = \min\{2M_3 + 175, M_3 + 375\}, G_4 = 800 - 4.5S$$

Since G_1 is a function of a single random variable M_1, its moments can be point-estimated using Eqs. (3.35a–c) in the same chapter. For a five-point estimate, using the inverse Rosenblatt transformation expressed as

$$T^{-1}(u_i) = F^{-1}[\Phi(u_i)]$$

where F is the cumulative distribution function of M_1 and Φ is the standard normal probability, the five estimating points of Eq. (3.35a–c) in the chapter just mentioned can be easily transformed into the original space as, $T^{-1}(u_{2-}) = 129.157$, $T^{-1}(u_{1-}) = 161.576$, $T^{-1}(u_0) = 197.787$, $T^{-1}(u_{1+}) = 242.113$, and $T^{-1}(u_{2+}) = 302.886$. Substituting these into

Eq. (3.35a–c) with the corresponding weights listed in Eq. (3.42a–e) in the same chapter, the first four moments of G_1 are approximately obtained

$$\mu_1 = 564.489\ tm, \sigma_1 = 44.164\ tm, \alpha_{31} = -0.479, \alpha_{41} = 2.943,$$

Similarly, the first four moments of G_2, G_3, G_4 are

$$\mu_2 = 553.979\ tm, \sigma_2 = 33.259\ tm, \alpha_{32} = -1.463, \alpha_{42} = 3.785,$$
$$\mu_3 = 564.489\ tm, \sigma_3 = 44.164\ tm, \alpha_{33} = -0.479, \alpha_{43} = 2.943,$$
$$\mu_4 = 575.000\ tm, \sigma_4 = 90.000\ tm, \alpha_{34} = -1.264, \alpha_{44} = 5.968,$$

Then using Eqs. (3.68a–d) from the same chapter, the first four moments of G^* are obtained approximately as: $\mu_G = 532.958$, $\sigma_G = 114.485$, $\alpha_{3G} = -0.704$, $\alpha_{4G} = 4.098$.

Finally, with the first-two moments of the performance function G^*, the second-moment reliability index is obtained as $\beta_{2M} = 4.655$. With the first three moments of the performance function, Eq. (4.27) in the chapter on Direct Methods of Moment gives the third-moment reliability index as $\beta_{3M} = 3.554$ with corresponding failure probability of $P_F = 1.971 \times 10^{-4}$. With the first four moments of the performance function, Eq. (4.55a) in that same chapter gives the fourth-moment reliability index as $\beta_{4M} = 3.265$ with corresponding probability of failure of $P_F = 5.47 \times 10^{-4}$.

Using Monte Carlo simulations (MCS) with 1 000 000 samples, the probability of failure for this system is estimated to be 5.34×10^{-4} with corresponding reliability index of $\beta = 3.272$. The coefficient of variation (COV) of this MCS estimate is 4.32%. In this case, only 20 function calls are used (with a total of 20 estimating points for all the variables). For this example, it can be observed that both the results of the fourth-moment approximations are in close agreement with the MCS results, whereas the third-moment reliability index errs about 8.6%, and the second-moment approximation is grossly in error (42% over-estimation in the reliability index).

Example 8.22   Consider Examples 8.3 and 8.15 again, for this example, the corresponding G_i of Eq. (3.67a) from the chapter on Moment Evaluation can be obtained as follows:

$$G_\mu = 1250\ \text{ft-kip}; G_1 = \min\{917 + M_1, 250 + 2M_1\}; G_2 = \min\{917 + M_2, 250 + 2M_2, 84 + 3M_2\};$$
$$G_3 = \min\{1250, 750 + M_3, 250 + 2M_3, -750 + 4M_3\}; G_4 = 2000 - 15S_1; G_5 = \min\{1250, 2584 - 10S_2\}$$

Using the point estimate method with five estimating points, the first four moments of G_1, G_2, G_3, G_4, G_5 are obtained approximately as

$$\mu_1 = 1248.98\ \text{ft-kip}, \sigma_1 = 146.79\ \text{ft-kip}, \alpha_{31} = -0.293, \alpha_{41} = 2.767,$$
$$\mu_2 = 1248.98\ \text{ft-kip}, \sigma_2 = 146.79\ \text{ft-kip}, \alpha_{32} = 0.291, \alpha_{42} = 2.767,$$
$$\mu_3 = 1246.88\ \text{ft-kip}, \sigma_3 = 29.229\ \text{ft-kip}, \alpha_{33} = -9.265, \alpha_{43} = 88.842,$$
$$\mu_4 = 1250.00\ \text{ft-kip}, \sigma_4 = 225.00\ \text{ft-kip}, \alpha_{34} = -0.927, \alpha_{44} = 4.547,$$
$$\mu_5 = 1250.00\ \text{ft-kip}, \sigma_5 = 0.00\ \text{ft-kip}, \alpha_{35} = --, \alpha_{45} = --,$$

In this case, σ_5 is 0, whereas α_{35} and α_{45} cannot be obtained according to Eqs.(3.35a–c) in the same chapter. This is because G_5 is almost a constant and it has almost no influence on the results of σ_G, α_{3G}, α_{4G}. Any values of α_{35} and α_{45} can be used; e.g. $\alpha_{35} = 0$ and $\alpha_{45} = 3$, and then use Eq. (3.68a–d) as usual, or substitute G_5 as a constant into Eq. (3.67a), both from the same chapter. The results would remain the same.

Then using Eqs. (3.68a–d), the first four moments of G^* are approximately obtained as $\mu_G = 1244.85$, $\sigma_G = 307.523$, $\alpha_{3G} = -0.307$, and $\alpha_{4G} = 3.426$.

With these first four moments of the performance function G^*, the second-moment reliability index is obtained as $\beta_{2M} = 4.048$, whereas the third-moment reliability index of Eq. (4.27) from the chapter on Direct Methods of Moment is $\beta_{3M} = 3.428$ with corresponding failure probability of $P_F = 3.04 \times 10^{-4}$. The fourth-moment reliability index of Eq. (4.55a) from the same chapter gives $\beta_{4M} = 3.292$ with corresponding failure probability of $P_F = 4.97 \times 10^{-4}$. MCS with 1 000 000 samples gives a probability of failure for this system as 6.45×10^{-4} with corresponding reliability index of $\beta = 3.218$. The coefficient of variation of this MCS estimate is 3.94%. For this example, the third-moment reliability index is about 6.5% in error, whereas the results of the fourth-moment approximation is in very close agreement with the MCS results. Again the second-moment reliability index has a significant error of about 26% in the reliability index.

The reliability analyses for this example are further extended assuming different types of distribution of the random variables. Assuming all the member strengths and loads are Weibull random variables, the results of the moment method and of the MCS with 1 000 000 samples are summarised in column (2) of Table 8.4. Similarly, the results for Gamma, Gumbel, and normal distributed random variables are summarised in columns (3), (4), and (5), respectively, in Table 8.4. From Table 8.4, it can be observed that irrespective of the types of

Table 8.4 Computational results for Example 8.22 with different types of PDF.

(1)	Weibull (2)	Gamma (3)	Gumbel (4)	Normal (5)
μ_G	1242.12	1244.58	1244.41	1243.58
σ_G	317.852	308.823	302.810	311.894
α_{3G}	−0.216	−0.207	−0.310	−0.054
α_{4G}	3.199	3.138	3.765	3.038
β_{2M}	3.908 (16%)	4.030 (15%)	4.110 (30%)	3.987 (5%)
β_{3M}	3.466 (2.57%)	3.577 (1.76%)	3.463 (9.9%)	3.858 (1.69%)
P_F	2.641×10^{-4}	1.738×10^{-4}	2.671×10^{-4}	5.716×10^{-5}
β_{4M}	3.390 (0.33%)	3.400 (3.27%)	3.196(1.43%)	3.806 (0.32%)
P_F	3.495×10^{-4}	3.369×10^{-4}	7.145×10^{-4}	5.591×10^{-5}
β_{MCS}	3.379	3.515	3.151	3.794
P_F	3.63×10^{-4}	2.2×10^{-4}	8.14×10^{-4}	7.4×10^{-5}
COV of P_F	5.2%	6.7%	3.5%	11.6%

Note: Percentage errors in the reliability index relative to MCS are in parentheses.

distribution, both the third- and fourth-moment reliability indices are in close agreement with the MCS results. The second-moment reliability indices, however, consistently contain significant errors.

Finally, this example is again further extended to examine the applicability (and limitation) to problems with extremely small probability of failure. For this purpose, the mean loads are assumed to be $\mu_{S1} = 35$ kip and $\mu_{S2} = 75$ kip. Using the seven-point estimates described in the chapter on Moment Evaluation, the first four moments of G^* are approximately obtained as $\mu_G = 1470.13$, $\sigma_G = 260.32$, $\alpha_{3G} = -0.096$, and $\alpha_{4G} = 3.186$. With these first four moments of the system performance function, the second-moment reliability index is obtained as $\beta_{2M} = 5.647$, the third-moment reliability index is found to be $\beta_{3M} = 5.198$ with $P_F = 1.007 \times 10^{-7}$, and the fourth-moment reliability index is obtained as $\beta_{4M} = 4.708$ with $P_F = 1.251 \times 10^{-6}$. Using MCS with 30 000 000 samples, the probability of failure for this system is obtained as 5.333×10^{-6} with corresponding reliability index of $\beta = 4.403$. The COV of this MCS estimate is 7.91%. For this example, the third- and fourth-moment reliability indices contain errors of about 18.1% and 5.6%, respectively, whereas the second-moment reliability index over-estimates the correct value by about 28%. It is observed that for this case involving very small failure probability, i.e. $P_F < 10^{-6}$, the accuracy of the fourth-moment reliability index decreases.

Example 8.23   A Beam-Cable System

Consider again the elasto-plastic beam-cable system Example 8.16.

Using Eqs. (3.67a–c) and Eq. (3.68a–d) from the chapter on Moment Evaluation and the five-point estimates, the first four moments of $G*$ are obtained as $\mu_G = 293.432$, $\sigma_G = 76.352$, $\alpha_{3G} = -0.574$, and $\alpha_{4G} = 3.265$. The second-moment reliability index is obtained as $\beta_{2M} = 3.843$, whereas the third-moment reliability index of Eq. (4.27) in the chapter on Direct Methods of Moment is $\beta_{3M} = 3.01$ with corresponding failure probability of $P_F = 1.31 \times 10^{-3}$. It should be noted that Eq. (4.55a) from this same chapter cannot give reasonable results since Eq. (4.55a) is essentially derived from the Type IV of the complete expression of the fourth-moment pseudo normal transformation (see Table 7.9 in the chapter on Transformation of Non-Normal Variables to Independent Normal Variables). While the combination of the first four moments belongs to the Type I of the transformation, based on which the fourth-moment reliability index is obtained as $\beta_{4M} = 3.093$ with $P_F = 9.919 \times 10^{-4}$.

MCS with 500 000 samples, gives the probability of failure for this system as 5.26×10^{-4} with corresponding reliability index of $\beta = 3.276$. The COV of this MCS estimate is 6.16%. It can be observed that the second-moment reliability index contains a significant number of errors (17%) and the third-moment reliability index underestimates the reliability index by 8.1%, whereas the fourth-moment reliability index has about 5.6% in error which is still large.

Using Eq. (3.66a–c) from the chapter on Moment Evaluation, the five-point estimates for the first four moments are obtained as $\mu_G = 297.270$, $\sigma_G = 78.780$, $\alpha_{3G} = -0.2089$, $\alpha_{4G} = 3.2546$. With these more accurate first four moments of the performance function, the second-moment reliability index is obtained as $\beta_{2M} = 3.773$, and the third- and fourth-moment reliability indices are obtained as $\beta_{3M} = 3.373$ and $\beta_{4M} = 3.264$. Clearly, with these more accurate first four moments, the third- and fourth-moment approximations are now in closer agreement with the MCS results. This means that if the first four moments are accurately obtained, the reliability of a system can be computed with no significant error. As illustrated in this example, the approximation of the system performance function with Eq. (3.67a) and the moments generated with Eq. (3.68a–d) from the same chapter may, in rare cases, contain significant errors. In this case, Eq. (3.66a–c) may be required to obtain more accurate results of the moments.

Example 8.24 2-Story 1-Bay Truss Structure

The example is an elasto-plastic truss structure with 2 stories and 1 bay as shown in Figure 8.31, which is also a series system. The statistics of the member strengths and loads are as follows: mean values are of $\mu_{T1} = \mu_{T2} = 90$ kip, $\mu_{T3} = 9$ kip, $\mu_{T4} = \mu_{T5} = 48$ kip, $\mu_{T6} = \mu_{T7} = 21$ kip, $\mu_{T8} = 15$ kip, $\mu_{T9} = \mu_{T10} = 30$ kip, $\mu_{F1} = 11$ kip, $\mu_{F2} = 3.6$ kip; and coefficients of variation are of $V_{T1} = ... = V_{T10} = 0.15$, $V_{F1} = 0.3$, $V_{F2} = 0.2$. The performance functions corresponding to the eight most likely failure modes are given in Eqs. (8.49a)–(8.49i) (Ono et al. 1990), with the respective FORM reliability indices listed in parentheses showing that none of the modes are significantly dominant.

$$g_1 = 0.7071T_4 + 0.7071T_5 - 2.2F_1 \quad (\beta_F = 3.409) \tag{8.49a}$$

$$g_2 = T_6 + 0.7071T_{10} - 1.2F_1 - F_2 \quad (\beta_F = 3.497) \tag{8.49b}$$

$$g_3 = T_3 + 0.7071T_5 + 0.7071T_{10} - 2.2F_1 \quad (\beta_F = 3.264) \tag{8.49c}$$

$$g_4 = T_8 + 0.7071T_{10} - 1.2F_1 \quad (\beta_F = 3.333) \tag{8.49d}$$

$$g_5 = T_6 + T_7 - 1.2F_1 \quad (\beta_F = 3.814) \tag{8.49e}$$

$$g_6 = T_3 + 0.7071T_5 - 1.2F_1 - F_2 \quad (\beta_F = 3.484) \tag{8.49f}$$

$$g_7 = 0.7071T_9 + 0.7071T_{10} - 1.2F_1 \quad (\beta_F = 3.846) \tag{8.49h}$$

$$g_8 = T_1 + 0.7071T_5 - 3.4F_1 - F_2 \quad (\beta_F = 3.793) \tag{8.49i}$$

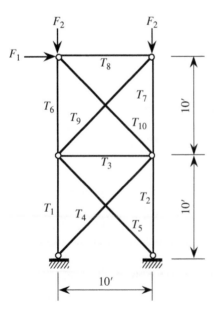

Figure 8.31 Two-story one-bay truss.

Using the five-point estimates, the first four moments of G^* are approximately obtained as $\mu_G = 22.316$, $\sigma_G = 5.379$, $\alpha_{3G} = -0.408$, and $\alpha_{4G} = 3.566$. With these first four moments of the system performance function, the moment-based reliability indices are obtained as $\beta_{2M} = 4.148$, $\beta_{3M} = 3.357$, and $\beta_{4M} = 3.246$. Using MCS with 1 000 000 samples, the probability of failure for this system is obtained as 7.86×10^{-4} with corresponding reliability index of $\beta = 3.161$. The COV of this MCS estimate is 3.57%. For this example, the third-moment reliability index errs about 6.2%, whereas the fourth-moment reliability index errs about 2.7%. The second-moment reliability index has a significant error of about 31%.

Example 8.25 A Brittle Systems

Consider again Examples 8.6 and 8.18. The first four moments of G^* are approximately $\mu_G = 993.3599$, $\sigma_G = 316.836$, $\alpha_{3G} = -0.275$, $\alpha_{4G} = 3.109$. The second-moment reliability index is obtained as $\beta_{2M} = 3.135$, and the third-moment reliability index is $\beta_{3M} = 2.789$. Again, Eq. (4.55a) from the chapter on Direct Methods of Moment cannot give reasonable results since Eq. (4.55a) is essentially derived from the Type IV of the complete expression of the fourth-moment pseudo normal transformation (see Table 7.9 in the chapter on Transformation of Non-Normal Variables to Independent Normal Variables). While the combination of the first four moments belongs to the Type III of the transformation based on which the fourth-moment reliability index is obtained as $\beta_{4M} = 2.818$ with $P_F = 2.413 \times 10^{-3}$.

MCS with 500 000 samples yield the probability of failure for this system as 2.506×10^{-3} with corresponding reliability index of $\beta = 2.806$. The COV of this MCS estimate is 2.82%. It can be observed that both the results of the third- and fourth-moment approximations are in close agreement with the MCS results, whereas the second-moment approximation overestimated the reliability index by 11.7%. These first four moments define a PDF using the cubic normal distribution (see Appendix C.3) which is also depicted in Figure 8.28, it can be observed that the histograms of the system performance function can be generally approached by the PDF defined by the first four moments.

Example 8.26 System Reliability of a Brittle Truss Structure
The example considers the brittle truss with one story and one bay in Example 8.7 as shown in Figure 8.11. The member strength T_i and loads F_i are independent lognormal random variables and the statistics of them are as follows: mean values are $\mu_{T1} = \mu_{T2} = 40\,t$, $\mu_{T3} = 10\,t$, $\mu_{T4} = \mu_{T5} = 20\,t$, $\mu_{F1} = 7\,t$, and $\mu_{F2} = 2\,t$; standard deviations are $\sigma_{T1} = \sigma_{T2} = 6\,t$, $\sigma_{T3} = 1.5\,t$, $\sigma_{T4} = \sigma_{T5} = 3\,t$, $\sigma_{F1} = 2.1\,t$, and $\sigma_{F2} = 0.6\,t$. There are five failure modes of this truss structure with respective performance functions listed as

$$g_1 = \max\left[T_1 - 1.127F_1 + 0.826F_2, \min\left(T_2 - \frac{4}{3}F_1, T_3 - F_1 + \frac{3}{4}F_2, T_4 - \frac{5}{4}F_2, T_5 - \frac{5}{3}F_1 + \frac{5}{4}F_2\right)\right]$$

(8.50a)

$$g_2 = \max\left[T_2 - 0.206F_1 - 0.826F_2, \min\left(T_1 - \frac{4}{3}F_1, T_3 - \frac{3}{4}F_2, T_4 - \frac{5}{3}F_1 - \frac{5}{4}F_2, T_5 - \frac{5}{4}F_2\right)\right]$$

(8.50b)

$$g_3 = \max\left[T_3 - 0.155F_1 + 0.13F_2, \min\left(T_1 - \frac{4}{3}F_1 + F_2, T_2 - F_2, T_4 - \frac{5}{3}F_1\right)\right] \quad (8.50c)$$

$$g_4 = \max\left[T_4 - 1.409F_1 - 0.217F_2, \min\left(T_1 - F_2, T_2 - \frac{4}{3}F_1 - F_2, T_3 - F_1, T_5 - \frac{5}{3}F_1\right)\right]$$
$$(8.50d)$$

$$g_4 = \max\left[T_5 - 0.258F_1 + 0.217F_2, \min\left(T_1 - \frac{4}{3}F_1 + F_2, T_2 - F_2, T_4 - \frac{5}{3}F_1\right)\right] \quad (8.50e)$$

The performance function of the system can be defined as the minimum of the above; i.e.

$$G(\mathbf{X}) = \min\{g_1, g_2, g_3, g_4, g_5\} \quad (8.50f)$$

Using the point estimate in the chapter on Moment Evaluation for Performance Functions, the first four moments of $G(\mathbf{X})$ are approximately $\mu_G = 8.972$, $\sigma_G = 3.732$, $\alpha_{3G} = -0.165$, and $\alpha_{4G} = 3.752$. The moment based reliability indices are obtained as $\beta_{2M} = 2.404$, $\beta_{3M} = 2.281$, and $\beta_{4M} = 2.211$.

Using MCS with 500 000 samples, the probability of failure for this system is 1.39×10^{-2} with a corresponding reliability index of $\beta = 2.199$. It can be observed that both the results of the third- and fourth-moment approximations are in close agreement with the MCS results, whereas the second-moment approximation overestimated the reliability index by 9.3%.

8.5 System Reliability Assessment of Ductile Frame Structures Using Methods of Moment

8.5.1 Challenges on System Reliability of Ductile Frames

The evaluation of system reliability for ductile frame structures has been an active area of research in the past 50 years. Several commonly used assumptions are summarised in the following (e.g. Ditlevsen and Bjerager 1986; Zhao and Ono 1998; Mahadevan and Raghothamachar 2000).

1) The constitutive relationship of the frame members is assumed elastic-perfectly plastic. The failure of a section means the imposition of a plastic hinge and an artificial moment equal to the plastic moment capacity of the section is applied at the hinge.
2) The frame member does not lose stability until it reaches the limit state of bending capacity.
3) Structural uncertainties are represented by considering only moment capacities of the frame members as random variables.
4) The effects of axial forces on the reduction of moment capacities of the frame members are neglected. Geometrical second-order and shear effects are also neglected.

Using these assumptions, based on the upper-bound theorem of plasticity (Livesley 1975), failure of ductile frame structure is defined as the formation of a kinematically admissible mechanism due to the formation of plastic hinges at a certain number of sections. Because of the uncertainties in the moment capacities of the structures and the load vector, frame structures may fail in different modes, and the general definition of performance function for ductile frame is the minimum of the performance functions corresponding to all the potential failure modes as

$$G(\mathbf{M}, \mathbf{P}) = \min\left[G_1(\mathbf{M}, \mathbf{P}), G_2(\mathbf{M}, \mathbf{P}), ..., G_n(\mathbf{M}, \mathbf{P})\right] \tag{8.51}$$

where $\mathbf{M}$ is the vector of moment capacities, $\mathbf{P}$ is the load vector, and $G_i(\mathbf{M}, \mathbf{P})$ is the performance function corresponding to ith failure mode.

To obtain the performance function defined by Eq. (8.51), the potential failure modes should be identified. However, there are generally an astronomically large number of potential failure paths in large-scale structures, and in most cases, only a small fraction of them contributes significantly to the overall failure probability. These may be referred to as *significant failure sequences*, and the estimation of system failure probability based on these sequences is expected to be close to the true answer. Several different approaches have been developed to identify the significant failure modes (or sequences) of a ductile frame structure, such as incremental load approach (e.g. Moses 1982; Feng 1988), β-unzipping method (e.g. Thoft-Christensen and Murotsu 1986; Thoft-Christensen 1990; Daghigh and Makouie 2003; Yang et al. 2014), truncated enumeration method (e.g. Melchers and Tang 1984; Nafday 1987), branch and bound method (e.g. Murotsu et al. 1984; Mahadevan and Raghothamachar 2000; Lee and Song 2011), stable configuration approach (Ang and Ma 1981; Bennett and Ang 1986), and mathematical programming techniques (Zimmerman et al. 1993), etc. Although some of these methods present elegant approaches for identifying the significant failure modes, they tend to be uneconomical or even unresolvable when applied to large structures.

If the potential failure modes are known or can be identified, computational procedures can be used for system reliability estimation such as bounding techniques (e.g. Cornell 1967; Ditlevsen 1979b; Feng 1989; Song 1992; Penmetsa and Grandhi 2002; Zhao et al. 2007), the PNET (Ang and Ma 1981), and direct or smart Monte Carlo simulations (Melchers 1994). However, the calculation of the failure probability is generally difficult for a system because of the large number of potential failure modes for most of practical structures, the difficulty in obtaining the sensitivity of the performance function, and the complications of in computing mutual correlation matrix and joint failure probability matrix among failure modes as described in Section 8.3.

Considering the difficulties in both failure mode identification and failure probability computation when using traditional methods, we will first define a performance function independent of failure modes and compute the corresponding probability of failure.

8.5.2 Performance Function Independent of Failure Modes

In order to define a failure mode-dependent performance function, one can first observe the limit state of function under only one load P. Because the structure will become a kinematically admissible mechanism when the load P is equal to the ultimate load, the performance function can be described as

$$G(\mathbf{M}, P) = u_p(\mathbf{M}) - P \tag{8.52}$$

where $u_p(\mathbf{M})$ is the ultimate load. The value of $u_p(\mathbf{M})$ is dependent on $\mathbf{M}$, and a different failure mode may happen with different components of $\mathbf{M}$.

In limit analysis, the ultimate load is generally described as the production of load P and a load factor, therefore, Eq. (8.52) can be described as follows:

$$G(\mathbf{M}, P) = \lambda(\mathbf{M}, P)P - P \tag{8.53}$$

where $\lambda(\mathbf{M}, P)$ is the load factor.

In Eq. (8.53), for a specific load P with certain direction, the amount of P does not influence the shape of the limit state surface $G(\mathbf{M}, \mathbf{P}) = 0$. Therefore, the same reliability analysis results will be obtained if it is written as follows:

$$G(\mathbf{M}, P) = \lambda(\mathbf{M}, P) - 1 \tag{8.54}$$

For multiple loads, the limit state function is defined in a load space, and it cannot be dealt with directly as Eq. (8.54). One can divide the load space into various load paths and consider one load path; for example, for a frame structure with two loads, $\mathbf{P} = [P_1, P_2]$. Considering a load path $\mathbf{P}(\theta_i)$ where θ_i is defined as $\tan\theta_i = P_2/P_1$, since the limit analysis is based on the principal of the proportional loading, the structure will become a kinematically admissible mechanism when the load increased along this load path. The utmost value of P_1 or P_2 can be described as the production of the load factor and P_1 or P_2 itself. The performance function can be written as follows, as already described:

$$G[\mathbf{M}, \mathbf{P}(\theta_i)] = \lambda[\mathbf{M}, \mathbf{P}(\theta_i)]P_1 - P_1 \text{ or } G[\mathbf{M}, \mathbf{P}(\theta_i)] = \lambda[\mathbf{M}, \mathbf{P}(\theta_i)]P_2 - P_2 \tag{8.55}$$

where $\lambda[\mathbf{M}, \mathbf{P}(\theta_i)]$ is the load factor and $\mathbf{P}(\theta_i)$ is the load path.

In Eq. (8.55), for a specific load path, the amount of P_1 or P_2 does not influence the shape of the limit state surface $G[\mathbf{M}, \mathbf{P}(\theta_i)] = 0$. Therefore, the same reliability analysis results will be obtained if it is written as follows:

$$G[\mathbf{M}, \mathbf{P}(\theta_i)] = \lambda[\mathbf{M}, \mathbf{P}(\theta_i)] - 1 \tag{8.56}$$

Different failure modes will happen with different $\mathbf{M}$ and different load path $\mathbf{P}(\theta_i)$, Eq. (8.56) however always holds true with any load path. With given $(\mathbf{M}, \mathbf{P})$, the loading path is certain and only one value of $\lambda(\mathbf{M}, \mathbf{P})$ can be obtained, therefore, one needs not to specify the loading path and Eq. (8.56) is written as

$$G(\mathbf{M}, \mathbf{P}) = \lambda(\mathbf{M}, \mathbf{P}) - 1 \tag{8.57}$$

Equation (8.57) is the general form of performance function of ductile frame structures (Zhao and Ono 1998). Because the ultimate load is obtained as the minimum load of formulation a kinematically admissible mechanism, limit state surface expressed in Eqs. (8.51) and (8.57) are the same, although their performance functions are different.

Using the performance function expressed in Eq. (8.57), the failure probability can be estimated from first- or second-order reliability method (FORM/SORM) and the response surface approach (RSA) without the identification procedure of failure modes and the correlation computation among the large number of failure modes (Zhao and Ono 1998; Leu and Yang 2000). However, the procedures need iteration and sometimes have local convergence problems; it is therefore very necessary to check the results of different fitting points. Besides, it becomes difficult to deal with complex large scale structures, especially with uniform distributed loads due to the strong-nonlinearity of the performance function. In order to avoid iteration and overcome the fitting point dependence described above, we will use the methods of moment described in the chapter on Direct Methods of Moment, in which, the first four moments of the performance function will be utilised to compute the probability of failure without derivation-based iteration.

A previous study (Zhao and Ono 1998) has revealed that the failure mode independent performance function in Eq. (8.57) was essentially the inner connotative surface of the limit state surfaces of all the potential failure modes of the structural system. The failure mode independent performance function in Eq. (8.57) is generally implicit, and for practical structures it may be strongly nonlinear and the distribution of the performance function may have strong non-normality. Since the methods of moment are good for weak non-normality of the distributions of the performance functions, to improve the computational efficiency of methods of moment in application to the system reliability analysis of frame structures, it is necessary to have a failure mode-independent performance function with distributions of weak non-normality.

It has been pointed out that the Box-Cox transformation (Box and Cox 1964; Zhang and Pandey 2013) is useful to bring the distributions of the transformed random variables or performance functions closer to the normality. Since the load factor $\lambda(\mathbf{M}, \mathbf{P})$ is greater than 0, according to the Box-Cox transformation, the performance function expressed in Eq. (8.57) can be equivalently expressed as:

$$G^*(\mathbf{M}, \mathbf{P}) = \begin{cases} \dfrac{[\lambda(\mathbf{M}, \mathbf{P})]^q - 1}{q}, & q \neq 0 \\ \ln[\lambda(\mathbf{M}, \mathbf{P})], & q = 0 \end{cases} \tag{8.58}$$

where q is an undetermined coefficient.

For simplicity, let $q = 0$, the performance function for the system reliability analysis of ductile frame structures can be expressed as

$$G^*(\mathbf{M}, \mathbf{P}) = \ln[\lambda(\mathbf{M}, \mathbf{P})] \tag{8.59}$$

Although the performance function of Eq. (8.59) has a different form than that of Eq. (8.57), they correspond to the same limit state physically. Since Eq. (8.59) can be considered as a logarithmic transformation of Eq. (8.57), it can reduce the variation of the performance function through reducing the measure scale. That is, the random fluctuations in Eq. (8.59) are generally much weaker than that in Eq. (8.57). In addition, the transformation is generally an effective way to make distributions less skewed, as will be demonstrated in the following numerical examples. Eq. (8.59) can therefore be used as the failure mode-independent performance function with weak non-normality for system reliability evaluation of ductile frame structures.

8.5.3 Limit Analysis

To determine $\lambda(\mathbf{M}, \mathbf{P})$ in Eq. (8.59), structural limit analysis is conducted using Compact Procedure (Aoyama 1988). In the procedure, the equilibrium equation is taken as the object function and the ultimate strength is taken as the limit condition. The mechanisms can be identified from the structural analysis when the total stiffness matric becomes singular. The limit analysis is defined as a problem to obtain the maximum load factor that satisfies the equilibrium equation and the limit condition using the linear programing method. The equilibrium equation and the limit condition are described as:

$$\lambda(\mathbf{M}, \mathbf{P})\mathbf{P} = \mathbf{HR} \tag{8.60}$$

$$-\mathbf{R}^l \le \mathbf{R} \le \mathbf{R}^u \tag{8.61}$$

where $\mathbf{H}$ is the coefficient matrix of the equilibrium equation; $\mathbf{R}$ is vector of member moment; and $\mathbf{R}^u$ and $\mathbf{R}^l$ denote the upper and lower bounds (i.e. utmost strength) of plastic limit moments, respectively.

If the ductile frame structures are only subjected to concentrated loads, the Gauss-Jordan method (Livesley 1975) can be applied to solve Eq. (8.60), and the main procedure is referred to as Compact Procedure (CP) including the following steps (Aoyama 1988; Zhao and Ono 1998):

1) Divide $\mathbf{R}$ into fundamental variables and non-fundamental variables based on the contents of $\mathbf{H}$. Specifically, finding the element h_{ij} with the largest absolute value of each row in the matrix $\mathbf{H}$ and making its value equal to 1, other elements' values in the jth column are assigned as 0. Then the vector of $\mathbf{R}$ can be divided into fundamental variables correspond to h_{ij} and non-fundamental variables.
2) Change the fundamental variables, and increase the load factor until the ultimate value (moment capacity) of a fundamental variable is reached.
3) Exchange the fundamental variables and the non-fundamental variables to further increase the load factor.
4) Repeat the previous steps until the load factor $\lambda(\mathbf{M}, \mathbf{P})$ reaches its maximum and the load factor $\lambda(\mathbf{M}, \mathbf{P})$ in Eq. (8.59) can then be obtained.

If there are uniform loads acting on the member of structures, the maximum value of moment may appear in a previously unknown position of the member. It thus becomes more difficult to obtain the load factor than that for structures with only concentrated loads. In order to solve this problem, the following steps (Yuan 2007) are suggested for CP:

1) For each member with uniform load q and length of l, assume a specific section which has a distance of αl from one side, where α is a position parameter and its initial value is $1/2$. Then according to the linear programming equation in CP, the limit load and internal moments can be obtained.
2) Suppose the maximum moment of section is R^e_{max} and its corresponding position is $\alpha_{max}l$, R^e_{max} can then be obtained based on the equilibrium equation, where $\alpha_{max} = 1/2 + (R_1 + R_2)/(ql^2)$, and R_1 and R_2 are the internal moments of both ends, respectively.
3) Check the limit condition $-R^e_u \le R^e_{max} \le R^e_u$, where R^e_u is the plastic limit moment of the member. Replace α by α_{max} until all the members satisfy the limit condition.

8.5.4 Methods of Moment for System Reliability of Ductile Frames

The procedure of the moment method is completely the same as described in the chapters on Moment Evaluation and Direct Methods of Moment. First, the hybrid dimension-reduction based point estimate method is utilised to evaluate the first four moments of the failure mode independent performance function with weak non-normality for system reliability evaluation of ductile frame structures, i.e. Eq. (8.59). According to Eq. (3.97), Eq. (8.59) can be approximated by

$$G(\mathbf{M},\mathbf{P}) \cong G_1^*(\mathbf{M}_S,\mathbf{P}_S) + G_2^*(\mathbf{M}_D,\mathbf{P}_D) - G(\mu_\mathbf{M},\mu_\mathbf{P}) \tag{8.62}$$

where

$$G_1^*(\mathbf{M}_S,\mathbf{P}_S) = \ln\left[\lambda\left(\mathbf{M}_S,\mathbf{P}_S,\mu_{\mathbf{M}_D},\mu_{\mathbf{P}_D}\right)\right] \tag{8.63}$$

$$G_2^*(\mathbf{M}_D,\mathbf{P}_D) = \ln\left[\lambda\left(\mathbf{M}_D,\mathbf{P}_D,\mu_{\mathbf{M}_I},\mu_{\mathbf{P}_I}\right)\right] \tag{8.64}$$

$$G(\mu_\mathbf{M},\mu_\mathbf{P}) = \ln\left[\lambda(\mu_1,\mu_2,...,\mu_n)\right] \tag{8.65}$$

where $G(\mu_\mathbf{M},\mu_\mathbf{P})$ represents the value of $G(\mathbf{M}, \mathbf{P})$ from which all the random variables take their mean values; $G_1^*(\mathbf{M}_S, \mathbf{P}_S)$ is the performance function of type 1, which means only dull variables $(\mathbf{M}_D, \mathbf{P}_D)$ take their mean values; and $G_2^*(\mathbf{M}_D, \mathbf{P}_D)$ is the performance function of type 2, in which only sensitive variables $(\mathbf{M}_I, \mathbf{P}_I)$ take their mean values.

The first four moments $G_1^*(\mathbf{M}_S, \mathbf{P}_S)$ is then estimated by using bivariate-dimension reduction method as described in Section 3.5.3, i.e. Eqs. (3.75)–(3.85d) in the chapter on Moment Evaluation; the univariate-dimension reduction method as described in Section 3.5.2, i.e. Eqs. (3.67a)–(3.68d) and (3.35a–c), is used to obtain the first four moments of $G_2^*(\mathbf{M}_D, \mathbf{P}_D)$; and finally using Eqs. (3.67a)–(3.68d) (all from the same chapter) again, the first four moments of the performance function of Eq. (8.61) can be obtained.

After the first four moments of Eq. (8.61) have been determined, the third-moment method or the fourth-moment method as described in the chapter on Direct Methods of Moment can be utilised to evaluate the probability of failure.

Example 8.27 Investigations on Fluctuations of the Performance Functions Eq. (8.59) and the Normality of their Distributions

Consider a one-story one-bay frame structure as shown in Figure 8.32, where P is a lateral load, and M is the moment capacity of frame members.

Based on Eqs. (8.57) and (8.59), the failure mode-independent performance functions for this example can be, respectively, written as follows

$$G(\mathbf{M},\mathbf{P}) = \lambda(M,P) - 1 = \frac{8M}{9P} - 1 \tag{8.66a}$$

Figure 8.32 A one-story one-bay frame structure subjected to one lateral load for Example 8.27.

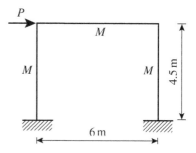

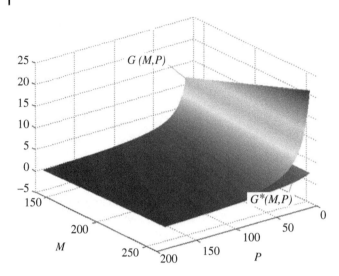

Figure 8.33 The limit state surfaces of the performance functions for Example 8.27.

$$G^*(\mathbf{M}, \mathbf{P}) = \ln[\lambda(M,P)] = \ln M - \ln P + \ln\left(\frac{8}{9}\right) \tag{8.66b}$$

where P and M = independent lognormal random variables with mean and standard deviations of $\mu_P = 100$ kN, $\mu_M = 200$ Mpa, $\sigma_P = 30$ kN, and $\sigma_M = 20$ Mpa.

The limit state surfaces corresponding to the performance functions in Eqs. (8.66a) and (8.66b) are illustrated in Figure 8.33. It can be observed that after the Box-Cox transformation, the performance function of Eq. (8.66a) with strong fluctuations turns to be a performance function of Eq. (8.66b) with weak fluctuations.

The first four moments of Eqs. (8.66a) and (8.66b) can be easily obtained and are shown in Table 8.5. The PDFs of Eqs. (8.66a) and (8.66b) approximated by the cubic normal distribution are then depicted in Figure 8.34. It can be observed from Table 8.5 and Figure 8.34 that after the Box-Cox transformation, the distribution of the performance function of Eq. (8.66a) with strong non-normality (skewness = 0.986 and kurtosis = 4.779) turns to be the distribution of the performance function of Eq. (8.66b) being normally distributed (skewness = 0.0 and kurtosis = 3.0).

In summary, after the Box-Cox transformation, the transformed failure mode-independent performance function for system reliability evaluation of ductile frame structures turns to be a weakly fluctuations function and a normally distributed random variable for this example.

Table 8.5 The first four moments of the performance functions for Example 8.27.

Performance function	Mean	Standard deviation	Skewness	Kurtosis
Eq. (8.66a)	0.938	0.616	0.986	4.779
Eq. (8.66b)	0.613	0.310	0.0	3.0

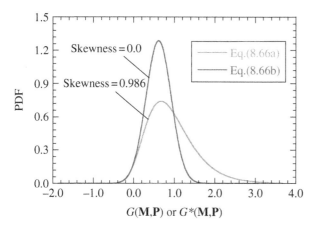

Figure 8.34 The PDFs of the performance functions for Example 8.27.

Example 8.28 System Reliability of a Six-Story Two-Bay Ductile Frame Structure under Lateral Concentrated Loads and Vertical Uniform Loads

The second example is a six-story two-bay plane building frame subjected to lateral concentrated loads and vertical uniform loads as shown in Figure 8.35, with the probabilistic information of member strength (R_b and R_c) and loads (L_1, L_2, and L_3) listed in Table 8.6.

According to Eq. (8.59), the proposed failure mode independent performance functions for this example can be expressed as

$$G^*(\mathbf{M}, \mathbf{P}) = \ln\left[\lambda(R_b, R_c, L_1, L_2, L_3)\right] \tag{8.67}$$

The sensitive indices of each variable are determined by using Eq. (3.93) from the chapter on Moment Evaluation with the seven-point estimate and are listed in Table 8.7.

According to Eq. (8.62), the performance function of Eq. (8.67) can be approximated by

$$G^*(\mathbf{M}, \mathbf{P}) = G_1^*(R_b, L_2, L_3) + G_2^*(R_c, L_1) - 0.616 \tag{8.68}$$

where

$$G_1^*(R_b, L_2, L_3) = \ln\left[\lambda\left(R_b, L_2, L_3, \mu_{L_1}, \mu_{R_c}\right)\right] \tag{8.69}$$

$$G_2^*(L_1, R_c) = \ln\left[\lambda\left(R_c, L_1, \mu_{R_b}, \mu_{L_2}, \mu_{L_3}\right)\right] \tag{8.70}$$

Using bivariate-dimension reduction based point estimate method, i.e. Eqs. (3.75a)–(3.85d) in the chapter just mentioned, the first four moments of G_1^* can be obtained as $\mu_{G_1^*} = 0.611$, $\sigma_{G_1^*} = 0.139$, $\alpha_{3G_1^*} = -0.030$, and $\alpha_{4G_1^*} = 2.981$, in which the load factor are computed using the compact procedure as given in Appendix I. Using Eqs. (3.67a)–(3.68d) and (3.35a–c) as before, the first four moments of G_2^* can be obtained as $\mu_{G_2^*} = 0.616$, $\sigma_{G_2^*} = 0.023$, $\alpha_{3G_2^*} = -0.048$, and $\alpha_{4G_2^*} = 3.008$.

Using Eqs. (3.67a)–(3.68d) again, the first four moments of the performance function $G^*(\mathbf{M}, \mathbf{P})$ are approximately $\mu_G = 0.611$, $\sigma_G = 0.141$, $\alpha_{3G} = -0.029$, $\alpha_{4G} = 2.981$. With these

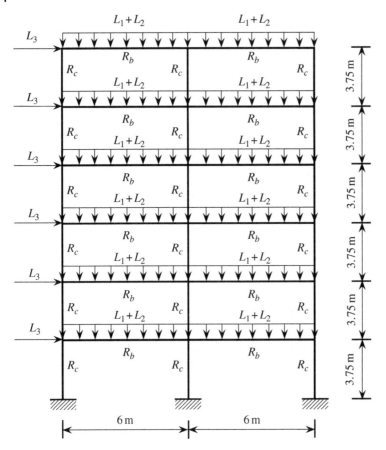

Figure 8.35 A six-story two-bay frame structure for Example 8.28.

Table 8.6 Random variables for Example 8.28.

Basic variables	Description	Distribution	Mean	Coefficient of variation (COV)
R_b	Moment capacity of beams	Lognormal	150 kNm	0.05
R_c	Moment capacity of columns	Lognormal	200 kNm	0.05
L_1	Dead load	Lognormal	20 kN/m	0.05
L_2	Live load	Gumbel	12 kN/m	0.30
L_3	Lateral load	Weibull	12 kN	0.40

first four moments of the proposed failure mode-independent performance function for the system reliability evaluation of the ductile frame structure, the moment-based reliability indices are $\beta_{3M} = 4.251$ with $P_f = 10.64 \times 10^{-6}$, and $\beta_{4M} = 4.305$ with $P_f = 8.332 \times 10^{-6}$. The total number of CP for structural limit analysis is 182 in the proposed method. For comparison, the probability of failure is also conducted with subset simulation (SS) (Au and

Table 8.7 Results of sensitivity analysis for Example 8.28.

Basic variables	Standard deviation, σ_{Gi}	Sensitive index, S_i (%)	Types of basic variables
R_b	0.639	20.1	Sensitive
R_c	0.058	1.8	Dull
L_1	0.314	9.9	Dull
L_2	1.167	36.7	Sensitive
L_3	1.002	31.5	Sensitive

Table 8.8 Results changed with the COV of the lateral load for Example 8.28.

COV of variable L_3	The proposed method (Runs of CP = 182)					SS	MCS
	μ_G	σ_G	α_{3G}	α_{4G}	β_{3M}	β_{SS}(Runs of CP)	β_{MCS}
0.6	0.613	0.206	0.067	3.223	3.079	2.855 (13 000)	3.052
0.8	0.617	0.255	0.158	3.316	2.569	2.452 (8400)	2.534
1.0	0.623	0.307	0.271	3.475	2.212	2.142 (4200)	2.234

Beck 2001a; Li and Cao 2016), which is instrumental in improving the computational efficiency of MCS. Here, the conditional probability of SS is 0.1 and the number of samples in each simulation level $N = 5000$ is used. The probability of failure is obtained as 9.261×10^{-5} with corresponding reliability index of 4.282, in which the total number of CP for structural limit analysis in SS is 23 000.

It can be observed that: (i) the reliability results of the proposed method and SS are in close agreement with the exact values; and (ii) the computational effort of the proposed method is about 0.72% of that of SS.

If the probabilistic information of the live load L_2 is changed as a Gumbel random variable with mean of 12 kN and COV = 0.6, 0.8, and 1.0, while the other random variables remain the same as shown in Table 8.6, the same procedures are carried out and the results of the proposed method, SS, and MCS are listed in Table 8.8. Likewise, the results by MCS (5×10^5 runs) are taken as the exact values, the COVs of the MCS estimate are 2.82%, 2.46%, and 1.78%, respectively. Again, it can be found that the proposed method can keep a very good trade-off of accuracy and efficiency for the system reliability assessment of ductile frame structures even when large coefficients of variation of original random variables are considered.

Example 8.29 System Reliability of a Six-Story Three-Bay Ductile Frame Structure under Lateral Concentrated Loads

The third example considers a frame structure with six stories and three bays as shown in Figure 8.36, and the probabilistic information of 28 random variables including member strength (M_1, ..., M_{22}) and loads (S_1, ..., S_6) are listed in Table 8.9.

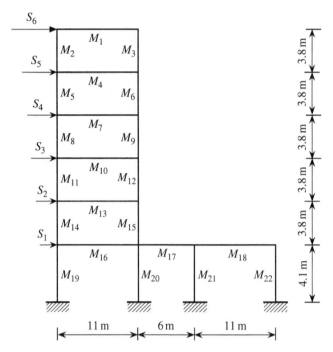

Figure 8.36 A six-story three-bay frame structure for Example 8.29.

Table 8.9 Random variables for Example 8.29.

Variable	Description	Distribution	Mean	COV
M_1, M_4, M_7, M_{17}, M_{18}	Moment capacity of beams	Lognormal	908 kNm	0.1
M_2, M_3, M_5, M_6, M_8, M_9, M_{21}, M_{22}	Moment capacity of columns	Lognormal	1452 kNm	0.1
M_{10}, M_{13}, M_{16}	Moment capacity of beams	Lognormal	1034 kNm	0.1
M_{11}, M_{12}, M_{14}, M_{15}, M_{19}, M_{20}	Moment capacity of columns	Lognormal	1628 kNm	0.1
S_1	Lateral load	Gumbel	25 kN	0.5
S_2	Lateral load	Gumbel	50 kN	0.5
S_3	Lateral load	Gumbel	75 kN	0.5
S_4	Lateral load	Gumbel	100 kN	0.5
S_5	Lateral load	Gumbel	125 kN	0.5
S_6	Lateral load	Gumbel	150 kN	0.5

According to Eq. (8.59), the proposed failure mode independent performance functions for this example can be expressed as

$$Z = G^*(\mathbf{M}, \mathbf{P}) = \ln\left[\lambda(M_1, ..., M_{22}, S_1, ..., S_6)\right] \tag{8.71}$$

The sensitive indices of each variable are determined by using Eq. (3.93) from the chapter on Moment Evaluation with the seven-point estimate and are listed in Table 8.10.

According to Eq. (8.62), the performance function of Eq. (8.71) can be approximated by

$$G^*(\mathbf{M}, \mathbf{P}) = G_1^*(S_3, S_4, S_5, S_6) + G_2^*(\mathbf{M}, S_1, S_2) - 0.812 \tag{8.72}$$

where

$$G_1^*(S_3, S_4, S_5, S_6) = \ln\left[\lambda\left(S_3, S_4, S_5, S_6, \boldsymbol{\mu}_{(\mathbf{M}, S_1, S_2)}\right)\right] \tag{8.73}$$

$$G_2^*(\mathbf{M}, S_1, S_2) = \ln\left[\lambda\left(\mathbf{M}, S_1, S_2, \mu_{S_4}, \mu_{S_5}, \mu_{S_6}\right)\right] \tag{8.74}$$

Using the hybrid dimension-reduction based point estimate method and the CP for structural limit analysis, the first four moments of Eq. (8.72) are summarised in Table 8.11, together with the corresponding third- and fourth-moment reliability indices. For

Table 8.10 Results of sensitivity analysis for Example 8.29.

Basic variables	Standard deviation, σ_{Gi}	Sensitive index, S_i (%)	Types of basic variables
$M_2, M_3, M_8, M_9, M_{11}, M_{12}, M_{16}\,M_{17}, M_{18}, M_{19},$ $M_{20}, M_{21}, M_{22}, S_1$	0	0	Dull
M_5, M_6	0.014	0.18	Dull
M_1, M_4	0.147	1.89	Dull
M_{14}, M_{15}	0.187	2.39	Dull
M_7	0.200	2.57	Dull
S_2	0.234	3.00	Dull
M_{10}, M_{13}	0.237	3.04	Dull
S_3	0.656	8.41	Sensitive
S_4	1.224	15.68	Sensitive
S_5	1.847	23.67	Sensitive
S_6	2.473	31.69	Sensitive

Table 8.11 Results changed with the COV of the lateral load for Example 8.29.

The proposed method (Runs of CP = 440)					SS	MCS
μ_G	σ_G	α_{3G}	α_{4G}	β_{3M}	β_{SS}(Runs of CP)	β_{MCS}
0.651	0.264	0.178	3.054	2.642	2.630 (8400)	2.601 (2×10^5)

comparison, the results of SS are also shown in Table 8.11, in which the conditional probability of 0.1 and the number of samples in each simulation level $N = 3000$ is used. The total number of CP for structural limit analysis in SS is 8400. Here, results of MCS with 2×10^5 runs are taken as the exact values, and the COVs of the MCS estimate are 3.28%. Again, the proposed method is very efficient and provides almost the same results with MCS and SS.

8.6   Summary

The concept of system reliability is described in this chapter. In order to improve the narrow bounds of the failure probability of a series structural system, a point estimation method is introduced for calculating the joint probability of every pair of failure modes of the system. Since the number of potential failure modes for most practical structures is very large, the determination of the mutual correlations and the joint failure probability matrices is quite cumbersome and difficult. A moment-based method for assessing the system reliability of series and non-series structures is investigated, with emphasis on series systems. The reliability indices (and associated failure probabilities) are directly calculated based on the first few moments of the system performance function of a structure. This does not require the reliability analysis of the individual failure modes, nor the iterative computation of derivatives, nor the computation of the mutual correlations among the failure modes, and does not require any design points. Thus, the moment method should be more effective for the system reliability evaluation of complex structures than currently available computational (non-MCS) methods.

As a practical topic, the evaluation of system reliability for ductile frame structures using methods of moment is introduced in the final part of this chapter. The difficulty in both failure mode identification and failure probability computation is still hard to overcome when using the traditional method for the system reliability assessment of ductile frame structures. To avoid the difficulties, a failure mode-independent performance function with weak non-normality is introduced. The methods of moment are then utilised to estimate the failure probability of ductile frame structures corresponding to the general performance function. From the numerical examples, it can be concluded that the introduced methodology successfully overcomes the two difficulties including the identification of significant failure modes and the computation of overall failure probabilities contributed from these significant modes in system reliability evaluation of ductile frame structures.

9

Determination of Load and Resistance Factors by Methods of Moment

9.1 Introduction

As the performance of a structure must be achieved under uncertainties, probabilistic analysis is thus necessary for reliability-based structural design. If the required safety factors are predetermined on the basis of specified probability-based requirements, reliability-based design may be accomplished through the adoption of appropriate deterministic design criteria, e.g. the use of traditional safety factors.

For obvious reasons, design criteria should be as simple as possible; moreover, they should be developed in a form that is familiar to the practical engineers. This can be accomplished through the use of load amplification factors and resistance reduction factors, known as the Load and Resistance Factors Design (LRFD) format (Ellingwood et al. 1982; Galambos and Ellingwood 1982; Ang and Tang 1984). In other words, the representative design loads are amplified by appropriate load factors and the nominal resistances are reduced by corresponding resistance factors. Structural safety is assured if the factored resistance is at least equal to the factored loads. The appropriate load and resistance factors must be developed in order to make the designed engineering structures achieve a prescribed level of reliability.

In principal, the load and resistance factors are determined using an acceptable reliability analysis method, such as first-order reliability-method (FORM) and Monte Carlo simulation (MCS). At the present stage, the load and resistance factors are prescribed in design codes, e.g. AIJ (2002). 'Recommendations for Limit State Design of Buildings' recommends several sets of load and resistance factors for target reliability levels of $b_T = 1, 2, 3, 4$. In general, the practicing engineers only use the load and resistance factors recommended in design codes without performing complicated reliability analysis in engineering designs. However, with the trend towards the performance design, there will be necessities for engineers to determine the load and resistance factors by themselves in order to conduct more flexible structural design. For example, one needs to determine load and resistance factors with different target reliability levels given by the codes or different uncertain characteristics from the assumption used in the codes even with the same reliability level. In such cases, it is expected that the design code recommend not only specific values of load and resistance factors, but also suitable and simple methods to determine these factors.

Structural Reliability: Approaches from Perspectives of Statistical Moments, First Edition.
Yan-Gang Zhao and Zhao-Hui Lu.
© 2021 John Wiley & Sons Ltd. Published 2021 by John Wiley & Sons Ltd.

The determination of load and resistance factors based on FORM is not suitable to be recommended as a simple method to practical engineers, not only because of the complexity to practical engineers, but also of some encountered problems such as the convergence in derivative-based iterations and multiple design points (Der Kiureghian and Dakessian 1998; Barranco-Cicilia et al. 2009). AIJ (2002) has provided a simple method, in which all the random variables are assumed to have known probability density functions (PDFs) and are required to transform into lognormal random variables.

In this chapter, the basic principle of the load and resistance factor is first reviewed, and the principle for the determination of these factors is discussed using methods of moment. Methods for the load and resistance factors estimation are introduced using the first few moments of the basic random variables, and simple formulae are adopted for the target mean resistance to avoid iterative computations. Since the introduced method is based on the first few moments of basic random variables, the load and resistances factors can be determined even when the probability distributions of the random variables are unknown.

9.2 Load and Resistance Factors

9.2.1 Basic Concept

Consider the simplest case which includes the resistance and only one load effect, the LRFD format may be expressed as follows.

$$\phi R_n \geq \gamma S_n \tag{9.1}$$

where ϕ = the resistance factor; γ = the load factor to be applied to load effect S; R_n = nominal value of the resistance; and S_n = presentative value of the load effect S.

In reliability-based structural design, the resistance factor ϕ and the load factor γ should be determined to achieve a specified level of reliability. That is, the design format, Eq. (9.1), should be equivalent to the following formula in terms of probability terms.

$$G(\mathbf{X}) = R - S \tag{9.2}$$

$$P_f \leq P_{fT} \text{ or } \beta \geq \beta_T \tag{9.3}$$

where R and S are the random variables representing the uncertainty in the resistance and load effect, respectively; P_f and β are the probability of failure and the reliability index corresponding to the performance function Eq. (9.2), respectively; P_{fT} and β_T are the acceptable probability of failure and target reliability index, respectively. Equation (9.3) implies that the failure probability P_f corresponding to the performance function in Eq. (9.2) should be less than a specified acceptable level and that the reliability index β should be larger than a specified target level.

9.2.2 Determination of LRFs by Second-Moment Method

If R and S are mutually independent normal random variables, the second-moment method is applicable and the design formula Eq. (9.3) becomes

$$\beta_{2M} \geq \beta_T \tag{9.4}$$

where

$$\beta_{2M} = \frac{\mu_G}{\sigma_G} \tag{9.5a}$$

$$\mu_G = \mu_R - \mu_S \tag{9.5b}$$

$$\sigma_G = \sqrt{\sigma_R^2 + \sigma_S^2} \tag{9.5c}$$

where β_{2M} is the second-moment reliability index; μ_G and σ_G are the mean value and standard deviation of the performance function G, respectively; μ_R and σ_R are the mean value and standard deviation of R, respectively; and μ_S and σ_S, are the mean value and standard deviation of S, respectively.

Substituting Eqs. (9.5a) to (9.5c) into Eq. (9.4) yields,

$$\frac{\mu_R - \mu_S}{\sqrt{\sigma_R^2 + \sigma_S^2}} \geq \beta_T \tag{9.6}$$

that is

$$\mu_R - \mu_S \geq \beta_T \sqrt{\sigma_R^2 + \sigma_S^2} = \beta_T \frac{\sigma_R^2 + \sigma_S^2}{\sqrt{\sigma_R^2 + \sigma_S^2}} = \frac{\beta_T \sigma_R^2}{\sqrt{\sigma_R^2 + \sigma_S^2}} + \frac{\beta_T \sigma_S^2}{\sqrt{\sigma_R^2 + \sigma_S^2}}$$

then, we have

$$\mu_R - \frac{\beta_T \sigma_R^2}{\sqrt{\sigma_R^2 + \sigma_S^2}} \geq \mu_S + \frac{\beta_T \sigma_S^2}{\sqrt{\sigma_R^2 + \sigma_S^2}}$$

or

$$\mu_R \left(1 - \frac{\sigma_R}{\mu_R} \frac{\sigma_R \beta_T}{\sqrt{\sigma_R^2 + \sigma_S^2}} \right) \geq \mu_S \left(1 + \frac{\sigma_S}{\mu_S} \frac{\sigma_S \beta_T}{\sqrt{\sigma_R^2 + \sigma_S^2}} \right)$$

Denote

$$\alpha_R = \frac{\sigma_R}{\sqrt{\sigma_R^2 + \sigma_S^2}}, \alpha_S = \frac{\sigma_S}{\sqrt{\sigma_R^2 + \sigma_S^2}} \tag{9.7}$$

as sensitivity factors for R and S, respectively, Eq. (9.6) becomes

$$\frac{\mu_R}{R_n}(1 - \alpha_R V_R \beta_T) R_n \geq \frac{\mu_S}{S_n}(1 + \alpha_S V_S \beta_T) S_n \tag{9.8}$$

where V_R and V_S are the coefficient of variation of R and S, respectively.

Comparing Eq. (9.8) with Eq. (9.1), the load and resistance factors may be expressed as,

$$\phi = (1 - \alpha_R V_R \beta_T) \frac{\mu_R}{R_n} \tag{9.9a}$$

$$\gamma = (1 + \alpha_S V_S \beta_T) \frac{\mu_S}{S_n} \tag{9.9b}$$

In general, for the case of multiple-load effects, the performance function and the LRFD format may be expressed as the follows.

$$G(\mathbf{X}) = R - \sum S_i \tag{9.10a}$$

$$\phi R_n \geq \Sigma \gamma_i S_{ni} \tag{9.10b}$$

If R and S_i are mutually independent normal random variables, the load and resistance factors may be expressed as,

$$\phi = (1 - \alpha_R V_R \beta_T) \frac{\mu_R}{R_n} \tag{9.11a}$$

$$\gamma_i = (1 + \alpha_{Si} V_{Si} \beta_T) \frac{\mu_{Si}}{S_{ni}} \tag{9.11b}$$

where V_R and V_{Si} are the coefficient of variation, respectively, of R and S_i; and α_R and α_{Si} are the sensitivity factors for R and S_i, respectively, which are given by

$$\alpha_R = \frac{\sigma_R}{\sigma_G}, \alpha_{Si} = \frac{\sigma_{Si}}{\sigma_G} \tag{9.12a}$$

$$\sigma_G = \sqrt{\sigma_R^2 + \sum \sigma_{Si}^2} \tag{9.12b}$$

It should be noted that in order to determine the load and resistance factors using Eqs. (9.11a) and (9.11b), the mean resistance and its standard deviation that meet with the target reliability level (referred to as *target mean resistance*, hereafter) should be predetermined. Therefore, the iterative computation is generally required to determine the load and resistance factors.

Example 9.1   Determine the Target Mean Resistance in Eq. (9.9a)

When the limit state is satisfied with the prescribed reliability level, β_T, Eq. (9.6) becomes

$$\mu_{RT} - \mu_S = \beta_T \sqrt{\sigma_R^2 + \sigma_S^2}$$

Then, we have

$$\mu_{RT}^2 - 2\mu_{RT}\mu_S + \mu_S^2 = \beta_T^2 \sigma_{RT}^2 + \beta_T^2 \sigma_S^2$$

That is

$$\mu_{RT}^2 \left(1 - \beta_T^2 V_{RT}^2\right) - 2\mu_{RT}\mu_S + \mu_S^2 \left(1 - \beta_T^2 V_S^2\right) = 0$$

The target mean resistance, μ_{RT}, can be expressed as

$$\mu_{RT} = \frac{\mu_S + \mu_S \beta_T \sqrt{V_S^2 + V_R^2 - \beta_T^2 V_S^2 V_R^2}}{\left(1 - \beta_T^2 V_R^2\right)} \tag{9.13}$$

In general, for multiple load effects, the mean resistance can be expressed as

$$\mu_{RT} = \frac{\sum \mu_{Si} + \beta_T \sqrt{\sum \sigma_{Si}^2 \left(1 - \beta_T^2 V_R^2\right) + V_R^2 \left(\sum \mu_{Si}\right)^2}}{\left(1 - \beta_T^2 V_R^2\right)} \tag{9.14}$$

One can observe that for independent normal random variables, the explicit expression of target mean resistance in Eqs. (9.13) and (9.14) can be easily obtained. However, for general cases, such as for non-normal random variables, the explicit expression of target mean resistance cannot be derived in a similar manner and iteration computation is generally required.

9.2.3 Determination of LRFs Under Lognormal Assumption

Consider Eq. (9.2) again, if R and S are mutually independent lognormal random variables with parameters λ_R, ζ_R and λ_S, ζ_S, according to Example 2.2 in the chapter on Fundamentals of Structural Reliability Theory the reliability index can be accurately given as

$$\beta = \frac{\lambda_R - \lambda_S}{\sqrt{\zeta_R^2 + \zeta_S^2}} \tag{9.15}$$

where λ_X and ζ_X are the mean and standard deviation of $\ln X$, respectively.

Substituting the Eq. (9.15) into Eq. (9.3), yields

$$\frac{\lambda_R - \lambda_S}{\sqrt{\zeta_R^2 + \zeta_S^2}} \geq \beta_T \tag{9.16a}$$

that is

$$\lambda_R - \lambda_S \geq \beta_T \sqrt{\zeta_R^2 + \zeta_S^2} = \beta_T \frac{\zeta_R^2 + \zeta_S^2}{\sqrt{\zeta_R^2 + \zeta_S^2}} = \frac{\beta_T \zeta_R^2}{\sqrt{\zeta_R^2 + \zeta_S^2}} + \frac{\beta_T \zeta_S^2}{\sqrt{\zeta_R^2 + \zeta_S^2}}$$

then we have

$$\lambda_R - \frac{\beta_T \zeta_R^2}{\sqrt{\zeta_R^2 + \zeta_S^2}} \geq \lambda_S + \frac{\beta_T \zeta_S^2}{\sqrt{\zeta_R^2 + \zeta_S^2}} \tag{9.16b}$$

Denote

$$\alpha_R = \frac{\zeta_R}{\sqrt{\zeta_R^2 + \zeta_S^2}}, \alpha_S = \frac{\zeta_S}{\sqrt{\zeta_R^2 + \zeta_S^2}} \tag{9.16c}$$

as sensitivity factors, respectively, for R and S, and note that

$$\lambda_R = \ln \frac{\mu_R}{\sqrt{1 + V_R^2}}, \lambda_S = \ln \frac{\mu_S}{\sqrt{1 + V_S^2}}$$

Equation (9.16b) becomes

$$\ln \frac{\mu_R}{\sqrt{1+V_R^2}} - \zeta_R \alpha_R \beta_T \geq \ln \frac{\mu_S}{\sqrt{1+V_S^2}} + \zeta_S \alpha_S \beta_T$$

that is

$$\ln \left[\frac{\mu_R}{\sqrt{1+V_R^2}} \exp\left(-\zeta_R \alpha_R \beta_T\right) \right] \geq \ln \left[\frac{\mu_S}{\sqrt{1+V_S^2}} \exp\left(\zeta_S \alpha_S \beta_T\right) \right]$$

or

$$\frac{\mu_R}{\sqrt{1+V_R^2}} \exp\left(-\zeta_R \alpha_R \beta_T\right) \geq \frac{\mu_S}{\sqrt{1+V_S^2}} \exp\left(\zeta_S \alpha_S \beta_T\right) \tag{9.17}$$

Comparing Eq. (9.17) with Eq. (9.1), the load and resistance factors may be expressed as,

$$\phi = \frac{1}{\sqrt{1+V_R^2}} \exp\left(-\zeta_R \alpha_R \beta_T\right) \frac{\mu_R}{R_n} \tag{9.18a}$$

$$\gamma = \frac{1}{\sqrt{1+V_S^2}} \exp\left(\zeta_S \alpha_S \beta_T\right) \frac{\mu_S}{S_n} \tag{9.18b}$$

Under the lognormal assumption, Eqs. (9.18a) and (9.18b) can be derived accurately since R/S is a lognormal random variable. However, for the case of multiple load effects, explicit expressions for load and resistance factors cannot be derived accurately even when all the basic random variables are independent.

Example 9.2 Determine the Target Mean Resistance in Eq. (9.18a)

When the limit state is satisfied with the prescribed reliability level, β_T, Eq. (9.16a) becomes

$$\lambda_{RT} = \lambda_S + \beta_T \sqrt{\zeta_R^2 + \zeta_S^2}$$

Then, the target mean resistance, μ_{RT}, can be expressed as

$$\mu_{RT} = \sqrt{1+V_R^2} \exp\left(\lambda_S + \beta_T \sqrt{\zeta_R^2 + \zeta_S^2}\right) \tag{9.19}$$

9.2.4 Determination of LRFs Using FORM

When R and S_i are non-normal random variables, the reliability index expressed in Eq. (9.5a) is not accurate. The reliability index can be obtained by the FORM. The design format can be expressed as

$$R^* \geq \Sigma S_i^* \tag{9.20}$$

and the load and resistance factors can be obtained as

$$\phi = \frac{R^*}{R_n}, \gamma_i = \frac{S_i^*}{S_{ni}} \tag{9.21}$$

where R^* and S_i^* are the values, respectively, of variable R and S_i at the design point of FORM in the original space.

According to Eq. (5.13a) in the chapter on Methods of Moment Based on First- and Second-Order Transformation, the design point in the standard normal space are given by

$$u_R^* = -\alpha_R\beta_T, u_{Si}^* = \alpha_{Si}\beta_T, i = 1, 2, ..., n \tag{9.22}$$

where α_R and α_{Si} are the sensitivity factors, respectively, for R and S_i,

$$\alpha_R = \frac{\frac{\partial G}{\partial u_R}\big|_{\mathbf{u}^*}}{\sqrt{\left(\frac{\partial G}{\partial u_R}\big|_{\mathbf{u}^*}\right)^2 + \sum_{i=1}^{n}\left(\frac{\partial G}{\partial u_{Si}}\big|_{\mathbf{u}^*}\right)^2}}, \alpha_{Si} = \frac{-\frac{\partial G}{\partial u_{Si}}\big|_{\mathbf{u}^*}}{\sqrt{\left(\frac{\partial G}{\partial u_R}\big|_{\mathbf{u}^*}\right)^2 + \sum_{i=1}^{n}\left(\frac{\partial G}{\partial u_{Si}}\big|_{\mathbf{u}^*}\right)^2}}, i = 1, 2, ...n$$

where $\mathbf{u}^*$ is the design point in the standard normal space obtained using derivative based iteration; u_R^* and u_{Si}^* are elements of $\mathbf{u}^*$ corresponding to R and S_i, respectively; and the explicit expressions of u_R^* and u_{Si}^* are not generally available.

Using the inverse Rosenblatt transformation, one has

$$R^* = F^{-1}\left[\Phi\left(u_R^*\right)\right], S_i^* = F^{-1}\left[\Phi\left(u_{Si}^*\right)\right] \tag{9.23}$$

And the load and resistance factors can be expressed as

$$\phi = \frac{F^{-1}[\Phi(-\alpha_R\beta_T)]}{R_n}, \gamma_i = \frac{F^{-1}[\Phi(\alpha_{Si}\beta_T)]}{S_{ni}} \tag{9.24}$$

Equation (9.24) is the general expressions for load and resistance factors by FORM.

In particular, when R and S_i are independent normal random variables, Eqs. (9.9a-b) or Eqs.(9.11a-b) can be readily obtained.

When R and S_i are independent lognormal random variables, the inverse Rosenblatt transformation in Eq. (9.23) can be expressed as

$$R^* = \exp[\zeta_R(-\alpha_R\beta_T) + \lambda_R] = \exp(\lambda_R)\exp(-\zeta_R\alpha_R\beta_T)$$
$$S_i^* = \exp[\zeta_{Si}(\alpha_{Si}\beta_T) + \lambda_{Si}] = \exp(\lambda_{Si})\exp(\zeta_{Si}\alpha_{Si}\beta_T)$$

Since

$$\lambda = \ln\frac{\mu}{\sqrt{1+V^2}}$$

Then one obtains

$$\phi = \frac{1}{\sqrt{1+V_R^2}}\exp\left(-\zeta_R\alpha_R\beta_T\right)\frac{\mu_R}{R_n} \tag{9.25a}$$

$$\gamma_i = \frac{1}{\sqrt{1+V_{Si}^2}} \exp\left(\zeta_{Si}\alpha_{Si}\beta_T\right)\frac{\mu_{Si}}{S_{ni}} \quad (9.25b)$$

When R and S_i are independent 3P-lognormal random variables, the inverse Rosenblatt transformation in Eq. (9.23) can be expressed as

$$R^* = \frac{3\sigma_R}{\alpha_{3R}}\left[\exp\left(\frac{\alpha_{3R}}{3}\alpha_R\beta_T - \frac{\alpha_{3R}^2}{18}\right)-1\right]+\mu_R$$

$$S_i^* = \frac{3\sigma_{Si}}{\alpha_{3Si}}\left[\exp\left(-\frac{\alpha_{3Si}}{3}\alpha_{Si}\beta_T - \frac{\alpha_{3Si}^2}{18}\right)-1\right]+\mu_{Si}$$

Then one obtains

$$\phi = \frac{3\sigma_R}{R_n\alpha_{3R}}\left[\exp\left(\frac{\alpha_{3R}}{3}\alpha_R\beta_T - \frac{\alpha_{3R}^2}{18}\right)-1\right]+\frac{\mu_R}{R_n} \quad (9.26a)$$

$$\gamma_i = \frac{3\sigma_{Si}}{S_{ni}\alpha_{3Si}}\left[\exp\left(-\frac{\alpha_{3Si}}{3}\alpha_{Si}\beta_T - \frac{\alpha_{3Si}^2}{18}\right)-1\right]+\frac{\mu_{Si}}{S_{ni}} \quad (9.26b)$$

where α_R and α_{Si} are the sensitivity factors in the standard normal space obtained using derivative based iteration, and explicit expressions of α_R and α_{Si} are not available. Needless to say, α_R and α_{Si} are not in the form of Eq. (9.16c). The flow chart is illustrated in Figure 9.1 to determine the load and resistance factors by FORM.

Example 9.3 Consider the statically indeterminate beam in Figure 9.2, which has been considered by Recommendations for Limit State Design of Buildings (AIJ 2002). The beam is loaded by three uniformly distributed loads, i.e. the dead load (D), live load (L), and snow load (S), in which the snow load is the dominating and time-dependent load. The probabilistic properties in member strength and loads are listed in Table 9.1. The design working life is assumed for 50 years.

The limit state function is expressed as

$$G(\mathbf{X}) = M_P - (M_D + M_L + M_S)$$

where M_P is the resistance; and $M_D = (Dl^2)/16$, $M_L = (Ll^2)/16$, and $M_S = (Sl^2)/16$ are the load effects of D, L, and S, respectively.

Determine the load and resistance factors for the performance function above, in order to achieve a reliability of $\beta_T = 2.0$.

Because S is a Gumbel random variable, the probability distribution of the maximum S during 50 years is also the Gumbel distribution (Melchers 1999). The values of mean, mean/nominal, coefficient of variation, and skewness corresponding to the maximum snow load during 50 years are readily obtained as: $\mu_{S50} = 2.595\mu_D$, $\mu_{S50}/S_n = 0.972$, $V_{S50} = 0.169$, and $\alpha_{3S50} = 1.140$.

Step 1; assume the initial mean value of resistance.

$$\mu_{Mp}^{(0)} = \mu_{M_D} + \mu_{M_L} + \mu_{M_S} = 3.895\mu_{M_D}$$

Figure 9.1 Flow chart for determining the load and resistance factors by FORM.

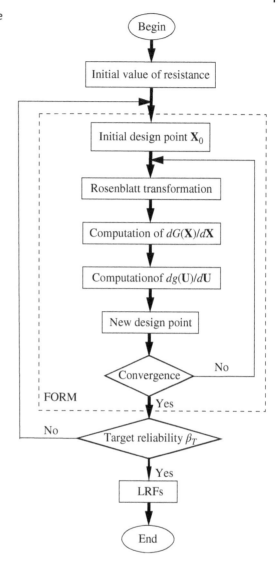

Figure 9.2 A statically indeterminate beam.

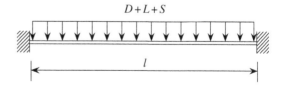

Table 9.1 Basic random variables for Example 9.3.

R or S_i	CDF	Mean/nominal	Mean	V_i	α_{3i}
D	Normal	1.0	μ_D	0.1	0.0
L	Lognormal	0.45	$0.3\mu_D$	0.4	1.264
S	Gumbel	0.47	$1.25\mu_D$	0.35	1.140
M_p	Lognormal	1.0	μ_{Mp}	0.1	0.301

First iteration:

Step 2; assume the mean value as an initial checking point.

$$r^{(0)} = \mu_{M_P} = 3.895\mu_{M_D}, \quad d^{(0)} = \mu_{M_D} = 1.0\mu_{M_D}, \quad l^{(0)} = \mu_{M_L} = 0.3\mu_{M_D}, \quad s^{(0)} = \mu_{M_S}$$

$$= 2.595\mu_{M_D}$$

That is, $\mathbf{x}^{(0)} = \{3.895, 1.0, 0.3, 2.595\}\mu_{M_D}$

Step 3; using the Rosenblatt transformation, obtain the corresponding checking point in the u-space,

$$u_R^{(0)} = \Phi^{-1}\left[F\left(r^{(0)}\right)\right] = \frac{\ln r^{(0)} - \lambda_R}{\zeta_R} = 0.04988\mu_{M_D}$$

$$u_D^{(0)} = \Phi^{-1}\left[F\left(d^{(0)}\right)\right] = \frac{d^{(0)} - \mu_D}{\sigma_D} = 0$$

$$u_L^{(0)} = \Phi^{-1}\left[F\left(l^{(0)}\right)\right] = \frac{\ln l^{(0)} - \lambda_L}{\zeta_L} = 0.1926\mu_{M_D}$$

$$u_S^{(0)} = \Phi^{-1}\left[F\left(s^{(0)}\right)\right] = 0.1773\mu_{M_D}$$

That is, $\mathbf{u}^{(0)} = \{0.04988, 0, 0.1926, 0.1773\}\mu_{M_D}$.

The first order reliability index at the current stage is given as

$$\beta_F^{(0)} = \sqrt{\mathbf{u}^{(0)^T}\mathbf{u}^{(0)}} = 0.2665$$

Step 4; determine the Jacobian Matrix evaluated at the checking point $\mathbf{u}^{(0)}$.

Since R, D, L, and S are independent,

$$\frac{\partial R}{\partial u_D} = \frac{\partial R}{\partial u_L} = \frac{\partial R}{\partial u_S} = 0, \quad \frac{\partial R}{\partial u_R} = 0.3885, \quad \frac{\partial D}{\partial u_R} = \frac{\partial D}{\partial u_L} = \frac{\partial D}{\partial u_S} = 0, \quad \frac{\partial D}{\partial u_D} = 0.1$$

$$\frac{\partial L}{\partial u_D} = \frac{\partial L}{\partial u_R} = \frac{\partial L}{\partial u_S} = 0, \quad \frac{\partial L}{\partial u_L} = 0.1156, \quad \frac{\partial S}{\partial u_D} = \frac{\partial S}{\partial u_R} = \frac{\partial S}{\partial u_L} = 0, \quad \frac{\partial S}{\partial u_S} = 0.4193$$

That is,

$$\mathbf{J} = \left.\frac{\partial \mathbf{x}}{\partial \mathbf{u}}\right|_{\mathbf{x}^{(0)}} = \begin{bmatrix} 0.3885 & 0 & 0 & 0 \\ 0 & 0.1 & 0 & 0 \\ 0 & 0 & 0.1156 & 0 \\ 0 & 0 & 0 & 0.4193 \end{bmatrix}$$

Step 5; evaluate the performance function and gradient vector at $\mathbf{u}_0$.

$$g\left(\mathbf{u}^{(0)}\right) = G\left(\mathbf{x}^{(0)}\right) = 0, \nabla G(\mathbf{x}_0) = \{1 \ -1 \ -1 \ -1\}^T$$

$$\nabla g(\mathbf{u}_0) = \mathbf{J}^T \nabla G(\mathbf{x}_0) = \begin{bmatrix} 0.3885 & 0 & 0 & 0 \\ 0 & 0.1 & 0 & 0 \\ 0 & 0 & 0.1156 & 0 \\ 0 & 0 & 0 & 0.4193 \end{bmatrix} \begin{Bmatrix} 1 \\ -1 \\ -1 \\ -1 \end{Bmatrix} = \begin{Bmatrix} 0.3885 \\ -0.1 \\ -0.1156 \\ -0.4193 \end{Bmatrix}$$

Step 6; obtain a new checking point.

$$\mathbf{u}^{(1)} = \frac{1}{\nabla^T g(\mathbf{u}_0)\nabla g(\mathbf{u}_0)} \left[\nabla^T g(\mathbf{u}_0)\mathbf{u}_0 - g(\mathbf{u}_0)\right]\nabla g(\mathbf{u}_0)$$

$$= \frac{1}{\begin{Bmatrix} 0.3885 \\ -0.1 \\ -0.1156 \\ -0.4193 \end{Bmatrix}^T \begin{Bmatrix} 0.3885 \\ -0.1 \\ -0.1156 \\ -0.4193 \end{Bmatrix}} \left[\begin{Bmatrix} 0.3885 \\ -0.1 \\ -0.1156 \\ -0.4193 \end{Bmatrix}^T \begin{Bmatrix} 0.04988 \\ 0 \\ 0.1926 \\ 0.1773 \end{Bmatrix} \mu_{M_D} - 0\right]$$

$$\begin{Bmatrix} 0.3885 \\ -0.1 \\ -0.1156 \\ -0.4193 \end{Bmatrix} = \begin{Bmatrix} -0.08571 \\ 0.02206 \\ 0.02550 \\ 0.09251 \end{Bmatrix} \mu_{M_D}$$

and in the space of the original variables, the checking point is

$$\mathbf{x}^{(1)} = \mathbf{x}^{(0)} + \mathbf{J}\left(\mathbf{u}^{(1)} - \mathbf{u}^{(0)}\right)$$

$$= \begin{Bmatrix} 3.895 \\ 1.0 \\ 0.3 \\ 2.595 \end{Bmatrix} \mu_{M_D} + \begin{bmatrix} 0.3885 & 0 & 0 & 0 \\ 0 & 0.1 & 0 & 0 \\ 0 & 0 & 0.1156 & 0 \\ 0 & 0 & 0 & 0.4193 \end{bmatrix} \left(\begin{Bmatrix} -0.08571 \\ 0.02206 \\ 0.02550 \\ 0.09251 \end{Bmatrix} \mu_{M_D}\right.$$

$$\left. - \begin{Bmatrix} 0.04988 \\ 0 \\ 0.1926 \\ 0.1773 \end{Bmatrix} \mu_{M_D}\right) = \begin{Bmatrix} 3.8423 \\ 1.0022 \\ 0.2807 \\ 2.5594 \end{Bmatrix} \mu_{M_D}$$

Step 7; calculate the reliability index.

$$\beta_F^{(1)} = \sqrt{\mathbf{u}^{(1)^T}\mathbf{u}^{(1)}} = \sqrt{\left\{\begin{array}{c} -0.08571 \\ 0.02206 \\ 0.02550 \\ 0.09251 \end{array}\right\}^T \left\{\begin{array}{c} -0.08571 \\ 0.02206 \\ 0.02550 \\ 0.09251 \end{array}\right\}} = 0.1305$$

Step 8; the relative difference between $\beta_F^{(1)}$ and $\beta_F^{(0)}$ is given as

$$\varepsilon = \frac{\left|\beta_F^{(1)} - \beta_F^{(0)}\right|}{\beta_F^{(1)}} = \frac{|0.1305 - 0.2665|}{0.1305} = 1.042$$

Since the difference ε is too large, repeat steps 3 through 7 until the convergence is achieved. The results of the first iteration are listed in Table 9.2, and the result of the first order reliability index is determined as $\beta_1 = 0.12916$.

Step 9; the relative difference between the first iteration β_1 and the target reliability index β_T is given as

$$\eta = \frac{|\beta_1 - \beta_T|}{\beta_T} = \frac{|0.129164 - 2|}{0.129164} = 0.9354$$

Since the difference is still too large, repeat steps 2 through 8 using the below $\mu_R^{(1)}$ as the new initial mean value of resistance.

$$\mu_{M_P}^{(1)} = \mu_{M_P}^{(0)} + (\beta_T - \beta_1) \times \sqrt{\sum \sigma_{x_i}^2} = 4.7155\mu_{M_D}$$

Table 9.2 Results of the first iteration.

No.	X	x*/ m_{MD}	u*/μ_{MD}	J	$\nabla G(x)$	$\nabla G_u(u)$	u*	β_F	ε
1	R	3.895	0.04988	diag	1	0.3885	−0.08571	0.1305	1.042
	D	1	0	(0.3885, 0.1,	−1	−0.1	0.02206		
	L	0.3	0.1926	0.1156, 0.4193)	−1	−0.1156	0.0255		
	S	2.595	0.1773		−1	−0.4193	0.09251		
2	R	3.8423	−0.08664	diag	1	0.3833	−0.08568	0.12917	0.0107
	D	1.0022	0.02206	(0.3833, 0.1,	−1	−0.1	0.02235		
	L	0.2807	0.01987	0.1081, 0.4066)	−1	−0.1081	0.02417		
	S	2.5594	0.09119		−1	−0.4066	0.09088		
3	R	3.8427	−0.08567	diag	1	0.3833	−0.08568	0.12916	0.000005
	D	1.0022	0.02235	(0.3833,0.1,	−1	−0.1	0.02235		
	L	0.2811	0.02417	0.1083, 0.4065)	−1	−0.1083	0.02421		
	S	2.5593	0.09088		−1	−0.4065	0.09087		

Table 9.3 The whole iterations of Example 9.3.

Iterations	$\mu_R^{(0)}$	FORM Iteration	$(R, D, L, S)/\mu_{MD}$	u^*/μ_{MD}	β_F	$\mu_R^{(1)}$
1	3.895	1	(3.895, 1.000, 0.300, 2.595)	(−0.086, 0.022, 0.026, 0.093)	0.131	4.716
		2	(3.842, 1.002, 0.281, 2.559)	(−0.086, 0.022, 0.024, 0.091)	0.129	
		3	(3.843, 1.002, 0.281, 2.559)	(−0.086, 0.022, 0.024, 0.091)	0.129	
2	4.716	1	(4.716, 1.000, 0.300, 2.595)	(−1.000, 0.213, 0.246, 0.891)	1.378	5.010
		2	(4.222, 1.021, 0.306, 2.894)	(−0.809, 0.192, 0.227, 1.018)	1.334	
		3	(4.327, 1.019, 0.304, 3.004)	(−0.783, 0.181, 0.212, 1.037)	1.329	
		4	(4.340, 1.018, 0.302, 3.019)	(−0.780, 0.180, 0.210, 1.040)	1.329	
...	...	...				...
12	5.311	1	(5.311, 1.000, 0.300, 2.595)	(−1.640, 0.310, 0.358, 1.299)	2.145	5.311
		2	(4.415, 1.031, 0.319, 3.065)	(−1.178, 0.267, 0.329, 1.590)	2.023	
		3	(4.69, 1.027, 0.316, 3.347)	(−1.093, 0.234, 0.284, 1.634)	1.999	
		4	(4.738, 1.023, 0.311, 3.404)	(−1.079, 0.228, 0.273, 1.644)	1.999	
13	5.311	1	(5.311, 1.000, 0.300, 2.595)	(−1.641, 0.310, 0.358, 1.299)	2.146	5.312
		2	(4.415, 1.031, 0.319, 3.065)	(−1.178, 0.267, 0.329, 1.590)	2.024	
		3	(4.690, 1.027, 0.316, 3.347)	(−1.093, 0.234, 0.285, 1.634)	2.000	
		4	(4.739, 1.023, 0.311, 3.405)	(−1.079, 0.228, 0.273, 1.645)	1.999	

The entire iterations are summarised in Table 9.3.

The relative difference between the result of the 13th iteration β_{13} and the target reliability index β_T is given as

$$\eta = \frac{|\beta_{13} - \beta_T|}{\beta_T} = \frac{|1.99923 - 2.0|}{2.0} = 0.000385$$

The convergence has been achieved.

Step 10; determine the load and resistance factors for the performance function.

Using the results in Table 9.3, the direction cosines at the design point in standard normal space $(-1.079, 0.228, 0.273, 1.645)$, are obtained as

$$\alpha = \frac{\nabla G(\mathbf{u})}{\sqrt{\nabla^T G(\mathbf{u}) \nabla G(\mathbf{u})}} = \left\{ \begin{array}{c} 0.5398 \\ -0.1142 \\ -0.1367 \\ -0.8228 \end{array} \right\}^T$$

Then, the sensitivity factors are obtained as

$$\alpha_{M_P} = 0.5398, \alpha_{M_D} = 0.1142, \alpha_{M_L} = 0.1367, \alpha_{M_S} = 0.8228$$

Therefore, the load and resistance factors can be obtained as

$$\phi = \frac{M_P^*}{M_{Pn}} = \frac{F^{-1}[\Phi(-\alpha_{M_P}\beta_T)]}{\mu_{M_P}} \cdot \frac{\mu_{M_P}}{M_{Pn}} = \frac{4.7448}{5.3112} \cdot 1.0 = 0.893$$

$$\gamma_{M_D} = \frac{M_D^*}{M_{Dn}} = \frac{F^{-1}[\Phi(\alpha_{M_D}\beta_T)]}{\mu_{M_D}} \cdot \frac{\mu_{M_D}}{M_{Dn}} = \frac{1.0228}{1.0} \cdot 1.0 = 1.023$$

$$\gamma_{M_L} = \frac{M_L^*}{M_{Ln}} = \frac{F^{-1}[\Phi(\alpha_{M_L}\beta_T)]}{\mu_{M_L}} \cdot \frac{\mu_{M_L}}{M_{Ln}} = \frac{0.3095}{0.3} \cdot 0.45 = 0.464$$

$$\gamma_{M_S} = \frac{M_S^*}{M_{Sn}} = \frac{F^{-1}[\Phi(\alpha_{M_S}\beta_T)]}{\mu_{M_S}} \cdot \frac{\mu_{M_S}}{M_{Sn}} = \frac{3.4137}{2.595} \cdot 0.972 = 1.279$$

The LRFD format and the target mean resistance for this example using FORM are obtained as

$$0.89 M_{Pn} \geq 1.02 M_{Dn} + 0.46 M_{Ln} + 1.28 M_{Sn}$$

$$\mu_{Mp} \geq 5.31 \mu_{M_D}$$

where $M_{Dn} = (D_n l^2)/16$, $M_{Ln} = (L_n l^2)/16$, and $M_{Sn} = (S_n l^2)/16$.

It can be observed from Figure 9.1 and Example 9.3 that to determine the load and resistance factors using FORM, the 'design point' must be determined and derivative-based iterations are necessary. As described in Section 9.1, there will be necessities for engineers to determine the load and resistance factors by themselves in order for more flexible structural design to accomplish performance design. Furthermore, all methods discussed assumed the basic random variables have known PDFs. There are practical needs for methods to determine the LRFs for random variables with unknown PDFs.

Since the first few moments of a random variable is much easier to be obtained than its PDF, the next sections will use the methods of moment to determine the load and resistance factors.

9.2.5 An Approximate Method for the Determination of LRFs

When R and S_i are independent lognormal random variables, it has been shown that the load and resistance factors can be simply expressed by Eqs. (9.25a) and (9.25b). However, the sensitivity factors α_R and α_{Si} have to be determined from the design points in FORM. For practical use, Mori (2002) developed empirical formulas to estimate the sensitivity factors as follows

$$\alpha_R = \frac{\zeta_R}{\sqrt{\zeta_R^2 + \sum (c_{Si}\zeta_{Si})^2}} \cdot \omega \tag{9.27a}$$

$$\alpha_{Si} = \frac{\zeta_{Si}}{\sqrt{\zeta_R^2 + \sum (c_{Si}\zeta_{Si})^2}} \cdot \omega \tag{9.27b}$$

where ω is a modification factor, and generally taken as 1.05; and c_{Si} is given by

$$c_{Si} = a_{Si} / \sum a_{Si} \tag{9.28}$$

where

$$a_{Si} = \exp\left(\lambda_{Si} + \frac{1}{2}\zeta_{Si}^2\right) \cdot \frac{\mu_{Si}}{S_{ni}} \tag{9.29}$$

If the independent S_i are not lognormal random variables, it is necessary to approximate them as lognormal random variables when using the formulas above. The normalised mean value of $\ln \widetilde{S}_i$ [mean value of $\ln\left(\widetilde{S}_i/\mu_{S_i}\right)$, i.e. $\widetilde{\lambda}_{S_i}$ $(\widetilde{\mu}_{\ln S_i})$]; standard deviation, i.e. $\widetilde{\zeta}_{S_i}$ $(\widetilde{\sigma}_{\ln S_i})$; and coefficient of variation $\widetilde{S}_i$, i.e. $\widetilde{V}_{S_i}$, are determined to satisfy the following conditions.

$$F_{Si}(s_{50}) = G_{Si}(s_{50}) \tag{9.30a}$$
$$F_{Si}(s_{99}) = G_{Si}(s_{99}) \tag{9.30b}$$

in which $F_{Si}(s)$ and $G_{Si}(s)$ are cumulative distribution functions (CDFs) of S_i and $\widetilde{S}_i$, respectively; and s_{50} and s_{99} are the value of satisfying $F_{Si}(s) = 0.5$ and $F_{Si}(s) = 0.99$ (0.999 if S_i is described by Type II extreme value distribution and $\beta_T \geq 2.5$), respectively.

For Type I extreme value distribution, $\widetilde{\lambda}_{S_i}$ and $\widetilde{\zeta}_{S_i}$ can be explicitly expressed as

$$\widetilde{\lambda}_{S_i} = \widetilde{\mu}_{\ln S_i} = \ln\left(1 - 0.164 V_{S_i}\right), \widetilde{\zeta}_{S_i} = \widetilde{\sigma}_{\ln S_i} = 0.430 \cdot \ln\left(\frac{1 + 3.14 V_{S_i}}{1 - 0.164 V_{S_i}}\right)$$

For Type II extreme value distribution, $\widetilde{\lambda}_{S_i}$ and $\widetilde{\zeta}_{S_i}$ can also be explicitly expressed as

1) $\beta_T \leq 2.5$

$$\widetilde{\lambda}_{S_i} = \widetilde{\mu}_{\ln S_i} = \frac{0.367}{k} - \ln\left[\Gamma\left(1 - \frac{1}{k}\right)\right], \widetilde{\zeta}_{S_i} = \widetilde{\sigma}_{\ln S_i} = \frac{1.82}{k}$$

2) $\beta_T > 2.5$

$$\widetilde{\lambda}_{S_i} = \widetilde{\mu}_{\ln S_i} = -\frac{0.170}{k} - \ln\left[\Gamma\left(1 - \frac{1}{k}\right)\right], \widetilde{\zeta}_{S_i} = \widetilde{\sigma}_{\ln S_i} = \frac{2.29}{k}$$

where k is the statistical characteristic coefficient of earthquake disaster for different regions, which can be found in AIJ (2015).

For the other distributions, the following regression formulae can be used.

$$\tilde{\lambda}_{S_i} = \tilde{\mu}_{\ln Si} = e_0 + e_1 V_{Si} + e_2 V_{Si}^2 + e_3 V_{Si}^3 \tag{9.31}$$

$$\tilde{\zeta}_{S_i} = \tilde{\sigma}_{\ln Si} = s_0 + s_1 V_{Si} + s_2 V_{Si}^2 + s_3 V_{Si}^3 \tag{9.32}$$

$$\tilde{V}_{Si} = \sqrt{\exp \tilde{\zeta}_{S_i}^{\,2} - 1} \tag{9.33}$$

where V_{Si} is the COV of annual maximum value of load effect before it is approximated to be lognormal random variate, and e_j and s_j are provided by Mori (2002) based on the probability distribution of the annual maximum value. The reference period of serviceability limit state design and ultimate limit state design are assumed to be 1 and 50 years, respectively.

In summary, when S_i are not lognormal random variables, the load and resistance factors can be determined by Eqs. (9.25a–9.29) only if λ_{S_i}, ζ_{S_i}, and V_{S_i} are replaced by $\tilde{\lambda}_{S_i}$, $\tilde{\zeta}_{S_i}$, and $\tilde{V}_{S_i}$, respectively. Detailed procedure can be found in Mori (2002) and AIJ (2002). A revised version of the method that is applicable for higher reliability can also be found in Mori et al. (2017) and Mori et al. (2019).

9.3 Load and Resistance Factors by Third-Moment Method

9.3.1 Determination of LRFs Using Third-Moment Method

For a performance function $G(\mathbf{X})$, as shown previously in the chapter on Direct Methods of Moment, the third-moment reliability index can be obtained as (Zhao and Lu 2011)

$$\beta_{3M} = -\frac{\alpha_{3G}}{6} - \frac{3}{\alpha_{3G}} \ln \left(1 - \frac{1}{3} \alpha_{3G} \beta_{2M} \right) \tag{9.34}$$

where

$$\beta_{2M} = \frac{\mu_G}{\sigma_G} = \frac{\mu_R - \Sigma \mu_{Si}}{\sigma_G} \tag{9.35a}$$

$$\sigma_G = \sqrt{\sigma_R^2 + \Sigma \sigma_{Si}^2} \tag{9.35b}$$

$$\alpha_{3G} = \frac{1}{\sigma_G^3} \left(\alpha_{3R} \sigma_R^3 - \Sigma \alpha_{3i} \sigma_{Si}^3 \right) \tag{9.35c}$$

where α_{3G} is the skewness of $G(\mathbf{X})$; α_{3R} and α_{3Si} are the skewness of R and S_i, respectively; and β_{2M} and β_{3M} are the second- and third-moment reliability index, respectively.

Substituting Eq. (9.34) into the design format in Eq. (9.3) yields,

$$\beta_{3M} \geq \beta_T \tag{9.36}$$

That is

$$-\frac{\alpha_{3G}}{6} - \frac{3}{\alpha_{3G}} \ln \left(1 - \frac{1}{3} \alpha_{3G} \beta_{2M} \right) \geq \beta_T$$

Reorganising the equation above gives

$$\beta_{2M} \geq \frac{3}{\alpha_{3G}}\left\{1 - \exp\left[\frac{\alpha_{3G}}{3}\left(-\beta_T - \frac{\alpha_{3G}}{6}\right)\right]\right\} \tag{9.37}$$

Denoting the right side of Eq. (9.37) as β_{2T}, one obtains

$$\beta_{2M} \geq \beta_{2T} \tag{9.38}$$

$$\beta_{2T} = \frac{3}{\alpha_{3G}}\left\{1 - \exp\left[\frac{\alpha_{3G}}{3}\left(-\beta_T - \frac{\alpha_{3G}}{6}\right)\right]\right\} \tag{9.39}$$

Equation (9.38) is as same as Eq. (9.3). It means that, if the second moment reliability index β_{2M} is at least equal to β_{2T}, the reliability index β will be equal to or larger than the target reliability index β_T, and the required reliability is satisfied. Therefore, β_{2T} can be considered a target value of β_{2M}, and hereafter is denoted as the *target second-moment reliability index*.

Since Eq. (9.38) is similar to Eq. (9.4) except that the right side is β_{2T}, the load and resistance factors corresponding to Eq. (9.38) can be easily obtained by substituting β_T to the right side of Eqs. (9.11a) and (9.11b) with β_{2T}

$$\phi = (1 - \alpha_R V_R \beta_{2T})\frac{\mu_R}{R_n} \tag{9.40a}$$

$$\gamma_i = (1 + \alpha_{Si} V_{Si} \beta_{2T})\frac{\mu_{Si}}{S_{ni}} \tag{9.40b}$$

The variation of the target second-moment reliability index β_{2T} is shown in Figure 9.3 with respect to the target reliability index β_T. From Figure 9.3, one can observe that β_{2T} is larger than β_T for negative α_{3G} and smaller for positive α_{3G}. For $\alpha_{3G} = 0$ and $\beta_{2T} = \beta_T$, Eq. (9.38) becomes exactly the same as Eq. (9.4), and the load and resistance factors can be determined using Eqs. (9.11a) and (9.11b).

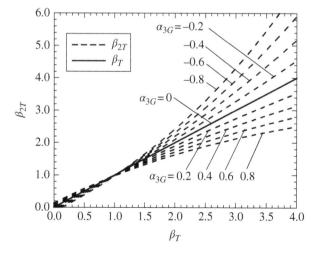

Figure 9.3 Target second-moment reliability index.

It should be noted that the load and resistance factors based on the third-moment method could also be achieved by using the other format of third-moment reliability indices in Eq. (9.36), of which details can be found in Zhao and Lu (2006) and Lu et al. (2010).

9.3.2 Estimation of the Mean Value of Resistance

9.3.2.1 Iteration Computation

Since the load and resistance factors are determined when the reliability index is equal to the target reliability index, the mean value of resistance should also be determined under this condition (hereafter referred to as the *target mean resistance*). The target mean resistance is computed using the following (Takada 2001).

$$\mu_{Rk} = \mu_{Rk-1} + (\beta_T - \beta_{k-1})\sigma_G \tag{9.41}$$

where μ_{Rk} and μ_{Rk-1} are values of the kth and $(k-1)$th iteration value for the mean value of resistance; and β_{k-1} is the value of the $(k-1)$th iteration for the third moment reliability index.

The procedure to determine the load and resistance factors using Eq. (9.41) are provided as the following:

1) Assume $\mu_{R0} = \Sigma \mu_{Si}$.
2) Calculate μ_G, σ_G, and α_{3G} using Eqs. (9.35a) to (9.35c) and determine β_{2M} with the aid of Eq. (9.35a).
3) Calculate β_{3M} using Eq. (9.34).
4) Calculate μ_{Rk} using Eq. (9.41).
5) Repeat the computation process of steps (2)–(4) until $|\beta_T - \beta_{k-1}| < 0.0001$, and then the target mean resistance is determined.
6) Determine the load and resistance factors using Eqs. (9.40a) and (9.40b).

Example 9.4 Consider the limit state function in Example 9.3, determine the load and resistance factors using the third-moment method to achieve a reliability of $\beta = 2.0$.

Step 1; assume the initial mean value of resistance.

$$\mu_{Mp0} = \Sigma \mu_{MSi} = 1.0\mu_{M_D} + 0.3\mu_{M_D} + 2.595\mu_{M_D} = 3.895\mu_{M_D}, \mu_{M_D0} = (\mu_D l^2)/16$$

Step 2; calculate μ_G, σ_G, and α_{3G} using Eqs. (9.35a) to (9.35c) and determine β_{2M} using Eq. (9.35a).

$$\mu_G = \mu_R - \Sigma \mu_{Si} = \mu_{Mp0} - \Sigma \mu_{MSi} = 0, \sigma_G = \sqrt{\sigma_R^2 + \Sigma \sigma_{Si}^2} = 0.607\mu_{M_D}$$

$$\alpha_{3G} = \frac{1}{\sigma_G^3}\left(\alpha_{3R}\sigma_R^3 - \Sigma \alpha_{3i}\sigma_{Si}^3\right) = -0.360, \beta_{2M} = \frac{\mu_G}{\sigma_G} = \frac{\mu_R - \Sigma \mu_{Si}}{\sigma_G} = 0$$

Table 9.4 Iteration computation of Example 9.4.

Iteration No.	$\mu_{Mp\,(k)}$	μ_G	σ_G	β_{2M}	α_{3G}	β_{3M}	$\mu_{Mp\,(k+1)}$	ε
1	$3.895\,\mu_{M_D}$	0	$0.607\,\mu_{M_D}$	0	-0.360	0.060	$5.073\,\mu_{M_D}$	1.940
2	$5.073\,\mu_{M_D}$	$1.178\,\mu_{M_D}$	$0.689\,\mu_{M_D}$	1.710	-0.181	1.658	$5.308\,\mu_{M_D}$	0.342
3	$5.308\,\mu_{M_D}$	$1.413\,\mu_{M_D}$	$0.706\,\mu_{M_D}$	2.001	-0.152	1.932	$5.356\,\mu_{M_D}$	0.068
4	$5.356\,\mu_{M_D}$	$1.461\,\mu_{M_D}$	$0.710\,\mu_{M_D}$	2.059	-0.146	1.987	$5.366\,\mu_{M_D}$	0.013
5	$5.366\,\mu_{M_D}$	$1.471\,\mu_{M_D}$	$0.710\,\mu_{M_D}$	2.07	-0.145	1.997	$5.367\,\mu_{M_D}$	0.003
6	$5.367\,\mu_{M_D}$	$1.472\,\mu_{M_D}$	$0.711\,\mu_{M_D}$	2.073	-0.144	2.000	$5.368\,\mu_{M_D}$	5×10^{-4}

Step 3; calculate β_{3M} using Eq. (9.34).

$$\beta_{3M0} = -\frac{\alpha_{3G}}{6} - \frac{3}{\alpha_{3G}}\ln\left(1 - \frac{1}{3}\alpha_{3G}\beta_{2M}\right) = 0.06003$$

Step 4; calculate μ_{Mp1} using Eq. (9.41).

$$\mu_{Mp1} = \mu_{Mp0} + (\beta_T - \beta_{3M})\sigma_G = 5.073$$

Step 5; the difference between β_T and β_{3M} is given as

$$\varepsilon = |\beta_T - \beta_{3M}| = |2.0 - 0.06003| = 1.93997 > 0.0001$$

Since the difference ε is too large, the above computation process should be repeated until $|\beta_T - \beta_{k-1}| < 0.001$. The entire iterations are summarised in Table 9.4.

One can observe that the convergence is achieved after six iteration computations. The target mean resistance obtained by the third-moment method is $\mu_{MpT} = 5.368\mu_{M_D}$.

Calculate α_{Mp} and α_{MSi} using Eq. (9.12a), and determine β_{2T} with the aid of Eq. (9.39).

$$\alpha_{Mp} = \sigma_{Mp}/\sigma_G = 0.755, \alpha_{M_D} = \sigma_{M_D}/\sigma_G = 0.141, \alpha_{M_L} = \sigma_{M_L}/\sigma_G = 0.169$$

$$\alpha_{MS50} = \sigma_{MS50}/\sigma_G = 0.617, \quad \beta_{2T} = \frac{3}{\alpha_{3G}}\left\{1 - \exp\left[\frac{\alpha_{3G}}{3}\left(-\beta_T - \frac{\alpha_{3G}}{6}\right)\right]\right\} = 2.073$$

Step 6; determine the load and resistance factors using Eqs. (9.40a) and (9.40b).

$$\phi = \mu_{M_P}(1 - \alpha_{M_P}V_{M_P}\beta_{2T})/R_n = 0.843$$
$$\gamma_{M_D} = \mu_{M_D}(1 + \alpha_{M_D}V_{M_D}\beta_{2T})/D_n = 1.029$$
$$\gamma_{M_L} = \mu_{M_L}(1 + \alpha_{M_L}V_{M_L}\beta_{2T})/L_n = 0.513$$
$$\gamma_{MS50} = \mu_{MS50}(1 + \alpha_{MS50}V_{MS50}\beta_{2T})/S_n = 1.182$$

The LRFD format and the target mean resistances using the third-moment method are obtained as

$$0.84 M_{Pn} \geq 1.03 M_{Dn} + 0.51 M_{Ln} + 1.18 M_{Sn}$$

$$\mu_{Mp} \geq 5.37 \mu_{MD}$$

where $M_{Dn} = (D_n l^2)/16$, $M_{Ln} = (L_n l^2)/16$, and $M_{Sn} = (S_n l^2)/16$.

9.3.2.2 A Simple Formula for Approximating the Iteration Computation

The above determination of target mean resistance requires iterations and is inconvenient for users or designers, for whom the computation procedure should be as simple and accurate as possible. In the following, a simple formula is presented for approximating the iteration computation.

At the limit state, Eq. (9.35a) and Eq. (9.38) give

$$\mu_R = \Sigma \mu_{Si} + \beta_{2T} \sigma_G \tag{9.42}$$

For Eq. (9.42), μ_{Si} is known and μ_R is to be determined. The unknown values of σ_G and β_{2T} are the functions of μ_R. Thus, in order to obtain the target mean resistance, one has to assume an initial value of the mean resistance μ_{R0}. Note that Eq. (9.42) can be expressed as

$$\mu_R = \Sigma \mu_{Si} + \beta_{2T} \sigma_G = \Sigma \mu_{Si} + \frac{3}{\alpha_{3G}} \left\{ 1 - \exp\left[\frac{\alpha_{3G}}{3} \left(-\beta_T - \frac{\alpha_{3G}}{6} \right) \right] \right\}$$

$$\times \left(\sqrt{1 + (\mu_R V_R)^2 / \Sigma \sigma_{Si}^2} \right) \sqrt{\Sigma \sigma_{Si}^2} \tag{9.43}$$

Since Eq. (9.34) is derived for $|\alpha_{3G}| < 1$ (Zhao and Ang 2003), and μ_R will become larger as β_T increases, the following approximation can be derived through trial and error.

$$\frac{3}{\alpha_{3G}} \left\{ 1 - \exp\left[\frac{\alpha_{3G}}{3} \left(-\beta_T - \frac{\alpha_{3G}}{6} \right) \right] \right\} \times \left(\sqrt{1 + (\mu_R V_R)^2 / \Sigma \sigma_{Si}^2} \right) \approx \sqrt{\beta_T^{3.5}} \tag{9.44}$$

Thus, the initial value μ_{R0} can be assumed as

$$\mu_{R_0} = \Sigma \mu_{Si} + \sqrt{\beta_T^{3.5} \Sigma \sigma_{Si}^2} \tag{9.45}$$

Through this discussion, a simple approximation formula for the iterative computation of the target mean resistance is expressed as

$$\mu_{RT} = \Sigma \mu_{Si} + \beta_{2T_0} \sigma_{G_0} \tag{9.46}$$

where μ_{RT} = the target mean resistance; σ_{G0} = the standard deviation of $G(\mathbf{X})$; and β_{2T0} = the target second-moment reliability index, are obtained using μ_{R0}.

The procedure to determine the load and resistance factors using the present simple approximation are as follows:

1) Calculate μ_{R0} using the Eq. (9.45).
2) Calculate σ_{G0}, α_{3G0}, and β_{2T0} using Eqs. (9.35b) and (9.35c), Eq. (9.39), respectively, and determine μ_{RT} with the aid of Eq. (9.46).
3) Calculate σ_G, α_{3G}, and β_{2T} using Eqs. (9.35b) and (9.35c), and Eq. (9.39), respectively, and calculate α_R and α_{Si} with the aid of Eq. (9.12a).
4) Determine the load and resistance factors using Eqs. (9.40a) and (9.40b).

Example 9.5 Consider the limit state function in Example 9.3 and apply the third-moment method to determine the target mean resistance using simple formulas, and determine the load and resistance factors to achieve a target reliability of $\beta_T = 2.0$.

According to Eq. (9.45)

$$\mu_{Mp0} = \Sigma \mu_{M_{Si}} + \sqrt{\beta_T^{3.5} \Sigma \sigma_{M_{Si}}^2} = 5.461 \mu_{M_D}, \mu_{Mp0} = (\mu_D l^2)/16$$

σ_{G0}, α_{3G0}, and β_{2T0} can be obtained using Eq. (9.35b), Eq. (9.35c), and Eq. (9.39), respectively,

$$\sigma_{G0} = \sqrt{\sigma_{M_{p0}}^2 + \Sigma \sigma_{M_{Si}}^2} = 0.718 \mu_{M_D}, \alpha_{3G0} = \left(\alpha_{3M_p} \sigma_{M_{p0}}^3 - \Sigma \alpha_{3MS_i} \sigma_{MS_i}^3\right)/\sigma_{G0}^3 = -0.133$$

$$\beta_{2T0} = \frac{3}{\alpha_{3G0}} \left\{ 1 - \exp\left[\frac{\alpha_{3G0}}{3}\left(-\beta_T - \frac{\alpha_{3G0}}{6}\right)\right] \right\} = 2.067$$

The target mean resistance μ_{MpT} can be estimated with the aid of Eq. (9.46)

$$\mu_{MpT} = \Sigma \mu_{M_{Si}} + \beta_{2T_0} \sigma_{G_0} = 5.379 \mu_{M_D}$$

Then σ_G, α_{3G}, and β_{2T} can be obtained as

$$\sigma_G = 0.711 \mu_{M_D}, \alpha_{3G} = -0.143, \beta_{2T} = 2.072$$

Considering $-1.0 < \alpha_{3G} = -0.143 < 0.386$, it is in the applicable range of the third-moment method.

Calculate α_{Mp} and α_{MSi} with the aid of Eq. (9.12a)

$$\alpha_{M_p} = \sigma_{M_p}/\sigma_G = 0.756, \alpha_{M_D} = \sigma_{M_D}/\sigma_G = 0.141$$

$$\alpha_{M_L} = \sigma_{M_L}/\sigma_G = 0.169, \alpha_{M_{S50}} = \sigma_{M_{S50}}/\sigma_G = 0.617$$

Determine the load and resistance factors using Eqs. (9.40a) and (9.40b)

$$\phi = \mu_{M_p}\left(1 - \alpha_{M_p} V_{M_p} \beta_{2T}\right)/R_n = 0.843, \gamma_{M_D} = \mu_{M_D}(1 + \alpha_{M_D} V_{M_D} \beta_{2T})/D_n = 1.029$$

$$\gamma_{ML} = \mu_{M_L}(1 + \alpha_{M_L} V_{M_L} \beta_{2T})/L_n = 0.513, \gamma_{M_{S50}} = \mu_{M_{S50}}(1 + \alpha_{M_{S50}} V_{M_{S50}} \beta_{2T})/S_n = 1.182$$

The LRFD format and the target mean resistance for this example using the third-moment method are obtained as

$$0.84 M_{Pn} \geq 1.03 M_{Dn} + 0.51 M_{Ln} + 1.18 M_{Sn}$$

$$\mu_{Mp} \geq 5.37 \mu_{M_D}$$

where $M_{Dn} = (D_n l^2)/16$, $M_{Ln} = (L_n l^2)/16$, and $M_{Sn} = (S_n l^2)/16$.

For the same problem using FORM, the LRFD format and the target mean resistance are

$$0.89M_{Pn} \geq 1.02M_{Dn} + 0.46M_{Ln} + 1.28M_{Sn}$$

$$\mu_{Mp} \geq 5.31\mu_{M_D}$$

The LRFD format for this example using the practical method (Mori, 2002; AIJ, 2002) is

$$0.88M_{Pn} \geq 1.02M_{Dn} + 0.49M_{Ln} + 1.28M_{Sn}$$

$$\mu_{Mp} \geq 5.40\mu_{M_D}$$

From this example, one can see that although the load and resistance factors obtained using the third-moment method are different from those obtained using FORM, the designed resistances under specific design conditions of the third-moment method are quite close to those of FORM.

Example 9.6 Consider the following performance function

$$G(X) = R - (D + L + E) \tag{9.47}$$

where R is the resistance; D denotes the dead load effect; L denotes the live load effect, and E is the maximum earthquake load effect during 50 years.

The CDFs of D, L, and E are unknown, the known information are listed in Table 9.5.

Determine the load and resistance factors for the performance function of Eq. (9.47), in order to achieve a reliability of $\beta = 2.4$.

According to Eq. (9.45)

$$\mu_{R0} = \Sigma\mu_{Si} + \sqrt{\beta_T^{3.5}\Sigma\sigma_{Si}^2} = 36.598\mu_D$$

σ_{G0}, α_{3G0}, and β_{2T0} can be obtained using Eqs. (9.35b) and (9.35c), and Eq. (9.39), respectively,

$$\sigma_{G0} = \sqrt{\sigma_R^2 + \Sigma\sigma_{Si}^2} = 12.761\mu_D, \quad \alpha_{3G0} = \left(\alpha_{3R}\sigma_{R0}^3 - \Sigma\alpha_{3Si}\sigma_{Si}^3\right)/\sigma_{G0}^3 = -0.215$$

$$\beta_{2T0} = \frac{3}{\alpha_{3G0}}\left\{1 - \exp\left[\frac{\alpha_{3G0}}{3}\left(-\beta_T - \frac{\alpha_{3G0}}{6}\right)\right]\right\} = 2.577$$

Table 9.5 Basic random variables for Example 9.6.

R or S_i	CDFs	μ_i/S_{ni} or μ_R/R_n	Mean	V_i	α_{3i}
R	Unknown	1.10	μ_R	0.3	0.927
D	Unknown	1.0	μ_D	0.1	0.0
L	Unknown	0.45	$0.5\mu_D$	0.4	1.264
E	Unknown	0.16	$5\mu_{Mp}$	1.3	6.097

The target mean resistance μ_{RT} can be estimated with the aid of Eq. (9.46)

$$\mu_{RT} = \Sigma\mu_{Si} + \beta_{2T_0}\sigma_{G_0} = 39.380\mu_D$$

Then σ_G, α_{3G}, and β_{2T} can be obtained as

$$\sigma_G = 13.486\mu_D, \alpha_{3G} = -0.059, \beta_{2T} = 2.448$$

Considering $-0.982 < \alpha_{3G} = -0.059 < 0.327$, it is in the applicable range of the third-moment method.

Calculate α_R and α_{Si} with the aid of Eq. (9.12a)

$$\alpha_R = \sigma_R/\sigma_G = 0.876, \alpha_D = \sigma_D/\sigma_G = 0.007$$
$$\alpha_L = \sigma_L/\sigma_G = 0.015, \alpha_E = \sigma_E/\sigma_G = 0.482$$

Determine the load and resistance factors using Eqs. (9.40a-b)

$$\phi = \mu_R(1 - \alpha_R V_R \beta_{2T})/R_n = 0.392, \gamma_D = \mu_D(1 + \alpha_D V_D \beta_{2T})/D_n = 1.002$$
$$\gamma_L = \mu_L(1 + \alpha_L V_L \beta_{2T})/L_n = 0.457, \gamma_E = \mu_E(1 + \alpha_E V_E \beta_{2T})/S_n = 0.405$$

The LRFD format and the target mean resistance for this example using the third-moment are obtained as

$$0.39R_n \geq 1.0D_n + 0.46L_n + 0.41E_n$$

$$\mu_R \geq 39.74\mu_D$$

Example 9.7 Consider the following nonlinear performance function of the fully plastic flexural capacity of a steel beam section

$$G(X) = YZ - M \tag{9.48}$$

where Y = the yield strength of steel, a lognormal variable; Z = section modulus of the section, a lognormal variable; and M = the applied bending moment at the pertinent section, a Gumbel variable.

Determine the mean design section for the performance function of Eq. (9.48), in order to achieve a target reliability of $\beta_T = 2.5$.

As this is a design problem, the purpose is to determine the appropriate μ_Z for any given μ_M to satisfy the required reliability. With $\mu_Y = 40$ ksi, the coefficients of variation are $V_Y = 0.125$, $V_Z = 0.05$, and $V_M = 0.20$ for Y, Z, and M, respectively.

First calculate the value of μ_{Z_0}

$$\mu_G = \mu_Y\mu_{Z_0} - \mu_M = \sqrt{\beta_T^{3.5}\sigma_M^2}, \mu_{Z_0} = \left(\sqrt{\beta_T^{3.5}\sigma_M^2} + \mu_M\right)/\mu_Y = 4.985 \times 10^{-2}\mu_M$$

Let $R = YZ$, then

$$\sigma_{R_0} = \sigma_{YZ_0} = \sqrt{\left(\mu_Y\mu_{Z_0}\right)^2\left[\left(1 + V_Y^2\right)\left(1 + V_Z^2\right) - 1\right]} = 0.269\mu_M$$

Therefore

$$\sigma_{G_0} = \sqrt{\sigma_{R0}^2 + \sigma_M^2} = 0.335\mu_M$$

The skewness of Y, Z, and M are readily obtained as

$$\alpha_{3Y} = 0.377, \alpha_{3Z} = 0.150, \alpha_{3M} = 1.14$$

The skewness of R can be obtained by

$$\alpha_{3R_0} = \alpha_{3YZ_0} = \left[\left(\alpha_{3Y}V_Y^3 + 3V_Y^2 + 1\right)\left(\alpha_{3Z}V_Z^3 + 3V_Z^2 + 1\right)\right.$$
$$\left. - 3\left(V_Y^2 + 1\right)\left(V_{Z_0}^2 + 1\right) + 2\right]/V_{YZ_0}^3 = 0.4068$$

Thus

$$\alpha_{3G_0} = \left[\alpha_{3R_0}\sigma_{R_0}^3 - \alpha_{3M}\sigma_M^3\right]/\sigma_{G_0}^3 = -0.0324$$

At the limit-state, the appropriate μ_{ZT} is obtained as

$$\mu_{ZT} = \left\{\mu_M + \left[\frac{3}{\alpha_{3G0}}\left\{1 - \exp\left[\frac{\alpha_{3G0}}{3}\left(-\beta_T - \frac{\alpha_{3G0}}{6}\right)\right]\right\}\right]\sigma_{G_0}\right\}/\mu_Y = 0.0462\mu_M$$

At the limit state, the design result of μ_{ZT} using FORM is obtained as $\mu_{ZT} = 0.0466\,\mu$. The relative difference of the results between the two methods is less than 1.0%.

From the above numerical examples, one can see that the third-moment method requires neither the iterative computation of derivatives, nor any design points. The designers or users can easily perform the reliability-based design with the aid of the third-moment method. Apparently, if the first three moments of the basic random variables are known, and using the third-moment method, the reliability-based design can be accomplished even when the probability distributions of the basic random variables are unknown.

9.4 General Expressions of Load and Resistance Factors Using Methods of Moment

The LRF formula we discussed can be extended to a general case of using the first several moments of $Z = G(\mathbf{X})$. Let

$$Z_S = \frac{Z - \mu_G}{\sigma_G}$$

Suppose that the relationship between the above standardised variable Z_S and the standard normal variable U can be expressed as the following functions of the first several moments of $Z = G(\mathbf{X})$,

$$U = S(Z_S, \mathbf{M}) \tag{9.49a}$$

$$Z_S = S^{-1}(U, \mathbf{M}) \tag{9.49b}$$

where $\mathbf{M}$ is a vector denoting the first several moments of $Z = G(\mathbf{X})$ and S^{-1} is the inverse function of S.

According to Eq. (4.12) in the chapter on Direct Methods of Moment, the moment-based reliability index is expressed as

$$\beta = -\Phi^{-1}(P_F) = -S(-\beta_{2M}, \mathbf{M}) \tag{9.50}$$

where β_{2M} is the second-moment reliability index.

Substituting Eq. (9.50) into the design format in Eq. (9.3) yields

$$\beta = -S(-\beta_{2M}, \mathbf{M}) \geq \beta_T \tag{9.51a}$$

From which

$$\beta_{2M} \geq -S^{-1}(-\beta_T, \mathbf{M}) \tag{9.51b}$$

Denoting the right side of Eq. (9.51b) as β_{2T}, one obtains

$$\beta_{2M} \geq \beta_{2T} \tag{9.52a}$$

$$\beta_{2T} = -S^{-1}(-\beta_T, \mathbf{M}) \tag{9.52b}$$

Equation (9.52a) is the same as Eq. (9.3). This implies that if the second-moment reliability index β_{2M} is equal to or larger than β_{2T}, the reliability index β will greater than or equal to the target reliability index β_T, and thus satisfy the required reliability. Therefore, β_{2T} is denoted as the target second-moment reliability index.

If the relationship between the standardised variable Z_S and the standard normal variable U is expressed as follows

$$Z_S = \sum_{i=0}^{n} a_i U^i \tag{9.53}$$

Then, β_{2T} can be expressed as

$$\beta_{2T} = -\sum_{i=0}^{n} a_i(-\beta_T)^i \tag{9.54}$$

Since Eq. (9.52a) is similar to Eq. (9.4) except that the right side is β_{2T}, the load and resistance factors corresponding to Eq. (9.52a) can be easily obtained by substituting β_T into the right side of Eq. (9.4) with β_{2T}. The design formula then becomes

$$\mu_R(1 - \alpha_R V_R \beta_{2T}) \geq \Sigma \mu_{Si}(1 + \alpha_{Si} V_{Si} \beta_{2T}) \tag{9.55}$$

and the load and resistance factors are obtained as

$$\phi = (1 - \alpha_R V_R \beta_{2T}) \frac{\mu_R}{R_n} \tag{9.56a}$$

$$\gamma_i = (1 + \alpha_{Si} V_{Si} \beta_{2T}) \frac{\mu_{Si}}{S_{ni}} \tag{9.56b}$$

where α_R and α_{Si} are the same as Eq. (9.7); V_R and $V_{Si} =$ the coefficient of variation, respectively, of R and S_i; α_R and $\alpha_{Si} =$ the sensitivity factors, respectively, for R and S_i; and $\beta_{2T} =$ the target second moment reliability index calculated from Eq. (9.54).

Since the formula is based on the first few moments of the load and the resistances, the LRFs can be determined even when the distributions of the random variables are unknown.

9.5 Determination of Load and Resistance Factors Using Fourth-Moment Method

9.5.1 Basic Formulas

The standardised variable Z_S can be expressed as a polynomial function of the standard normal variable U as following (Zhao and Lu 2007a)

$$Z_S = -l_1 + k_1 U + l_1 U^2 + k_2 U^3 \tag{9.57}$$

where the coefficients l_1, k_1, and k_2 are given as:

$$l_1 = \frac{\alpha_{3G}}{6(1 + 6l_2)}, l_2 = \frac{1}{36}\left(\sqrt{6\alpha_{4G} - 8\alpha_{3G}^2 - 14} - 2\right) \tag{9.58}$$

$$k_1 = \frac{1 - 3l_2}{(1 + l_1^2 - l_2^2)}, k_2 = \frac{l_2}{(1 + l_1^2 + 12l_2^2)} \tag{9.59}$$

where α_{4G} = the 4th dimensionless central moment, i.e., the kurtosis of $G(X)$, which is calculated from

$$\alpha_{4G} = \frac{1}{\sigma_G^4}\left(\alpha_{4R}\sigma_R^4 + 6\sigma_R^2\Sigma_{i=1}^{n}\sigma_{Si}^2 + \Sigma_{i=1}^{n}\alpha_{4Si}\sigma_{Si}^4 + 6\Sigma_{i=1}^{n-1}\Sigma_{j>i}^{n}\sigma_{Si}^2\sigma_{Sj}^2\right) \tag{9.60}$$

where α_{4R} and α_{4Si} are the kurtosis of R and S_i, respectively.

In this section, Eq. (9.57) is used to obtain the target second-moment reliability index, which is given as

$$\beta_{2T} = l_1 + k_1\beta_T - l_1\beta_T^2 + k_2\beta_T^3 \tag{9.61}$$

Especially when $\alpha_{4G} = 3$ and α_{3G} is small enough, one has $l_2 = k_2 = 0$, $k_1 = 1$, $l_1 = \alpha_{3G}/6$. Eq. (9.61) becomes

$$\beta_{2T} = \beta_T - \frac{1}{6}\alpha_{3G}\left(\beta_T^2 - 1\right) \tag{9.62}$$

which is essentially the same as the target second-moment reliability index obtained by the third-moment method.

For $\alpha_{4G} = 3$ and $\alpha_{3G} = 0$, Eq. (9.61) becomes $\beta_{2T} = \beta_T$, which is exactly the same as Eq. (9.4), and the load and resistance factors can be determined using Eqs. (9.9a and b).

The variation of the target second moment reliability index β_{2T} is shown in Figure 9.4a with respect to the target reliability index β_T for the case of $\alpha_{3G} = 0$, and in Figure 9.4b–d for the cases of $\alpha_{4G} = 2.8$, $\alpha_{4G} = 3.0$ and $\alpha_{4G} = 3.2$, respectively. It can be observed that compared with β_T, β_{2T} is generally larger for negative α_{3G} and smaller for positive α_{3G}. One can also observe that β_{2T} is generally larger than β_T for $\alpha_{4G} > 3.0$ but smaller for positive $\alpha_{4G} < 3.0$.

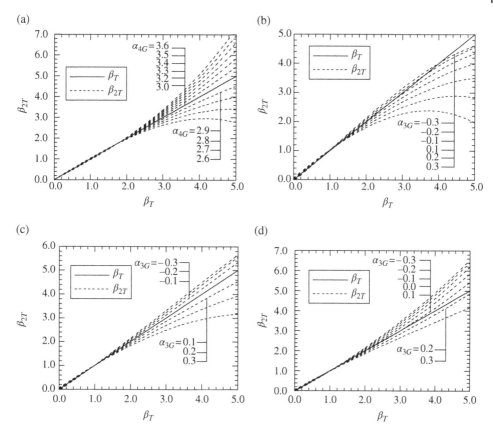

Figure 9.4 Variation of target second-moment reliability index with respect to α_{3G} and α_{4G}. (a) $\alpha_{3G} = 0$. (b) $\alpha_{4G} = 2.8$. (c) $\alpha_{4G} = 3.0$. (d) $\alpha_{4G} = 3.2$.

9.5.2 Determination of the Mean Value of Resistance

9.5.2.1 The Iteration Method

Since the load and resistance factors are determined when the reliability index is equal to the target reliability index, the mean value of the resistance should be determined under this condition (hereafter referred to as the *target mean resistance*). Generally, the target mean resistance is computed using the following iteration equation.

$$\mu_{Rk} = \mu_{Rk-1} + (\beta_T - \beta_{k-1})\sigma_G \tag{9.63}$$

where μ_{Rk} and μ_{Rk-1} are the values of the kth and (k–1)th iteration for the mean value of resistance; and β_{k-1} is the value of the (k–1)th iteration for the third- or fourth-moment reliability index.

9.5.2.2 Simple Formulae for the Target Mean Resistance

The following simple formula is utilised to avoid the iterative computations of the target mean resistance

$$\mu_{RT} = \Sigma\mu_{Si} + \beta_{2T_0}\sigma_{G_0} \qquad (9.64)$$

where μ_{RT} = the target mean resistance; σ_{G0} = the standard deviation of $G(\mathbf{X})$; and β_{2T0} = the target second-moment reliability index, which are obtained using μ_{R0}.

μ_{R0} is given by the following equation, which is obtained from trial and error (Lu et al. 2010)

$$\mu_{R_0} = \Sigma\mu_{Si} + \sqrt{\beta_T^{3.3}\Sigma\sigma_{Si}^2} \qquad (9.65)$$

The steps for determining the load and resistance factors using the fourth-moment method are as follows:

1) Calculate μ_{R0} using the Eq. (9.65).
2) Calculate σ_{G0}, α_{3G0}, α_{4G0}, and β_{2T0} using Eqs. (9.35b) and (9.35c), Eq. (9.60), and Eq. (9.61), respectively, and determine μ_{RT} with the aid of Eq. (9.64).
3) Calculate σ_G, α_{3G}, α_{4G}, and β_{2T} using Eqs. (9.35b) and (9.35c), Eq. (9.60), and Eq. (9.61), respectively, and calculate α_R and α_{Si} with the aid of Eq. (9.7).
4) Determine the load and resistance factors using Eqs. (9.56a) and (9.56b).

9.5.2.3 The Efficiency of the Simple Formula

In order to investigate the efficiency of the simple formula for target mean resistance, consider the following performance function

$$G(\mathbf{X}) = R - (D + L + S) \qquad (9.66)$$

where R = resistance, with unknown PDF and $\mu_R/R_n = 1.1$, $V = 0.15$, $\alpha_{3R} = 0.453$, $\alpha_{4R} = 3.368$;

D = dead load, with unknown PDF and $\mu_D/D_n = 1$, $V = 0.1$, $\alpha_{3D} = 0.0$, $\alpha_{4D} = 3.0$;
L = live load, with unknown PDF and $\mu_L/L_n = 0.45$, $V = 0.4$, $\alpha_{3L} = 1.264$, $\alpha_{4L} = 5.969$; and
S = snow load, with unknown PDF and $\mu_S/S_n = 0.47$, $V = 0.25$, $\alpha_{3S} = 1.140$, $\alpha_{4S} = 5.4$.

Consider the mean value of D, L with $\mu_D = 1.0$, $\mu_L/\mu_D = 0.5$, the load and resistance factors obtained using the simple formula are illustrated in Figure 9.5a–c, compared with the corresponding factors obtained using iterative computations from the fourth-moment method for $\beta_T = 2.0$, 3.0, and 4.0. The target mean resistances obtained using the simple formula and those obtained using iterative calculations are illustrated in Figure 9.5d. One can observe from Figure 9.5 that the load and resistance factors and the target mean resistances obtained by the two methods are essentially the same for a given target reliability index.

Example 9.8 Consider the following performance function

$$G(\mathbf{X}) = R - (D + L) \qquad (9.67)$$

where R = resistance, a lognormal variable with $\mu_R/R_n = 1.1$, COV = 0.15;

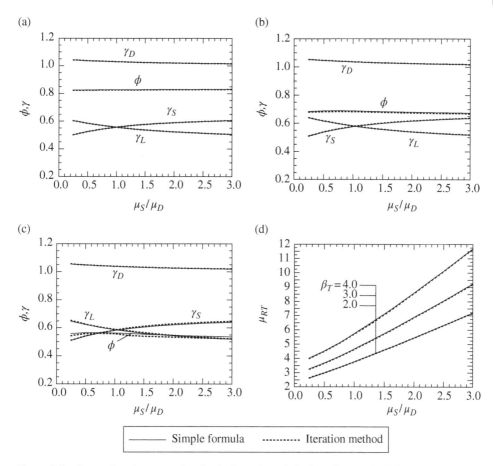

Figure 9.5 Comparison between the simple formula and the iteration method (fourth-moment method). (a) Load and resistance factors (β_T = 2.0); (b) Load and resistance factors (β_T = 3.0). (c) Load and resistance factors (β_T = 4.0). (d) Target mean resistances.

D = dead load, a normal variable with $\mu_D/D_n = 1.0$, COV = 0.1; and L = live load, a Weibull variable with $\mu_L/L_n = 0.45$, COV = 0.4.

The skewness are 0.453, 0, and 0.2768 for R, D, and L, respectively, and the kurtosis are 3.368, 3, and 2.78 for R, D, and L, respectively.

The load and resistance factors obtained using the methods of moments are illustrated in Figure 9.6a,b for $\beta_T = 2$ and $\beta_T = 3$, respectively, in comparison with corresponding factors obtained using FORM. The target mean resistances obtained using the methods of moments and those obtained by FORM are illustrated in Figure 9.7a,b for $\beta_T = 2$ and $\beta_T = 3$, respectively. From Figures 9.6 and 9.7, one can observe that although the load and resistance factors obtained by methods of moment are different from those obtained by FORM, the target mean resistance are essentially the same as those obtained by FORM. In other words,

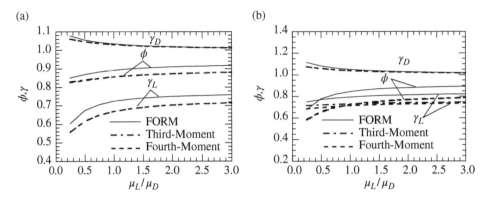

Figure 9.6 Load and resistance factors for Example 9.8. (a) $\beta_T = 2$. (b) $\beta_T = 3$.

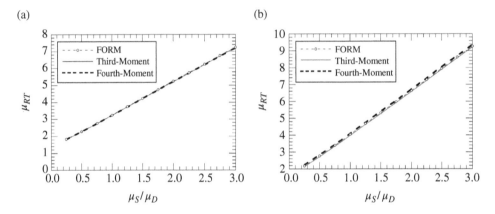

Figure 9.7 Target mean resistance for Example 9.8. (a) $\beta_T = 2$. (b) $\beta_T = 3$.

FORM and methods of moment give the same design results even when the load and resistance factors of the methods are different.

Example 9.9 Consider the following performance function

$$G(\mathbf{X}) = R - (D + L + S) \tag{9.68}$$

where R = resistance with unknown CDF and $\mu_R/R_n = 1.1$, COV = 0.15, $\alpha_{3R} = 0.453$, $\alpha_{4R} = 3.368$; D = dead load, a normal variable with $\mu_D/D_n = 1.0$, COV = 0.1, $\alpha_{3D} = 0$, $\alpha_{4D} = 3$; L = live load with unknown CDF and $\mu_L/L_n = 0.45$, $\mu_L/\mu_D = 0.5$, COV = 0.4, $\alpha_{3L} = 1.264$, $\alpha_{4L} = 5.969$; and S = snow load which is the main load, a Gumbel variable with $\mu_S/S_n = 0.45$, COV − 0.4, $\alpha_{3S} = 1.14$, $\alpha_{4S} = 5.4$.

Since the CDFs of R and L are unknown, the FORM is not applicable. Here, the LRFs are obtained using the methods of moment.

The load and resistance factors obtained using the methods of moments are illustrated in Figure 9.8a,b for $\beta_T = 2$ and $\beta_T = 3$, respectively. The target mean resistances obtained using the simple formula are illustrated in Figure 9.9a,b for $\beta_T = 2$ and $\beta_T = 3$, respectively. From Figures 9.8 and 9.9, one can observe that the results of the load and resistance factors and the target mean resistances have good agreements for the third- and fourth-moment methods.

From Figure 9.8a, one can observe that all results of the load and resistance factors obtained by the third- and fourth-moment methods have good agreements for $\beta_T = 2$. From Figure 9.8b, one can observe that all results of the load factors obtained by the third- and fourth-moment methods have good agreements, but the resistance factors obtained by the

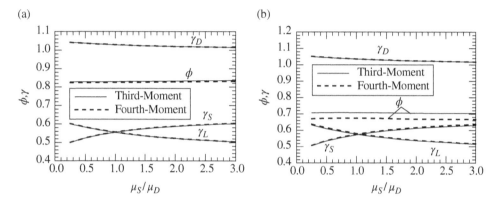

Figure 9.8 Load and resistance factors for Example 9.9. (a) $\beta_T = 2$. (b) $\beta_T = 3$.

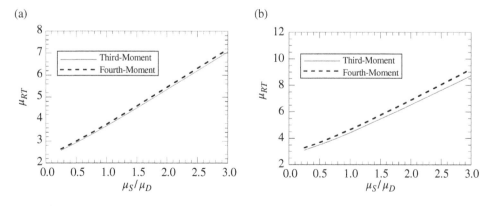

Figure 9.9 Target mean resistance for Example 9.9. (a) $\beta_T = 2$. (b) $\beta_T = 3$.

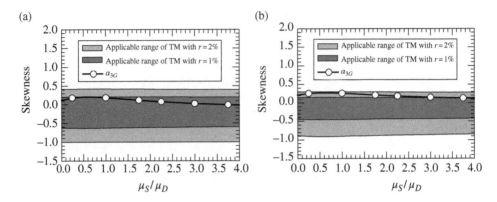

Figure 9.10 Investigation on the application of third-moment method. (a) β_T = 2. (b) β_T = 3.

two methods have visible difference for $\beta_T = 3$. From Figure 9.9, one can observe that the target mean resistances obtained by the third- and fourth-moment methods are almost the same for $\beta_T = 2$ and have visible differences for $\beta_T = 3$. This can be attributed to the fact that the third-moment method in the latter case produces relatively larger error as only the first three moments of the performance function are used. The skewness for the cases above are depicted in Figure 9.10a,b, for $\beta_T = 2$ and $\beta_T = 3$, respectively. In the same figures, the application range of third-moment method in terms of skewness are also presented using Eq. (4.26a) in the chapter on Direct Methods of Moment.

From Figure 9.10a,b, one can observe that the skewness of the performance function for $\beta_T = 2$ is within the application range of third-moment method with $r = 1\%$ while that for $\beta_T = 3$ is beyond. This explains why there are visible differences between the target mean resistance obtained by the third- and the fourth-moment methods.

9.6 Summary

This chapter presented a simple method to determine the load and resistance factors using the methods of moment. Derivative-based iteration, which is necessary in FORM, is no longer required. For this reason, the methods of moment are easier to apply. Although the load and resistance factors obtained by the methods of moment are different from those obtained by FORM for some cases, the target mean resistance obtained by both methods are essentially the same. Furthermore, since the methods of moment are based on the first few moments of the basic variables of loads and the resistance, instead of their CDF/PDFs, the LRFs can be determined even when the distributions of the random variables are unknown.

10

Methods of Moment for Time-Variant Reliability

10.1 Introduction

Until now, all the contents in previous chapters are related to time-invariant problems in structural reliability. However, in practical engineering there are many issues that are time-variant, which generally include two kinds. One is dynamic reliability problem, in which the uncertainty in dynamic response has to be included, and the response is generally treated by a random process rather than a random variable. For convenience, the basic theory of random process is outlined in Appendix D. Another issue is dealing with the problems where the resistance and load effects are time-dependent.

In dynamic reliability problems, one main concern is to calculate the probabilities that a structural response exceeds the critical threshold levels in the given time interval $[0, T]$. It is considered that failure occurs when the critical threshold value is exceeded at its first time interval and this kind of analysis falls into the category of the first passage problem. This chapter will describe the application of methods of moment for estimating the first passage probability of structural responses.

Meanwhile, during the service life of civil infrastructures, their performance, functionality, and serviceability are essentially time-dependent reliability problems, as both structural resistance and load effect are changed with time. The moment-based approach is introduced to compute the time-dependent failure probability of structures with complicated, multi-dimensional and implicit performance functions considering the correlation among random variables.

10.2 Simulating Stationary Non-Gaussian Process Using the Fourth-Moment Transformation

10.2.1 Brief Review on Simulating Stationary Non-Gaussian Process

The response of structures under random excitation is difficult to determine analytically. As such, Monte Carlo simulation (MCS) is often applied, where the simulation of random excitations is the first step. Since the random excitations are usually non-Gaussian in practice, methods for simulating the stationary non-Gaussian processes have been developed by

Structural Reliability: Approaches from Perspectives of Statistical Moments, First Edition.
Yan-Gang Zhao and Zhao-Hui Lu.
© 2021 John Wiley & Sons Ltd. Published 2021 by John Wiley & Sons Ltd.

using spectral representation method (SRM, see Appendix D) (e.g. Shinozuka and Jan 1972; Shinozuka and Deodatis 1991) and transformation models from Gaussian to non-Gaussian processes (e.g. Grigoriu 1998; Shields and Deodatis 2013; Gurley et al. 1996; Yang and Gurley 2015). The transformation from Gaussian to non-Gaussian processes can be realised in two ways: one is to utilise the marginal probability density functions (MPDFs) and the correlation function matrix (CFM) or power spectral density matrix (PSDM) of target non-Gaussian processes (Grigoriu 1995, 1998; Yamazaki and Shinozuka 1988; Deodatis and Micaletti 2001; Bocchini and Deodatis 2008; Masters and Gurley 2003; Shields et al. 2011; Shields and Deodatis 2013); and the other is to utilise the first four moments and CFM or PSDM of the target non-Gaussian processes, where the Hermite polynomial model (HPM) is widely used (Gurley et al. 1996; Gurley 1997; Gurley and Kareem 1997; Gurley et al. 1997; Gurley and Kareem 1998a; Gurley and Kareem 1998b; Yang and Gurley 2015). This section will discuss the simulation using SRM and the first four moments and CFM or PSDM.

In essential, the HPM mentioned is essentially a fourth-moment transformation model. In the following sections, based on the fourth-moment transformation model, the marginal PDF of a non-Gaussian process represented by a function of PDF of a standard normal distribution and the complete transformation model of CFM from non-Gaussian to Gaussian processes with clear applicable range are introduced. An efficient procedure for simulating stationary non-Gaussian processes is then presented by using SRM and the fourth-moment transformation model.

10.2.2 Transformation for Marginal Probability Distributions

Without loss of generality, a non-Gaussian process $X_i(t)$ can be standardised as

$$X_{is}(t) = \left[X_i(t) - \mu_{X_i}\right]/\sigma_{X_i} \tag{10.1}$$

where μ_{X_i} and σ_{X_i} are mean and standard deviation of $X_i(t)$; and $X_{is}(t)$ is the standardised non-Gaussian process with zero-mean and unit standard deviation.

According to the fourth-moment transformation model (Fleishman 1978), $X_{is}(t)$ can be expressed as

$$X_{is}(t) = S^{-1}[U_i(t)] = a_{1i} + a_{2i}U_i(t) + a_{3i}[U_i(t)]^2 + a_{4i}[U_i(t)]^3 \tag{10.2}$$

where $U_i(t)$ is the standard Gaussian random processes; $S^{-1}[\cdot]$ is the translation function; and the coefficients of a_{1i}, a_{2i}, a_{3i}, and a_{4i} can be determined by making the first four moments of the left side of Eq. (10.2) equal to those of the right side (see Section 7.4.3 in the chapter on Transformation of Non-Normal Variables to Independent Normal Variables).

Based on the translation process theory (Grigoriu 1984), the marginal cumulative distribution function (CDF) of $X_i(t)$ can be expressed as

$$F[X_i(t)] = \Phi[U_i(t)] \tag{10.3}$$

where $F(\cdot)$ and $\Phi(\cdot)$ are the marginal CDFs of $X_i(t)$ and $U_i(t)$, respectively.

Taking the derivative of the left and right sides of Eq. (10.3) with respect to $X_i(t)$, the marginal PDF of $X_i(t)$, $f(\cdot)$, is formulated as

Table 10.1 Expression of $U_i(t)$ in terms of $X_{is}(t)$.

Parameter			Range of $X_i(t)$	Normal transformation $U_i(t)$	Type
$a_{4i} < 0$			$J_{2i}^* < X_i(t) < J_{1i}^*$	$-2r_i \cos[(\theta_i + \pi)/3] - a_i/3$	I
$a_{4i} > 0$	$p_i < 0$	$a_{3X_i} \geq 0$	$J_{1i}^* < X_i(t) < J_{2i}^*$	$2r_i \cos(\theta_i/3) - a_i/3$	II
			$X_i(t) \geq J_{2i}^*$	$\sqrt[3]{A_i} + \sqrt[3]{B_i} - a_i/3$	
		$a_{3X_i} < 0$	$J_1^* < X_i(t) < J_{2i}^*$	$-2r_i \cos[(\theta_i - \pi)/3] - a_i/3$	III
			$X_i(t) \leq J_{1i}^*$	$\sqrt[3]{A_i} + \sqrt[3]{B_i} - a_i/3$	
	$p_i \geq 0$			$\sqrt[3]{A_i} + \sqrt[3]{B_i} - a_i/3$	IV
$a_{4i} = 0$		$a_{3X_i} \neq 0$	$a_{2i}^2 + 4a_{3i}[a_{3i} + X_{is}(t)] \geq 0$	$\left\{ -a_{2i} + \sqrt{a_{2i}^2 + 4a_{3i}[a_{3i} + X_{is}(t)]} \right\}/2a_{3i}$	V
		$a_{3X_i} = 0$		$X_{is}(t)$	VI

$$f[X_i(t)] = \frac{\phi[U_i(t)]}{\sigma_{X_i}\{3a_{4i}[U_i(t)]^2 + 2a_{3i}U_i(t) + a_{2i}\}} \tag{10.4}$$

where $U_i(t)$ is expressed in terms of $X_i(t)$ and is listed in Table 10.1, which is essentially the same as that in Table 7.9 in the chapter just mentioned.

The related parameters in Table 10.1 are given as follows

$$A_i = -\frac{q_i}{2} + \sqrt{\Delta_i}, B_i = -\frac{q_i}{2} - \sqrt{\Delta_i}, \theta_i = \arccos\left(\frac{-q_i}{2r_i^3}\right), r_i = \sqrt{-\frac{p_i}{3}}, a_i = \frac{a_{3i}}{a_{4i}}, c_i = \frac{a_{2i}}{a_{4i}} \tag{10.5}$$

$$\Delta_i = \left(\frac{p_i}{3}\right)^3 + \left(\frac{q_i}{2}\right)^2, p_i = \frac{3a_{2i}a_{4i} - a_{3i}^2}{3a_{4i}^2}, q_i = \frac{2}{27}a_i^3 - \frac{a_i c_i}{3} - a_i - \frac{X_{is}(t)}{a_{4i}} \tag{10.6}$$

$$J_{1i}^* = \sigma_{X_i} a_{4i}\left(-2r_i^3 + \frac{2}{27}a_i^3 - \frac{a_i c_i}{3} - a_i\right) + \mu_{X_i}, J_{2i}^* = \sigma_{X_i} a_{4i}\left(2r_i^3 + \frac{2}{27}a_i^3 - \frac{a_i c_i}{3} - a_i\right) + \mu_{X_i} \tag{10.7}$$

Since the derivation of the marginal probability distributions is based on the fourth-moment transformation model, it is referred to as the *Cubic normal probability model*. Studies have shown that the cubic normal distribution has rich flexibility and the whole applicable range of the cubic normal distribution covers a large area in the skewness–kurtosis plane (Zhao et al. 2018b). Therefore, it has the capacity to model the marginal PDFs of strongly non-Gaussian processes.

10.2.3 Transformation for Correlation Functions

Let $X_s(t) = \{X_{1s}(t), ..., X_{is}(t), ..., X_{js}(t), ..., X_{ns}(t)\}$ be a correlated multivariate stationary non-Gaussian vector process and $U(t) = \{U_1(t), ..., U_i(t), ..., U_j(t), ..., U_n(t)\}$ be a correlated multivariate stationary Gaussian vector process. The correlation function $R_{X_{is}X_{js}}(\tau)$ of the non-Gaussian processes $X_{is}(t)$ and $X_{js}(t+\tau)$ can be determined using (Grigoriu 1998):

$$R_{X_{is}X_{js}}(\tau) = E\big[X_{is}(t)\cdot X_{js}(t+\tau)\big] = E\big\{S^{-1}[U_i(t)]\cdot S^{-1}[U_j(t+\tau)]\big\}$$

$$= \int_{-\infty}^{\infty}\int_{-\infty}^{\infty} S^{-1}[u_i(t)]\cdot S^{-1}[u_j(t+\tau)]\cdot\Phi\Big[u_i, u_j; \rho_{U_iU_j}(\tau)\Big]\, du_i du_j \tag{10.8}$$

where $\Phi\Big[u_i, u_j; \rho_{U_iU_j}(\tau)\Big]$ is the joint PDF of a bivariate standard Gaussian vector process with Gaussian correlation coefficient function $\rho_{u_iu_j}(\tau)$ of $U_i(t)$ and $U_j(t+\tau)$ for time lag τ.

In non-Gaussian simulation, $R_{X_{is}X_{js}}(\tau)$ is prescribed and it is necessary to invert Eq. (10.8) to determine the underlying Gaussian correlation function $\rho_{U_iU_j}(\tau)$ using numerical iteration for fixed values of the time lag τ. Thus, the determination of $\rho_{U_iU_j}(\tau)$ is cumbersome. However, if the fourth-moment transformation model is used as the translation function, the Gaussian to non-Gaussian correlation relationship can be determined without numerical integration. For Gaussian processes $U_i(t)$ and $U_j(t+\tau)$ of $U(t)$ with $1 \le i, j \le n$, their corresponding non-Gaussian processes $X_{is}(t)$ and $X_{js}(t+\tau)$ can be expressed as:

$$X_{is}(t) = S^{-1}[U_i(t)] = a_{1i} + a_{2i}U_i(t) + a_{3i}[U_i(t)]^2 + a_{4i}[U_i(t)]^3 \tag{10.9a}$$

$$X_{js}(t+\tau) = S^{-1}[U_j(t+\tau)] = a_{1j} + a_{2j}U_j(t+\tau) + a_{3j}\big[U_j(t+\tau)\big]^2 + a_{4j}\big[U_j(t+\tau)\big]^3 \tag{10.9b}$$

Since each $X_{is}(t)$ and $X_{js}(t+\tau)$ is zero-mean and unit standard deviation, the correlation function $R_{X_{is}X_{js}}(\tau)$ is reduced to the non-Gaussian correlation coefficient function $\rho_{X_{is}X_{js}}(\tau)$:

$$\rho_{X_{is}X_{js}}(\tau) = E\big[X_{is}(t)\cdot X_{js}(t+\tau)\big] = (a_{1i}, a_{2i}, a_{3i}, a_{4i})\mathbf{R}(a_{1j}, a_{2j}, a_{3j}, a_{4j})^T \tag{10.10}$$

where

$$\mathbf{R} = E\Big\{\big\{1, U_i(t), [U_i(t)]^2, [U_i(t)]^3\big\}^T \cdot \big\{1, U_j(t+\tau), [U_j(t+\tau)]^2, [U_j(t+\tau)]^3\big\}\Big\}$$

$$= \begin{bmatrix} 1 & 0 & 1 & 0 \\ 0 & \rho_{U_iU_j}(\tau) & 0 & 3\rho_{U_iU_j}(\tau) \\ 1 & 0 & 2\big[\rho_{U_iU_j}(\tau)\big]^2 + 1 & 0 \\ 0 & 3\rho_{U_iU_j}(\tau) & 0 & 6\big[\rho_{U_iU_j}(\tau)\big]^3 + 9\rho_{U_iU_j}(\tau) \end{bmatrix} \tag{10.11}$$

Substituting Eq. (10.11) into Eq. (10.10) leads to:

$$\rho_{X_{is}X_{js}}(\tau) = (a_{2i}a_{2j} + 3a_{4i}a_{2j} + 3a_{2i}a_{4j} + 9a_{4i}a_{4j})\rho_{U_iU_j}(\tau) + 2a_{3i}a_{3j}\big[\rho_{U_iU_j}(\tau)\big]^2$$

$$+ 6a_{4i}a_{4j}\big[\rho_{U_iU_j}(\tau)\big]^3 \tag{10.12}$$

When $i = j$, $\rho_{X_{is}X_{js}}(\tau)$ reduces to auto-correlation coefficient function $\rho_{X_{is}X_{is}}(\tau)$ of $X_{is}(t)$ and is one of the diagonal terms of the correlation coefficient function matrix (CCFM); when $i \ne$

j, $\rho_{X_{is}X_{js}}(\tau)$ is the cross-correlation coefficient function of $X_{is}(t)$ and $X_{js}(t+\tau)$, and is an off-diagonal term of CCFM.

It can be observed that $\rho_{U_iU_j}(\tau)$ can be determined from solving Eq. (10.12). It is worth noting that the valid solution of $\rho_{U_iU_j}(\tau)$ should be restricted by the following conditions to satisfy the definition of the correlation coefficient:

$$-1 \le \rho_{U_iU_j}(\tau) \le 1, \rho_{X_{is}X_{js}}(\tau)\cdot\rho_{U_iU_j}(\tau) \ge 0, \text{and } \left|\rho_{U_iU_j}(\tau)\right| \ge \left|\rho_{X_{is}X_{js}}(\tau)\right| \tag{10.13}$$

Following Eqs. (10.12) and (10.13), the solutions of $\rho_{U_iU_j}(\tau)$ and the applicable bound of original correlation coefficient $\rho_{X_{is}X_{js}}(\tau)$, i.e. $\left[\min \rho_{X_{is}X_{js}}(\tau), \max \rho_{X_{is}X_{js}}(\tau)\right]$, to ensure the transformation executable, can be obtained and are summarised in Table 10.2. It is essentially the same as Algorithm 7.1 except that the former uses functions of random processes for the the correlation coefficients between two time (t_1 and t_1+t), while the latter uses constant correlation coefficients between random variables.

Table 10.2 Expressions of $\rho_{U_iU_j}(\tau)$ and the applicable bound of $\rho_{X_{is}X_{js}}(\tau)$.

---

1) For $a_{4i}a_{4j} < 0$

$$\rho_{U_iU_j}(\tau) = -2r\cos\left(\frac{\theta + \pi}{3}\right) - \frac{t_2}{3}$$

$$\min \rho_{X_{is}X_{js}}(\tau) = \begin{cases} 6a_{4i}a_{4j}\cdot h\left[\rho_{U_iU_j-1}(\tau)\right], 6a_{4i}a_{4j}\cdot h\left[\rho_{U_iU_j-1}(\tau)\right] > -1 \text{ and } \rho_{U_iU_j-1}(\tau) > -1 \\ 6a_{4i}a_{4j}\cdot h(-1), \quad \text{otherwise} \end{cases}$$

$$\max \rho_{X_{is}X_{js}}(\tau) = \begin{cases} 6a_{4i}a_{4j}\cdot h\left[\rho_{U_iU_j-2}(\tau)\right], 6a_{4i}a_{4j}\cdot h\left[\rho_{U_iU_j-2}(\tau)\right] < 1 \text{ and } \rho_{U_iU_j-2}(\tau) < 1 \\ 6a_{4i}a_{4j}\cdot h(1), \quad \text{otherwise} \end{cases}$$

in which

$$\rho_{U_iU_j-1}(\tau) = \left(-t_2 - \sqrt{t_2^2 - 3t_1}\right)/3, \rho_{U_iU_j-2}(\tau) = \left(-t_2 + \sqrt{t_2^2 - 3t_1}\right)/3$$

$$h(y) = y^3 + t_2y^2 + t_1y, t_1 = \frac{a_{2i}a_{2j} + 3a_{4i}a_{2j} + 3a_{2i}a_{4j} + 9a_{4i}a_{4j}}{6a_{4i}a_{4j}}, t_2 = \frac{a_{3i}a_{3j}}{3a_{4i}a_{4j}}$$

$$r = \sqrt{-\frac{p}{3}}, \theta = \arccos\left(-\frac{q}{2r^3}\right), p = t_1 - \frac{t_2^2}{3}, q = \frac{2t_2^3}{27} - \frac{t_1t_2}{3} - \frac{\rho_{X_{is}X_{js}}(\tau)}{6a_{4i}a_{4j}}$$

2) For $a_{4i}a_{4j} > 0$ and $D = 4\left(t_2^2 - 3t_1\right) \le 0$

$$\rho_{U_iU_j}(\tau) = \sqrt[3]{A} + \sqrt[3]{B} - \frac{t_2}{3}$$

$$\min \rho_{X_{is}X_{js}}(\tau) = 6a_{4i}a_{4j}\cdot h(-1), \max \rho_{X_{is}X_{js}}(\tau) = 6a_{4i}a_{4j}\cdot h(1)$$

where

$$A = -\frac{q}{2} + \sqrt{\Delta}, B = -\frac{q}{2} - \sqrt{\Delta}, \Delta = \left(\frac{p}{3}\right)^3 + \left(\frac{q}{2}\right)^2$$

(Continued)

Table 10.2 (Continued)

3) For $a_{4i}a_{4j} > 0$, $D = 4(t_2^2 - 3t_1) > 0$, $t_2 > 0$, and $\Delta \geq 0$

$$\rho_{U_i U_j}(\tau) = \sqrt[3]{A} + \sqrt[3]{B} - \tfrac{t_2}{3}$$

$$\min \rho_{X_{is} X_{js}}(\tau) = \begin{cases} 6a_{4i}a_{4j} \cdot h\left[\rho_{U_i U_j - 2}(\tau)\right], & 6a_{4i}a_{4j} \cdot h\left[\rho_{U_i U_j - 2}(\tau)\right] > -1 \text{ and } \rho_{U_i U_j - 2}(\tau) > -1 \\ 6a_{4i}a_{4j} \cdot h(-1), & \text{otherwise} \end{cases}$$

$$\max \rho_{X_{is} X_{js}}(\tau) = 6a_{4i}a_{4j} \cdot h(1)$$

4) For $a_{4i}a_{4j} > 0$, $D = 4(t_2^2 - 3t_1) > 0$, $t_2 > 0$, and $\Delta < 0$

$$\rho_{U_i U_j}(\tau) = 2r \cos\left(\frac{\theta}{3}\right) - \frac{t_2}{3}$$

$$\min \rho_{X_{is} X_{js}}(\tau) = \begin{cases} 6a_{4i}a_{4j} \cdot h\left[\rho_{U_i U_j - 2}(\tau)\right], & 6a_{4i}a_{4j} \cdot h\left[\rho_{U_i U_j - 2}(\tau)\right] > -1 \text{ and } \rho_{U_i U_j - 2}(\tau) > -1 \\ 6a_{4i}a_{4j} \cdot h(-1), & \text{otherwise} \end{cases}$$

$$\max \rho_{X_{is} X_{js}}(\tau) = 6a_{4i}a_{4j} \cdot h(1)$$

5) For $a_{4i}a_{4j} > 0$, $D = 4(t_2^2 - 3t_1) > 0$, $t_2 < 0$, and $\Delta \geq 0$

$$\rho_{U_i U_j}(\tau) = \sqrt[3]{A} + \sqrt[3]{B} - \frac{t_2}{3}$$

$$\min \rho_{X_{is} X_{js}}(\tau) = 6a_{4i}a_{4j} \cdot h(-1)$$

$$\max \rho_{X_{is} X_{js}}(\tau) = \begin{cases} 6a_{4i}a_{4j} \cdot h\left[\rho_{U_i U_j - 1}(\tau)\right], & 6d_i d_j \cdot h\left[\rho_{U_i U_j - 1}(\tau)\right] < 1 \text{ and } \rho_{U_i U_j - 1}(\tau) < 1 \\ 6a_{4i}a_{4j} \cdot h(1), & \text{otherwise} \end{cases}$$

6) For $a_{4i}a_{4j} > 0$, $D = 4(t_2^2 - 3t_1) > 0$, $t_2 < 0$, and $\Delta < 0$

$$\rho_{U_i U_j}(\tau) = -2r \cos\left(\frac{\theta - \pi}{3}\right) - \frac{t_2}{3}$$

$$\min \rho_{X_{is} X_{js}}(\tau) = 6a_{4i}a_{4j} \cdot h(-1)$$

$$\max \rho_{X_{is} X_{js}}(\tau) = \begin{cases} 6a_{4i}a_{4j} \cdot h\left[\rho_{U_i U_j - 1}(\tau)\right], & 6d_i d_j \cdot h\left[\rho_{U_i U_j - 1}(\tau)\right] < 1 \text{ and } \rho_{U_i U_j - 1}(\tau) < 1 \\ 6a_{4i}a_{4j} \cdot h(1), & \text{otherwise} \end{cases}$$

7) For $a_{4i}a_{4j} = 0$ and $a_{3i}a_{3j} > 0$

$$\rho_{U_i U_j}(\tau) = \frac{\sqrt{\left(a_{2i}a_{2j} + 3a_{2i}a_{4j} + 3a_{4i}a_{2j}\right)^2 + 8a_{3i}a_{3j}\rho_{X_{is} X_{js}}(\tau)} - \left(a_{2i}a_{2j} + 3a_{2i}a_{4j} + 3a_{4i}a_{2j}\right)}{4a_{3i}a_{3j}}$$

$$\min \rho_{X_{is} X_{js}}(\tau) = \begin{cases} h_2\left[\rho_{U_i U_j - \text{sym}}(\tau)\right], & h_2\left[\rho_{U_i U_j - \text{sym}}(\tau)\right] > -1 \text{ and } \rho_{U_i U_j - \text{sym}}(\tau) > -1 \\ h_2(-1), & \text{otherwise} \end{cases}$$

$$\max \rho_{X_{is} X_{js}}(\tau) = h_2(1)$$

where

$$\rho_{U_i U_j - \text{sym}}(\tau) = -\frac{a_{2i}a_{2j} + 3a_{2i}a_{4j} + 3a_{4i}a_{4j}}{4a_{3i}a_{3j}}, \quad h_2(y) = \left(a_{2i}a_{2j} + 3a_{4i}a_{2j} + 3a_{2i}a_{4j}\right) \cdot y + 2a_{3i}a_{3j}y^2$$

Table 10.2 (Continued)

8) For $a_{4i}a_{4j} = 0$ and $a_{3i}a_{3j} < 0$

$$\rho_{U_iU_j}(\tau) = \frac{\sqrt{\left(a_{2i}a_{2j} + 3a_{2i}a_{4j} + 3a_{4i}a_{2j}\right)^2 + 8a_{3i}a_{3j}\rho_{X_{is}X_{js}}(\tau)} - \left(a_{2i}a_{2j} + 3a_{2i}a_{4j} + 3a_{4i}a_{2j}\right)}{4a_{3i}a_{3j}}$$

$$\min \rho_{X_{is}X_{js}}(\tau) = h_2(-1)$$

$$\max \rho_{X_{is}X_{js}}(\tau) = \begin{cases} h_2\left[\rho_{U_iU_j-\mathrm{sym}}(\tau)\right], h_2\left[\rho_{U_iU_j-\mathrm{sym}}(\tau)\right] < 1 \text{ and } \rho_{U_iU_j-\mathrm{sym}}(\tau) < 1 \\ h_2(1), \qquad \text{otherwise} \end{cases}$$

9) For $a_{4i}a_{4j} = 0$ and $a_{3i}a_{3j} = 0$

$$\rho_{U_iU_j}(\tau) = \frac{\rho_{X_{is}X_{js}}(\tau)}{a_{2i}a_{2j} + 3a_{2i}a_{4j} + 3a_{4i}a_{2j}}$$

$$\min \rho_{X_{is}X_{js}}(\tau) = -\left(a_{2i}a_{2j} + 3a_{2i}a_{4j} + 3a_{4i}a_{2j}\right), \ \max \rho_{X_{is}X_{js}}(\tau) = a_{2i}a_{2j} + 3a_{2i}a_{4j} + 3a_{4i}a_{2j}$$

10.2.4 Methods to Deal with the Incompatibility

When simulating stationary non-Gaussian processes using the fourth-moment transformation model, the potential incompatibility of the transformation from Gaussian to non-Gaussian processes needs to be investigated. There are two types of the incompatibility (Deodatis and Micaletti 2001), which are discussed separately in the following.

The first type of incompatibility arises when the target-Gaussian CFM is outside its limited range as listed in Table 10.2, or the combination of the skewness and kurtosis falls out of the applicable range of fourth-moment transformation model (see Figure 7.7 in the chapter on Transformation of Non-Normal Variables to Independent Normal Variables). In such cases, the translation from Gaussian to non-Gaussian processes can no longer be fulfilled by the fourth-moment transformation model.

The second type of incompatibility refers to the case when the translated Gaussian CFM is non-positive definite, which is due to numerical disturbance. Once this type of incompatibility occurs, the PSDM obtained from CFM via Wiener-Khintchine transformation is also non-positive definite and cannot be decomposed by Cholesky method, which interrupts the numerical simulation in the SRM (see Appendix D). To address such incompatibility, Shields and Deodatis (2013) presented an iterative methodology to estimate positive definite Gaussian PSDM. However, this method needs iterations and is unavoidably time-consuming. Under this circumstance, a recently developed eigenvalue decomposition method (Ji et al. 2018) is adopted. Through this method, the underlying Gaussian PSDM can be decomposed into the following form:

$$\mathbf{S_G}(\omega) = \sum_{i=1}^{n} v_i(\omega) \cdot \boldsymbol{\psi_i}(\omega) \cdot \boldsymbol{\psi_i^*}(\omega)^T \tag{10.14}$$

where $v_i(\omega)$ and $\boldsymbol{\psi_i}(\omega)$ are the ith eigenvalue and corresponding eigenvector of $\mathbf{S_G}(\omega)$; and $\boldsymbol{\psi_i^*}(\omega)$ denotes the complex conjugate of $\boldsymbol{\psi_i}(\omega)$.

To make $\mathbf{S_G}(\omega)$ positive definite, which requires all of the eigenvalues positive, the small negative eigenvalues in $\mathbf{v}(\omega) = \{v_1(\omega), v_2(\omega), \ldots, v_n(\omega)\}$ are substituted by small positive values, e.g. 0.0001. The modified positive definite $\mathbf{S_G}(\omega)$, $\mathbf{S'_G}(\omega)$ is then expressed as

$$\mathbf{S'_G}(\omega) = \sum_{i=1}^{n} v'_i(\omega) \cdot \mathbf{\psi_i}(\omega) \cdot \mathbf{\psi_i^*}(\omega)^T \tag{10.15}$$

where $v'_i(\omega)$ denotes the modified ith eigenvalue.

10.2.5 Scheme of Simulating Stationary Non-Gaussian Random Processes

Based on the fourth-moment transformation model, a scheme for simulating stationary non-Gaussian random processes developed from Yang and Gurley (2015) is described as follows:

1) Initialize the probabilistic and spectral models. The probabilistic model is prescribed by specifying the first four statistical moments μ_{X_i}, σ_{X_i}, α_{3X_i}, and α_{4X_i} of each component $X_i(t)$. The target PSDM $S_X(\omega)$ is prescribed by specifying each term $S_{X_iX_j}(\omega)$. Without loss of generality, each term is normalised

$$S_{X_{is}X_{js}}(\omega) = \frac{S_{X_iX_j}(\omega)}{\sqrt{\int_{-\infty}^{\infty} S_{X_iX_i}(\omega)d\omega \int_{-\infty}^{\infty} S_{X_jX_j}(\omega)d\omega}} \tag{10.16}$$

2) The coefficients of a_{1i}, a_{2i}, a_{3i}, and a_{4i} in Eq. (10.2) can be determined using the moment matching method only if the first four moments, i.e. μ_{X_i}, σ_{X_i}, α_{3X_i}, and α_{4X_i} of each component $X_i(t)$ are specified.
3) Compute each term of the CCFM by inverse Wiener-Khintchine relationship

$$\rho_{X_{is}X_{js}}(\tau) = R_{X_{is}X_{js}}(\tau) = \int_{-\infty}^{\infty} S_{X_{is}X_{js}}(\omega)e^{i\omega\tau}d\omega \tag{10.17}$$

4) Transform $\rho_{X_{is}X_{js}}(\tau)$ into $\rho_{U_iU_j}(\tau)$ using Table 10.2. Check whether $\rho_{X_{is}X_{js}}(\tau)$ falls into the region of $\left[\min \rho_{X_{is}X_{js}}(\tau),\ \max \rho_{X_{is}X_{js}}(\tau)\right]$ for all τ. If yes, then go to step 5. If not, then Type I incompatibility is encountered, and the transformation method cannot be utilised.
5) Calculate each Gaussian spectral term $S_{U_iU_j}(\omega)$ of Gaussian PSDM $S_U(\omega)$ by Wiener-Khintchine relationship:

$$S_{U_iU_j}(\omega) = \frac{1}{2\pi} \int_{-\infty}^{\infty} \rho_{U_iU_j}(\tau)e^{-i\omega\tau}d\tau \tag{10.18}$$

6) Perform Cholesky decomposition of Gaussian PSDM $S_U(\omega)$

$$S_U(\omega) = H(\omega)H^*(\omega)^T \tag{10.19}$$

where $H(\omega)$ is the Cholesky decomposition of $S_U(\omega)$. $H^*(\omega)^T$ is the transpose conjugate of $H(\omega)$. The Cholesky decomposition of the Gaussian PSDM $S_U(\omega)$ requires that $S_U(\omega)$ be positive definite. It is necessary to check whether $S_U(\omega)$ is positive definite for all ω. If yes, then get $H(\omega)$ and go to step 7. If not, then Type II incompatibility is encountered, and the eigenvalue decomposition method, Eqs. (10.14) and (10.15) have to be used.

7) Generate Gaussian time histories $U(t)$ using the SRM.
8) Obtain each non-Gaussian $X_{is}(t)$ via Eq. (10.2), and apply the desired translation and dilation $X_i(t) = \mu_{X_i} + \sigma_{X_i} X_{is}(t)$.

Example 10.1 Statistical Response Analysis of the Time-Domain Solution of a Tension Leg Platform

The transformation model is applied to simulate data resulting from a time-domain solution of a simple numerical model of a tension leg platform (TLP) subject to irregular seas, as shown in Figure 10.1 (Choi and Sweetman 2010). Surge response time histories have been simulated using a simplified two-dimensional nonlinear numerical model, which approximates the Snorre TLP in the Norwegian sector of the North Sea.

Simulations are for a significant wave height of 10.3 m with a spectral peak period of 15.5 seconds. Wind force and current velocity are held constant at 3×10^7 N and -0.1 m/s, respectively. The time-step of integration is 0.05 seconds, and the total time duration for the long time-histories is 90 minutes (the natural period is around 72 seconds). The wave is simulated from a JONSWAP spectrum, which is formulated as (Sarpkaya and Issacson 1981)

$$S_j(\omega) = \frac{\alpha g^2}{\omega^5} \exp\left[-\frac{5}{4}\left(\frac{\omega_p}{\omega}\right)^4 \right] \gamma^r \tag{10.20}$$

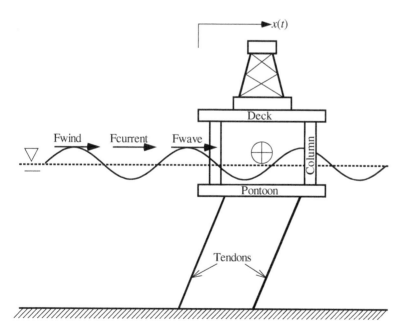

Figure 10.1 Tension leg platform. *Source*: Modified from Choi, M. and Sweetman, B. (2010). The Hermite moment model for highly skewed response with application to tension leg. J. Offshore Mech. Arct. Eng., 132 (2): 021602.

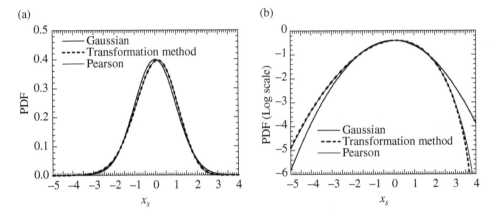

Figure 10.2 Comparison of simulated PDFs for the response of the TLP obtained by different methods. (a) In ordinary coordinate. (b) In logarithmic coordinate.

$$r = \exp\left[-\frac{\left(\omega - \omega_p\right)^2}{2\sigma^2\omega_p^2}\right] \tag{10.21}$$

$$\alpha = 0.076\left(\frac{U_{10}^2}{F_S g}\right)^{0.22}, \omega_p = 22\left(\frac{g^2}{U_{10}F_S}\right)^{1/3}, \gamma = 3.3, \sigma = \begin{cases} 0.07, & \omega \leq \omega_p \\ 0.09, & \omega > \omega_p \end{cases} \tag{10.22}$$

where U_{10} is the wind speed at a height of 10 m above the sea surface and F_S is the distance over which the wind blows with constant velocity. For time duration, 0–4700 seconds, the skewness and kurtosis of the resulting response are −0.2402 and 2.9885, respectively.

The PDFs simulated by the introduced model, and the Pearson system (the benchmark), as well as the PDF of the Gaussian process (Gaussian), are plotted in Figure 10.2a,b. It can be observed that the simulated PDFs obtained by the introduced model fit the benchmark PDFs very well for all values of x_s, while the PDFs of the Gaussian processes have great differences from the benchmark PDFs obtained by the Pearson system for the responses.

Example 10.2 Non-Gaussian Process Simulation

This example considers the simulation of a non-Gaussian vector process $X(t) = \{x_1(t), x_2(t), x_3(t)\}$ with known MPDFs and PSDM, which is revised from the numerical studies of Shields and Deodatis (2013) and Yang and Gurley (2015). The MPDFs of $x_1(t)$ and $x_2(t)$ are assumed to be shifted Weibull distribution as shown in Eq. (10.23), and the MPDF of $x_3(t)$ is assumed to be shifted lognormal distribution as given by Eq. (10.24).

$$f(x) = k_w \theta^{-k_w} x^{-k_w - 1} \exp\left[-\left(\frac{x + \mu}{\theta_w}\right)^{k_w}\right], \quad x > -\mu \tag{10.23}$$

where k_w is the shape parameter, θ_w is the scale parameter, and μ is the mean value.

$$f(x) = \frac{1}{\sqrt{2\pi}\sigma_N x} \exp\left\{ -\frac{[\ln(x+\mu) - \mu_N]^2}{2\sigma_N^2} \right\}, \quad x > -\mu \tag{10.24}$$

where σ_N and μ_N are given by

$$\sigma_N^2 = \ln\left(1 + \sigma^2/\mu^2\right), \mu_N = \ln\mu - \sigma_N^2/2 \tag{10.25}$$

The parameters of the MPDFs of $\mathbf{X}(t)$ are shown in Table 10.3, and the first four moments of $x_i(t)$ ($i = 1, ..., 3$) can be readily calculated and are listed in Table 10.3.

The diagonal terms of the PSDM are assumed as follows:

$$S_{Nii}^T(\omega) = \alpha(1 + \beta \cdot \omega)^{-5/3}, i = 1, 2, 3 \tag{10.26}$$

where α and β are deterministic parameters. For $i = 1$, $\alpha = 2.34$, and $\beta = 6.19$; for $i = 2$, $\alpha = 2.61$, and $\beta = 6.98$; and $\alpha = 7.73$ and $\beta = 21.8$ for $i = 3$. The off-diagonal terms are defined as (Deodatis 1996):

Table 10.3 Distributions and the first four moments of $x_i(t)$ (i = 1, 2, 3) for Example 10.2.

				The first four central moments		
i	Distribution	Parameters	Mean	Standard deviation	Skewness	Kurtosis
1	Shifted Weibull	$k_w = 1, \theta_w = 1, \mu = 1$	1.000	1.000	2.000	9.000
2	Shifted Weibull	$k_w = 1.258, \theta_w = 1.344, \mu = 1.25$	1.250	1.000	1.415	5.737
3	Shifted lognormal	$\mu = 1.8, \sigma = 1.0$	1.800	1.000	1.830	9.550

Table 10.4 Investigation of the first type of incompatibility due to the limited range of the target non-Gaussian CFM.

		Extremum of $\xi_{ij}(\tau)$		
Components	Limited range of $\xi_{ij}(\tau)$	Minimum	Maximum	Type
1,1	(−0.6655, 1.0000)	−0.0028	1.0000	II
2,2	(−0.7868, 1.0000)	−0.0031	1.0000	II
3,3	(−0.8030, 1.0000)	−0.0084	1.0000	IV
1,2	(−0.7277, 0.9948)	−0.0029	0.9170	II
1,3	(−0.7349, 0.9915)	−0.0048	0.5230	IV
2,3	(−0.7860, 0.9910)	−0.0050	0.5465	IV

$$S_{Nij}^{T}(\omega) = \sqrt{S_{Nii}^{T}(\omega) S_{Njj}^{T}(\omega)} \gamma_{Nij}^{T}(\omega), \quad i \neq j \tag{10.27}$$

where $\gamma_{Nij}^{T}(\omega)$ is the coherence function, which is assumed as

$$\gamma_{N12}^{T}(\omega) = \exp(-0.1757 \cdot \omega), \gamma_{N13}^{T}(\omega) = \exp(-3.478 \cdot \omega), \gamma_{N23}^{T}(\omega) = \exp(-3.292 \cdot \omega) \tag{10.28}$$

As described in Section 10.2.5, the first step of the simulation of $\mathbf{X}(t)$ is to check the first type of incompatibility. It can be readily found that the combination of the given skewness and kurtosis are in the applicable range of the fourth-moment transformation model. The limited ranges of the target non-Gaussian CFM can be obtained from Table 10.2 by using the skewness and kurtosis of the non-Gaussian processes, which are listed in Table 10.4, together with extremums of all elements of the target non-Gaussian CFM. As can be observed from Table 10.4, the extremums of all elements of the target non-Gaussian CFM fall into in the limited ranges of the proposed model, which means the first type of incompatibility does not occur in this example and the proposed model can be applied to translate the Gaussian CFM from the target non-Gaussian CFM.

Since the Gaussian PSDM is no longer positive definite in some certain frequency domains after translation, the second type of incompatibility occurs in this example. The eigenvalue decomposition method mentioned in Section 10.2.4, i.e. Eqs. (10.14) and (10.15), is used to modify the Gaussian PSDM. The accuracy of the modified Gaussian PSDM can be verified by comparing it with the unmodified one. The elements of the relative difference matric ε can be determined by (Shields and Deodatis 2013):

$$\varepsilon_{ij} = 100 \sqrt{\frac{\sum_{l=0}^{N-1} \left[S'_{Gij}(\omega_l) - S_{Gij}(\omega_l) \right]^2}{\sum_{l=0}^{N-1} \left[S_{Gij}(\omega_l) \right]^2}} \tag{10.29}$$

$$\omega_{ml} = l\Delta\omega, l = 0, 1, ..., N-1 \tag{10.30}$$

where $S'_{Gij}(\omega)$ and $S_{Gij}(\omega)$ denote the elements of the modified and unmodified Gaussian PSDM, respectively. With the aid of Eqs. (10.29) and (10.30), ε for this example is obtained as:

$$\varepsilon = \begin{bmatrix} 0.862\% & 0.674\% & 0.591\% \\ 0.674\% & 1.032\% & 0.937\% \\ 0.591\% & 0.937\% & 1.215\% \end{bmatrix} \tag{10.31}$$

As can be observed from Eq. (10.31), the relative differences between the modified and unmodified Gaussian PSDM are very small (less than 1.3% for all the elements), which means that the eigenvalue decomposition method has no great influence on the PSDM of the Gaussian vector process to be simulated.

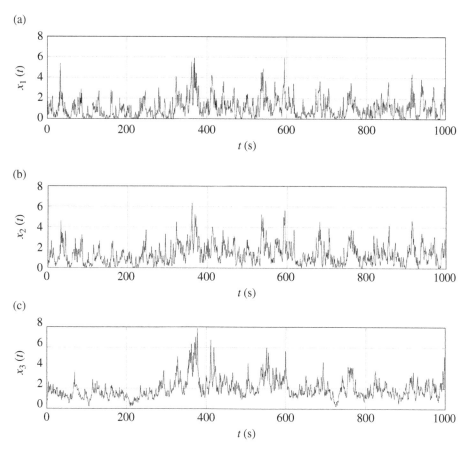

Figure 10.3 Simulated $x_i(t)$ (i = 1,2,3) in order from top to bottom. a $x_1(t)$; b $x_2(t)$; c $x_3(t)$.

After overcoming the second type of incompatibility, the stationary Gaussian process $U(t)$ is generated by SRM and then translated into the target process $X(t)$ based on the fourth-moment transformation. The simulated non-Gaussian samples are plotted in Figure 10.3a–c, which shows that $x_1(t)$ and $x_2(t)$ have similar changing tendency but they are inconsistent with $x_3(t)$. The similarity of change tendency meets the need of coherence relationships among them and demonstrates the accuracy of the simulation method based on the fourth-moment transformation model.

The PSDM and CFM estimated from the simulated samples are plotted in Figure 10.4a–l. The target non-Gaussian PSDM and CFM can be obtained from Eqs. (10.26–28), which are taken as the exact values for comparison. It can be observed from Figure 10.4 that the simulated PSDM and CFM obtained by using the fourth-moment transformation model are in good agreement with the target values for all the cases, which proves the accuracy of the fourth-moment transformation model in simulating stationary non-Gaussian processes.

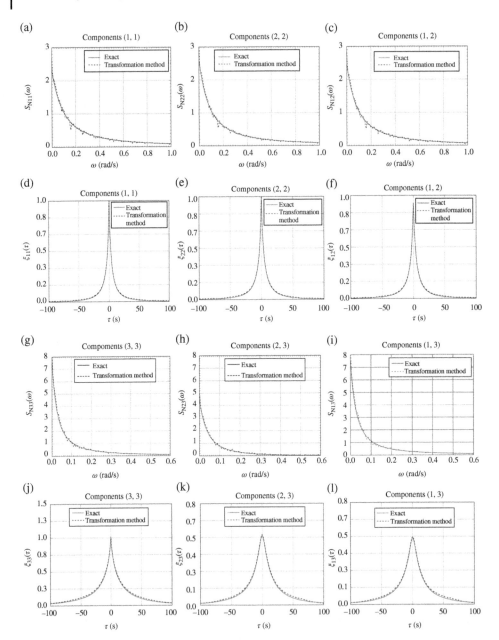

Figure 10.4 Comparison between simulated and exact PSDM or CFM. (a) PSDs not involved with $x_3(t)$-components (1,1). (b) PSDs not involved with $x_3(t)$-components (2,2). (c) PSDs not involved with $x_3(t)$-components (1,2). (d) CFs not involved with $x_3(t)$-components (1,1). (e) CFs not involved with $x_3(t)$-components (2,2). (f) CFs not involved with $x_3(t)$-components (1,2). (g) PSDs involved with $x_3(t)$-components (3,3). (h) PSDs involved with $x_3(t)$-components (2,3). (i) PSDs involved with $x_3(t)$-components (1,3). (j) CFs involved with $x_3(t)$-components (3,3). (k) CFs involved with $x_3(t)$-components (2,3). (l) CFs involved with $x_3(t)$-components (1,3).

Example 10.3   Simulation of Measured Wind Pressure Coefficients

The aim of this example is to simulate time histories of the wind pressure coefficient (Cp) that have the same first four statistical moments, CFM or PSDM behaviour as those of experimental Cp data obtained from wind tunnel tests conducted at the University of Western Ontario (UWO) (Ho et al. 2005; Pierre et al. 2005). The test model is an industrial building, which has full-scale size of 19.05 m × 12.2 m × 3.66 m with a roof slope of 1:12 and is assumed to be located in suburban terrain. The model scale is 1:100, with 335 taps (marked by '+') distributed on the rooftop as shown in Figure 10.5. The sampling frequency was 500 Hz with a sampling time of 100 seconds. The tests were conducted in suburban terrain with roughness length of about 0.3 m, under a reference mean wind speed of 13.7 m/s at the equivalent of 10 m above the ground, which corresponds to a mean wind speed of 6.1 m/s at the roof height (3.66 m). Only the wind direction of 315° is considered in this study.

For brevity, four representative taps, i.e. Tap 1, Tap 2, Tap 3, and Tap 4, are selected for investigations. Tap 1 is chosen because the wind pressure at this tap shows strong non-Gaussianity. Tap 2 and Tap 3 are chosen since they are next to Tap 1 and the wind pressure at these taps is different from Tap 1 in their higher moments. Tap 4 is chosen because the wind pressure here is a hardening process. Table 10.5 presents the first four moments obtained from the measure data for each tap.

The original Cp data is first standardised before simulation. Similar to Example 10.2, the first step of simulating the Cp time histories is to check the first type of incompatibility. It can be readily found that the combination of the obtained skewness and kurtosis are in the applicable range of the fourth-moment transformation model. The limited ranges of the target non-Gaussian CFM obtained from Table 10.2 by using the skewness and kurtosis and extremums of all elements of the target CFM of the standardised Cp time histories are listed

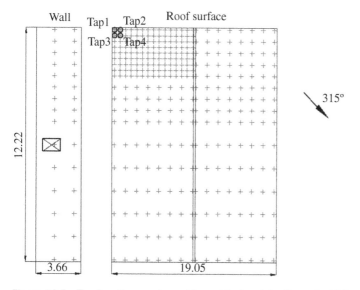

Figure 10.5 Tap locations and panel layout (unit: m) for Example 10.3.

Table 10.5 Statistical moments of original and simulated Cp data for Example 10.3.

Tap	Moments of original data				Moments of simulated data			
	Mean	Standard deviation	Skewness	Kurtosis	Mean	Standard deviation	Skewness	Kurtosis
1	0.8656	0.8851	2.1041	10.7527	0.8688	0.8993	2.1131	11.1235
2	1.6674	1.5036	1.5344	6.2847	1.6798	1.5316	1.4423	5.7379
3	1.6965	1.3192	1.2825	5.0956	1.7102	1.3524	1.2616	5.0694
4	−0.5542	0.0890	−0.0040	2.9736	−0.5542	0.0873	−0.0031	2.9301

Table 10.6 Investigation of the first type of incompatibility due to the limited range of the target non-Gaussian CFM in Example 10.3.

Components	Limited range of $\xi_{ij}(\tau)$	Extremum of $\xi_{ij}(\tau)$		Type
		Minimum	Maximum	
Tap 1, Tap 1	(−0.7252, 1.0000)	−0.0370	1.0000	IV
Tap 2, Tap 2	(−0.7590, 1.0000)	−0.0450	1.0000	II
Tap 3, Tap 3	(−0.8078, 1.0000)	−0.0302	1.0000	II
Tap 4, Tap 4	(−1.0000, 1.0000)	−0.0350	1.0000	IV
Tap 1, Tap 2	(−0.7357, 0.9931)	−0.0346	0.5661	II
Tap 1, Tap 3	(−0.7582, 0.9880)	−0.0254	0.5091	I
Tap 1, Tap 4	(−0.9198, 0.9191)	−0.0313	0.2497	I
Tap 2, Tap 3	(−0.7837, 0.9989)	−0.0307	0.3389	I
Tap 2, Tap 4	(−0.9380, 0.9373)	−0.0365	0.2532	I
Tap 3, Tap 4	(−0.9510, 0.9504)	−0.0417	0.3058	IV

in Table 10.6, which shows that the first type of incompatibility does not occur. It is found that the second type of incompatibility still exists, which makes the Gaussian PSDM unable to be decomposed by Cholesky method. The eigenvalue decomposition method is again used to modify the Gaussian PSDM. Based on Eqs. (10.29) and (10.30), ε for this example is calculated as

$$\varepsilon = \begin{bmatrix} 0.167\% & 0.345\% & 0.183\% & 0.194\% \\ 0.345\% & 0.181\% & 0.215\% & 0.261\% \\ 0.183\% & 0.215\% & 0.175\% & 0.176\% \\ 0.194\% & 0.261\% & 0.176\% & 0.356\% \end{bmatrix} \tag{10.32}$$

Equation (10.32) shows that the largest value of the elements of ε is 0.356%, which is relatively small and thus proves the accuracy of eigenvalue decomposition method. After the incompatibility is checked and dealt with, the Cp time histories simulated by fourth-moment transformation model are illustrated in Figure 10.6a–d, and together with the

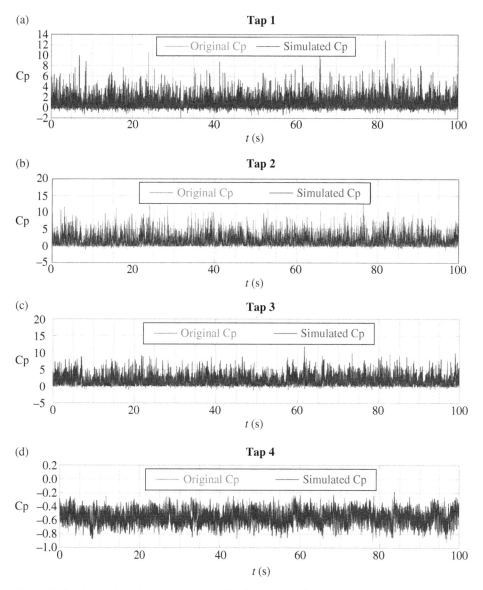

Figure 10.6 Comparison between measured and simulated Cp data for Example 10.3. (a) Cp time history of Tap1. (b) Cp time history of Tap 2. (c) Cp time history of Tap 3. (d) Cp time history of Tap 4.

original data. It can be observed that the simulated time histories closely match original time histories with respect to the magnitude and frequency of the peaks. The simulated first four moments for each tap are also presented in Table 10.5, showing good agreement as expected.

Comparisons between the exact and simulated PSDs and correlation functions for some representative components are respectively depicted in Figure 10.7a–l, which show that the simulated auto-PSD and auto-correlation function of the Cp data of Taps 1–4 obtained by using the transformation model matches the exact ones very well.

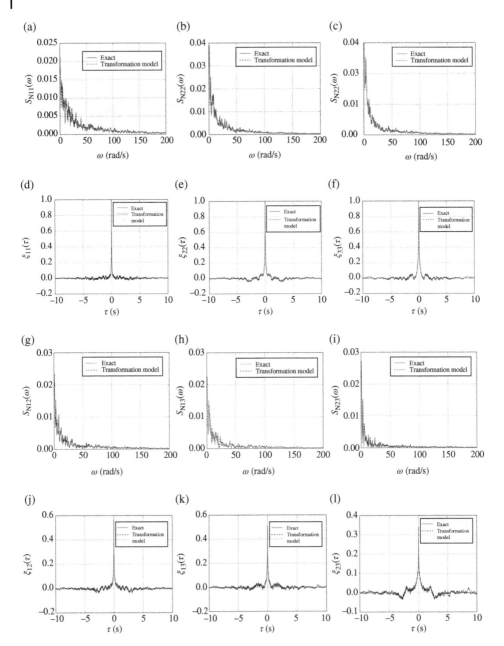

Figure 10.7 Comparison between simulated and exact PSDs or correlation functions for Example 10.3. (a) Auto-PSD (Tap 1, Tap 1). (b) Auto-PSD (Tap 2, Tap 2). (c) Auto-PSD (Tap 3, Tap 3). (d) Auto-correlation function (Tap 1, Tap 1). (e) Auto-correlation function (Tap 2, Tap 2). (f) Auto-correlation function (Tap 3, Tap 3). (g) Cross-PSD (Tap 1, Tap 2). (h) Cross-PSD (Tap 1, Tap 3). (i) Cross-PSD (Tap 2, Tap 3). (j) Cross-correlation function (Tap 1, Tap 2). (k) Cross-correlation function (Tap 1, Tap 3). (l) Cross-correlation function (Tap 2, Tap 3).

10.3 First Passage Probability Assessment of Stationary Non-Gaussian Processes Using the Fourth-Moment Transformation

10.3.1 Brief Review on First Passage Probability

The failure of a random dynamic system is usually defined based on the exceedance of a predefined critical threshold level. In many engineering applications, it is important to calculate the probabilities that a structural response exceeds the critical threshold levels in the given time interval [0, *T*]. The critical threshold level could be the yield stress, strain, or displacement. Failure occurs when the critical threshold value is exceeded at its first time and this kind of analysis falls into the category of the first passage problem (Rice 1944; Rice 1945; Coleman 1959; Lin 1970; Vanmarcke 1975).

Generally, there are no simple analytical solutions existing for the first passage problems (Rice 1944; Rice 1945; Coleman 1959; Lin 1970; Crandall 1970; Vanmarcke 1975; Naess 1990; Barbato and Conte 2011; Ghazizadeh et al. 2011). Many numerical studies have been devoted to the computation of the first passage probability (Au and Beck 2001b; Naess and Gaidai 2009; Naess et al. 2010), and MCS is the most general and robust, although it is computationally expensive for low probability events. In most cases, the computation of the first passage probability is often based on the up-crossing approach which assumes that the crossings are independent and the responses are Gaussian.

The well-known Gaussian model is generally inappropriate for calculating the first passage probability of a structure when structural behaviour is non-linear, or the excitation is non-Gaussian (e.g. wind and wave loads). Structural responses generally cannot be modelled as Gaussian stochastic processes for these cases. This is important particularly for non-Gaussian responses whose tail behaviours are different from Gaussian responses. Series distribution methods (Grigoriu 1984; Ochi 1986) can be used to transform known statistical results (such as mean up-crossing rates and extreme values of a Gaussian process) into those of a non-Gaussian process by finding a simple functional transformation of the equivalent Gaussian statistics. For these cases, various non-linear models have been formulated through series approximation to map a non-Gaussian response process into a standard Gaussian one. The polynomials using response moments are used to form non-Gaussian response contributions and to realise the transformation between non-Gaussian process and Gaussian process (Winterstein 1988; Jensen 1994; Puig and Akian 2004; Winterstein and Kashef 2000; He and Zhao 2007; Winterstein and Mackenzie 2011; Zhao et al. 2019). The approach based on the fourth-moment transformation will be introduced in this section.

10.3.2 Formulation of the First Passage Probability of Stationary Non-Gaussian Structural Responses

In the structural dynamical reliability analysis, one often assumes that the crossings of a random response follow the Poisson distribution and the first passage probability, $P_f(T)$ is given by (Coleman 1959)

$$P_f(T) = 1 - \exp\{-E[N^+(T)]\} \tag{10.33}$$

where $E[N^+(T)]$ is the mean number of crossings of $X(t)$ exceeding the critical level x during the interval $[0, T]$ for regular streams of up-crossings. In the stationary case, one has $E[N^+(T)] = v^+T$, where v^+ denotes the mean up-crossing rate and can be estimated by Rice formula (Rice 1944) as

$$v^+(t) = \int_0^\infty \dot{x} f_{X\dot{X}}(x, \dot{x}, t) d\dot{X} \tag{10.34}$$

where $f_{X\dot{X}}(x, \dot{x}, t)$ is the joint PDF of $X(t)$ and its derivative $\dot{X}(t)$ at time t. For a stationary standard Gaussian process $U(t)$ at critical level u, Eq. (10.34) can be written as

$$v^+ = v_0 \exp\left(-\frac{u^2}{2}\right), v_0 = \frac{\sigma_{\dot{U}}}{2\pi} \tag{10.35}$$

where v_0 is the mean rate of response cycles; $\sigma_{\dot{U}}$ is the standard deviation of $\dot{U}(t)$; and $\dot{U}(t)$ is the derivative of $U(t)$ at time t. For a linear oscillator, v_0 can be written as

$$v_0 = \frac{\omega_0}{2\pi} \tag{10.36}$$

where ω_0 is the undamped natural frequency of a dynamical system.

This approach has been proven to be exact for very high critical levels (Cramér 1966). If $X(t)$ is narrow-banded or the critical threshold level is low, the clumping effect of the up-crossing events should be considered to modify the mean level up-crossing rates in Eq. (10.34) (He 2009; He 2015). Since the exceedances of the critical threshold level tend to occur in clumps, the clump size $\langle cs \rangle$ is proposed to improve Eq. (10.33) (Langley 1988). When the stationary start condition and the clump size are considered, Eq. (10.33) can be further expressed as Eqs. (10.37–10.40) for stationary Gaussian processes (Ditlevsen 1986; Langley 1988; He and Zhao 2007).

$$P_f(T) = 1 - [1 - P_f(0)] \exp\left\{-\frac{v^+ T}{\langle cs \rangle \cdot [1 - P_f(0)]}\right\} \tag{10.37}$$

where

$$P_f(0) = 1 - \Phi[u(x)], \quad v^+ = v_0 \exp\left[-\frac{u^2(x)}{2}\right] \tag{10.38}$$

$$\langle cs \rangle = 1 + \sum_{n=2}^\infty P(cs \geq n), P(cs \geq n) = 2\left\{1 - \Phi\left[\frac{u(x) \cdot (1 - R_U(\tau_n))}{\sqrt{1 - R_U^2(\tau_n)}}\right]\right\} \tag{10.39}$$

$$R_U(\tau) = \exp[-\zeta\omega_0\tau][\cos\omega_d\tau + (\zeta\omega_0/\omega_d)\sin\omega_d\tau], \omega_d = \omega_0\sqrt{1 - \zeta^2} \tag{10.40}$$

where $P_f(0)$ is the probability of instantaneous failure at $T = 0$; $R_U(\tau)$ is the autocorrelation function of the standard Gaussian process $U(t)$; $\tau_n = t_n - t_1$ is a suitable choice of τ at which $R_U(\tau)$ displays its nth distinguishable peak (including the peak $\tau = 0$); ζ is damp ratio; ω_0 is the undamped natural frequency of a dynamic system; $\Phi[\cdot]$ is the CDF of standard normal variables; and $u(x)$ is the equivalent Gaussian fractile.

For high critical level, Eq. (10.37) can be improved as (Ditlevsen 1986),

$$P_f(T) = 1 - [1 - P_f(0)] \exp\left\{ -\frac{v^+ T}{[1 - P_f(0)]} \right\} \tag{10.41}$$

When the structural responses are stationary non-Gaussian processes, Eqs. (10.37–10.39) can still be utilised to compute the first passage probability, except that the equivalent Gaussian fractile $u(x)$ and the autocorrelation function $R_U(\tau)$ will be obtained from the stationary non-Gaussian response using the fourth-moment transformation model. The detailed procedure for the transformations of marginal probability distributions and correlation functions between stationary non-Gaussian processes and Gaussian ones has been discussed in Section 10.2.2 and 10.2.3, respectively.

10.3.3 Computational Procedure for the First Passage Probability of Stationary Non-Gaussian Structural Responses

The procedure for calculating the first passage probability of a stationary non-Gaussian process $X_i(t)$ based on the fourth-moment transformation is described as follows (Zhao et al. 2019):

1) Estimate the first four moments (μ_{X_i}, σ_{X_i}, α_{3X_i}, and α_{4X_i}) and autocorrelation function $R_{X_i}(\tau)$ of a stationary non-Gaussian structural response;
2) Determine the four polynomial coefficients (a_{1i}, a_{2i}, a_{3i}, and a_{4i}) of Eq. (10.2) using the obtained first four moments;
3) Compute the autocorrelation function $R_{U_i}(\tau)$ according to $R_{X_i}(\tau)$ and Table 10.2.
4) Determine the critical threshold level x_i, and compute the equivalent Gaussian fractile u (x_i) using Table 10.1.
5) Compute the first passage probability of a stationary non-Gaussian structural response using Eqs. (10.37)–(10.39).

Example 10.4 First Passage Probabilities of Structural Responses

This example considers a linear single-degree-of-freedom (SDOF) structure excited by a quadratic forcing function $F(t)$ as shown in Figure 10.8. Specifically, it is assumed that the forcing function is $F(t) = \alpha_1 V(t) + \alpha_2 V^2(t)$, where $V(t)$ is a stationary zero-mean Gaussian process, and α_1 and α_2 are constants. This example is chosen primarily for its simplicity, but a similar forcing function could serve as a model for dynamic wind pressure, although with parameters different from those chosen here.

The structural response to $F(t)$ is determined by the following equation

$$\ddot{X}(t) + 2\omega_0 \zeta \dot{X}(t) + \omega_0^2 X(t) = F(t) \tag{10.42}$$

In this example, it is assumed that $\alpha_1 = \alpha_2 = \omega_0^2$ and $\omega_0 = 1.0\,\text{rad/s}$. For simplicity, the variance spectrum $S_V(\omega)$ of $V(t)$ is 0.5 m^2s between $\omega_l = 0.5\,\text{rad/s}$ and $\omega_u = 1.5\,\text{rad/s}$, in which zero is outside this interval for $\omega \geq 0$ and $S_V(-\omega) = S_V(\omega)$. It follows that Var $[X(t)] = 1$.

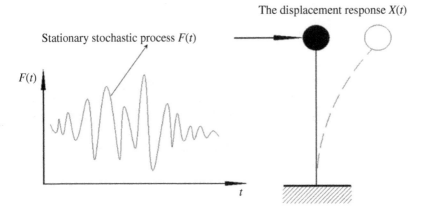

Figure 10.8 A SDOF structure excited by a quadratic forcing function $F(t)$ for Example 10.4.

The first four moments of the quadratic part of stationary structural responses $X(t)$, i.e. the responses to $\alpha_2 V^2(t)$ determined under the damping ratio (ζ) being equal to 0.1, are obtained as $\mu_X = 1.0$, $\sigma_X = 1.62$, $\alpha_{3X} = 1.7$, and $\alpha_{4X} = 9.26$, respectively (Naess 1987) ('softening' responses, i.e. $\alpha_{4X} > 3$). For this example, the correlation function $R_X(\tau)$ of the quadratic part of stationary structural responses $X(t)$ is approximated by $R_U(\tau)$ as given by Eq. (10.40), since the differences between $R_X(\tau)$ and $R_U(\tau)$ are very small and the times that display their distinguishable peaks are almost the same (He and Zhao 2007). Therefore, $R_U(\tau)$ can be used to calculate the mean clump size of $X(t)$, that is Eq. (10.39) is used.

The first passage probabilities of stationary non-Gaussian structural responses for critical threshold level $x = 4.5\sigma_X$ and $x = 5\sigma_X$ obtained by using the introduced transformation method are shown in Figure 10.9a,b, respectively, compared with the results obtained by assuming Gaussian model and the MCS method (Au and Beck 2001b; Bayer and Bucher 1999). From Figure 10.9a,b:

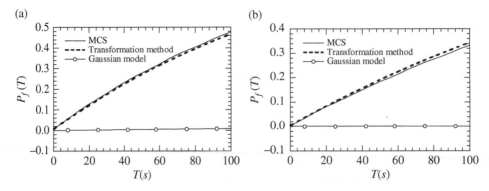

Figure 10.9 The first passage probabilities of structural responses in Example 10.4. (a) $x = 4.5\sigma_X$; (b) $x = 5\sigma_X$.

1) The results of Gaussian model have the greatest differences from those of MCS method (with 100 000 samples) and are too conservative.
2) The introduced transformation method performs better than the Gaussian model, and the obtained results are in close agreement with those by MCS method during the whole interval *T*. Therefore, for 'softening' responses the introduced procedure is more suitable for engineering applications.

10.4 Time-Dependent Structural Reliability Analysis Considering Correlated Random Variables

10.4.1 Brief Review on Time-Dependent Structural Reliability Methods

Another time-variant reliability problem is reliability evaluation with time-dependent resistance and load effects. In practical engineering, structural resistance may deteriorate due to ageing under the aggressive service environment, and the occurrence and intensity of loads may change with time due to varying service and environmental demands. Therefore, time-dependent reliability problems for structural safety and serviceability during the entire service life have to be solved.

Many studies (Oswald and Schueller 1984; Madsen and Tvedt 1990; Mori and Elling-wood 1993b; Hong 2000; Li et al. 2015; Wang et al. 2016; Melchers and Beck 2018) have been conducted on the approximated methodologies to determine the time-dependent probability of failure. The methodology proposed by Mori and Ellingwood (1993b) was one of the first attempts to assess time-dependent reliability of structures considering both the randomness of resistance and the stochastic nature of load (Li et al. 2015), and was widely adopted by the succeeding researches (Enright and Frangopol 1998; Hong 2000; Bhargava et al. 2011; Li et al. 2015; Wang et al. 2017). Mori and Ellingwood (1993b) treated the occurrence of rare events such as extra-ordinary live loads during the service life of the structure as stationary Poisson point process consisting of a sequence of identically distributed and statistically independent intensities with a constant mean occurrence rate (Li et al. 2015) (i.e. the live load intensity is treated as the time-dependent random variable). However, non-stationary load models are more realistic in some cases, such as an increasing number of trucks and relative frequency of heavy vehicles (Nowak et al. 2010) and increasing hurricane wind speeds (Knutson et al. 2010; Li et al. 2016). An improved method was developed by Li et al. (2015) for evaluating time-dependent reliability of ageing structures, in which non-stationary load processes, consisting of time-dependent mean occurrence rate and time-varying CDF of load intensity, are taken into account. Since the CDFs of the load intensity/effects are used in the formulas of these methodologies, the formulas are only applicable for the cases that the live load intensity/effect can be separated from the limit state functions, i.e. the performance functions should be explicit. However, the effect of live load on a structure is generally determined by finite element method (FEM), and thus the performance function is implicit.

In this section, the time-dependent failure probability of structures with complicated, multi-dimensional and implicit performance functions considering the correlation among random variables is mathematically formulated, which is then evaluated by using moment-based approach (Lu et al. 2019).

10.4.2 Formulation of Time-Dependent Failure Probability

In order to compute the reliability of a structure over time interval (0, *T*], the time-varying performance function should be established first and can be expressed as

$$G(t) = R(t) - S(t) \tag{10.43}$$

As the structural resistance deteriorates with time, the time-dependent degraded resistance may be given by (Mori and Ellingwood 1993b)

$$R(t) = R_0 \cdot g(t) \tag{10.44}$$

where R_0 is the original (undegraded) resistance, and $g(t)$ is the time-dependent degradation function of structural resistance determined by the environmental stressor.

Under the assumptions that the duration of live load event is generally very short and the live load intensity is treated as constant during a load event, the live load can be modelled as a sequence of randomly occurring pulses with random intensity S_i, $i = 1, 2, ..., n$, at time t_i with a small duration τ during time interval (0, *T*] (Mori and Ellingwood 1993b). The duration is sufficiently small and it can be assumed that the resistance remains the same during the load event. If n independent events occur within time interval (0, *T*] at deterministic times t_i, $i = 1, 2, ..., n$, the structural time-dependent failure probability can be expressed as (Mori and Ellingwood 1993b; Li et al. 2015)

$$P_f(T) = 1 - P[R(t_1) > S_1 \cap R(t_2) > S_2 \cap \cdots \cap R(t_n) > S_n] \tag{10.45}$$

The occurrence of live load effects can be modelled as a stationary Poisson point process with a constant mean occurrence rate, namely, each live load effect is statistically independent and identically distributed (the time of occurrence follows a uniform distribution). Based on the forgoing description of a live load event, assuming all random variables are mutually independent, the time-dependent failure probability of a structure within time interval (0, *T*] under dead and live loads can be expressed as (Mori and Ellingwood 1993b)

$$P_f(T) = 1 - \int_0^\infty \int_0^\infty \exp\left\{ -\lambda_{S_Q} \cdot \int_0^T [1 - F_{S_Q}(r_0 \cdot g(t) - s_D)] dt \right\} f_{R_0}(r_0) f_{S_D}(s_D) dr_0 ds_D \tag{10.46}$$

where λ_{S_Q} is the mean occurrence rate of live load; S_D is the dead load effect; $F_{S_Q}(\cdot)$ is the CDF of live load effect; and $f_{S_D}(\cdot)$ and $f_{R_0}(\cdot)$ are the PDFs of dead load effect and initial resistance, respectively.

If the occurrence of live load effects is modelled as a non-stationary Poisson point process with a time-variant mean occurrence rate and time-dependent CDF of intensity, namely, each live load effect is statistically independent and diversely distributed (the time of occurrence does no longer follow a uniform distribution). Eq. (10.46) can be rewritten as (Li et al. 2015)

$$P_f(T) = 1 - \int_0^\infty \int_0^\infty \exp\left\{ -\int_0^T \lambda_{S_Q}(t) \cdot \left[1 - F_{S_Q}(r_0 \cdot g(t) - s_D, t) \right] dt \right\} f_{R_0}(r_0) f_{S_D}(s_D) dr_0 ds_D.$$

(10.47)

Equations (10.46) and (10.47) are widely adopted within the assessment of the time-dependent reliability of deteriorating structures (Enright and Frangopol 1998; Hong 2000; Bhargava et al. 2011; Wang et al. 2017). However, there still exist several limitations within these approaches and further studies are needed. For instance, in Eqs. (10.46) and (10.47), the time-dependent degraded resistance is simplified as the product of original resistance and degradation function, which is generally empirical or experimental. The CDFs of the live load intensity/effect used in Eqs. (10.46) and (10.47) restrict the application of these approaches for implicit performance functions. Furthermore, the correlations between R_0, S_D, and S_Q are not considered in Eqs. (10.46) and (10.47). Therefore, an efficient approach should be developed to compute the multi-dimensional integration over time domain and random-variate space.

In engineering application, the resistance and load effect may be expressed as functions of structural parameters, loads, and environmental actions parameters, among others. For such complicated cases, the parameters used in the reliability assessment can be divided into two categories: one is the time-independent random variables that remain the same during the given period $(0, T]$ such as parameters of dead load, original material strength, and initial geometry variables; and the other is the time-dependent random variables (Hong 2000) that at arbitrary time instants are randomly generated including in-service loads and environmental actions, corrosion rate, and so on. Letting $\mathbf{X}$ and $\mathbf{Y}(t)$ denote the vectors of time-independent and time-dependent random variables, respectively, the general time-dependent performance function of a structure can then be expressed as $G[\mathbf{X}, \mathbf{Y}(t), t]$, in which $G[\mathbf{X}, \mathbf{Y}(t), t] > 0$ represents the safe domain.

When $\mathbf{X}$ and $\mathbf{Y}(t)$ are mutually independent, the probability of failure within a given time interval $(0, T]$ can then be formulated as (Hong 2000)

$$P_f(T) = 1 - \int_{\Omega_{\mathbf{X}}} P_{\mathbf{Y}|\mathbf{X}}(\mathbf{x}, T) \cdot f_{\mathbf{X}}(\mathbf{x}) d\mathbf{x}$$

(10.48)

where $\Omega_{\mathbf{X}}$ is the domain region of $\mathbf{X}$; $\mathbf{x}$ is the sample of $\mathbf{X}$; $f_{\mathbf{X}}(\mathbf{x})$ is the joint PDF of $\mathbf{X}$; and $P_{\mathbf{Y}|\mathbf{X}}(\mathbf{x}, T)$ denotes the probability of survival during the period $(0, T]$ conditioned on the value of the random vector $\mathbf{X} = \mathbf{x}$.

Assuming that the occurrence in time of $\mathbf{Y}(t)$ is modelled as Poisson point process, the derived expression of $P_{\mathbf{Y}|\mathbf{X}}(\mathbf{x}, T)$ is given by

$$P_{\mathbf{Y}|\mathbf{X}}(\mathbf{x}, T) = \exp\left[-\int_0^T \lambda_{\mathbf{Y}}(t) \cdot P_{\mathbf{Y}|\mathbf{X},t}(\mathbf{x}, t) dt \right]$$

(10.49)

in which $\lambda_{\mathbf{Y}}(t)$ is the mean occurrence rate of the time-dependent random variable combination, e.g. the live load combination; and $P_{\mathbf{Y}|\mathbf{X},t}(\mathbf{x}, t)$ represents the probability of failure at time t conditioned on $\mathbf{x}$, which can be written as

$$P_{Y|X,t}(\mathbf{x}, t) = P\{G[\mathbf{x}, \mathbf{Y}(t), t] \leq 0\} = \int_{G(\mathbf{x},\mathbf{Y}(t),t) \leq 0} f_{\mathbf{Y}}(\mathbf{y}, t)d\mathbf{y} \qquad (10.50)$$

where $G[\mathbf{x}, \mathbf{Y}(t), t] \leq 0$ is the domain region of $\mathbf{Y}(t)$; and $f_{\mathbf{Y}}(\mathbf{y}, t)$ denotes the joint PDF of $\mathbf{Y}(t)$ at time t. Note that once $\mathbf{x}$ and t are determined, $P_{Y|X,t}(\mathbf{x}, t)$ in Eq. (10.49) can be assessed using traditional structural reliability analysis methods.

Assembling Eqs. (10.48)–(10.50), the problem of the time-dependent probability of failure within time interval $(0, T]$ can then be formulated mathematically as follows

$$P_f(T) = 1 - \int_{\Omega_{\mathbf{X}}} \exp\left\{ -\int_0^T \lambda_{\mathbf{Y}}(t) \cdot \left[\int_{G(\mathbf{x},\mathbf{Y}(t),t) \leq 0} f_{\mathbf{Y}}(\mathbf{y}, t)d\mathbf{y} \right] dt \right\} f_{\mathbf{X}}(\mathbf{x})d\mathbf{x} \qquad (10.51)$$

When the correlation between $\mathbf{X}$ and $\mathbf{Y}(t)$ at the specified time point is considered, the transformation of $\mathbf{X}$ and $\mathbf{Y}(t)$ into independent standard normal space is first conducted using the inverse normal transformation, e.g. Rosenblatt transformation (Hohenbichler and Rackwitz 1981), Nataf transformation (Liu and Der Kiureghian 1986), third-moment transformation (Lu et al. 2017a), or fourth-moment transformation (Lu et al. 2020), and then Eq. (10.51) can be rewritten as.

$$P_f(T) = 1 - \int_{\Omega_{\mathbf{U}_{\mathbf{X}}}} \exp\left\{ -\int_0^T \lambda_{\mathbf{Y}}(t) \cdot \left[\int_{G[T^{-1}(\mathbf{u}_{\mathbf{X}},\mathbf{U}_{\mathbf{Y}}),t] \leq 0} f_{\mathbf{U}_{\mathbf{Y}}}(\mathbf{u}_{\mathbf{Y}}, t)d\mathbf{u}_{\mathbf{Y}} \right] dt \right\} f_{\mathbf{U}_{\mathbf{X}}}(\mathbf{u}_{\mathbf{X}})d\mathbf{u}_{\mathbf{X}}$$

$$(10.52)$$

where $T^{-1}(\cdot)$ is the inverse normal transformation; $\Omega_{\mathbf{U}_{\mathbf{X}}}$ is the domain region of $\mathbf{X}$ in standard normal space; $\mathbf{U}_{\mathbf{X}}$ and $\mathbf{U}_{\mathbf{Y}}$ are independent standard normal random vectors, respectively; and $\mathbf{u}_{\mathbf{X}}$ and $\mathbf{u}_{\mathbf{Y}}$ are the samples of $\mathbf{U}_{\mathbf{X}}$ and $\mathbf{U}_{\mathbf{Y}}$, respectively.

10.4.3 Fast Integration Algorithms for the Time-Dependent Failure Probability

From Eq. (10.51) or Eq. (10.52), it can be observed that the time-dependent reliability analysis involves a multi-dimensional integral over time domain and random-variate space. When the performance function is complicated, implicit, or involves multiple random variables, the computation of the multi-dimensional integral by direct integration is almost impossible. Therefore, it is necessary to develop an efficient approach for evaluating the time-dependent failure probability. In this section, Gauss-Legendre quadrature and point-estimate method are utilised to transform the integrals with respect to time domain and random-variate space to a form of dual summation. The time-dependent failure probability is then expressed as a function of a series of probabilities of failure with time-dependent random variables at specified time points conditioned on the estimating points of the time-independent random variables. The description of the fast integrations algorithms for Eq. (10.51) is described in detail as follows, and the solution of Eq. (10.52) can be disposed in a similar way.

10.4.3.1 Gauss-Legendre Quadrature for Disassembling the Integral with Respect to Time Domain

Letting $t = T \cdot \tau/2 + T/2$, the integral interval of time t, $(0, T]$, in Eq. (10.49) can be transformed into $(-1, 1]$ for τ, which leads to

$$P_{Y|X}(\mathbf{x}, T) = \exp\left[-\frac{T}{2}\int_{-1}^{1} \lambda_Y\left(\frac{T}{2}\tau + \frac{T}{2}\right)\cdot P_{Y|X,t}\left(\mathbf{x}, \frac{T}{2}\tau + \frac{T}{2}\right)d\tau\right] \tag{10.53}$$

Using the Gauss-Legendre quadrature (Abramowitz and Stegun 1972), Eq. (10.53) can be rewritten as

$$P_{Y|X}(\mathbf{x}, T) \approx \exp\left[-\frac{T}{2}\cdot\sum_{k=1}^{m_T} \lambda_Y\left(\frac{T}{2}\tau_k + \frac{T}{2}\right)\cdot P_{Y|X,t}\left(\mathbf{x}, \frac{T}{2}\tau_k + \frac{T}{2}\right)\cdot w_k\right] \tag{10.54}$$

where τ_k and w_k are the abscissae and the corresponding weights for Gauss quadrature with weight function 1, respectively; and m_T is the number of the abscissae. The abscissae τ_k is defined as the roots of the m_Tth-order Legendre polynomial, which may be expressed using Rodrigues' formula (Abramowitz and Stegun 1972):

$$P_{m_T}(\tau) = \frac{1}{2^{m_T}m_T!}\cdot\frac{d^{m_T}}{d\tau^{m_T}}\left(\tau^2 - 1\right)^{m_T} \tag{10.55a}$$

in which $d^{m_T}/d\tau^{m_T}$ denotes the m_Tth-order derivative about τ and $m_T!$ is the factorial of m_T. The weights w_k are obtained from the following function (Abramowitz and Stegun 1972):

$$w_k = \frac{2}{\left(1 - \tau_k^2\right)\left[P'_{m_T}(\tau_k)\right]^2} \tag{10.55b}$$

where $P'_{m_T}(\tau_k)$ is the derivative of $P_{m_T}(\tau)$ at τ_k.

For $m_T = 3$, 4, and 5, the abscissae and their corresponding weights calculated using Eqs. (10.55a) and (10.55b) are listed in Table 10.7.

10.4.3.2 Point-Estimate Method for Disassembling the Integral Corresponding to Random Variables

Substituting Eq. (10.54) into Eq. (10.49), one obtains

$$P_f(T) = 1 - \int_{\Omega_X} \exp\left[-\frac{T}{2}\sum_{k=1}^{m_T}\lambda_Y\left(\frac{T}{2}\tau_k + \frac{T}{2}\right)\cdot P_{Y|X,t}\left(\mathbf{x}, \frac{T}{2}\tau_k + \frac{T}{2}\right)\cdot w_k\right]\cdot$$
$$f_X(\mathbf{x})d\mathbf{x} = 1 - E\left[P_{Y|X}(\mathbf{X}, T)\right] \tag{10.56a}$$

Table 10.7 The abscissae τ_k and the corresponding weights w_k of Gauss-Legendre quadrature for the number of the abscissae m_T = 3, 4, and 5.

Number of abscissae, m_T	3		4		5		
Abscissae, τ_k	0	$\pm\sqrt{\dfrac{3}{5}}$	$\pm\sqrt{\dfrac{3}{7}-\dfrac{2}{7}\sqrt{\dfrac{6}{5}}}$	$\pm\sqrt{\dfrac{3}{7}+\dfrac{2}{7}\sqrt{\dfrac{6}{5}}}$	0	$\pm\dfrac{1}{3}\sqrt{5-2\sqrt{\dfrac{10}{7}}}$	$\pm\dfrac{1}{3}\sqrt{5+2\sqrt{\dfrac{10}{7}}}$
Weights, w_k	$\dfrac{8}{9}$	$\dfrac{5}{9}$	$\dfrac{18+\sqrt{30}}{36}$	$\dfrac{18-\sqrt{30}}{36}$	$\dfrac{128}{225}$	$\dfrac{322+13\sqrt{70}}{900}$	$\dfrac{322-13\sqrt{70}}{900}$

where $E(\cdot)$ denotes the expectation; and $P_{Y|X}(\mathbf{X}, T)$ is

$$P_{Y|X}(\mathbf{X}, T) = \exp\left[-\frac{T}{2} \cdot \sum_{k=1}^{m_T} \lambda_Y\left(\frac{T}{2}\tau_k + \frac{T}{2}\right) \cdot P_{Y|X,t}\left(\mathbf{X}, \frac{T}{2}\tau_k + \frac{T}{2}\right) \cdot w_k\right] \quad (10.56b)$$

Equation (10.56a) shows that the key issue of evaluating the time-dependent failure probability is to determine the mean of $P_{Y|X}(\mathbf{X}, T)$, which can be obtained by the point-estimate method in standard normal space (Zhao and Ono 2000a) as described in the chapter on Moment Evaluation for Performance Functions.

Based on the inverse normal transformation (Hohenbichler and Rackwitz 1981; Liu and Der Kiureghian 1986; Lu et al. 2017a; Lu et al. 2020), $P_{Y|X}(\mathbf{X}, T)$ can be rewritten as

$$P_{Y|X}(\mathbf{X}, T) = \exp\left\{-\frac{T}{2} \cdot \sum_{k=1}^{m_T} \lambda_Y\left(\frac{T}{2}\tau_k + \frac{T}{2}\right) \cdot P_{Y|X,t}\left[T^{-1}(\mathbf{U_X}), \frac{T}{2}\tau_k + \frac{T}{2}\right] \cdot w_k\right\}$$
$$(10.57)$$

in which $\mathbf{U_X}$ is n_X-dimensional independent standard normal random vector; and n_X is the number of the random variables in $\mathbf{X}$. Once T is obtained and m_T, τ_k, and w_k are calculated using Eqs. (10.55a) and (10.55b) or Table 10.7, Eq. (10.57) is a function associated with the parameter $\mathbf{U_X}$, which is denoted as $h(\mathbf{U_X})$ for brevity.

To reduce the computation burden, the bivariate dimension-reduction method (Xu and Rahman 2004) can be adopted, and the function $h(\mathbf{U_X})$ can be approximated by

$$h(\mathbf{U_X}) \approx h_2 - (n_X - 2)h_1 + \frac{(n_X - 1)(n_X - 2)}{2}h_0 \quad (10.58)$$

where h_2 is a summation of $n_X(n_X - 1)/2$ two-dimensional functions; h_1 is a summation of n_X one-dimensional functions; and h_0 is a constant. h_2, h_1, and h_0 are, respectively, expressed as

$$h_2 = \sum_{i<j} h(0, \cdots, U_i, \cdots, U_j, \cdots, 0) = \sum_{i<j} h_{ij}(U_i, U_j) \quad (10.59a)$$

$$h_1 = \sum_{i=1}^{n_X} h(0, \cdots, U_i, \cdots, 0) = \sum_{i=1}^{n_X} h_i(U_i) \quad (10.59b)$$

$$h_0 = h(0, \cdots, 0, \cdots, 0) \quad (10.59c)$$

in which $h_{ij}(\cdot)$ is a two-dimensional function, $i, j = 1, 2, \ldots, n_X$ and $i < j$; and $h_i(\cdot)$ is a one-dimensional function.

The mean of $P_{Y|X}(\mathbf{X}, T)$ can then be computed as

$$E[P_{Y|X}(\mathbf{X}, T)] = E[h(\mathbf{U_X})] = \sum_{i<j} \mu_{ij} - (n_X - 2) \sum_{i=1}^{n_X} \mu_i + \frac{(n_X - 1)(n_X - 2)}{2}h_0 \quad (10.60)$$

where μ_{ij} and μ_i are the mean of $h_{ij}(\cdot)$ and $h_i(\cdot)$, respectively. According to point-estimate method in standard normal space (Zhao and Ono 2000a), μ_{ij} and μ_i can be estimated as

$$\mu_{ij} = \sum_{r=1}^{m_X} \sum_{k=1}^{m_X} p_r p_k h_{ij}(u_{ir}, u_{jk}) \tag{10.61a}$$

$$\mu_i = \sum_{r=1}^{m_X} p_r h_i(u_{ir}) \tag{10.61b}$$

where u_{ir}, u_{jk}, p_r, and p_k are estimating points and the corresponding weights in standard normal space, respectively; and m_X is the number of estimating points in one-dimensional case. The estimating points and their corresponding weights can be readily from Table 3.3 in the chapter on Moment Evaluation for Performance Functions.

10.4.3.3 Conditional Probability of Failure at the Given Time Point

Through the preceding approaches, the computation of time-dependent failure probability is transformed into the calculation of the time-invariant probability $P_{Y|X,t}[T^{-1}(u_X), t]$ (i.e. $P\{G[T^{-1}(u_X), Y, t] \leq 0\}$), in which $u_X = (0, ..., u_i, ..., 0)$ or $(0, ..., u_i, ..., u_j, ..., 0)$ and $t = (\tau_k + 1) \cdot T/2$. The time-invariant probability can then be evaluated FORM, SORM, methods of moment, and simulation methods. Substituting all the results of $P_{Y|X,t}[T^{-1}(u_X), t]$ into Eqs. (10.61a) and (10.61b), (10.60), and (10.56a), the time-dependent failure probability of a structure during the time interval (0, T] can be obtained.

In summary, the computation procedure of the developed method for evaluating the time-dependent failure probability within a given period (0, T] is described as follows, and the flowchart is illustrated in Figure 10.10a, in which the computation procedure of μ_i is displayed in Figure 10.10b and the computation procedure of μ_{ij} is similar to that of μ_i.

1) Based on the Gauss-Legendre quadrature, the time-dependent probability of failure, P_f (T), is assessed using Eq. (10.56a), in which the abscissas, τ_k, their corresponding weights, w_k, can be determined from Eqs. (10.55a) and (10.55b) given the number of abscissa, m_T.
2) Based on the inverse normal transformation and the correlation coefficients among time-independent random variables, we can transform $P_{Y|X}[X, T]$ into $h(U_X)$.
3) Using bivariate dimension-reduction integration method, $h(U_X)$ can be approximated by Eq. (10.58).
4) Utilising point-estimate method to obtain the mean of $P_{Y|X}[X, T]$, i.e. $h(U_X)$, in which the estimating points in independent standard normal space, u_r, their corresponding weights, p_r, can be determined from Table 3.3 in the chapter on Moment Evaluation once the number of estimating point, m_X, is selected.
5) Calculate μ_i and μ_{ij}. First, assess $P_{Y|X,t}[T^{-1}(u_X), t]$, in which $u_X = (0, ..., u_i, ..., 0)$ or $(0, ..., u_i, ..., u_j, ..., 0)$ and $t = (\tau_k + 1) \cdot T/2$, by employing traditional techniques for structural reliability; then, compute $P_{Y|X}[T^{-1}(u_X), T]$, i.e. $h(u_X)$, using Eq. (10.57); finally, derive μ_i and μ_{ij} by Eqs. (10.61a) and (10.61b).
6) Obtain $P_f(T)$ with the aid of Eqs. (10.56a) and (10.60).

(a) (b)

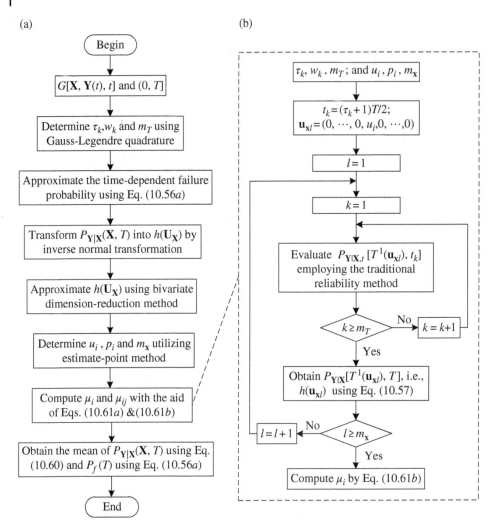

Figure 10.10 Flowchart of the moment-based approach for time-dependent reliability assessment. (a) Main flow chart; (b) Computation procedure of μ_i.

Example 10.5   Illustration of the Computational Procedure of the Moment-Based Approach

This example considers a simple performance function, which has been studied by Mori and Ellingwood (1993b) and Li et al. (2015) and is formulated as

$$G(\mathbf{X}, \mathbf{Y}(t), t) = R_0 \cdot (1 - 0.005t) - S_D - S_Q \tag{10.62}$$

where R_0 is the original resistance and follows a lognormal distribution with the mean of $1.15R_n$ (R_n is the nominal resistance) and coefficient of variation (COV) of 0.15; S_D is the dead load effect and is assumed to be deterministic as $1.00S_{Dn}$ (S_{Dn} is the nominal value);

and S_Q is the live load effect and follows a Gumbel distribution with the mean of $0.4S_{Qn}$ (S_{Qn} is the nominal value) and COV of 0.5. The combination of nominal live load effect and dead load effect need to satisfy the following design equation:

$$0.9R_n = 1.4S_{Dn} + 1.7S_{Qn} \tag{10.63}$$

For simplicity, it is assumed that $S_{Dn} = S_{Qn}$, and R_0 and S_Q are mutually independent. The service life is considered as 60 years. The mean occurrence rate of live load, λ_{S_Q}, is 0.5 per year. To elaborate the moment-based approach, the detailed procedure for evaluating the time-dependent probability of failure within the time interval (0, 60] is described as follows.

For this example, the time-independent random vector $\mathbf{X}$ includes R_0 only, and the time-dependent random vector $\mathbf{Y}(t)$ includes S_Q only. According to Eq. (10.51), the time-dependent probability of failure with the period (0, 60] can be formulated as

$$P_f(60) = 1 - \int_{-\infty}^{+\infty} \exp\left\{ -\int_0^{60} 0.5 \cdot \left[\int_{G(r_0, S_Q, t) \leq 0} f_{S_Q}(s_Q) ds_Q \right] dt \right\} f_{R_0}(r_0) dr_0 \tag{10.64}$$

Based on Eq. (10.56a) and the four abscissa estimate of Gauss-Legendre quadrature, Eq. (10.64) can be computed as

$$P_f(60) = 1 - \int_{-\infty}^{+\infty} \exp\left[-15 \times \sum_{k=1}^{4} P_{S_Q|R_0,t}(r_0, t_k) \times w_k \right] f_{R_0}(r_0) dr_0 = 1 - E\left[P_{S_Q|R_0}(R_0, 60) \right] \tag{10.65}$$

in which τ_k and w_k are obtained from Table 10.7 and $t_k = T \cdot \tau_k/2 + T/2$ as: $t_1 = 4.165911$, $w_1 = 0.3478548$; $t_2 = 19.80057$, $w_2 = 0.6521452$; $t_3 = 40.19943$, $w_3 = 0.6521452$; $t_4 = 55.834089$, and $w_4 = 0.3478548$.

Using the inverse Rosenblatt transformation (Hohenbichler and Rackwitz 1981), $P_{S_Q|R_0}(R_0, 60)$ can be rewritten as

$$P_{S_Q|R_0}(R_0, 60) = \exp\left\{ -15 \times \sum_{k=1}^{4} P_{S_Q|R_0,t}\left[T^{-1}(U_{R_0}), t_k \right] \times w_k \right\} \approx h(U_{R_0}) \tag{10.66}$$

where U_{R_0} is independent standard normal random variable.

Using the seven-point estimate method in standard normal space, the mean of $P_{S_Q|R_0}(R_0, 60)$ can be computed as

$$E\left[P_{S_Q|R_0}(R_0, 60) \right] = E[h(U_{R_0})] = \sum_{i=1}^{7} \exp\left\{ -15 \times \sum_{k=1}^{4} P_{S_Q|R_0,t}\left[T^{-1}(u_i), t_k \right] \times w_k \right\} \times p_i \tag{10.67}$$

in which u_i and p_i are readily obtained from Table 3.3 (in the chapter previously mentioned), as: $u_1 = -u_7 = -3.7504397$, $p_1 = p_7 = 5.4826886 \times 10^{-4}$; $u_2 = -u_6 = -2.3667594$, $p_2 = p_6 = 3.0757124 \times 10^{-2}$; $u_3 = -u_5 = -1.1544054$, $p_3 = p_5 = 0.24012318$; $u_4 = 0$, and $p_4 = 0.45714286$.

Since u_i and t_k are determinate through the above approaches, the failure probability $P_{S_Q|R_0,t}[T^{-1}(u_i), t_k]$ can be computed using methods of moment and is listed in Table 10.8.

Table 10.8 Conditional probability of failure at the given time point (T = 60 years).

	The failure probability, $P_{S_Q \mid R_0, t}\left[T^{-1}(u_i), t_k\right]$			
	Abscissa, t_1	Abscissa, t_2	Abscissa, t_3	Abscissa, t_4
Estimating point, u_1	3.485×10^{-3}	1.067×10^{-2}	4.533×10^{-2}	1.328×10^{-1}
Estimating point, u_2	1.391×10^{-4}	5.526×10^{-4}	3.339×10^{-3}	1.320×10^{-2}
Estimating point, u_3	4.526×10^{-6}	2.364×10^{-5}	2.043×10^{-4}	1.067×10^{-3}
Estimating point, u_4	9.244×10^{-8}	6.587×10^{-7}	8.540×10^{-6}	6.085×10^{-5}
Estimating point, u_5	9.088×10^{-10}	9.367×10^{-9}	1.965×10^{-7}	2.025×10^{-6}
Estimating point, u_6	2.773×10^{-12}	4.539×10^{-11}	1.741×10^{-9}	2.849×10^{-8}
Estimating point, u_7	8.882×10^{-16}	2.820×10^{-14}	2.491×10^{-12}	7.739×10^{-11}

Substituting the results of $P_{S_Q \mid R_0, t}[T^{-1}(u_i), t_k]$ into Eqs. (10.67) and (10.65), the time-dependent probability of failure, $P_f(60)$, can be obtained as

$$P_f(60) = 5.592 \times 10^{-3} \tag{10.68}$$

According to the relationship between reliability index and probability of failure, the corresponding reliability index is 2.537, while the result of MCS with 14 900 000 deterministic runs (COV of MCS is 1.9%) is 2.537, respectively. It can be observed that: (i) The result obtained from the moment-based approach agrees well with that of MCS; and (ii) The moment-based approach method is very efficient since only 168 times of deterministic runs are required.

The time-dependent reliability indices within different periods $(0, T]$ obtained from the moment-based approach (168 deterministic runs for each period), and MCS with deterministic runs approximately ranged from 5 960 000 to 193 000 000 (COVs of MCS are less than 3.0%) are depicted in Figure 10.11. It indicates that the results obtained from the moment-based approach are in good agreement with those obtained by MCS in the entire investigation range.

Example 10.6 Time-Dependent Reliability Assessment of a Reinforced Concrete Beam

This example assesses the reliability of a simply supported reinforced concrete beam under bending moment, which has been analysed by Val (2007). The beam is subjected to uniformly distributed loads including dead and live loads and chloride attacking as shown in Figure 10.12a,b. The flexural performance function is

$$G(\mathbf{X}, \mathbf{Y}(t), t) = \alpha \cdot \min\left\{ A_s(t) f_y \left[d - \frac{A_s(t) f_y}{1.7 f_c b} \right], \frac{1}{3} f_c b d^2 \right\} - \frac{(D + Q) L^2}{8} \tag{10.69}$$

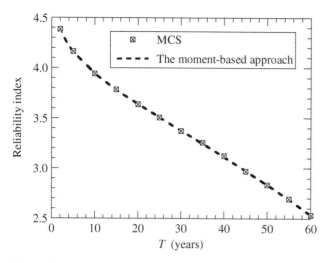

Figure 10.11 Time-dependent reliability indices obtained by different methods for Example 10.5.

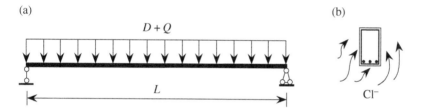

Figure 10.12 Beam subjected to uniform loads and chloride attacking for Example 10.6. (a) A simple beam. (b) Chloride ion penetration.

where α is the model uncertainty of the resistance; f_y is the yield strength of reinforcing steel; f_c is the compressive strength of concrete; b is the cross-sectional width ($b = 0.35$ m); d is the effective depth of cross-section; L is the span of beam ($L = 10$ m); D and Q are the dead and live loads, respectively; and $A_s(t)$ is the total remaining cross-sectional area of longitudinal bars at time t, which is given as

$$A_s(t) = n_0 \frac{\pi D_0^2}{4} - \sum_{i=1}^{n_0} A_{p,i}(t) \tag{10.70}$$

where n_0 is the number of reinforcing bars ($n_0 = 9$); D_0 is the initial diameter of reinforcing bar ($D_0 = 25.4$ mm); and $A_{p,i}(t)$ is the cross-sectional area of pitting corrosion in the ith reinforcing bar at time t, which can be determined by the hemispherical model (Val 2007) shown in Figure 10.13a–c, formulated as

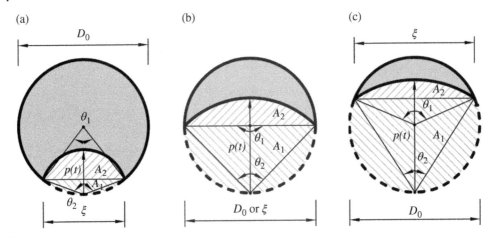

Figure 10.13 Pitting corrosion with hemispherical shape used in Example 10.6. (a) $0 < p(t) < D_0/\sqrt{2}$; (b) $p(t) = D_0/\sqrt{2}$; (c) $D_0/\sqrt{2} < p(t) < D_0$.

$$A_{p,i}(t) = \begin{cases} A_1 + A_2, & p(t) \le \dfrac{D_0}{\sqrt{2}} \\ \dfrac{\pi D_0^2}{4} - A_1 + A_2, & \dfrac{D_0}{\sqrt{2}} < p(t) \le D_0 \\ \dfrac{\pi D_0^2}{4}, & p(t) > D_0 \end{cases} \qquad (10.71)$$

in which

$$A_1 = \frac{1}{2}\left[\theta_1\left(\frac{D_0}{2}\right)^2 - \xi\left|\frac{D_0}{2} - \frac{p(t)^2}{D_0}\right|\right], A_2 = \frac{1}{2}\left[\theta_2 p(t)^2 - \xi\frac{p(t)^2}{D_0}\right] \qquad (10.72a)$$

$$\xi = 2p(t)\sqrt{1 - \left[\frac{p(t)}{D_0}\right]^2}, \theta_1 = 2\arcsin\frac{\xi}{D_0}, \theta_2 = 2\arcsin\frac{\xi}{2p(t)} \qquad (10.72b)$$

where $p(t)$ is the depth of a pit after t years since corrosion initiation and can be evaluated by

$$p(t) = 0.0116 i_{corr} t\kappa \qquad (10.73)$$

where i_{corr} denotes the corrosion current density and is assumed as $1\,\mu A/cm^2$ for illustration; and κ is the ratio between the maximum pit depth and the average corrosion depth.

It is assumed that the corrosion of reinforcements is initiated at time $t = 0$. The live load process is described by a Poisson point process and the mean occurrence rate is considered to be time-varying ($\lambda(t) = 1 + 0.01\,t$ year^{-1}). α, κ, f_y, f_c, d, and D are assumed as time-independent random variables, and Q is assumed as time-dependent random variable. All random variables are considered mutually independent and their statistical properties are listed in Table 10.9.

The time-dependent reliability indices within different periods $(0, T]$ can be readily obtained using the moment-based approach (8120 deterministic runs for each period)

Table 10.9 Statistical properties of random variables used in Case 2 of Example 10.6.

Variable	Mean	COV	Distribution
Model uncertainty, α	1.1	0.12	Normal
Parameter of pitting corrosion, κ	11.1	0.12	Gumbel
Steel strength, f_y (MPa)	490	0.10	Lognormal
Concrete compressive strength, f_c (MPa)	26.2	0.18	Lognormal
Effective depth of cross section, d (mm)	710	0.02	Normal
Dead load, D (kN/m)	21	0.10	Normal
Live load, $Q(t)$ (kN/m)	17.5	0.50	Gamma

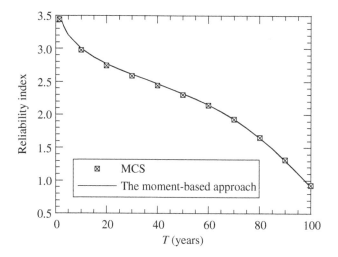

Figure 10.14 Time-dependent reliability indices by different methods for Case2 of Example 10.6.

and are depicted in Figure 10.14. For comparison, the MCS results with deterministic runs ranged from 2 130 000 to 18 100 000 (COVs of MCS are less than 2.0%) are also shown in Figure 10.14. It can be observed that for the complex performance function involving multiple random variables, the moment-based approach can still provide quite close results with those obtained from MCS.

Example 10.7 Time-Dependent Reliability Assessment of a Three-Bay Six-Story Frame Structure

Time-dependent reliability of a three-bay six-story steel frame structure under lateral loads is assessed in this example as shown in Figure 10.15 (He et al. 2014). The serviceability limit state is investigated. The failure criterion with respect to serviceability is reached when the

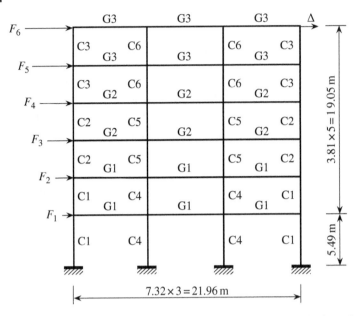

Figure 10.15 The three-bay six-story steel frame structure under lateral loads for Example 10.7.

horizontal top floor displacement exceeds a threshold value. The occurrence of lateral loads F_i $(i = 1, 2, ..., 6)$ is modelled as non-stationary Poisson point process with time-dependent CDF of intensity and time-variant mean occurrence rate $\lambda(t) = 1 + 0.01\,t\,\mathrm{yr}^{-1}$. Accordingly, the limit state function is

$$G(\mathbf{X}, \mathbf{Y}(t), t) = \Delta_{\lim} - \Delta(\mathbf{X}, \mathbf{Y}(t), t) \tag{10.74}$$

where $\Delta_{\lim}$ is the threshold value ($\Delta_{\lim} = 0.077$ m); and $\Delta(\mathbf{X}, \mathbf{Y}(t), t)$ is the horizontal top floor displacement, which can be obtained by FEM.

Here, it is assumed that the deterioration of the structure is described by the degradation of elastic modulus, and the time-dependent elastic modulus $E(t)$ is given by (Fan et al. 2017)

$$E(t) = E_0 \cdot (1 - \eta_1 \cdot t^{\eta_2}) \tag{10.75}$$

in which E_0 is the initial elastic modulus; and η_1 and η_2 are dimensionless parameters with $\eta_1 = 0.007$ and $\eta_2 = 0.68$. All elements have the same Young's modulus. The cross-sections of the frame are given in Table 10.10. The statistical properties of the random variables

Table 10.10 Sections (AISC) of frame members within Example 10.7.

Story	Exterior column	Interior column	Girder
1, 2	C1: W14 × 159	C4: W27 × 161	G1: W24 × 94
3, 4	C2: W14 × 132	C5: W27 × 114	G2: W24 × 76
5, 6	C3: W14 × 99	C6: W27 × 84	G3: W24 × 55

Table 10.11 Statistical properties of random variables used in Example 10.7.

Variable	Mean	COV	Distribution
Initial elastic modulus, E_0 (GPa)	200	0.10	Lognormal
Lateral load, F_6 (kN)	$120 \times (1 + 0.002\,t)$	0.40	Gumbel
Lateral load, F_5 (kN)	$101 \times (1 + 0.002\,t)$	0.40	Gumbel
Lateral load, F_4 (kN)	$83 \times (1 + 0.002\,t)$	0.40	Gumbel
Lateral load, F_3 (kN)	$64 \times (1 + 0.002\,t)$	0.40	Gumbel
Lateral load, F_2 (kN)	$45 \times (1 + 0.002\,t)$	0.40	Gumbel
Lateral load, F_1 (kN)	$27 \times (1 + 0.002\,t)$	0.40	Gumbel

considered in the reliability assessment are listed in Table 10.11. In this example, E_0 is independent of F_i ($i = 1, 2, ..., 6$). The correlation coefficient ρ among the lateral loads F_i is considered.

The moment-based approach is adopted to compute the time-dependent probability of failure. Four abscissa estimate of Gauss-Legendre quadrature and five-point estimate method are used to disassemble the time-dependent failure probability into the conditional probability of failure at the given time point. The time-invariant probability is obtained from the fourth-moment method as described in the chapters on Moment Evaluation and Direct Methods of Moment, in which the first four moments are evaluated using five-point estimate method with the aid of bivariate dimension-reduction method. Accordingly, the time-dependent reliability indices under different correlation coefficients ρ (i.e. $\rho = 0.6$ and 0.9) are calculated as shown in Figure 10.16, in which 8120 times of deterministic structural analyses for each period are required. To verify accuracy of the moment-based approach, the time-dependent reliability indices during the investigated periods are

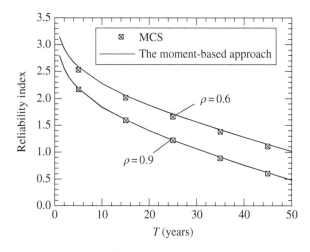

Figure 10.16 Time-dependent reliability indices with different correlation coefficients for Example 10.7.

calculated using MCS with deterministic runs ranging from 225 000 to 999 000 (COVs of MCS are less than 3.0%) and are also shown in Figure 10.16. It can be concluded from Figure 10.16 that: (i) the moment-based approach provides almost the identical results with MCS; and (ii) the correlation coefficient among the time-dependent random variables has a significant effect on the time-dependent structural reliability, and the time-dependent reliability index decrease as the increase of the positive correlation within the investigated time period $(0, T]$.

10.5 Summary

In this chapter, the time-invariant problems in structural reliability are discussed. Simulating stationary non-Gaussian vector process using the fourth-moment transformation model is first described, in which the translations of the MPDFs and CFM between a stationary non-Gaussian process and a Gaussian process are clarified. The transformation is then applied to evaluate the first passage probability of stationary non-Gaussian structural responses in structural dynamical reliability. Finally, a moment-based approach is introduced for time-dependent static structural reliability analysis considering correlated random variables, in which the integral of time dependent failure probability with respect to time domain and random-variate space are estimated by Gauss-Legendre quadrature and point-estimate method based on bivariate dimension-reduction integration, respectively. Accordingly, the time-dependent probability of failure within a given period is decomposed into a series of probabilities of failure with time-dependent random variables at specified time points conditioned on the estimating points of the time-independent random variables, which can be evaluated from state-of-the-art techniques for reliability assessment.

11

Methods of Moment for Structural Reliability with Hierarchical Modelling of Uncertainty

11.1 Introduction

Hierarchical models are rather common in decision-making under uncertainty. They arise when there is a 'correct' or 'ideal' (first-level) uncertainty model about a phenomenon of interest, but the model itself is uncertain. The model uncertainty is often called second-level uncertainty. Both the first-level uncertainty and the uncertainty of the first-level uncertainty model considered together make up a hierarchical problem. In the present case, it has two levels, but it is not difficult to conceive of hierarchical models with three, or four, or even more levels. Here, however, we shall only be concerned with two-level hierarchical models.

Broadly speaking, the uncertainty is categorised as aleatory uncertainty (objective uncertainty) and epistemic uncertainty (subjective uncertainty) (Ang 1970; Chen and Wan 2019). The former is caused by inherent variability, therefore it can be measured but not reduced, while the latter one is due to imperfect modelling, simplification, and limited availability of database, thus it can hardly be quantified (Melchers 1999). To facilitate structural reliability analysis, Der Kiureghian and Ditlevsen (2009) summarised the sources and characteristics of aleatory and epistemic uncertainties in practical structural reliability analysis, and pointed out that ignoring these uncertainties may lead to an incorrect estimation of failure probability. In the probability theory, the probability density function (PDF) or cumulative distribution function (CDF) of a fundamental random variable can be considered as the quantitative description of aleatory uncertainty, and one type of epistemic uncertainty can be considered as the uncertainty of this PDF or CDF (Chen and Wan 2019). It is necessary to find an effective way to capture and measure the probability of structural failure when considering the uncertainty in the probability distribution.

Reconsider the fundamental reliability problem

$$P_f = \int_{G(\mathbf{X}) \leq 0} f_{\mathbf{X}}(\mathbf{x})d\mathbf{x} \tag{11.1}$$

where P_f is the probability of structural failure; $\mathbf{X}$ is a random vector representing uncertain quantities such as loads, material properties, geometric dimensions, and boundary conditions; $f_{\mathbf{X}}(\mathbf{x})$ is the joint PDF of $\mathbf{X}$; and $G(\mathbf{X})$ is the limit state function or performance function.

The uncertainties contained directly in the components of $\mathbf{X}$ are the first-level uncertainty. If only the first-level uncertainties are considered, the PDFs of the basic random

Structural Reliability: Approaches from Perspectives of Statistical Moments, First Edition.
Yan-Gang Zhao and Zhao-Hui Lu.
© 2021 John Wiley & Sons Ltd. Published 2021 by John Wiley & Sons Ltd.

variables are generally assumed known, and distribution parameters of their PDFs are usually assumed to be given or certain. However, in practical engineering, one is faced with the problem of imperfect states of knowledge about such distributions. For example, the distribution parameters of the basic random variables such as mean and standard deviation involved in loads, environmental actions including chloride, temperature, oxygen, carbonation, moisture, and structural resistance are estimated from statistical data of limited sample sizes, and these distribution parameters may change as the amount of corresponding statistical data increases. All this results in uncertainties in the distribution types and their parameters. These distribution uncertainties associated with the basic random variables in **X**, which belong to second-level uncertainties, thus lead to uncertainty in the calculated failure probability and in the associated reliability index (Der Kiureghian 1989; Zhao et al. 2018c). It is therefore necessary to consider both levels of uncertainties simultaneously, which makes up a hierarchical model as just described.

Two typical types of hierarchical models will be discussed in structural reliability. One is structural reliability with consideration of the uncertainties in distribution parameters, and another one is the dynamic reliability considering uncertainties contained in structural parameters. Methods of moment will be applied to these two types of structural reliability problems to account for hierarchical modelling of uncertainty.

11.2 Formulation Description of the Structural Reliability with Hierarchical Modelling of Uncertainty

The formulation description of the structural reliability with hierarchical modelling of uncertainty is first illustrated for structural reliability problems to account for uncertainties in the distribution parameters. Assume that the uncertainties in the distribution parameters, such as the mean and standard deviation of the basic random variables in **X**, the distribution parameters are treated as a random vector **Θ**, whereby the joint probability density function $f_{\mathbf{X}}(\mathbf{x})$ becomes a conditional distribution function $f_{\mathbf{X,\Theta}}(\mathbf{x}, \boldsymbol{\theta})$. Therefore, the conditional probability of failure becomes

$$P_f(\boldsymbol{\Theta}) = \int_{G(\mathbf{X},\boldsymbol{\Theta}) \leq 0} f_{\mathbf{X,\Theta}}(\mathbf{x}, \boldsymbol{\theta}) d\mathbf{x} \tag{11.2}$$

where $G(\mathbf{X}, \boldsymbol{\Theta})$ expresses the performance function; $f_{\mathbf{X,\Theta}}(\mathbf{x}, \boldsymbol{\theta})$ is the joint PDF of **X** and **Θ**; and the conditional failure probability $P_f(\boldsymbol{\Theta})$ is a function of the distribution parameters **Θ**. Because the distribution parameters **Θ** are uncertain, the conditional failure probability $P_f(\boldsymbol{\Theta})$ is also uncertain. The corresponding conditional reliability index $\beta(\boldsymbol{\Theta})$, is also uncertain and is given by

$$\beta(\boldsymbol{\Theta}) = \Phi^{-1}[1 - P_f(\boldsymbol{\Theta})] \tag{11.3}$$

where $\Phi^{-1}(\cdot)$ denotes the inverse of the CDF of a standard normal random variable. Because $P_f(\boldsymbol{\Theta})$ and $\beta(\boldsymbol{\Theta})$ are random variables, they have PDFs as well as statistical moments, such as their mean, standard deviation, skewness, and kurtosis.

The most interest in practical engineering for this problem is the overall probability of failure with consideration of both the two levels of uncertainties in $\mathbf{X}$ and $\mathbf{\Theta}$. For random vector $\mathbf{X}$, whose joint PDF includes uncertain parameters $\mathbf{\Theta}$, the overall probability of failure is then defined as the expectation of the conditional failure probability $P_f(\mathbf{\Theta})$ over the outcome space of the uncertain parameters $\mathbf{\Theta}$, which can be formulated as

$$P_F = \int_{G(\mathbf{X},\mathbf{\Theta}) \leq 0} f_{\mathbf{X},\mathbf{\Theta}}(\mathbf{x}, \mathbf{\theta})d\mathbf{x}d\mathbf{\theta} \tag{11.4}$$

In most cases, Eq. (11.4) cannot be directly solved because of the difficulty in determining the explicit expression of the performance function $G(\mathbf{X}, \mathbf{\Theta})$ and the joint PDF $f_{\mathbf{X},\mathbf{\Theta}}(\mathbf{x}, \mathbf{\theta})$. This is because $\mathbf{\Theta}$ represents the distribution parameters of $\mathbf{X}$, and $\mathbf{X}$ is a function of $\mathbf{\Theta}$. However, the conditional failure probability of the structural system for given distribution parameter values $\mathbf{\Theta} = \mathbf{\theta}$ can be readily evaluated using state-of-the-art techniques such as the first- and second-order reliability methods, methods of moment, and simulation methods (Zhao and Ono 2001; Ang and Tang 1984; Ditlevsen and Madsen 1996; Choi et al. 2007). Therefore, the overall probability of failure incorporating the uncertainties in distribution parameters can be formulated generally as

$$P_F = \int_{\mathbf{\theta}} P_f(\mathbf{\Theta})f_{\mathbf{\Theta}}(\mathbf{\theta})d\mathbf{\theta} \tag{11.5}$$

where $P_f(\mathbf{\theta})$ is the conditional probability of failure for a given $\mathbf{\Theta} = \mathbf{\theta}$ (which can be evaluated from state-of-the-art techniques), and $f_{\mathbf{\Theta}}(\mathbf{\theta})$ is the joint PDF of $\mathbf{\Theta}$.

The overall probability of failure in Eq. (11.5) is essentially the mean of the conditional failure probability $P_f(\mathbf{\Theta})$ when the parameter uncertainties are considered for the basic random variables in $\mathbf{X}$. An advanced first-order second-moment method, which was developed from the first-order reliability method by introducing an auxiliary variable has been proposed by Wen and Chen (1987) and Zhao and Jiang (1992) for solving Eq. (11.5), in which the effect of distribution parameter uncertainties was discussed on the overall probability of failure. Hong (1996) proposed a procedure to evaluate the overall probability of failure by using the point-estimate method. Der Kiureghian (2008) derived a simple approximate formula by using the first-order approximation method to compute the mean of the conditional reliability index.

Another point of interest in practical engineering of this problem is to evaluate the uncertainty in $P_f(\mathbf{\Theta})$ or $\beta(\mathbf{\Theta})$ since both of them are functions of random variables $\mathbf{\Theta}$. For the sake of transparency in communicating risk, it is necessary to determine not only the mean value but also the quantile or even the probability distribution of the conditional failure probability $P_f(\mathbf{\Theta})$, or the corresponding conditional reliability index $\beta(\mathbf{\Theta})$. Der Kiureghian and Ditlevsen (2009) obtained the probability distributions of the conditional reliability index $\beta(\mathbf{\Theta})$ and the corresponding conditional probability of failure when the explicit PDF of $\beta(\mathbf{\Theta})$ could be easily determined. However, the explicit PDF of the conditional reliability index β $(\mathbf{\Theta})$ generally cannot be obtained in engineering practice. Ang and De Leon (2005) therefore utilised MCS to obtain both the mean and quantile of the conditional failure probability $P_f(\mathbf{\Theta})$.

In the following, the application of first-order reliability method (FORM) to the overall probability of failure with consideration of both the two levels of uncertainties will first be introduced; methods of moment and point-estimate method from previous chapters will then be applied to evaluate the overall structural reliability under such a condition. Other methods for evaluating the quantile or even the distribution of the conditional failure probability $P_f(\Theta)$, or the corresponding conditional reliability index $\beta(\Theta)$ using point-estimate methods are also introduced in Section 11.4.

11.3 Overall Probability of Failure Due to Hierarchical Modelling of Uncertainty

11.3.1 Evaluating Overall Probability of Failure Based on FORM

As described in Section 11.2, Eq. (11.5) represents the overall probability of failure due to hierarchical modelling of uncertainty. Since the performance function in Eq. (11.5) is unknown, it cannot be directly solved by the methods of moment and FORM. Instead, one might use the Monte Carlo simulation (MCS). Generating N sets of Θ according to its probability density function and using available methods for each set of Θ evaluating $P_f(\Theta)$, the overall probability can be calculated as

$$P_F = \frac{1}{N}\sum_{k=1}^{N} P_f(\theta_k) \tag{11.6}$$

where θ_k is a vector of the kth set of sample; and $P_f(\theta_k)$ is the conditional probability of failure for given $\Theta = \theta_k$, $k = 1, ..., N$.

Another alternative is to use the so-called second-moment method (Wen and Chen 1987). By expanding $P_f(\Theta)$ with respect to Θ, at the mean value of Θ, the probability of failure, P_F, can be estimated as the following equation where the second order derivatives of $P_f(\Theta)$ are required (Wen and Chen 1987)

$$P_F = P_f(\mu_\theta) + \frac{1}{2}\sum_i\sum_j \frac{\partial^2 P_f(\theta)}{\partial\theta_i\partial\theta_j}\rho_{ij}\sigma_i\sigma_j \tag{11.7}$$

The accuracy of Eq. (11.7) obviously depends on the variability and smoothness of $P_f(\Theta)$ with respect to Θ.

In order to solve Eq. (11.5) using the methods of moment and FORM, a performance function corresponding to Eq. (11.5) is needed. For this purpose, an auxiliary variable Θ_{n+1} can be introduced which has an arbitrary probability distribution but satisfies

$$F_{\Theta_{n+1}}(0\mid\theta) = P_f(\theta) \tag{11.8}$$

then we can easily obtain

$$P_F = \int_\Theta P_f(\theta)f(\theta)d\theta$$
$$= \int_\Theta\left[\int_{\Theta_{n+1}\leq 0} f_{\Theta_{n+1}}(\theta_{n+1}\mid\theta)d\theta_{n+1}\right]f(\theta)d\theta \tag{11.9}$$
$$= \int_{\Theta_{n+1}\leq 0}\left[\int_\Theta f_{\Theta_{n+1}}(\theta_{n+1}\mid\theta)f(\theta)d\theta\right]d\theta_{n+1}$$

Comparing Eq. (11.9) with Eq. (11.2), one can obtain the performance function can be obtained as

$$G(\Theta, \Theta_{n+1}) = \Theta_{n+1} \tag{11.10}$$

Since both Eqs. (11.4) and (11.5) are different expressions for the same problem, we have

$$P_f(\Theta) = \int_{G(\mathbf{X}|\Theta) \leq 0} f_{\mathbf{X}|\Theta}(\mathbf{x}|\theta) dx \tag{11.11}$$

Assume $z = G(\mathbf{x}|\theta)$, where z is an expression of the actual performance function and a function of $\mathbf{X}$, so we can obtain

$$F_Z(z) = \int_{G(\mathbf{X}|\Theta) \leq z} f_{\mathbf{X}|\Theta}(\mathbf{x}|\theta) dx \tag{11.12}$$

For $z = 0$

$$F_Z(z = 0) = P_f(\theta) \tag{11.13}$$

This means that although Eq. (11.10) might be different from the actual performance function (which is difficult to determine in practice), the limit state surfaces coincide for both performance functions in Eq. (11.13) and Eq. (11.8).

Since the reliability is not dependent on the probability distribution of Θ_{n+1} in Eq. (11.10), we can select an arbitrary distribution of Θ_{n+1} as long as it satisfies Eq. (11.8). For example, if the normal distribution is selected with the unit standard deviation for Θ_{n+1}, the mean value of Θ_{n+1} can be derived from Eq. (11.8) as

$$\mu_{\Theta_{n+1}} = \Phi^{-1}[P_f(\theta)] \tag{11.14}$$

Then the following performance function that is proposed by Wen and Chen (1987) can be obtained

$$G_U(\mathbf{U}, U_{n+1}) = U_{n+1} - \Phi^{-1}\{P_f[T^{-1}(\mathbf{u})]\} \tag{11.15}$$

where U_{n+1} is an auxiliary standard normal variable independent of $\mathbf{U}$; and $T^{-1}(\mathbf{u})$ is the inverse Rosenblatt transformation. Neither the distribution of the auxiliary variable, Θ_{n+1}, nor the Rosenblatt transformation from Θ_{n+1} to U_{n+1} is required for reliability analysis using Eqs. (11.10) or (11.15) as the performance function, so the difficulties described can thus be avoided.

Furthermore, we can select different distributions of Θ_{n+1} to obtain many forms of performance functions. If distributions of Θ_{n+1} satisfy Eq. (11.8), the limit state expressed by the selected performance function will coincide with the actual limit state. For example, if Θ_{n+1} fits with the Laplace's distribution, and assuming the probability density function is

$$f_{\Theta_{n+1}}(\Theta_{n+1}) = 0.5 \exp[|\Theta_{n+1} - a|]$$

the performance function in $\mathbf{U}$-space is

$$G_U(\mathbf{U}, U_{n+1}) = \ln[2\Phi(u_{n+1})] - \ln\{2P_f[T^{-1}(\mathbf{u})]\}$$

The general form of the performance function in $\mathbf{U}$-space can be expressed as

$$G_U(\mathbf{U}, U_{n+1}) = g[\Phi(u_{n+1})] - g\{P_f[T^{-1}(\mathbf{u})]\}$$

where g is a determinate function. It can be easily seen that the methods described above yield the same limit surface and the same design point as those of Eq. (11.15).

To graphically interpret the performance function in Eq. (11.10) (Zhao et al. 2014), assume that the performance function in Eq. (11.1) includes only two random variables, $G(\mathbf{X}) = G(x_1, x_2)$. Without loss of generality, suppose that Θ is the only random variable used to represent the model uncertainties. Then, the performance function considering the model uncertainty becomes $G(\mathbf{X}) = G(x_1, x_2, \theta)$, and the corresponding limit state surface defined in standard normal space (u_1, u_2, u_θ) is shown in Figure 11.1, where point A is the design point having minimum distance from the limit state surface to the initial point in the standard normal space. For the limit state probability $P_f(\theta)$ for given θ, the overall probability of failure in Eq. (11.4) can be solved as

$$P_F = \int_\theta P_f(\theta)d\theta \tag{11.16}$$

When the auxiliary variable, θ_{n+1}, that satisfies Eq. (11.8) is introduced, $\theta_{n+1} = 0$ represents a curve through point A such that each point on the curve has the minimum distance from the curve $G(\mathbf{X}) = G(x_1, x_2 \mid \theta)$ to the axis u_θ. For this case, the method reduces a three-dimensional problem to a two-dimensional one by utilising the characteristic that the limit state probability, with θ given, is easy to obtain. In general, for an $(n + m)$-dimensional problem, $G(\mathbf{X}, \mathbf{\Theta}) = 0$, where the dimension of $\mathbf{X}$ is m and that of $\mathbf{\Theta}$ is n, by utilising the characteristic that the limit state probability, with part of the variable $\mathbf{\Theta}$ given, is easy to obtain, the reliability problem can be reduced to an $n + 1$-dimensional one, and the design point can be searched for an $(n + 1)$-dimensional subsurface $G(\mathbf{\Theta}, \theta_{n+1}) = \theta_{n+1} = 0$ instead of the original limit surface $G(\mathbf{X}, \mathbf{\Theta}) = 0$.

The performance function expressed in Eq. (11.10) or (11.15) enables the evaluation of structural reliability considering the parameter uncertainties to be performed using FORM.

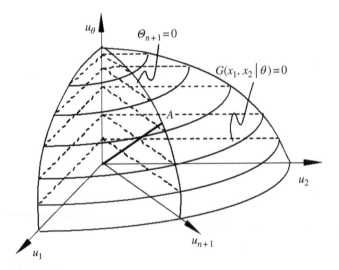

Figure 11.1 Graphical interpretation for the auxiliary variable.

As an important step of FORM, the gradients of the performance function, G, i.e. the inverse of the Jacobian matrix, is expressed as

$$[J_{n+1}]^{-1} = \left[\frac{\partial \boldsymbol{\theta}}{\partial \mathbf{u}}\right] = \begin{bmatrix} J_n^{-1} & \{0\}_n \\ \left\{\dfrac{\partial \theta_{n+1}}{\partial u_i}\right\}^T_{i \leq n} & \dfrac{\partial \theta_{n+1}}{\partial u_{n+1}} \end{bmatrix} \tag{11.17}$$

where

$$\left\{\frac{\partial \theta_{n+1}}{\partial u_i}\right\}_{i \leq n} = -[J_n^{-1}]^T \left\{\frac{\partial P_f(\boldsymbol{\theta})}{\partial \theta_i}\right\}_{i \leq n} \tag{11.18}$$

$$\frac{\partial \theta_{n+1}}{\partial u_{n+1}} = \phi(u_{n+1}) \tag{11.19}$$

where

$$u_{n+1} = \Phi^{-1}[P_f(\boldsymbol{\theta})] \tag{11.20}$$

where $\Phi(\cdot)$ and $\phi(\cdot)$ are the CDF and PDF of standard normal random variables, respectively; and $P_f(\boldsymbol{\theta})$ is the conditional probability of failure for given $\boldsymbol{\Theta} = \boldsymbol{\theta}$.

It is not always easy to obtain these gradients, especially for non-linear structural performance functions. The response surface approach (RSA) has been introduced (Yao and Wen 1996; Zhao et al. 1999) to avoid the required sensitivity analysis in FORM, where the response surface function is expressed as a polynomial of basic random variables in the original space. For simplification, the performance function, shown in Eq. (11.15), is approximated by the following second-order polynomial of standard normal random variables, in which mixed terms are neglected.

$$G(\mathbf{U}) = a_0 + U_{n+1} + \sum_{j=1}^{n} \gamma_j U_j + \sum_{j=1}^{n} \lambda_j U_j^2 \tag{11.21}$$

Here, a_0, γ_j, and λ_j are $(2n+1)$ regression coefficients with j ranging from 1 to n. If the practical performance function $G(\mathbf{U})$ in Eq. (11.15) is fitted by $G(\mathbf{U})$ in Eq. (11.21) at fitting points in the vicinity of the design point, regression coefficients a_0, γ_j, and λ_j can be determined from linear equations of a_0, γ_j, and λ_j obtained at each fitting point.

11.3.2 Evaluating Overall Probability of Failure Based on Methods of Moment

In this section, methods of moment will be used to obtain the overall probability of failure corresponding to the performance functions in Eqs. (11.10) or (11.15). If the first three or four moments of $G(\mathbf{X})$ in Eqs. (11.10) or (11.15) are obtained, reliability analysis can be conducted by approximating the distribution of a specific random variable with its known first three or four moments, as have been discussed in the chapter on Direct Methods of Moment.

For the performance function in Eq. (11.15), it can be approximated by

$$Z = G'(\mathbf{U}) = \sum_{i=1}^{n} (G_i - G_\mu) + G_\mu + U_{n+1} \tag{11.22}$$

where

$$G_\mu = -\Phi^{-1}\{P_f[T^{-1}(\boldsymbol{\mu})]\} \tag{11.23a}$$

$$G_i = -\Phi^{-1}\{P_f[T^{-1}(\mu_1, ..., u_i, ..., \mu_n)]\} \tag{11.23b}$$

Therefore, the first four moments of Eq. (11.22) can be expressed as

$$\mu_G = \sum_{i=1}^{n}(\mu_{G_i} - G_\mu) + G_\mu \tag{11.24a}$$

$$\sigma_G^2 = 1 + \sum_{i=1}^{n}\sigma_{G_i}^2 \tag{11.24b}$$

$$\alpha_{3G}\sigma_G^3 = \sum_{i=1}^{n}\alpha_{3G_i}\sigma_{G_i}^3 \tag{11.24c}$$

$$\alpha_{4G}\sigma_G^4 = 3 + \sum_{i=1}^{n}\alpha_{4G_i}\sigma_{G_i}^4 + 6\sum_{i=1}^{n-1}\sum_{j>i}^{n}\sigma_{G_i}^2\sigma_{G_j}^2 + 6\sum_{i=1}^{n}\sigma_{G_i}^2 \tag{11.24d}$$

where the first four moments of G_i can be obtained according to Eqs. (3.68a)–(3.68d) in the chapter on Moment Evaluation.

11.3.3 Evaluating Overall Probability of Failure Based on Direct Point Estimates Method

The overall failure probability in Eq. (11.5) can also be considered as the mean value of $P_f(\boldsymbol{\theta})$ with rspect to $f(\boldsymbol{\theta})$. Beside the FORM and the methods of moment, the probability of failure in Eq. (11.5) can be directly calculated using point estimated method (Zhao et al. 2018c) as described in the chapter on Method Evaluation. Rewrite Eq. (11.5) in standard u-space, the probability of failure considering parameter uncertainties is expressed as

$$P_F = \int_{\mathbf{u}} P_f[T^{-1}(\mathbf{u})] \, \phi(\mathbf{u})d\mathbf{u} \tag{11.25}$$

where $T^{-1}(\mathbf{u})$ denotes inverse Rosenblatt transformation, and $\phi(\mathbf{u})$ denotes the PDF of standard normal variables.

Using the univariate-dimension reduction based point estimate method in the chapter on Moment Evaluation, P_F can be point estimated as

$$P_F = \sum_{i=1}^{n}\mu_i - (n-1)P_f(\boldsymbol{\mu}) \tag{11.26}$$

where $\boldsymbol{\mu}$ represents the vector in which all the random variables take their mean values, and μ_i is the mean value of

$$P_{fi} = P_f[\mathbf{X}_i] = P_f[T^{-1}(\mathbf{U}_i)] \tag{11.27}$$

where $\mathbf{X}_i = [\mu_1, ..., \mu_{i-1}, x_i, \mu_{i+1}, ..., \mu_n]^T$; $\mathbf{U}_i = [u_{\mu 1}, ..., u_{\mu i-1}, u_i, u_{\mu i+1}, ..., u_{\mu n}]^T$; and $u_{\mu k}, k = 1, ..., n$ except i, is the kth value of u_μ, which is the vector in u-space corresponding to $\boldsymbol{\mu}$.

Since P_{fi} is a function of only one standard normal random variable u_i, its mean value can be point-estimated from

$$\mu_i = \sum_{k=1}^{m} P_k P_f \left[T^{-1}(u_{ik}) \right] \tag{11.28}$$

The estimating points u_{ik} and their corresponding weights P_k can be found in Table 3.3 in the chapter on Moment Evaluation.

If the bivariate-dimension reduction based point estimate method is used, P_F can be point estimated as

$$P_F = \sum_{i<j} \mu_{ij} - (n-2) \sum_{i<j} \mu_i + \frac{(n-1)(n-2)}{2} P_f(\boldsymbol{\mu}) \tag{11.29}$$

where μ_i is determined by Eq. (11.28), and μ_{ij} is the mean value of

$$P_{fij} = P_f \left[\mu_1, ..., X_i, ..., X_j, ..., \mu_n \right] = P_{fij}(X_i, X_j) \tag{11.30}$$

Since P_{fij} is a function of only two standard normal random variables u_i and u_j, its mean value can be point-estimated from

$$\mu_{ij} = \sum_{k_1=1}^{m} \sum_{k_2=1}^{m} P_{k_1} P_{k_2} \{ P_{fij} \left[T^{-1}(u_{i,k_1}), T^{-1}(u_{j,k_2}) \right] \}^k \tag{11.31}$$

where the estimating points u_{i,k_1}, u_{j,k_2} and their corresponding weights P_{k_1}, P_{k_2} can be found in Table 3.3 in the chapter on Moment Evaluation.

Example 11.1 Structural Reliability Including Uncertainties in Distribution Parameters

This example considers a bar subjected to tension stress, which has been investigated by Zhang and Mahadevan (2003). The bar fails when the applied load exceeds the tensile strength of the bar, and the performance function is simply expressed as

$$G(\mathbf{X}) = R - S \tag{11.32}$$

where R is the resistance of the bar and S is the applied load.

Assume that both R and S are independent normal variables with mean and standard deviation of μ_R, μ_S, σ_R, and σ_S, respectively. For this example, the reliability index is available as a closed-form equation in terms of the distribution parameters

$$\beta = \frac{\mu_R - \mu_S}{\sqrt{\sigma_R^2 + \sigma_S^2}} \tag{11.33}$$

and the failure probability is

$$P_f = \Phi(-\beta) \tag{11.34}$$

If the distribution parameters μ_R, μ_S, σ_R, and σ_S are assumed as random variables and their probabilistic information is listed in Table 11.1, the overall failure probability can be formulated by

Table 11.1 Probabilistic information of the distribution parameters in Example 11.1.

Parameters	Mean	Standard deviation	Distribution
$\mu_R\ (\theta_1)$	50	2	Lognormal distribution
$\mu_s\ (\theta_2)$	35	2	Lognormal distribution
$\sigma_s\ (\theta_3)$	5	1	Lognormal distribution
$\sigma_R\ (\theta_4)$	7	1.4	Lognormal distribution

$$P_F = \int_\theta P_f(\mathbf{\theta}) f_\Theta(\mathbf{\theta}) d\mathbf{\theta} = \int_{\mu_R,\mu_s,\sigma_R,\sigma_s} P_f(\mu_R,\mu_s,\sigma_R,\sigma_s) f(\mu_R,\mu_s,\sigma_R,\sigma_s) d\mu_R d\mu_s d\sigma_R d\sigma_s \quad (11.35)$$

in which

$$P_f(\mathbf{\theta}) = P_f(\mu_R,\mu_s,\sigma_R,\sigma_s) = \Phi\left(-\frac{\mu_R - \mu_S}{\sqrt{\sigma_R^2 + \sigma_S^2}}\right) \quad (11.36)$$

If the methods of moment are used, the detailed procedure is described as follows. According to Eq. (11.15), the performance function is expressed as

$$G_U(\mathbf{U}, U_{n+1}) = U_{n+1} - \Phi^{-1}\{P_f[T^{-1}(\mathbf{u})]\} \quad (11.37)$$

where $P_f(\cdot)$ is shown in Eq. (11.36).

Using univariate-dimension reduction based point estimate method, Eq. (11.37) can be approximated by

$$Z = G'(\mathbf{U}) = \sum_{i=1}^{4}(G_i - G_\mu) + G_\mu + U_{n+1} \quad (11.38)$$

where

$$G_\mu = -\Phi^{-1}\{P_f[T^{-1}(\mathbf{\mu})]\} = \frac{\mu_{\theta_1} - \mu_{\theta_2}}{\sqrt{\left(\mu_{\theta_3}\right)^2 + \left(\mu_{\theta_4}\right)^2}} = \frac{50 - 35}{\sqrt{7^2 + 5^2}} = 1.7437$$

$$G_1 = -\Phi^{-1}\{P_f[T^{-1}(u_1, ..., \mu_i, ..., \mu_n)]\} = \frac{\theta_1 - \mu_{\theta_2}}{\sqrt{\left(\mu_{\theta_3}\right)^2 + \left(\mu_{\theta_4}\right)^2}} = \frac{\theta_1 - 35}{\sqrt{74}}$$

$$G_2 = \frac{50 - \theta_2}{\sqrt{74}}, G_3 = \frac{15}{\sqrt{\theta_3^2 + 25}}, G_4 = \frac{15}{\sqrt{49 + \theta_4^2}}$$

Using five-point estimate in standardised normal space, the first four moments of G_1, are readily obtained as

$$\mu_{G_1} = \sum_{k=1}^{5} P_k G_1[T^{-1}(u_k)] = P_{2-} G_1[T^{-1}(u_{2-})] + P_{1-} G_1[T^{-1}(u_{1-})] + P_0 G_1[T^{-1}(u_0)]$$

$$+ P_{1+} G_1[T^{-1}(u_{1+})] + P_{2+} G_1[T^{-1}(u_{2+})]$$

$$= 0.011257 \times 1.1121 + 0.22208 \times 1.4326 + 0.5333 \times 1.7391 + 0.22208 \times 2.0626 + 0.011257 \times 2.4419$$

$$= 1.7437$$

$$\sigma_{G_1} = \sqrt{\sum_{k=1}^{5} P_k \left\{ G_1[T^{-1}(u_k)] - \mu_{G_1} \right\}^2} = 0.2325$$

$$\alpha_{3G_1} = \frac{\sum_{k=1}^{5} P_k \cdot \left\{ G_1[T^{-1}(u_k)] - \mu_{G_1} \right\}^3}{\sigma_{G_1}^3} = 0.1201$$

$$\alpha_{4G_1} = \frac{\sum_{k=1}^{5} P_k \cdot \left\{ G_1[T^{-1}(u_k)] - \mu_{G_1} \right\}^4}{\sigma_{G_1}^4} = 3.0256$$

Similarly, the first four moments of G_2, G_3, and G_4 are respectively, obtained as

$$\mu_{G_2} = 1.7437, \sigma_{G_2} = 0.2325, \alpha_{3G_2} = -0.1716, \alpha_{4G_2} = 3.0524$$

$$\mu_{G_3} = 1.7441, \sigma_{G_3} = 0.1147, \alpha_{3G_3} = -0.5471, \alpha_{4G_3} = 3.3181$$

$$\mu_{G_4} = 1.7653, \sigma_{G_4} = 0.2228, \alpha_{3G_4} = -0.0291, \alpha_{4G_4} = 2.8192$$

According to Eqs. (11.24a) to (11.24d), the first four moments of the performance function of Eq. (11.38) are approximately obtained as

$$\mu_G = 1.766, \quad \sigma_G = 1.082, \quad \alpha_{3G} = -0.0014, \quad \alpha_{4G} = 2.9999$$

Using Eqs. (4.3), (4.27), and (4.55a) from the chapter on Direct Methods of Moment, the moment-based reliability indices are as follows:

$$\beta_{2M} = \beta_{3M} = \beta_{4M} = 1.632 \text{ with } P_F = 5.134 \times 10^{-2}$$

If the direct point-estimate method is utilised, according to Eq. (11.26), the overall failure probability can be approximated as

$$P_F = \sum_{i=1}^{4} \mu_i - (4-1)P_f(\boldsymbol{\mu}) = \sum_{i=1}^{4} \mu_i - 3P_f(\boldsymbol{\mu}) \tag{11.39}$$

where

$$P_f(\boldsymbol{\mu}) = \Phi\left(-\frac{50 - 35}{\sqrt{7^2 + 5^2}} \right) = 0.0406$$

and μ_i is the mean value of $P_f(\theta_i)$, and for this example $P_f(\theta_i)$ can be expressed as the following equations.

$$P_f(\theta_1) = \Phi\left(-\frac{\theta_1 - 35}{\sqrt{74}} \right), \quad P_f(\theta_2) = \Phi\left(-\frac{50 - \theta_2}{\sqrt{74}} \right), \quad P_f(\theta_3) = \Phi\left(-\frac{15}{\sqrt{\theta_3^2 + 5^2}} \right),$$

$$P_f(\theta_4) = \Phi\left(-\frac{15}{\sqrt{7^2 + \theta_4^2}} \right)$$

Using five-points estimate, the estimating points of θ_1 (i.e., μ_R) in original normal space are given as follows.

$$\mu_{R2-} = 44.5669, \quad \mu_{R1-} = 47.3241, \quad \mu_{R0} = 49.96, \quad \mu_{R1+} = 52.7428, \quad \mu_{R2+} = 56.0059$$

Therefore, the mean of $P_f(\theta_1)$, μ_1, is readily obtained as

$$\mu_1 = \sum_{k=1}^{5} P_k P_f(\mu_{Rk}) = P_{2-}P_f(\mu_{R2-}) + P_{1-}P_f(\mu_{R1-}) + P_0 P_f(\mu_{R0}) + P_{1+}P_f(\mu_{R1+}) + P_{2+}P_f(\mu_{R2+})$$

$$= 0.011257 \times 1.3304 \times 10^{-1} + 0.22208 \times 7.5979 \times 10^{-2} + 0.5333 \times 4.1012 \times 10^{-2}$$
$$+ 0.22208 \times 1.9577 \times 10^{-2} + 0.011257 \times 7.3054 \times 10^{-2} = 4.4673 \times 10^{-2}$$

Similarly, the mean of $P_f(\theta_2)$, $P_f(\theta_3)$, and $P_f(\theta_4)$ are respectively obtained as $\mu_2 = 4.4774 \times 10^{-2}$, $\mu_3 = 4.2448 \times 10^{-2}$, and $\mu_4 = 4.1595 \times 10^{-2}$.

According to Eq. (11.39), the overall probability of failure is obtained as

$$P_F = \sum_{i=1}^{4} \mu_i - 3P_f(\boldsymbol{\mu}) = \mu_1 + \mu_2 + \mu_3 + \mu_4 - 3P_f(\boldsymbol{\mu})$$

$$= 4.4673 \times 10^{-2} + 4.4774 \times 10^{-2} + 4.2448 \times 10^{-2} + 4.1595 \times 10^{-2} - 3 \times 0.0406 = 5.168 \times 10^{-2}$$

Using an MCS with 1 000 000 samples, the overall probability of failure (Eq. 11.6), i.e. the mean of the conditional failure probability (Eq. 11.36), is obtained as 5.123×10^{-2}. It can be observed that the probability of failure is close to the result of MCS for the methods of moment and the direct point estimate method, in which the reliability analysis can be easily conducted without shortcomings associated with design points, and the methods do not require either iteration or computation of derivatives.

Furthermore, assume that μ_R varies from 30 to 70, the results of FORM, methods of moment, and point-estimate method considering the parameters uncertainties are shown in Figure 11.2, in which the results of MCS with 100 000 samples are utilised as a benchmark. The results for given parameters, i.e. without considering the parameters uncertainties are also illustrated in Figure 11.2. From Figure 11.2, it can be observed that: (i) all the

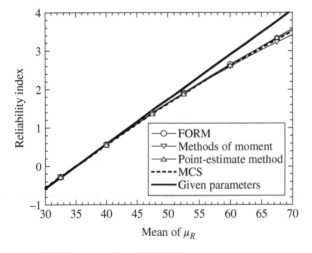

Figure 11.2 Reliability function for Example 11.1.

three methods, i.e. FORM, the methods of moment, and point estimate method provide almost the same results with that of MCS. (ii) As the reliability index gradually increases, the results considering parameter uncertainties become smaller when compared to those with given parameters. This implies that neglecting the effects of parameter uncertainties would overestimate structural reliability, especially for high-safety structures.

11.4 The Quantile of the Conditional Failure Probability

When second-level uncertainties are considered, the effects of these uncertainties will lead to uncertainty in the calculated failure probability and in the safety index. In this light, the conditional failure probability $P_f(\Theta)$ and the conditional reliability index $\beta(\Theta)$ become random variables.

In order to quantitatively estimate the uncertainty in the failure probability $P_f(\Theta)$ and the reliability index $\beta(\Theta)$ due to the second-level uncertainties, the quantile of $P_f(\Theta)$ is generally necessary. For this purpose, the distribution of $P_f(\Theta)$ needs to be determined. Although the distribution of $P_f(\Theta)$ is generally approximated by histogram obtained by MCS (Ang and De Leon 2005), a large number of samples are required. In this section, the quantile of $P_f(\Theta)$ will be approximated by methods of moment (Zhao et al. 2018c).

Since $P_f(\Theta)$ is a monotonic function of $\beta(\Theta)$, the same values of quantile of $P_f(\Theta)$ or $\beta(\Theta)$ can be obtained using the distribution of $P_f(\Theta)$ or $\beta(\Theta)$. Since the variability in $\beta(\Theta)$ is much smaller than that in $P_f(\Theta)$, we will approximate the distribution of $\beta(\Theta)$ rather than that of $P_f(\Theta)$.

Using the bivariate dimension-reduction based point estimate method in the chapter on Moment Evaluation. The function $\beta(\Theta)$ can then be approximated by $\beta^*(\Theta)$ as follows:

$$\beta(\Theta) \cong \beta^*(\Theta) = \beta^*\left[T^{-1}(\mathbf{U})\right] = \sum_{i<j}\beta_{ij} - (n-2)\sum_{i=1}^{n}\beta_i + \frac{(n-1)(n-2)}{2}\beta_0 \quad (11.40)$$

where

$$\beta_{i,j} = \beta\left[\mu_1, ..., T^{-1}(U_i), ..., T^{-1}(U_j), ..., \mu_n\right] = \beta_{ij}\left(U_i, U_j\right) \quad (11.41a)$$

$$\beta_i = \beta\left[\mu_1, ..., T^{-1}(U_i), ..., \mu_n\right] = \beta_i(U_i) \quad (11.41b)$$

$$\beta_0 = \beta(\mu_1, ..., \mu_i, ..., \mu_n) \quad (11.41c)$$

where $\beta_{i,j}$ is a two-dimensional function, $i, j = 1, 2, ..., n$ and $i < j$; β_i is a one-dimensional function; and β_0 is a constant.

Therefore, using the inverse Rosenblatt transformation, the kth raw moments of $\beta(\Theta)$, $\mu_{k\beta}$, can be formulated approximately as

$$\mu_{k\beta} = E\left\{[\beta(\Theta)]^k\right\} \cong E\left\{[\beta^*(\Theta)]^k\right\} = E\left\{\{\beta^*\left[T^{-1}(\mathbf{U})\right]\}^k\right\}$$

$$\cong \sum_{i<j}\mu_{k-\beta_{ij}} - (n-2)\sum_{i=1}^{n}\mu_{k-\beta_i} + \frac{(n-1)(n-2)}{2}\beta_0^k \quad (11.42)$$

where

$$\beta_0^k = [\beta(\mu_1, \ldots, \mu_i, \ldots, \mu_n)]^k \tag{11.43}$$

$$\mu_{k-\beta_i} = \int_{-\infty}^{\infty} \{\beta_i(u_i)\}^k \phi(u_i) du_i \tag{11.44a}$$

$$\mu_{k-\beta_{ij}} = \int_{-\infty}^{\infty} \int_{-\infty}^{\infty} \left\{\beta_{ij}(u_i, u_j)\right\}^k \phi(u_i)\phi(u_j) du_i du_j \tag{11.44b}$$

Using the point-estimate method, the one-dimensional integral in Eq. (11.44a) can be estimated as follows:

$$\mu_{k-\beta_i} = \sum_{r=1}^{m} P_r \{\beta_i[(u_{ir})]\}^k \tag{11.45}$$

Similarly, the two-dimensional integral in Eq. (11.44b) can be estimated as

$$\mu_{k-\beta_{ij}} = \sum_{r_1=1}^{m} \sum_{r_2=1}^{m} P_{r_1} P_{r_2} \left\{\beta_{ij}[(u_{ir1}, u_{jr2})]\right\}^k \tag{11.46}$$

The estimating points and the corresponding weights can be found in Table 3.3 in the chapter on Moment Evaluation.

Finally, the mean, standard deviation, and skewness of the conditional reliability index $\beta(\Theta)$ can be estimated using Eqs. (3.85a–d) also in the chapter just mentioned.

After the first four moments of $\beta(\Theta)$ are obtained, the distribution of $\beta(\Theta)$ can be approximated by using the square normal distributions (Zhao and Ono 2000b; Zhao et al. 2001) (see Appendix B.3) or the cubic normal distribution (Zhao and Lu 2007a, 2008; Zhao et al. 2018b) (see Appendix C.3). For the square normal distribution, when the skewness is small, the PDF and CDF of this distribution are respectively expressed as

$$f(\beta) = \frac{3\phi\left[\frac{1}{\alpha_{3\beta}}\left(\sqrt{9 + \alpha_{3\beta}^2 + 6\alpha_{3\beta}\frac{\beta - \mu_\beta}{\sigma_\beta}} - 3\right)\right]}{\sigma_\beta\sqrt{9 + \alpha_{3\beta}^2 + 6\alpha_{3\beta}\frac{\beta - \mu_\beta}{\sigma_\beta}}} \tag{11.47}$$

$$F(\beta) = \Phi\left[\frac{1}{\alpha_{3\beta}}\left(\sqrt{9 + \alpha_{3\beta}^2 + 6\alpha_{3\beta}\frac{\beta - \mu_\beta}{\sigma_\beta}} - 3\right)\right] \tag{11.48}$$

For the confidence level α of $P_f(\Theta)$, the fractile of $\beta(\Theta)$ is $1 - \alpha$. The corresponding quantile to the confidence level α can be determined by the following equation.

$$F[\beta(\alpha)] = \Phi\left[\frac{1}{\alpha_{3\beta}}\left(\sqrt{9 + \alpha_{3\beta}^2 + 6\alpha_{3\beta}\frac{\beta(\alpha) - \mu_\beta}{\sigma_\beta}} - 3\right)\right] = 1 - \alpha \tag{11.49}$$

Therefore, the quantile corresponding to the confidence level α is given as

$$\beta(\alpha) = \mu_\beta + \sigma_\beta\left\{-\frac{\alpha_{3\beta}}{6} + \Phi^{-1}(1-\alpha) + \frac{\alpha_{3\beta}}{6}\left[\Phi^{-1}(1-\alpha)\right]^2\right\} \tag{11.50}$$

and the corresponding failure probability of the confidence level $(1 - \alpha)$ is given as

$$P_f(1-\alpha) = \Phi[-\beta(\alpha)] \tag{11.51}$$

When using the simplification of cubic normal distribution (Appendix C.3), the PDF and CDF of this distribution are, respectively, expressed as

$$f(\beta) = \frac{\phi(u)}{\sigma(k_1 + 2l_1 u + 3k_2 u^2)} \tag{11.52}$$

$$F(\beta) = \Phi(u) \tag{11.53}$$

where

$$u = -\frac{\sqrt[3]{2}p}{\sqrt[3]{-q+\Delta}} + \frac{\sqrt[3]{-q+\Delta}}{\sqrt[3]{2}} - \frac{l_1}{3k_2} \tag{11.54}$$

where the coefficients l_1, k_1, k_2, Δ, p, and q are defined as

$$\Delta = \sqrt{q^2 + 4p^3}, \quad p = \frac{3k_1 k_2 - l_1^2}{9k_2^2}, \quad q = \frac{2l_1^3 - 9k_1 k_2 l_1 + 27k_2^2[-l_1 - (\beta-\mu_\beta)/\sigma_\beta]}{27k_2^3} \tag{11.55a}$$

$$l_1 = \frac{\alpha_{3\beta}}{6(1+6l_2)}, \quad k_1 = \frac{1-3l_2}{(1+l_1^2-l_2^2)}, \quad k_2 = \frac{l_2}{(1+l_1^2+12l_2^2)}, \quad l_2 = \frac{\sqrt{6\alpha_{4\beta} - 8\alpha_{3\beta}^2 - 14} - 2}{36} \tag{11.55b}$$

In theory, the α-percentile value (in %) of the conditional reliability index $\beta(\Theta)$, namely, β_α, can be obtained from the following formula

$$F(\beta_\alpha) = \Phi(u) = \alpha \tag{11.56}$$

Since

$$\beta_s = \frac{\beta(\Theta) - \mu_\beta}{\sigma_\beta} = S^{-1}(u) = -l_1 + k_1 u + l_1 u^2 + k_2 u^3 \tag{11.57}$$

Then the solution of β_α can be derived as

$$\beta_\alpha = \mu_\beta + \sigma_\beta \cdot S^{-1}[\Phi^{-1}(\alpha)] \tag{11.58}$$

Therefore, the $(1-\alpha)$-percentile value (in %) of the conditional failure probability can be expressed as

$$P(1-\alpha) = \Phi[-\beta_\alpha] \tag{11.59}$$

Example 11.2 The Conditional Reliability Index with an Explicit Expression

For the same Example 11.1, we will approximate the distribution of $\beta(\Theta)$ rather than that of $P_f(\Theta)$ since the variability in $\beta(\Theta)$ is much smaller than that in $P_f(\Theta)$. According to

Eq. (11.33), when considering the uncertainties of the distribution parameters, $\beta(\Theta)$ is expressed as

$$\beta(\Theta) = \frac{\mu_R - \mu_S}{\sqrt{\sigma_R^2 + \sigma_S^2}} \tag{11.60}$$

Using the bivariate-dimension reduction based point estimate method, the first three moments of $\beta(\Theta)$ are readily obtained as, $\mu_\beta = 1.7674$, $\sigma_\beta = 0.4268$, and $\alpha_{3\beta} = 0.3999$, respectively.

After the first three moments of $\beta(\Theta)$ have been determined, the probability distribution of $\beta(\Theta)$ can be approximated by using the square normal distributions. According to Eq. (11.47), the PDF of this distribution is expressed as

$$f(\beta) = \frac{7.03\phi\left[2.5\sqrt{9.08 + 5.62(\beta - 1.77)} - 7.47\right]}{\sqrt{9.08 + 5.62(\beta - 1.77)}} \tag{11.61}$$

The histogram of the conditional reliability index $\beta(\Theta)$ obtained by using the MCS with 1 000 000 samples is shown in Figure 11.3, together with the PDF curve (denoted as the thick solid line) obtained from Eq. (11.61). It can be seen in Figure 11.3 that the histogram of the conditional reliability index $\beta(\Theta)$ is well behaved and is approximated well by the PDF of the square normal distribution determined by using its first three moments. The histogram of the conditional failure probability $P_f(\Theta)$ obtained by using the MCS with 1 000 000 samples is shown in Figure 11.4. It can be observed in Figure 11.4 that the histogram of $P_f(\Theta)$ is skewed to the right and is truncated when $P_f(\Theta)$ tends to zero, which is difficult to approximate by well-known distributions.

According to Eq. (11.50), the 10% and 5% fractiles of $\beta(\Theta)$ can be obtained as β $(0.1) = 1.239$ and $\beta(0.05) = 1.114$, respectively, which are also shown in Figure 11.3. Then, according to Eq. (11.51), the corresponding 90% and 95% confidence levels of $P_f(\Theta)$ are readily obtained as $P_f(0.9) = 0.1077$ and $P_f(0.95) = 0.1332$, respectively.

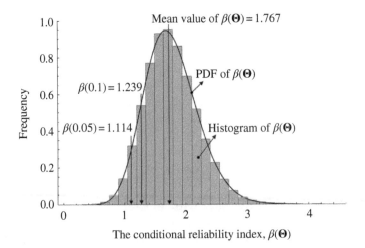

Figure 11.3 Histogram and PDF curve of the conditional reliability index for Example 11.2.

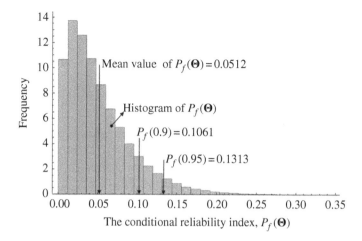

Figure 11.4 Histogram of the conditional failure probability for Example 11.2.

It should be noted that the mean value of $\beta(\Theta)$ is 1.767 and its corresponding failure probability is 0.0386, which is not equal to the mean of $P_f(\Theta)$ ($= 0.0152$).

Using the MCS with 1 000 000 samples, the 90% and 95% confidence levels of $P_f(\Theta)$ are easily obtained as $P_f(0.9) = 0.1061$ and $P_f(0.95) = 0.1313$, respectively, which are also shown in Figure 11.4.

From the discussion above, one may conclude that although it is very simple to determine the quantile of $P_f(\Theta)$ by utilising the square normal distribution for approximating the conditional reliability index, the results obtained from methods of moments are almost the same as those obtained by MCS method.

Example 11.3   The Conditional Reliability Index with an Implicit Expression

The second example considers an existing reinforced concrete T-beam bridge as shown in Figure 11.5a, which has been investigated by Wang et al. (2015). The bridge consists of a 19.5-m simply supported span and five beams spaced equally at 1.6-m intervals. The cross-section of one of these girders is shown in Figure 11.5b and the reinforcement

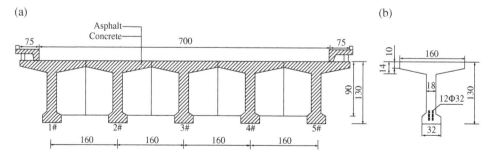

Figure 11.5 A reinforced concrete bridge (cm). (a) Cross-section of the RC bridge. (b) Cross-section of T-girder.

diameter is 32 mm. The thickness of the wearing surface (asphalt) is 5.0 cm. The initial yield strength of the reinforcement is 280 MPa.

Considering the randomness of the resistance and load effects, the flexural limit state function in the midspan cross-section can be written as

$$Z = G(\mathbf{X}) = K_P \cdot f_g A_g \left(h_0 - \frac{f_g A_g}{2 f_c \cdot b_i} \right) - S_G - S_Q \tag{11.62}$$

where K_P is normally distributed with a mean of 1.098 and a standard deviation of 0.078. Furthermore, f_g is the yield strength of the corroded reinforcement: $f_g = f_{g0}(1 - \eta)$, where η is the steel strength loss, and f_{g0} is the initial yield strength of the reinforcement. In addition, A_g is the area of reinforcement, f_c is the concrete strength, S_G is the effect of the dead load, S_Q is the effect of a live load, $h_0 = 1.1$ m, and $b_i = 1.6$ m.

The concrete strength f_c (MPa), steel strength loss η (%), dead load effect S_G (kN·m), and the live load effect S_Q (kN·m) are assumed as random variables and their probabilistic information are listed in Table 11.2.

The reliability analysis of the performance function in Eq. (11.62) can be readily evaluated using state-of-the-art techniques. Here, FORM is utilised and the reliability index is readily obtained as 5.341, with corresponding failure probability of 4.614×10^{-8}.

In this example, the distribution parameters (i.e. mean and standard deviation) of the four random variables, i.e. μ_{fc}, μ_{η}, μ_{SG}, μ_{SQ}, σ_{fc}, σ_{η}, σ_{SG}, and σ_{SQ} are assumed to be random variables, and their probabilistic information is listed in Table 11.3. Estimating the mean value and quantile of the conditional failure probability is described below.

Table 11.2 Probabilistic information about the random variables in Example 11.3.

Variable	Distribution	Mean	Standard deviation
Concrete strength, f_c (MPa)	Normal	20	1.45
Steel strength loss, η	Normal	22	2.04
Dead load effect, S_G (kN·m)	Normal	613.18	26.45
Live load effect, S_Q (kN·m)	Gumbel	535.42	84.08

Table 11.3 Probabilistic information about the distribution parameters in Example 11.3.

Variable	Distribution	Mean	Standard deviation
μ_{fc}	Normal	21.34	1.72
μ_{η}	Normal	22.08	2.56
μ_{SG}	Normal	597.03	28.36
μ_{SQ}	Normal	610	7.14
σ_{fc}	Lognormal	1.51	0.072
σ_{η}	Lognormal	2.67	0.12
σ_{SG}	Normal	25.32	2.16
σ_{SQ}	Normal	96.23	4.68

According to Eq. (11.4), the overall failure probability can be formulated as

$$P_F = \int_{G(\mathbf{X},\Theta) \leq 0} f_{\mathbf{X},\Theta}(\mathbf{x},\theta)d\mathbf{x}d\theta = \int_{\theta} P_f(\Theta)f_{\Theta}(\theta)d\theta \qquad (11.63)$$

According to Eq. (11.26), the conditional failure probability $P_f(\Theta)$ can be approximated as

$$P_f(\Theta) \cong P_f{}^*(\Theta) = \sum_{i=1}^{8} P_{fi} - 7P_f(\boldsymbol{\mu}) \qquad (11.64)$$

where

$$P_{f1} = P_f\left(\mu_{fc}\right), P_{f2} = P_f\left(\mu_\eta\right), P_{f3} = P_f(\mu_{SG}), P_{f4} = P_f\left(\mu_{SQ}\right),$$

$$P_{f5} = P_f\left(\sigma_{fc}\right), P_{f6} = P_f\left(\sigma_\eta\right), P_{f7} = P_f(\sigma_{SG}), P_{f8} = P_f(\sigma_{SQ})$$

Although explicit expressions for P_{fi} and $P_f(\boldsymbol{\mu})$ are not available, they can be easily estimated by using FORM. Because $P_f(\boldsymbol{\mu})$ is a function of the means of all eight random variables, we replace the original mean and standard deviation of the four random variables in Table 11.2 are replaced by means of these parameters as given in Table 11.3, whereby $P_f(\boldsymbol{\mu})$ can then be easily obtained as 5.919×10^{-7} by using FORM.

Using a seven-point estimate in standard normal space, the estimating points of P_{f1}, i.e. $P_f(\mu_{fc})$ in original space, can be obtained as follows with the aid of an inverse Rosenblatt transformation:

$$\mu_{fc1} = 14.889, \mu_{fc2} = 17.269, \mu_{fc3} = 19.354, \mu_{fc4} = 21.340, \mu_{fc5} = 23.326, \mu_{fc6} = 25.411, \mu_{fc7} = 27.791$$

Similar to estimating $P_f(\boldsymbol{\mu})$, $P_f(\mu_{fci})$, $i = 1, ..., 7$, can be readily evaluated using FORM. Using the point-estimate method, the mean of $P_f(\mu_{fc})$ or P_{f1}, $\mu_{P_{f1}}$, is readily obtained as

$$\mu_{P_{f1}} = \sum_{k=1}^{7} P_k P_f\left(\mu_{fck}\right) = 5.961 \times 10^{-7}$$

Similarly, the means of $P_f(\mu_\eta)$ or P_{f2}, $P_f(\mu_{SG})$ or P_{f3}, $P_f(\mu_{SQ})$ or P_{f4}, $P_f(\sigma_{fc})$ or P_{f5}, $P_f(\sigma_\eta)$ or P_{f6}, $P_f(\sigma_{SG})$ or P_{f7}, and $P_f(\sigma_{SQ})$ or P_{f8} are obtained as $\mu_{P_{f2}} = 8.783 \times 10^{-7}$, $\mu_{P_{f3}} = 6.347 \times 10^{-7}$, $\mu_{P_{f4}} = 5.945 \times 10^{-7}$, $\mu_{P_{f5}} = 5.919 \times 10^{-7}$, $\mu_{P_{f6}} = 5.928 \times 10^{-7}$, $\mu_{P_{f7}} = 5.921 \times 10^{-7}$, and $\mu_{P_{f8}} = 6.635 \times 10^{-7}$, respectively.

Therefore, according to Eq. (11.64), the overall probability of failure, i.e. the mean of the conditional failure probability, is readily estimated as

$$P_F = E[P_f(\Theta)] \cong \sum_{i=1}^{8} \mu_{P_{fi}} - 7P_f(\boldsymbol{\mu}) = 1.001 \times 10^{-6}$$

Using the MCS with 1 000 000 samples, the overall probability of failure, i.e. the mean of the conditional failure probability, is obtained as 1.052×10^{-6}. It can be observed that the result obtained by using the direct point estimate method is almost the same as that of MCS. Again, the mean of the conditional failure probability when considering the parameter uncertainties (1.001×10^{-6}) is larger than the failure probability without considering the parameter uncertainties (4.614×10^{-8}). That again shows that neglecting parameter uncertainties will lead to the structural reliability being overestimated.

Then the probability distribution is estimated for the conditional reliability index and the quantile of conditional failure probability.

According to Eq. (11.40), the conditional reliability index $\beta(\Theta)$ can be approximated as

$$\beta(\Theta) \cong \beta^*(\Theta) = \beta^*\left[T^{-1}(\mathbf{U})\right] = \sum_{i<j}\beta_{i,j} - 6\sum_{i=1}^{8}\beta_i + 21\beta_0 \tag{11.65}$$

where

$$\beta_{1,2} = \beta\left(\mu_{f_c}, \mu_\eta\right), \beta_{1,3} = \beta\left(\mu_{f_c}, \mu_{SG}\right), \beta_{1,4} = \beta\left(\mu_{f_c}, \mu_{SQ}\right), \beta_{1,5} = \beta\left(\mu_{f_c}, \sigma_{f_c}\right), \beta_{1,6} = \beta\left(\mu_{f_c}, \sigma_\eta\right),$$
$$\beta_{1,7} = \beta\left(\mu_{f_c}, \sigma_{SG}\right), \beta_{1,8} = \beta\left(\mu_{f_c}, \sigma_{SQ}\right), \beta_{2,3} = \beta(\mu_\eta, \mu_{SG}), \beta_{2,4} = \beta(\mu_\eta, \mu_{SQ}), \beta_{2,5} = \beta(\mu_\eta, \sigma_{f_c}),$$
$$\beta_{2,6} = \beta(\mu_\eta, \sigma_\eta), \beta_{2,7} = \beta(\mu_\eta, \sigma_{SG}), \beta_{2,8} = \beta(\mu_\eta, \sigma_{SQ}), \beta_{3,4} = \beta(\mu_{SG}, \mu_{SQ}), \beta_{3,5} = \beta(\mu_{SG}, \sigma_{f_c}),$$
$$\beta_{3,6} = \beta(\mu_{SG}, \sigma_\eta), \beta_{3,7} = \beta(\mu_{SG}, \sigma_{SG}), \beta_{3,8} = \beta(\mu_{SG}, \sigma_{SQ}), \beta_{4,5} = \beta(\mu_{SQ}, \sigma_{f_c}), \beta_{4,6} = \beta(\mu_{SQ}, \sigma_\eta),$$
$$\beta_{4,7} = \beta(\mu_{SQ}, \sigma_{SG}), \beta_{4,8} = \beta(\mu_{SQ}, \sigma_{SQ}), \beta_{5,6} = \beta(\sigma_{f_c}, \sigma_\eta), \beta_{5,7} = \beta(\sigma_{f_c}, \sigma_{SG}), \beta_{5,8} = \beta(\sigma_{f_c}, \sigma_{SQ}),$$
$$\beta_{6,7} = \beta(\sigma_\eta, \sigma_{SG}), \beta_{6,8} = \beta(\sigma_\eta, \sigma_{SQ}), \beta_{7,8} = \beta(\sigma_{SG}, \sigma_{SQ}) \tag{11.66}$$

$$\beta_1 = \beta\left(\mu_{fc}\right), \beta_2 = \beta(\mu_\eta), \beta_3 = \beta(\mu_{SG}), \beta_4 = \beta(\mu_{SQ})$$
$$\beta_5 = \beta(\sigma_{fc}), \beta_6 = \beta(\sigma_\eta), \beta_7 = \beta(\sigma_{SG}), \beta_8 = \beta(\sigma_{SQ}) \tag{11.67}$$
$$\beta_0 = \Phi^{-1}\left[1 - P_f(\mathbf{\mu})\right] = 4.858$$

According to the bivariate dimension-reduction based point estimate method, i.e. Eqs. (11.40)–(11.46), in which the estimation of the reliability indices for determining $\mu_{k-\beta_i}$ and $\mu_{k-\beta_{i,j}}$ in Eqs. (11.44) and (11.45) is evaluated from Eq. (11.62) using FORM, the first three moments of $\beta(\Theta)$ can be obtained as, $\mu_\beta = 4.856$, $\sigma_\beta = 0.218$, and $\alpha_{3\beta} = -0.023$, respectively. Substituting the obtained first three moments of $\beta(\Theta)$ into Eq. (11.47), the PDF of the conditional reliability index $\beta(\Theta)$ is expressed as

$$f_\beta[\beta(\Theta)] = \frac{13.76\phi\left[-43.48\left(\sqrt{9 - 0.63[\beta(\Theta) - 4.86]} - 3\right)\right]}{\sqrt{9 - 0.63[\beta(\Theta) - 4.86]}} \tag{11.68}$$

The histogram of the conditional reliability index $\beta(\Theta)$ obtained by using the MCS with 1 000 000 samples is shown in Figure 11.6, together with the PDF curve (denoted as the thick solid line) obtained from Eq. (11.68). It can be observed from Figure 11.6 that the histogram of the conditional reliability index $\beta(\Theta)$ is well behaved and can be approximated well by the PDF of the square normal distribution determined by using its first three moments. The histogram of the conditional failure probability $P_f(\Theta)$ obtained by using the MCS with 1 000 000 samples is shown in Figure 11.7 where it can be observed that the histogram of $P_f(\Theta)$ is skewed to the right and is truncated when $P_f(\Theta)$ tends to zero, which is difficult to approximate by well-known distributions.

According to Eq. (11.50), the 10% and 5% fractiles of $\beta(\Theta)$ can be obtained as β $(0.1) = 4.576$ and $\beta(0.05) = 4.496$, respectively, which are also shown in Figure 11.6. Then, according to Eq. (11.51), the corresponding 90% and 95% confidence levels of $P_f(\Theta)$ are

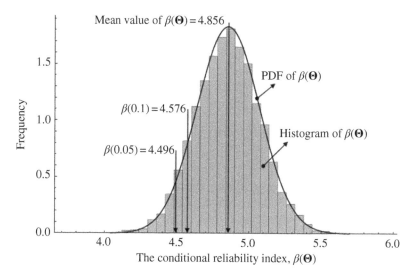

Figure 11.6 Histogram and PDF curve of the conditional reliability index for Example 11.3.

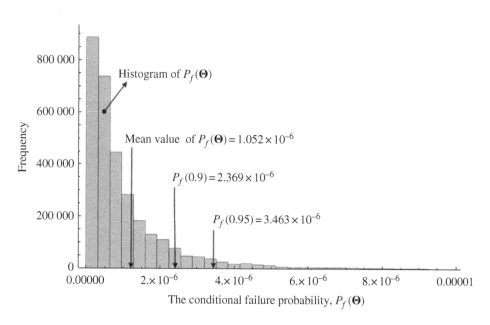

Figure 11.7 Histogram of the conditional failure probability for Example 11.3.

readily obtained as $P_f(0.9) = 2.370 \times 10^{-6}$ and $P_f(0.95) = 3.462 \times 10^{-6}$, respectively. Using MCS, the 90% and 95% confidence levels of $P_f(\Theta)$ are easily obtained as $P_f(0.9) = 2.369 \times 10^{-6}$ and $P_f(0.95) = 3.463 \times 10^{-6}$, respectively, which are also shown in Figure 11.7. It should be noted that the failure probability corresponding to the mean of $\beta(\Theta)$ is obtained as: $\Phi(-4.856) = 5.989 \times 10^{-7}$, which is not equal to the mean of $P_f(\Theta)$ (= 1.052×10^{-6}).

From this discussion, it can be concluded again that although it is very simple to determine the quantile of $P_f(\Theta)$ by utilising the square normal distribution for the conditional reliability index, the results estimated by the methods of moment above are almost the same as those obtained by MCS method.

11.5 Application to Structural Dynamic Reliability Considering Parameters Uncertainties

Another typical problem of hierarchical models encountered in practical engineering is the dynamic reliability considering uncertainties contained in input parameters. For structural dynamic reliability, the problem is generally described as a structural system with parameters $\mathbf{X}_1$ under the action of time-varying loadings $\mathbf{S}(t, \mathbf{X}_2)$, where $\mathbf{S}$ is a vector random process of time, and the process is specified by a set of parameters $\mathbf{X}_2$. Let $\mathbf{X}$ denotes all parameters in the structure and loadings, where $\mathbf{X}$ is a vector of quantities that are uncertain but time invariant. The performance function must include a vector of random response variables, $\mathbf{R}$, because $\mathbf{R}$ is dependent not only on $\mathbf{X}$ but also on the random process, $\mathbf{S}$. The probability of failure can be similarly formulated by (Zhao et al. 1999)

$$P_F = \int_{G[\mathbf{X},\mathbf{R}(t)] \leq 0} f_{\mathbf{X},\mathbf{R}}(\mathbf{x}, \mathbf{r}) d\mathbf{x} d\mathbf{r} \tag{11.69}$$

Formally, Eq. (11.69) can be solved by the methods of moment and FORM. The required information includes the known form of the performance function, G; the probability distributions of both system parameters, $\mathbf{X}$, and response variables, $\mathbf{R}$; and even the joint PDF, $f_{\mathbf{X},\mathbf{R}}(\mathbf{x}, \mathbf{r})$. The method would be very complicated because of the number of response variables and the difficulty in determining the distribution of response variables of interest, and furthermore, the joint PDF, $f_{\mathbf{X},\mathbf{R}}(\mathbf{x}, \mathbf{r})$, since $\mathbf{R}$ is dependent on $\mathbf{X}$.

On the other hand, if the failure probability of the structural system for given parameter values of $\mathbf{X} = \mathbf{x}$ can be evaluated from state-of-the-art techniques, e.g. those based on random process and vibration methods, then overall reliability is formulated by

$$P_F = \int_{\mathbf{X}} P_f(\mathbf{x}) f_{\mathbf{X}}(\mathbf{x}) d\mathbf{x} \tag{11.70}$$

where $P_f(\mathbf{x})$ is the conditional probability of failure for given $\mathbf{X} = \mathbf{x}$, which is evaluated from state-of-the-art techniques; and $f(\mathbf{x})$ is the joint PDF of $\mathbf{X}$.

It can be observed that both Eqs. (11.69) and (11.70) are in the same forms as those of Eqs. (11.4) and (11.5), and therefore the methods described in previous sections can be directly applied to compute the overall probability of failure and to estimate the distribution of the conditional failure probability.

Example 11.4 Consider an example of a linear single-degree-of-freedom system under dynamic excitation. If the excitation is modelled as Gaussian white noise, for a simple limit state of a given threshold level of displacement being exceeded, failure probability with deterministic system parameters is

$$P_f = 1 - \exp\left\{ -fD\exp\left[-\frac{1}{2}\left(\frac{d}{\sigma_r}\right)^2 \right] \right\} \tag{11.71}$$

where

$$\sigma_r = \sqrt{\frac{\pi S}{4(2\pi f)^3 \zeta}} \tag{11.72}$$

where S is the spectra intensity; D is the duration of the excitation; f is the natural frequency; ζ is the damping ratio; and d is a threshold of response.

When the uncertainties in the parameters S, D, f, and ζ are considered and they are treated as random variables, the problem of dynamic reliability analysis considering parameter uncertainties can be expressed as

$$P_F = \int_{S,D,f,\zeta} P_f(S,D,f,\zeta)f(S,D,f,\zeta)dSdDdfd\zeta \tag{11.73}$$

The probabilistic properties of these four uncertain parameters are listed in Table 11.4. For a threshold of $d = 0.35$, consider only uncertainties contained in the duration, D, and damping ratio, ζ. According to Eq. (11.15) and direct methods of moment described in the chapters on Moment Evaluation and Direct Methods of Moment, the first four moments of the performance function are approximately obtained as $\mu_G = 3.599$, $\sigma_G = 1.5662$, $\alpha_{3G} = 0.1276$, and $\alpha_{4G} = 3.117$. The 4M reliability index is readily obtained $\beta_{4M} = 2.372$.

Using MCS with 5 000 000 samples, the first four moments are obtained as $\mu_G = 3.605$, $\sigma_G = 1.5723$, $\alpha_{3G} = 0.1133$, and $\alpha_{4G} = 3.0797$; and the probability of failure is obtained as 0.00933 with a corresponding reliability index of $\beta_{MCS} = 2.352$. One can observe that the probability of failure obtained using the direct methods of moment are close to that of MCS.

The results of different threshold levels are shown in Figure 11.8, in which the result of MCS with 10 000 samples is illustrated; the result using the second moment method is obtained from Eq. (11.7). From Figure 11.8, it can be observed that the curves obtained by FORM and methods of moment are approximately equal to that obtained by the MCS method, whereas that of the second moment method is considerably different. It can also be observed that the degree of the slope of the curve obtained considering parameter uncertainties is much smaller than that with given parameters; i.e. the limit state in the reliability analysis considering parameter uncertainties is not as dominant as in that with given parameters.

From Figure 11.8, it can also be observed that for this example, when the reliability index is high, the results considering parameter uncertainties are lower than those with given parameters. On the other hand, when the reliability index is negative (in case of dangerous structures or those subjected to large intensity but low occurrence rate load), results considering parameter uncertainties are higher than those with given parameters. Therefore, neglecting effects of parameter uncertainties would overestimate structural reliability of high-safety structures and underestimate those of low-safety structures.

Table 11.4 Probabilistic properties of uncertain parameters.

Parameter	Distribution	Mean	COV
S	Type II extreme value	0.25	0.60
D	Lognormal	10 sec	0.3
f	Normal	2 Hz	0.1
ζ	Lognormal	0.02	0.4

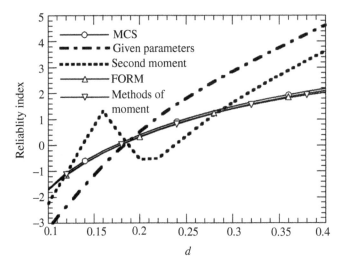

Figure 11.8 Reliability function for Example 11.4.

Example 11.5 A Nonlinear SDOF System under Dynamic Excitation

Consider an example of a nonlinear SDOF system under dynamic excitation. The excitation is modelled as Gaussian white noise, nonlinearity coefficient ε, the intensity S_0, the duration D of the excitation and the natural frequency v, damping ratio ξ are uncertain and modelled as random variables as shown in Table 11.5. For a simple limit state of a given threshold level of the displacement (d) being exceeded, the failure probability with deterministic system parameters is

$$P_f = 1 - exp\left\{ -vDexp\left[-\frac{1}{2}\left(\frac{d}{\sigma_y}\right)^2 \right] \right\} \tag{11.74}$$

where

$$\sigma_y = \sqrt{\frac{\pi S_0}{2(2\pi v)^3 \xi}\left(1 - 3\varepsilon\frac{\pi S_0}{2(2\pi v)^3 \xi} \right)} \tag{11.75}$$

Table 11.5 Statistical model of each uncertain parameter.

Parameter	Distribution	Mean	COV
ε	Lognormal	0.1	0.2
v	Normal	2 Hz	0.1
D	Lognormal	11 sec	0.3
S_0	Type II extreme value	0.25 m^2/s^3	0.6
ξ	Lognormal	0.03	0.3

When the uncertainties in the parameters S_0, v, ξ, ε, and D are considered, the problem of dynamic reliability analysis considering parameter uncertainties can be expressed as

$$P_F = \int_{S_0,v,\xi,\varepsilon,D} P_f(S_0,v,\xi,\varepsilon,D)f(S_0,v,\xi,\varepsilon,D)dS_0 dv d\xi d\varepsilon dD \tag{11.76}$$

In order to investigate the influence of the parameter uncertainties on the overall reliability, three cases are considered.

Case 1: Considering the uncertainties of excitation parameters D, S_0, and ξ.
Case 2: Considering the uncertainties of structural parameters v and ε.
Case 3: Considering uncertainties of all the parameters.

The variation of overall reliability indices with respect to different threshold levels are shown in Figure 11.9a,b, where the results of MCS with 1 000 000 samples are illustrated with solid lines, and those obtained by the direct point estimate method are depicted with dashed lines. The result with given parameters (the uncertainties in the parameters being not considered, that is, the mean value of parameters is taken) is also presented with solid lines. From Figure 11.9a,b it can be observed that:

1) The results obtained by the direct point estimate method are in close agreement with those obtained by the MCS method for all three cases.

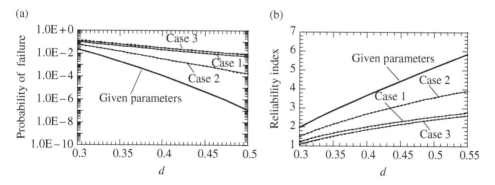

Figure 11.9 Overall reliability vs thresholds for Example 11.5. (a) Probability of failure. (b) Reliability index.

2) The degree of the slope of the curve obtained considering parameter uncertainties is much smaller than that with given parameters; i.e. the threshold in the reliability analysis considering parameter uncertainties is not so sensitive as in that with given parameters.
3) The effect of uncertainties in excitation parameters on the overall reliability index is greater than that of uncertainties in structural parameters.
4) For this example, results considering parameter uncertainties are far smaller than those with given parameters. Therefore, neglecting effects of parameter uncertainties would overestimate structural reliability of high-safety structures.

According to Eqs. (11.3) and (11.74), the conditional reliability index $\beta(\mathbf{X})$ can be expressed as

$$\beta(\mathbf{X}) = \Phi^{-1}(1 - P_f) = \Phi^{-1}\left\{ \exp\left\{ -vD\exp\left[-\frac{1}{2}\left(\frac{d}{\sigma_y}\right)^2 \right] \right\} \right\} \tag{11.77}$$

For a threshold of $d = 0.3$, using the bivariate dimension-reduction based point estimate method, the first four moments of $\beta(\mathbf{X})$ are obtained as $\mu_\beta = 2.154$, $\sigma_\beta = 1.694$, $\alpha_{3\beta} = -0.266$, and $\alpha_{4\beta} = 3.419$.

According to Eq. (11.55b), the polynomial coefficients of the third-order polynomial transformation are obtained as $l_1 = 0.041$, $k_1 = 0.962$, and $k_2 = 0.012$. According to Eq. (11.52), the PDF of the conditional reliability index $\beta(\mathbf{X})$ based on the cubic normal distribution is expressed as

$$f(\beta) = \frac{\phi(u)}{1.694(0.962 - 0.082u + 0.036u^2)} \tag{11.78}$$

The histogram of the conditional reliability index $\beta(\mathbf{X})$ obtained by using the MCS with 1 000 000 samples is shown in Figure 11.10, together with the PDF curve (denoted as the thick solid line) obtained from Eq. (11.78). It can be observed from Figure 11.10 that the histogram of the conditional reliability index $\beta(\mathbf{X})$ is well behaved and can be approximated well by the PDF of the cubic normal distribution determined by using its first four moments. The histogram of the conditional failure probability $P_f(\mathbf{X})$ obtained by using the MCS with 1 000 000 samples is shown in Figure 11.11. It can be observed in Figure 11.11 that the histogram of $P_f(\mathbf{X})$ is skewed to the right and is truncated when $P_f(\mathbf{X})$ tends to zero, which is difficult to approximate by well-known distributions.

It can also be observed from Figure 11.11 that the histogram of $P_f(\mathbf{X})$ has an increase in frequency just below 1.0. The reason lies in the fact that when the conditional reliability index is in the range of -1.65 to 0.52, the corresponding range of the conditional failure probability is 0.30 to 0.95. The samples of the conditional failure probability in this range are small, and its horizontal axis range is large, thus the frequency density of the conditional failure probability within this range is close to zero. While the conditional reliability index is less than -1.65, the corresponding conditional failure probability is close to 1.0. Although the number of samples of the conditional failure probability in this range is also small, its horizontal axis range is even smaller, so the frequency density of the conditional failure probability within this range has an increase. Therefore, in Figure 11.11, the histogram

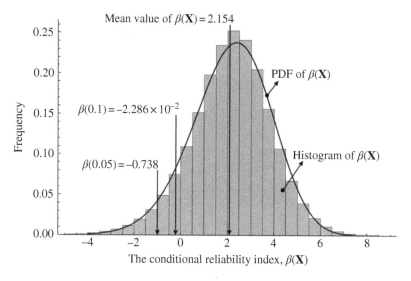

Figure 11.10 Histogram and PDF curve of the conditional reliability index for Example 11.5.

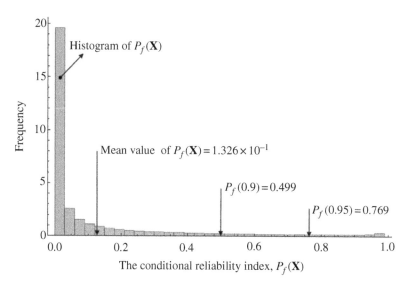

Figure 11.11 Histogram of the conditional failure probability for Example 11.5.

of $P_f(\mathbf{X})$ has an increase in frequency just below 1.0, and this phenomenon occurs when the conditional failure probability has larger values. And in this situation, the frequency distribution histogram of the conditional failure probability is almost impossible to effectively fit by using the existing well-known distribution forms.

According to Eq. (11.58), the 10% and 5% fractiles of $\beta(\mathbf{X})$ can be obtained as β (0.1) $= -2.286 \times 10^{-2}$ and $\beta(0.05) = -0.738$, respectively, which are also shown in Figure 11.10. Then, according to Eq. (11.59), the corresponding 90% and 95% confidence levels of $P_f(\mathbf{X})$ are readily obtained as $P_f(0.9) = 0.509$ and $P_f(0.95) = 0.769$, respectively. Using MCS, the 90% and 95% confidence levels of $P_f(\mathbf{X})$ are easily obtained as $P_f(0.9) = 0.499$ and $P_f(0.95) = 0.769$, respectively, which are also shown in Figure 11.11. It should be noted that the failure probability corresponding to the mean of $\beta(\mathbf{X})$ is obtained as: $\Phi(-2.154) = 1.562 \times 10^{-2}$, which is not equal to the mean of $P_f(\mathbf{X})$ ($= 1.326 \times 10^{-1}$).

From this discussion, it can be concluded that although it is very simple to determine the quantile of $P_f(\mathbf{X})$ by utilising the cubic normal distribution for the conditional reliability index, the results estimated by the methods of moment above are almost the same as those obtained by MCS method.

11.6 Summary

In this chapter, two typical problems of hierarchical models are discussed for structural reliability, where one is structural reliability analysis considering the uncertainties in distribution parameters, and the other is the dynamic reliability considering uncertainties contained in input parameters. The application of FORM, methods of moment, and point-estimate method are introduced to compute the overall probability of failure with consideration of both the two levels of uncertainties. The point-estimate method is also introduced for evaluating the quantile of the conditional failure probability. It can be concluded that neglecting effects of second-level uncertainty overestimates structural reliability of high-safety structures and underestimates that of low-safety structures.

12

Structural Reliability Analysis Based on Linear Moments

12.1 Introduction

Structural reliability analysis accounting for random variables with unknown distributions has been discussed in the chapter on Structural Reliability Assessment Based on the Information of Moments of Random Variables, in which the x–u and u–x transformations are achieved through using the third- or fourth-moment transformation technique when the first four central moments (C-moments) and correlation coefficient matrix of the basic random variables are known.

Whether the first four C-moments from samples is accurate or not will have great effect on the accuracy of the transformations. Studies have shown that the numerical values of sample C-moments, particularly when the sample size is small, may be different from those of the probability distributions from which the sample was drawn (Hosking 1990). Recently, other types of moments such as the linear moments (L-moments) (Hosking 1990; Hosking and Wallis 1997; Chen and Tung 2003; Gao et al. 2017), fractional moments (Inverardi and Tagliani 2003; Zhang and Pandey 2013), and probability weight moments (Greenwood et al. 1979; Deng and Pandey 2008) have been applied in structural reliability. In this chapter, structural reliability analysis based on the first few L-moments will be introduced. Following the definition of L-moments of random variables, we will first present the x–u and u–x transformations based on the first few L-moments and correlation coefficient matrix of the basic random variables, and structural reliability analysis method based on the first few L-moments and correlation coefficient matrix will then be discussed.

12.2 Definition of L-Moments

Let X be a random variable, and let $X_{1:r} \le X_{2:r} \le \dots \le X_{r:r}$ be the order statistics of an observation of size r drawn from the distribution of X. Hosking (1990) defined the L-moments of X to be the quantities:

$$\lambda_r = r^{-1} \sum_{j=0}^{r-1} (-1)^k \binom{r-1}{k} E(X_{r-k:r}), \quad r = 1, 2, \dots \tag{12.1}$$

Structural Reliability: Approaches from Perspectives of Statistical Moments, First Edition.
Yan-Gang Zhao and Zhao-Hui Lu.
© 2021 John Wiley & Sons Ltd. Published 2021 by John Wiley & Sons Ltd.

where λ_r is a linear function of the expected order statistics; $X_{r-k:r}$ denotes the $r-k$th smallest observation from the size r; and $E(X_{r-k:r})$ is the expectation of order statistic $X_{r-k:r}$, which may be written as (David 1981)

$$E(X_{r-k:r}) = \frac{r!}{(r-k-1)!k!} \int x[F(x)]^{r-k-1}[1-F(x)]^k dF(x) \tag{12.2}$$

where $F(x)$ is the cumulative distribution function (CDF) of X.

For common uses in engineering practice, the first four L-moments can be expressed as

$$\lambda_1 = E(X_{1:1}) = \int_0^1 x dF(x) \tag{12.3}$$

$$\lambda_2 = \frac{1}{2}E(X_{2:2} - X_{1:2}) = \int_0^1 x[2F(x) - 1]dF(x) \tag{12.4}$$

$$\lambda_3 = \frac{1}{3}E(X_{3:3} - 2X_{2:3} + X_{1:3}) = \int_0^1 x\left[6F^2(x) - 6F(x) + 1\right]dF(x) \tag{12.5}$$

$$\lambda_4 = \frac{1}{4}E(X_{4:4} - 3X_{3:4} + 3X_{2:4} - X_{1:4}) = \int_0^1 x\left[20F^3(x) - 30F^2(x) + 12F(x) - 1\right]dF(x) \tag{12.6}$$

where $X_{1:1}$ denotes the single observations, and is a measure of the location of the distribution; $X_{2:2} - X_{1:2}$ denotes the difference between the two observations, and is a measure of the scale of the distribution; $X_{3:3} - 2X_{2:3} + X_{1:3}$ denotes the central second difference among the three observations, and is a measure of the skewness of the distribution; and $X_{4:4} - 3X_{3:4} + 3X_{2:4} - X_{1:4}$ denotes the central third difference among the four observations, and is a measure of the kurtosis of the distribution (Hosking and Wallis 1997).

The computation of λ_r can be simplified by relating $E(X_{r-k:r})$ to the probability weighted moments (PWM) (Greenwood et al. 1979) defined as

$$M_{l,r,k} = E\left\{X^l[F(x)]^r[1-F(x)]^k\right\} = \int_0^1 x^l[F(x)]^r[1-F(x)]^k dF(x) \tag{12.7}$$

in which l, r, and k are integer numbers.

In terms of a particular type of the PWM we have, $\eta_r = M_{1,\,r,\,0}$, which can be defined as

$$\eta_r = M_{1,r,0} = E\{X[F(x)]^r\} = \int_0^1 x[F(x)]^r dF(x) \tag{12.8}$$

Thus, the first four L-moments can also be expressed as

$$\lambda_1 = \eta_0 \tag{12.9}$$

$$\lambda_2 = 2\eta_1 - \eta_0 \tag{12.10}$$

$$\lambda_3 = 6\eta_2 - 6\eta_1 + \eta_0 \tag{12.11}$$

$$\lambda_4 = 20\eta_3 - 30\eta_2 + 12\eta_1 - \eta_0 \tag{12.12}$$

where λ_1 and λ_2 may be regarded as measures of location and scale, respectively. It is often convenient to standardise the higher moments λ_r, $r > 3$, so that they are independent of the units of measurement of X. We define, therefore, the L-moment ratios of X to be the quantities (Hosking 1990):

$$\tau_r = \lambda_r/\lambda_2, r = 3, 4, ... \tag{12.13a}$$

in which

$$\lambda_2 \geq 0 \tag{12.13b}$$

where the L-moment ratios τ_3 and τ_4 are dimensionless analogues of λ_3 and λ_4, respectively, and are plausible measures of skewness and kurtosis. The constraints on τ_3 and τ_4 are (Hosking 1990)

$$-1 < \tau_3 < 1, \quad \frac{5\tau_3^2 - 1}{4} \leq \tau_4 < 1 \tag{12.13c}$$

According to Eqs. (12.3)–(12.6) or Eqs. (12.8)–(12.12), the first four L-moments can be obtained when the probability distribution of the random variable is known. Table 12.1 gives the first four L-moments and C-moments for some commonly used distributions.

In practice, L-moments are usually estimated from random samples drawn from an unknown distribution. Let $x_1, x_2, ..., x_n$ be the samples and $x_{1:n} \leq x_{2:n} \leq ... \leq x_{n:n}$ be the samples in order, and the first four sample L-moments, i.e. l_1, l_2, l_3, and l_4, can be given by (e.g. MacKenzie and Winterstein 2011)

$$l_1 = \frac{1}{n} \sum_{j=1}^{n} x_{j:n} \tag{12.14}$$

$$l_2 = \frac{2}{n} \sum_{j=2}^{n} \frac{(j-1)}{(n-1)} x_{j:n} - \frac{1}{n} \sum_{j=1}^{n} x_{j:n} \tag{12.15}$$

$$l_3 = \frac{6}{n} \sum_{j=3}^{n} \frac{(j-1)(j-2)}{(n-1)(n-2)} x_{j:n} - \frac{6}{n} \sum_{j=2}^{n} \frac{(j-1)}{(n-1)} x_{j:n} + \frac{1}{n} \sum_{j=1}^{n} x_{j:n} \tag{12.16}$$

$$l_4 = \frac{20}{n} \sum_{j=4}^{n} \frac{(j-1)(j-2)(j-3)}{(n-1)(n-2)(n-3)} x_{j:n} - \frac{30}{n} \sum_{j=3}^{n} \frac{(j-1)(j-2)}{(n-1)(n-2)} x_{j:n}$$
$$+ \frac{12}{n} \sum_{j=2}^{n} \frac{(j-1)}{(n-1)} x_{j:n} - \frac{1}{n} \sum_{j=1}^{n} x_{j:n} \tag{12.17}$$

12.3 Structural Reliability Analysis Based on the First Three L-Moments

12.3.1 Transformation for Independent Random Variables

12.3.1.1 The u–x Transformation

By using the second-order polynomial transformation technique (Zhao and Ono 2000b), a univariate random variable X can be approximated by a second-order polynomial of a standard normal random variable as

$$X = S^{-1}(U) = a + bU + cU^2 \tag{12.18}$$

Table 12.1 C-moments and L-moments of some common distributions.

Distribution	Normal	Lognormal	Gumbel	Exponential
$F(X)$ or $X(F)$	$F = \Phi\left(\dfrac{X-\mu}{\sigma}\right)$	$F(x) = \Phi\left[\dfrac{\ln(x-\xi) - \lambda}{\zeta}\right]$	$x = \xi - \alpha \ln(-\ln F)$	$X = -\dfrac{1}{\lambda}\ln(1-F)$
L-moments	$\lambda_1 = \mu$ $\lambda_2 = \sigma/\sqrt{\mu}$ $\tau_3 = 0$ $\tau_4 = 0.1226$	$\lambda_1 = \xi + \exp\left(\lambda + \dfrac{\zeta^2}{2}\right)$ $\lambda_2 = \exp\left(\lambda + \dfrac{\zeta^2}{2}\right) erf\left(\tfrac{\zeta}{2}\right)$ $\tau_3 = \dfrac{6}{\sqrt{\pi}\,erf\left(\tfrac{\zeta}{2}\right)} \int_0^{\tfrac{\zeta}{2}} erf\left(\dfrac{x}{\sqrt{3}}\right) \exp\left(-x^2\right) dx$	$\lambda_1 = \xi + \gamma\alpha$ $\lambda_2 = \alpha \ln 2$ $\tau_3 = 0.1669$ $\tau_4 = 0.1504$	$\lambda_1 = 1/\lambda$ $\lambda_2 = 1/(2\lambda)$ $\tau_3 = 1/3$ $\tau_4 = 1/6$
C-moments	$\mu = \mu$ $\sigma^2 = \sigma^2$ $\alpha_3 = 0$ $\alpha_4 = 3$	$\mu = \exp\left[\dfrac{1}{2}\zeta^2 + \lambda\right]$ $\sigma = \mu\sqrt{\exp\left(\zeta^2\right) - 1}$ $\alpha_3 = V(V^2+3)$ $\alpha_4 = V^8 + 6V^6 + 15V^4 + 16V^2 + 3$	$\mu = \xi + \gamma\alpha\sigma^2 = \pi^2\alpha^2/6$ $\alpha_3 \approx 1.14$ $\alpha_4 = 5.4$	$\mu = 1/\lambda$ $\sigma^2 = 1/\lambda^2$ $\alpha_3 = 2$ $\alpha_4 = 9$

where X is an arbitrary random variable; U is the standard normal random variable; and a, b, c are polynomial coefficients that can be determined by making the first three L-moments of $S^{-1}(U)$ to be equal to those of X.

According to Eq. (12.8) and assuming $F(x) = \Phi(u)$, η_m can be expressed as (Chen and Tung 2003)

$$
\begin{aligned}
\eta_m\left[S^{-1}(U)\right] = \eta_m(X) &= \int_0^1 xF^m(x)dF(x) \\
&= \int_{-\infty}^{\infty} (a + bu + cu^2)\Phi^m(u)\phi(u)du, \qquad m = 0, 1, 2 \\
&= aC_{m,0} + bC_{m,1} + cC_{m,2}
\end{aligned}
\tag{12.19}
$$

where $m = 0, 1, 2$, and $C_{m, n}$ is defined as

$$
C_{m,n} = \int_{-\infty}^{\infty} u^n \Phi^m(u)\phi(u)du, \qquad m, n = 0, 1, 2
\tag{12.20}
$$

Then

$$
C_{0,0} = 1, C_{0,1} = 0, C_{0,2} = 1
\tag{12.21a}
$$

$$
C_{1,0} = \frac{1}{2}, C_{1,1} = \frac{1}{2\sqrt{\pi}}, C_{1,2} = \frac{1}{2}
\tag{12.21b}
$$

$$
C_{2,0} = \frac{1}{3}, C_{2,1} = \frac{1}{2\sqrt{\pi}}, C_{2,2} = \frac{\sqrt{3} + 2\pi}{6\pi}
\tag{12.21c}
$$

According to Eq. (12.19) and Eqs. (12.9)–(12.11), the first three L-moments of $S^{-1}(U)$ can be obtained as

$$
\lambda_{1S} = \eta_0\left[S^{-1}(U)\right] = a + c
\tag{12.22a}
$$

$$
\lambda_{2S} = 2\eta_1\left[S^{-1}(U)\right] - \eta_0\left[S^{-1}(U)\right] = \frac{b}{\sqrt{\pi}}
\tag{12.22b}
$$

$$
\lambda_{3S} = 6\eta_2\left[S^{-1}(U)\right] - 6\eta_1\left[S^{-1}(U)\right] + \eta_0\left[S^{-1}(U)\right] = \frac{\sqrt{3}c}{\pi}
\tag{12.22c}
$$

Making the first three L-moments of $S^{-1}(U)$ to be equal to those of X leads to

$$
\lambda_{1S} = a + c = \lambda_{1X}
\tag{12.23a}
$$

$$
\lambda_{2S} = \frac{b}{\sqrt{\pi}} = \lambda_{2X}
\tag{12.23b}
$$

$$
\lambda_{3S} = \frac{\sqrt{3}c}{\pi} = \lambda_{3X}
\tag{12.23c}
$$

Thus, explicit expressions for the polynomial coefficients a, b, and c can be obtained as

$$
a = \lambda_{1X} - \frac{\pi}{\sqrt{3}}\lambda_{3X}
\tag{12.24a}
$$

$$
b = \sqrt{\pi}\lambda_{2X}
\tag{12.24b}
$$

$$
c = \frac{\pi}{\sqrt{3}}\lambda_{3X}
\tag{12.24c}
$$

As can be observed from Eq. (12.18) and Eqs. (12.24a)–(12.24c), the second-order polynomial transformation technique based on the first three L-moments has no limitations on the L-skewness. Particularly, if $\lambda_{1X} = \lambda_{3X} = 0$ and $\lambda_{2X} = 1/\sqrt{\pi}$, then $a = c = 0$ and $b = 1$, the u–x transformation function will then reduce to $x = u$.

12.3.1.2 The x–u Transformation

From Eq. (12.18), the x–u transformation is readily obtained as

i) when $\lambda_{3X} \neq 0$

$$U = \frac{-\sqrt{3}\lambda_{2X} + \sqrt{3\lambda_{2X}^2 + 4\sqrt{3}\lambda_{3X}(X - \lambda_{1X}) + 4\pi\lambda_{3X}^2}}{2\sqrt{\pi}\lambda_{3X}} \tag{12.25}$$

ii) when $\lambda_{3X} = 0$, and $\lambda_{2X} \neq 0$

$$U = \frac{X - \lambda_{1X}}{\sqrt{\pi}\lambda_{2X}} \tag{12.26}$$

12.3.2 Transformation for Correlated Random Variables

12.3.2.1 The u–x Transformation for Correlated Random Variables

For two correlated non-normal random variables X_i and X_j, if their first three L-moments and the correlation coefficient (ρ_{ij}) are known, X_i and X_j may be expressed as

$$X_i = a_i + b_i Z_i + c_i Z_i^2 = (a_i, b_i, c_i) \cdot \left(1, Z_i, Z_i^2\right)^T \tag{12.27a}$$

$$X_j = a_j + b_j Z_j + c_j Z_j^2 = (a_j, b_j, c_j) \cdot \left(1, Z_j, Z_j^2\right)^T \tag{12.27b}$$

where Z_i and Z_j are two correlated standard normal variables with the correlation coefficient of ρ_{0ij}. According to Eqs. (12.24a)–(12.24c), the coefficients of a_i, b_i, and c_i in Eq. (12.27a) and a_j, b_j, and c_j in Eq. (12.27b) can be determined as

$$a_i = \lambda_{1X_i} - \frac{\pi}{\sqrt{3}}\lambda_{3X_i}, b_i = \sqrt{\pi}\lambda_{2X_i}, c_i = \frac{\pi}{\sqrt{3}}\lambda_{3X_i} \tag{12.28a}$$

$$a_j = \lambda_{1X_j} - \frac{\pi}{\sqrt{3}}\lambda_{3X_j}, b_j = \sqrt{\pi}\lambda_{2X_j}, c_j = \frac{\pi}{\sqrt{3}}\lambda_{3X_j} \tag{12.28b}$$

where λ_{1X_i}, λ_{2X_i}, λ_{3X_i}, and λ_{1X_j}, λ_{2X_j}, λ_{3X_j} are the first three L-moments of X_i and X_j, respectively.

The correlation coefficient between Z_i and Z_j, i.e. ρ_{0ij}, can be derived according to the definition of correlation coefficient, which can be described as follows:

$$\rho_{ij} = \frac{Cov(X_i X_j)}{\sqrt{D(X_i)D(X_j)}} = \frac{E(X_i X_j) - E(X_i)E(X_j)}{\sigma_{X_i}\sigma_{X_j}} = \frac{E(X_i X_j) - \lambda_{1X_i}\lambda_{1X_j}}{\sigma_{X_i}\sigma_{X_j}} \tag{12.29}$$

where $D(X_i)$ and $D(X_j)$ are the variance of X_i and X_j, respectively; σ_{X_i} and σ_{X_j} are the standard deviation of X_i and X_j, respectively; and $E(X_i X_j)$ can be derived as

$$E\left(X_i X_j\right) = E\left[\left(a_i, b_i, c_i\right)\left(1, Z_i, Z_i^2\right)^T \cdot \left(1, Z_j, Z_j^2\right)\left(a_j, b_j, c_j\right)^T\right]$$

$$= \left(a_i, b_i, c_i\right)\begin{bmatrix} 1 & 0 & 1 \\ 0 & \rho_{0ij} & 0 \\ 1 & 0 & 2\rho_{0ij}^2 + 1 \end{bmatrix}\left(a_j, b_j, c_j\right)^T \tag{12.30}$$

Substituting Eqs. (12.30), (12.28a), and (12.28b) into Eq. (12.29), the relationship between ρ_{ij} and ρ_{0ij} can be expressed as

$$\rho_{ij} = \frac{2c_i c_j \rho_{0ij}^2 + b_i b_j \rho_{0ij}}{\sigma_{X_i} \sigma_{X_j}} \tag{12.31}$$

that is

$$\rho_{ij} = \frac{2\pi^2 \lambda_{3X_i} \lambda_{3X_j} \rho_{0ij}^2 + 3\pi \lambda_{2X_i} \lambda_{2X_j} \rho_{0ij}}{3\sigma_{X_i} \sigma_{X_j}} \tag{12.32}$$

where ρ_{0ij} can be determined from solving Eq. (12.32). It is worth noting that the valid solution of ρ_{0ij} should be restricted by the following conditions to satisfy the definition of the correlation coefficient:

$$-1 \le \rho_{0ij} \le 1, \, -1 \le \rho_{ij} \le 1, \, \rho_{ij} \cdot \rho_{0ij} \ge 0, \, \frac{\partial \rho_{ij}}{\partial \rho_{0ij}} \ge 0, \, \text{and} \, \left|\rho_{0ij}\right| \ge \left|\rho_{ij}\right| \tag{12.33}$$

With Eqs. (12.32) and (12.33), the expressions of the correlation coefficient ρ_{0ij} and the upper and lower bounds of original correlation coefficient ρ_{ij} to ensure the transformation executable are summarised in Table 12.2. The derivation of Table 12.2 is illustrated in Example 12.1.

Each $\rho_{ij\text{-min}}$ and $\rho_{ij\text{-max}}$ form the lower and upper initial correlation matrix (initial correlation matrix) $\boldsymbol{\rho}_{\min}$ and $\boldsymbol{\rho}_{\max}$, respectively. Then the boundary Θ of initial correlation matrix $\boldsymbol{\rho}$ (to make sure that every element of the matrix can be conducted for the transformation) can be described as

Table 12.2 Expressions of ρ_{0ij} and applicable range of ρ_{ij} for the transformation based on the first three L-moments.

Conditions	ρ_{0ij}	Range of A and B	Lower and upper value of ρ_{ij}
$A > 0$	$\rho_{0ij} = \frac{-B + \sqrt{B^2 + 4A\rho_{ij}}}{2A}$	$B^2 - 4A \ge 0$	$\rho_{ij\text{-min}} = \max\{-1, A - B\}$ $\rho_{ij\text{-max}} = \min\{1, A + B\}$
		$B^2 - 4A < 0$	$\rho_{ij\text{-min}} = \max\{-B^2/(4A), A - B\}$ $\rho_{ij\text{-max}} = \min\{1, A + B\}$
$A < 0$	$\rho_{0ij} = \frac{-B + \sqrt{B^2 + 4A\rho_{ij}}}{2A}$	$B^2 + 4A \ge 0$	$\rho_{ij\text{-min}} = \max\{-1, A - B\}$ $\rho_{ij\text{-max}} = \min\{1, A + B\}$
		$B^2 + 4A < 0$	$\rho_{ij\text{-min}} = \max\{-1, A - B\}$ $\rho_{ij\text{-max}} = \min\{-B^2/(4A), A + B\}$
$A = 0$	$\rho_{0ij} = \frac{\rho_{ij}}{B}$	---	$\rho_{ij\text{-min}} = \max\{-1, -B\}$ $\rho_{ij\text{-max}} = \min\{1, B\}$

Note: $A = 2c_i c_j / \sigma_i \sigma_j$ and $B = b_i b_j / \sigma_i \sigma_j > 0$.

$$\Theta = \left[\max\left(\rho_{ij-\min} \right), \ \min\left(\rho_{ij-\max} \right) \right] \tag{12.34}$$

The preceding procedure can be readily extended to n-dimension correlated non-normal random vector if we know the first three L-moments, standard deviations, and correlation matrix. The polynomial coefficients can be determined using Eqs. (12.24a)–(12.24c), for any two correlated non-normal random variables, and the corresponding equivalent correlation coefficients of standard normal variables can be determined from Table 12.2. The equivalent correlation matrix of standard normal variables, $\mathbf{C_Z}$, can then be expressed as Eq. (7.105) from the chapter on Transformation of Non-Normal Variables to Independent Normal Variables.

With the aid of Cholesky decomposition, the correlated standard normal random vector $\mathbf{Z}$ can then be transformed into the independent standard normal vector $\mathbf{U}$, i.e.

$$\mathbf{Z} = \mathbf{L_0 U} \tag{12.35}$$

where $\mathbf{L_0}$ is a lower triangular matrix obtained from Cholesky decomposition, which can be expressed as Eq. (7.108) in the chapter just mentioned.

For the cases of original correlation matrix with very small eigenvalues, the equivalent correlation matrix might become a non-positive semidefinite matrix. The method for solving the problem (Ji et al. 2018) mentioned in Section 7.5.2 also in the chapter on Transformation of Non-Normal Variables to Independent Normal Variables can be adopted to make Cholesky decomposition ready.

According to Eqs. (12.35), Z_i can then be expressed as:

$$Z_i = \sum_{k=1}^{i} l_{ik} U_k, \quad i = 1, 2, \cdots, n \tag{12.36}$$

Substituting Eq. (12.36) into Eq. (12.27a), the u–x transformation for correlated random variables using the second-order polynomial transformation is expressed as:

$$X_i = a_i + b_i \sum_{k=1}^{i} l_{ik} U_k + c_i \left(\sum_{k=1}^{i} l_{ik} U_k \right)^2, \quad i = 1, 2, \cdots, n \tag{12.37}$$

12.3.2.2 The x–u Transformation for Correlated Random Variables

In order to achieve the normal transformation, the correlated non-normal random vector $\mathbf{X}$ is first transformed into the correlated standard normal random vector $\mathbf{Z}$, and $\mathbf{Z}$ is then transformed into the independent standard normal random vector $\mathbf{U}$.

According to Eqs. (12.25) and (12.26), the X–Z transformation can be given by

$$Z_i = S(X_i) = \begin{cases} \dfrac{-\sqrt{3}\lambda_{2X_i} + \sqrt{3\lambda_{2X_i}^2 + 4\sqrt{3}\lambda_{3X_i}(X_i - \lambda_{1X_i}) + 4\pi\lambda_{3X_i}^2}}{2\sqrt{\pi}\lambda_{3X}}, & c_i \neq 0 \\[4mm] \dfrac{X_i - \lambda_{1X_i}}{\sqrt{\pi}\lambda_{2X_i}}, & c_i = 0 \end{cases}$$

$$\tag{12.38}$$

Then, the correlated standard normal random vector $\mathbf{Z}$ can be transformed into independent standard normal vector $\mathbf{U}$,

$$\mathbf{U} = \mathbf{L}_0^{-1}\mathbf{Z} \tag{12.39}$$

where $\mathbf{L}_0^{-1}$ is the inverse matrix of $\mathbf{L}_0$, which is a lower triangular matrix as shown in Eq. (7.111) in the chapter on Transformation of Non-Normal Variables to Independent Normal Variables.

According to Eq. (12.39), U_i is expressed as:

$$U_i = \sum_{k=1}^{i} h_{ik} Z_k, \quad i = 1, 2, \cdots, n \tag{12.40}$$

Substituting Eq. (12.38) into Eq. (12.40), the X–U transformation can be expressed as:

$$U_i = \sum_{k=1}^{i} h_{ik} S(X_k), \quad i = 1, 2, \cdots, n \tag{12.41}$$

where $S(\cdot)$ is given by Eq. (12.38); and h_{ik} is the elements of $\mathbf{L}_0^{-1}$.

12.3.3 Reliability Analysis Using the First Three L-Moments and Correlation Matrix

The computation procedure of first-order reliability method (FORM) using the first three L-moments and correlation matrix is similar to that of first-order third-moment method (FOTM) in the third-moment pseudo standard normal space as described in Section 6.4.2 in the chapter on Structural Reliability Assessment, where the third-moment pseudo transformations based on the first three central moments are replaced by the transformations based on the first three L-moments, and the element of Jacobian matrix can be derived from Eq. (12.37), i.e.

$$\frac{\partial x_i}{\partial u_j} = l_{ij}\left(b_i + 2c_i \sum_{k=1}^{i} l_{ik} u_k\right), \quad i, j = 1, 2, ..., n \tag{12.42}$$

Example 12.1 The General Formulae of Equivalent Correlation Coefficients and Applicable Range for Original Correlation Coefficient

For the propose of illustration, rewritten Eq. (12.31) as

$$\rho_{ij} = A\rho_{0ij}^2 + B\rho_{0ij}, \quad (B > 0) \tag{12.43}$$

where

$$A = \frac{2c_i c_j}{\sigma_{Xi}\sigma_{Xj}}, \quad B = \frac{b_i b_j}{\sigma_{Xi}\sigma_{Xj}} \tag{12.44}$$

According to Eqs. (12.24a)–(12.24c), b_i (or b_j) will be constant greater than 0, therefore $B > 0$; while the sign of A is determined by the sign of $c_i c_j$ (i.e. $\lambda_{3Xi}\lambda_{3Xj}$).

For $A \neq 0$, Eq. (12.43) is a quadratic equation about ρ_{0ij} and can be equivalently expressed as:

$$\frac{\rho_{ij}}{A} = \rho_{0ij}^2 + \frac{B}{A}\rho_{0ij} \quad (B > 0) \tag{12.45}$$

For convenience of expression, the right side of Eq. (12.45) is expressed as $h(\rho_{0ij})$, i.e.

$$\frac{\rho_{ij}}{A} = h\left(\rho_{0ij}\right) = \rho_{0ij}^2 + \frac{B}{A}\rho_{0ij} \tag{12.46}$$

Then, the extreme value of $h(\rho_{0ij})$ can be determined as

$$h_{ext} = h\left(\frac{-B}{2A}\right) = \frac{-B^2}{4A^2} \tag{12.47}$$

1) For $A > 0$, Eq. (12.47) has two real roots $\rho_{0ij-1} = 0$ and $\rho_{0ij-2} = -B/A < 0$. Thus, the shape of $h(\rho_{0ij})$ is depicted for $A > 0$ in Figure 12.1, where the solid line denotes the region satisfying the condition of $\rho_{ij} \cdot \rho_{0ij} \geq 0$ and ρ_{0ij} is a monotonic function of ρ_{ij}.

From Figure 12.1, the equivalent correlation coefficient ρ_{0ij} for $A > 0$ can be given as:

$$\rho_{0ij} = \frac{-B + \sqrt{B^2 + 4A\rho_{ij}}}{2A} \tag{12.48}$$

To ensure ρ_{0ij} given by Eq. (12.48) to satisfy the definition of correlation coefficient, i.e. $-1 \leq \rho_{0ij} \leq 1$, there should exist application bounds for ρ_{ij}, i.e. $\rho_{ij} \in [\rho_{ij-min}, \rho_{ij-max}]$, where ρ_{ij-min} and ρ_{ij-max} are the lower and upper bounds, respectively.

For $h_{ext} \leq -1/A$ ($B^2 \geq 4A$), i.e. the interval, $(-\infty, 0]$ of ρ_{0ij} can be divided into two sub-intervals by the abscissas corresponding to $\rho_{ij} = -1$, as shown in Figure 12.1a. When $\rho_{0ij} = -1$ locates in interval I, the lower bound of $\rho_{ij-min} = A - B$ ensures $\rho_{0ij} \geq -1$. When $\rho_{0ij} = -1$ locates in interval II, the lower bound is $\rho_{ij-min} = -1$. The lower bound for $B^2 \geq 4A$ can thus be given as $\rho_{ij-min} = \max\{-1, A - B\}$.

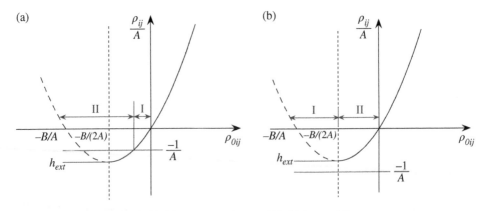

Figure 12.1 The shape of $h(\rho_{0ij})$ for $A > 0$. (a) $h_{ext} \leq -1/A$; (h) $h_{ext} > -1/A$.

For $h_{ext} > -1/A$ ($B^2 - 4A < 0$), i.e. the interval, $(-\infty, 0]$ of ρ_{0ij} can be divided into two subinterval by $\rho_{0ij} = -B/2A$, as shown in Figure 12.1b. When $\rho_{0ij} = -1$ locates in interval I, the lower bound is $\rho_{ij-min} = -B^2/4A$. When $\rho_{0ij} = -1$ locates in interval II, the lower bound of $\rho_{ij-min} = A - B$ ensures $\rho_{0ij} \geq -1$. The lower bound for $B^2 - 4A < 0$ can therefore be given as $\rho_{ij-min} = \max\{-B^2/4A, A-B\}$.

From this discussion, for $A > 0$, the suitable maximum original correlation coefficient, i.e. the lower bound of ρ_{ij}, can be summarised as:

$$\rho_{ij-min} = \begin{cases} \max\{-1, A-B\}, & B^2 - 4A \geq 0 \\ \max\{-B^2/(4A), A-B\}, & B^2 - 4A < 0 \end{cases} \tag{12.49}$$

Substituting Eq. (12.44) into Eq. (12.49), the lower bound of ρ_{ij} can be rewritten as

$$\rho_{ij-min} = \begin{cases} \max\{-1, (2c_ic_j - b_ib_j)/(\sigma_i\sigma_j)\}, & c_i, c_j \in \Omega \\ \max\{-b_i^2 b_j^2/(\sigma_i\sigma_j c_ic_j), (2c_ic_j - b_ib_j)/(\sigma_i\sigma_j)\}, & \text{otherwise} \end{cases} \tag{12.50}$$

where

$$\Omega = \left\{ c_i, c_j \,\middle|\, c_ic_j \leq b_i^2 b_j^2/(8\sigma_i\sigma_j) \right\} \tag{12.51}$$

Similarly, the upper bound of ρ_{ij}, for $A > 0$, can also be determined as:

$$\rho_{ij-max} = \min\{1, A+B\} \tag{12.52}$$

that is

$$\rho_{ij-max} = \min\{1, (2c_ic_j + b_ib_j)/(\sigma_i\sigma_j)\} \tag{12.53}$$

2) Similarly for $A < 0$, ρ_{0ij} is also given by Eq. (12.48), and the application bound of ρ_{ij} is expressed as:

$$\rho_{ij-max} = \begin{cases} \min\{1, A+B\}, & B^2 + 4A \geq 0 \\ \min\{-B^2/(4A), A+B\}, & \text{otherwise} \end{cases} \tag{12.54}$$

$$\rho_{ij-min} = \max\{-1, A-B\} \tag{12.55}$$

3) For $A = 0$, Eq. (12.43) reduces to a linear equation of ρ_{0ij}:

$$\rho_{0ij} = \rho_{ij}/B \tag{12.56}$$

And the application bounds of ρ_{ij} can be determined as follow:

$$\rho_{ij-max} = \min\{1, B\} \tag{12.57}$$

$$\rho_{ij-min} = \max\{-1, -B\} \tag{12.58}$$

Example 12.2 Illustration of the Computational Procedure Using a Simple Performance Function

Consider the following performance function

Table 12.3 Probability distributions and correlation matrix of random variables for Example 12.2.

Variables	Distribution	Mean	Coefficient of variation	Correlation matrix
X_1	Lognormal	1000	0.2	$\begin{pmatrix} 1.0 & 0.2 & 0.2 \\ 0.2 & 1.0 & 0.2 \\ 0.2 & 0.2 & 1.0 \end{pmatrix}$
X_2	Lognormal	2×10^{10}	0.05	
X_3	Lognormal	3.9025×10^{-5}	0.1	

$$G(\mathbf{X}) = 0.02 - \frac{8X_1}{X_2 X_3} \tag{12.59}$$

where X_1, X_2, and X_3 are random variables, and their probability distributions and correlation matrix are listed in Table 12.3.

From Table 12.1, the first three L-moments of X_1, X_2, and X_3 can be readily obtained as

$$\lambda_{1X_1} = 1000, \quad \lambda_{2X_1} = 111.369, \quad \lambda_{3X_1} = 10.753$$
$$\lambda_{1X_2} = 2 \times 10^{10}, \quad \lambda_{2X_2} = 5.637 \times 10^8, \quad \lambda_{3X_2} = 1.376 \times 10^7$$
$$\lambda_{1X_3} = 3.9025 \times 10^{-5}, \quad \lambda_{2X_3} = 2.194 \times 10^{-6}, \quad \lambda_{3X_3} = 1.069 \times 10^{-7}$$

Using Eqs. (12.24a)–(12.24c), the corresponding polynomial coefficients can be obtained as

$$a_1 = 980.496, \quad b_1 = 197.396, \quad c_1 = 19.504$$
$$a_2 = 2 \times 10^{10}, \quad b_2 = 9.992 \times 10^8, \quad c_2 = 2.496 \times 10^7$$
$$a_3 = 3.883 \times 10^{-5}, \quad b_3 = 3.890 \times 10^{-6}, \quad c_3 = 1.939 \times 10^{-7}$$

According to Eq. (12.44), the parameter matrix $\mathbf{A}$ can be calculated as

$$\mathbf{A} = \begin{bmatrix} 0.0190199 & 0.00486773 & 0.00969043 \\ 0.00486773 & 0.00124579 & 0.00248006 \\ 0.00969043 & 0.00248006 & 0.00493717 \end{bmatrix}$$

As can be observed, all the elements of $\mathbf{A}$ are greater than 0, and then using Eqs. (12.50) and (12.53), the lower and upper bounds of ρ_{ij} can be rewritten as

$$\rho_{ij-min} = \begin{bmatrix} -0.955114 & -0.981258 & -0.973818 \\ -0.981258 & -0.997019 & -0.993135 \\ -0.973818 & -0.993135 & -0.988036 \end{bmatrix},$$

$$\rho_{ij-max} = \begin{bmatrix} 0.993153 & 0.990993 & 0.993198 \\ 0.990993 & 0.999511 & 0.998095 \\ 0.993198 & 0.998095 & 0.99791 \end{bmatrix}$$

The applicable range of initial correlation coefficients matrix can be readily obtained as $\Theta = [-0.955114, 0.990993]$, which implies that the initial correlation matrix given in this case is applicable for the third L-moments transformation method. It can be observed that the applicable range of original correlation coefficient is large enough and therefore is satisfied for general engineering problems.

Using Eq. (12.48), the equivalent correlation matrix can be readily obtained as

$$\mathbf{C_Z} = \begin{bmatrix} 1.0000 & 0.2019 & 0.2020 \\ 0.2019 & 1.0000 & 0.2005 \\ 0.2020 & 0.2005 & 1.0000 \end{bmatrix}$$

Then, the Cholesky decomposition results of the equivalent correlation matrix can then be determined as

$$\mathbf{L_0} = \begin{bmatrix} 1.0 & 0 & 0 \\ 0.2019 & 0.9794 & 0 \\ 0.2020 & 0.1631 & 0.9657 \end{bmatrix}, \quad \mathbf{L_0^{-1}} = \begin{bmatrix} 1.0 & 0 & 0 \\ -0.2061 & 1.0210 & 0 \\ -0.1744 & -0.1724 & 1.0355 \end{bmatrix}$$

Assume initial checking point as:

$$x_0 = (x_1, x_2, x_3)^T = \left(1000, 2 \times 10^{10}, 3.9025 \times 10^{-5}\right)^T$$

Using Eq. (12.41), the corresponding checking point in standard normal space can be readily given as $\mathbf{u_0} = (0.0979, 0.0053, 0.0301)^T$, then the initial reliability index β_0 at $\mathbf{u_0}$ can be obtained as $\beta_0 = \sqrt{\mathbf{u_0^T u_0}} = 0.1026$.

The Jacobian matrix evaluated at the checking point $\mathbf{u_0}$ using Eq. (12.42) is determined as:

$$\mathbf{J} = \begin{pmatrix} 201.214 & 0 & 0 \\ 2.0193 \times 10^8 & 9.7979 \times 10^8 & 0 \\ 7.8954 \times 10^{-7} & 6.3739 \times 10^{-7} & 3.7740 \times 10^{-6} \end{pmatrix}$$

The performance function and gradient vector at $\mathbf{u_0}$ (using Eq. (6.29) from the chapter on Structural Reliability Assessment) are obtained as:

$$g(\mathbf{u_0}) = G(\mathbf{x_0}) = 0.00975, \quad \nabla g(\mathbf{u_0}) = \begin{pmatrix} -1.752 \times 10^{-3} \\ 6.695 \times 10^{-4} \\ 9.912 \times 10^{-4} \end{pmatrix}$$

New checking point $\mathbf{u_1}$ is obtained (using Eq. (6.30) also from the chapter on Structural Reliability Assessment) as:

$$\mathbf{u_1} = \frac{1}{\nabla^T g(\mathbf{u_0}) \cdot \nabla g(\mathbf{u_0})} \left[\nabla^T g(\mathbf{u_0}) \cdot \mathbf{u_0} - g(\mathbf{u_0})\right] \nabla g(\mathbf{u_0}) = \begin{pmatrix} 3.8499 \\ -1.4716 \\ -2.1787 \end{pmatrix}$$

The corresponding reliability index is calculated as $\beta = \sqrt{\mathbf{u_1^T u_1}} = 4.66195$. Because the absolute difference between β and β_0 is

$$\varepsilon_r = \frac{|\beta - \beta_0|}{\beta} = 0.9780 > 1.0 \times 10^{-6}$$

The checking point in original space is determined as:

$$\mathbf{x}_1 = \mathbf{x}_0 + \mathbf{J}(\mathbf{u}_1 - \mathbf{u}_0) = \left(1754.95, 1.93105 \times 10^{10},\ 0.0000327\right)^T$$

The above procedure is repeated using $\mathbf{x}_1$ as the new checking point until convergence is achieved. The convergence is achieved in five steps with $\beta_{\text{L-M}} = 3.395$ at the design point in standard normal space $\mathbf{u}^* = (2.70419, -1.15433, -1.69832)^T$ corresponding to the design point in original space $\mathbf{x}^* = (1626.92, 1.93994 \times 10^{10}, 3.341643 \times 10^{-5})^T$.

As for comparison, reliability analysis in which the x–u and u–x transformations are achieved using the first three C-moments, is conducted for the same example. Similarly, the reliability index $\beta_{\text{C-M}}$ can be obtained as 3.373 after the convergence is achieved ($\varepsilon_r \leq 1.0 \times 10^{-6}$) in five steps. The last design point in standard normal space $\mathbf{u}^* = (2.70197, -1.13515, -1.67008)^T$ corresponding to the design point in original space $\mathbf{x}^* = (1663.32, 1.94175 \times 10^{10}, 3.42644 \times 10^{-5})^T$.

According to the full known distribution information and initial correlation matrix of random variables, the failure probability is obtained as 3.52×10^{-4} (coefficient of variation of MCS is 3.77%) using MCS procedure (sample size $= 2 \times 10^6$) with the corresponding reliability index of 3.388. It can be observed that both the results of FORM based on the L-moments and C-moments are in close agreement with that of MCS.

Furthermore, suppose that the coefficient of variation of x_1 is 0.8, and then the skewness of x_1 turns to be 2.924, which is out of the scope of the third-moment transformation based on C-moments. In this case, the boundary of initial correlation matrix checked as $\Theta = [-0.545, 0.857]$, and reliability analysis based on L-moments can therefore be conducted. The obtained reliability index $\beta_{\text{L-M}} = 1.266$ converges to $\mathbf{u}^* = (1.24335, -0.133144, -0.199123)^T$ corresponding to $\mathbf{x}^* = (1964.38.76, 2.0117 \times 10^{10}, 3.9059 \times 10^{-5})^T$, after the convergence is achieved ($\varepsilon_r \leq 1.0 \times 10^{-6}$) in five steps. Using MCS, the failure probability is obtained as 9.261×10^{-2} (the coefficient of variation of MCS is 0.989%) with the corresponding reliability index of 1.324. We can observe that FORM based on L-moments provides comparable result compared with MCS, while FORM based on C-moments cannot be conducted for this case.

Figure 12.2 presents the reliability indices with respect to coefficients of variation of x_1 obtained using FORM based on the L-moments and C-moments, together with those of MCS. It can be observed from Figure 12.2 that both the results of FORM based on the L-moments and FORM based on C-moments are in good agreement with those obtained by MCS when the coefficients of variation of x_1 varies from 0.2 to 0.783. When the coefficients of variation of x_1 is greater than 0.783, FORM based on L-moments provides comparable results compared with MCS, while FORM based on C-moments cannot be conducted since the skewness of x_1 is beyond the application range of third-moment transformation based on C-moments.

Example 12.3 Reliability Analysis Using the First Three L-Moments for Implicit Performance Functions

This example considers a three-bay five-story frame structure subjected to lateral loads as shown in Figure 12.3. The total height of the frame is $H = 21.0\,\text{m}$. In this example,

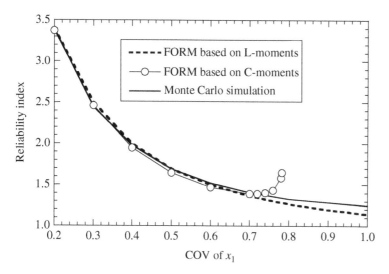

Figure 12.2 The reliability indexes with respect to coefficients of variation of x_1.

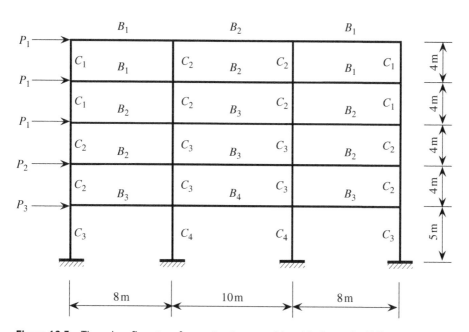

Figure 12.3 Three-bay five-story frame structure considered in Example 12.3.

21 random variables are considered. The random variables associated with the frame elements are shown in Table 12.4, and their probabilistic information is listed in Table 12.5. The implicit performance function is formulated as (Chakraborty and Chowdhury 2016):

$$G(\mathbf{X}) = u_{\text{lim}} - u(\mathbf{X}) \tag{12.60}$$

Table 12.4 Characteristics of the frame elements in Example 12.3.

Element	Young's modulus	Moment of inertia	Cross-sectional area
B_1	E_1	I_1	A_1
B_2	E_1	I_2	A_2
B_3	E_1	I_3	A_3
B_4	E_1	I_4	A_4
C_1	E_2	I_5	A_5
C_2	E_2	I_6	A_6
C_3	E_2	I_7	A_7
C_4	E_2	I_8	A_8

Table 12.5 Probabilistic information of the basic random variables for Example 12.3.

Variable	Distribution	Mean	C.O.V	λ_1	λ_2	τ_3
					First three moments	
P_1	Gumbel	135.0 kN	0.2	135.0 kN	14.592	0.1669
P_2	Gumbel	90.0 kN	0.2	90.0 kN	9.728	0.1669
P_3	Gumbel	70.0 kN	0.2	70.0 kN	7.56622	0.1669
E_1	Lognormal	2.1×10^7 kN/m^2	0.05	2.1×10^7 kN/m^2	591 906	0.024411
E_2	Lognormal	2.4×10^7 kN/m^2	0.05	2.4×10^7 kN/m^2	676 464	0.024411
I_1	Lognormal	8.11×10^{-3} m^4	0.1	8.11×10^{-3} m^4	4.6×10^{-4}	0.048712
I_2	Lognormal	0.011 m^4	0.1	0.011 m^4	6.2×10^{-4}	0.048712
I_3	Lognormal	0.0213 m^4	0.1	0.0213 m^4	0.0012	0.048712
I_4	Lognormal	0.0295 m^4	0.1	0.0295 m^4	0.0017	0.048712
I_5	Lognormal	0.0108 m^4	0.1	0.0108 m^4	6.1×10^{-4}	0.048712
I_6	Lognormal	0.0141 m^4	0.1	0.0141 m^4	7.9×10^{-4}	0.048712
I_7	Lognormal	0.0232 m^4	0.1	0.0232 m^4	1.3×10^{-3}	0.048712
I_8	Lognormal	0.0259 m^4	0.1	0.0259 m^4	1.5×10^{-3}	0.048712
A_1	Lognormal	0.312 m^2	0.1	0.312 m^2	0.0175	0.048712
A_2	Lognormal	0.372 m^2	0.1	0.372 m^2	0.0209	0.048712
A_3	Lognormal	0.505 m^2	0.1	0.505 m^2	0.0284	0.048712
A_4	Lognormal	0.557 m^2	0.1	0.557 m^2	0.0313	0.048712
A_5	Lognormal	0.253 m^2	0.1	0.253 m^2	0.0142	0.048712
A_6	Lognormal	0.291 m^2	0.1	0.291 m^2	0.0164	0.048712
A_7	Lognormal	0.372 m^2	0.1	0.372 m^2	0.0209	0.048712
A_8	Lognormal	0.418 m^2	0.1	0.418 m^2	0.0235	0.048712

where $u_{\lim} = 0.02$ m, and $u(\mathbf{X})$ is the top-floor displacement determined from finite element analysis. The correlation matrix of basic random variables is given as:

$$
\mathbf{C_X} = \begin{array}{c} P_1 \\ P_2 \\ P_3 \\ E_1 \\ E_2 \\ I_1 \\ \vdots \\ A_8 \end{array}
\begin{pmatrix}
1 & & & & & & & \\
0.5 & 1 & & & & & & \\
0.5 & 0.5 & 1 & & & & & \\
0 & 0 & 0 & 1 & & & & \\
0 & 0 & 0 & 0.9 & 1 & & & \\
0 & 0 & 0 & 0 & 0.13 & 1 & & \\
\vdots & \vdots & \vdots & \vdots & \vdots & \ddots & \ddots & \\
0 & 0 & 0 & 0 & 0.13 & \cdots & 0.13 & 1
\end{pmatrix}
\tag{12.61}
$$

The first three L-moments can be computed from their CDFs of the random variables and are also listed in Table 12.5. The u–x and x–u transformations of the random variables are then performed using Eqs. (12.37) and (12.41), respectively.

For the implicit performance function here, the central difference method as shown in Eq. (5.19) in the chapter on Methods of Moment Based on First- and Second-Order Transformation is used to determine the gradient vector $\nabla G(\mathbf{x}_0)$. After the convergence is achieved ($\varepsilon_r \leq 1.0 \times 10^{-4}$) in five steps, the reliability index can be readily obtained as $\beta_{L-M} = 2.4227$. The failure probability P_f obtained from MCS based on Nataf transformation with 10^6 samples is 0.011764 (the COV of the MCS result is 0.92%), and the corresponding reliability index is 2.2648. It can be observed that the result of FORM based on the first three L-moments is close to that of MCS.

12.4 Structural Reliability Analysis Based on the First Four L-Moments

12.4.1 Transformation for Independent Random Variables

12.4.1.1 The *u–x* Transformation Using the First Four L-Moments
For a random variable, when the first four L-moments are known, the u–x transformation can be realised using the third-order polynomial transformation (Fleishman 1978)

$$X = S^{-1}(U) = a + bU + cU^2 + dU^3 \tag{12.62}$$

Explicit and simple expressions have been developed by Tung (1999) for the polynomial coefficients in Eq. (12.62).

$$a = \lambda_{1X} - 1.81379937\lambda_{3X} \tag{12.63a}$$

$$b = 2.25518617\lambda_{2X} - 3.93740250\lambda_{4X} \tag{12.63b}$$

$$c = 1.81379937\lambda_{3X} \tag{12.63c}$$

$$d = -0.19309293\lambda_{2X} + 1.57496 1\lambda_{4X} \tag{12.63d}$$

It can be observed that the polynomial coefficients are simple explicit functions of the L-moments, which are much simpler and easier than those based on C-moments where the coefficients have to be obtained by nonlinear equations as described in Section 7.4.3 in the chapter on Transformation of Non-Normal Variables to Independent Normal Variables. This can be considered as one advantage of the transformation based on the first four L-moments.

12.4.1.2 The *x–u* Transformation Using the First Four L-Moments

Similar to the derivation of complete expressions of the *x–u* transformation/third-order polynomial normal transformation (TPNT) based on the first four C-moments as described in Section 7.4.3 as noted in the previous paragraph, the complete monotonic expressions of the *x–u* transformation (TPNT) under different combinations of the first four L-moments of random variables, i.e. $U = S(X)$, can be readily obtained according to the Cardano formula (Zwillinger 2018) and are summarised in Table 12.6 (Zhao et al. 2020). The parameters in Table 12.6 are given by

$$h = \frac{2c^3}{27d^3} - \frac{bc}{3d^2} + \frac{a}{d} - \frac{X}{d}, \quad A = \left(-\frac{c^2 - 3bd}{9d^2}\right)^3 + \frac{h^2}{4} \tag{12.64a}$$

$$\alpha = \arccos\left(-\frac{h}{2\sqrt{(c^2 - 3bd)^3/(9d^2)^3}}\right) \tag{12.64b}$$

$$Q_1^* = d\left(-2\left|\frac{\Delta_0}{9d^2}\right|^{3/2} + \frac{2c^3}{27d^3} - \frac{bc}{3d^2} + \frac{a}{d}\right), \quad Q_2^* = d\left(2\left|\frac{\Delta_0}{9d^2}\right|^{3/2} + \frac{2c^3}{27d^3} - \frac{bc}{3d^2} + \frac{a}{d}\right) \tag{12.64c}$$

$$Q_0^* = -\frac{b^2 - 4ac}{4c}, \quad \Delta_0 = c^2 - 3bd \tag{12.64d}$$

From Table 12.6, it can be found that the discriminant can be directly expressed by L-skewness τ_3 and L-kurtosis τ_4 rather than the coefficients of *a*, *b*, *c*, and *d* in Eq. (12.62). While the discriminant for the complete expressions of TPNT based on C-moments cannot be directly expressed by the skewness and kurtosis as shown in Table 7.9 in the chapter on Transformation of Non-Normal Variables to Independent Normal Variables. This is another advantage of TPNT based on L-moments over that of C-moments.

Using Table 12.6, the *x–u* transformation is easy to accomplish. For the six types, Figure 12.4a–f shows the unsuitable and suitable roots of the monotonic *x–u* transformation based on the first four L-moments, in which the suitable ones are illustrated in solid lines and the dashed lines represent the unsuitable ones. As can be observed from Figure 12.4a–f, for each type of the TPNT based on L-moments, the *x–u* transformation is monotone by excluding the improper value of *U*.

12.4.1.3 Boundary Among Each Expression of the TPNT in the $\tau_3 - \tau_4$ Plane

According to Hosking and Wallis (1997), L-moment ratios should be satisfied with $|\tau_r| < 1$ for all $r \geq 3$, and the relationship between τ_3 and τ_4 is given by

$$\frac{1}{4}\left(5\tau_3^2 - 1\right) \leq \tau_4 < 1 \tag{12.65}$$

Table 12.6 The monotonic x–u transformation (TPNT) based on L-moments $S(X)$.

L-moments ratio	x–u expression	Range of X	Type
$\tau_3^2 \leq -5.65487\tau_4^2 + 3.93218\tau_4 - 0.397092$, $\tau_4 \neq 0.1226$	$\sqrt[3]{-h/2 - \sqrt{A}} + \sqrt[3]{-h/2 + \sqrt{A}} - c/3d$	$(-\infty, +\infty)$	1
$\tau_3^2 > -5.65487\tau_4^2 + 3.93218\tau_4 - 0.397092$, $\tau_4 > 0.1226$, $\tau_3 \geq 0$	$2\sqrt{\Delta_0/(9d^2)}\cos(\alpha/3) - c/3d$	$Q_1^* < X < Q_2^*$	2
	$\sqrt[3]{-h/2 - \sqrt{A}} + \sqrt[3]{-h/2 + \sqrt{A}} - c/3d$	$X \geq Q_2^*$	
$\tau_3 < 0$	$\sqrt[3]{-h/2 - \sqrt{A}} + \sqrt[3]{-h/2 + \sqrt{A}} - c/3d$	$X \leq Q_1^*$	3
	$-2\sqrt{\Delta_0/(9d^2)}\cos[(\alpha - \pi)/3] - c/3d$	$Q_1^* < X < Q_2^*$	
$\tau_3^2 > -5.65487\tau_4^2 + 3.93218\tau_4 - 0.397092$, $\tau_4 < 0.1226$	$-2\sqrt{\Delta_0/(9d^2)}\cos[(\alpha + \pi)/3] - c/3d$	$Q_2^* \leq X \leq Q_1^*$	4
$\tau_4 = 0.1226$, $\tau_3 > 0$	$\left(\sqrt{b^2 - 4b(a-X)} - b\right)/2c$	$X \geq Q_0^*$	5
$\tau_3 < 0$	$\left(\sqrt{b^2 - 4c(a-X)} - b\right)/2c$	$X \leq Q_0^*$	
$\tau_3 = 0$	$(X-a)/b$	$(-\infty, +\infty)$	6

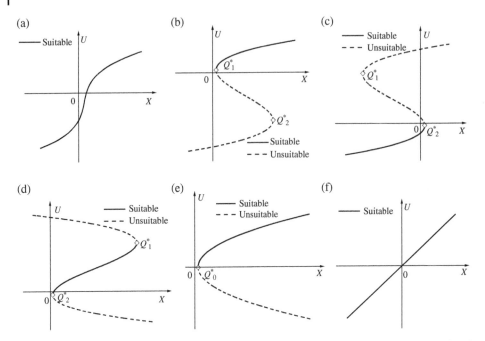

Figure 12.4 Illustration of the monotonicity of the six types of the TPNT based on the first four L-moments. (a) Type 1. (b) Type 2. (c) Type 3. (d) Type 4. (e) Type 5. (f) Type 6.

Table 12.6 indicates that there are three boundaries among each expression of the x–u transformation formulated by the relationship between τ_3 and τ_4 as follows:

Boundary 1: The boundary for $\Delta_0 = c^2 - 3bd = 0$.

When $\Delta_0 = 0$, the relationship between τ_3 and τ_4 can be expressed as the following, according to Eqs. (12.68b)–(12.68d):

$$\tau_3^2 = -5.65487\tau_4^2 + 3.93218\tau_4 - 0.397092 \tag{12.66}$$

Boundary 2: The boundary when $\tau_4 = 0.1226$.
Boundary 3: The boundary when $\tau_3 = 0$.

According to the three boundaries and Eq. (12.65), the boundary lines are depicted in Figure 12.5 for the different types of the TPNT based on L-moments in the τ_3–τ_4 plane, where the shaded region is the non-applicable area of the transformation. It can be observed from Figure 12.5 that:

1) Type 1 is an ellipse in the middle of the region.
2) The range of Types 2 and 3 are symmetric with respect to the line of $\tau_3 = 0$, and they are in the upper right and upper left of Type 1, respectively.
3) Type 4 is at the bottom of the region. The area of Type 1, 2, 3, and 4 have little difference, which shows that these types of the TPNT based on L-moments are important and each of them cannot be ignored in practical engineering problems.

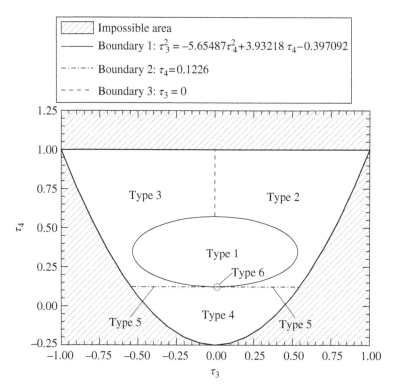

Figure 12.5 Boundaries of particular types in the polynomial normal transformation based on L-moments.

4) Types 5 and 6 are degenerated forms of the polynomial normal transformation for $d = 0$ ($\tau_4 = 0.1226$), which can be represented by a line ($\tau_4 = 0.1226$ and $\tau_3 \neq 0$) and a point ($\tau_4 = 0.1226$ and $\tau_3 = 0$).

We can observe from Eqs. (12.63a)–(12.63d) that the coefficients of the TPNT based on L-moments do not have any limitations. Therefore, τ_3 and τ_4 can be theoretically taken at any value within the boundaries given by Hosking and Wallis (1997). This means that the TPNT based on L-moments can be implemented within the definition range of τ_3 and τ_4. Compared with the TPNT based on C-moments, which has limitations for the pairs of skewness and kurtosis to determine of the polynomial coefficients of a, b, c, and d (see Section 7.4.3 as previously mentioned), this is another advantage of the TPNT based on L-moments.

The applicable range of the TPNT based on L-moments in the $\tau_3 - \tau_4$ plane is illustrated in Figure 12.6, together with the $\tau_3 - \tau_4$ relationship of some commonly used distributions. As can be observed from Figure 12.6, the Lognormal, Weibull, Gamma, and Beta (mean = 0.1) distribution are each represented by a curve, and the Normal, Gumbel, and Exponential distribution are represented by a point. Figure 12.6 reveals that:

1) The range of the TPNT based on L-moments covers a large area, and the relationships between τ_3 and τ_4 for selected commonly-used distributions are within the applicable

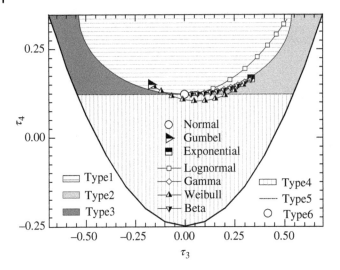

Figure 12.6 Some commonly used distributions in the applicable ranges of the polynomial normal transformation based on L-moments.

range of the TPNT based on L-moments. This implies that the TPNT based on L-moments is generally applicable for reliability engineering.

2) In practical engineering, the pairs of τ_3 and τ_4 of the commonly-used distributions almost fall into the area of Type 1 of the TPNT based on L-moments, such as Gumbel and Lognormal distributions, and Weibull distribution with coefficient of variation (COV) = 0.1. A small number of commonly used distributions fall into the area of Type 2 and 4, for example, Beta, Gamma, and Exponential distributions, and Weibull distribution with COV = 0.8–1.0 belong to Type 2, while Weibull distribution with COV = 0.2–0.7 belongs to Type 4. Normal distribution with $\tau_3 = 0$ and $\tau_4 = 0.1226$ is the Type 6 of TPNT based on L-moments. This implies that the complete expressions of the TPNT based on L-moments are necessary for practical engineering.

Example 12.4   Statistical Uncertainty of the TPNT Based on L-Moments

To investigate the statistical uncertainty of the TPNT based on L-moments due to a limited sample size, the exceedance probability (P_e) as formulated in Eq. (12.67) is estimated using simulated data with a sample size from 10^3 to 10^6.

$$P_e(X = x) = 1 - F[x] = 1 - \Phi[U(x)] \tag{12.67}$$

where x is the threshold and assumed to be $\mu + 3\sigma$, of which μ and σ are the mean and standard deviation of the random variable X, respectively; and $U(x)$ is the x–u transformation achieved by the TPNT. For illustration, four cases are considered in this example: X is assumed as a Gamma distribution (mean = 10, COV = 0.1); a Lognormal distribution (mean = 10, COV = 0.2); a Gumbel distribution (mean = 10, COV = 0.3); and a Weibull distribution (mean = 10, COV = 0.4).

The sample data is generated by using MCS. For each sample, the generation of the first 1000 samples will be used in the first simulation, the generation of the first 2000 samples will be used in the following simulation, and using the entire samples in previous simulations plus an extra 1000 samples until the total data of 10^6 in each successive simulation. Using the sample data, the first four L-moments and C-moments of X can be obtained, and $P_e(x)$ can be computed using Eq. (12.67). For the purpose of comparison, the exact $P_e(x)$ is obtained by Rosenblatt transformation using the given distributions. The variations of $P_e(x)$ $(x = \mu + 3\sigma)$ with respect to the sample size are depicted in Figures 12.7a–d for the four assumed distributions, respectively. It can be observed from Figures 12.7a–d that: (i) The fluctuations of $P_e(x)$ obtained from the TPNT based on C-moments are obviously larger than those from L-moments, especially when the sample size is small; (ii) $P_e(x)$ obtained by TPNT based on L-moments is closer to the exact values. This example shows that L-moments perform better with smaller sample sizes than C-moments, which has been pointed out by Hosking (1990).

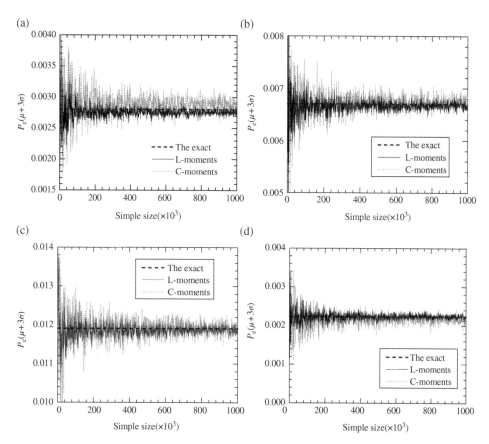

Figure 12.7 Variations of P_e $(\mu + 3\sigma)$ with respect to sample size in Example 12.4. (a) Gamma distribution. (b) Lognormal distribution. (c) Gumbel distribution. (d) Weibull distribution.

Example 12.5 Investigation on the x–u Transformation Using the First Four L-Moments

For a random variable X with a known PDF, $f(X)$, using the complete monotonic expressions of TPNT based on L-moments, the x–u transformation can be achieved for random variables with all practical combinations of the first four L-moments. To demonstrate its accuracy and efficiency, the second example considers three random variables following Weibull, Lognormal, and Gamma distributions. The distribution parameters and the first four L-moments of these random variables are listed in Table 12.7.

Figures 12.8a–c illustrate the changes of the x–u transformation for three non-normal random variables, which are obtained from different methods including the Rosenblatt transformation (exact), the TPNT based on C-moments (Zhao et al. 2018a), and the TPNT based on L-moments. Table 12.7 and Figure 12.8a–c reveal the followings:

1) The TPNT based on C-moments can provide good results for the Lognormal and Gamma variables (Figure 12.8b,c). However, for the Weibull variables with COV = 0.2, there are large differences between the results obtained by the TPNT based on C-moments and the exact ones when the value of X is far from the mean value (see Figure 12.8a). This can be attributed to the fact that the solved U from the TPNT based on C-moments (Zhao et al. 2018a) are complex numbers for $X > 1.42$ of the Weibull variables with COV of 0.2, where only real parts of complex numbers are taken for the transformation.
2) The results of the TPNT based on L-moments are in good agreement with those of the Rosenblatt transformation for all the three cases considered in the entire investigation range, and the TPNT based on L-moments performs better than the TPNT based on C-moments, especially for the Weibull variable with COV of 0.2 (Figure 12.8a), when $X = 1.53$, the exact value of U is 3.27, while those obtained by the TPNT based on L-moments and C-moments are 3.24 and 3.86, respectively.
3) The x–u transformations considered in this example include three types of the TPNT based on L-moments, i.e. Type 1, 2, and 4, which again demonstrates the necessity of the definition of the complete monotonic expression of the TPNT based on L-moments.

12.4.2 Transformation for Correlated Random Variables

Assume that X_i and X_j are correlated non-normal random variables and the correlation coefficient between X_i and X_j is ρ_{ij}, Z_i and Z_j are correlated standard normal random variables the correlation coefficient between Z_i and Z_j is ρ_{0ij}.

The third-order polynomial normal transformation is suggested by Fleishman (1978), in which a non-normal random variable can be expressed as

$$X_i = S^{-1}(Z_i) = a_i + b_i Z_i + c_i Z_i^2 + d_i Z_i^3 \tag{12.68}$$

where a_{0i}, a_{1i}, a_{2i}, and a_{3i} are the polynomial coefficients and are expressed as

$$a_i = \lambda_{1X_i} - 1.81379937\lambda_{3X_i} \tag{12.69a}$$

$$b_i = 2.25518617\lambda_{2X_i} - 3.93740250\lambda_{4X_i} \tag{12.69b}$$

Table 12.7 Probability distributions and their statistical parameters of random variables in Example 12.5.

| Distribution | Mean | COV | First four L-moments | | | | Parameters of the transformation based on L-moments | | | | Type |
			λ_1	λ_2	λ_3	λ_4	a	b	c	d	
Weibull	1	0.2	1	0.1127	−0.0072	0.0133	1.0130	1.0753	−0.0130	−0.0009	4
Lognormal	1	0.2	1	0.1114	0.0108	0.0145	0.9805	1.0706	0.0195	0.0013	1
Gamma	1	0.3	1	0.1674	0.0164	0.0210	0.9702	1.0449	0.0298	0.0008	2

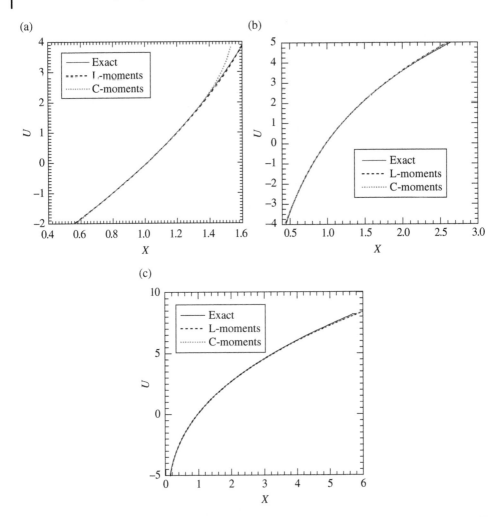

Figure 12.8 Comparison of x–u transformation with different methods for Example 12.5. (a) Weibull (COV = 0.2). (b) Lognormal (COV = 0.2). (c) Gamma (COV = 0.3).

$$c_i = 1.81379937\lambda_{3X_i} \tag{12.69c}$$

$$d_i = -0.19309293\lambda_{2X_i} + 1.574961\lambda_{4X_i} \tag{12.69d}$$

The correlation coefficient is defined by the covariance, and expressed as

$$\rho_{ij} = \frac{Cov(X_iX_j)}{\sqrt{D(X_i)D(X_j)}} = \frac{E(X_iX_j) - E(X_i)E(X_j)}{\sqrt{D(X_i)D(X_j)}} = \frac{E(X_iX_j) - \mu_{X_i}\mu_{X_j}}{\sigma_{X_i}\sigma_{X_j}} \tag{12.70}$$

in which $Cov(X_iX_j)$ = covariance; $E(\cdot)$ = expectation; $D(\cdot)$ = variance; μ_{X_i} and μ_{X_j} are the mean of X_i and X_j; and σ_{X_i} and σ_{X_j} are the standard deviation of X_i and X_j.

According to Eq. (12.68), $E(X_i X_j)$ in Eq. (12.70) can be expressed as

$$
\begin{aligned}
E(X_i X_j) &= E\left[(a_i, b_i, c_i, d_i)\left(1, Z_i, Z_i^2, Z_i^3\right)^T \left(1, Z_j, Z_j^2, Z_j^3\right)\left(a_j, b_j, c_j, d_j\right)^T\right] \\
&= (a_i, b_i, c_i, d_i) E\left[\left(1, Z_i, Z_i^2, Z_i^3\right)^T \left(1, Z_j, Z_j^2, Z_j^3\right)\right]\left(a_j, b_j, c_j, d_j\right)^T \qquad (12.71) \\
&= (a_i, b_i, c_i, d_i)\mathbf{R}\left(a_j, b_j, c_j, d_j\right)^T
\end{aligned}
$$

where $\mathbf{R}$ can be expressed as:

$$
\mathbf{R} = \begin{bmatrix}
1 & 0 & 1 & 0 \\
0 & \rho_{0ij} & 0 & 3\rho_{0ij} \\
1 & 0 & 2\rho_{0ij}^2 + 1 & 0 \\
0 & 3\rho_{0ij} & 0 & 6\rho_{0ij}^3 + 9\rho_{0ij}
\end{bmatrix} \qquad (12.72)
$$

Substituting Eq. (12.72) into Eq. (12.71), one can obtain

$$
E(X_i X_j) = \mu_{X_i}\mu_{X_j} + \left(b_i b_j + 3 d_i b_j + 3 b_i d_j + 9 d_i d_j\right)\cdot\rho_{0ij} + 2 c_i c_j \rho_{0ij}^2 + 6 d_i d_j \rho_{0ij}^3 \qquad (12.73)
$$

According to Eq. (12.70) and Eq. (12.73), ρ_{0ij} can be determined as

$$
\rho_{ij} = \frac{\left(b_i b_j + 3 d_i b_j + 3 b_i d_j + 9 d_i d_j\right)\cdot\rho_{0ij} + 2 c_i c_j \rho_{0ij}^2 + 6 d_i d_j \rho_{0ij}^3}{\sigma_{X_i}\sigma_{X_j}} \qquad (12.74)
$$

In order to satisfy the definition of correlation coefficient, ρ_{0ij} in Eq. (12.74) should satisfy the following conditions:

$$
-1 \le \rho_{0ij} \le 1, \quad \rho_{ij}\cdot\rho_{0ij} \ge 0 \quad \text{and} \quad \left|\rho_{0ij}\right| \ge \left|\rho_{ij}\right| \qquad (12.75)
$$

According to the Cardano formula (Zwillinger 2018), together with Eqs. (12.74) and (12.75), the equivalent correlation coefficients and bounds of the original correlation coefficients of the transformation can be determined and are summarised in Table 12.8. The parameters in Table 12.8 are given by

$$
A = \left(-\frac{c^2 - 3bd}{9d^2}\right)^3 + \frac{h^2}{4}, \quad h = \frac{2c^3}{27d^3} - \frac{bc}{3d^2} - \frac{\rho_{ij}}{d}, t = \frac{c}{d} \qquad (12.76a)
$$

$$
\Delta_0 = c^2 - 3bd, \alpha = \arccos\left(-\frac{h}{2\sqrt{\Delta_0^3/(9d^2)^3}}\right), f(x) = bx + cx^2 + dx^3 \qquad (12.76b)
$$

$$
\rho_{0ij-1} = \frac{-c - \sqrt{c^2 - 3bd}}{3d}, \rho_{0ij-2} = \frac{-c + \sqrt{c^2 - 3bd}}{3d}, \rho_{0ij-cri} = -\frac{b}{2c} \qquad (12.76c)
$$

$$
b = \frac{b_i b_j + 3 d_i b_j + 3 b_i d_j + 9 d_i d_j}{\sigma_{X_i}\sigma_{X_j}}, c = \frac{2 c_i c_j}{\sigma_{X_i}\sigma_{X_j}}, \quad d = \frac{6 d_i d_j}{\sigma_{X_i}\sigma_{X_j}} \qquad (12.76d)
$$

From the preceding discussion, any two arbitrary correlated variables with known first four L-moments, standard deviation, and correlation coefficient can be transformed into two correlated standard normal variables.

Table 12.3 The expression of ρ_{0ij}.

L-moments ratio and parameters	Expression of ρ_{0ij}	ρ_{ij-max}	ρ_{ij-min}
$(\tau_{4X_i}-0.1226)$, $(\tau_{4X_j}-0.1226)>0$; $\Delta_0 \leq 0$	$\sqrt[3]{-h/2-\sqrt{A}}+\sqrt[3]{-h/2+\sqrt{A}}-t/3$	$\min\{f(1),\,1\}$	$\max\{f(-1),\,-1\}$
$(\tau_{4X_i}-0.1226)$, $(\tau_{4X_j}-0.1226)>0$; $\Delta_0>0$; $\tau_{3X_i}\tau_{3X_j}\geq 0$; $A>0$	$\sqrt[3]{-h/2-\sqrt{A}}+\sqrt[3]{-h/2+\sqrt{A}}-t/3$	$\min\{f(1),\,1\}$	$\begin{cases} f(\rho_{0ij-2}), & f(\rho_{0ij-2})\geq -1 \\ \max\{f(-1),\,-1\}\cdot f(\rho_{0ij-2}), & f(\rho_{0ij-2})<-1 \end{cases}$
$A \leq 0$	$-2\sqrt{\Delta_0/(9d^2)}\cos(\alpha/3)-t/3$		$\begin{cases} f(\rho_{0ij-2}), & f(\rho_{0ij-2})\geq -1 \\ \max\{f(-1),\,-1\}\cdot f(\rho_{0ij-2}), & f(\rho_{0ij-2})<-1 \end{cases}$
$\tau_{3X_i}\tau_{3X_j}<0$; $A>0$	$\sqrt[3]{-h/2-\sqrt{A}}+\sqrt[3]{-h/2+\sqrt{A}}-t/3$	$\begin{cases} f(\rho_{0ij-1}), & f(\rho_{0ij-1})\leq 1 \\ \min\{f(1),1\}\cdot f(\rho_{0ij-1}), & f(\rho_{0ij-1})>1 \end{cases}$	$\max\{f(-1),\,-1\}$
$A \leq 0$	$-2\sqrt{\Delta_0/(9d^2)}\cos[(\alpha-\pi)/3]-t/3$	$\begin{cases} f(\rho_{0ij-1}), & f(\rho_{0ij-1})\leq 1 \\ \min\{f(1),1\}\cdot f(\rho_{0ij-1}), & f(\rho_{0ij-1})>1 \end{cases}$	
$(\tau_{4X_i}-0.1226)$, $(\tau_{4X_j}-0.1226)<0$; $\Delta_0 \geq 0$	$-2\sqrt{\Delta_0/(9d^2)}\cos[(\alpha+\pi)/3]-t/3$	$\begin{cases} f(\rho_{0ij-2}), & f(\rho_{0ij-2})\leq 1 \\ \min\{f(1),1\}\cdot f(\rho_{0ij-2}), & f(\rho_{0ij-2})>1 \end{cases}$	$\begin{cases} f(\rho_{0ij-1}), & f(\rho_{0ij-1})\geq -1 \\ \max\{f(-1),\,-1\}\cdot f(\rho_{0ij-1}), & f(\rho_{0ij-1})<-1 \end{cases}$
$\tau_{3X_i}\tau_{3X_j}>0$	$\left(-b+\sqrt{b^2+4c\rho_{ij}}\right)/2c$	$\min\{f(1),\,1\}$	$\begin{cases} \max\{f(-1),\,f(\rho_{0ij-crt})\}\cdot f(\rho_{0ij-crt}), & f(\rho_{0ij-crt})\geq -1 \\ \max\{f(-1),\,-1\}, & f(\rho_{0ij-crt})<-1 \end{cases}$
$\tau_{3X_i}\tau_{3X_j}<0$	$\left(-b+\sqrt{b^2+4c\rho_{ij}}\right)/2c$	$\begin{cases} \min\{f(1),\,f(\rho_{0ij-crt})\}\cdot f(\rho_{0ij-crt})\leq 1 \\ \min\{f(1),\,1\}, & f(\rho_{0ij-crt})>1 \end{cases}$	$\max\{f(-1),\,-1\}$
$\tau_{3X_i}\tau_{3X_j}=0$	ρ_{ij}/b	$\min\{f(1),\,1\}$	$\max\{f(-1),\,-1\}$

Similarly, the preceding procedure can be easily extended to n-variables with known L-moments and correlation matrix. The polynomial coefficients of each variable can be obtained by Eqs. (12.69a)–(12.69d), and for any two correlated variables, the equivalent correlation coefficients of standard normal variables can be determined from Table 12.8. The equivalent correlation matrix of standard normal variables, $\mathbf{C_Z}$, and the lower triangular matrix $\mathbf{L_0}$ obtained from Cholesky decomposition of $\mathbf{C_Z}$ can then be obtained. Their forms are the same as Eqs. (7.105) and (7.108), respectively in the chapter on Transformation of Non-Normal Variables to Independent Normal Variables. For the cases of original correlation matrix with very small eigenvalues, the equivalent correlation matrix might become a non-positive semidefinite matrix. The method for solving the problem (Ji et al. 2018) mentioned in Section 7.5.2 in the same chapter can be adopted to make Cholesky decomposition ready.

From Eqs. (12.68), (12.36), and Table 12.8, the u–x transformations can be expressed as

$$X_i = a_i + b_i \sum_{k=1}^{i} l_{ik}U_k + c_i \left(\sum_{k=1}^{i} l_{ik}U_k \right)^2 + d_i \left(\sum_{k=1}^{i} l_{ik}U_k \right)^3, \quad (i = 1, 2, \cdots, n)$$

(12.77)

where l_{ik} is the tth row kth column element of matrix $\mathbf{L_0}$ as shown in Eq. (7.108) as noted in the previous paragraph.

From Eq. (12.40), and Table 12.6, the x–u transformation can be expressed as

$$U_i = \sum_{k=1}^{i} h_{ik}Z_k = \sum_{k=1}^{i} h_{ik}S(X_k), \quad (i = 1, 2, \cdots, n)$$

(12.78)

where h_{ik} is the tth row kth column element of matrix $\mathbf{L_0}^{-1}$ as shown in Eq. (7.111) in the chapter just mentioned, and $S(X)$ is given in Table 12.6.

12.4.3 Reliability Analysis Using the First Four L-Moments and Correlation Matrix

The computation procedure of FORM using the first four L-moments and correlation matrix is similar to that described in Section 12.3.3, where the third L-moment transformations are replaced by the fourth L-moment transformations.

Example 12.6 Structural Reliability Analysis Using TPNT Based on the First Four L-Moments

This example considers a 72-bar spatial truss structure under lateral loads (Xu et al. 2017), as shown in Figure 12.9. The cross-sectional area of each element A is 34.849 cm^2. The elastic modulus E and the applied wind loads F_1–F_8 on nodes 5, 8, 9, 12, 13, 16, 17, 20 are random variables, of which the distribution and the first four L-moments for the random variables are listed in Table 12.9.

The finite-element analysis of the space truss is carried out within the linear-elastic and small- deformation range, and the implicit performance function is expressed as

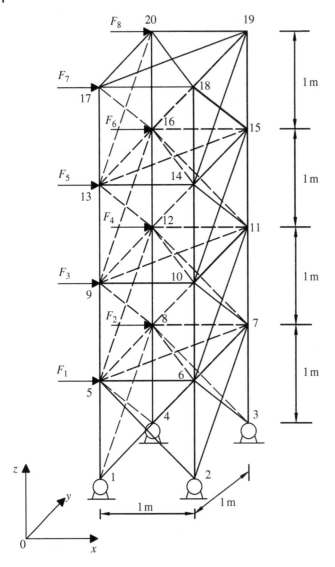

Figure 12.9 Spatial truss structure for Example 12.6.

$$Z = G(\mathbf{X}) = 0.0012 - \max_{5 \le i \le 20} |u_i(\mathbf{X})| \tag{12.79}$$

where the horizontal displacement of node i is denoted by u_i for $i = 5, 6, ..., 20$ (along x direction).

The probability of failure obtained by MCS is 5.99×10^{-5} (the COV of the MCS result is 4.09%) and the corresponding reliability index is $\beta_{MCS} = 3.8465$, which are employed as the benchmark for comparison.

Using the CDFs/PDFs of random variables, the FORM reliability index (β_{FORM}) is obtained as 3.8876. Using the first four L-moments of random variables, the FORM

Table 12.9 Random variables in Example 12.6.

Variable	Distribution	First four L-moments				Parameters of the normal transformation				Type
		λ_1	λ_2	λ_3	λ_4	a	b	c	d	
E	Normal	2.1×10^{11}	5.92×10^9	0	7.26×10^8	2.1×10^{11}	1.05×10^{10}	0	−18.90	6
F_1	Weibull	200 000	22 538.2	−1436.74	2655.26	202 606	40 372.9	−2606	−170.0	4
F_2	Weibull	200 000	22 538.2	−1436.74	2655.26	202 606	40 372.9	−2606	−170.0	4
F_3	Weibull	300 000	33 807.2	−2155.11	3982.89	303 909	60 559.4	−3909	−255.0	4
F_4	Weibull	300 000	33 807.2	−2155.11	3982.89	303 909	60 559.4	−3909	−255.0	4
F_5	Lognormal	400 000	44 547.7	4301.23	5788.11	392 198	77 673.2	7801.6	514.2	1
F_6	Lognormal	400 000	44 547.7	4301.23	5788.11	392 198	77 673.2	7801.6	514.2	1
F_7	Lognormal	500 000	55 684.6	5376.54	7235.14	490 248	97 091.5	9752.0	642.75	1
F_8	Lognormal	500 000	55 684.6	5376.54	7235.14	490 248	97 091.5	9753.0	642.75	1

reliability index $\beta_{L\text{-FORM}}$ can be calculated as 3.9058. Similarly, the FORM reliability index by using the first four C-moments ($\beta_{C\text{-FORM}}$) is calculated as 3.9102. It can be concluded that for this example the FORM reliability index obtained using the first four L-moments is almost in agreement with that of using the CDFs/PDFs, and it is closer to MCS result than that obtained by using the first four C-moments.

Example 12.7 Structural Reliability Analysis Including Correlated Random Variables Using TPNT Based on the First Four L-Moments

Consider again the implicit performance function of Eq. (12.60). The probabilistic information is the same with that of Example 12.3 including the probability distributions and the correlation coefficients among the random variables. The first four L-moments computed from the CDFs of the basic random variables are shown in Table 12.10.

Table 12.10 Probabilistic information of the basic random variables for Example 12.7.

Variable	Distribution	Mean	COV	λ_1	λ_2	τ_3	τ_4
					First four L-moments		
P_1	Gumbel	135.0 kN	0.2	135.0 kN	14.592	0.1669	0.1504
P_2	Gumbel	90.0 kN	0.2	90.0 kN	9.728	0.1669	0.1504
P_3	Gumbel	70.0 kN	0.2	70.0 kN	7.56622	0.1669	0.1504
E_1	Lognormal	2.1×10^7 kN/m²	0.05	2.1×10^7 kN/m²	591 906	0.024412	0.1231
E_2	Lognormal	2.4×10^7 kN/m²	0.05	2.4×10^7 kN/m²	676 464	0.024412	0.1231
I_1	Lognormal	8.11×10^{-3} m⁴	0.1	8.11×10^{-3} m⁴	4.6×10^{-4}	0.048712	0.1245
I_2	Lognormal	0.011 m⁴	0.1	0.011 m⁴	6.2×10^{-4}	0.048712	0.1245
I_3	Lognormal	0.0213 m⁴	0.1	0.0213 m⁴	0.0012	0.048712	0.1245
I_4	Lognormal	0.0295 m⁴	0.1	0.0295 m⁴	0.0017	0.048712	0.1245
I_5	Lognormal	0.0108 m⁴	0.1	0.0108 m⁴	6.1×10^{-4}	0.048712	0.1245
I_6	Lognormal	0.0141 m⁴	0.1	0.0141 m⁴	7.9×10^{-4}	0.048712	0.1245
I_7	Lognormal	0.0232 m⁴	0.1	0.0232 m⁴	1.3×10^{-3}	0.048712	0.1245
I_8	Lognormal	0.0259 m⁴	0.1	0.0259 m⁴	1.5×10^{-3}	0.048712	0.1245
A_1	Lognormal	0.312 m²	0.1	0.312 m²	0.0175	0.048712	0.1245
A_2	Lognormal	0.372 m²	0.1	0.372 m²	0.0209	0.048712	0.1245
A_3	Lognormal	0.505 m²	0.1	0.505 m²	0.0284	0.048712	0.1245
A_4	Lognormal	0.557 m²	0.1	0.557 m²	0.0313	0.048712	0.1245
A_5	Lognormal	0.253 m²	0.1	0.253 m²	0.0142	0.048712	0.1245
A_6	Lognormal	0.291 m²	0.1	0.291 m²	0.0164	0.048712	0.1245
A_7	Lognormal	0.372 m²	0.1	0.372 m²	0.0209	0.048712	0.1245
A_8	Lognormal	0.418 m²	0.1	0.418 m²	0.0235	0.048712	0.1245

Table 12.11 Comparison of results by different methods for Example 12.7.

Reliability method	Transformation	Reliability index
FORM	Third-moment transformation using L-moments	2.4227
	Fourth-moment transformation using L-moments	2.3441
MCS	Nataf transformation	2.273

The reliability index obtained from FORM based on the third- and fourth-moment transformation using L-moments are listed in Table 12.11, together with that from MCS. Since the marginal PDFs and the correlation matrix of the basic random variables are known, MCS based on Nataf transformation can be conducted and the result is taken as exact value. The failure probability P_f from MCS based on Nataf transformation with 10^6 samples is 0.011764 (the COV of the MCS result is 0.92%), and the corresponding reliability index is also listed in Table 12.11. As can be observed from Table 12.11, FORM using the first four L-moments provides almost the same result as MCS, and it gives better result than FORM using the first three L-moments.

12.5 Summary

In this chapter, the reliability evaluation using the first few L-moments of random variables is discussed. The first part of this chapter introduces the definition of the first four L-moments and the computation of the first four L-moments from the probability distributions or statistical data of random variables. The second- and third-order polynomial normal transformation techniques are then investigated using the first three and four L-moments, respectively. Explicit third- and fourth-moment standardisation functions using L-moments are described. Using these methods, the normal transformation for random variables (independent or correlated) using their L-moments can be achieved without using the Rosenblatt or Nataf transformations. Through the numerical examples presented, FORM based on the transformation techniques using L-moments is demonstrated to be sufficiently accurate in structural reliability assessment. If one neglects the drawbacks that the definitions of L-moments are not familiar with practical engineering, the transformation based on L-moments is a viable alternative in FORM for structural reliability, since the polynomial coefficients can be explicitly expressed by using the L-moments, and the applicable range of these transformation is broader than that based on C-moments.

13

Methods of Moment with Box-Cox Transformation

13.1 Introduction

As described in chapters on Moment Evaluation and Direct Methods of Moment, the direct methods of moment mainly include two steps for structural reliability analysis: (i) the evaluation of the first four moments (i.e. mean, standard deviation, skewness, and kurtosis) of performance functions; and (ii) the estimation of failure probability using the third- or fourth-moment reliability indices, which are derived from the assumption that the unknown distributions of performance functions follow one of flexible three- or four-parameter distributions, e.g. 3P lognormal distribution (Tichy 1994), the square normal distribution (Zhao et al. 2001), Pearson system (Stuart and Ord 1987), cubic normal distribution (Zhao and Lu 2008; Zhao et al. 2018b), and so on. The direct methods of moment require neither iterations nor computation of derivatives of the performance function (Zhao and Ono 2001; Xu and Kong 2019), so are therefore convenient for finite-element based structural reliability analysis. However, if the pairs of the skewness and kurtosis of the performance functions are far from those of a normal random variable, it will be difficult to find a solution in which we can have confidence (Pearson et al. 1979). In other words, the direct methods of moment might not be applicable to problems with extremely strong non-normality (Zhao and Ono 2004), which make it necessary to make these performance functions close to normally distributed.

One option for this purpose is to utilise the Box-Cox transformation (Box and Cox 1964), which provides a family of power transformations and has been used to transform statistical data of random variables closer to be normally distributed and widely accepted and extensively applied in many fields of natural and social sciences (Harold et al. 1979; Osborne 2010; Zhang and Yang 2017). In this chapter, the application of Box-Cox transformation for dealing with the performance functions for structural reliability will be investigated, and a criterion is introduced to determine the optimal Box-Cox transformation parameter that makes the transformed performance function approximately normally distributed. Since the transformed performance function approaches normally distributed from the perspective of its first four moments, the failure probability can be accurately evaluated using the simple third-moment reliability index as expressed in Eq. (4.27) in the chapter on Direct Methods of Moment. Hereafter, the resulted method for structural reliability analysis is referred to as the methods of moment combined with Box-Cox transformation.

Structural Reliability: Approaches from Perspectives of Statistical Moments, First Edition.
Yan-Gang Zhao and Zhao-Hui Lu.
© 2021 John Wiley & Sons Ltd. Published 2021 by John Wiley & Sons Ltd.

13.2 Methods of Moment with Box-Cox Transformation

13.2.1 Criterion for Determining the Box-Cox Transformation Parameter

Without loss of generality, the performance function $G(\mathbf{X})$ in structural reliability analysis can be expressed as:

$$G(\mathbf{X}) = R(\mathbf{X}) - S(\mathbf{X}) \tag{13.1}$$

where $R(\mathbf{X})$ and $S(\mathbf{X})$ represent structural resistance and load effect, respectively. In engineering applications, the resistance $R(\mathbf{X})$ and load effect $S(\mathbf{X})$ are generally positive. If $R(\mathbf{X}) > 0$ and $S(\mathbf{X}) > 0$, the performance function $G(\mathbf{X})$ can be equivalently reformulated as:

$$g(\mathbf{X}) = \frac{R(\mathbf{X})}{S(\mathbf{X})} - 1 = \lambda(\mathbf{X}) - 1 \tag{13.2}$$

where $\lambda(\mathbf{X})$ denotes the proportion of $R(\mathbf{X})$ to $S(\mathbf{X})$, which is also a positive function of $\mathbf{X}$.

Based on the Box-Cox transformation (Box and Cox 1964), the performance function expressed in Eq. (13.2) can be equivalently expressed as:

$$\tilde{g}(\mathbf{X}) = \begin{cases} \dfrac{\lambda(\mathbf{X})^q - 1}{q}, q \neq 0 \\ \ln\left[\lambda(\mathbf{X})\right], \quad q = 0 \end{cases} \tag{13.3}$$

where $\tilde{g}(\mathbf{X})$ is the transformed performance function; and q is the Box-Cox transformation parameter, which can be any real number.

To apply the Box-Cox transformation, the selection of the parameter q is very important. In classical applications of this transformation, the parameter q is generally determined using the maximum likelihood estimation (MLE) (Box and Cox 1964), which involves maximising the log-likelihood function (Murphy and Vaart 2000) by means of the observations. Therefore, MLE may be only applicable to simulation-based reliability method, but it is not easy to determine the size of samples. In this section, to improve the methods of moment for structural reliability analysis, the parameter q is selected based on the skewness and kurtosis of the transformed performance function expressed in Eq. (13.3) considering the difficulty in the calculation of moments higher than the fourth order. Therefore, the optimal Box-Cox transformation parameter q is determined by minimising the distance between the pair of skewness and kurtosis of the transformed performance function, i.e. $(\alpha_{3\tilde{g}}, \alpha_{4\tilde{g}})$ and that of a normal random variable, i.e. $(0, 3)$, which can be equivalently formulated as (Cai et al. 2020):

$$\begin{cases} \text{Find:} \quad q \\ \text{Minimize:} \ d_N^2 = \alpha_{3\tilde{g}}^2 + \left(\alpha_{4\tilde{g}} - 3\right)^2 \end{cases} \tag{13.4}$$

where $\alpha_{3\tilde{g}}$ and $\alpha_{4\tilde{g}}$ denote the skewness and kurtosis of the transformed performance function $\tilde{g}(\mathbf{X})$, respectively; and d_N is the distance between the pair of skewness and kurtosis of the transformed performance function and those of a normal random variable. Equation (13.4) is a one-dimensional unconstrained optimization problem, which can be easily solved with the simplex search method (Nelder and Mead 1965; Mehta and Dasgupta 2012).

It can be observed from Eq. (13.4) that the computation of the Box-Cox transformation parameter q involves the evaluation of the skewness and kurtosis of the transformed performance function, which can be evaluated using the point-estimate method as described in the chapter on Moment Evaluation. Because the transformed performance function can be approximated as a normal random variable from the perspective of the first four moments, the failure probability can be accurately evaluated using the third-moment reliability index as expressed by Eq. (4.27) in the chapter on Direct Methods of Moment.

13.2.2 Procedure of the Methods of Moment with Box-Cox Transformation for Structural Reliability

The computational procedure of the methods of moment combined with Box-Cox transformation for structural reliability analysis can be summarised as follows:

1) Generate the estimate-points and corresponding weights in independent standard normal space by means of the point estimate method (details see the chapter on Moment Evaluation);
2) Calculate and compose the estimate points and weights in independent standard normal space according to the moment estimation formula, i.e. direct point estimate method or point-estimate method with bivariate dimension-reduction method;
3) Transform these estimate points into original space using the inverse normal transformation;
4) Evaluate the values of function $\lambda(\mathbf{X})$ at each estimate-point by deterministic performance function evaluation or structural analysis;
5) Determine the Box-Cox transformation parameter q using Eq. (13.4) and simplex search method; and
6) Obtain the first three central moments of transformed performance function $\tilde{g}(\mathbf{X})$, and the third-moment reliability index and the corresponding failure probability using Eq. (4.27) in the chapter on Direct Methods of Moment.

From the computational procedure described above, it can be observed that the calculation of the Box-Cox transformation parameter q can be straightforwardly solved without additional performance function evaluations. Therefore, the total computational effort of the methods of moment combined with Box-Cox transformation is equivalent with that of the direct methods of moment. If the first four moments of the transformed performance function are evaluated using point-estimate method combined with bivariate dimension-reduction method, the total number of deterministic performance function evaluations or structural analysis (for a n-dimensional random vector using m estimate points) is $n(n-1)(m^2-2m+1)/2 + (m-1)n + 1$. For the generally used five-point scheme, the total number of function evaluations is $8n^2 - 4n + 1$.

Example 13.1 Reliability Analysis for Nonnormally Distributed Performance Functions and Small Failure Probability

This example considers the following simple performance function, which is a fundamental reliability model that has been used in many situations:

$$G(\mathbf{X}) = R - S \tag{13.5}$$

where R is resistance and S is load effect. Assume that R and S are mutually independent random variables with means and coefficient of variations (COVs) of μ_R, V_R and μ_S, V_S, respectively.

Because only two random variables are involved, the exact failure probability can be determined by directly numerical integral as $P_f = \int_{-\infty}^{+\infty} F_R(s) f_S(s) ds$, in which $F_R(\cdot)$ denotes the cumulative distribution function (CDF) of R and $f_S(\cdot)$ is the PDF of S.

Since $R > 0$ and $S > 0$, Eq. (13.5) can be rewritten as

$$G(\mathbf{X}) = R/S - 1 = \lambda(\mathbf{X}) - 1 \tag{13.6}$$

The following two cases are investigated.

Case 1 investigates the applicability of the methods of moment combined with Box-Cox transformation for structural reliability analysis involving nonnormally distributed performance functions. Assume that R is a lognormal random variable with $\mu_R = 200$ and $V_R = 0.25$, and S is a lognormal random variable with μ_S and V_S listed in Table 13.1. The skewness and kurtosis of the original performance function, i.e. Eq. (13.5), and those of the transformed performance function, i.e. Eq. (13.6), are also listed in Table 13.1, together with the Box-Cox transformation parameter q and the reliability indices obtained from the direct methods of moment, the methods of moment with Box-Cox transformation, and the numerical integral. Herein, the moments of the transformed performance function are evaluated by direct point estimate method with five-point estimate, i.e. the first four central moments are calculated using Eqs. (3.66a–c) in the chapter on Moment Evaluation with $m = 5$; and the moments of the original performance function are calculated using the definitions of the statistical moments, which are exact values. It can be observed from Table 13.1 that: (i) the direct methods of moment may provide inaccurate results when the skewness or kurtosis of the performance function is far away from that of a normal random variable; and (ii) the results obtained from the methods of moment with Box-Cox transformation are identical with the exact ones because the transformed performance function is close to be normally distributed.

Table 13.1 The results by different methods for Example 13.1 (Case 1).

		Direct methods of moment			Methods of moment with Box-Cox transformation				
μ_S	V_S	α_{3G}	α_{4G}	β	q	α_{3g}	α_{4g}	β_{3M}	Exact
32	0.50	0.615	3.916	3.911 (8.85%)	5.96×10^{-9}	1.72×10^{-10}	3.0	3.593	3.593
27	0.75	0.468	4.079	3.277 (6.27%)	9.03×10^{-9}	2.89×10^{-9}	3.0	3.083	3.083
23	1.00	0.282	4.881	2.850 (0.18%)	2.98×10^{-9}	2.33×10^{-9}	3.0	2.855	2.855
20	1.25	0.038	7.297	2.571 (6.17%)	1.04×10^{-8}	1.53×10^{-8}	3.0	2.741	2.741
18	1.50	−0.323	14.131	2.385 (10.5%)	-5.96×10^{-9}	-7.15×10^{-9}	3.0	2.665	2.665

Note: Value in parentheses is the relative error compared with the exact one.

To present the procedure in detail for the methods of moment with Box-Cox transformation, it is assumed that both R and S are lognormal random variables with $\mu_R = 200$, $V_R = 0.25$, and $\mu_S = 32$, $V_S = 0.5$.

According to the point-estimate method presented in the chapter on Moment Evaluation, if the direct point estimate method is used and the number of estimate points for each dimension is five (the total number of estimate points is $5^2 = 25$), the estimate points and corresponding weights in independent standard normal space (i.e. **U**-space) can be obtained using Table 3.3 in the chapter just mentioned, and which are listed in Table 13.2. Then, the estimate-points in the original space (i.e. **X**-space) can be obtained using the inverse normal transformation (here Rosenblatt transformation is used), which

Table 13.2 The estimate-points and corresponding weights for Example 13.1.

i	Estimate-points in U-space and weights			Estimate-points in X-space		Value of $\lambda(X)$	Values of $\tilde{g}(X)$
	u_1	u_2	P	r	s		
1	0.0	0.0	0.284444	194.0285	28.62167	6.779077	1.91384
2	1.355626	0.0	0.11844	270.9104	28.62167	9.465221	2.24762
3	2.856970	0.0	0.006004	392.0738	28.62167	13.69850	2.61729
4	−1.35563	0.0	0.118440	138.9650	28.62167	4.855236	1.58006
5	−2.85697	0.0	0.006004	96.02033	28.62167	3.354812	1.21040
6	0.0	1.355626	0.118440	194.0285	54.30063	3.573227	1.27347
7	1.355626	1.355626	0.049318	270.9104	54.30063	4.989085	1.60725
8	2.856970	1.355626	0.00250	392.0738	54.30063	7.220429	1.97691
9	−1.35563	1.355626	0.049318	138.9650	54.30063	2.559178	0.93969
10	−2.85697	1.355626	0.002500	96.02033	54.30063	1.768310	0.57002
11	0.0	2.856970	0.006004	194.0285	110.3593	1.758152	0.56426
12	1.355626	2.856970	0.002500	270.9104	110.3593	2.454803	0.89805
13	2.856970	2.856970	0.000127	392.0738	110.3593	3.552702	1.26771
14	−1.35563	2.856970	0.002500	138.9650	110.3593	1.259204	0.23048
15	−2.85697	2.856970	0.000127	96.02033	110.3593	0.870070	−0.13918
16	0.0	−1.35563	0.118440	194.0285	15.08638	12.86117	2.55421
17	1.355626	−1.35563	0.049318	270.9104	15.08638	17.95728	2.88800
18	2.856970	−1.35563	0.002500	392.0738	15.08638	25.98859	3.25766
19	−1.35563	−1.35563	0.049318	138.9650	15.08638	9.211285	2.22043
20	−2.85697	−1.35563	0.002500	96.02033	15.08638	6.364703	1.85077
21	0.0	−2.85697	0.006004	194.0285	7.423024	26.13874	3.26342
22	1.355626	−2.85697	0.002500	270.9104	7.423024	36.49596	3.59720
23	2.856970	−2.85697	0.000127	392.0738	7.423024	52.81861	3.96686
24	−1.35563	−2.85697	0.002500	138.9650	7.423024	18.72080	2.92964
25	−2.85697	−2.85697	0.000127	96.02033	7.423024	12.93547	2.55997

are also given in Table 13.2. Using these estimate-points, the values of function $\lambda(\mathbf{X})$ at each point can be determined, and are also listed in Table 13.2. With the aid of the estimate weights P and the values of $\lambda(\mathbf{X})$, the first four moments of the transformed performance function can be formulated as the function of only the Box-Cox transformation parameter q. Based on Eq. (13.4) and the simplex search method, the minimum of d_N is obtained as 9.4398×10^{-18} at $q = 5.9605 \times 10^{-9}$. Therefore, with the aid of the Box-Cox transformation, the performance function can be equivalently expressed as:

$$\tilde{g}(\mathbf{X}) = \frac{(R/S)^{5.9605 \times 10^{-9}} - 1}{5.9605 \times 10^{-9}} \tag{13.7}$$

Substituting the estimate-points into Eq. (13.7), the values of the transformed performance function can be determined as listed in Table 13.2. Then, the first four moments of the transformed performance function (i.e. mean, standard deviation, skewness, and kurtosis) can be obtained as 1.91384, 0.532699, 1.716332×10^{-10}, and 3.0, respectively. With the aid of the third-order moment reliability index as given by Eq. (4.27) as previously mentioned, the reliability index can be obtained as 3.592726, which is almost the same with the exact one (= 3.593) from the numerical integral.

Case 2 investigates the accuracy of the methods of moment with Box-Cox transformation for structural reliability analysis involving small failure probabilities. Assume that R is a Gamma random variable with $V_R = 0.15$ and μ_R in the range of 200–1000; and S is a log-normal random variable with $\mu_S = 100$ and $V_S = 0.25$ and 0.40. Similarly, the moments of the transformed performance function are calculated using direct point estimate method with five-point estimate and the moments of the original performance function are calculated using the definitions of the statistical moments. The failure probabilities obtained from the methods of moment with Box-Cox transformation, the direct methods of moment, and numerical integral are presented and compared in Figure 13.1. It can be observed from Figure 13.1 that: (i) the direct methods of moment may provide inaccurate result when the failure probability is relatively small; and (ii) the results obtained from the methods of moment with Box-Cox transformation are in excellent agreement with the exact ones for the entire investigation range.

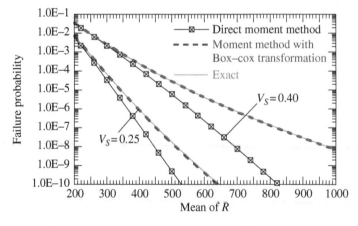

Figure 13.1 Failure probabilities obtained from different methods for Example 13.1 (Case 2).

Example 13.2 Insensitivity to Different Formulations of Equivalent Limit States

The second example discusses the insensitivity to different formulations of equivalent limit states using a steel bar subjected to an axial load N. The diameter of the bar is d_0, and the yield point is f_y. Based on the design criterion assumption, the reliability margin can be expressed in the following four ways (Tichy 1994).

$$\text{Case 1; axial load criterion} : G(\mathbf{X}) = \frac{\pi d_0^2 f_y}{4} - N \tag{13.8a}$$

$$\text{Case 2; cross-sectionarea criterion} : G(\mathbf{X}) = \frac{\pi d_0^2}{4} - \frac{N}{f_y} \tag{13.8b}$$

$$\text{Case 3; diameter criterion} : G(\mathbf{X}) = d_0 - \sqrt{\frac{4N}{\pi f_y}} \tag{13.8c}$$

$$\text{Case 4; stress criterion} : G(\mathbf{X}) = f_y - \frac{4N}{\pi d_0^2} \tag{13.8d}$$

where d_0, f_y, and N are mutually independent random variables; and their statistical distributions are given in Table 13.3.

The results obtained from the methods of moment with Box-Cox transformation, including the Box-Cox transformation parameter q, the first four moments, and the reliability index, are listed in Table 13.4. Herein, the first four moments of the transformed performance function are evaluated using the direct point estimate method with five-point

Table 13.3 Statistical distribution of random variables in Example 12.2.

Variable	Distribution	Mean	COV
d_0 (mm)	Lognormal	40.0	0.10
f_y (kN/mm^2)	Lognormal	0.40	0.15
N (kN)	Lognormal	120.0	0.70

Table 13.4 Results of the methods of moment combined with Box-Cox transformation for Example 13.2.

	The methods of moment combined with Box-Cox transformation						
Case	q	$\mu_{\tilde{g}}$	$\sigma_{\tilde{g}}$	$\alpha_{3\tilde{g}}$	$\alpha_{4\tilde{g}}$	β_{3M}	MCS
Case 1	7.45×10^{-9}	1.611	0.679	1.389×10^{-9}	3.00	2.3727	2.3726
Case 2	7.45×10^{-9}	1.611	0.679	1.389×10^{-9}	3.00	2.3727	2.3726
Case 3	2.61×10^{-8}	0.805	0.339	1.200×10^{-8}	3.00	2.3727	2.3726
Case 4	7.45×10^{-9}	1.611	0.679	1.389×10^{-8}	3.00	2.3727	2.3726

estimate. The total number of performance function evaluations is 125. For comparison, the 'exact' failure probabilities are calculated using Monte Carlo simulation (MCS) with 1.0×10^8 samples (the samples of random variables used in each case are identical and the COVs of failure probability for Case 1–4 are 1.06%, 1.06%, 1.06%, and 1.06%, respectively) and the corresponding reliability indices are also presented in Table 13.4. It can be observed from Table 13.4 that although the first four moments of the transformed performance function may be different (Case 3 is different from other cases), the reliability indices obtained from the methods of moment with Box-Cox transformation are identical for different formulations and in excellent agreement with the MCS results. Therefore, it can be concluded that the methods of moment with Box-Cox transformation is insensitive to the different mathematical formulations for equivalent limit states.

Example 13.3 Reliability Analysis for Performance Functions with Correlated Random Variables

To investigate the applicability of the methods of moment with Box-Cox transformation for structural reliability analysis with correlated random variables, the following example considers a slope as shown in Figure 13.2. The performance function for the stability of this slope can be expressed as follows (Low 2007; Lu et al. 2017a).

$$G(\mathbf{X}) = \frac{cA + N \tan\phi}{W\left(\sin\psi_p + \alpha\cos\psi_p\right) + V\cos\psi_p - T\sin\theta} - 1 \tag{13.9}$$

where

$$A = (H - z)/\sin\psi_p \tag{13.10a}$$

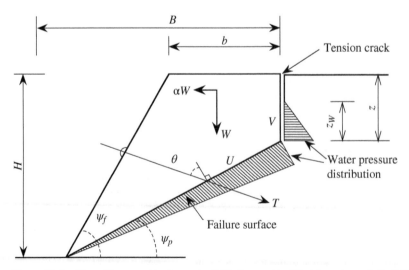

Figure 13.2 Slope illustrated in Example 13.3. *Source:* Modified from Low, B. K. (2007). Reliability analysis of rock slopes involving correlated nonnormals. Int. J. Rock. Mech. Min. Sci., 44(6), 922–935.

$$N = W\left(\cos\psi_p - a\sin\psi_p\right) - U - V\sin\psi_p + T\cos\theta \tag{13.10b}$$

$$W = \frac{1}{2}\gamma H^2\left[\left(1 - \frac{z^2}{H^2}\right)\cot\psi_p - \cot\psi_f\right] \tag{13.10c}$$

$$U = \frac{1}{2}\gamma_w \cdot r \cdot z \cdot A \tag{13.10d}$$

$$V = \frac{1}{2}\gamma_w r^2 z^2 \tag{13.10e}$$

in which $\gamma = 2.6\times10^4\,\text{N/m}^3$; $\gamma_w = 1.0\times10^4\,\text{N/m}^3$; $\psi_f = 50°$; $\psi_p = 35°$; $T = 0$; $\theta = 0$; and $H = 60\text{m}$. Note that c, ϕ, r, z, and a are random variables and their statistical distributions are listed in Table 13.5; and all four variables are generally correlated in engineering application. Assume that the correlation coefficients between c and ϕ, r and z are $\rho_{c,\phi}$ and $\rho_{r,z}$, respectively. All the other random variables are mutually independent.

The results obtained from the methods of moment combined with Box-Cox transformation are listed in Table 13.6 including the Box-Cox parameter, first four moments, and reliability index, together with the reliability index obtained from MCS with 1.0×10^6 samples (the largest COV of failure probability is 3.39%). It can be observed from Table 13.6 that (i) the reliability indices obtained from the methods of moment with Box-Cox

Table 13.5 Statistical information of random variables for Example 13.3.

Variable	Distribution	Mean	Standard deviation
c	Lognormal	140 kPa	28 kPa
ϕ	Lognormal	35°	5°
r	Truncated Exponential (0, 1.0)[a]	0.344	0.263
z	Lognormal	14 m^2	3 m^2
a	Truncated Exponential (0, 0.16)[b]	0.055	0.042

Note: [a]The distribution of r is truncated from the exponential distribution with mean equal 0.5; [b]The distribution of a is truncated from the exponential distribution with mean equal 0.08.

Table 13.6 The results of the methods of moment combined with Box-Cox transformation for Example 13.3.

Correlation coefficient		methods of moment with Box-Cox transformation						MCS
$\rho_{c,\phi}$	$\rho_{r,z}$	q	$\mu_{\hat{g}}$	$\sigma_{\hat{g}}$	$\alpha_{3\hat{g}}$	$\alpha_{4\hat{g}}$	β_{3M}	
0.0	0.0	0.411	0.507	0.214	0.010	3.145	2.377 (0.34%)	2.369
−0.25	−0.50	0.174	0.491	0.169	−0.021	3.124	2.873 (0.59%)	2.890
−0.50	−0.25	0.592	0.543	0.190	−0.007	3.134	2.847 (0.07%)	2.849
−0.50	−0.50	0.235	0.501	0.157	−0.054	3.190	3.107 (0.80%)	3.132

Note: Value in parentheses is the relative difference compared with the MCS results.

transformation are in excellent agreement with the MCS results; and (ii) the correlation coefficient of random variables has significant influence on the reliability index.

Example 13.4   Reliability Analysis for Strongly Nonlinear Performance Functions

To explore the applicability of the methods of moment with Box-Cox transformation for structural reliability analysis involving strongly nonlinear performance function, this example investigates a two-degree-of-freedom primary-secondary system subjected to white noise base excitation with intensity S_0 as shown in Figure 13.3, which has been investigated by Der Kiureghian and De Stefano (1991), Bourinet et al. (2011), and Sundar and Shields (2016). The system is characterised by the masses m_p and m_s, spring stiffnesses k_p and k_s, natural frequencies $\omega_p = (k_p/m_p)^{1/2}$ and $\omega_s = (k_s/m_s)^{1/2}$, and damping rates ζ_p and ζ_s, where the subscripts p and s refer to the primary and secondary oscillators, respectively. Due to the tuning and interaction effects, the peak response of the secondary oscillator is strongly sensitive to the properties of the system and it is thus worthwhile to investigate the influence of uncertainty within the system parameters on the reliability of the secondary oscillator. If the force capacity of the secondary spring is F_s, the reliability against failure of the secondary spring can be evaluated by the following performance function (Der Kiureghian and De Stefano 1991):

$$G(\mathbf{X}) = F_s - 3k_s \cdot \sqrt{\frac{\pi S_0}{4\zeta_s \omega_s^3} \left[\frac{\zeta_a \zeta_s}{\zeta_p \zeta_s \left(4\zeta_a^2 + \theta_{sr}^2\right) + \gamma \zeta_a^2} \frac{\left(\zeta_p \omega_p^3 + \zeta_s \omega_s^3\right)\omega_p}{4\zeta_a \omega_a^4} \right]} \tag{13.11}$$

where $\gamma = m_s/m_p$ denotes the mass ratio; $\omega_a = (\omega_p + \omega_s)/2$ and $\zeta_a = (\zeta_p + \zeta_s)/2$ represent the average frequency and damping ratio, respectively; and $\theta_{sr} = (\omega_p = \omega_s)/\omega_a$ is a tuning parameter. The statistical distribution properties of the random variables are summarised in Table 13.7.

The reliability indices obtained from the methods of moment with Box-Cox transformation, FORM, and SORM are presented in Figure 13.4, together with the 'exact' values evaluated from MCS with 5.0×10^6 (the largest COV of failure probability is 1.76%). It can be observed from Figure 13.4 that: (i) the reliability index provided by FORM significantly underestimates the reliability because the performance function is strongly nonlinear; (ii) the SORM result gives a significant improvement compared with that of FORM; and (iii) the reliability indices obtained from SORM and the methods of moment with Box-Cox transformation are in excellent agreement with the MCS results for the entire investigation range.

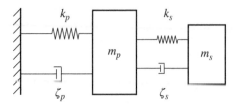

Figure 13.3 The two-degrees-of-freedom system for Example 13.4.

Table 13.7 Statistical information of random variables in Example 13.4.

Variable	Distribution	Mean	COV
m_p	Lognormal	1.00	0.05
m_s	Lognormal	0.01	0.05
k_p	Lognormal	1.00	0.10
k_s	Lognormal	0.01	0.10
ζ_p	Lognormal	0.05	0.15
ζ_s	Lognormal	0.02	0.15
F_s	Lognormal	20–30	0.40
S_0	Lognormal	100.0	0.25

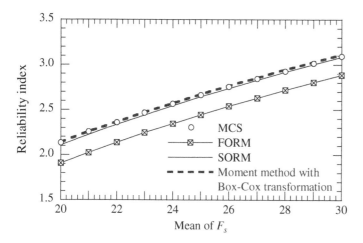

Figure 13.4 Comparison of reliability indices by different methods for Example 13.4.

Example 13.5   Reliability Assessment Integrated with Finite Element Analysis for an Implicit Performance Function with Small Failure Probability

To examine the accuracy and efficiency of the methods of moment with Box-Cox transformation for structural reliability analysis with a large number of random variables and implicit performance function, a seven-story three-bay reinforced concrete (RC) nonlinear frame structure shown in Figure 13.5a is analysed. The total height of this frame structure is $H = 24.3$ m, wherein the height of the first story is 4.5 m, the individual heights of the second story to seventh story are 3.3 m, and the span of each bay is 6.0 m. The sections and arrangement of the reinforcements of column and beam members are shown in Figures 13.5b–d. In this example, the serviceability of this RC frame structure under lateral loads is investigated.

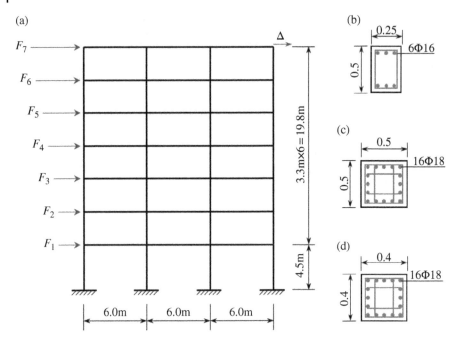

Figure 13.5 RC frame structure considered in Example 13.5 (unit: m).

Assume that the failure is reached when the horizontal displacement of the upper floor exceeds the allowable limit Δ_{lim}. The performance function is implicitly expressed as:

$$G(\mathbf{X}) = \Delta_{\text{lim}} - \Delta(\mathbf{X}) \tag{13.12}$$

where Δ_{lim} is the allowable horizontal displacement; and $\Delta(\mathbf{X})$ is the horizontal displacement of the upper floor, which can be determined from the nonlinear finite element analysis of this frame structure.

The finite element model of this nonlinear RC frame structure is built with OpenSees software. The inelastic fibre elements are used to simulate the sections of the column and beam members, which are composed of the steel bars and concrete. The uniaxial material 'Concrete02' is used to simulate the unconfined and confined concrete, and material 'Steel02' is applied to simulate the longitudinal steel bar. All lateral loads and the properties of concrete and steel are assumed to be mutually independent random variables, and their statistical distributions are given in Table 13.8.

With the aid of the methods of moment combined with Box-Cox transformation, the failure probabilities can be obtained and are depicted in Figure 13.6. In this example, if the moments of the transformed performance function are determined by direct point estimate method, the computation is prohibitively large due to the horizontal displacement $\Delta(\mathbf{X})$ involving 20 random variables. To improve the computational efficiency, the moments of the transformed performance function are evaluated using five-point estimate method combined with the bivariate dimension-reduction method, i.e. the first four moments are obtained by Eqs. (3.78)–(3.85d) in the chapter on Moment Evaluation with $m = 5$.

Table 13.8 Statistical distribution of random variables in Example 13.5.

Variable	Distribution	Mean	COV
f_{pc-c}, confined concrete compressive strength (MPa)	Lognormal	33.8	0.15
f_{pc-u}, unconfined concrete compressive strength (MPa)	Lognormal	28.2	0.15
ε_{c0-c}, confined concrete strain at maximum strength	Lognormal	0.0024	0.15
ε_{c0-u}, unconfined concrete strain at maximum strength	Lognormal	0.002	0.15
f_{cu-c}, confined concrete crushing strength (MPa)	Lognormal	6.77	0.15
f_{cu-u}, unconfined concrete crushing strength (MPa)	Lognormal	5.64	0.15
ε_{cu-c}, confined concrete strain at crushing strength	Lognormal	0.021	0.15
ε_{cu-u}, unconfined concrete strain at crushing strength	Lognormal	0.004	0.15
f_t, concrete tensile strength (MPa)	Lognormal	2.82	0.15
E_t, concrete tension softening stiffness (MPa)	Lognormal	4.14×10^3	0.15
F_y, steel yield strength (MPa)	Lognormal	452.0	0.05
E_0, steel initial elastic tangent (MPa)	Lognormal	2.10×10^5	0.05
b, steel strain-hardening ratio	Lognormal	0.01	0.05
F_1–F_2, lateral load (kN)	Gumbel	30.0	0.35
F_3, lateral load (kN)	Gumbel	45.0	0.35
F_4, lateral load (kN)	Gumbel	65.0	0.35
F_5, lateral load (kN)	Gumbel	83.0	0.35
F_6, lateral load (kN)	Gumbel	90.0	0.35
F_7, lateral load (kN)	Gumbel	95.0	0.35

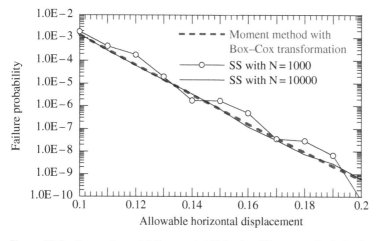

Figure 13.6 Comparison of failure probabilities by different methods for Example 13.5.

The total number of finite element analysis is 3121. Because the performance function involves the computation of nonlinear finite element model, the 'exact' results are difficult to obtain by MCS. For comparison, the failure probabilities are also conducted with subset simulation (SS) (Au and Beck 2001a; Li and Cao 2016), which is instrumental in improving the computational efficiency of MCS. Here, the conditional probability of SS is 0.1 and the number of samples in each simulation level $N = 1000$ and $N = 10\,000$ are used. The total number of finite element analysis in SS is varying from 2800 ($\Delta_{\text{lim}} = 0.1$ m) to 10 900 ($\Delta_{\text{lim}} = 0.2$ m) for $N = 1000$, and 28 000 ($\Delta_{\text{lim}} = 0.1$ m) to 99 000 ($\Delta_{\text{lim}} = 0.2$ m) for $N = 10\,000$, respectively. The failure probabilities obtained from SS are also depicted in Figure 13.6.

It can be observed from Figure 13.6 that: (i) the methods of moment with Box-Cox transformation and SS with $N = 10\,000$ provide almost identical results for the entire investigation range; and (ii) the result obtained from SS with $N = 1000$ is unstable and may have large difference compared with that obtained from SS with $N = 10\,000$ for some cases.

From the preceding discussion, very little difference can also be observed for the required computational effort between the methods of moment with Box-Cox transformation and the SS with $N = 1000$ for large failure probabilities, while the methods of moment with Box-Cox transformation is much more efficient for small failure probabilities. The number of finite element analysis required for the methods of moment with Box-Cox transformation is just 3.15–11.15% of that of SS with $N = 10\,000$. In summary, the methods of moment with Box-Cox transformation is more efficient than SS without loss of accuracy in this example.

Example 13.6 Dynamic Reliability Assessment of a Nonlinear Single-Degree-of-Freedom (SDOF) System under Stochastic Excitations

To investigate the applicability of the methods of moment combined with Box-Cox transformation for dynamic reliability assessment of nonlinear structural systems under stochastic excitations, a Single-Degree-of-Freedom (SDOF) hysteretic oscillator under stochastic seismic excitations is considered in this example, which has been investigated by Xu and Dang (2019). This dynamic system is expressed as:

$$m\ddot{u}(t) + c\dot{u}(t) + k[\alpha u(t) + (1-\alpha)z(t)] = -m\ddot{u}_g(t) \tag{13.13}$$

where $\ddot{u}(t)$, $\dot{u}(t)$, and $u(t)$ are the acceleration, velocity, and displacement of the oscillator, respectively; $m = 2.2 \times 10^5$ kg and $k = 3.0 \times 10^7$ N/m are the mass and stiffness, respectively; and the Rayleigh's damping $c = am + bk$ is considered, in which $a = 0.1998$ Hz and $b = 0.0048$ sec. The hysteretic displacement $z(t)$ obeys the extended Bouc-Wen hysteretic law (Baber and Wen 1981; Baber and Noori 1985):

$$\dot{z}(t) = \frac{A\dot{u}(t) - v(t)\left[\kappa|\dot{u}(t)|z(t)|^{q-1}z(t) + \gamma\dot{u}(t)|z(t)|^q\right]}{\eta(t)} \tag{13.14a}$$

where

$$v(t) = 1 + \delta_v \varepsilon(t), \eta(t) = 1 + \delta_\eta \varepsilon(t), \varepsilon(t) = \int_0^t z(\tau)\dot{u}(\tau)d\tau \tag{13.14b}$$

in which $\alpha = 0.1$, $A = 1.0$, $q = 1.0$, $\kappa = 80.0$, $\gamma = 40.0$, $\delta_v = 2000.0$, and $\delta_\eta = 2000.0$.

The nonstationary ground acceleration $\ddot{u}_g(t)$ is simulated using the spectral representation method (Shinozuka and Deodatis 1991; Chen et al. 2017) as:

$$\ddot{u}_g(t) = g(t) \cdot \sqrt{2} \sum_{k=0}^{N-1} \sqrt{2S(\omega_k)\Delta\omega} \cos(\omega_k t + \Phi_k) \tag{13.15}$$

where $\Delta\omega = \omega_u/N$, in which $\omega_u = 100.0$ rad/s is the truncated frequency; $\omega_k = k \cdot \Delta\omega$ ($k = 0$, 1, ..., N–1), in which $N = 1000$ is adopted in this example; $\Phi_0, \Phi_1, ..., \Phi_{N-1}$ are independent uniformly distributed random variables over the interval $[0, 2\pi]$; $S(\omega)$ is the power spectral density function; and $g(t)$ is the envelope function. $S(\omega)$ and $g(t)$ are assumed as follows (Clough and Penzien 2003).

$$S(\omega) = \frac{\omega_g^4 + 4\zeta_g^2 \omega_g^2 \omega^2}{\left(\omega^2 - \omega_g^2\right)^2 + 4\zeta_g^2 \omega_g^2 \omega^2} \cdot \frac{\omega^2}{\left(\omega^2 - \omega_f^2\right)^2 + 4\zeta_f^2 \omega_f^2 \omega^2} \cdot S_0 \tag{13.16a}$$

$$g(t) = \begin{cases} (t/t_1)^2, & 0 \le t < t_1 \\ 1.0, & t_1 \le t < t_2 \\ \exp[-c(t - t_2)], & t_2 \le t < T \\ 0.0, & t \ge T \end{cases} \tag{13.16b}$$

in which $\omega_g = 11.42$ rad/s; $\zeta_g = 0.80$; $\omega_f = 0.1\omega_g$; $\zeta_f = \zeta_g$; $S_0 = 79.73$ cm^2/s^3; $t_1 = 0.8$ s; $t_2 = 7.0$ s; $T = 17$ s; and $c = 0.35$.

The limit state function can be expressed as (Xu and Dang 2019)

$$G(\mathbf{X}) = u_{\lim} - \max_{t \in [0, T]}\{|u(\mathbf{X}, t)|\} \tag{13.17}$$

where $u_{\lim} = 0.01$ m is the threshold, and $\mathbf{X} = [\Phi_0, \Phi_1, ..., \Phi_{N-1}]^T$. A total of 1000 random variables are involved in this example.

In this example, to verify whether the Box-Cox transformation can be applied to normalise the performance function involving stochastic processes or not, the statistical moments of the transformed performance function are determined using MCS with 5000 samples. The Box-Cox transformation parameter q, the first four moments of the transformed performance function, and the 3 M reliability indices are obtained as $q = 0.174$, $\mu_{\tilde{g}} = 0.876$, $\sigma_{\tilde{g}} = 0.256$, $\alpha_{3\tilde{g}} = 2.406 \times 10^{-3}$, $\alpha_{4\tilde{g}} = 2.928$, and $\beta_{3M} = 3.413$, respectively. The failure probability obtained by MCS with 2.0×10^6 samples is 2.49×10^{-4}, and the corresponding reliability index is 3.482, in which the COV of the failure probability is obtained as 4.48%. Since the moments of the transformed performance function is evaluated using MCS, the kurtosis of the transformed performance function slightly deviates from that of a normal random variable. Although the transformed performance function is not perfectly normally distributed, the reliability result provided by the methods of moment combined with Box-Cox transformation is accurate enough (the relative errors of 3 M reliability index is 1.98%). That is, the methods of moment combined with Box-Cox transformation can be applied for dynamic reliability assessment of nonlinear structural systems under stochastic excitations.

13.3 Summary

In this chapter, the methods of moment with Box-Cox transformation are described for structural reliability analysis. It can be concluded that: (i) with the aid of the Box-Cox transformation, the transformed performance function can be approximated as normally distributed from the perspective of the first four moments; (ii) the methods of moment with successful Box-Cox transformation have no shortcoming of the sensitivity to the different mathematical formulas of equivalent limit states; and (iii) the methods of moment with Box-Cox transformation can provide sufficiently accurate results even when the structural reliability analysis involves nonnormally distributed, nonlinear, or implicit performance function with small failure probability, correlated random variables or stochastic processes.

Appendix A

Basic Probability Theory

In this appendix, we summarise the basic probability theory that is essential for structural reliability methods described in this book, with emphasis on the concept and utility of the first four moments of random variables. For details of probability theory, the readers can refer to many textbooks, some of which are listed in the references (e.g. Abramowitz and Stegun 1972; Ang and Tang 1975; Ang and Tang 2006; Johnson and Kotz 1969; Johnson and Kotz 1970a; Johnson and Kotz 1970b; Johnson and Kotz 1972; Stuart and Ord 1987).

A.1 Events and Probability

A.1.1 Introduction

It may be recognised from the chapter on Measures of Structural Safety that when speaking of probability, we are referring to the occurrence of an event relative to other events. In other words, there is more than one possibility, since otherwise the problem would be deterministic. For quantitative purposes, therefore, probability can be considered as a numerical measure of the likelihood of occurrence of an event relative to a set of alternative events.

In calculating the probability that an event occurs, a basis for assigning probability measures to the various possible outcomes is necessary. The assignment may be based on prior conditions, or the results of empirical observation, or both. The usefulness of a calculated probability will depend on the appropriateness of the basis for its determination. The empirical relative frequency basis must rely on a large amount of observational data. When data is limited, the relative frequency by itself may have limited usefulness.

A.1.2 Events and Their Combinations

Three special events are defined as follows.

- **Impossible event:** denoted as ϕ, is the event with no sample point. It is therefore an empty set in a sample.
- **Certain event:** denoted as S, is the event containing all the sample points in a sample space, that is, it is the sample space itself.
- **Complementary event $\overline{E}$:** for an event E in a sample space S, the complementary event contains all the sample points in S that are not in E.

Structural Reliability: Approaches from Perspectives of Statistical Moments, First Edition.
Yan-Gang Zhao and Zhao-Hui Lu.
© 2021 John Wiley & Sons Ltd. Published 2021 by John Wiley & Sons Ltd.

(a)   (b)   (c)

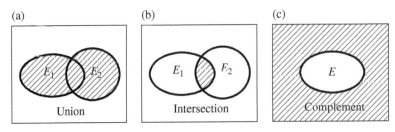

Figure A.1 Venn diagrams of union and intersection of events E_1 and E_2, and complementary event $\overline{E}$. (a) $E_1 \cup E_2$; (b) $E_1 \cap E_2$; (c) $\overline{E}$.

In many practical problems, the event of interest may be some combination of other events. There are two basic ways that events may be combined or derived from other events; by union or intersection. Consider two events, E_1 and E_2:

- The union of E_1 and E_2, denoted $E_1 \cup E_2$, is an event that means the occurrence of E_1 or E_2, or both. In other words, $E_1 \cup E_2$ is the subset of sample points that belong to E_1 or E_2.
- The intersection of E_1 and E_2, denoted $E_1 \cap E_2$, is an event that means the joint occurrence of E_1 and E_2. In other words, $E_1 \cap E_2$ is the subset of sample points that belong to both E_1 and E_2.

The union, intersection and complement are shown in Figure A.1a–c.

Mutually exclusive events: If the occurrence of one event precludes the occurrence of another event, then the two events are mutually exclusive. This means that the corresponding subsets will have no overlap; that is, the subsets are disjointed. The intersection of two mutually exclusive events therefore, is an impossible event as shown in Figure A.2.

A.1.3 Mathematical Operations of Sets

The notations to designate sets and the associated operations are as follows:

$\cup$ = the union
$\cap$ = the intersection
$\supset$ = contains
$\subset$ = belongs to, or is contained in
$\overline{E}$ = the complement of E

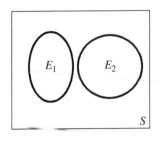

Figure A.2 Exclusive events.

With these notations, the mathematical rules governing the operations of sets are described in the following:

Commutative Rule: The union and intersection of sets are commutative, that is, for two sets A and B

$$A \cup B = B \cup A$$
$$A \cap B = B \cap A$$

Associative Rule: The union and intersection of sets are associative, that is, for three sets A, B, and C

(a)   (b)   (c)

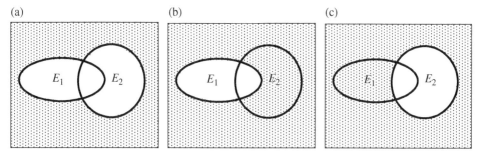

Figure A.3 Venn diagrams showing De Morgan's rule. (a) $\overline{E_1 \cup E_2}$; (b) $\overline{E_1}$; (c) $\overline{E_2}$.

$$(A \cup B) \cup C = A \cup (B \cup C)$$
$$(A \cap B) \cap C = A \cap (B \cap C)$$

Distributive rule: The union and intersection of sets are distributive, that is, for three sets A, B, and C

$$(A \cup B) \cap C = (A \cap C) \cup (B \cap C)$$
$$(A \cap B) \cup C = (A \cup C) \cap (B \cup C)$$

De Morgan's rule: The complement of the union of two sets E_1 and E_2 is the same as the intersection of their complements, i.e.,

$$\overline{E_1 \cup E_2} = \overline{E_1} \cap \overline{E_2}$$

The general validity of this relationship can be shown with the Venn diagrams in Figure A.3a–c. In more general terms, the De Morgan's rule is

$$\overline{E_1 \cup E_2 \cup ... \cup E_k} = \overline{E_1} \cap \overline{E_2} \cap ... \cap \overline{E_k}$$

A.1.4 Mathematics of Probability

1) Basic axioms of probability
   a) The probability $P(E)$ of an event E is a real non-negative number;

   $$0 \le P(E) \le 1$$

   b) The probability $P(S)$ of the certain event S is 1.0 and the probability of the impossible event is 0, i.e.

   $$P(S) = 1, \quad P(\phi) = 0$$

   c) Addition rule: The probability that either or both of two events E_1 and E_2 occur is

   $$P(E_1 \cup E_2) = P(E_1) + P(E_2) - P(E_1 \cap E_2) \tag{A.1a}$$

for two mutually exclusive events

$$P(E_1 \cup E_2) = P(E_1) + P(E_2) \qquad\qquad (A.1b)$$

2) Conditional probability and multiplication rule:

The probability of an event may depend on the occurrence of another event. If this dependence exists, the associated probability is a conditional probability. The conditional probability of E_1 assuming E_2 has occurred (the conditional probability of E_1 given E_2) denoted $P(E_1 \mid E_2)$, can be given as

$$P(E_1 \mid E_2) = \frac{P(E_1 E_2)}{P(E_2)} \qquad\qquad (A.2)$$

The multiplication rule is expressed as

$$P(E_1 E_2) = P(E_1 \mid E_2)P(E_2) = P(E_2 \mid E_1)P(E_1) \qquad\qquad (A.3)$$

3) Theorem of total probability:

Sometimes the probability of an event A cannot be determined directly. However, its occurrence is always accompanied by the occurrence of other events E_i, $i = 1, 2, ..., n$, so that the probability of A will depend on which of the events E_i has occurred. In such cases, the probability of A will be an expected probability. Such problems require the theorem of total probability.

Formally, consider n mutually exclusive and collective exhaustive events $E_1, E_2, ..., E_n$; that is $E_1 \cup E_2 \cup \cdots \cup E_n = S$. Then, the *total probability theorem* is expressed as

$$P(A) = \sum_{i=1}^{n} P(A \mid E_i)P(E_i) \qquad\qquad (A.4)$$

The total probability theorem is illustrated in Figure A.4.

4) Bayes' Theorem:

In the situation underlying the total probability theorem, if the event A occurred, what is the probability that a particular E_i also occurred? This may be considered as a reverse probability.

Applying Eq. (A.3) to the joint event AE_i, we have

$$P(A \mid E_i)P(E_i) = P(E_i \mid A)P(A)$$

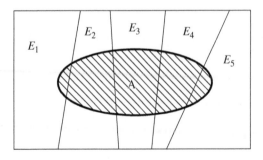

Figure A.4 Illustration of the theorem of total probability.

Therefore, we obtain the desired probability

$$P(E_i \mid A) = \frac{P(A \mid E_i)P(E_i)}{P(A)} \tag{A.5a}$$

which is known as the Bayes' theorem. If $P(A)$ is expanded using the total probability theorem, then, we have

$$P(E_i \mid A) = \frac{P(A \mid E_i)P(E_i)}{\sum\limits_{i=1}^{n} P(A \mid E_i)P(E_i)} \tag{A.5b}$$

A.2 Random Variables and Their Distributions

A *random variable* is a mathematical vehicle for representing an event in analytical form. In contrast to a deterministic variable that can assume definite values, a *random variable*, which is usually denoted with a capital letter, may assume a numerical value only with an associated probability or probability measure. That is to say, if X is defined as a *random variable*, then $X = x$, or $X < x$, $a < X < b$ represents an event, all of these events occurred with a probability. The rule for describing the probability measures associated with all the values of a random variable is a probability distribution.

If X is a random variable, its probability distribution can always be described by its cumulative distribution function (CDF), which is

$$F_X(x) = P(X \le x) \quad \text{for all } x \tag{A.6}$$

Here X is a discrete random variable only if certain discrete values of x have positive probabilities. Alternatively, X is a continuous random variable if probability measures are defined for any value of x.

For a discrete random variable X, its probability distribution may also be described in terms of a probability mass function (PMF), which is simply a function expressing $p_X(x_i) = P(X = x_i)$ for all x_i. Therefore, if X is a discrete random variable with PMF $p_X(x_i)$, its cumulative distribution function is

$$F_X(x) = P(X \le x) = \sum_{\text{all } x_i \le x} P(X = x_i) = \sum_{\text{all } x_i \le x} p_X(x_i) \tag{A.7}$$

However, if X is continuous, probabilities are associated with intervals on the real line (since events are defined as intervals on the real line); consequently, at a specific value of X, such as $X = x$, only the density of probability is defined. Thus, for a continuous random variable, the probability law may also be described in terms of a probability density function (PDF), so that if $f_X(x)$ is the PDF of X, the probability of X in the interval (a, b) is

$$P(a < X \le b) = \int_a^b f_X(x)dx \tag{A.8}$$

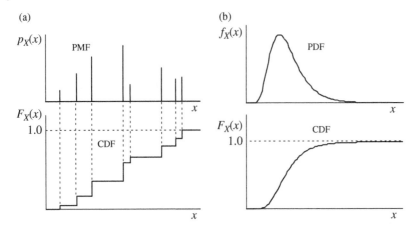

Figure A.5 Probability distributions of discrete and continuous random variable. (a) Discrete; (b) Continuous.

It follows then that the corresponding CDF is

$$F_X(x) = P(X \leq x) = \int_{-\infty}^{x} f_X(t)dt \tag{A.9}$$

Accordingly, if $F_X(x)$ has a first derivative, then, from Eq. (A.9),

$$f_X(x) = \frac{dF_X(x)}{dx} \tag{A.10}$$

We might reiterate that $f(x)$ is not a probability; however, $f_X(x)dx = P(x < X \leq x + dx)$ is the probability that values of X will be in the interval $(x, x + dx)$.

Figure A.5a,b presents graphic examples of legitimate probability distributions.

It should be emphasised that any function used to represent the probability distribution of a random variable must necessarily satisfy the axioms of probability. For this reason, the function must be nonnegative and the probabilities associated with all possible values of the random variable must add up to 1.0. In other words, if $F_X(x)$ is the CDF of X, then it must have the following properties:

a) $F_X(-\infty) = 0$, $F_X(+\infty) = 1$;
b) $F_X(x) \geq 0$ and is non-decreasing with x; and
c) It is continuous with x.

Conversely, any function possessing these properties is a bona fide CDF. By virtue of these properties and Eqs. (A.7) through (A.10), the PMF and PDF are nonnegative functions of x, whereas the probabilities of a PMF add up to 1.0, and the total area under a PDF is also equal to 1.0. We observe that we can write Eq. (A.8) as

$$P(a < X \leq b) = F_X(b) - F_X(a) \tag{A.11}$$

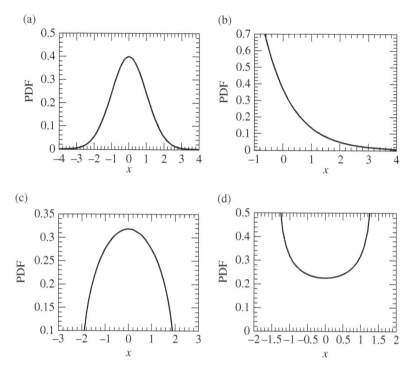

Figure A.6   Shape types of PDF. (a) Bell-shaped type; (b) J-shaped type; (c) n-shaped type; (d) U-shaped type.

Sometimes, the random variables may be classified in terms of the shape of their PDFs. For example, as illustrated in Figure A.6a–d, random variables may be classified as bell-shaped type, J-type, n-type and U-type.

A.3   Main Descriptors of a Random Variable

The probabilistic characteristics of a random variable would be described completely if the form of the distribution function (or equivalently its PDF or mass function) and the associated parameters are specified. In practice, however, the form of the distribution function may not be known. Consequently, approximate description of a random variable is necessary. The probabilistic characteristics of a random variable may be described approximately in terms of certain key quantities or main descriptors of the random variable. The most important of these are the measures of location and dispersion of the random variable, and the measures of skewness and kurtosis may also be important and useful when the underlying distribution is known to be non-normal.

Moreover, even when the distribution function is known, the principal quantities remain useful, because they convey information on the properties of the random variable that are of first importance in practical applications. The parameters of the distribution may also be derived as functions of these quantities or may be the parameters themselves.

A.3.1 Measures of Location

There are three measures of location in common use; the means, the median, and the mode. We consider them in turn.

The mean, is perhaps the most generally used statistical measure, and in fact is far older than the science of statistics itself. Since there is a range of possible values of a random variable, we would naturally be interested in its average value. Because the different values of the random variable are associated with different probabilities or probability densities, the 'weighted' average would be of special interest; this is known as the mean value or the expected value of the random variable. If X is a discrete random variable with PMF $p_X(x_i)$, its mean value, denoted $E(X)$ or μ_X, is

$$\mu_X = E(X) = \sum_{all\ x_i} x_i p_x(x_i) \tag{A.12}$$

Similarly, for a continuous random variable X with PDF $f_X(x)$, the mean value is

$$\mu_X = E(X) = \int_{-\infty}^{\infty} x f_X(x) dx \tag{A.13}$$

Figure A.7 presents graphic examples of PDFs with different mean values.

The mode: If the PDF of a random variable has a local maximum at a value x_l, i.e. if $f(x)$ is greater at x_l than at neighbouring values below and above x_l, there is said to be a mode of the distribution at x_l. For a continuous random variable, the first derivation of the PDF at x_l is equal to 0, i.e.

$$\left.\frac{df}{dx}\right|_{x=x_l} = 0$$

If there is only one such modal value, the distribution is called unimodal, and if there are several, the distribution is multimodal. For a unimodal distribution, the mode x_l is the most

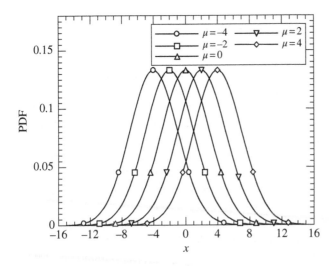

Figure A.7 PDFs with different means.

probable value of the variable, that is, it is the value of the random variable with the largest probability or the highest probability density.

The median is the value of a random variable at which the values above and below it are equally probable; that is, if x_m is the median of X, then

$$F_X(x_m) = \int_{-\infty}^{x_m} f(x)dx = 0.5 \tag{A.14}$$

In general, the mean, median, and mode of a random variable are different, especially if the density function is not symmetric as shown in Figure A.8. However, if the PDF is symmetric and unimodal, these three quantities coincide as shown in Figure A.9.

A.3.2 Measures of Dispersion

After the mean value, the next most important quantity of a random variable is the measure of dispersion or variability, that is, the quantity that gives a measure of how closely the values of the variate are clustered or conversely, how widely they are spread around the central value. If the deviation is taken with respect to the mean value, then a suitable average measure of dispersion is the variance, which is defined as

$$Var(X) = E\left[(X - \mu_X)^2\right] \tag{A.15}$$

in which $\mu_X = E(X)$ is the mean value of X.

For a discrete random variable X with PMF $p_X(x_i)$, the variance of X is

$$Var(X) = \sum_{all\ x_i} (x - \mu_X)^2 p_X(x_i) \tag{A.16}$$

For a continuous random variable X with PDF $f_X(x)$, the variance of X is

$$Var(X) = \int_{-\infty}^{\infty} (x - \mu_X)^2 f_X(x)dx \tag{A.17}$$

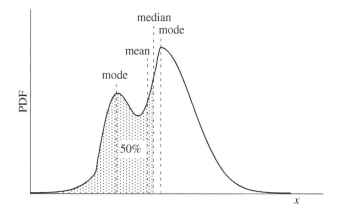

Figure A.8 Mean, median, and mode of a variable with asymmetric PDF.

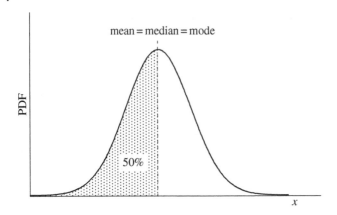

Figure A.9 Mean, median, and mode of a variable with symmetric and unimodal PDF.

Expanding Eq. (A.16), we have

$$Var(X) = E\left[X^2 - 2\mu_X X + \mu_X^2\right] = E(X^2) - 2\mu_X E(X) + \mu_X^2 \tag{A.18}$$

Thus, a useful relation for the variance is

$$Var(X) = E(X^2) - \mu_X^2 \tag{A.19}$$

In Eq. (A.19), the term $E(X^2)$ is known as the mean-square value of X.

Dimensionally, a more convenient measure of dispersion is the square root of the variance, or the standard deviation σ; that is,

$$\sigma_X = \sqrt{Var(X)} \tag{A.20a}$$

Figure A.10a presents graphic examples of PDFs with the same mean and different standard deviation σ. From Figure A.10a, one can see that the larger the standard deviation σ, the wider the values of the variable are spread around the mean value.

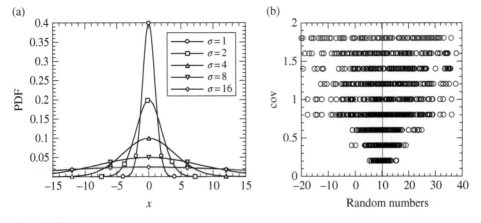

Figure A.10 PDFs and random numbers with different standard deviations and the same mean. (a) PDFs; (b) Random numbers.

It is hard to say, solely on the basis of the variance or standard deviation, whether the dispersion is large or small; for this purpose, the measure of dispersion relative to the mean value is more useful. For this reason, the coefficient of variation (COV)

$$V_X = \frac{\sigma_X}{\mu_X} \qquad \text{(A.20b)}$$

is often a preferred and convenient non-dimensional measure of dispersion or variability.

Figure A.10b presents graphic examples of random numbers with the same mean and different COV. From Figure A.10b, one can see that the larger the standard coefficients of variation, the wider the values of the variable are spread around the mean value.

A.3.3 Measures of Asymmetry

Another useful property of a random variable is the symmetry or lack of symmetry of its probability distribution, and its associated degree and direction of asymmetry. A measure of this asymmetry is skewness. Consider the third-order central moment of a random variable X, which is defined as (refer to Section A.4)

$$E\left[(X - \mu_X)^3\right] = \sum_{all \ x_i} (x - \mu_X)^3 p_X(x_i) \quad \text{for discrete } X$$

$$E\left[(X - \mu_X)^3\right] = \int_{-\infty}^{\infty} (x - \mu_X)^3 f_X(x) dx \quad \text{for continuous } X$$

Observe that $E[(X - \mu_X)^3]$ is zero if the probability distribution is symmetric about μ_X; otherwise it may be positive or negative. It will be positive if the values of X that are greater than μ_X are more widely dispersed than the dispersion of $X < \mu_X$. On the other hand, it will be negative if the reverse situation is true.

A convenient non-dimensional measure of the asymmetry above is skewness, which is defined as the third central moment of a random variable divided by its third power of standard deviation,

$$\alpha_3 = \frac{E\left[(X - \mu_X)^3\right]}{\sigma_X^3} \qquad \text{(A.21)}$$

Figure A.11a,b illustrates the PDFs with positive and negative skewness coefficient in which all the distributions have the same mean value of 0 and same standard deviation of 1. From Figure A.11a,b, one can see that the larger the absolute skewness coefficient, the stronger the asymmetry of the PDF.

Figure A.12a,b presents graphic examples of random numbers with different skewness but the same mean and COV. From Figure A.12a,b, one can see that the larger the skewness, the stronger the asymmetry of the random numbers.

A.3.4 Measures of Sharpness

Another useful property of a random variable is the sharpness or peakedness of its probability distribution; a measure of this sharpness is kurtosis. Consider the fourth-order central moment of a random variable X, which is defined as (refer to Section A.4)

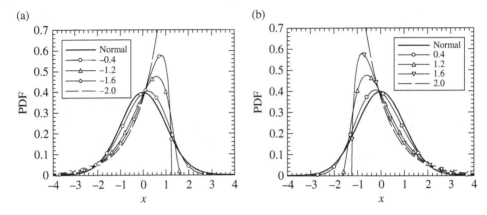

Figure A.11 PDFs with negative and positive skewness. (a) Negative skewness; (b) Positive skewness.

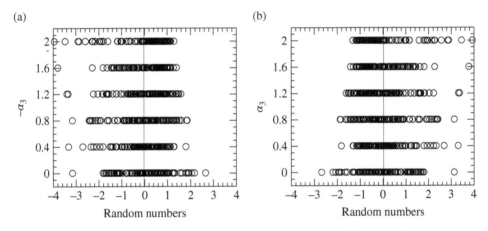

Figure A.12 Random numbers with negative and positive skewness. (a) Negative skewness; (b) Positive skewness.

$$E\left[(X-\mu_X)^4\right] = \sum_{all\,x_i}(x-\mu_X)^4 p_X(x_i) \quad \text{for discrete } X$$

$$E\left[(X-\mu_X)^4\right] = \int_{-\infty}^{\infty}(x-\mu_X)^4 f_X(x)dx \quad \text{for continuous } X$$

Observe that $E[(X-\mu_X)^4]$ is always positive. A convenient non-dimensional measure of sharpness is the kurtosis, which is defined as the fourth-order central moment of a random variable divided by its fourth power of standard deviation,

$$\alpha_4 = \frac{E\left[(X-\mu_X)^4\right]}{\sigma_X^4} \tag{A.22}$$

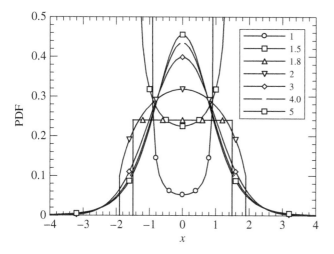

Figure A.13 The PDFs with different kurtosis.

Figure A.13 illustrates the PDFs with different kurtosis in which all the distributions have the same mean value of 0, the same standard deviation of 1, and the same skewness of 0. One can see that the larger the kurtosis, the sharper the PDF. When $\alpha_4 = 1.8$, the distribution is a uniform distribution as shown in Section A.6.2. The minimum value of kurtosis is 1 since the kurtosis should satisfy the following equation (Stuart and Ord 1987).

$$\alpha_4 \geq \alpha_3^2 + 1 \tag{A.23}$$

From Figure A.13, one can see that when $\alpha_4 < 1.8$, the distribution is U type.

In order to understand the skewness and kurtosis more comprehensively, Figure A.14a–d illustrates the PDFs with different skewness and kurtosis in which all the distributions have the same mean value of 0, and the same standard deviation of 1. In Figure A.14a,b, the PDFs are with the same skewness but different kurtosis, and in Figure A.14c,d the PDFs are with the same kurtosis but different skewness.

A.4 Moments and Cumulants

A.4.1 Moments

The expected value of X^r for any real number r is termed the *rth moment about zero*. The words 'about zero' are commonly omitted, unless they are needed to ensure clarity. Here, the rth moment is denoted by the symbols μ_{rX}. When $r = 1$, the first moment is the mean value, which is commonly denoted as μ_X as described in Section A.3.1.

The *rth moment about a*, where a is a constant, is the expected value of $(X - a)^r$. The special case, $a = E(X) = \mu_X$ is by far the most frequently encountered. The corresponding moments are called *central moments*, and here we denote the rth central moment by M_r.

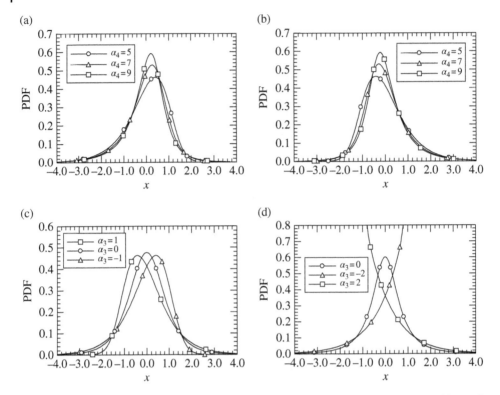

Figure A.14 The PDFs with some different skewness and kurtosis. (a) $\alpha_3 = -1$; (b) $\alpha_3 = 1$; (c) $\alpha_4 = 5$; (d) $\alpha_4 = 9$.

Thus

$$\mu_r = E(X^r) \tag{A.24a}$$

$$M_r = E[(X - \mu_X)^r] \tag{A.24b}$$

Observe that the first central moment M_1 must be zero. The second central moment M_2 is the variance of X which is commonly written as $Var(X)$ as described in Section A.3.2. The third and fourth central moments, M_3 and M_4, have been used to define the skewness and kurtosis, respectively, as described in Sections A.3.3 and A.3.4.

The *rth moment ratio*, sometimes called the *rth dimensionless central moment*, is defined as

$$\alpha_{rX} = \frac{M_{rX}}{\sigma_X^r} = \frac{E[(X - \mu_X)^r]}{\sigma_X^r} \quad \text{for} \quad r > 2 \tag{A.25}$$

Since moment ratios are indices of the shape of a distribution, they are also called *shape factors*. Observe that the third moment ratio α_{3X} is skewness as described in Section A.3.3 and the fourth moment ratio α_{4X} is the kurtosis as described in Section A.3.4.

Note that the moment ratios have the same value for any linear function $(a + bX)$ with $b > 0$. If $b < 0$, the absolute values are not altered but ratios of odd order have their signs reversed. Partially for this reason, the description of a random variable X is often simplified by introducing the *standardised random variable*

$$X_s = \frac{X - \mu_X}{\sigma_X} \qquad (\text{A.26})$$

Its mean is equal to 0 and the standard deviation is 1. Its shape factors are the same as those of X.

It is often convenient to calculate the central moments M_r from the moments about zero, and though less often, vice-versa. Formulas by which this can be done are

$$M_r = E[(X - \mu)^r] = \sum_{j=0}^{r} (-1)^j \binom{r}{j} \mu^j E(X^{r-j}) \qquad (\text{A.27})$$

$$\mu_r = E[X^r] = \sum_{j=0}^{r} \binom{r}{j} \mu^j E\left[(X - \mu)^{r-j}\right] \qquad (\text{A.28})$$

where

$$\binom{r}{j} = \frac{r!}{(r-j)! j!} \qquad (\text{A.29})$$

is the *Binomial coefficient*, it is also known as the number of combinations of j objects chosen from a set of r.

In particular,

$$
\begin{aligned}
M_2 &= \sigma^2 = E(X^2) - \mu^2 \\
M_3 &= \alpha_3 \sigma^3 = E(X^3) - 3E(X^2)\mu + 2\mu^3 \\
M_4 &= \alpha_4 \sigma^4 = E(X^4) - 4E(X^3)\mu + 6E(X^2)\mu^2 - 3\mu^4
\end{aligned} \qquad (\text{A.30})
$$

or

$$
\begin{aligned}
M_2 &= \sigma^2 = \mu_2 - \mu^2 \\
M_3 &= \alpha_3 \sigma^3 = \mu_3 - 3\mu_2\mu + 2\mu^3 \\
M_4 &= \alpha_4 \sigma^4 = \mu_4 - 4\mu_3\mu + 6\mu_2\mu^2 - 3\mu^4
\end{aligned}
$$

For the inverse calculation

$$
\begin{aligned}
\mu_2 &= E[X^2] = M_2 + \mu^2 \\
\mu_3 &= E[X^3] = M_3 + 3M_2\mu + \mu^3 \\
\mu_4 &= E[X^4] = M_4 + 4M_3\mu + 6M_2\mu^2 + \mu^4
\end{aligned}
$$

or

$$
\begin{aligned}
\mu_2 &= E[X^2] = \sigma^2 + \mu^2 \\
\mu_3 &= E[X^3] = \alpha_3\sigma^3 + 3\sigma^2\mu + \mu^3 \\
\mu_4 &= E[X^4] = \alpha_4\sigma^4 + 4\alpha_3\sigma^3\mu + 6\sigma^2\mu^2 + \mu^4
\end{aligned} \qquad (\text{A.31})
$$

A.4.2 Moment and Cumulant Generating Functions

The moment-generating function of a random variable X, denoted as $\phi_X(t)$, is defined as the expected value of e^{tX}; that is

$$\phi_X(t) = E(e^{tX}) \tag{A.32}$$

where t is an auxiliary (deterministic) variable.

Since

$$e^{tX} = \sum_{r=0}^{\infty} \frac{1}{r!} t^r x^r$$

the moment generating function can be expressed as

$$\phi_X(t) = E(e^{tX}) = E\left(\sum_{r=0}^{\infty} \frac{1}{r!} t^r x^r\right) = \sum_{r=0}^{\infty} \frac{1}{r!} t^r E(x^r) = \sum_{r=0}^{\infty} \frac{1}{r!} t^r \mu_r$$

Therefore, the rth moment of X can be obtained as the coefficient of $t^r/r!$ in the Taylor series expansion of its moment generating function $\phi_X(t)$.

Note that

$$\frac{d\phi_X(t)}{dt} = E(Xe^{tX})$$

Therefore

$$\frac{d\phi_X(t)}{dt}\Big|_{t=0} = E(X) = \mu \tag{A.33}$$

and in general,

$$\frac{d^r \phi_X(t)}{dt^r}\Big|_{t=0} = E(X^r) = \mu_r \tag{A.34}$$

This is to say, the rth moment of a random variable is given by the rth derivative of its moment generating function at $t = 0$.

Since

$$E[t(X - \mu)] = e^{-t\mu} E(e^{tX})$$

It follows that the coefficient in Taylor series of

$$e^{-t\mu} \phi_X(t) = e^{-t\mu} E(e^{tX}) \tag{A.35}$$

is the rth central moment of X. The function of Eq. (A.35) is therefore called the *central moment generating function* of X.

$$M_r = E[(X - \mu)^r] = \frac{d^r [e^{-t\mu} \phi(t)]}{dt^r}\Big|_{t=0} \tag{A.36}$$

The logarithm of the moment generating function of X is the *cumulant generating function* of X.

$$\psi(t) = \ln[\phi_X(t)] \tag{A.37}$$

If the moment generating function exists, so does the cumulant generating function. The coefficient of $t^r/r!$ in the Taylor series expansion of the cumulant generating function $\psi(t)$ at $t = 0$ is the *rth cumulant* of X, and is denoted by the symbols $k_r(X)$ or, when no confusion is likely to arise, simply, k_r. Evidently k_r is a function of the moments of X.

$$k_r = \frac{d^r[\psi(t)]}{dt^r}\bigg|_{t=0} \tag{A.38}$$

Since, for any constant a,

$$\phi_{X+a}(t) = E\left[e^{t(X+a)}\right] = e^{ta}\phi_X(t)$$

It follows that

$$\psi_{X+a}(t) = \ln[\phi_{X+a}(t)] = ta + \psi_X(t)$$

Hence for $r \geq 2$, the coefficients of $t^r/r!$ in the Taylor series expansion of left and right sides of the equation above are the same; that is

$$k_r(X+a) = k_r(X) \quad \text{for } r \geq 2 \tag{A.39a}$$

While

$$k_1(X+a) = k_1(X) + a \tag{A.39b}$$

Thus, the cumulants (for $r \geq 2$) are not affected by adding a constant to X, but k_1 is changed by the addition of the same constant.

Sometimes we may need to have the relationships between cumulants and moments. For central moments, the formulas are expressed as

$$k_1 = 0, \quad k_2 = M_2, \quad k_3 = M_3, \quad k_4 = M_4 - 3M_2^2 \tag{A.40}$$

and

$$M_1 = 0, \quad M_2 = k_2, \quad M_3 = k_3, \quad M_4 = k_4 + 3k_2^2 \tag{A.41}$$

For some distributions, the moment generating function $\phi_X(t) = E(e^{tX})$ may be infinite. However, the function (of t) $E(e^{itX})$, where $i = \sqrt{-1}$, always exists and is finite. This is called the *characteristic function* of the distribution. It has properties similar to the moment generating function.

$$\phi_X(it) = E\left(e^{itX}\right) \tag{A.42}$$

In terms of the characteristic function, the rth moment of X is given by

$$\mu_r = E(X^r) = \frac{1}{i^r}\frac{d^r\phi_X(it)}{dt^r}\bigg|_{t=0} \tag{A.43}$$

whereas the special relation for the variance is

$$Var(X) = E(X^2) = -\frac{d^2\phi_X(it)}{dt^2}\bigg|_{t=0} \tag{A.44}$$

A.5 Normal and Lognormal Distributions

A.5.1 The Normal Distribution

Perhaps the best-known and most widely used probability distribution is the normal distribution, also known as the Gaussian distribution. The normal distribution has a PDF given by

$$f_X(x) = \frac{1}{\sigma\sqrt{2\pi}} \exp\left[-\frac{1}{2}\left(\frac{x-\mu}{\sigma}\right)^2\right] \quad, -\infty < x < \infty \tag{A.45}$$

where μ and σ are the parameters of the distribution, which are also the mean and standard deviation, respectively, of the variate. A short notation for this distribution is $N(\mu, \sigma)$, which we shall adopt. The random variable with normal distribution is called a normal random variable.

The standard normal distribution. A Gaussian distribution with parameters $\mu = 0$ and $\sigma = 1.0$ is known as the *standard normal distribution* and is denoted appropriately as $N(0, 1)$. The random variable with the standard normal distribution is called a *standard normal variable*. Because of its wide usage, special notations $\phi(u)$ and $\Phi(u)$ are commonly used to designate the PDF and CDF, respectively, of a standard normal variable U; that is, $\phi(u) = f_U(u)$ and $\Phi(u) = F_U(u)$, where U has $N(0, 1)$ distribution. The PDF of U is expressed as

$$\phi(u) = \frac{1}{\sqrt{2\pi}} \exp\left(-\frac{1}{2}u^2\right) \quad, -\infty < x < \infty \tag{A.46}$$

$\phi(u)$ is symmetrical about $u = 0$.

The rth moment of U is expressed as

$$\mu_r = E(U^r) = \frac{1}{\sqrt{2\pi}} \int_{-\infty}^{\infty} u^r e^{-\frac{1}{2}u^2} du \tag{A.47}$$

If r is odd,

$$\mu_r = 0 \tag{A.48}$$

If r is even, μ_r is obtained as

$$\mu_r = (r-1)!! = (r-1)(r-3)...3\cdot 1 \tag{A.49}$$

Particularly, the first four moments of U are

$$\mu = 0$$
$$\sigma = 1$$
$$\alpha_3 = 0$$
$$\alpha_4 = 3$$

Obviously, a standard normal variable can be regarded as a standardised variable of a normal variable

$$U = \frac{X - \mu}{\sigma} \tag{A.50}$$

where U has $N(0, 1)$ distribution and X has $N(\mu, \sigma)$ distribution.

Comparing Eq. (A.46) with Eq. (A.45), one can easily understand that

$$f_X(x) = \frac{1}{\sigma} \phi\left(\frac{x - \mu}{\sigma}\right) \tag{A.51}$$

Referring to the definition of the CDF of a random variable, the probability of $p = P(U \le a)$ is denoted as

$$p = P(U \le a) = \Phi(a) = \int_{-\infty}^{a} \phi(u)du \tag{A.52}$$

which is the shaded area in Figure A.15.

Conversely, the value of a standard normal variable at a cumulative probability p would be denoted as

$$u_p = \Phi^{-1}(p) \tag{A.53}$$

This notation is used through this book.

The CDF of $N(0, 1)$ that is, $\Phi(u)$, is tabulated widely as tables of normal probabilities in many textbooks. Nowadays, $\Phi(u)$ has been included as a common function in many disposal software such as MS Excel.

Suppose a normal variable X with distribution $N(\mu, \sigma)$, the CDF of X is expressed as

$$F_X(x) = P(X \le x) = \int_{-\infty}^{x} f_X(t)dt$$

Substituting Eqs. (A.50) and (A.51) into the equation above, it follows that,

$$F_X(x) = \int_{-\infty}^{x} \frac{1}{\sigma} \phi\left(\frac{t - \mu}{\sigma}\right) dt = \int_{-\infty}^{\frac{x - \mu}{\sigma}} \phi(u)du$$

That is,

$$F_X(x) = \Phi\left(\frac{x - \mu}{\sigma}\right) \tag{A.54}$$

Using Eq. (A.54), the probability that X falls in an interval (a, b) is

$$P(a < X \le b) = F_X(b) - F_X(a) = \Phi\left(\frac{b - \mu}{\sigma}\right) - \Phi\left(\frac{a - \mu}{\sigma}\right) \tag{A.55}$$

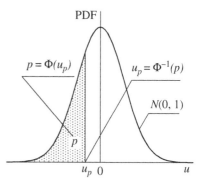

Figure A.15 The standard normal density function.

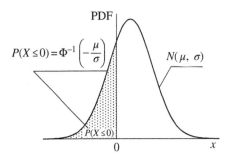

Figure A.16 Probability for $X \le 0$.

Especially, the probability for $X \le 0$ is given by

$$P(X \le 0) = \Phi\left(-\frac{\mu}{\sigma}\right) \tag{A.56}$$

as illustrated in Figure A.16.

Some typical areas covered by the PDF of the standard normal random variable is shown in Figure A.17a–d, where one can see that the area covered within ±1, ±2, ±3, and ±4 are 68.26%, 95.45%, 99.73%, and 99.99%, respectively. The same probabilities are also valid for any general normal distributions, i.e. $N(\mu, \sigma)$, with the same ± number of σ's about the mean μ.

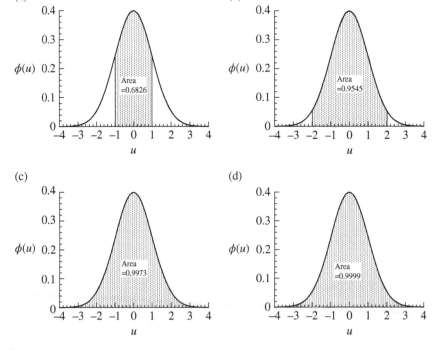

Figure A.17 Areas under the PDF of standard normal variable. (a) $-1 < u < 1$; (b) $-2 < u < 2$; (c) $-3 < u < 3$; (d) $-4 < u < 4$.

A.5.2 The Logarithmic Normal Distribution

A random variable X has a logarithmic normal (or simply lognormal) probability distribution if $\ln X$ (the natural logarithm of X) is normal. The PDF of X is

$$f_X(x) = \frac{1}{\sqrt{2\pi}\zeta x}\exp\left[-\frac{1}{2}\left(\frac{\ln x - \lambda}{\zeta}\right)^2\right], \qquad 0 < x < \infty \tag{A.57}$$

where $\lambda = E(\ln X) = \mu_{\ln X}$ and $\zeta = \sqrt{Var(\ln X)} = \sigma_{\ln X}$ are, respectively, the mean and standard deviation of $\ln X$, and the parameters of the distribution. The PDF of a lognormal variable is illustrated in Figure A.18 for various values of ζ. Although there is no definition for $\ln X$ at $x = 0$, one may easily derived that the limit value of $f_X(x)$ when $x \to 0$ is 0, that is

$$\lim_{x \to 0} f_X(x) = \lim_{x \to +\infty} f_X(x) = 0$$

From the definition of the mode, one can obtain the mode of lognormal variable as

$$mode(X) = Exp(\lambda - \zeta)$$

when ζ is large, the mode will be close to zero such as the case of $\zeta = 2$ and $\zeta = 3$ in Figure A.18.

The relationship between a lognormal variable X and the standard normal variable U is expressed as

$$U = \frac{\ln X - \lambda}{\zeta} \tag{A.58}$$

Eq. (A.58) follows that

$$f_X(x) = \frac{1}{\zeta x}\phi\left(\frac{\ln x - \lambda}{\zeta}\right) \tag{A.59}$$

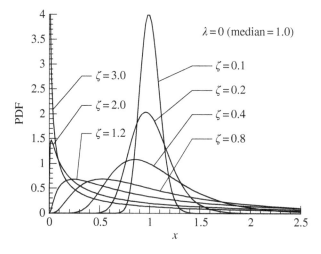

Figure A.18 Lognormal density functions.

and

$$F_X(x) = \Phi\left(\frac{\ln x - \lambda}{\zeta}\right) \tag{A.60}$$

On the basis of Eq. (A.60), the probability that X will assume values in an interval (a, b) is

$$P(a < X \le b) = F_X(b) - F_X(a) = \Phi\left(\frac{\ln b - \lambda}{\zeta}\right) - \Phi\left(\frac{\ln a - \lambda}{\zeta}\right) \tag{A.61}$$

In view of this convenient facility for calculating probability of lognormal variables and also because the values of the random variable are always positive, the lognormal distribution may be useful in those applications where the values of the variable are known to be strictly positive; for example, the strength and the fatigue life of material, the intensity of rainfall, the time for project completion, and the volume of air traffic.

We observe from Eq. (A.61) that the probability is a function of the parameters λ and ζ. These parameters are related to the mean μ and standard deviation σ of the variable as follows.

$$\mu = E(X) = \exp\left[\frac{1}{2}\zeta^2 + \lambda\right] \tag{A.62}$$

$$\sigma = \mu\sqrt{\exp\left(\zeta^2\right) - 1} \tag{A.63}$$

Equations (A.62) and (A.63) are used to obtain the mean and deviation of a lognormal variable X with parameters of λ and ζ.

The shape factors, coefficient of skewness α_3 and coefficient of kurtosis α_4 are obtained as

$$\alpha_3 = (\omega - 1)^{1/2}(\omega + 2) \tag{A.64}$$

$$\alpha_4 = \omega^4 + 2\omega^3 + 3\omega^2 - 3 \tag{A.65}$$

where $\omega = \exp(\zeta^2)$.

Since $\omega = V^2 + 1$, Eqs. (A.64) and (A.65) are also given as

$$\alpha_3 = V(V^2 + 3) \tag{A.66a}$$

$$\alpha_4 = V^8 + 6V^6 + 15V^4 + 16V^2 + 3 \tag{A.66b}$$

The shape factors α_3 and α_4 changed with V are listed in Table A.1.

From Eqs. (A.66a) and (A.66b), one may see that α_3 and α_4 can be simply approximated by the following equations for small V, e.g., $V < 0.3$.

$$\alpha_3 = 3V \tag{A.67a}$$

$$\alpha_4 = 3 + 16V^2 \tag{A.67b}$$

From Eqs. (A.63), one obtains

$$\zeta^2 = \ln\left(1 + \frac{\sigma^2}{\mu^2}\right) \tag{A.68}$$

Table A.1 Parameters and moments of lognormal distribution.

V	ζ	$\lambda - \ln\mu$	α_3	α_4
0.1	0.066	−0.002	0.301	3.162
0.2	0.131	−0.009	0.608	3.664
0.3	0.193	−0.019	0.927	4.566
0.4	0.254	−0.032	1.264	5.969
0.5	0.311	−0.048	1.625	8.035
0.6	0.365	−0.067	2.016	11
0.7	0.416	−0.087	2.443	15.21
0.8	0.464	−0.107	2.912	21.12
0.9	0.508	−0.129	3.429	29.42
1.0	0.549	−0.151	4	41

and from Eq. (A.62), we have

$$\lambda = \ln\mu - \frac{1}{2}\zeta^2 \tag{A.69a}$$

Thus we also have

$$\lambda = \ln \frac{\mu}{\sqrt{1 + V^2}} \tag{A.69b}$$

Equations (A.68) and (A.69b) are used to obtain the parameters of λ and ζ of a lognormal variable X with known mean μ and standard deviation σ.

If σ/μ is not large, say $\sigma/\mu < 0.3$, then

$$\ln\left(1 + \frac{\sigma^2}{\mu^2}\right) \approx \frac{\sigma^2}{\mu^2}$$

In such cases, therefore

$$\zeta \approx \frac{\sigma}{\mu} = V \tag{A.70}$$

The median is often used to designate the central value of a lognormal variable. If x_m is the median, then by the definition of the median,

$$P(X \le x_m) = 0.5$$

Thus,

$$\frac{\ln x_m - \lambda}{\zeta} = \Phi^{-1}(0.5) = 0$$

Therefore, in terms of the median, the parameter λ is

$$\lambda = \ln x_m \tag{A.71}$$

Conversely,

$$x_m = e^{\lambda} = \frac{\mu}{\sqrt{1 + V^2}} \tag{A.72}$$

This means that the median of a lognormal variable is always less than its mean value; that is $x_m < \mu$.

A.6 Commonly Used Distributions

A.6.1 Introduction

Any function possessing all the properties cited in Section A.2 can be used to describe the probability distribution of a random variable. However, there are a number of discrete and continuous functions that are particularly useful because of one or more of the following reasons: (i) The function is the result of an underlying physical process and is derived on the basis of certain physically reasonable assumptions; (ii) the function is the result of some limiting process, and (iii) it is widely known and the necessary statistical information including probability tables is available widely. The most commonly used probability distributions in structural reliability theory and the normal and lognormal distribution have been discussed in detail in Section A.5. Several other commonly used probability distribution functions are presented and their special properties will be described in this section. Comprehensive accounts of all the major theoretical distributions can be found in Johnson and Kotz (1969, 1970a,b, 1972) and Patil et al. (1985a,b,c).

A.6.2 Rectangular Distribution

Rectangular distribution is also called the uniform distribution. If the random variable is uniformly distributed between a and b, then the PDF and CDF of X are given as

$$f_X(x) = \frac{1}{b-a}, \quad a \leq x \leq b \tag{A.73a}$$

$$F_X(x) = \frac{x-a}{b-a}, \quad a \leq x \leq b \tag{A.73b}$$

Graphically, the PDF and CDF of the rectangular distribution would appear as shown in Figure A.19.

In particular, when $a = 0, b = 1$, the standard form of Eqs. (A.73a) and (A.73b) are given as

$$f_X(x) = 1, \quad 0 \leq x \leq 1 \tag{A.74a}$$

$$F_X(x) = x, \quad 0 \leq x \leq 1 \tag{A.74b}$$

The first four moments of X are given as

$$\mu_X = \frac{1}{2}(a + b) \tag{A.75a}$$

$$\sigma_X = \frac{1}{12}(a - b)^2 \tag{A.75b}$$

$$\alpha_{3X} = 0 \tag{A.75c}$$

(a)

$f_X(x)$

PDF

a b x

(b)

$F_X(x)$

1.0

CDF

a b x

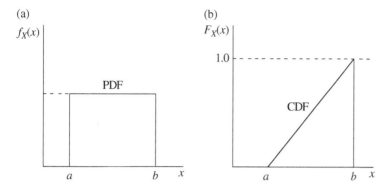

Figure A.19   PDF and CDF of rectangular distribution.

$$\alpha_{4X} = \frac{9}{5} = 1.8 \tag{A.75d}$$

A.6.3 Bernoulli Sequences and the Binomial Distribution

Bernoulli sequence is used to model the potential occurrence or recurrence of an event in a sequence of repeated 'trials.' If the probability of occurrence of event in each trial is p (and probability of non-occurrence is $1 - p$), then the probability of exactly x occurrences among n trials in a Bernoulli sequence is given by the binomial PMF as follows

$$P(X = x) = \binom{n}{x} p^x (1-p)^{n-x}, \qquad x = 1, 2, \cdots, n \tag{A.76}$$

where n and p are parameters and

$$\binom{n}{x} = \frac{n!}{(n-x)!x!}$$

is the binomial coefficient.

The first four moments are

$$\mu = np \tag{A.77a}$$

$$\sigma^2 = np(1-p) \tag{A.77b}$$

$$\alpha_3 = \frac{1-2p}{\sqrt{np(1-p)}} \tag{A.77c}$$

$$\alpha_4 = \frac{1-6p(1-p)}{np(1-p)} + 3 \tag{A.77d}$$

A.6.4 The Geometric Distribution

In a Bernoulli sequence, the number of trials until a specified event occurs for the first time is governed by the geometric distribution. We observe that if the first occurrence of the event is realised on the tth trial, then there must be no occurrence of this event in any of the prior $(t - 1)$ trials. Therefore, if T is the appropriate random variable,

$$P(T = t) = pq^{t-1} \quad t = 1, 2, \ldots \tag{A.78}$$

which is known as the geometric distribution.

The return period: In a time (or space) problem that can be modelled as a Bernoulli sequence, the number of time (or space) intervals until the first occurrence of an event is called the *first occurrence time*. The mean recurrence time is popularly known in engineering as the average *return period*.

The first four moments are

$$\mu = E(T) = \frac{1}{p} \tag{A.79a}$$

$$\sigma = \frac{\sqrt{q}}{p} \tag{A.79b}$$

$$\alpha_3 = \frac{1+q}{\sqrt{q}} \tag{A.79c}$$

$$\alpha_4 = \frac{1 + 7q + q^2}{q} \tag{A.79d}$$

A.6.5 The Poisson Process and Poisson Distribution

Many physical problems of interest to engineers involve the possible occurrences of events at any point in time and/or space. The occurrences of the event may be modelled with a *Poisson sequence* or *Poisson process*. The PMF of Poisson process is given as

$$P(X = x) = \frac{\lambda^x}{x!} e^{-\lambda} \quad x = 0, 1, 2, \ldots; \lambda > 0 \tag{A.80}$$

where $\lambda = vt$, v is the mean occurrence rate, that is, the average number of occurrence of the event per unit time (or space) interval.

Graphically, the PMF of the Poisson distribution would appear as shown in Figure A.20. The first four moments are

$$\mu = \lambda \tag{A.81a}$$

$$\sigma = \sqrt{\lambda} \tag{A.81b}$$

$$\alpha_3 = \frac{1}{\sqrt{\lambda}} \tag{A.81c}$$

$$\alpha_4 = \frac{1}{\lambda} + 3 \tag{A.81d}$$

A.6.6 The Exponential Distribution

The random variable X has an exponential distribution if it has a PDF of form:

$$f_X(x) = \lambda e^{-\lambda x}, \quad x \geq 0 \tag{A.82a}$$

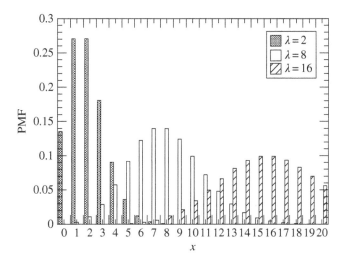

Figure A.20 The PMF for Poisson distribution.

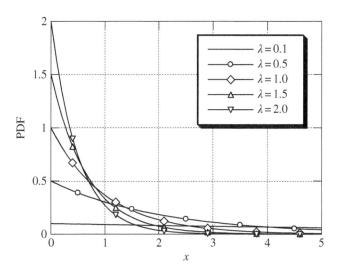

Figure A.21 PDF of the exponential distribution.

Graphically, the PDF of the exponential distribution would appear as shown in Figure A.21.

The corresponding CDF of X is

$$F_X(x) = 1 - e^{-\lambda x} \quad x \geq 0 \tag{A.82b}$$

The first four moments are

$$\mu = \frac{1}{\lambda}, \quad \sigma = \frac{1}{\lambda}, \quad \alpha_3 = 2, \quad \alpha_4 = 9 \tag{A.83}$$

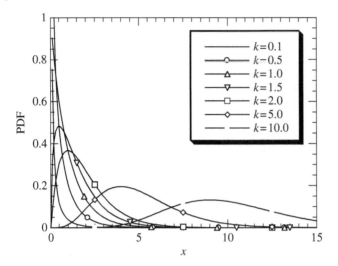

Figure A.22 PDF of the standard Gamma distribution.

A.6.7 The Gamma Distribution

The random variable X has a Gamma distribution if it has a PDF of form:

$$f_X(x) = \frac{v(vx)^{k-1}e^{-vx}}{\Gamma(k)} \quad x \geq 0 \tag{A.84a}$$

where v and k are parameters.

When $v = 1$, the distribution is standard Gamma distribution

$$f_X(x) = \frac{x^{k-1}e^{-x}}{\Gamma(k)} \quad x \geq 0 \tag{A.84b}$$

Graphically, the PDF of the standard Gamma would appear as shown in Figure A.22. The first four moments are

$$\mu = \frac{k}{v} \quad \sigma = \frac{\sqrt{k}}{v}, \quad \alpha_3 = \frac{2}{\sqrt{k}}, \quad \alpha_4 = 3 + \frac{6}{k} \tag{A.85a}$$

Since the COV of X is expressed as $V = 1/\sqrt{k}$, the skewness and kurtosis can be expressed as

$$\alpha_3 = 2V, \quad \alpha_4 = 3 + 6V^2 \tag{A.85b}$$

A.7 Extreme Value Distributions

A.7.1 Introduction

The extremes of natural phenomena are often of special interest and importance in engineering. The statistics of such extremes are of special significance to many problems involving natural hazards, such as the largest earthquake intensity expected at a building site in

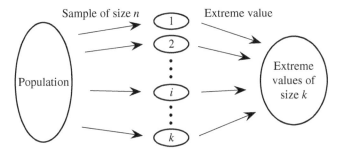

Figure A.23   Concept of extreme value distribution.

the working life of the building. Consider a random variable or population with known probability distribution. A sample of size n from this population will have a largest and smallest value, and these extreme values will also have their respective distributions which are related to the distribution of the initial random variable or population. That is, when we speak of extreme values, we are considering the largest and smallest values from a sample of size n within a known population as shown in Figure A.23.

In order to understand extreme value distribution comprehensively, consider a standard normal random variable, and generate a population of random numbers with size 10 000 000, the histogram is shown in Figure A.24a. From the population, taking 100 000 samples of size 100, the smallest values of each sample of size 100 compose a new population of smallest values of size 100 000. Then the histogram of the smallest values is shown in Figure A.24b, and the first four statistical moments are given as

$$\mu_{min} = -2.509, \quad \sigma_{min} = 0.429, \quad \alpha_{3\,min} = -0.645, \quad \alpha_{4\,min} = 3.733$$

Similarly, the largest values of each sample of size 100 compose a new population of largest values of size 100 000. The histogram of the largest values is shown in Figure A.24c, and the first four statistical moments are given as

$$\mu_{max} = 2.509, \quad \sigma_{max} = 0.428, \quad \alpha_{3\,max} = 0.677, \quad \alpha_{4\,max} = 3.838$$

One can see that the absolute of the mean values, standard deviations, skewness, and kurtosis of the smallest and the largest value distributions are the same. The mean values and skewness of the smallest value distribution are negative, while those of the largest value

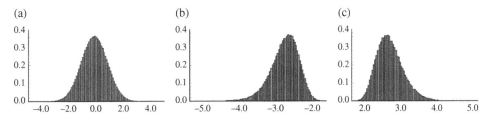

Figure A.24   Extreme value distribution. (a) Normal population. (b) Smallest of each 100 numbers. (c) Largest of each 100 numbers.

(a)

(b)

(c)

(d)

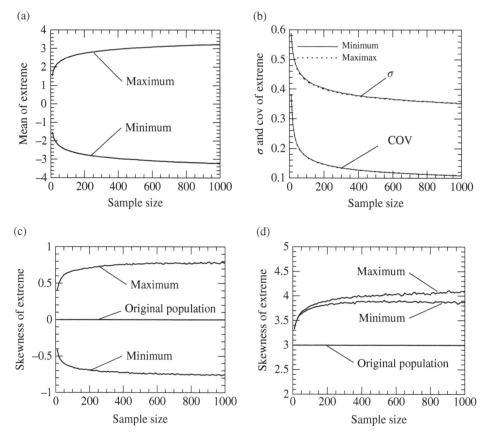

Figure A.25 Variations of the first four moments respect to the sample size. (a) Mean. (b) Standard deviation. (c) Skewness. (d) Kurtosis.

distribution are positive. That is to say, the two distributions have the same shape but located axisymmetrically.

Using the same population, the variations of the first four statistical moments with respect to the sample size are depicted in Figure A.25a–d. From Figure A.25a–d, one can see that the absolute mean values of both the largest and smallest increase with the increase of the sample size, while the standard deviations and the coefficients of variation of both the distribution decrease with the increase of the sample size. Both the skewness and kurtosis of the largest and the smallest value distributions increase with the increase of the sample size. The differences in kurtosis between the largest and the smallest value distribution are due to the relative size of the population size and the sample size. With the increase of the relative population size, the difference will be reduced.

A.7.2 The Asymptotic Distributions

It has been shown that the distribution of an extreme value converges asymptotically in distribution as the sample size n increases. According to Gumbel (1958), there are three types of asymptotic distribution depending on the tail behaviour of the initial PDFs as follows.

Type I: The double exponential form.
Type II: The single exponential form.
Type III: The exponential form with an upper or lower bound.

In engineering, these asymptotic distributions are generally simply called *extreme value distributions*. Extreme value distributions are generally considered to comprise the three following families:

1) Type I

$$\text{Largest}: F_X(x) = \exp\left[-\exp\left(\frac{\zeta-x}{\theta}\right)\right], \quad x > \zeta; \text{ otherwise } 0 \tag{A.86a}$$

$$\text{Smallest}: F_X(x) = 1-\exp\left[-\exp\left(\frac{x-\zeta}{\theta}\right)\right], \quad x < \zeta; \text{ otherwise } 1 \tag{A.86b}$$

2) Type II

$$\text{Largest}: F_X(x) = \exp\left[-\left(\frac{x-\zeta}{\theta}\right)^{-k}\right], \quad x > \zeta; \text{ otherwise } 0 \tag{A.87a}$$

$$\text{Smallest}: F_X(x) = 1-\exp\left[-\left(\frac{\zeta-x}{\theta}\right)^{-k}\right], \quad x < \zeta; \text{ otherwise } 1 \tag{A.87b}$$

3) Type III

$$\text{Largest}: F_X(x) = \exp\left[-\left(\frac{\zeta-x}{\theta}\right)^{k}\right], \quad x < \zeta; \text{ otherwise } 0 \tag{A.88a}$$

$$\text{Smallest}: F_X(x) = 1-\exp\left[-\left(\frac{x-\zeta}{\theta}\right)^{k}\right], \quad x > \zeta; \text{ otherwise } 1 \tag{A.88b}$$

where ζ, $\theta(>0)$ and $k(>0)$ are parameters.

Recently, Kanda (1981, 1994) added Type IV, the exponential form with an upper and lower bound, which is shown as follows.

4) Type IV

$$F_X(x) = \exp\left[-\left(\frac{\zeta-x}{\theta(x-\varepsilon)}\right)^{k}\right], \quad \varepsilon < x < \zeta \tag{A.89}$$

where ζ is the upper limit and ε is the lower limit.

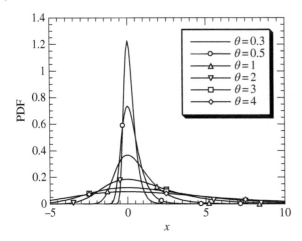

Figure A.26 PDFs of the Gumbel distribution (with $\zeta = 0$).

A.7.3 The Gumbel Distribution

Of the extreme value distributions above, Type I largest is by far the one most commonly referred to in discussions of extreme value distributions. Indeed, some authors call Type I 'the extreme value distribution'. Types II and III can be transformed to Type I by the following simple transformations respectively

$$Z = \ln (X - \zeta); \quad Z = -\ln (\zeta - X)$$

Type I largest distribution is also named *Gumbel distribution* in engineering. From Eq. (A.86a), the PDF of Gumbel distribution is obtained as

$$f_X(x) = \frac{1}{\theta} \exp \left(\frac{\zeta - x}{\theta} \right) \exp \left[- \exp \left(\frac{\zeta - x}{\theta} \right) \right] \qquad \text{(A.90a)}$$

The standard form is

$$f_X(x) = \exp \left[-x - \exp \left(-x \right) \right] \qquad \text{(A.90b)}$$

Graphically, the PDF of the Gumbel would appear as shown in Figure A.26.
The first four moments are

$$\mu = \zeta + 0.5772\theta, \ \sigma = 1.2825\,\theta, \ \alpha_3 = 1.1395, \ \alpha_4 = 5.4 \qquad \text{(A.91)}$$

Note that ζ and θ are purely location and scale parameters, respectively.

Related Distribution: If X is a Gumbel distributed variable with parameter ζ and θ, then $Y = \exp[-(X - \zeta)/\theta]$ is a exponentially distributed variable with PDF of $f_Y(y) = \exp(-y)$.

Example A.1 The Derivation of Eq. (A.91)

The moment generating function of a random variable X of Gumbel distribution is

$$\phi(t) = e^{t\zeta}\Gamma(1 - \theta t) \qquad (\theta \, |t| < 1)$$

and the cumulant generating function is

$$\Psi(t) = \ln[\phi(t)] = \zeta t + \ln[\Gamma(1 - \theta t)]$$

The cumulants of X are

$$K_1 = \left.\frac{d[\Psi(t)]}{dt}\right|_{t=0} = \zeta + \theta\psi(1) = \zeta + \gamma\theta = \zeta + 0.5772\theta$$

$$K_r = \left.\frac{d^r[\Psi(t)]}{dt^r}\right|_{t=0} = (-\theta)^r\psi^{(r-1)}(1) = \theta^r(r-1)!\sum_{i=0}^{\infty}\frac{1}{(i+1)^r} \quad (r \geq 2)$$

where $\psi(t) = \dfrac{d\{\ln[\Gamma(t)]\}}{dt} = \dfrac{\Gamma'(t)}{\Gamma(t)}$ is a polygamma function and

$$K_3 = 2.40410\theta^2$$

$$K_4 = \frac{1}{15}\pi^4\theta^4 = 6.49390\theta^2$$

$$\psi^{(k)}(z) = \sum_{i=0}^{\infty}\frac{(-1)^{k+1}k!}{(z+i)^{k+1}} \quad (r \geq 2)$$

Therefore,

$$K_2 = \frac{1}{6}\pi^2\theta^2 = 1.64490\theta^2, \quad K_3 = 2.40410\theta^2, \quad K_4 = \frac{1}{15}\pi^4\theta^4 = 6.49390\theta^2$$

According to Eq. (A.41), the first moments expressed in Eq. (A.91) can be obtained.

A.7.4 The Frechet Distribution

The *Type-II Largest*, which is also named *Frechet distribution*, is also commonly used in engineering. When $\zeta = 0$, the PDF of Frechet distribution has the form:

$$f_X(x) = \frac{k}{\theta}\left(\frac{x}{\theta}\right)^{-(k+1)}\exp\left[-\left(\frac{x}{\theta}\right)^{-k}\right], \qquad x > 0 \tag{A.92}$$

Graphically, the PDF of the Frechet distribution would appear as shown in Figure A.27. The rth moment of a random variable X of Frechet distribution is

$$\mu_r = E(X^r) = \theta^r\Gamma\left(1 - \frac{r}{k}\right)$$

For μ_r to exist, $k > r$ is necessary.

Using the equation above, the mean, the second, third, and fourth central moments of X are expressed as

$$\mu = \theta\Gamma\left(1 - \frac{1}{k}\right) \tag{A.93a}$$

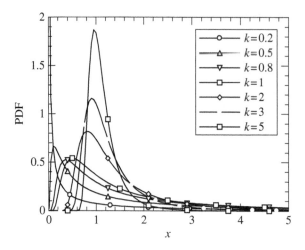

Figure A.27 PDF of the Frechet distribution ($\theta = 1$).

$$\sigma^2 = \theta^2\left[\Gamma\left(1 - \frac{2}{k}\right) - \Gamma^2\left(1 - \frac{1}{k}\right)\right] \tag{A.93b}$$

$$\alpha_3 = \frac{1}{\sigma^3}\theta^3\left[\Gamma\left(1 - \frac{3}{k}\right) - 3\Gamma\left(1 - \frac{1}{k}\right)\Gamma\left(1 - \frac{2}{k}\right) + \Gamma^3\left(1 - \frac{1}{k}\right)\right] \tag{A.93c}$$

$$\alpha_4 = \frac{1}{\sigma^4}\theta^4\left[\Gamma\left(1 - \frac{4}{k}\right) - 4\Gamma\left(1 - \frac{1}{k}\right)\Gamma\left(1 - \frac{3}{k}\right) + 6\Gamma^2\left(1 - \frac{1}{k}\right)\Gamma\left(1 - \frac{2}{k}\right) - 3\Gamma^4\left(1 - \frac{1}{k}\right)\right] \tag{A.93d}$$

For a known COV, $V = \sigma/\mu$, the parameter k is obtained from the following equation

$$\Gamma\left(1 - \frac{2}{k}\right) = [1 + V^2]\Gamma^2\left(1 - \frac{1}{k}\right)$$

After obtaining k, then θ is obtained as

$$\theta = \mu/\Gamma\left(1 - \frac{1}{k}\right)$$

The parameters k and μ/θ, and skewness and kurtosis changed with the COVs, V, are listed in Table A.2. As can be observed from Table A.2, the skewness and kurtosis do not exist as V becomes larger than 0.7 and 0.5, respectively.

Related Distribution: There is a power transformation of any variable with a Frechet distribution, which produces an exponentially distributed variable. That is, if X is a Frechet distributed variable with parameters θ and k, then $Y = (X/\theta)^{-k}$ is an exponentially distributed variable with PDF of

$$f_Y(y) = \exp(-y)$$

Table A.2 Parameters and moments of Frechet distribution.

V	k	θ/μ	α_3	α_4
0.1	13.62	0.954	1.662	8.732
0.2	7.263	0.908	2.353	16.43
0.3	5.184	0.865	3.353	40.58
0.4	4.173	0.825	5.007	301.6
0.5	3.586	0.79	8.423	—
0.6	3.21	0.759	20.32	—
0.7	2.953	0.733	—	—
0.8	2.769	0.711	—	—
0.9	2.633	0.692	—	—
1	2.53	0.676	—	—

A.7.5 The Weibull Distribution

The *Type-III Smallest*, which is also named *Weibull distribution*, is also commonly used in engineering. When $\zeta = 0$, the PDF of Weibull distribution has the form:

$$f_X(x) = k\theta^{-k}x^{k-1}\exp\left[-\left(\frac{x}{\theta}\right)^k\right], \quad x > 0 \tag{A.94a}$$

where θ and k are parameters.

The distribution is also known as the Type-III Smallest distribution. The standard density function is

$$f_X(x) = kx^{k-1}\exp\left(-x^k\right), \quad x > 0 \tag{A.94b}$$

Graphically, the PDF of the standard Weibull would appear as shown in Figure A.28. The rth moment of a random variable X of Weibull distribution is

$$\mu_r = E(X^r) = \theta^r\Gamma\left(1 + \frac{r}{k}\right) \tag{A.95}$$

Using Eq. (A.30), the mean, the second, third, and fourth central moments of X are expressed as

$$\mu = \theta\Gamma\left(1 + \frac{1}{k}\right) \tag{A.96a}$$

$$\sigma^2 = \theta^2\left[\Gamma\left(1 + \frac{2}{k}\right) - \Gamma^2\left(1 + \frac{1}{k}\right)\right] \tag{A.96b}$$

$$\alpha_3 = \frac{1}{\sigma^3}\theta^3\left[\Gamma\left(1 + \frac{3}{k}\right) - 3\Gamma\left(1 + \frac{1}{k}\right)\Gamma\left(1 + \frac{2}{k}\right) + \Gamma^3\left(1 + \frac{1}{k}\right)\right] \tag{A.96c}$$

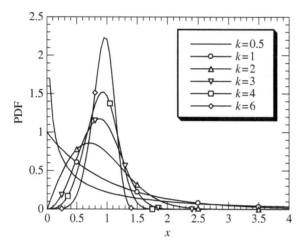

Figure A.28 PDF of the standard Weibull distribution.

$$\alpha_4 = \frac{1}{\sigma^4}\theta^4\left[\Gamma\left(1 + \frac{4}{k}\right) - 4\Gamma\left(1 + \frac{1}{k}\right)\Gamma\left(1 + \frac{3}{k}\right) + 6\Gamma^2\left(1 + \frac{1}{k}\right)\Gamma\left(1 + \frac{2}{k}\right) - 3\Gamma^4\left(1 + \frac{1}{k}\right)\right]$$

(A.96d)

For a known COV, $V = \sigma/\mu$, the parameter k is obtained from the following equation,

$$\Gamma\left(1 + \frac{2}{k}\right) = [1 + V^2]\Gamma^2\left(1 + \frac{1}{k}\right)$$

After obtaining k, then θ is obtained as

$$\theta = \mu/\Gamma\left(1 + \frac{1}{k}\right)$$

The parameters k and μ/θ, and skewness and kurtosis changed with the COVs, V, are listed in Table A.3.

Related Distribution: There is a power transformation of any variable with a Weibull distribution which produces an exponentially distributed variable. That is, if X is a Weibull distributed variable with parameter θ and k, then $Y = (X/\theta)^k$ is an exponentially distributed variable with PDF of $f_Y(y) = \exp(-y)$.

A.8 Multiple Random Variables

A.8.1 Joint and Conditional Probability Distribution

The concept of a random variable and its probability distribution can be extended to two or more random variables. In order to numerically identify events that are the results of two or more physical processes, the events in a sample space may be mapped into two or more dimensions of the real space, implicitly this requires two or more random variables.

Table A.3 Parameters and moments of Weibull distribution.

v	k	θ/μ	α_3	α_4
0.1	12.15	1.043	−0.715	3.78
0.2	5.797	1.08	−0.352	3.004
0.3	3.714	1.108	−0.026	2.723
0.4	2.696	1.125	0.2768	2.788
0.5	2.101	1.129	0.5664	3.131
0.6	1.717	1.122	0.8496	3.732
0.7	1.451	1.103	1.1313	4.594
0.8	1.258	1.075	1.4152	5.737
0.9	1.113	1.04	1.7041	7.192
1	1	1	2	9

Considering the case of two random variables, X and Y, the probabilities for all possible pairs of x and y may be described with the joint distribution function of random variables X and Y, defined as

$$F_{X,Y}(x,y) = P(X \leq x, Y \leq y) \tag{A.97}$$

which is the cumulative probability of the joint occurrences of the events identified by $X \leq x$, $Y \leq y$. In order to comply with the axioms of probability, the joint distribution function must satisfy the follows:

a) $F_{X,Y}(-\infty, -\infty) = 0$, $F_{X,Y}(\infty, \infty) = 1$
b) $F_{X,Y}(-\infty, y) = 0$, $F_{X,Y}(\infty, y) = F_Y(y)$; $F_{X,Y}(x, -\infty) = 0$, $F_{X,Y}(x, \infty) = F_X(x)$
c) $F_{X,Y}(x, y)$ is nonnegative, and a non-decreasing function of x and y.

If the random variables X and Y are discrete, the probability distribution may also be described with the joint probability mass function (PMF), which is simply

$$p_{X,Y}(x,y) = P(X = x, Y = y) \tag{A.98}$$

Then the distribution function becomes

$$F_{X,Y}(x,y) = \sum_{\{x_i \leq x, y_i \leq y\}} p_{X,Y}(x_i, y_i) \tag{A.99}$$

which is simply the sum of probabilities associated with all point pairs (x_i, y_i) in the subset $\{x_i \leq x, y_i \leq y\}$.

The probability of $(X = x)$ may depend on the values of Y, or vice versa. Accordingly, we have the conditional probability mass function

$$p_{X|Y}(x \mid y) = P(X = x \mid Y = y) = \frac{p_{X,Y}(x,y)}{p_Y(y)} \quad \text{if } p_Y(y) \neq 0 \tag{A.100a}$$

Similarly

$$p_{Y|X}(y \mid x) = \frac{p_{X,Y}(x,y)}{p_X(x)} \quad \text{if } p_X(x) \neq 0 \tag{A.100b}$$

The PMF of the individual random variables may be obtained from the joint PMF; applying the theorem of total probability, we obtain the marginal PMF of X as

$$p_X(x) = \sum_{\text{all } y_i} P(X = x | Y = y_i) P(Y = y_i) = \sum_{\text{all } Y_i} P(X = x, Y = y_i) = \sum_{\text{all } p_i} p_{x,y}(x, y_i) \tag{A.101a}$$

By the same token,

$$p_X(x) = \sum_{\text{all } y_i} p_{X,Y}(x, y_i) \tag{A.101b}$$

If the random variables X and Y are statistically independent,

$$p_{X|Y}(x \mid y) = p_X(x), \quad p_{Y|X}(y \mid x) = p_Y(y)$$

then, Eq. (A.98) becomes

$$p_{X,Y}(x,y) = p_X(x)p_Y(y) \tag{A.102}$$

If the random variable X and Y are continuous, the probability distribution may also be described with the joint PDF and CDF

$$F_{X,Y}(x,y) = \int_{-\infty}^{x} \int_{-\infty}^{y} f_{X,Y}(x,y) dx dy \tag{A.103a}$$

$$f_{X,Y}(x,y) = \frac{\partial^2 F_{X,Y}(x,y)}{\partial x \partial y} \tag{A.103b}$$

Analogous to Eq. (A.100a), the conditional density function of X given Y, is

$$f_{X|Y}(x \mid y) = \frac{f_{X,Y}(x,y)}{f_Y(y)} \tag{A.104}$$

Therefore, in general,

$$f_{X,Y}(x,y) = f_{X|Y}(x \mid y)f_Y(y) \tag{A.105a}$$

or

$$f_{X,Y}(x,y) = f_{Y|X}(y \mid x)f_X(x) \tag{A.105b}$$

However, if X and Y are statistically independent, that is, $f_{X|Y}(x|y) = f_X(x)$ and $f_{Y|X}(y|x) = f_Y(y)$, then

$$f_{X,Y}(x,y) = f_X(x)f_Y(y) \tag{A.105c}$$

Applying the total probability theorem, we obtain the marginal density functions

$$f_X(x) = \int_{-\infty}^{\infty} f_{X,Y}(x,y) dy \tag{A.106a}$$

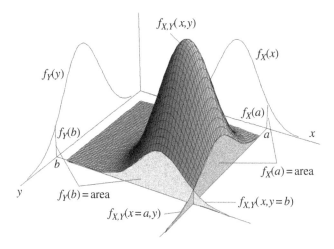

Figure A.29 Joint and marginal PDF of continuous random variables X and Y.

$$f_Y(y) = \int_{-\infty}^{\infty} f_{X,Y}(x,y)dx \qquad \text{(A.106b)}$$

The characteristics of a joint density function for two random variables X and Y, and the associated marginal densities, are portrayed in Figure A.29.

A simple example of joint probability density is two dimensional normal distribution.

$$f_{X,Y}(x,y) = \frac{1}{2\pi\sigma_X\sigma_Y\sqrt{1-\rho^2}} \cdot \exp\left\{\frac{-1}{2(1-\rho^2)}\left[\left(\frac{x-\mu_X}{\sigma_X}\right)^2 - 2\rho\left(\frac{x-\mu_X}{\sigma_X}\right)\left(\frac{y-\mu_Y}{\sigma_Y}\right) + \left(\frac{y-\mu_Y}{\sigma_Y}\right)^2\right]\right\}$$

$$\text{(A.107)}$$

where ρ is the correlation coefficient between X and Y.

A.8.2 Covariance and Correlation

The joint second moments of X and Y is

$$E(XY) = \int_{-\infty}^{\infty}\int_{-\infty}^{\infty} xy f_{X,Y}(x,y)dxdy \qquad \text{(A.108a)}$$

and if X and Y are statistically independent, the equation above becomes

$$E(XY) = \int_{-\infty}^{\infty}\int_{-\infty}^{\infty} xy f_X(x)f_Y(y)dxdy = E(X)E(Y) \qquad \text{(A.108b)}$$

The joint second moment about the means μ_X and μ_Y is the covariance of X and Y, that is,

$$\text{Cov}(X,Y) = E[(X-\mu_X)(Y-\mu_Y)] = E(XY) - E(X)E(Y) \qquad \text{(A.109)}$$

In view of Eq. (A.109), $\text{Cov}(X, Y) = 0$ if X and Y are statistically independent.

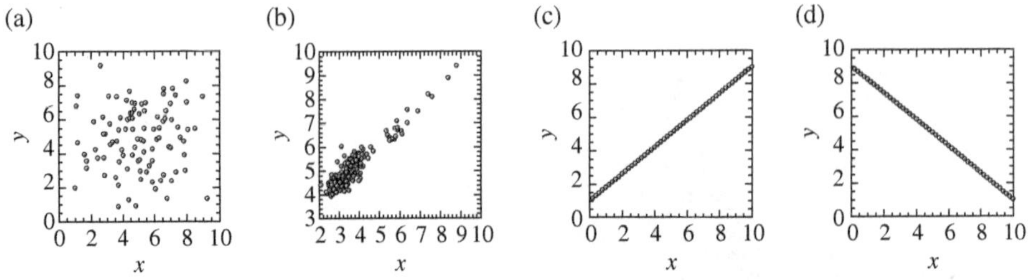

(a)

(b)

(c)

(d)

Figure A.30 Significance of correlation coefficient ρ. (a) $\rho = 0$; (b) $0 < \rho < 1$; (c) $\rho = 1$; (d) $\rho = -1$.

The $Cov(X,Y)$ is a measure of the degree of interrelationship between the variables X and Y. For this purpose, it is preferable to use the normalised covariance or correlation coefficient, which is defined as

$$\rho = \frac{Cov(X,Y)}{\sigma_X \sigma_Y}$$ (A.110a)

The values of ρ range between -1 and 1; that is,

$$-1 \leq \rho \leq 1$$ (A.110b)

When $\rho = 0$, values of X and Y may appear as in Figure A.30a, when $\rho = -1$ and 1, X, and Y are linearly related as shown in Figure A.30c,d, respectively. For intermediate values of ρ, values of X and Y would appear as in Figure A.30b, and the scatter decreased as ρ increases.

Example A.2 (Adapted from Ang and Tang 1975).

A cantilever beam is subjected to two random loads P_1 and P_2 as shown in Figure A.31, which are statistically independent with means and standard deviations μ_1, σ_1 and μ_2, σ_2 respectively.

The shear force Q and bending moment M at the fixed support are

$$Q = P_1 + P_2, \quad M = aP_1 + 2aP_2$$

which are also random variables with means and variances as follows

$$\mu_Q = \mu_1 + \mu_2, \quad \mu_M = a\mu_1 + 2a\mu_2$$
$$\sigma_Q^2 = \sigma_1^2 + \sigma_2^2, \sigma_M^2 = a^2\left(\sigma_1^2 + 4\sigma_2^2\right)$$

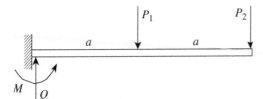

Figure A.31 A cantilever beam. *Source:* Adapted from Ang, A. H-S. and Tang, W.H. (1975) Probability Concepts in Engineering Planning and Design, Vol. I: Basic Principles, New York: John Wiley & Sons.

Although P_1 and P_2 are statistically independent, Q and M will be correlated. The correlation can be evaluated as follows

$$E(QM) = E[(P_1 + P_2)(aP_1 + 2aP_2)] = a(\sigma_1^2 + 2\sigma_2^2) + \mu_Q\mu_M$$

Therefore

$$Cov(Q, M) = E(QM) - \mu_Q\mu_M = a(\sigma_1^2 + 2\sigma_2^2)$$

The corresponding correlation coefficient is

$$\rho_{Q,M} = \frac{Cov(Q, M)}{\sigma_Q\sigma_M} = \frac{(\sigma_1^2 + 2\sigma_2^2)}{\sqrt{(\sigma_1^2 + \sigma_2^2)(\sigma_1^2 + 4\sigma_2^2)}}$$

Hence, if $\sigma_1 = \sigma_2$

$$\rho_{Q,M} = \frac{3}{\sqrt{10}} = 0.948$$

indicating strong correlation between the shear and moment at the support. This correlation arises because Q and M are functions of the same load P_1 and P_2.

A.9 Functions of Random Variables

A.9.1 Function of a Single Random Variable

Engineering problems often involve the evaluation of functional relations between a dependent variable and one or more basic variables. If any of the basic variables are random, the dependent variable will likewise be random, and its probability distribution, as well as its moments, will be functionally related to and may be derived from those of the basic random variables.

Consider the function of a single random variable

$$Y = g(X) \tag{A.111a}$$

This means that when $Y = y$, $X = x = g^{-1}(y)$ where g^{-1} is the inverse function of g. Assume for the moment that $g(x)$ is a monotonically increasing function of x with a unique inverse $g^{-1}(y)$ (as shown in Figure A.32a,b), then

$$P(Y \le y) = P[g(X) \le y] = P\left[X \le g^{-1}(y)\right]$$

That is, the CDF of Y is

$$F_Y(y) = F_X\left[g^{-1}(y)\right] \tag{A.111b}$$

For discrete X, since

$$P(Y = y) = P(X = x) = P\left[X = g^{-1}(y)\right]$$

The PMF of Y is

$$p_Y(y) = p_X\left[g^{-1}(y)\right] \tag{A.112}$$

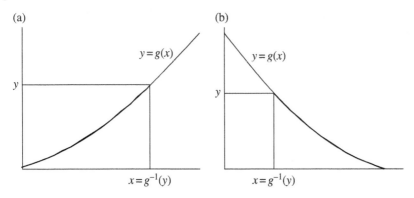

Figure A.32 Monotony of function $Y = g(X)$. (a) Monotonically increasing. (b) Monotonically decreasing.

whereas for continuous X, the PDF of Y is

$$f_Y(y) = \frac{dF_Y(y)}{dy} = f_X\left[g^{-1}(y)\right]\frac{dg^{-1}(y)}{dy} \tag{A.113a}$$

When Y decreases with increasing X (as shown in Figure A.32b), since

$$P(Y \le y) = P[g(X) \le y] = P\left[X \ge g^{-1}(y)\right]$$

then

$$F_Y(y) = 1 - F_X\left[g^{-1}(y)\right] \tag{A.113b}$$

$$f_Y(y) = -f_X\left[g^{-1}(y)\right]\frac{dg^{-1}(y)}{dy} \tag{A.113c}$$

Equations (A.113a) and (A.113c) can be summarised as

$$f_Y(y) = f_X\left[g^{-1}(y)\right]\left|\frac{dg^{-1}(y)}{dy}\right| \tag{A.113d}$$

In particular, for $Y = -X$, $f_Y(y) = f_X(-y)$.

The equations above have many applications. Suppose that X is normal variable with parameters μ and σ.

Let $Y = (X - \mu)/\sigma$, then,

$$x = \sigma y + \mu \quad \text{and} \quad \frac{dx}{dy} = \sigma$$

Thus Eq. (A.113a) yields

$$f_Y(y) = \frac{1}{\sqrt{2\pi}\sigma}\exp\left[-\frac{1}{2}\left(\frac{\sigma y + \mu - \mu}{\sigma}\right)^2\right]\sigma = \frac{1}{\sqrt{2\pi}}\exp\left[-\frac{1}{2}y^2\right]$$

Therefore Y is a standard normal variable with PDF $N(0, 1)$.

Let $Y = \exp(X)$, then,

$$x = \ln y \quad \text{and} \quad \frac{dx}{dy} = \frac{1}{y}$$

Thus Eq. (A.113a) yields

$$f_Y(y) = \frac{1}{\sqrt{2\pi}\sigma y} \exp\left[-\frac{1}{2}\left(\frac{\ln y - \mu}{\sigma} \right)^2 \right]$$

Obviously Y is a lognormal variable.

Example A.3

Assuming that X is a lognormal random variable with parameters $\lambda = \mu_{\ln X}$, $\zeta = \sigma_{\ln X}$, determine the PDF of $Y = X^n$.

According to the definition of lognormal variable, we have

$$f_X(x) = \frac{1}{\sqrt{2\pi}\zeta x} \exp\left[-\frac{1}{2}\left(\frac{\ln x - \lambda}{\zeta} \right)^2 \right] \quad \text{and} \quad x = y^{\frac{1}{n}}, \quad \frac{dx}{dy} = \frac{1}{n}y^{\frac{1-n}{n}}$$

Since Y increases with increasing X for $n > 0$ and decreases with increasing X for $n < 0$, Eqs. (A.113a) and (A.113b) yields

$$f_Y(y) = \begin{cases} \dfrac{1}{n}y^{\frac{1}{n}-1}f_X\left(y^{\frac{1}{n}}\right), & n > 0 \\[2mm] -\dfrac{1}{n}y^{\frac{1}{n}-1}f_X\left(y^{\frac{1}{n}}\right), & n < 0 \end{cases}$$

That is,

$$f_Y(y) = \frac{1}{\sqrt{2\pi}|n|\zeta y} \exp\left[-\frac{1}{2}\left(\frac{\ln y - n\lambda}{n\zeta} \right)^2 \right]$$

One can see that Y is also a lognormal random variable with parameter $n\lambda = \mu_{\ln Y}$, $n\zeta = \sigma_{\ln Y}$. An example of PDF of $Y = X^n$ for a lognormal X with $\lambda = 0.5$ and $\zeta = 0.4$ is shown in Figure A.33.

Similarly, one can obtain the PDF of $Y = X^n$ when X is a normal random variable with mean μ and standard deviation σ. Since X is defined in the range of $(-\infty, \infty)$, we only consider positive n. Directly using Eq. (A.89c), one obtains that,

$$f_Y(y) = \frac{1}{n\sqrt{2\pi}\sigma_X}\left| y^{\frac{1-n}{n}} \right| \exp\left[-\frac{1}{2}\left(\frac{y^{\frac{1}{n}} - \mu_X}{\sigma_X} \right)^2 \right], \quad n > 0$$

Obviously, Y is defined in the range of $(-\infty, \infty)$, except $y = 0$ for odd n and $(0, \infty)$ for even n.

Examples of PDF of $Y = X^n$ for a normal X with $N(1, 0.5)$, $N(1, 0.2)$ and $N(0, 1)$ are shown in Figure A.34a–c.

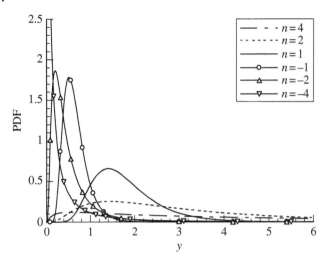

Figure A.33 PDF of $Y = X^n$ for a lognormal X with $\lambda = 0.5$ and $\zeta = 0.4$.

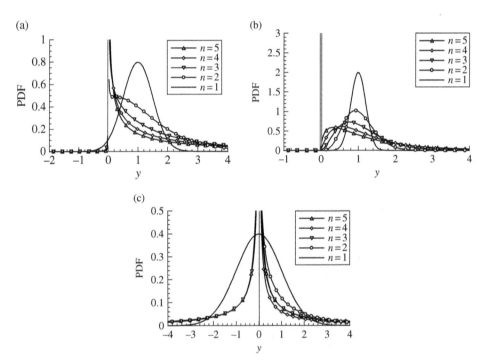

Figure A.34 PDF of $Y = X^n$ for a normal X. (a) X: $N(1,0.5)$; (b) X: $N(1,0.2)$; (c) X: $N(0,1)$.

A.9.2 Function of Multiple Random Variables

A.9.2.1 Generalisation of Probability Distributions of Function of Multiple Random Variables

Consider the function of two random variables X and Y,

$$Z = g(X, Y) \tag{A.114}$$

Since

$$P(Z \le z) = P[g(x, y) \le z]$$

For continuous X, Y, the CDF of Z is

$$F_Z(z) = \int\int_{g(X,Y) \le z} f_{X,Y}(x,y)dxdy = \int_{-\infty}^{\infty}\int_{-\infty}^{z} f_{X,Y}(g^{-1},y)\left|\frac{\partial g^{-1}}{\partial z}\right|dzdy \tag{A.115a}$$

Then, the PDF of Z is

$$f_Z(z) = \int_{-\infty}^{\infty} f_{X,Y}(g^{-1},y)\left|\frac{\partial g^{-1}}{\partial z}\right|dy \tag{A.115b}$$

For the sum of two random variable, say $Z = g(X, Y) = X + Y$, we have

$$X = g^{-1}(Z, Y) = Z - Y \text{ and } \frac{\partial x}{\partial z} = 1$$

$$f_Z(z) = \int_{-\infty}^{\infty} f_{X,Y}(z-y,y)dy$$

An application is shown in Example 2.6, where the distribution of the sum of two normal random variables are derived.

For the product of two random variables, say $Z = g(X, Y) = XY$, we have

$$X = g^{-1}(Z, Y) = \frac{Z}{Y} \text{ and } \frac{\partial x}{\partial z} = \frac{1}{y}$$

Thus

$$f_Z(z) = \int_{-\infty}^{\infty} \left|\frac{1}{y}\right| f_{X,Y}\left(\frac{z}{y},y\right)dy$$

Similarly, for the quotient of two random variables, for example $Z = g(X, Y) = X/Y$, the PDF of Z would be

$$f_Z(z) = \int_{-\infty}^{\infty} |y| f_{X,Y}(zy,y)dy$$

The method described above for a function of two variables can be generalised to derive the distribution of a function of n random variables, briefly, if

$$Z = g(X_1, X_2, \cdots, X_n) \tag{A.116}$$

Then the CDF is

$$F_Z(z) = \int\limits_{g(X_1,\cdots X_n) \le z} \cdots \int f_{X_1,\cdots X_n}(x_1, x_2, \cdots, x_n \, dx_1 dx_2 \cdots dx_n)$$

$$= \int_{-\infty}^{\infty} \cdots \int_{-\infty}^{\infty} \int_{-\infty}^{z} f_{X_1,\cdots X_n}\left(g^{-1}, x_2, \cdots, x_n\right) \left|\frac{\partial g^{-1}}{\partial z}\right| dz dx_2 \cdots dx_n$$

(A.117)

where $g^{-1} = g^{-1}(z, x_2, \cdots, x_n)$,

Then the PDF of Z is

$$f_Z(z) = \int_{-\infty}^{\infty} \cdots \int_{-\infty}^{\infty} f_{X_1,\cdots X_n}\left(g^{-1}, x_2, \cdots, x_n\right) \left|\frac{\partial g^{-1}}{\partial z}\right| dx_2 \cdots dx_n$$

(A.118)

A.9.2.2 Sum (and Difference) of Normal Variables

If X and Y are statistically independent normal variables with means and standard deviations μ_X, σ_X and μ_Y, σ_Y, respectively; the PDF of Z is (Ang and Tang 2006)

$$f_Z(z) = \frac{1}{\sqrt{2\pi}\sqrt{\sigma_X^2 + \sigma_Y^2}} \exp\left[-\frac{1}{2}\left(\frac{z - (\mu_X + \mu_Y)}{\sqrt{(\sigma_X^2 + \sigma_Y^2)}}\right)^2\right]$$

Therefore, Z is also a normal random variable with mean and variance $\mu_Z = \mu_X + \mu_Y$, $\sigma_Z^2 = \sigma_X^2 + \sigma_Y^2$.

Similarly, $Z = X - Y$ is also a normal random variable with mean and variance $\mu_Z = \mu_X - \mu_Y$, $\sigma_Z^2 = \sigma_X^2 + \sigma_Y^2$.

Based on these results, it can be shown inductively that if

$$Z = \sum_{i=1}^{n} a_i X_i$$

(A.119)

where a_i are constants, and X_i are statistical independent normal variables $N(\mu_{X_i}, \sigma_{X_i})$, then Z is also Gaussian with mean and variance

$$\mu_Z = \sum_{i=1}^{n} a_i \mu_{X_i}$$

(A.120)

$$\sigma_Z^2 = \sum_{i=1}^{n} a_i^2 \sigma_{X_i}^2$$

(A.121)

In other words, any linear function of independent normal variables is also a normal variable. The derivation of distributions of sum of dependent normal variables can be found in Example 2.6, from which one can observe that any linear function of dependent normal variables is also a normal random variable.

A.9.2.3 Products and Quotients of Random Variables

By virtue of the result for sums and difference of normal random variables, it follows that the product and quotient of statistically independent lognormal random variable is also a lognormal random variable. Suppose

$$Z = \prod_{i=1}^{n} X_i \tag{A.122}$$

where X_i are statistically independent lognormal variables with respective parameters $\lambda_{X_i}, \zeta_{X_i}$, then

$$\ln Z = \sum_{i=1}^{n} \ln X_i$$

Since each $\ln X_i$ is normal, it follows that $\ln Z$ is also normal with mean and variance, according to Eqs. (A.120) and (A.121), as follows:

$$\lambda_Z = \mu_{\ln Z} = \sum_{i=1}^{n} \lambda_{X_i} \tag{A.123}$$

$$\zeta_Z^2 = \sigma_{\ln Z}^2 = \sum_{i=1}^{n} \zeta_{X_i}^2 \tag{A.124}$$

Hence Z is lognormal with the above parameters λ_Z and ζ_Z.

A.9.2.4 The Central Limit Theorem

One of the most significant theorems in probability theory is that pertaining to the limiting distribution of a sum of random variables known as the *central limit theorem*. Stated loosely, the theorem says that the sum of a large number of individual random components, none of which is dominant, tends to the normal distribution as the number of components (regardless of their initial distributions) increases without limit. That is, regardless of the distributions of X_i, the sum

$$Z = \sum_{i=1}^{n} X_i$$

will approach a normal distribution as $n \to \infty$.

Therefore, if a physical process is the result of the totality of a large number of individual effects, then according to the central limit theorem, the process would tend to be a normal distribution.

By virtue of the central limit theorem, the product of a large number of independent factors (none of which dominates the product) will tend to the lognormal distribution. That is, regardless of the distributions of X_i, the product

$$Z = \prod_{i=1}^{n} X_i$$

will approach a lognormal distribution as $n \to \infty$.

A.10 Summary

In this chapter, the basic theory of statistics and applied probability, which will constitute the fundamentals of the structural reliability theory, is introduced. The concepts of moments, especially the first four moments, of a random variable, are introduced and illustrated in detail. This can help to understand the methods of moments for structural reliability of this book. Furthermore, the commonly used distributions are introduced with emphasis on the first four moments corresponding to each distribution, which are essentially useful in the methods of moment of this book.

Appendix B

Three-Parameter Distributions

B.1 Introduction

Determination of the probability distributions of the basic random variables is essential for the accurate evaluation of the reliability of a structure. Often, the method for determining the required distribution is to fit the histogram of the statistical data of a variable with a candidate distribution (Ang and Tang 1975), and apply statistical goodness-of-fit tests.

Two-parameter distributions such as the normal, lognormal, Weibull, and Gamma distributions are often selected as the candidate distribution, in which the parameters of the distribution are generally evaluated from the mean value and standard deviation of the available data. After the two parameters are determined, the distribution form and any higher-order moments, such as skewness, may be evaluated; quite often, these higher-order moments may not be consistent with those of the available data. This is illustrated with the following: The two histograms shown in Figure B.1 represent the observed variabilities in the properties of H-shape structural steel (Ono et al. 1986). Figure B.1a shows the histogram of the section area and Figure B.1b shows the histogram of the residual stress at the flange. From Figure B.1a, one can see that the coefficient of variation (COV) of the section area is small, 0.0514, whereas the skewness is large 0.7085. Conversely, the COV of the residual stress from Figure B.1b is large, 0.7492, whereas the skewness is 0.823. The skewness of the normal, lognormal, Weibull, and Gamma distributions that have the same mean value and standard deviation as the data in Figure B.1a can be obtained as, respectively, 0.00, 0.1555, −0.9121, and 0.1024. Clearly, none of these two-parameter distributions has skewness that is consistent with the skewness of the data, which is 0.7085. Similarly, with respect to Figure B.1b, the skewness of these same four distributions can be determined as 0.00, 1.7819, 0.6834, and 1.4984, respectively. Again, none of these matches the skewness of the data, which is 0.823 in this case.

As illustrated with the data of Figure B.1, two-parameter distributions may not be appropriate when the skewness of the statistical data is important and must be reflected in the distribution.

For this purpose, three-parameter (3P) distributions are required. In this Appendix, the 3P distributions such as the 3P lognormal distribution (Stuart and Ord 1987; Tichy 1994), and the square normal distribution (Zhao et al. 2001) are introduced, in which the three parameters can be directly defined in terms of the mean value, standard deviation, and

Structural Reliability: Approaches from Perspectives of Statistical Moments, First Edition.
Yan-Gang Zhao and Zhao-Hui Lu.
© 2021 John Wiley & Sons Ltd. Published 2021 by John Wiley & Sons Ltd.

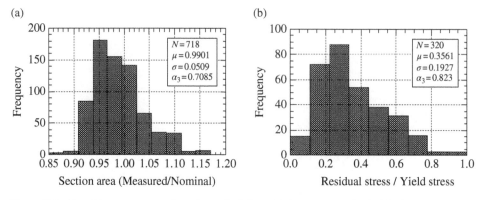

Figure B.1 Two histogram examples of practical data. (a) Section area. (b) Residual stress.

skewness. Other 3P distributions such as 3P Gamma distribution (Zhao and Ang 2002) can be referred to Stuart and Ord (1987).

B.2 The 3P Lognormal Distribution

B.2.1 Definition of the Distribution

For a standardised variable

$$X_s = \frac{X - \mu_X}{\sigma_X}$$

the random variable X obeys the 3P lognormal distribution when the distribution of

$$Y = \ln|X_s - u_b| \tag{B.1}$$

is normal. Here, μ_X and σ_X are the mean and standard deviation of X, respectively; and u_b is the standardised bound of the distribution.

The probability density function (PDF) is defined as (Stuart and Ord 1987; Tichy 1994)

$$f(x_s) = \frac{1}{\sqrt{2\pi \ln(A)}|x_s - u_b|} \exp\left\{ -\frac{1}{2\ln(A)}\left[\ln\left(\sqrt{A}\frac{|x_s - u_b|}{|u_b|}\right)\right]^2 \right\} \tag{B.2a}$$

or

$$f(x_s) = \frac{\text{sign}(\alpha_3)}{\sqrt{2\pi \ln(A)}(x_s - u_b)} \exp\left\{ -\frac{1}{2\ln(A)}\left[\ln\left(\frac{\sqrt{A}(u_b - x_s)}{u_b}\right)\right]^2 \right\} \tag{B.2b}$$

where the parameters A and u_b are given by (Zhao and Ono 2001)

$$A = 1 + \frac{1}{u_b^2}, u_b = (a + b)^{\frac{1}{3}} + (a - b)^{\frac{1}{3}} - \frac{1}{\alpha_{3X}} \tag{B.3a}$$

$$a = -\frac{1}{\alpha_3}\left(\frac{1}{2} + \frac{1}{\alpha_{3X}^2}\right), b = \frac{1}{2\alpha_{3X}^2}\sqrt{\alpha_{3X}^2 + 4} \qquad (B.3b)$$

in which α_{3X} is the third dimensionless central moment, i.e. the skewness of X.

Since A and u_b in Eq. (B.1) only depend on α_{3X}, the distribution is determined by the three parameters, μ_X, σ_X, and α_{3X}.

The third and fourth dimensionless central moments are given as

$$\alpha_{3X} = -\frac{1}{u_b}\left(3 + \frac{1}{u_b^2}\right) \qquad (B.4a)$$

$$\alpha_{4X} = A^4 + 2A^3 + 3A^2 - 3 \qquad (B.4b)$$

The relationship between the standard normal random variable U and X_s can be given as (Zhao and Ono 2001)

$$U = \frac{\text{sign}(\alpha_{3G})}{\sqrt{\ln(A)}} \ln\left[\sqrt{A}\left(1 - \frac{X_s}{u_b}\right)\right] \qquad (B.5a)$$

and

$$X_s = u_b\left\{1 - \frac{1}{\sqrt{A}}\exp\left[\text{sign}(\alpha_{3G})\sqrt{\ln(A)}U\right]\right\} \qquad (B.5b)$$

Example B.1 Derivation of the Distribution

When $X_s > u_b$, Eq. (B.1) becomes

$$Y = \ln(X_s - u_b)$$

Let $Z = X_s - u_b$, then

$$\mu_{Xs} = \mu_Z + u_b = 0 \qquad (B.6a)$$

$$\sigma_{Xs} = \sigma_Z = 1 \qquad (B.6b)$$

$$\alpha_{3Xs} = \alpha_{3X} = \alpha_{3Z} \qquad (B.6c)$$

Since $Y = \ln Z$ is normal, supposing the mean and standard deviation of Y are λ and ζ, respectively, then, according to Eqs. (A.62) through (A.64), we have

$$\mu_Z = \exp\left(\frac{1}{2}\zeta^2 + \lambda\right), \quad \sigma_Z = \mu_Z\sqrt{\exp(\zeta^2) - 1}, \quad \alpha_{3Z} = \left[\exp(\zeta^2) - 1\right]^{1/2}\left[\exp(\zeta^2) + 2\right]$$

or

$$\mu_Z = \sqrt{A}\exp\lambda \qquad (B.7a)$$

$$\sigma_Z = \mu_Z\sqrt{A - 1} \qquad (B.7b)$$

$$\alpha_{3Z} = \sqrt{A - 1}(A + 2) \qquad (B.7c)$$

where $A = \exp(\zeta^2)$.

From Eqs. (B.6a–b) and (B.7a–b), one obtains that

$$u_b = -\mu_Z = \frac{-1}{\sqrt{A-1}}, \quad A = 1 + \frac{1}{u_b^2} \tag{B.8}$$

Then Eq. (B.6c) becomes

$$-\frac{1}{u_b}\left(\frac{1}{u_b^2} + 3\right) = \alpha_{3X} \tag{B.9a}$$

Equation (B.9a) can be rewritten as

$$\alpha_{3X} u_b^3 + 3u_b^3 + 1 = 0 \tag{B.9b}$$

The solution of the equation above is given as Eqs. (B.3a) and (B.3b). Using u_b obtained from Eq. (B.8), λ and ζ are given as

$$\zeta = \sqrt{\ln A} \tag{B.10a}$$

$$\lambda = \ln \frac{-u_b}{\sqrt{A}} \tag{B.10b}$$

Then, the relationship between the standard normal random variable U and X_s can be given as

$$U = \frac{\ln Z - \lambda}{\zeta} = \frac{\ln(X_s - u_b) - \ln \frac{-u_b}{\sqrt{A}}}{\sqrt{\ln A}}$$

That is

$$U = \frac{1}{\sqrt{\ln(A)}} \ln\left[\sqrt{A}\left(1 - \frac{X_s}{u_b}\right)\right]$$

Then, the PDF of X_s can be easily given as

$$f_{Xs}(x_s) = \frac{1}{\sqrt{2\pi \ln(A)}(x_s - u_b)} \exp\left[-\frac{1}{2\ln(A)}\left[\ln\left(\sqrt{A}\left(1 - \frac{x_s}{u_b}\right)\right)\right]^2\right]$$

From Eq. (B.9a), one can see that in this case $\alpha_{3X} > 0$ and $u_b < 0$

When $X_s < u_b$, Eq. (B.1) becomes

$$Y = \ln(u_b - X_s)$$

Let $Z = u_b - X_s$, then

$$\mu_{Xs} = u_b - \mu_Z = 0 \tag{B.11a}$$
$$\sigma_{Xs} = \sigma_Z = 1 \tag{B.11b}$$
$$\alpha_{3Xs} = \alpha_{3X} = -\alpha_{3Z} \tag{B.11c}$$

From Eqs. (B.6a) to (B.6b) and (B.11a) to (B.11b), one obtains that

$$u_b = \mu_Z = \frac{1}{\sqrt{A-1}}, \quad A = 1 + \frac{1}{u_b^2} \tag{B.12}$$

Then, Eq. (B.6c) becomes

$$\frac{1}{u_b}\left(\frac{1}{u_b^2}+3\right) = -\alpha_{3X} \tag{B.13}$$

The solution of the equation above is given as Eqs. (B.3a) and (B.3b).
Using u_b obtained from Eq. (B.12), λ and ζ are given as

$$\zeta = \sqrt{\ln A} \tag{B.14a}$$

$$\lambda = \ln\frac{u_b}{\sqrt{A}} \tag{B.14b}$$

Then, the relationship between the standard normal random variable U and Z can be given as

$$U = \frac{\ln Z - \lambda}{\zeta}$$

and the PDF of Z can be easily given as

$$f_z(z) = \frac{1}{\sqrt{2\pi}z\zeta}\exp\left[-\frac{1}{2}\left(\frac{\ln z - \lambda}{\zeta}\right)\right]$$

Since $X_s = u_b - Z$ decreases with Z increases, according to Eq. (A.113b), we have

$$F_{Xs}(x_s) = 1 - F_Z[u_b - x_s]$$

Then

$$U = \Phi^{-1}[F_{Xs}(x_s)] = \Phi^{-1}[1 - F_Z(u_b - x_s)] = -\Phi^{-1}[F_Z(u_b - x_s)]$$

That is

$$U = -\frac{\ln(u_b - X_s) - \lambda}{\zeta} = \frac{\ln(u_b - X_s) - \ln\frac{u_b}{\sqrt{A}}}{\sqrt{\ln A}}$$

Therefore

$$U = \frac{-1}{\sqrt{\ln(A)}}\ln\left[\sqrt{A}\left(1 - \frac{X_s}{u_b}\right)\right]$$

According to Eq. (A.113c), the PDF is given as

$$f_{Xs}(x_s) = \frac{-1}{\sqrt{2\pi\ln(A)}(x_s - u_b)}\exp\left[-\frac{1}{2\ln(A)}\left[\ln\left(\sqrt{A}\left(1 - \frac{X_s}{u_b}\right)\right)\right]^2\right]$$

From Eq. (B.13), one can see that in this case $\alpha_{3X} < 0$ and $u_b > 0$.
Then, Eq. (B.5a) will be obtained.

B.2.2 Simplification of the Distribution

One may see that the expression of the 3P lognormal distribution is quite complicated. Since sometimes the distribution is only applied in engineering when the absolute value of α_{3X} is relatively small, as shown in Section 4.2 of the chapter on Direct Methods of Moment, the distribution will be simplified under the condition of small skewness of the random variable.

For $-1 < \alpha_{3X} < 1$, Eq. (B.3a) can be simplified as the following approximations for u_b with errors less than 3% (Zhao and Ang 2003). Other approximations may be found in Ugata and Moriyama (1996) and Ugata (2000).

$$u_b = -\frac{3}{\alpha_{3X}} \tag{B.15}$$

Then, the relationship between the standard normal random variable U and the standardised random variable Xs can be expressed as follows:

$$U = \frac{1}{6}\alpha_{3X} + \frac{3}{\alpha_{3X}} \ln\left[1 + \frac{1}{3}\alpha_{3X}X_s\right] \tag{B.16a}$$

$$X_s = \frac{3}{\alpha_{3X}}\left[\exp\left(\frac{1}{3}\alpha_{3X}U - \frac{1}{18}\alpha_{3X}^2\right) - 1\right] \tag{B.16b}$$

The simplified PDF is expressed as

$$f(x_s) = \frac{1}{\sqrt{2\pi}\left(1 + \frac{1}{3}\alpha_{3X}x_s\right)} \exp\left[-\frac{1}{2}\left[\frac{1}{6}\alpha_{3X} + \frac{3}{\alpha_{3X}} \ln\left(1 + \frac{1}{3}\alpha_{3X}x_s\right)\right]^2\right] \tag{B.17}$$

To examine the accuracy of the approximation expressed in Eq. (B.15), the values of the standardised bound u_b are depicted in Figure B.2, where the solid lines indicate the exact values of u_b obtained from Eq. (B.3a) and the dashed lines indicate those obtained with Eq. (B.15). From Figure B.2, one can see that although Eq. (B.15) is much simpler than

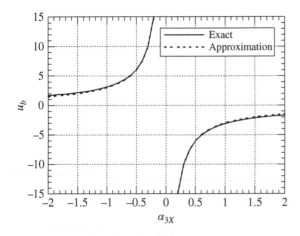

Figure B.2 Variations of the standardised bound.

Eq. (B.3a), the u_b results obtained with Eq. (B.15) agree well with those obtained from Eq. (B.3a).

Example B.2 The Derivation of Eqs. (B.15) and (B.16a) to (B.16b)

For a random variable X of the 3P lognormal distribution, we introduce a new random variable Z

$$Z = |X_s - u_b|$$

According to the definition of the distribution, since the logarithm of Z, $Y = \ln Z$ obeys normal distribution, Z obeys lognormal distribution. Suppose parameters of the lognormal distribution are λ and ζ, and then the skewness α_{3Z} is given by the following equation according to Eq. (A.64).

$$\alpha_{3Z} = \left[\exp\left(\zeta^2\right) + 2 \right] \sqrt{\exp\left(\zeta^2\right) - 1} \qquad (B.18)$$

When the absolute value of α_{3Z} is less than 1, the absolute value of ζ is less than 0.34, and the following approximation applies with an error less than 3%.

$$\sqrt{\exp\left(\zeta^2\right) - 1} = \zeta$$

Then, one obtains

$$\alpha_{3Z} = 3\zeta + \zeta^3$$

For small ζ (e.g. $\zeta < 0.34$), this equation becomes

$$\alpha_{3Z} = 3\zeta \qquad (B.19)$$

The relationship between α_3 and ζ is graphically shown in Figure B.3, from which one may see that although Eq. (B.19) is much simpler than Eq. (B.18), they give good agreements when the absolute value of α_{3Z} is less than 1.

According to Eqs. (B.6c) and (B.11c)

$$\alpha_{3Z} = \text{sign}(\alpha_{3X})\alpha_{3X}$$

Figure B.3 Relationship between α_3 and ζ.

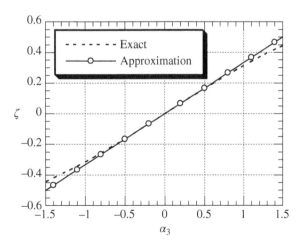

Comparing Eq. (B.8) with Eq. (B.12), one can easily see that

$$u_b = -\frac{\text{sign}(\alpha_{3X})}{\zeta} = -\frac{3 \cdot \text{sign}(\alpha_{3X})}{\alpha_{3Z}} = -\frac{3}{\alpha_{3X}} \tag{B.20}$$

That is, Eq. (B.15) will be obtained.

Since

$$A = \exp\left(\zeta^2\right)$$

Substituting A into Eqs. (B.5a) and (B.5b), one obtains Eqs. (B.16a) to (B.16b).

B.3 Square Normal Distribution

B.3.1 Definition of the Distribution

For a standardised variable

$$X_s = \frac{X - \mu_X}{\sigma_X}$$

The random variable X obeys the square normal distribution (Zhao et al. 2001) when X_s can be expressed by the following second-order polynomial of standard normal random variable

$$X_s = -\lambda + \sqrt{1 - 2\lambda^2}U + \lambda U^2 \tag{B.21}$$

The PDF and CDF of X_s are given as

$$f(x_s) = \frac{\phi\left[\frac{1}{2\lambda}\left(\sqrt{1 + 2\lambda^2 + 4\lambda x_s} - \sqrt{1 - 2\lambda^2}\right)\right]}{\sqrt{1 + 2\lambda^2 + 4\lambda x_s}} \tag{B.22}$$

$$F(x_s) = \Phi\left[\frac{1}{2\lambda}\left(\sqrt{1 + 2\lambda^2 + 4\lambda x_s} - \sqrt{1 - 2\lambda^2}\right)\right] \tag{B.23}$$

in which Φ and ϕ are the CDF and PDF of a standard normal random variable, and μ_X, σ_X and λ are the three parameters of the distribution. The random variable x is defined in the following ranges:

$$-\infty \le \frac{X - \mu_X}{\sigma_X} \le -\frac{1}{4\lambda} - \frac{1}{2}\lambda \quad \text{for } \lambda < 0 \tag{B.24a}$$

$$-\frac{1}{4\lambda} - \frac{1}{2}\lambda \le \frac{X - \mu_X}{\sigma_X} \le \infty \quad \text{for } \lambda > 0 \tag{B.24b}$$

The first four central moments of X are obtained as:

$$E[X] = \mu_X \tag{B.25a}$$

$$E\left[(X - \mu_X)^2\right] = \sigma_X^2 \tag{B.25b}$$

$$E\left[\left(\frac{X - \mu_X}{\sigma_X}\right)^3\right] = 6\lambda - 4\lambda^3 \tag{B.25c}$$

$$E\left[\left(\frac{X - \mu_X}{\sigma_X}\right)^4\right] = 3 + 48\lambda^2 - 48\lambda^4 \tag{B.25d}$$

Using Eq. (B.25c), parameter λ can be easily determined as

$$\lambda = \text{sign}(\alpha_{3X})\sqrt{2}\cos\left[\frac{\pi + |\theta|}{3}\right] \tag{B.26a}$$

$$\theta = \arctan\left(\frac{\sqrt{8 - \alpha_{3X}^2}}{\alpha_{3X}}\right) \tag{B.26b}$$

in which α_{3X} is third dimensionless central moment, i.e. the skewness of X.

The relationship between the standard normal random variable U and X_s can be given by Eq. (B.21) and the following equation as (Zhao and Ono 2000b)

$$U = \frac{1}{2\lambda}\left(\sqrt{1 + 2\lambda^2 + 4\lambda X_s} - \sqrt{1 - 2\lambda^2}\right) \tag{B.27}$$

The distribution is referred to as the 'square normal' distribution since Eq. (B.21) is a quadratic form of the standard normal variable.

From Eq. (B.26b), in order to make Eq. (B.26a) operable, α_{3X} should be limited in the range of

$$-2\sqrt{2} \leq \alpha_{3X} \leq 2\sqrt{2} \tag{B.28}$$

The limitation is not strict for general engineering use (Zhao and Ono 2000b).

Particularly, when α_{3X} approaches 0, $|\theta|$ approaches $\pi/2$, and λ approaches 0. The limit of Eq. (B.23) can be obtained as

$$\lim_{\lambda \to 0}[F(x_s)] = \Phi\left\{\lim_{\lambda \to +\infty}\left[\frac{1}{2\lambda}\sqrt{1 + 2\lambda^2 + 4\lambda x_s} - \sqrt{1 - 2\lambda^2}\right]\right\} = \Phi(x_s) \tag{B.29}$$

That is to say, the distribution approaches normal distribution when the skewness approaches 0.

The PDFs of the standard three-parameter distribution for $\alpha_{3X} > 0$ and $\alpha_{3X} < 0$ are shown in Figure B.4a,b. When $\alpha_{3X} = 0$, the PDF degenerates as that of the standard normal

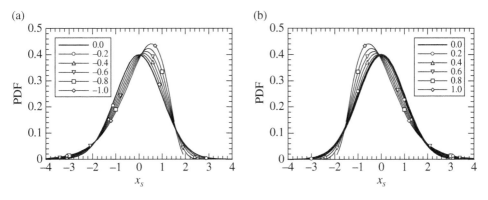

Figure B.4 The PDF of the square normal distribution. (a) $\alpha_{3X} < 0$, (b) $\alpha_{3X} > 0$.

distribution and is depicted as a solid line in Figure B.4. From Figure B.4, one can see that the distribution obviously reflects the characteristics of skewness. As investigated in the ensuing sections, the distribution satisfies the requirements of flexibility, simplicity, and generality described previously.

B.3.2 Simplification of the Distribution

For small α_{3X}, using Maclaurin expansion of θ, the parameter λ expressed in Eq. (B.26a) can be simply approximated as (Zhao et al. 2001):

$$\lambda = \frac{1}{6}\alpha_{3X} \tag{B.30}$$

By substituting Eq. (B.30) into the equations in the preceding section, a simple three-parameter distribution directly defined in terms of mean value μ_X, standard deviation σ_X, and skewness α_{3X} of a random variable can be obtained.

CDF:

$$F(x_s) = \Phi\left[\frac{1}{\alpha_{3X}}\left(\sqrt{9 + \alpha_{3X}^2 + 6\alpha_{3X}x_s} - 3\right)\right] \tag{B.31a}$$

PDF:

$$f(x_s) = \frac{3\phi\left[\frac{1}{\alpha_{3X}}\left(\sqrt{9 + \alpha_{3X}^2 + 6\alpha_{3X}x_s} - 3\right)\right]}{\sqrt{9 + \alpha_{3X}^2 + 6\alpha_{3X}x_s}} \tag{B.31b}$$

Definition range of X_s:

$$-\infty \le X_s \le -\frac{3}{2\alpha_{3X}} - \frac{1}{12}\alpha_{3X} \text{ for } \alpha_{3X} < 0 \tag{B.32a}$$

$$-\frac{3}{2\alpha_{3X}} - \frac{1}{12}\alpha_{3X} \le X_s \le \infty \text{ for } \alpha_{3X} > 0 \tag{B.32b}$$

Relationship between X_s and U:

$$U = \frac{1}{\alpha_{3X}}\left(\sqrt{9 + \alpha_{3X}^2 + 6\alpha_{3X}X_s} - 3\right) \tag{B.33a}$$

$$X = -\frac{1}{6}\alpha_{3X} + U + \frac{1}{6}\alpha_{3X}U^2 \tag{B.33b}$$

The first four moments:

$$E[X] = \mu_X \tag{B.34a}$$

$$E\left[(X - \mu_X)^2\right] = \sigma_X^2 \tag{B.34b}$$

$$E\left[\left(\frac{X - \mu_X}{\sigma_X}\right)^3\right] = \alpha_{3X} - \frac{1}{54}\alpha_{3X}^3 \tag{B.34c}$$

$$E\left[\left(\frac{X - \mu_X}{\sigma_X}\right)^4\right] = 3 + \frac{4}{3}\alpha_{3X}^2 - \frac{1}{27}\alpha_{3X}^4 \tag{B.34d}$$

When $|\alpha_{3X}| \leq 1$, since the last term in Eqs. (B.34c) and (B.34d) can be ignored with an error smaller than 2%, one can rewrite Eqs. (B.34c) and (B.34d) simply as:

$$E\left[\left(\frac{X - \mu_X}{\sigma_X}\right)^3\right] = \alpha_{3X} \tag{B.34e}$$

$$E\left[\left(\frac{x - \mu_X}{\sigma_X}\right)^4\right] = 3 + \frac{4}{3}\alpha_{3X}^2 \tag{B.34f}$$

From Eqs. (B.34a), (B.34b), and (B.34e), one can understand that the mean value, standard deviation, and skewness of x are equal to the three parameters μ_X, σ_X, and α_{3X}, respectively, of the distribution.

Since Eqs. (B.31a) and (B.31b) are very simple and are defined directly in terms of the mean value, standard deviation and skewness of a random variable, the distribution has more applications in structural reliability.

B.4 Comparison of the 3P Distributions

The PDFs comparison of different 3P distributions are depicted in Figure B.5a–d for skewness of $\alpha_{3X} = 0.4, 0.8, 1.2, 1.6, 2.0$, in which we include the square normal and 3P lognormal distributions, along with the 3P Gamma distribution (Zhao and Ang 2002). From Figure B.5, one can see that for small α_{3X}, e.g. $\alpha_{3X} = 0.4, 0.8$, there is no significant difference among the PDFs of the three different distributions.

The relationship between α_{3X} and α_{4X} for the three 3P distributions are shown in Figure B.6, from which one can observe that all the three distributions approach the normal distribution when α_{3X} approaches 0. For small skewness, e.g. $\alpha_{3X} < 1$, the three lines in Figure B.6 are quite close; this is the reason why there is no significant difference among the PDFs of the three distributions as described above. One may also see that the exponential point is in the 3P Gamma line because the exponential distribution is a special form of the Gamma distribution.

B.5 Applications of the 3P Distributions

The applications of the 3P distributions for the derivation of the third-moment reliability index have been described in Section 4.2 in the chapter on Direct Methods of Moment. Here, other applications of the 3P distributions such as in statistical data analysis and representations of one- and two-parameter distributions will be introduced.

B.5.1 Statistical Data Analysis

The 3P distributions are often appropriate for fitting statistical data of a random variable. Here, we only investigate the application of square normal distributions. The application of 3P lognormal and other 3P distributions can be similarly conducted. Consider the measured data of H-shape structural steel described earlier. The histogram of the section area is shown

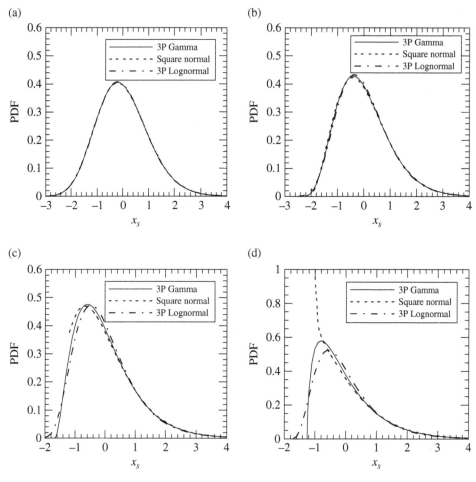

Figure B.5 Comparisons of 3P Distributions. (a) $\alpha_{3X} = 0.4$, (b) $\alpha_{3X} = 0.8$, (c) $\alpha_{3X} = 1.2$, (d) $\alpha_{3X} = 1.6$.

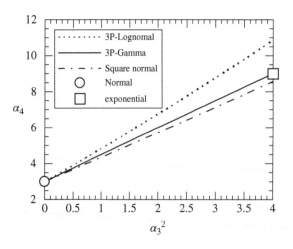

Figure B.6 Relationship between α_3 and α_4 of the 3P distributions.

(a)   (b)

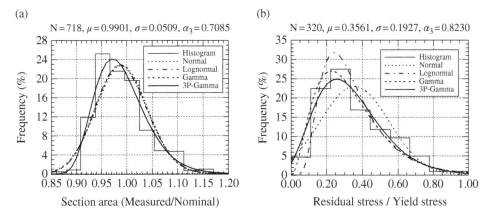

Figure B.7 Fitting results for data of H-shape structural steel. (a) Section area. (b) Residual stress.

in Figure B.7a, in which the PDFs of the normal and lognormal distributions, with the same mean value and standard deviation as the data, and the PDF of the square normal distribution whose mean value, standard deviation, and skewness are equal to those of the data, are depicted, revealing the following:

1) The normal distribution is clearly not appropriate, as its skewness is zero, whereas the skewness of the data is 0.7085.
2) The skewness of the lognormal distribution is a function of the COV. In the present case, the COV of the lognormal distribution is 0.051, thus, its skewness is 0.1555, which is much smaller than that of the data (0.7085).
3) The skewness of the square normal distribution can be equal to that of the statistical data, and thus can fit the histogram much better than the normal or the lognormal distribution.

Results of the chi-square tests of the three distributions are listed in Table B.1, in which the goodness-of-fit tests were obtained using the following equation (Ang and Tang 1975):

$$T = \sum_{i=1}^{k} \frac{(O_i - E_i)^2}{E_i} \tag{B.35}$$

where O_i and E_i are the observed and theoretical frequencies, respectively; k is the number of intervals used; and T is a measure of the respective goodness-of-fit. From Table B.1, one can see that the goodness-of-fit of the square normal distribution is $T = 24.8$, which is much smaller than those of the normal with $T = 91.8$ or the lognormal with $T = 70.8$.

Similarly, the histogram of measured residual stress and the PDFs of the three distributions are shown in Figure B.7b; the corresponding results of the chi-square tests are listed in Table B.2. From Figure B.7b, one can observe that the lognormal distribution with a skewness of 1.7819 is much larger than that of the data of 0.7492. Clearly, therefore, the square normal distribution fits the histogram much better than the normal or lognormal distribution. Additionally, from Table B.2, the goodness-of-fit tests verify that the square normal

Table B.1 Results of χ^2 test for section area.

Intervals	Freq.	Predicted frequency			Goodness-of-fit test		
		Nor.	Log.	Square nor.	Nor.	Log.	Square nor.
< 0.908	9	38.7	35.3	18.0	22.8	19.6	4.46
0.908–0.938	85	69.3	72.2	83.7	3.54	2.27	0.02
0.938–0.967	181	123	129	154	26.8	20.8	4.61
0.967–0.996	155	160	162	168	0.13	0.28	1.03
0.996–1.025	141	150	146	132	0.54	0.17	0.63
1.025–1.054	66	102	97.5	82.6	12.9	10.2	3.36
1.054–1.083	35	50.5	49.1	44.1	4.77	4.06	1.87
1.083–1.113	34	18.2	19.3	20.8	13.6	11.2	8.38
> 1.113	5	4.75	5.95	14.5	6.67	2.28	0.43
Sum	718	718	718	718	91.8	70.8	24.8

Note: Freq. = Frequency; Nor. = Normal; Log. = Lognormal; Square nor. = Square normal.

Table B.2 Results of χ^2 test for residual stress.

Intervals	Freq.	Predicted frequency			Goodness-of-fit test		
		Nor.	Log.	Square nor.	Nor.	Log.	Square nor.
< 0.111	15	32.6	6.54	23.2	9.48	10.9	2.88
0.111–0.222	72	45.4	73.2	61.0	15.6	0.02	1.97
0.222–0.333	88	67.0	95.9	78.1	6.57	0.66	1.25
0.333–0.444	54	71.6	66.0	66.5	4.33	2.17	2.35
0.444–0.555	38	55.4	37.1	44.2	5.44	0.02	0.87
0.555–0.667	31	31.0	19.5	24.9	0.00	6.75	1.51
0.667–0.778	16	12.5	10.1	12.4	0.96	3.40	1.03
0.778–0.889	3	3.67	5.29	5.68	0.12	0.99	1.27
> 0.889	3	0.91	6.32	4.00	4.81	1.75	0.25
Sum	320	320	320	320	47.3	26.7	13.4

Note: Freq. = Frequency; Nor. = Normal; Log. = Lognormal; Square nor. = Square normal.

distribution has a better fit with $T = 13.4$, than the normal with $T = 47.3$, or the lognormal with $T = 26.7$.

These examples clearly demonstrate the advantages of the square normal distribution for fitting statistical data with significant skewness.

B.5.2 Representations of One- and Two-Parameter Distributions

The square normal distribution, as defined in Eqs. (B.31a) and (B.31b) can be used to represent or approximate any two-parameter distributions by equating the respective three moments. This is illustrated with the Gamma, Weibull, lognormal, and Exponential distributions, all of which are one- or two-parameter distributions. Figures B.8a–c show the PDFs of the above distributions depicted as solid lines; in these same figures the respective square normal distributions with the same first three moments as those of the corresponding one- or two-parameter distributions, are depicted as dashed lines. In these figures, all the two-parameter distributions are shown with mean values of $\mu = 25, 30, 35,$ and 40, and coefficients of variation $V = 0.1, 0.2, 0.3,$ and 0.4.

Figures B.8a–c show that the dashed lines coincide closely with the solid lines, demonstrating the flexibility of the square normal distribution for representing two-parameter distributions. This flexibility can be useful in the structural reliability analysis as described in the chapter on Direct Methods of Moment.

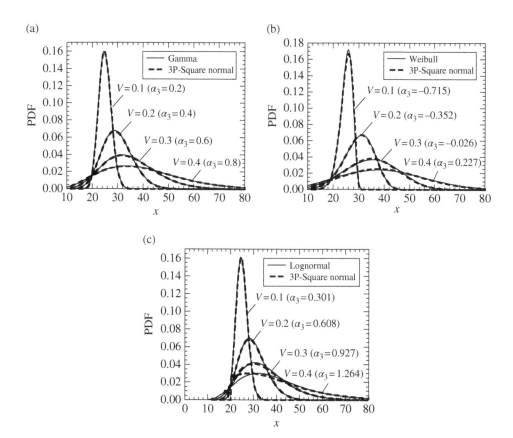

Figure B.8 PDF comparisons with some two-parameter distributions. (a) Gamma distribution. (b) Weibull distribution. (c) Lognormal distribution.

B.5.3 Distributions of Some Random Variables Used in Structural Reliability

As an application in presenting the distributions of some random variables used in structural reliability, the uncertainties included in some properties of structural steel are analysed using the statistical data collected by Ono et al. (1986). The data was collected from 1030 papers and reports published in the journals and reports of the following institutions during the past 30 years:

Architecture Institute of Japan (AIJ);
Japanese Society of Civil Engineers (JSCE);
Japanese Society of Steel Construction (JSSC);
American Society of Civil Engineers (ASCE); and
Welding Research Council, among others.

Therefore, the data have quite good generality to reflect the uncertainties of the basic variables. The statistical results for some basic variables are listed in Table B.3. In Table B.3, the yield stress, ultimate stress, Young's modulus, and elongation correspond to SS41 material (JIS-G-3101 1976) without distinction of the section types. The section area and thickness correspond to rolled and welded H-shape steel without distinction of the type of steel; the residual stress corresponds to the part of the flange of the welded H-shape steel, and Poisson's ratio is for all types of steel and all types of sections because the amount of data are very limited.

Table B.3 Distribution of some basic random variables.

Basic random variables	N	k	Moments			Goodness-of-fit test		
			μ	σ	α_3	Normal	Lognormal	Square normal
(1)	(2)	(3)	(4)	(5)	(6)	(7)	(8)	(9)
Yield stress (t/cm^2)	2195	8	3.055	0.364	0.512	72.11	18.33	13.43
Ultimate stress (t/cm^2)	1932	7	4.549	0.317	0.153	74.25	49.21	55.11
Young's modulus (10^3 t/cm^2)	626	6	2.082	0.096	0.163	20.79	25.44	27.01
Elongation (%)	1572	6	28.2	5.216	0.491	61.26	7.237	10.51
Poisson's ratio	165	6	0.283	0.029	0.639	31.51	16.05	12.85
Section area[a]	718	9	0.990	0.051	0.709	91.80	70.76	23.86
Residual stress[b]	320	9	0.356	0.193	0.823	47.32	26.71	13.86
Thickness[a]	884	7	0.986	0.045	0.649	115.7	84.84	40.87

[a]Measured value/nominal value; [b]Measured residual stress/measured yield stress.

Since these variables were generally treated as normal or lognormal distributions (Ono et al. 1986), in order to investigate which distribution fits the statistical data better, the results of the chi-square test for normal, lognormal, and the square normal distributions are listed in Table B.3. In Table B.3, the goodness of the fit test are obtained from Eq. (B.35), and the smallest goodness of fit among the three distributions are underlined. From Table B.3, one can see that:

1) For Young's modulus, the goodness-of-fit of the normal distribution (20.79) is smaller than those of lognormal (25.44) and the square normal (27.01) distributions, while for elongation, the goodness-of-fit of the lognormal distribution (7.237) is smaller than those of normal (61.26) and the square normal (10.51) distributions. That is to say, the normal and lognormal distributions are suitable to Young's modulus and elongation, respectively.

2) For yield stress, Poisson's ratio, section area, and residual stress, the goodness-of-fit test of the square normal distribution (13.43, 12.85, 23.86, 13.86) are much smaller than those of the normal (72.11, 31.51, 91.80, 47.32) and lognormal (18.33, 16.05, 70.76, 26.71) distributions. It means that the square normal distribution is more suitable for these variables than normal and lognormal distributions.

3) For ultimate stress and thickness, although the goodness-of-fit test of the square normal distribution is relatively small among the three distributions, the goodness-of-fit test is still quite large (55.11 and 40.87). This may be because the first three moments are not enough to express the probability characteristics of the random variables. The use of higher-order moments may be required.

B.6 Summary

In this Appendix, the 3P distributions such as the 3P lognormal distribution and the square normal distribution are introduced, in which the three parameters can be directly defined in terms of the mean value, standard deviation, and skewness. Besides being quite simple and flexible for fitting statistical data of basic random variables, the 3P distributions are used for the derivation of the third-moment reliability index in the chapter on Direct Methods of Moment, and are also convenient for performing normal transformations needed in structural reliability analysis as described in the chapter on Methods of Moment Based on First- and Second-Order Transformation.

Appendix C

Four-Parameter Distributions

C.1 Introduction

We have shown in Appendix B that the two-parameter (2P) distributions such as the normal, lognormal, Gumbel, and Weibull distributions may not be appropriate when the skewness of the statistical data is important and must be reflected in the distribution. Additionally, the three-parameter (3P) distributions such as square normal distribution have been introduced as candidate distributions. The 3P distributions, which can effectively reflect the information of skewness as well as the mean value and standard deviation of statistical data, have more flexibility for fitting statistical data of basic random variables, and can more effectively fit the histograms of available data than 2P distributions.

If the 3P distributions are selected as the candidate distribution and the three parameters are determined, the distribution form and higher-order moments, such as kurtosis, may be evaluated. However, because the kurtosis of the 3P distributions is dependent on the skewness, it may not be consistent with those of the available data. This is illustrated with the two histograms shown in Figure C.1a,b, which represent the observed variability in the properties of H-shape structural steel. Figure C.1a shows the histogram of the thickness, in which the number of the data is 885 and the first four moments of the data are obtained as the mean value $\mu = 0.986$, the standard deviation $\sigma = 0.0457$, the skewness $\alpha_3 = 0.883$, and the kurtosis $\alpha_4 = 5.991$. Figure C.1b shows the histogram of the ultimate stress, in which the number of the data is 1932 and the first four moments of the data are obtained as $\mu = 4.549$, $\sigma = 0.317$, $\alpha_3 = 0.153$, and $\alpha_4 = 6.037$. The kurtosis of the square normal distribution that has the same mean value, standard deviation, and skewness as the data in Figure C.1a,b can be obtained as 4.04 and 3.03, respectively. Apparently, the kurtosis of the 3P distributions is too small to match those of the data for the two illustrated cases.

As illustrated with the data of Figure C.1, the 3P distributions are not flexible enough to reflect the kurtosis of statistical data of a random variable, and distributions that can be determined by effectively using the information of kurtosis as well as the mean value, standard deviation, and skewness of the statistical data are required.

For most practical purposes, it is sufficient to use four parameters that are determined by the first four moments of random variables. There are several existing systems of frequency

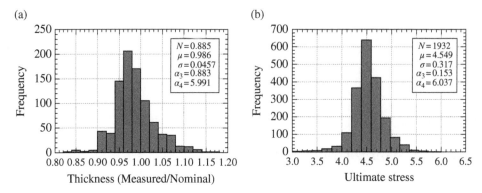

Figure C.1 Two histogram examples of practical data. (a) Thickness (Measured/Nominal). (b) Ultimate stress.

curves that have four parameters, such as the Pearson, Johnson, and Burr systems (Pearson et al. 1979; Slifker and Shapiro 1980; Stuart and Ord 1987), and Ramberg's Lambda distribution (Ramberg et al. 1979). Recently, the cubic normal distribution for practical engineering has been developed by Zhao and Lu (2007a, 2008) and Zhao et al. (2018b). The four parameters of each system are determined in terms of the mean value, standard deviation, skewness, and kurtosis of the sample data. The Johnson system has been investigated by Parkinson (1978) and Hong (1996), and the Lambda distribution has been investigated by Grigoriu (1983) and Zhao et al. (2006b). For the sake of simplicity, generality, and flexibility, the Pearson system and the cubic normal distribution will be introduced in this appendix.

C.2   The Pearson System

C.2.1   Definition of the System

This system was originated by Pearson (1895). For every member of the Pearson system, $f_Y(y)$, the PDF of a random variable Y satisfies a differential equation of form (Stuart and Ord 1987)

$$\frac{1}{f}\frac{df}{dy} = -\frac{e+y}{c_0 + c_1 y + c_2 y^2} \tag{C.1}$$

in which

$$c_0 = \frac{\mu_2(4\beta_2 - 3\beta_1)}{(10\beta_2 - 12\beta_1 - 18)} \tag{C.2a}$$

$$e = c_1 = \frac{\sqrt{\beta_1}\sqrt{\mu_2}(\beta_2 + 3)}{(10\beta_2 - 12\beta_1 - 18)} \tag{C.2b}$$

$$c_2 = \frac{(2\beta_2 - 3\beta_1 - 6)}{(10\beta_2 - 12\beta_1 - 18)} \tag{C.2c}$$

$\beta_1 = \alpha_3{}^2, \beta_2 = \alpha_4$ are the first- and second-moment ratios, and μ_2 is the second-order central moment, respectively, of Y.

Without loss of generality, we write

$$X = \frac{Y - \mu}{\sigma}$$

where μ and σ are the mean value and standard deviation of Y, respectively. Then, $\mu_2 = 1$, Eq. (C.1) can be rewritten as

$$\frac{1}{f}\frac{df}{dx} = -\frac{ax + b}{c + bx + dx^2} \tag{C.3}$$

where parameters a, b, c, d are expressed as

$$a = 10\alpha_4 - 12\alpha_3^2 - 18 \tag{C.4a}$$

$$b = \alpha_3(\alpha_4 + 3) \tag{C.4b}$$

$$c = 4\alpha_4 - 3\alpha_3^2 \tag{C.4c}$$

$$d = 2\alpha_4 - 3\alpha_3^2 - 6 \tag{C.4d}$$

Since for an arbitrary random variable (Stuart and Ord 1987),

$$\alpha_4 \geq \alpha_3^2 + 1 \tag{C.5a}$$

then

$$c = 4\alpha_4 - 3\alpha_3^2 \geq 4 \tag{C.5b}$$

always holds true.

Since the conditions

$$f(x) \geq 0 \tag{C.6a}$$

and

$$\int_{-\infty}^{\infty} f(x)dx = 1 \tag{C.6b}$$

must be satisfied, it follows from Eq. (C.3) that $f(x)$ must tend to zero as x tends to infinity, and so, also must df/dx. This may not be true of formal solutions of Eq. (C.3), but in such cases the condition Eq. (C.6a) is not satisfied and it is necessary to restrict the range of values of x to those for which Eq. (C.6a) is satisfied, and assign the value $f(x) = 0$ when x is outside this range.

C.2.2 Various Types of the PDF in the Pearson System

The shape of the curve representing the probability density function varies considerably with a, b, c, d, and various types of the PDF of x can be obtained as follows, according to the relative values of b and d. In the following, the distribution will be discussed according to whether b equals to 0 or not, i.e. the distribution is symmetry ($\alpha_3 = 0$) or asymmetry ($\alpha_3 \neq 0$).

1) $b \neq 0$ $(a_3 \neq 0)$

For $\Delta = b^2 - 4cd > 0$, there are two real roots for equation $c + bx + dx^2 = 0$:

$$r_1 = \frac{-b - \sqrt{\Delta}}{2d}, \quad r_2 = \frac{-b + \sqrt{\Delta}}{2d}$$

When $d < 0$, since $\Delta > b^2$, the two roots are of the opposite signs and $r_1 > 0$, $r_2 < 0$, for either $b > 0$ or $b < 0$, Eq. (C.3) is equivalent to

$$\frac{1}{f}\frac{df}{dx} = -\frac{ax + b}{d(r_1 - x)(x - r_2)} \tag{C.7a}$$

f can be given as

$$f(x) = K(x - r_2)^{\frac{-1}{\sqrt{\Delta}}(ar_2 + b)}(r_1 - x)^{\frac{1}{\sqrt{\Delta}}(ar_1 + b)} \tag{C.7b}$$

where K is determined by $F_x(+\infty) = 1$.

Equation (C.7b) is corresponding to the Pearson type I distribution. The range of x should be limited so that $r_1 - x > 0$ and $x - r_2 > 0$, which means

$$r_2 < x < r_1$$

The PDFs of type I distributions are shown in Figure C.2.

When $d = 0$, Eq. (C.3) degenerates as

$$\frac{1}{f}\frac{df}{dx} = -\frac{ax + b}{c + bx} \tag{C.8a}$$

f can be given as

$$f(x) = K(c + bx)^{(ac - b^2)/b^2} \exp\left[-\frac{ax}{b}\right] \tag{C.8b}$$

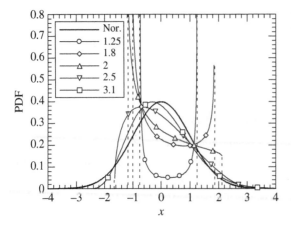

Figure C.2 The PDF of Type I distribution.

the range of x should be limited so that $c + bx > 0$, which means $x > -b/c$ for $b > 0$ and $x < -b/c$ for $b < 0$. Eq. (C.8b) is corresponding to the Pearson type III distribution.

When $d > 0$, since $\Delta < b^2$, the two roots are of the same sign and $r_1 < r_2$. If $b > 0$, then, $r_1 < r_2 < 0$, Eq. (C.3) is equivalent to

$$\frac{1}{f}\frac{df}{dx} = -\frac{ax + b}{d(x - r_1)(x - r_2)} \tag{C.9a}$$

f can be given as

$$f(x) = K(x - r_1)^{\frac{1}{\sqrt{\Delta}}(ar_1 + b)}(x - r_2)^{\frac{-1}{\sqrt{\Delta}}(ar_2 + b)} \tag{C.9b}$$

The range of x should be limited so that $x - r_1 > 0$ and $x - r_2 > 0$, which means

$$x > r_2$$

If $b < 0$, then, $0 < r_1 < r_2$, Eq. (C.3) is equivalent to

$$\frac{1}{f}\frac{df}{dx} = -\frac{ax + b}{d(r_1 - x)(r_2 - x)} \tag{C.10a}$$

f can be given as

$$f(x) = K(r_1 - x)^{\frac{1}{\sqrt{\Delta}}(ar_1 + b)}(r_2 - x)^{\frac{-1}{\sqrt{\Delta}}(ar_2 + b)} \tag{C.10b}$$

The range of x should be limited so that $r_1 - x > 0$ and $r_2 - x > 0$, which means

$$x < r_1$$

Equations (C.9b) and (C.10b) can be summarized as

$$f(x) = K|x - r_1|^{\frac{1}{\sqrt{\Delta}}(ar_1 + b)}|x - r_2|^{\frac{-1}{\sqrt{\Delta}}(ar_2 + b)}, \quad x < r_1 \text{ for } b < 0, x > r_2 \text{ for } b > 0 \tag{C.11}$$

Equation (C.11) corresponds to the Pearson type VI distribution.

For $\Delta = b^2 - 4cd = 0$, there is only one real root for equation $c + bx + dx^2 = 0$,

$$r_1 = r_2 = r_0 = -\frac{b}{2d}$$

Since $c > 0$, we have $d > 0$. Therefore, $r_0 > 0$ for $b < 0$ and $r_0 < 0$ for $b > 0$. Equation (C.3) degenerates as

$$\frac{1}{f}\frac{df}{dx} = -\frac{ax + b}{(x - r_0)^2}, \quad \text{for } b > 0 \tag{C.12a}$$

$$\frac{1}{f}\frac{df}{dx} = -\frac{ax + b}{(r_0 - x)^2}, \quad \text{for } b < 0 \tag{C.12b}$$

respectively, and f can be given as

$$f(x) = K(x - r_0)^{-\frac{a}{d}}\exp\left[\frac{ar_0 + b}{d(x - r_0)}\right], \quad x > r_0 \tag{C.13a}$$

$$f(x) = K(r_0 - x)^{-\frac{q}{d}} \exp\left[\frac{ar_0 + b}{d(x - r_0)}\right], \quad x < r_0 \tag{C.13b}$$

Equations (C.13a) and (C.13b) can be summarized as

$$f(x) = K|x - r_0|^{-\frac{q}{d}} \exp\left[\frac{ar_0 + b}{d(x - r_0)}\right], \quad x < r_0 \text{ for } b < 0, x > r_0 \text{ for } b > 0 \tag{C.14}$$

Equation (C.14) corresponds to the Pearson type V distribution.

For $\Delta = b^2 - 4cd < 0$, there is no real root for equation $c + bx + dx^2 = 0$, Eq. (C.3) is equivalent to

$$\frac{1}{f}\frac{df}{dx} = -\frac{ax + b}{(c - b^2/d/4) + d(x + b/d/2)^2} \tag{C.15a}$$

f can be given as

$$f(x) = K(c + bx + dx^2)^{-\frac{a}{2d}} \exp\left[\frac{ab - 2bd}{d\sqrt{-\Delta}} \tan^{-1}\left(\frac{b + 2dx}{\sqrt{-\Delta}}\right)\right] \tag{C.15b}$$

Since $c + bx + dx^2 > 0$ always holds true in this case, the range of x is $-\infty < x < +\infty$. Equation (C.15b) corresponds to the Pearson type IV distribution.

2) $b = 0$ ($\alpha_3 = 0$)

For $\Delta = b^2 - 4cd > 0$, which means $d < 0$, $\alpha_3 = 0$, $\alpha_4 < 3$, there are two real roots for equation $c + bx + dx^2 = 0$:

$$r_1 = -\frac{\sqrt{\Delta}}{2d} > 0, \quad r_2 = \frac{\sqrt{\Delta}}{2d} < 0$$

f can be given as

$$f(x) = K\left(-\frac{c}{d} - x^2\right)^{\frac{-a}{2d}} \tag{C.16}$$

The range of x should be limited so that $r_1 - x > 0$ and $x - r_2 > 0$, which means

$$\frac{\sqrt{\Delta}}{2d} < x < -\frac{\sqrt{\Delta}}{2d}$$

Equation (C.16) corresponds to the Pearson type II distribution, the PDFs of which are shown in Figure C.3.

In particular, when $\alpha_3 = 0$, $\alpha_4 = 1$, we have, $b = 0$, $a = -8$, $c = 4$, $d = -4$, $\sqrt{\Delta} = 8$

$$f(x) = 0.05232\frac{1}{1 - x^2}, \quad -1 < x < 1$$

When $\alpha_3 = 0$, $\alpha_4 = 1.5$, we have, $b = 0$, $a = -3$, $c = 6$, $d = -3$, $\sqrt{\Delta} = 6\sqrt{2}$

$$f(x) = 0.3184\frac{1}{\sqrt{2 - x^2}}, \quad -\sqrt{2} < x < \sqrt{2}$$

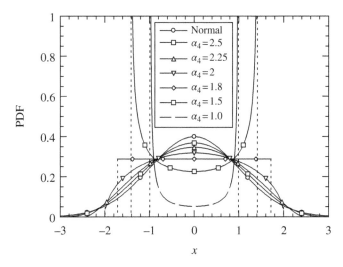

Figure C.3 The PDFs of Type II distributions.

When $\alpha_3 = 0$, $\alpha_4 = 1.8$, we have, $b = 0$, $a = 0$, $c = 7.2$, $d = -2.4$, $\sqrt{\Delta} = 4.8\sqrt{3}$

$$f(x) = \frac{1}{6}\sqrt{3}, \quad -\sqrt{3} < x < \sqrt{3}$$

When $\alpha_3 = 0$, $\alpha_4 = 2$, we have, $b = 0$, $a = 2$, $c = 8$, $d = -2$, $\sqrt{\Delta} = 8$

$$f(x) = 0.1592\sqrt{4 - x^2}, \quad -2 < x < 2$$

When $\alpha_3 = 0$, $\alpha_4 = 2.25$, we have, $b = 0$, $a = 4.5$, $c = 9$, $d = -1.5$, $\sqrt{\Delta} = 3\sqrt{6}$

$$f(x) = 0.02358\sqrt[3]{6 - x^2}, \quad -\sqrt{6} < x < \sqrt{6}$$

When $\alpha_3 = 0$, $\alpha_4 = 2.5$, we have, $b = 0$, $a = 7$, $c = 10$, $d = -1$, $\sqrt{\Delta} = 2\sqrt{10}$

$$f(x) = 0.0001164\left(10 - x^2\right)^{3.5}, \quad -\sqrt{10} < x < \sqrt{10}$$

For $\Delta = b^2 - 4cd = 0$, which means $d = 0$, $\alpha_3 = 0$, $\alpha_4 = 3$, Eq. (C.3) degenerates as

$$\frac{1}{f}\frac{df}{dx} = -\frac{ax}{c} \tag{C.17a}$$

f can be given as

$$f(x) = K \exp\left[-\frac{ax^2}{2c}\right] \tag{C.17b}$$

Equation (C.17b) corresponds to the PDF of a standard normal random variable. Since $a = c = 12$, Eq. (C.17b) can be expressed as

$$f(x) = \frac{1}{\sqrt{2\pi}} \exp\left(-\frac{x^2}{2}\right) \tag{C.17c}$$

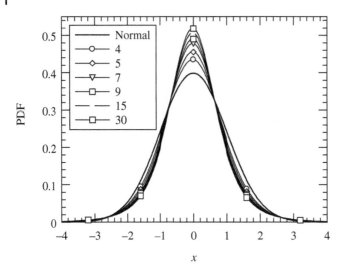

Figure C.4 The PDFs of Type VII distributions.

For $\Delta = b^2 - 4cd < 0$, which means $d > 0$, $\alpha_3 = 0$, $\alpha_4 > 3$, there is no real root for equation $c + bx + dx^2 = 0$.

f can be given as

$$f(x) = K\left(\frac{c}{d} + x^2\right)^{\frac{-a}{2d}}, \quad -\infty < x < \infty \tag{C.18}$$

Equation (C.18) corresponds to the Pearson type VII distribution, the PDFs of which are shown in Figure C.4.

In particular, when $\alpha_3 = 0$, $\alpha_4 = 4$, we have, $b = 0$, $a = 22$, $c = 16$, $d = 2$,

$$f(x) = 40320\frac{1}{(8 + x^2)^{5.5}}, \quad -\infty < x < \infty$$

When $\alpha_3 = 0$, $\alpha_4 = 5$, we have, $b = 0$, $a = 32$, $c = 20$, $d = 4$,

$$f(x) = 284.7\frac{1}{(5 + x^2)^4}, \quad -\infty < x < \infty$$

When $\alpha_3 = 0$, $\alpha_4 = 9$, we have, $b = 0$, $a = 72$, $c = 36$, $d = 12$,

$$f(x) = 13.232\frac{1}{(3 + x^2)^3}, \quad -\infty < x < \infty$$

The member of Pearson system is summarized in Table C.1 and the division of the $\alpha_3^2 - \alpha_4$ plane among the various types is exhibited in Figure C.5.

Table C.1 Various types of the PDF in Pearson system.

Parameter		PDF	Range	Type
$b \neq 0$	$\Delta > 0,\, d < 0$	$f(x) = K(x-r_2)^{\frac{-ar_2-b}{\sqrt{\Delta}}}(r_1-x)^{\frac{ar_1+b}{\sqrt{\Delta}}}$	$r_2 < x < r_1$	I
	$\Delta > 0,\, d = 0$	$f(x) = K(c+bx)^{\frac{ac-b^2}{b^2}}\exp\left[-\frac{ax}{b}\right]$	$x < -\dfrac{b}{c},\quad$ for $b<0$ $x > -\dfrac{b}{c},\quad$ for $b>0$	III
	$\Delta > 0,\, d > 0$	$f(x) = K\lvert x-r_1\rvert^{\frac{ar_1+b}{\sqrt{\Delta}}}\lvert x-r_2\rvert^{\frac{-ar_2-b}{\sqrt{\Delta}}}$	$x < r_1,\quad$ for $b<0$ $x > r_2,\quad$ for $b>0$	VI
	$\Delta = 0$	$f(x) = K\lvert x-r_0\rvert^{-\frac{a}{d}}\exp\left[\frac{ar_0+b}{d(x-r_0)}\right]$	$x < r_0,\quad$ for $b<0$ $x > r_0,\quad$ for $b>0$	V
	$\Delta < 0$	$f(x) = K\left(c+bx+dx^2\right)^{-\frac{a}{2d}}$ $\exp\left[\frac{ab-2bd}{d\sqrt{-\Delta}}\tan^{-1}\left(\frac{b+2dx}{\sqrt{-\Delta}}\right)\right]$	$-\infty < x < +\infty$	IV
$b = 0$	$\Delta > 0\ (d < 0)$	$f(x) = K\left(-\frac{c}{d}-x^2\right)^{\frac{-a}{2d}}$	$\dfrac{\sqrt{\Delta}}{2d} < x < -\dfrac{\sqrt{\Delta}}{2d}$	II
	$\Delta = 0\ (d = 0)$	$f(x) = \dfrac{1}{\sqrt{2\pi}}\exp\left[-\frac{x^2}{2}\right]$	$-\infty < x < +\infty$	N
	$\Delta < 0\ (d > 0)$	$f(x) = K\left(\frac{c}{d}+x^2\right)^{\frac{-a}{2d}}$	$-\infty < x < +\infty$	VII

Note: $\Delta = b^2 - 4cd$, $r_1 = \dfrac{-b-\sqrt{\Delta}}{2d}$, $r_2 = \dfrac{-b+\sqrt{\Delta}}{2d}$, $r_0 = \dfrac{-b}{2d}$, and K are determined by $F_X(\infty) = 1$.

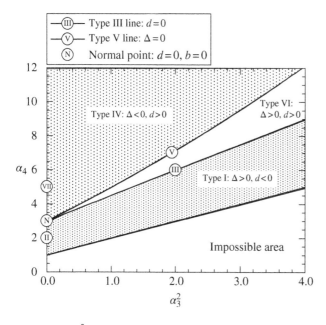

Figure C.5 $\alpha_3^2 - \alpha_4$ plane for Pearson system.

C.3 Cubic Normal Distribution

C.3.1 Definition of the Distribution

The distribution is defined on the base of the following polynomial normal transformation (Fleishman 1978; Hong and Lind 1996; Chen and Tung 2003; Zhao and Lu 2007a)

$$X = \frac{Y - \mu}{\sigma} = a_1 + a_2 U + a_3 U^2 + a_4 U^3 \tag{C.19}$$

The CDF and PDF corresponding to Eq. (C.19) are expressed as (Zhao and Lu 2008; Zhao et al. 2018b)

$$F(x) = \Phi(u) \tag{C.20a}$$

$$f(x) = \frac{\phi(u)}{\sigma(a_2 + 2a_3 u + 3a_4 u^2)} \tag{C.20b}$$

in which F, f, μ, and σ are the CDF, PDF, mean value, and standard deviation of Y, respectively; Φ and ϕ are the CDF and PDF of a standard normal random variable U; and a_1, a_2, a_3, and a_4 are deterministic coefficients. Since the distribution is defined by the third-order polynomial of standard normal random variables, it has been named cubic normal distribution (Zhao and Lu 2008).

The complete monotonic expressions of the inverse third-order polynomial transformation (Zhao et al. 2018a), i.e. x-u transformation, has been discussed in Section 7.4.3 (see Table 7.9) in the chapter on Transformation of Non-Normal Variables to Independent Normal Variables. The polynomial coefficients a_1, a_2, a_3, and a_4 can be determined by moment-matching method as has been discussed in the section just mentioned. Here, for convenience, a table to approximate the parameters of a_1, a_2, a_3, and a_4 is given in Table C.2. If the values of α_3 and α_4 are known, the parameters a_1, a_2, a_3, and a_4 can be determined from Table C.2 using the α_3 and α_4 values as entry points. One simply picks the values of a_1, a_2, a_3, and a_4 for which the α_3 and α_4 are closest to the desired values. If α_3 is negative, one uses its absolute value, and after finding the values of a_1, a_2, a_3, and a_4, changes the sign of a_1 and a_3 (the density with a skewness of $-\alpha_3$ is the mirror image of the density with a skewness of α_3).

C.3.2 Representative PDFs of the Distribution

Once the parameters are determined, the probability density curves can be plotted with the aid of Eqs. (C.20a) and (C.20b).

The representative standard PDFs of this distribution (each has a mean of zero and standard deviation of one) includes a wide range of curve shapes as illustrated by the density plots in Figure C.6. The densities are indexed by the values of the skewness α_3 and kurtosis α_4. In Figure C.6a–c, the skewness is fixed and three values of kurtosis are illustrated, while in Figure C.6d–f, the kurtosis is fixed and three values of skewness are illustrated. From Figure C.6, one can see that the distribution reflects the characteristics of the skewness and kurtosis quite well. One can also see that the left tail of PDF is long for negative α_3 and the right tail is long for positive α_3. This characteristic is especially important when the distribution is used as a fourth-moment reliability index as described in the chapter on Direct Methods of Moment.

Table C.2 The four parameters of a_1, a_2, a_3, and a_4 for given values of skewness α_3 and kurtosis α_4 (Cubic normal distribution).

	$\alpha_3 = 0.00$				$\alpha_3 = 0.05$				$\alpha_3 = 0.10$		
α_4	$a_1 = -a_3$	a_2	a_4	α_4	$a_1 = -a_3$	a_2	a_4	α_4	$a_1 = -a_3$	a_2	a_4
2.0	0.0	1.2210	−0.0802	2.0	−0.0138	1.2224	−0.0808	2.0	−0.0280	1.2268	−0.0828
2.2	0.0	1.1478	−0.0520	2.2	−0.0116	1.1487	−0.0524	2.2	−0.0233	1.1514	−0.0535
2.4	0.0	1.0972	−0.0335	2.4	−0.0103	1.0979	−0.0338	2.4	−0.0206	1.0998	−0.0346
2.6	0.0	1.0585	−0.0199	2.6	−0.0094	1.0590	−0.0201	2.6	−0.0189	1.0605	−0.0207
2.8	0.0	1.0269	−0.0090	2.8	−0.0088	1.0273	−0.0092	2.8	−0.0177	1.0285	−0.0097
3.0	0.0	1.0000	0.0000	3.0	−0.0083	1.0004	−0.0001	3.0	−0.0167	1.0014	−0.0006
3.2	0.0	0.9765	0.0078	3.2	−0.0080	0.9768	0.0076	3.2	−0.0160	0.9777	0.0073
3.4	0.0	0.9555	0.0146	3.4	−0.0076	0.9558	0.0145	3.4	−0.0153	0.9566	0.0142
3.6	0.0	0.9365	0.0207	3.6	−0.0074	0.9368	0.0206	3.6	−0.0148	0.9375	0.0203
3.8	0.0	0.9191	0.0263	3.8	−0.0072	0.9193	0.0262	3.8	−0.0143	0.9200	0.0259
4.0	0.0	0.9030	0.0314	4.0	−0.0070	0.9032	0.0313	4.0	−0.0139	0.9038	0.0310
4.2	0.0	0.8879	0.0361	4.2	−0.0068	0.8881	0.0360	4.2	−0.0136	0.8886	0.0358
4.4	0.0	0.8738	0.0404	4.4	−0.0066	0.8739	0.0404	4.4	−0.0132	0.8744	0.0402
4.6	0.0	0.8604	0.0445	4.6	−0.0065	0.8606	0.0445	4.6	−0.0129	0.8610	0.0443
4.8	0.0	0.8477	0.0484	4.8	−0.0063	0.8479	0.0483	4.8	−0.0127	0.8483	0.0482
5.0	0.0	0.8357	0.0521	5.0	−0.0062	0.8358	0.0520	5.0	−0.0124	0.8362	0.0518
5.2	0.0	0.8241	0.0555	5.2	−0.0061	0.8243	0.0555	5.2	−0.0122	0.8247	0.0553
5.4	0.0	0.8131	0.0588	5.4	−0.0060	0.8132	0.0588	5.4	−0.0120	0.8136	0.0586
5.6	0.0	0.8025	0.0620	5.6	−0.0059	0.8026	0.0619	5.6	−0.0118	0.8029	0.0618
5.8	0.0	0.7922	0.0650	5.8	−0.0058	0.7923	0.0650	5.8	−0.0116	0.7927	0.0648
6.0	0.0	0.7824	0.0679	6.0	−0.0057	0.7825	0.0679	6.0	−0.0114	0.7828	0.0677
6.2	0.0	0.7728	0.0707	6.2	−0.0056	0.7729	0.0707	6.2	−0.0113	0.7732	0.0705
6.4	0.0	0.7636	0.0734	6.4	−0.0056	0.7637	0.0733	6.4	−0.0111	0.7639	0.0732
6.6	0.0	0.7546	0.0760	6.6	−0.0055	0.7547	0.0759	6.6	−0.0110	0.7550	0.0758
6.8	0.0	0.7459	0.0785	6.8	−0.0054	0.7460	0.0785	6.8	−0.0109	0.7462	0.0784
7.0	0.0	0.7374	0.0809	7.0	−0.0054	0.7375	0.0809	7.0	−0.0107	0.7377	0.0808
7.2	0.0	0.7291	0.0833	7.2	−0.0053	0.7292	0.0833	7.2	−0.0106	0.7295	0.0832
7.4	0.0	0.7211	0.0856	7.4	−0.0052	0.7211	0.0855	7.4	−0.0105	0.7214	0.0854
7.6	0.0	0.7132	0.0878	7.6	−0.0052	0.7133	0.0878	7.6	−0.0104	0.7135	0.0877
7.8	0.0	0.7055	0.0900	7.8	−0.0051	0.7056	0.0899	7.8	−0.0103	0.7058	0.0899
8.0	0.0	0.6980	0.0921	8.0	−0.0051	0.6981	0.0920	8.0	−0.0102	0.6983	0.0920
8.2	0.0	0.6907	0.0941	8.2	−0.0050	0.6907	0.0941	8.2	−0.0101	0.6909	0.0940
8.4	0.0	0.6835	0.0961	8.4	−0.0050	0.6835	0.0961	8.4	−0.0100	0.6837	0.0960
8.6	0.0	0.6764	0.0981	8.6	−0.0049	0.6765	0.0981	8.6	−0.0099	0.6767	0.0980
8.8	0.0	0.6695	0.1000	8.8	−0.0049	0.6695	0.1000	8.8	−0.0098	0.6697	0.0999
9.0	0.0	0.6627	0.1019	9.0	−0.0049	0.6627	0.1019	9.0	−0.0097	0.6629	0.1018

(*Continued*)

Table C.2 (Continued)

	$\alpha_3 = 0.15$				$\alpha_3 = 0.20$				$\alpha_3 = 0.25$		
α_4	$a_1 = -a_3$	a_2	a_4	α_4	$a_1 = -a_3$	a_2	a_4	α_4	$a_1 = -a_3$	a_2	a_4
2.0	−0.0429	1.2343	−0.0863	2.0	−0.0590	1.2459	−0.0917	2.0	−0.0775	1.2634	−0.1000
2.2	−0.0354	1.1560	−0.0555	2.2	−0.0481	1.1627	−0.0585	2.2	−0.0617	1.1718	−0.0625
2.4	−0.0313	1.1032	−0.0360	2.4	−0.0422	1.1079	−0.0380	2.4	−0.0537	1.1143	−0.0407
2.6	−0.0286	1.0631	−0.0218	2.6	−0.0385	1.0668	−0.0233	2.6	−0.0487	1.0716	−0.0253
2.8	−0.0267	1.0306	−0.0106	2.8	−0.0358	1.0336	−0.0118	2.8	−0.0453	1.0375	−0.0134
3.0	−0.0252	1.0032	−0.0013	3.0	−0.0338	1.0057	−0.0023	3.0	−0.0426	1.0090	−0.0036
3.2	−0.0240	0.9792	0.0067	3.2	−0.0322	0.9814	0.0058	3.2	−0.0406	0.9842	0.0047
3.4	−0.0231	0.9580	0.0137	3.4	−0.0309	0.9599	0.0129	3.4	−0.0389	0.9623	0.0119
3.6	−0.0223	0.9387	0.0199	3.6	−0.0298	0.9404	0.0192	3.6	−0.0374	0.9426	0.0183
3.8	−0.0215	0.9211	0.0255	3.8	−0.0288	0.9226	0.0249	3.8	−0.0362	0.9246	0.0241
4.0	−0.0209	0.9048	0.0307	4.0	−0.0280	0.9062	0.0301	4.0	−0.0352	0.9080	0.0294
4.2	−0.0204	0.8896	0.0354	4.2	−0.0273	0.8908	0.0349	4.2	−0.0342	0.8925	0.0343
4.4	−0.0199	0.8753	0.0398	4.4	−0.0266	0.8765	0.0394	4.4	−0.0334	0.8780	0.0388
4.6	−0.0195	0.8618	0.0440	4.6	−0.0260	0.8629	0.0436	4.6	−0.0326	0.8644	0.0430
4.8	−0.0190	0.8491	0.0479	4.8	−0.0255	0.8501	0.0475	4.8	−0.0319	0.8514	0.0470
5.0	−0.0187	0.8369	0.0516	5.0	−0.0250	0.8379	0.0512	5.0	−0.0313	0.8391	0.0507
5.2	−0.0183	0.8253	0.0551	5.2	−0.0245	0.8262	0.0547	5.2	−0.0307	0.8274	0.0540
5.4	−0.0180	0.8142	0.0584	5.4	−0.0241	0.8151	0.0581	5.4	−0.0302	0.8162	0.0576
5.6	−0.0177	0.8035	0.0616	5.6	−0.0237	0.8043	0.0613	5.6	−0.0297	0.8054	0.0608
5.8	−0.0175	0.7932	0.0646	5.8	−0.0233	0.7940	0.0643	5.8	−0.0292	0.7950	0.0639
6.0	−0.0172	0.7833	0.0675	6.0	−0.0230	0.7841	0.0672	6.0	−0.0288	0.7850	0.0669
6.2	−0.0170	0.7737	0.0703	6.2	−0.0226	0.7744	0.0701	6.2	−0.0284	0.7754	0.0697
6.4	−0.0167	0.7644	0.0730	6.4	−0.0223	0.7651	0.0728	6.4	−0.0280	0.7660	0.0724
6.6	−0.0165	0.7554	0.0757	6.6	−0.0220	0.7561	0.0754	6.6	−0.0276	0.7569	0.0751
6.8	−0.0163	0.7467	0.0782	6.8	−0.0218	0.7473	0.0779	6.8	−0.0273	0.7481	0.0776
7.0	−0.0161	0.7382	0.0806	7.0	−0.0215	0.7388	0.0804	7.0	−0.0269	0.7396	0.0801
7.2	−0.0159	0.7299	0.0830	7.2	-0.0213	0.7305	0.0828	7.2	−0.0266	0.7312	0.0825
7.4	−0.0157	0.7218	0.0853	7.4	−0.0210	0.7224	0.0851	7.4	−0.0263	0.7231	0.0848
7.6	−0.0156	0.7139	0.0875	7.6	−0.0208	0.7145	0.0873	7.6	−0.0260	0.7152	0.0870
7.8	−0.0154	0.7062	0.0897	7.8	−0.0206	0.7067	0.0895	7.8	−0.0258	0.7074	0.0892
8.0	−0.0153	0.6987	0.0918	8.0	−0.0204	0.6992	0.0916	8.0	−0.0255	0.6999	0.0914
8.2	−0.0151	0.6913	0.0939	8.2	−0.0202	0.6918	0.0937	8.2	−0.0253	0.6924	0.0934
8.4	−0.0150	0.6841	0.0959	8.4	−0.0200	0.6846	0.0957	8.4	−0.0250	0.6852	0.0955
8.6	−0.0148	0.6770	0.0979	8.6	−0.0198	0.6775	0.0977	8.6	−0.0248	0.6781	0.0975
8.8	−0.0147	0.6701	0.0998	8.8	−0.0196	0.6705	0.0996	8.8	−0.0246	0.6711	0.0994
9.0	−0.0146	0.6633	0.1017	9.0	−0.0195	0.6637	0.1015	9.0	−0.0244	0.6643	0.1013

Table C.2 (Continued)

	$\alpha_3 = 0.30$				$\alpha_3 = 0.35$				$\alpha_3 = 0.40$		
α_4	$a_1 = -a_3$	a_2	a_4	α_4	$a_1 = -a_3$	a_2	a_4	α_4	$a_1 = -a_3$	a_2	a_4
2.2	−0.0766	1.1836	−0.0678	2.2	−0.0936	1.1989	−0.0749	2.2	−0.1136	1.2192	−0.0848
2.4	−0.0660	1.1225	−0.0443	2.4	−0.0793	1.1328	−0.0488	2.4	−0.0941	1.1455	−0.0545
2.6	−0.0595	1.0778	−0.0279	2.6	−0.0709	1.0855	−0.0312	2.6	−0.0832	1.0949	−0.0352
2.8	−0.0550	1.0425	−0.0154	2.8	−0.0653	1.0486	−0.0179	2.8	−0.0761	1.0560	−0.0210
3.0	−0.0517	1.0131	−0.0053	3.0	−0.0611	1.0182	−0.0074	3.0	−0.0710	1.0242	−0.0099
3.2	−0.0491	0.9878	0.0033	3.2	−0.0579	0.9921	0.0015	3.2	−0.0671	0.9972	−0.0006
3.4	−0.0470	0.9654	0.0107	3.4	−0.0554	0.9692	0.0092	3.4	−0.0640	0.9736	0.0074
3.6	−0.0452	0.9454	0.0172	3.6	−0.0532	0.9487	0.0159	3.6	−0.0614	0.9526	0.0143
3.8	−0.0437	0.9271	0.0231	3.8	−0.0514	0.9301	0.0219	3.8	−0.0592	0.9336	0.0205
4.0	−0.0424	0.9102	0.0285	4.0	−0.0498	0.9129	0.0274	4.0	−0.0574	0.9161	0.0262
4.2	−0.0413	0.8946	0.0335	4.2	−0.0484	0.8971	0.0325	4.2	−0.0557	0.9000	0.0313
4.4	−0.0402	0.8799	0.0380	4.4	−0.0472	0.8822	0.0371	4.4	−0.0543	0.8849	0.0361
4.6	−0.0393	0.8661	0.0423	4.6	−0.0461	0.8683	0.0415	4.6	−0.0530	0.8708	0.0405
4.8	−0.0384	0.8531	0.0463	4.8	−0.0451	0.8551	0.0455	4.8	−0.0518	0.8574	0.0446
5.0	−0.0377	0.8407	0.0501	5.0	−0.0441	0.8426	0.0494	5.0	−0.0507	0.8447	0.0485
5.2	−0.0370	0.8289	0.0537	5.2	−0.0433	0.8306	0.0530	5.2	−0.0497	0.8327	0.0522
5.4	−0.0363	0.8176	0.0571	5.4	−0.0425	0.8192	0.0564	5.4	−0.0488	0.8212	0.0557
5.6	−0.0357	0.8067	0.0603	5.6	−0.0418	0.8083	0.0597	5.6	−0.0480	0.8101	0.0590
5.8	−0.0351	0.7963	0.0634	5.8	−0.0411	0.7978	0.0629	5.8	−0.0472	0.7995	0.0622
6.0	−0.0346	0.7862	0.0664	6.0	−0.0405	0.7877	0.0659	6.0	−0.0465	0.7893	0.0652
6.2	−0.0341	0.7765	0.0693	6.2	−0.0399	0.7779	0.0687	6.2	−0.0458	0.7795	0.0681
6.4	−0.0336	0.7671	0.0720	6.4	−0.0394	0.7684	0.0715	6.4	−0.0451	0.7699	0.0709
6.6	−0.0332	0.7580	0.0747	6.6	−0.0388	0.7592	0.0742	6.6	−0.0445	0.7607	0.0736
6.8	−0.0328	0.7491	0.0772	6.8	−0.0383	0.7503	0.0768	6.8	−0.0439	0.7517	0.0762
7.0	−0.0324	0.7405	0.0797	7.0	−0.0379	0.7417	0.0793	7.0	−0.0434	0.7430	0.0788
7.2	−0.0320	0.7322	0.0821	7.2	−0.0374	0.7333	0.0817	7.2	−0.0429	0.7346	0.0812
7.4	−0.0316	0.7240	0.0844	7.4	−0.0370	0.7251	0.0840	7.4	−0.0424	0.7263	0.0836
7.6	−0.0313	0.7160	0.0867	7.6	−0.0366	0.7171	0.0863	7.6	−0.0419	0.7183	0.0858
7.8	−0.0310	0.7083	0.0889	7.8	−0.0362	0.7093	0.0885	7.8	−0.0415	0.7104	0.0881
8.0	−0.0307	0.7007	0.0911	8.0	−0.0358	0.7017	0.0907	8.0	−0.0411	0.7028	0.0902
8.2	−0.0304	0.6932	0.0931	8.2	−0.0355	0.6942	0.0928	8.2	−0.0406	0.6953	0.0924
8.4	−0.0301	0.6860	0.0952	8.4	−0.0351	0.6869	0.0948	8.4	−0.0403	0.6880	0.0944
8.6	−0.0298	0.6788	0.0972	8.6	−0.0348	0.6797	0.0968	8.6	−0.0399	0.6808	0.0964
8.8	−0.0295	0.6719	0.0991	8.8	−0.0345	0.6727	0.0988	8.8	−0.0395	0.6737	0.0984
9.0	−0.0293	0.6650	0.1010	9.0	−0.0342	0.6658	0.1007	9.0	−0.0392	0.6668	0.1003
9.2	−0.0290	0.6583	0.1029	9.2	−0.0339	0.6591	0.1026	9.2	−0.0388	0.6600	0.1022

(Continued)

Table C.2 (Continued)

	$\alpha_3 = 0.45$				$\alpha_3 = 0.50$				$\alpha_3 = 0.55$		
α_4	$a_1 = -a_3$	a_2	a_4	α_4	$a_1 = -a_3$	a_2	a_4	α_4	$a_1 = -a_3$	a_2	a_4
2.2	−0.1406	1.2523	−0.1014	2.4	−0.1311	1.1811	−0.0714	2.6	−0.1297	1.1370	−0.0543
2.4	−0.1110	1.1613	−0.0618	2.6	−0.1120	1.1202	−0.0465	2.8	−0.1147	1.0881	−0.0350
2.6	−0.0967	1.1064	−0.0402	2.8	−0.1005	1.0755	−0.0294	3.0	−0.1048	1.0500	−0.0208
2.8	−0.0877	1.0649	−0.0248	3.0	−0.0926	1.0400	−0.0165	3.2	−0.0977	1.0187	−0.0095
3.0	−0.0814	1.0314	−0.0129	3.2	−0.0868	1.0104	−0.0060	3.4	−0.0923	0.9920	−0.0002
3.2	−0.0767	1.0033	−0.0031	3.4	−0.0824	0.9850	0.0027	3.6	−0.0880	0.9687	0.0078
3.4	−0.0730	0.9789	0.0052	3.6	−0.0787	0.9625	0.0103	3.8	−0.0844	0.9478	0.0148
3.6	−0.0699	0.9572	0.0125	3.8	−0.0757	0.9424	0.0170	4.0	−0.0814	0.9290	0.0210
3.8	−0.0673	0.9377	0.0189	4.0	−0.0731	0.9241	0.0230	4.2	−0.0788	0.9116	0.0267
4.0	−0.0651	0.9198	0.0247	4.2	−0.0709	0.9072	0.0284	4.4	−0.0766	0.8956	0.0318
4.2	−0.0632	0.9034	0.0300	4.4	−0.0689	0.8916	0.0334	4.6	−0.0745	0.8806	0.0366
4.4	−0.0615	0.8880	0.0348	4.6	−0.0672	0.8769	0.0381	4.8	−0.0727	0.8666	0.0410
4.6	−0.0600	0.8736	0.0394	4.8	−0.0656	0.8631	0.0424	5.0	−0.0711	0.8533	0.0451
4.8	−0.0586	0.8601	0.0436	5.0	−0.0642	0.8501	0.0464	5.2	−0.0696	0.8407	0.0490
5.0	−0.0574	0.8473	0.0475	5.2	−0.0629	0.8377	0.0502	5.4	−0.0683	0.8288	0.0527
5.2	−0.0562	0.8351	0.0513	5.4	−0.0617	0.8259	0.0538	5.6	−0.0670	0.8173	0.0562
5.4	−0.0552	0.8234	0.0548	5.6	−0.0606	0.8146	0.0573	5.8	−0.0659	0.8063	0.0595
5.6	−0.0542	0.8122	0.0582	5.8	−0.0595	0.8038	0.0605	6.0	−0.0648	0.7958	0.0627
5.8	−0.0533	0.8015	0.0614	6.0	−0.0586	0.7934	0.0636	6.2	−0.0638	0.7856	0.0657
6.0	−0.0525	0.7912	0.0645	6.2	−0.0577	0.7833	0.0666	6.4	−0.0629	0.7758	0.0686
6.2	−0.0517	0.7813	0.0674	6.4	−0.0569	0.7736	0.0695	6.6	−0.0620	0.7664	0.0714
6.4	−0.0510	0.7717	0.0703	6.6	−0.0561	0.7643	0.0723	6.8	−0.0611	0.7572	0.0741
6.6	−0.0503	0.7624	0.0730	6.8	−0.0553	0.7552	0.0749	7.0	−0.0604	0.7483	0.0767
6.8	−0.0496	0.7533	0.0756	7.0	−0.0546	0.7463	0.0775	7.2	−0.0596	0.7396	0.0792
7.0	−0.0490	0.7446	0.0782	7.2	−0.0540	0.7377	0.0800	7.4	−0.0589	0.7312	0.0817
7.2	−0.0484	0.7361	0.0806	7.4	−0.0533	0.7294	0.0824	7.6	−0.0582	0.7230	0.0840
7.4	−0.0478	0.7278	0.0830	7.6	−0.0527	0.7212	0.0847	7.8	−0.0576	0.7150	0.0863
7.6	−0.0473	0.7197	0.0853	7.8	−0.0522	0.7133	0.0870	8.0	−0.0570	0.7072	0.0886
7.8	−0.0468	0.7118	0.0876	8.0	−0.0516	0.7055	0.0892	8.2	−0.0564	0.6995	0.0907
8.0	−0.0463	0.7041	0.0898	8.2	−0.0511	0.6980	0.0913	8.4	−0.0558	0.6921	0.0928
8.2	−0.0458	0.6965	0.0919	8.4	−0.0506	0.6905	0.0934	8.6	−0.0553	0.6848	0.0949
8.4	−0.0454	0.6892	0.0940	8.6	−0.0501	0.6833	0.0955	8.8	−0.0548	0.6776	0.0969
8.6	−0.0450	0.6819	0.0960	8.8	−0.0496	0.6762	0.0975	9.0	−0.0543	0.6706	0.0989
8.8	−0.0446	0.6749	0.0980	9.0	−0.0492	0.6692	0.0994	9.2	−0.0538	0.6637	0.1008
9.0	−0.0442	0.6679	0.0999	9.2	−0.0488	0.6624	0.1013	9.4	−0.0533	0.6570	0.1027
9.2	−0.0438	0.6611	0.1018	9.4	−0.0483	0.6556	0.1032	9.6	−0.0529	0.6503	0.1045

Table C.2 (Continued)

	$\alpha_3 = 0.60$				$\alpha_3 = 0.65$				$\alpha_3 = 0.70$		
α_4	$a_1 = -a_3$	a_2	a_4	α_4	$a_1 = -a_3$	a_2	a_4	α_4	$a_1 = -a_3$	a_2	a_4
2.6	−0.1512	1.1576	−0.0645	2.6	−0.1793	1.1848	−0.0790	2.6	−0.2462	1.2756	−0.1300
2.8	−0.1310	1.1032	−0.0420	2.8	−0.1503	1.1214	−0.0507	2.8	−0.1744	1.1436	−0.0622
3.0	−0.1183	1.0619	−0.0260	3.0	−0.1337	1.0759	−0.0324	3.0	−0.1517	1.0926	−0.0403
3.2	−0.1095	1.0284	−0.0137	3.2	−0.1225	1.0397	−0.0186	3.2	−0.1372	1.0530	−0.0246
3.4	−0.1029	1.0002	−0.0036	3.4	−0.1144	1.0096	−0.0077	3.4	−0.1271	1.0206	−0.0125
3.6	−0.0978	0.9757	0.0049	3.6	−0.1082	0.9838	0.0015	3.6	−0.1196	0.9931	−0.0025
3.8	−0.0936	0.9540	0.0122	3.8	−0.1033	0.9611	0.0093	3.8	−0.1136	0.9692	0.0059
4.0	−0.0901	0.9345	0.0188	4.0	−0.0991	0.9408	0.0162	4.0	−0.1088	0.9479	0.0132
4.2	−0.0871	0.9166	0.0246	4.2	−0.0957	0.9223	0.0223	4.2	−0.1047	0.9286	0.0197
4.4	−0.0844	0.9002	0.0300	4.4	−0.0927	0.9053	0.0279	4.4	−0.1012	0.9110	0.0256
4.6	−0.0821	0.8848	0.0349	4.6	−0.0900	0.8895	0.0330	4.6	−0.0982	0.8948	0.0309
4.8	−0.0801	0.8705	0.0395	4.8	−0.0877	0.8748	0.0377	4.8	−0.0955	0.8796	0.0358
5.0	−0.0782	0.8569	0.0437	5.0	−0.0856	0.8610	0.0421	5.0	−0.0932	0.8654	0.0403
5.2	−0.0766	0.8441	0.0477	5.2	−0.0837	0.8479	0.0462	5.2	−0.0910	0.8521	0.0445
5.4	−0.0750	0.8319	0.0515	5.4	−0.0820	0.8355	0.0501	5.4	−0.0891	0.8394	0.0485
5.6	−0.0736	0.8203	0.0550	5.6	−0.0804	0.8236	0.0537	5.6	−0.0873	0.8273	0.0523
5.8	−0.0723	0.8092	0.0584	5.8	−0.0789	0.8123	0.0572	5.8	−0.0857	0.8158	0.0558
6.0	−0.0711	0.7985	0.0616	6.0	−0.0776	0.8015	0.0605	6.0	−0.0842	0.8048	0.0592
6.2	−0.0700	0.7882	0.0647	6.2	−0.0763	0.7910	0.0636	6.2	−0.0828	0.7942	0.0624
6.4	−0.0690	0.7783	0.0677	6.4	−0.0751	0.7810	0.0666	6.4	−0.0815	0.7840	0.0655
6.6	−0.0680	0.7687	0.0705	6.6	−0.0740	0.7713	0.0695	6.6	−0.0803	0.7741	0.0684
6.8	−0.0670	0.7594	0.0733	6.8	−0.0730	0.7619	0.0723	6.8	−0.0791	0.7646	0.0712
7.0	−0.0662	0.7504	0.0759	7.0	−0.0720	0.7528	0.0750	7.0	−0.0780	0.7554	0.0740
7.2	−0.0653	0.7417	0.0784	7.2	−0.0711	0.7440	0.0776	7.2	−0.0770	0.7465	0.0766
7.4	−0.0645	0.7332	0.0809	7.4	−0.0702	0.7354	0.0801	7.4	−0.0760	0.7378	0.0791
7.6	−0.0638	0.7249	0.0833	7.6	−0.0694	0.7270	0.0825	7.6	−0.0751	0.7293	0.0816
7.8	−0.0631	0.7168	0.0856	7.8	−0.0686	0.7189	0.0848	7.8	−0.0743	0.7211	0.0840
8.0	−0.0624	0.7090	0.0879	8.0	−0.0679	0.7109	0.0871	8.0	−0.0734	0.7131	0.0863
8.2	−0.0617	0.7013	0.0901	8.2	−0.0672	0.7032	0.0893	8.2	−0.0726	0.7053	0.0885
8.4	−0.0611	0.6937	0.0922	8.4	−0.0665	0.6956	0.0915	8.4	−0.0719	0.6976	0.0907
8.6	−0.0605	0.6864	0.0943	8.6	−0.0658	0.6882	0.0936	8.6	−0.0712	0.6902	0.0928
8.8	−0.0599	0.6792	0.0963	8.8	−0.0652	0.6809	0.0956	8.8	−0.0705	0.6828	0.0949
9.0	−0.0594	0.6721	0.0983	9.0	−0.0646	0.6738	0.0976	9.0	−0.0698	0.6757	0.0969
9.2	−0.0589	0.6652	0.1002	9.2	−0.0640	0.6669	0.0996	9.2	−0.0692	0.6687	0.0989
9.4	−0.0584	0.6584	0.1021	9.4	−0.0634	0.6600	0.1015	9.4	−0.0685	0.6618	0.1008
9.6	−0.0579	0.6518	0.1039	9.6	−0.0629	0.6533	0.1033	9.6	−0.0680	0.6550	0.1027

(Continued)

Table C.2 (Continued)

	$\alpha_3 = 0.75$				$\alpha_3 = 0.80$				$\alpha_3 = 0.85$		
α_4	$a_1 = -a_3$	a_2	a_4	α_4	$a_1 = -a_3$	a_2	a_4	α_4	$a_1 = -a_3$	a_2	a_4
2.8	−0.2076	1.1733	−0.0791	3.0	−0.2021	1.1369	−0.0638	3.0	−0.2505	1.1780	−0.0897
3.0	−0.1736	1.1125	−0.0503	3.2	−0.1750	1.0872	−0.0412	3.2	−0.2013	1.1093	−0.0532
3.	−0.1544	1.0687	−0.0320	3.4	−0.1580	1.0483	−0.0252	3.4	−0.1778	1.0659	−0.0339
3.4	−0.1414	1.0334	−0.0182	3.6	−0.1461	1.0162	−0.0128	3.6	−0.1622	1.0307	−0.0195
3.6	−0.1320	1.0038	−0.0072	3.8	−0.1372	0.9889	−0.0026	3.8	−0.1510	1.0011	−0.0081
3.8	−0.1248	0.9784	0.0020	4.0	−0.1302	0.9651	0.0059	4.0	−0.1425	0.9755	0.0013
4.0	−0.1191	0.9559	0.0098	4.2	−0.1246	0.9439	0.0133	4.2	−0.1357	0.9530	0.0094
4.2	−0.1143	0.9358	0.0167	4.4	−0.1198	0.9247	0.0199	4.4	−0.1301	0.9329	0.0164
4.4	−0.1103	0.9175	0.0229	4.6	−0.1158	0.9072	0.0258	4.6	−0.1254	0.9145	0.0227
4.6	−0.1068	0.9006	0.0285	4.8	−0.1123	0.8910	0.0311	4.8	−0.1214	0.8977	0.0283
4.8	−0.1037	0.8850	0.0336	5.0	−0.1093	0.8759	0.0360	5.0	−0.1179	0.8820	0.0335
5.0	−0.1010	0.8704	0.0383	5.2	−0.1065	0.8618	0.0406	5.2	−0.1148	0.8674	0.0383
5.2	−0.0986	0.8567	0.0427	5.4	−0.1040	0.8484	0.0448	5.4	−0.1120	0.8537	0.0427
5.4	−0.0964	0.8437	0.0468	5.6	−0.1018	0.8358	0.0488	5.6	−0.1095	0.8407	0.0468
5.6	−0.0944	0.8314	0.0506	5.8	−0.0998	0.8238	0.0526	5.8	−0.1072	0.8284	0.0507
5.8	−0.0926	0.8196	0.0543	6.0	−0.0979	0.8123	0.0562	6.0	−0.1051	0.8166	0.0544
6.0	−0.0909	0.8084	0.0577	6.2	−0.0962	0.8013	0.0595	6.2	−0.1032	0.8054	0.0579
6.2	−0.0894	0.7976	0.0610	6.4	−0.0946	0.7908	0.0628	6.4	−0.1014	0.7947	0.0612
6.4	−0.0879	0.7872	0.0642	6.6	−0.0931	0.7806	0.0658	6.6	−0.0997	0.7843	0.0644
6.6	−0.0866	0.7772	0.0672	6.8	−0.0917	0.7708	0.0688	6.8	−0.0982	0.7743	0.0674
6.8	−0.0853	0.7676	0.0701	7.0	−0.090	0.7613	0.0716	7.0	−0.0967	0.7647	0.0703
7.0	−0.0841	0.7582	0.0728	7.2	−0.0891	0.7521	0.0743	7.2	−0.0954	0.7554	0.0731
7.2	−0.0830	0.7492	0.0755	7.4	−0.0880	0.7432	0.0770	7.4	−0.0941	0.7463	0.0757
7.4	−0.0819	0.7404	0.0781	7.6	−0.0869	0.7346	0.0795	7.6	−0.0929	0.7376	0.0783
7.6	−0.0809	0.7319	0.0806	7.8	−0.0858	0.7262	0.0820	7.8	−0.0917	0.7290	0.0808
7.8	−0.0800	0.7235	0.0830	8.0	−0.0848	0.7180	0.0844	8.0	−0.0907	0.7208	0.0833
8.0	−0.0791	0.7154	0.0853	8.2	−0.0839	0.7100	0.0867	8.2	−0.0896	0.7127	0.0856
8.2	−0.0782	0.7075	0.0876	8.4	−0.0830	0.7022	0.0889	8.4	−0.0886	0.7048	0.0879
8.4	−0.0774	0.6998	0.0898	8.6	−0.0821	0.6946	0.0911	8.6	−0.0877	0.6971	0.0901
8.6	−0.0766	0.6923	0.0920	8.8	−0.0813	0.6871	0.0932	8.8	−0.0868	0.6896	0.0923
8.8	−0.0758	0.6849	0.0941	9.0	−0.0805	0.6799	0.0953	9.0	−0.0859	0.6822	0.0944
9.0	−0.0751	0.6777	0.0961	9.2	−0.0797	0.6727	0.0973	9.2	−0.0851	0.6750	0.0964
9.2	−0.0744	0.6706	0.0981	9.4	−0.0790	0.6657	0.0993	9.4	−0.0843	0.6679	0.0984
9.4	−0.0737	0.6637	0.1001	9.6	−0.0783	0.6589	0.1012	9.6	−0.0836	0.6610	0.1004
9.6	−0.0731	0.6569	0.1020	9.8	−0.0776	0.6521	0.1031	9.8	−0.0828	0.6542	0.1023
9.8	−0.0725	0.6502	0.1038	10.0	−0.0770	0.6455	0.1049	10.0	−0.0821	0.6476	0.1041

Table C.2 (Continued)

	$\alpha_3 = 0.90$				$\alpha_3 = 0.95$				$\alpha_3 = 1.00$		
α_4	$a_1 = -a_3$	a_2	a_4	α_4	$a_1 = -a_3$	a_2	a_4	α_4	$a_1 = -a_3$	a_2	a_4
3.2	−0.2380	1.1372	−0.0705	3.4	−0.2362	1.1113	−0.0601	3.6	−0.2375	1.0904	−0.0524
3.4	−0.2027	1.0866	−0.0450	3.6	−0.2056	1.0674	−0.0384	3.8	−0.2096	1.0507	−0.0330
3.6	−0.1815	1.0476	−0.0278	3.8	−0.1860	1.0316	−0.0228	4.0	−0.1910	1.0175	−0.0186
3.8	−0.1670	1.0152	−0.0147	4.0	−0.1721	1.0014	−0.0106	4.2	−0.1776	0.9891	−0.0071
4.0	−0.1563	0.9875	−0.0041	4.2	−0.1617	0.9754	−0.0006	4.4	−0.1674	0.9643	0.0024
4.2	−0.1480	0.9634	0.0048	4.4	−0.1536	0.9525	0.0078	4.6	−0.1594	0.9424	0.0105
4.4	−0.1413	0.9420	0.0124	4.6	−0.1470	0.9320	0.0152	4.8	−0.1528	0.9227	0.0176
4.6	−0.1358	0.9227	0.0192	4.8	−0.1414	0.9134	0.0217	5.0	−0.1472	0.9048	0.0239
4.8	−0.1311	0.9051	0.0252	5.0	−0.1367	0.8963	0.0275	5.2	−0.1424	0.8882	0.0295
5.0	−0.1270	0.8888	0.0307	5.2	−0.1326	0.8806	0.0328	5.4	−0.1383	0.8729	0.0347
5.2	−0.1234	0.8737	0.0357	5.4	−0.1290	0.8658	0.0377	5.6	−0.1346	0.8585	0.0395
5.4	−0.1203	0.8595	0.0403	5.6	−0.1258	0.8520	0.0422	5.8	−0.1313	0.8450	0.0439
5.6	−0.1174	0.8461	0.0446	5.8	−0.1229	0.8389	0.0464	6.0	−0.1284	0.8322	0.0480
5.8	−0.1149	0.8334	0.0487	6.0	−0.1203	0.8266	0.0504	6.2	−0.1257	0.8201	0.0519
6.0	−0.1125	0.8214	0.0525	6.2	−0.1179	0.8148	0.0541	6.4	−0.1233	0.8085	0.0556
6.2	−0.1104	0.8099	0.0561	6.4	−0.1157	0.8035	0.0576	6.6	−0.1210	0.7974	0.0590
6.4	−0.1084	0.7989	0.0595	6.6	−0.1137	0.7927	0.0610	6.8	−0.1189	0.7868	0.0624
6.6	−0.1066	0.7883	0.0627	6.8	−0.1118	0.7823	0.0642	7.0	−0.1170	0.7766	0.0655
6.8	−0.1049	0.7782	0.0658	7.0	−0.1100	0.7723	0.0672	7.2	−0.1152	0.7667	0.0685
7.0	−0.1033	0.7683	0.0688	7.2	−0.1084	0.7626	0.0702	7.4	−0.1135	0.7572	0.0714
7.2	−0.1018	0.7589	0.0717	7.4	−0.1068	0.7533	0.0730	7.6	−0.1119	0.7480	0.0742
7.4	−0.1004	0.7497	0.0744	7.6	−0.1054	0.7443	0.0757	7.8	−0.1104	0.7391	0.0769
7.6	−0.0991	0.7408	0.0771	7.8	−0.1040	0.7355	0.0783	8.0	−0.1090	0.7304	0.0794
7.8	−0.0978	0.7321	0.0796	8.0	−0.1027	0.7269	0.0808	8.2	−0.1076	0.7220	0.0819
8.0	−0.0966	0.7237	0.0821	8.2	−0.1015	0.7186	0.0833	8.4	−0.1064	0.7138	0.0843
8.2	−0.0955	0.7155	0.0845	8.4	−0.1003	0.7106	0.0856	8.6	−0.1051	0.7058	0.0867
8.4	−0.0944	0.7076	0.0868	8.6	−0.0992	0.7027	0.0879	8.8	−0.1040	0.6980	0.0890
8.6	−0.0934	0.6998	0.0891	8.8	−0.0981	0.6950	0.0901	9.0	−0.1029	0.6903	0.0912
8.8	−0.0924	0.6922	0.0912	9.0	−0.0971	0.6874	0.0923	9.2	−0.1018	0.6829	0.0933
9.0	−0.0915	0.6847	0.0934	9.2	−0.0962	0.6801	0.0944	9.4	−0.1008	0.6756	0.0954
9.2	−0.0906	0.6774	0.0955	9.4	−0.0952	0.6729	0.0965	9.6	−0.0999	0.6685	0.0974
9.4	−0.0897	0.6703	0.0975	9.6	−0.0943	0.6658	0.0985	9.8	−0.0989	0.6615	0.0994
9.6	−0.0889	0.6633	0.0995	9.8	−0.0935	0.6589	0.1004	10.0	−0.0980	0.6546	0.1014
9.8	−0.0881	0.6565	0.1014	10.0	−0.0926	0.6521	0.1024	10.2	−0.0972	0.6479	0.1033
10.0	−0.0873	0.6497	0.1033	10.2	−0.0918	0.6454	0.1042	10.4	−0.0964	0.6413	0.1051
10.2	−0.0866	0.6431	0.1051	10.4	−0.0911	0.6389	0.1061	10.6	−0.0956	0.6348	0.1069

(Continued)

Table C.2 (Continued)

$\alpha_3 = 1.05$				$\alpha_3 = 1.10$				$\alpha_3 = 1.15$			
α_4	$a_1 = -a_3$	a_2	a_4	α_4	$a_1 = -a_3$	a_2	a_4	α_4	$a_1 = -a_3$	a_2	a_4
3.6	−0.2895	1.1224	−0.0765	3.8	−0.2874	1.0995	−0.0669	4.2	−0.2960	1.0670	−0.0565
3.8	−0.2407	1.0727	−0.0465	4.0	−0.2453	1.0574	−0.0417	4.4	−0.2578	1.0322	−0.0350
4.0	−0.2144	1.0361	−0.0286	4.2	−0.2201	1.0233	−0.0250	4.6	−0.2333	1.0016	−0.0196
4.2	−0.1965	1.0049	−0.0150	4.4	−0.2025	0.9937	−0.0120	4.8	−0.2158	0.9746	−0.0075
4.4	−0.1834	0.9780	−0.0041	4.6	−0.1895	0.9680	−0.0015	5.0	−0.2027	0.9506	0.0025
4.6	−0.1734	0.9543	0.0050	4.8	−0.1795	0.9452	0.0073	5.2	−0.1924	0.9293	0.0109
4.8	−0.1653	0.9332	0.0128	5.0	−0.1714	0.9248	0.0149	5.4	−0.1841	0.9100	0.0182
5.0	−0.1587	0.9142	0.0197	5.2	−0.1647	0.9063	0.0216	5.6	−0.1771	0.8923	0.0246
5.2	−0.1531	0.8968	0.0258	5.4	−0.1590	0.8893	0.0276	5.8	−0.1712	0.8761	0.0304
5.4	−0.1482	0.8807	0.0314	5.6	−0.1541	0.8736	0.0330	6.0	−0.1661	0.8610	0.0357
5.6	−0.1440	0.8657	0.0364	5.8	−0.1498	0.8590	0.0380	6.2	−0.1616	0.8469	0.0405
5.8	−0.1403	0.8516	0.0411	6.0	−0.1460	0.8452	0.0426	6.4	−0.1576	0.8335	0.0450
6.0	−0.1369	0.8384	0.0454	6.2	−0.1426	0.8322	0.0468	6.6	−0.1540	0.8210	0.0492
6.2	−0.1339	0.8259	0.0495	6.4	−0.1395	0.8199	0.0508	6.8	−0.1507	0.8090	0.0531
6.4	−0.1312	0.8140	0.0533	6.6	−0.1366	0.8082	0.0546	7.0	−0.1477	0.7976	0.0568
6.6	−0.1286	0.8026	0.0569	6.8	−0.1341	0.7970	0.0582	7.2	−0.1450	0.7867	0.0603
6.8	−0.1263	0.7917	0.0604	7.0	−0.1317	0.7862	0.0615	7.4	−0.1424	0.7762	0.0636
7.0	−0.1242	0.7812	0.0636	7.2	−0.1295	0.7759	0.0648	7.6	−0.1401	0.7661	0.0668
7.2	−0.1222	0.7711	0.0667	7.4	−0.1274	0.7660	0.0678	7.8	−0.1379	0.7564	0.0698
7.4	−0.1203	0.7614	0.0697	7.6	−0.1255	0.7564	0.0708	8.0	−0.1359	0.7470	0.0727
7.6	−0.1186	0.7520	0.0726	7.8	−0.1237	0.7471	0.0736	8.2	−0.1340	0.7379	0.0755
7.8	−0.1169	0.7429	0.0753	8.0	−0.1220	0.7381	0.0763	8.4	−0.1322	0.7291	0.0781
8.0	−0.1154	0.7341	0.0779	8.2	−0.1204	0.7294	0.0790	8.6	−0.1304	0.7206	0.0807
8.2	−0.1139	0.7255	0.0805	8.4	−0.1189	0.7209	0.0815	8.8	−0.1288	0.7122	0.0832
8.4	−0.1125	0.7172	0.0830	8.6	−0.1174	0.7127	0.0839	9.0	−0.1273	0.7041	0.0856
8.6	−0.1112	0.7091	0.0854	8.8	−0.1161	0.7046	0.0863	9.2	−0.1259	0.6962	0.0880
8.8	−0.1100	0.7012	0.0877	9.0	−0.1148	0.6968	0.0886	9.4	−0.1245	0.6885	0.0902
9.0	−0.1088	0.6934	0.0899	9.2	−0.1136	0.6891	0.0908	9.6	−0.1231	0.6810	0.0925
9.2	−0.1076	0.6859	0.0921	9.4	−0.1124	0.6816	0.0930	9.8	−0.1219	0.6736	0.0946
9.4	−0.1065	0.6785	0.0942	9.6	−0.1112	0.6743	0.0951	10.0	−0.1207	0.6664	0.0967
9.6	−0.1055	0.6713	0.0963	9.8	−0.1102	0.6671	0.0972	10.2	−0.1195	0.6593	0.0987
9.8	−0.1045	0.6642	0.0983	10.0	−0.1091	0.6601	0.0992	10.4	−0.1184	0.6524	0.1007
10.0	−0.1035	0.6573	0.1003	10.2	−0.1081	0.6532	0.1011	10.6	−0.1173	0.6456	0.1026
10.2	−0.1026	0.6505	0.1022	10.4	−0.1072	0.6465	0.1031	10.8	−0.1163	0.6390	0.1045
10.4	−0.1017	0.6438	0.1041	10.6	−0.1062	0.6399	0.1049	11.0	−0.1153	0.6324	0.1064
10.6	−0.1009	0.6372	0.1060	10.8	−0.1054	0.6334	0.1068	11.2	−0.1144	0.6260	0.1082

Table C.2 (Continued)

	$\alpha_3 = 1.30$				$\alpha_3 = 1.40$				$\alpha_3 = 1.50$		
α_4	$a_1 = -a_3$	a_2	a_4	α_4	$a_1 = -a_3$	a_2	a_4	α_4	$a_1 = -a_3$	a_2	a_4
4.6	−0.3132	1.0424	−0.0516	5.0	−0.3386	1.0218	−0.0509	5.4	−0.3772	1.0023	−0.0559
4.8	−0.2746	1.0122	−0.0313	5.2	−0.2959	0.9958	−0.0302	5.6	−0.3233	0.9814	−0.0318
5.0	−0.2491	0.9843	−0.0164	5.4	−0.2681	0.9703	−0.0152	5.8	−0.2911	0.9587	−0.0161
5.2	−0.2309	0.9590	−0.0046	5.6	−0.2482	0.9464	−0.0034	6.0	−0.2683	0.9362	−0.0037
5.4	−0.2171	0.9364	0.0051	5.8	−0.2332	0.9247	0.0064	6.2	−0.2514	0.9152	0.0064
5.6	−0.2063	0.9160	0.0133	6.0	−0.2215	0.9049	0.0147	6.4	−0.2383	0.8959	0.0150
5.8	−0.1975	0.8974	0.0205	6.2	−0.2120	0.8869	0.0219	6.6	−0.2277	0.8782	0.0223
6.0	−0.1902	0.8804	0.0268	6.4	−0.2041	0.8703	0.0282	6.8	−0.2190	0.8619	0.0288
6.2	−0.1839	0.8647	0.0325	6.6	−0.1973	0.8550	0.0339	7.0	−0.2116	0.8468	0.0346
6.4	−0.1785	0.8500	0.0377	6.8	−0.1915	0.8406	0.0391	7.2	−0.2052	0.8326	0.0398
6.6	−0.1737	0.8363	0.0425	7.0	−0.1864	0.8271	0.0438	7.4	−0.1996	0.8193	0.0446
6.8	−0.1695	0.8233	0.0469	7.2	−0.1818	0.8144	0.0482	7.6	−0.1947	0.8067	0.0490
7.0	−0.1656	0.8110	0.0510	7.4	−0.1777	0.8023	0.0524	7.8	−0.1902	0.7948	0.0532
7.2	−0.1622	0.7993	0.0549	7.6	−0.1740	0.7908	0.0562	8.0	−0.1862	0.7834	0.0570
7.4	−0.1590	0.7882	0.0585	7.8	−0.1706	0.7798	0.0598	8.2	−0.1825	0.7725	0.0607
7.6	−0.1561	0.7775	0.0620	8.0	−0.1675	0.7693	0.0633	8.4	−0.1792	0.7621	0.0641
7.8	−0.1534	0.7672	0.0653	8.2	−0.1646	0.7592	0.0665	8.6	−0.1761	0.7521	0.0674
8.0	−0.1509	0.7573	0.0684	8.4	−0.1619	0.7495	0.0696	8.8	−0.1732	0.7425	0.0705
8.2	−0.1486	0.7478	0.0714	8.6	−0.1594	0.7400	0.0726	9.0	−0.1705	0.7331	0.0735
8.4	−0.1464	0.7386	0.0742	8.8	−0.1571	0.7309	0.0755	9.2	−0.1680	0.7241	0.0764
8.6	−0.1443	0.7296	0.0770	9.0	−0.1549	0.7221	0.0782	9.4	−0.1657	0.7154	0.0791
8.8	−0.1424	0.7209	0.0796	9.2	−0.1528	0.7135	0.0808	9.6	−0.1635	0.7069	0.0817
9.0	−0.1406	0.7125	0.0822	9.4	−0.1509	0.7052	0.0834	9.8	−0.1614	0.6986	0.0843
9.2	−0.1389	0.7043	0.0847	9.6	−0.1491	0.6971	0.0858	10.0	−0.1594	0.6906	0.0867
9.4	−0.1373	0.6963	0.0871	9.8	−0.1473	0.6892	0.0882	10.2	−0.1576	0.6828	0.0891
9.6	−0.1357	0.6885	0.0894	10.0	−0.1457	0.6815	0.0905	10.4	−0.1558	0.6751	0.0914
9.8	−0.1342	0.6809	0.0916	10.2	−0.1441	0.6740	0.0928	10.6	−0.1541	0.6677	0.0937
10.0	−0.1328	0.6735	0.0938	10.4	−0.1426	0.6666	0.0949	10.8	−0.1525	0.6604	0.0958
10.2	−0.1315	0.6662	0.0959	10.6	−0.1411	0.6594	0.0971	11.0	−0.1509	0.6532	0.0979
10.4	−0.1302	0.6591	0.0980	10.8	−0.1397	0.6524	0.0991	11.2	−0.1495	0.6463	0.1000
10.6	−0.1289	0.6521	0.1000	11.0	−0.1384	0.6455	0.1011	11.4	−0.1480	0.6394	0.1020
10.8	−0.1277	0.6453	0.1020	11.2	−0.1372	0.6387	0.1031	11.6	−0.1467	0.6327	0.1039
11.0	−0.1266	0.6386	0.1039	11.4	−0.1359	0.6321	0.1050	11.8	−0.1454	0.6261	0.1059
11.2	−0.1255	0.6320	0.1058	11.6	−0.1348	0.6256	0.1069	12.0	−0.1441	0.6196	0.1077
11.4	−0.1244	0.6255	0.1076	11.8	−0.1336	0.6192	0.1087	12.2	−0.1429	0.6133	0.1095
11.6	−0.1234	0.6192	0.1094	12.0	−0.1325	0.6129	0.1105	12.4	−0.1417	0.6070	0.1113

(Continued)

Table C.2 (Continued)

	$\alpha_3 = 1.60$				$\alpha_3 = 1.70$				$\alpha_3 = 1.80$		
α_4	$a_1 = -a_3$	a_2	a_4	α_4	$a_1 = -a_3$	a_2	a_4	α_4	$a_1 = -a_3$	a_2	a_4
6.0	−0.3603	0.9665	−0.0370	6.4	−0.4200	0.9446	−0.0498	7.0	−0.4192	0.9155	−0.0386
6.2	−0.3202	0.9484	−0.0194	6.6	−0.3594	0.9366	−0.0259	7.2	−0.3662	0.9102	−0.0186
6.4	−0.2926	0.9279	−0.0059	6.8	−0.3236	0.9202	−0.0105	7.4	−0.3317	0.8958	−0.0042
6.6	−0.2726	0.9077	0.0050	7.0	−0.2983	0.9016	0.0017	7.6	−0.3071	0.8789	0.0072
6.8	−0.2573	0.8888	0.0140	7.2	−0.2796	0.8833	0.0116	7.8	−0.2888	0.8620	0.0166
7.0	−0.2452	0.8713	0.0217	7.4	−0.2651	0.8659	0.0199	8.0	−0.2746	0.8459	0.0246
7.2	−0.2353	0.8551	0.0284	7.6	−0.2535	0.8498	0.0271	8.2	−0.2632	0.8307	0.0314
7.4	−0.2270	0.8400	0.0344	7.8	−0.2440	0.8347	0.0334	8.4	−0.2536	0.8165	0.0375
7.6	−0.2199	0.8259	0.0398	8.0	−0.2359	0.8206	0.0391	8.6	−0.2456	0.8032	0.0429
7.8	−0.2137	0.8127	0.0447	8.2	−0.2289	0.8073	0.0442	8.8	−0.2386	0.7906	0.0478
8.0	−0.2083	0.8002	0.0493	8.4	−0.2228	0.7948	0.0489	9.0	−0.2324	0.7786	0.0524
8.2	−0.2034	0.7883	0.0535	8.6	−0.2174	0.7829	0.0532	9.2	−0.2269	0.7673	0.0565
8.4	−0.1990	0.7770	0.0574	8.8	−0.2125	0.7716	0.0573	9.4	−0.2220	0.7564	0.0605
8.6	−0.1950	0.7662	0.0611	9.0	−0.2081	0.7608	0.0610	9.6	−0.2176	0.7460	0.0641
8.8	−0.1913	0.7559	0.0646	9.2	−0.2041	0.7505	0.0646	9.8	−0.2135	0.7360	0.0676
9.0	−0.1880	0.7459	0.0679	9.4	−0.2004	0.7405	0.0680	10.0	−0.2097	0.7264	0.0709
9.2	−0.1849	0.7363	0.0710	9.6	−0.1970	0.7310	0.0712	10.2	−0.2063	0.7172	0.0740
9.4	−0.1820	0.7270	0.0740	9.8	−0.1939	0.7217	0.0742	10.4	−0.2031	0.7082	0.0770
9.6	−0.1793	0.7181	0.0769	10.0	−0.1909	0.7128	0.0771	10.6	−0.2001	0.6995	0.0798
9.8	−0.1768	0.7094	0.0797	10.2	−0.1882	0.7041	0.0799	10.8	−0.1973	0.6911	0.0826
10.0	−0.1744	0.7009	0.0823	10.4	−0.1856	0.6957	0.0826	11.0	−0.1946	0.6829	0.0852
10.2	−0.1721	0.6927	0.0849	10.6	−0.1832	0.6875	0.0852	11.2	−0.1921	0.6750	0.0877
10.4	−0.1700	0.6847	0.0873	10.8	−0.1809	0.6795	0.0877	11.4	−0.1898	0.6672	0.0902
10.6	−0.1680	0.6770	0.0897	11.0	−0.1787	0.6718	0.0901	11.6	−0.1876	0.6596	0.0925
10.8	−0.1661	0.6694	0.0920	11.2	−0.1767	0.6642	0.0924	11.8	−0.1855	0.6522	0.0948
11.0	−0.1643	0.6619	0.0943	11.4	−0.1747	0.6568	0.0947	12.0	−0.1834	0.6450	0.0970
11.2	−0.1626	0.6547	0.0965	11.6	−0.1729	0.6496	0.0969	12.2	−0.1815	0.6380	0.0992
11.4	−0.1609	0.6476	0.0986	11.8	−0.1711	0.6425	0.0990	12.4	−0.1797	0.6310	0.1013
11.6	−0.1593	0.6407	0.1006	12.0	−0.1694	0.6356	0.1011	12.6	−0.1779	0.6243	0.1033
11.8	−0.1578	0.6338	0.1027	12.2	−0.1678	0.6288	0.1031	12.8	−0.1763	0.6176	0.1053
12.0	−0.1563	0.6272	0.1046	12.4	−0.1662	0.6221	0.1051	13.0	−0.1747	0.6111	0.1072
12.2	−0.1549	0.6206	0.1065	12.6	−0.1647	0.6156	0.1070	13.2	−0.1731	0.6047	0.1091
12.4	−0.1536	0.6142	0.1084	12.8	−0.1633	0.6092	0.1088	13.4	−0.1716	0.5984	0.1110
12.6	−0.1523	0.6078	0.1102	13.0	−0.1619	0.6029	0.1107	13.6	−0.1702	0.5922	0.1128
12.8	−0.1511	0.6016	0.1120	13.2	−0.1605	0.5967	0.1125	13.8	−0.1688	0.5861	0.1145
13.0	−0.1499	0.5955	0.1137	13.4	−0.1592	0.5906	0.1142	14.0	−0.1675	0.5801	0.1163

Table C.2 (Continued)

	$\alpha_3 = 1.90$				$\alpha_3 = 2.00$		
α_4	$a_1 = -a_3$	a_2	a_4	α_4	$a_1 = -a_3$	a_2	a_4
7.6	−0.4336	0.8861	−0.0335	8.4	−0.4001	0.8616	−0.0126
7.8	−0.3795	0.8858	−0.0142	8.6	−0.3609	0.8539	0.0020
8.0	−0.3440	0.8740	−0.0001	8.8	−0.3337	0.8407	0.0134
8.2	−0.3188	0.8588	0.0111	9.0	−0.3137	0.8263	0.0227
8.4	−0.3002	0.8432	0.0203	9.2	−0.2985	0.8121	0.0305
8.6	−0.2857	0.8281	0.0280	9.4	−0.2862	0.7985	0.0371
8.8	−0.2740	0.8138	0.0347	9.6	−0.2761	0.7855	0.0430
9.0	−0.2643	0.8002	0.0406	9.8	−0.2675	0.7733	0.0483
9.2	−0.2561	0.7875	0.0459	10.0	−0.2600	0.7616	0.0531
9.4	−0.2489	0.7754	0.0508	10.2	−0.2535	0.7505	0.0575
9.6	−0.2426	0.7639	0.0552	10.4	−0.2477	0.7398	0.0615
9.8	−0.2370	0.7530	0.0593	10.6	−0.2424	0.7296	0.0653
10.0	−0.2320	0.7425	0.0631	10.8	−0.2377	0.7199	0.0689
10.2	−0.2274	0.7324	0.0667	11.0	−0.2334	0.7104	0.0723
10.4	−0.2232	0.7227	0.0701	11.2	−0.2294	0.7013	0.0755
10.6	−0.2194	0.7134	0.0734	11.4	−0.2257	0.6925	0.0785
10.8	−0.2158	0.7044	0.0764	11.6	−0.2223	0.6840	0.0814
11.0	−0.2125	0.6957	0.0794	11.8	−0.2191	0.6757	0.0841
11.2	−0.2094	0.6872	0.0822	12.0	−0.2161	0.6677	0.0868
11.4	−0.2065	0.6790	0.0848	12.2	−0.2133	0.6599	0.0894
11.6	−0.2038	0.6710	0.0874	12.4	−0.2106	0.6523	0.0918
11.8	−0.2013	0.6632	0.0899	12.6	−0.2081	0.6448	0.0942
12.0	−0.1988	0.6556	0.0923	12.8	−0.2057	0.6375	0.0965
12.2	−0.1965	0.6482	0.0947	13.0	−0.2034	0.6304	0.0987
12.4	−0.1944	0.6410	0.0969	13.2	−0.2013	0.6235	0.1009
12.6	−0.1923	0.6339	0.0991	13.4	−0.1992	0.6167	0.1030
12.8	−0.1903	0.6270	0.1012	13.6	−0.1973	0.6100	0.1051
13.0	−0.1884	0.6202	0.1033	13.8	−0.1954	0.6035	0.1070
13.2	−0.1866	0.6136	0.1053	14.0	−0.1936	0.5971	0.1090
13.4	−0.1849	0.6070	0.1072	14.2	−0.1919	0.5907	0.1109
13.6	−0.1832	0.6006	0.1092	14.4	−0.1902	0.5845	0.1127
13.8	−0.1816	0.5943	0.1110	14.6	−0.1886	0.5785	0.1145
14.0	−0.1801	0.5882	0.1128	14.8	−0.1871	0.5725	0.1163
14.2	−0.1786	0.5821	0.1146	15.0	−0.1856	0.5666	0.1180
14.4	−0.1772	0.5761	0.1164	15.2	−0.1842	0.5607	0.1197
14.6	−0.1758	0.5702	0.1181	15.4	−0.1828	0.5550	0.1214

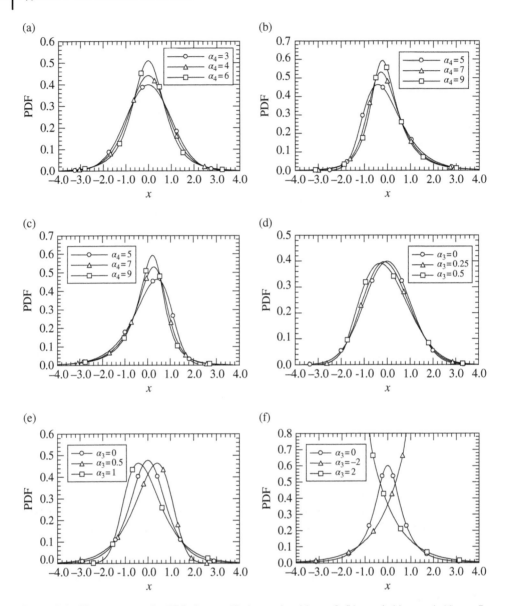

Figure C.6 The representative PDFs for specified α_3 and α_4. (a) $\alpha_3 = 0$; (b) $\alpha_3 = 1$; (c) $\alpha_3 = -1$; (d) $\alpha_4 = 3$; (e) $\alpha_4 = 5$; (f) $\alpha_4 = 9$.

C.3.3 Application in Data Analysis

The introduced cubic normal distribution can be applied to structural analysis, and the fourth-moment reliability index derived from the distribution has been thoroughly discussed in the chapter on Direct Methods of Moment. It can also be used to approximate

(a)

(b)

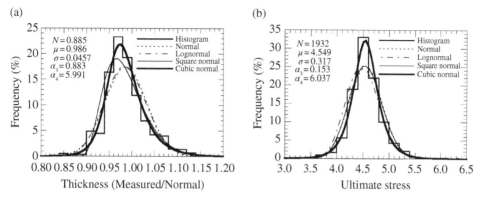

Figure C.7 Data Fitting for Thickness and Ultimate Stress. (a) Thickness (Measured/Nominal). (b) Ultimate stress.

one-, two-, and three-parameter distributions (Zhao and Lu 2008). Also, it can be used to structural reliability analysis including random variables with unknown distributions as described in the chapter on Structural Reliability Assessment. In this section, the application of the distribution to fitting statistical data of a random variable will be introduced.

Consider the measured data of H-shape structural steel described earlier. The fitting result of the histogram of the ratio between measured values and nominal values of the thickness is shown in Figure C.7a, in which the PDFs of the normal and lognormal distributions, with the same mean value and standard deviation as the data; the PDF of the square normal distribution whose mean value, standard deviation, and skewness are equal to those of the data; and the PDF of the cubic normal distribution whose mean value, standard deviation, skewness, and kurtosis are equal to those of the data, are depicted, revealing the following:

1) The PDFs of the normal distribution and lognormal distribution have the greatest differences from the histogram of the statistical data among the four distributions. Since the normal distribution is a symmetrical distribution with the skewness = 0.0 and the kurtosis = 3.0, respectively, it obviously cannot be used to fit the histogram that has such a large skewness (0.883) and kurtosis (5.991), respectively. Although the lognormal distribution can reflect skewness and kurtosis to some degree, the skewness and kurtosis of the lognormal distribution are functions of the coefficient of variation (COV). In the present case, the COV of the lognormal distribution is very small (0.0463), and thus its skewness and kurtosis are 0.139 and 3.03, respectively, which are too small to match those of the data.

2) Since the first three moments of the square normal distribution are equal to those of the data, it fits the histogram much better than the normal and lognormal distributions. However, the kurtosis of this distribution depends on the skewness. The kurtosis corresponding to the skewness of the data is obtained as 4.04, which is too small to match that of the data.

3) The first four moments of the cubic normal distribution can be equal to those of the data, and thus can fit the histogram much better than the normal, lognormal, and square normal distributions.

Results of the chi-square tests of the four distributions are listed in Table C.3, in which the goodness-of-fit tests were obtained using the following equation (Ang and Tang 2006)

$$T = \sum_{i=1}^{k} \frac{(O_i - E_i)^2}{E_i} \tag{C.21}$$

where O_i and E_i are the observed and theoretical frequencies, respectively; k is the number of intervals used; and T is a measure of the respective goodness-of-fit. From Table C.3, one can see that the goodness-of-fit of the introduced distribution is $T = 38.87$, which is much smaller than those of other distributions.

Similarly, the fitting result of the histogram of the ultimate stress is shown in Figure C.7b. From this figure, one can see that since the skewness of the data is quite small, the square normal distribution cannot show significant improvement upon the normal and lognormal distributions, whereas the cubic normal can effectively fit the histograms of the available data. Also, from Table C.4, the goodness-of-fit tests verify that the introduced distribution has the best fit with $T = 21.33$ among all the distributions.

Table C.3 Results of test for thickness.

| | | Predicted frequency | | | | Goodness-of-fit | | | |
| | | | | Square | Cubic | | | Square | Cubic |
Intervals	Freq.	Nor.	Log.	nor.	nor.	Nor.	Log.	nor.	nor.
< 0.9	11	26.5	22.8	1.4	14.3	9.07	6.11	65.83	0.76
0.90–0.92	43	39.3	39.5	29.7	25.4	0.35	0.31	5.96	12.20
0.92–0.94	39	73.2	76.3	97.1	66.8	15.98	18.23	34.76	11.57
0.94–0.96	145	113	117.9	153	138.3	9.06	6.23	0.42	0.32
0.96–0.98	206	144.3	147.7	166.4	188.6	26.38	23.01	9.42	1.61
0.98–1.00	170	152.7	151.9	145.3	169.5	1.96	2.16	4.20	0.00
1.00–1.02	105	133.8	129.7	109.6	116.7	6.20	4.70	0.19	1.17
1.02–10.4	61	97.2	92.9	74.5	70.8	13.48	10.95	2.45	1.36
1.04–1.06	37	58.4	56.4	46.9	40.7	7.84	6.67	2.09	0.34
1.06–1.08	35	29	29.3	27.8	23.1	1.24	1.11	1.86	6.13
1.08–1.10	13	12	13.1	15.7	13	0.08	0.00	0.46	0.00
1.10–1.12	12	4.1	5.1	8.6	7.4	15.22	9.34	1.34	2.86
> 1.12	8	1.5	2.4	9	10.4	28.17	13.07	0.11	0.55
Sum	885	885	885	885	885	135.03	101.89	129.10	38.87

Note: Freq. – Frequency; Nor. = Normal; Log. = Lognormal; Square nor. = Square normal; Cubic nor. = Cubic normal.

Table C.4 Results of test for ultimate stress.

Intervals	Freq.	Predicted frequency				Goodness-of-fit			
		Nor.	Log.	Square nor.	Cubic nor.	Nor.	Log.	Square nor.	Cubic nor.
< 3.8	32	17.5	10.4	12.1	27.4	12.01	44.86	32.73	0.77
3.8–4.0	32	63	57.1	59	45.7	15.25	11.03	12.36	4.11
4.0–4.2	108	181.3	189.7	187.7	130.8	29.64	35.19	33.84	3.97
4.2–4.4	365	354.9	377.8	371.6	345.4	0.29	0.43	0.12	1.11
4.4–4.6	638	472.8	480.4	477.9	589.5	57.72	51.70	53.63	3.99
4.6–4.8	424	428.6	410.9	415.3	459.1	0.05	0.42	0.18	2.68
4.8–5.0	193	264.4	247.6	252.2	204.8	19.28	12.04	13.90	0.68
5.0–5.2	82	110.9	109.4	110.1	78.1	7.53	6.86	7.17	0.19
5.2–5.4	40	31.6	36.7	35.5	30	2.23	0.30	0.57	3.33
> 5.4	18	7	12	10.6	21.2	17.29	3.00	5.17	0.48
Sum	1932	1932	1932	1932	1932	161.29	165.83	159.66	21.33

Note: Freq. = Frequency; Nor. = Normal; Log. = Lognormal; Square nor. = Square normal; Cubic nor. = Cubic normal.

From these examples, one can clearly see that since the first four moments of the cubic normal distribution are equal to those of the statistical data, it fits the histogram much better than the normal, the lognormal, and the square normal distribution.

C.4 Summary

In this Appendix, the 4P distributions such as the Pearson system and the cubic normal distribution are introduced, in which the four parameters can be directly defined in terms of the mean value, standard deviation, skewness, and kurtosis. Besides being quite simple and flexible for fitting statistical data of basic random variables, the 4P distributions are used for the derivation of the fourth-moment reliability index in the chapter on Direct Methods of Moment, and are convenient also for performing normal transformations needed in structural reliability analysis as described in the chapter on Transformation of Non-Normal Variables to Independent Normal Variables.

Appendix D

Basic Theory of Stochastic Process

Basic theory of stochastic process that is essential for time-dependent reliability method described in this book is summarised in this appendix, with emphasis on the concept and utility of the stationary process. For details of stochastic process theory, the readers can refer to many textbooks, some of which are listed in the references (Cramer and Leadbetter 1967; Papoulis 1984; Lutes and Sarkani 2004; Shueller and Shinozuka 1987; Li et al. 1993; Li and Chen 2009).

D.1   General Concept of Stochastic Process

A stochastic or random process $X(t)$ is defined as a parameterized family of random variables X with the parameter t. In the following, X may be interpreted as random excitation (or structural dynamical response) and t means time. In other words, for any fixed time, t_i, the corresponding value $X(t_i)$ is a random variable. In this point of view, a one-dimensional stochastic process may be understood as an extension of a random vector (Figure D.1a,b). Therefore, those basic concepts that hold for multidimensional random vectors are still true for one-dimensional stochastic processes.

A stochastic process can be classified into four broad categories, namely:

1) Discretely parameterized discrete stochastic processes;
2) Discretely parameterized continuous stochastic processes;
3) Continuously parameterized discrete stochastic processes; and
4) Continuously parameterized continuous stochastic processes.

The so-called counting process (the famous Poisson process is one of them) belongs to the third category, while the wind, earthquake, and structural dynamical responses under random excitations belong to the fourth category. In the following parts of this appendix, some useful but brief definitions related to the third and fourth category of stochastic processes are given.

Structural Reliability: Approaches from Perspectives of Statistical Moments, First Edition.
Yan-Gang Zhao and Zhao-Hui Lu.
© 2021 John Wiley & Sons Ltd. Published 2021 by John Wiley & Sons Ltd.

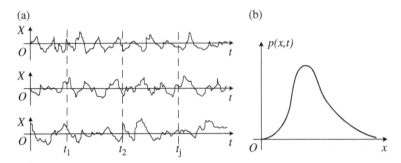

Figure D.1 A stochastic process and the one-dimensional PDF. (a) Stochastic samples. (b) One-dimensional PDF.

D.2 Time Domain Description of Stochastic Processes

D.2.1 Probability Distributions of Stochastic Processes

Assume that $\{X(t), t \in T\}$ is a stochastic process, where t is a time parameter that belongs to a set T (a time interval). To describe its probability properties, what first counts is the cumulative distribution function (CDF) of every one-time random variable (or the random variable at a certain time t, $t \in T$). This is denoted by

$$F(x,t) = P\{X(t) < x\}, \quad t \in T \tag{D.1}$$

and is called the one-dimensional distribution of $\{X(t), t \in T\}$.

It is obvious that the one-dimensional CDF alone is insufficient to describe a stochastic process. Then, it naturally comes into our minds to study how the random variables of a stochastic process at different time instants are correlated. For this purpose, we introduce

$$F(x_1, t_1; x_2, t_2) = P\{X(t_1) < x_1, X(t_2) < x_2\}, \quad t_1, t_2 \in T \tag{D.2}$$

as the two-dimensional CDF of $\{X(t), t \in T\}$.

In increasing order of completeness, for any finite $\{t_1, t_2, ...t_n \in T\}$, there is

$$F(x_1, t_1; x_2, t_2; ...; x_n, t_n) = P\{X(t_1) < x_1, X(t_2) < x_2, ...X(t_n) < x_n\} \tag{D.3}$$

which is called the n-dimensional CDF of $\{X(t), t \in T\}$.

For a stochastic process $\{X(t), t \in T\}$, its one-, two-, ..., and n-dimensional CDFs constitute its complete probabilistic structure. In fact, once this family of finite-dimensional CDFs is given, we can determine the correlation of any finite random variables of the stochastic process at different time instants. That is, the probabilistic structure of $\{X(t), t \in T\}$ is obtained.

Similar to the cases of a random variable and a random vector, for a stochastic process the finite-dimensional probability density functions (PDF) are defined by the derivatives of the corresponding CDFs; that is:

$$f(x,t) = \frac{\partial F(x,t)}{\partial x}$$

$$f(x_1, t_1; x_2, t_2) = \frac{\partial^2 F(x_1, t_1; x_2, t_2)}{\partial x_1 \partial x_2} \qquad \text{(D.4)}$$

$$\cdots$$

$$f(x_1, t_1; x_2, t_2; \ldots; x_n, t_n) = \frac{\partial^n F(x_1, t_1; x_2, t_2; \ldots; x_n, t_n)}{\partial x_1 \partial x_2 \ldots \partial x_n}$$

Certainly, this family can completely describe the probabilistic structures of a stochastic process as well.

The concepts discussed above can be extended to several jointly distributed stochastic processes. For example, for stochastic processes $X(t)$ and $Y(t')$, the joint $(n + m)$-dimensional CDF is

$$F(x_1, t_1; \ldots; x_n, t_n; y_1, t_1'; \ldots; y_m, t_m') = P\{X(t_1) < x_1, \ldots, X(t_n) < x_n, Y(t_1') < y_1, \ldots, Y(t_m') < y_m\}$$
$$\text{(D.5)}$$

and the corresponding $(n + m)$-dimensional PDF is

$$f(x_1, t_1; \ldots; x_n, t_n; y_1, t_1'; \ldots; y_m, t_m') = \frac{\partial^{n+m} F(x_1, t_1; \ldots; x_n, t_n; y_1, t_1'; \ldots; y_m, t_m')}{\partial x_1 \ldots \partial x_n \partial y_1 \ldots \partial y_m} \qquad \text{(D.6)}$$

D.2.2 Moment Functions of Stochastic Processes

Similar as a random variable, the expectation or mean of a one-dimensional stochastic process $X(t)$ is defined by

$$\mu_X(t) = E[X(t)] = \int_{-\infty}^{\infty} x(t) f(x,t) dx \qquad \text{(D.7)}$$

This equation represents the first origin moment of the random variable of the stochastic process at time t, and therefore $\mu_X(t)$ is generally a function of t. The mean square and variance of $X(t)$ can be given by

$$E[X^2(t)] = \int_{-\infty}^{\infty} x^2(t) f(x,t) dx \qquad \text{(D.8)}$$

$$\sigma_X^2(t) = E\{[X(t) - \mu_X(t)]^2\} = E[X^2(t)] - \mu_X^2(t) \qquad \text{(D.9)}$$

Unlike random variables, the correlation function is a very important statistical parameter for stochastic processes, which serves as a measure of interrelation of any two different states of stochastic processes. It quantifies how close the values of the random variables specified at two different time instants are in the sense of probability. There are auto- and cross-correlation functions according to whether the correlation information of one or between two stochastic processes needs to be characterised.

The *autocorrelation function* of $X(t)$ is defined for two random variables from the same process by

$$R_{XX}(t_1, t_2) = E[X(t_1)X(t_2)] = \int_{-\infty}^{\infty}\int_{-\infty}^{\infty} x(t_1)x(t_2)f(x_1, t_1; x_2, t_2)dx_1 dx_2 \qquad \text{(D.10)}$$

As $t_1 = t_2 = t$, Eq. (D.10) reduces to Eq. (D.8), i.e. the *autocorrelation function* of $X(t)$ becomes its mean square.

On the other hand, the *cross-correlation function* is assigned to those from two different processes. Suppose $X(t)$ and $Y(t)$ are two stochastic processes, and then the cross-correlation function is defined by

$$R_{XY}(t_1, t_2) = E[X(t_1)Y(t_2)] = \int_{-\infty}^{\infty}\int_{-\infty}^{\infty} x(t_1)y(t_2)f(x, t_1; y, t_2)dxdy \qquad \text{(D.11)}$$

where $f(x, t_1; y, t_2)$ is the joint PDF of $X(t_1)$ and $Y(t_2)$.

The cross-correlation function describes the interrelation between two stochastic processes in the time domain. In other words, it indicates the degree of probabilistic similarity between two stochastic processes at different time instants.

The normalised correlation functions are called as *correlation coefficient*. The *autocorrelation coefficient* is denoted by

$$\rho_{XX}(t_1, t_2) = \frac{R_{XX}(t_1, t_2) - \mu_X(t_1)\mu_X(t_2)}{\sigma_X(t_1)\sigma_X(t_2)} \qquad \text{(D.12)}$$

Correspondingly, the *cross-correlation coefficient* is

$$\rho_{XY}(t_1, t_2) = \frac{R_{XY}(t_1, t_2) - \mu_X(t_1)\mu_Y(t_2)}{\sigma_X(t_1)\sigma_Y(t_2)} \qquad \text{(D.13)}$$

As noted, the correlation is based on the second-origin moments of processes, whereas the following covariance is on the second central moments.

For a stochastic process $X(t)$, the *auto-covariance function* is defined as

$$\begin{aligned} B_{XX}(t_1, t_2) &= E\{[X(t_1) - \mu_X(t_1)][X(t_2) - \mu_X(t_2)]\} \\ &= \int_{-\infty}^{\infty}\int_{-\infty}^{\infty} [x(t_1) - \mu_X(t_1)][x(t_2) - \mu_X(t_2)]f(x_1, t_1; x_2, t_2)dx_1 dx_2 \end{aligned} \qquad \text{(D.14)}$$

When $t_1 = t_2 = t$, Eq. (D.14) reduces to Eq. (D.9), i.e. the *auto-covariance function* of $X(t)$ becomes its variance.

For two stochastic processes, the *cross-covariance function* is defined as

$$\begin{aligned} B_{XY}(t_1, t_2) &= E\{[X(t_1) - \mu_X(t_1)][Y(t_2) - \mu_Y(t_2)]\} \\ &= \int_{-\infty}^{\infty}\int_{-\infty}^{\infty} [x(t_1) - \mu_X(t_1)][y(t_2) - \mu_Y(t_2)]f(x, t_1; y, t_2)dxdy \end{aligned} \qquad \text{(D.15)}$$

It is easy to verify the following relationships:

$$B_{XX}(t_1, t_2) = R_{XX}(t_1, t_2) - \mu_X(t_1)\mu_X(t_2) \qquad \text{(D.16)}$$

$$B_{XY}(t_1, t_2) = R_{XY}(t_1, t_2) - \mu_X(t_1)\nu_Y(t_2) \qquad \text{(D.17)}$$

These two equations imply that, for stochastic processes with zero mean, the covariance function equals the correlation function.

D.2.3 Stationary and Nonstationary Process

A stochastic process $X(t)$ is called *strictly stationary*, if finite-dimensional distributions of time remain unchanged if the time scale is shifted by an arbitrary value τ.

$$f(x_1, t_1; x_2, t_2; ...; x_n, t_n) = f(x_1, t_1 + \tau; x_2, t_2 + \tau; ...; x_n, t_n + \tau) \tag{D.18}$$

In contrast, a stochastic process is called a *weakly stationary* process if its expectation is a constant and its autocorrelation function depends only on the time difference $\tau = t_2 - t_1$ (independent of t_1 and t_2). That is

$$\mu_X(t) = E[X(t)] = \mu_X = \text{constant} \tag{D.19}$$

and

$$R_{XX}(t_1, t_2) = R_{XX}(t_1, t_1 + \tau) = R_{XX}(\tau) \tag{D.20}$$

Generally speaking, a weakly stationary process is not necessarily a strictly stationary one, while a strictly stationary process must be a weakly stationary one. Only for a Gaussian process, if weakly stationary, it is also strictly stationary. In practical applications, the weak stationarity is much more widely used than the strict stationarity. Thus, the stationary processes mentioned below (also in the chapter on Methods of Moment for Time-Variant Reliability), unless specified otherwise, mean the weakly stationary ones.

The following properties hold true for the autocorrelation function:

a) It is symmetric;

$$R_X(\tau) = R_X(-\tau) \tag{D.21}$$

b) It is nonnegative definite

$$\sum_{i=1}^{n} \sum_{j=1}^{n} R_X(t_i - t_j) h(t_i) h^*(t_j) \geq 0 \tag{D.22}$$

where $h(t)$ is any arbitrary complex function and $h^*(t)$ is the complex conjugate;

c) It is bounded;

$$|R_X(\tau)| \leq R_X(0) \tag{D.23}$$

d) If $X(t)$ does not contain periodic components, then for a zero-mean stochastic process

$$\lim_{\tau \to \infty} R(\tau) = 0 \tag{D.24}$$

A typical autocorrelation function of a zero-mean stationary process is shown in Figure D.2.

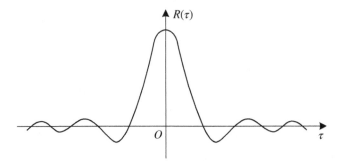

Figure D.2 Autocorrelation function of a zero-mean stationary process.

Notice that, for a stationary process, the mean square and variance of $X(t)$ become

$$E[X^2(t)] = R_{XX}(0) \tag{D.25a}$$

$$\sigma_X^2 = B_{XX}(0) = R_{XX}(0) - \mu_X^2 \tag{D.25b}$$

If two stochastic processes are jointly stationary, the cross-correlation function defined by Eq. (D.11) can be rewritten as

$$R_{XY}(\tau) = R_{XY}(t_2 - t_1) \tag{D.26}$$

If a stochastic process $X(t)$ cannot satisfy the conditions specified by Eq. (D.18) or Eqs. (D.19) and (D.20), it is referred to *nonstationary*.

D.2.4   Ergodicity of a Stochastic Process

A stationary stochastic process is said to be ergodic if the expectation and autocorrelation function of $X(t)$ defined by Eq. (D.7) and Eq. (D.10) are identical with those obtained from a single time history $X_i(t)$. That is

$$\mu_X(t) = E[X(t)] = \lim_{T \to \infty} \frac{1}{T} \int_0^T x_i^k(t) dt \tag{D.27}$$

and

$$R_{XX}(\tau) = E[X(t)X(t + \tau)] = \lim_{T \to \infty} \frac{1}{T} \int_0^T x_i(t)x_i(t + \tau) dt \tag{D.28}$$

D.3   Frequency Domain Description of Stochastic Processes

D.3.1   Power Spectral Density Function

If Fourier Transform (FT) is applied to the process $X(t)$, a new process $X(\omega)$ is obtained by means of

$$X(\omega) = \frac{1}{2\pi} \int_{-\infty}^{\infty} X(t) \exp(-i\omega t) dt \qquad \text{(D.29)}$$

The inverse transform yields

$$X(t) = \int_{-\infty}^{\infty} X(\omega) \exp(i\omega t) d\omega \qquad \text{(D.30)}$$

In view of Eq. (D.30), it becomes obvious that $X(t)$ may be considered a process which is composed of a sum of harmonic components with amplitudes $X(\omega)d\omega$. However, the above mentioned Fourier transforms do not exist for stationary processes, since these processes have infinite duration and are not absolutely integrable. If Fourier transform is applied to the autocorrelation function $R_{XX}(\tau)$, the power spectral density (PSD) $S_{XX}(\omega)$ is obtained

$$S_{XX}(\omega) = \frac{1}{2\pi} \int_{-\infty}^{\infty} R_{XX}(\tau) \exp(-i\omega\tau) d\tau \qquad \text{(D.31)}$$

Reversely

$$R_{XX}(\tau) = \int_{-\infty}^{\infty} S_{XX}(\omega) \exp(i\omega\tau) d\omega \qquad \text{(D.32)}$$

Eqs. (D.31) and (D.32) are denoted as Wiener-Khintchine-Theorem. The PSD may be interpreted as the distribution of energy over frequency. It exists if $R_{XX}(\tau)$ is absolutely integrable. This condition, which is met by most weakly stationary processes, exists. Noting that

$$R_{XX}(0) = E[X(t)X(t)] = \sigma_X^2 \qquad \text{(D.33)}$$

from Eq. (D.32) the following important relation is obtained

$$\sigma_X^2 = \int_{-\infty}^{\infty} S_{XX}(\omega) d\omega \qquad \text{(D.34)}$$

Eq. (D.34) is the basis for the Power Spectral Method (PSM) in stochastic structural dynamics.

A cross-spectral density of two stochastic processes $X(t)$ and $Y(t)$, which are jointly stationary in the wide sense, is defined by

$$S_{XY}(\omega) = \frac{1}{2\pi} \int_{-\infty}^{\infty} R_{XY}(\tau) \exp(-i\omega\tau) d\tau \qquad \text{(D.35)}$$

And the cross-correlation function obtained from

$$R_{XY}(\tau) = \int_{-\infty}^{\infty} S_{XY}(\omega) \exp(i\omega\tau) d\omega \qquad \text{(D.36)}$$

D.3.2 Wide- and Narrow-Band Processes

A stationary process is called to be narrow-banded if its spectral density is concentrated around a particular value e.g. $|\omega_0|$. Otherwise, it is considered a wide-band process. The

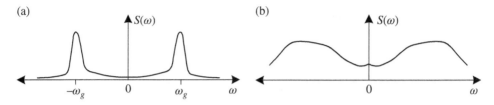

Figure D.3 Spectral density functions of stochastic processes. (a) Narrow-band process. (b) Wide-band process.

schematic representation of spectral density functions of a wide-band process is shown in Figure D.3.

Generally, the spectral moments, λ_i, which are defined as

$$\lambda_i = \int_{-\infty}^{\infty} |\omega^i| S(\omega) d\omega \tag{D.37}$$

are used to determine some measures of the band-width of stochastic process, i.e.

$$\gamma^2 = 1 - \frac{\lambda_2^2}{\lambda_0 \lambda_4} \tag{D.38}$$

For example, the value for γ approaches zero for narrow-band process and approaches one for wide-band processes.

D.4 Special Processes

D.4.1 White Noise Process

A weakly stationary stochastic process with a constant spectral density, S_0, is called a white noise, $W(t)$, defined by either one of the following equivalent properties:

$$S_{WW}(\omega) = S_0, \quad -\infty < \omega < \infty \tag{D.39}$$

$$R_{WW}(\tau) = 2\pi S_0 \delta(\tau) \tag{D.40}$$

where $\delta(\tau)$ is the Dirac delta function.

As can be seen from Eqs. (D.39) and (D.40), the mean square value of a white noise is unbounded. It is quite clear that a physical stochastic process cannot have an infinite average energy. Therefore the white-noise process has to be considered as mathematical idealizations. It should be noted that justifications are required to ensure that the results of the simplified analysis represent a close approximation of what would be obtained in case of physical excitations.

D.4.2 Markov Process

If the following relations between the conditional probability functions are defined such that

$$F(x_n, t_n | x_{n-1}, t_{n-1}; x_{n-2}, t_{n-2}; ...; x_1, t_1) = F(x_n, t_n | x_{n-1}, t_{n-1}), \quad t_1 < \cdots < t_{n-1} < t_n$$

(D.41)

a stochastic process is called Markovian.

As the state of the process at time $t = t_n$ depends only on the state at $t = t_{n-1}$, it is also called a 'one-step memory' stochastic process. The right hand side of Eq. (D.41) expresses the transition probabilitites. A discrete Markov process is completely defined by its initial state and the transition probabilities. For stationary properties, the latter suffices for complete characterisation.

For continuous valued stochastic processes $X(t)$ the Markovian properties are defined by

$$f(x_n, t_n | x_{n-1}, t_{n-1}; x_{n-2}, t_{n-2}; ...; x_1, t_1) = f(x_n, t_n | x_{n-1}, t_{n-1}), \quad t_1 < \cdots < t_{n-1} < t_n$$

(D.42)

It is found that the transition probabilities of a Markov process are governed by the well-known Fokker-Planck or diffusion equation, and the simplest, i.e. the one dimensional or scalar, Markov process is the so-called Wiener process.

D.4.3 Poisson Process

This process is memoryless, i.e. the probabilities of event arrivals are independent of each other. Given stationarity, the probability of event arrival in any time interval $t + dt$ is equal to νdt where ν is a positive constant. Furthermore, it is assumed that a simultaneous event arrival within $t + dt$ is negligibly small. Based on these assumptions, it can be shown that the probability of n event arrivals within a time period $(0, t)$ is:

$$P(n|\nu, t) = \frac{(\nu t)^2}{n!} \exp(-\nu t), \quad n = 0, 1, ..., \infty$$

(D.43)

D.4.4 Gaussian Process

A stochastic process $X(t)$ is said to be a Gaussian process if all joint probability density functions [as defined in Eq.(D.3)] are normally distributed. A Gaussian process is completely defined by its mean function $\mu_X(t)$ and covariance function $R_{XX}(t_1, t_2)$. For an arbitrary weakly stationary process, $\mu_X(t)$ is a constant and $\phi_{XX}(t_1, t_2)$ is a function of $\tau = t_2 - t_1$; again, for the special case of a Gaussian process, no distinction between the weakly and the strongly stationary cases is necessary since one implies the other. It is important to note that a linear transformation of a set of Gaussian variables produces a new set of Gaussian variables. Similarly, a linear operation, such as differentiation or intergration, on a Gaussian process results in another Gaussian process.

D.5 Spectral Representation Method

A zero-mean, stationary Gaussian vector process $\mathbf{U}(t) = \{u_1(t), u_2(t), ..., u_n(t)\}$ with prescribed PSDM $\mathbf{S}_G(\omega)$ (or equivalently CFM) can be simulated by the following series as $N \rightarrow \infty$ (Shinozuka and Jan 1972; Shinozuka and Deodatis 1991):

$$u_i(t) = 2\sqrt{\Delta\omega} \sum_{m=1}^{i} \sum_{l=1}^{N} |H_{im}(\omega_{ml})| \cdot \cos[\omega_{ml}t - \theta_{im}(\omega_{ml}) + \phi_{ml}], \quad i = 1, 2, ..., m$$

$$\tag{D.44}$$

$$\omega_{ml} = l\Delta\omega, l = 0, 1, ..., N - 1 \tag{D.45}$$

$$\Delta\omega = \frac{\omega_u}{N} \tag{D.46}$$

$$\theta_{im}(\omega_l) = \tan^{-1}\left(\frac{\Im[H_{im}(\omega_l)]}{\Re[H_{im}(\omega_l)]}\right) \tag{D.47}$$

where ω_u is the upper cutoff frequency beyond which the elements of the PSDM have negligibly small (or zero) values; $\Im$ and $\Re$ are the imaginary and real parts of a complex number, respectively; and ϕ_{ml} is m-sequences of independent uniformly distributed random phase angles over the interval $[0, 2\pi)$. Eq. (D.44) requires the PSDM being decomposed into the following form:

$$\mathbf{S}_G(\omega) = \mathbf{H}(\omega)\mathbf{H}^*(\omega)^T \tag{D.48}$$

where $\mathbf{H}^*(\omega)$ denotes the complex conjugate of $\mathbf{H}(\omega)$.

The simulated time histories are periodic with period:

$$T = \frac{2\pi}{\Delta\omega} \tag{D.49}$$

To avoid aliasing, the time step Δt separating the generated values of $u_i(t)$ must satisfy the following condition:

$$\Delta t \leq \frac{2\pi}{2\omega_u} \tag{D.50}$$

D.6 Summary

In this chapter, the basic theory of stochastic process, which constitutes the fundamentals of the time-dependent structural reliability theory, is introduced. The general concepts, the time and frequency domain descriptions of stochastic processes with emphasis on stationary processes are briefly described, and some special processes are also introduced, which are essentially useful for better understanding the time-dependent structural reliability methods described in this book.

References

Abramowitz, M. and Stegun, I.E. (1972). *Handbook of Mathematical Functions*. New York: Dover Publications.

Adduri, R.P., Penmetsa, R.C., and Grandhi, R.V. (2004). Estimation of structural system reliability using fast Fourier transforms, Proceedings of 10th AIAA/ISSMO *Multidisciplinary Analysis and Optimization Conference*, Albany, New York, AIAA, 2004-4341.

Adhikari, S. (2005). Asymptotic distribution method for structural reliability analysis in high dimensions. *Proc. R. Soc. A.* 461 (2062): 3141–3158.

AIJ (2002). *Recommendations for Limit State Design of Buildings*. Tokyo: Architectural Institute of Japan (AIJ) (in Japanese).

AIJ (2015). *Recommendations for Loads on Buildings*. Tokyo: Architectural Institute of Japan (AIJ) (in Japanese).

Ang, A.H.-S. (1970). Extended reliability basis of structural design under uncertainties. In: *Proceeding of the 9th Reliability and Maintainability Conference, Annals of Reliability and Maintainability*, 642–649. New York: IEEE.

Ang, A.H.-S. and Amin, M. (1968). Reliability of structures and structural systems. *J. Eng. Mech. Div.* 94 (2): 671–691.

Ang, A.H.-S. and Cornell, C.A. (1974). Reliability bases of structural safety and design. *J. Struct. Div.* 100 (9): 1755–1769.

Ang, A.H.-S. and De Leon, D. (2005). Modeling and analysis of uncertainties for risk-informed decisions in infrastructures engineering. *Struct. Infrastruct. Eng.* 1 (1): 19–31.

Ang, A.H.-S. and Ma, H.F. (1981). *On the reliability of structural systems*, Proceedings of the 3rd International Conference on Structural Safety and Reliability, Trondheim, Norway, pp. 295–314.

Ang, A.H.-S. and Tang, W.H. (1975). *Probability Concepts in Engineering Planning and Design, Vol. I: Basic Principles*. New York: Wiley.

Ang, A.H.-S. and Tang, W. (1984). *Probability Concepts in Engineering Planning and Design, Vol II: Decision, Risk, and Reliability*. New York: Wiley.

Ang, A.H.-S. and Tang, W. (2006). *Probability Concepts in Engineering: Emphasis on Applications to Civil and Environmental Engineering*. New York: Wiley.

Aoyama, H. (1988). *Structural Analysis by Method of Matrix*. Tokyo: Paifukan Press (in Japanese).

Au, S.K. and Beck, J.L. (2001a). Estimation of small failure probabilities in high dimensions by subset simulation. *Probab. Eng. Mech.* 16 (4): 263–277.

Structural Reliability: Approaches from Perspectives of Statistical Moments, First Edition.
Yan-Gang Zhao and Zhao-Hui Lu.
© 2021 John Wiley & Sons Ltd. Published 2021 by John Wiley & Sons Ltd.

Au, S.K. and Beck, J.L. (2001b). First excursion probabilities for linear systems by very efficient importance sampling. *Probab. Eng. Mech.* 16 (3): 193–207.

Au, S.K. and Beck, J.L. (2003). Subset simulation and its application to seismic risk based on dynamic analysis. *J. Eng. Mech.* 129 (8): 901–917.

Ayyub, B.M. and Haldar, A. (1984). Practical structural reliability techniques. *J. Struct. Eng.* 110 (8): 1707–1724.

Baber, T.T. and Noori, M.N. (1985). Random vibration of degrading pinching systems. *J. Eng. Mech.* 111 (8): 1010–1026.

Baber, T.T. and Wen, Y.K. (1981). Random vibrations of hysteretic degrading systems. *J. Eng. Mech.* 107 (6): 1069–1089.

Baecher, G.B. and Christian, J.T. (2003). *Reliability and Statistics in Geotechnical Engineering*. New York: Wiley.

Barbato, M. and Conte, J.P. (2011). Structural reliability applications of nonstationary spectral characteristics. *J. Eng. Mech.* 137 (137): 371–382.

Barranco-Cicilia, F., Castro-Prates de Lima, E., and Sudati-Sagrilo, L.V. (2009). Structural reliability analysis of limit state functions with multiple design points using evolutionary strategies. *Ingeniería Investigación y Tecnología* 10 (2): 87–97.

Bayer, V. and Bucher, C. (1999). Importance sampling for first passage problems of nonlinear structures. *Probab. Eng. Mech.* 14 (1–2): 27–32.

Benjamin, J.R. and Cornell, C.A. (1970). *Probability, Statistics, and Decision for Civil Engineers*. New York: McGraw-Hill.

Bennett, R.M. and Ang, A.H.-S. (1986). Formulation of structural system reliability. *J. Eng. Mech.* 112 (11): 1135–1151.

Bhargava, K., Mori, Y., and Ghosh, A.K. (2011). Time-dependent reliability of corrosion-affected RC beams. Part 2: estimation of time-dependent failure probability. *Nucl. Eng. Des.* 241 (5): 1385–1394.

Bjerager, P. (1991). Methods for structural reliability computation. In: *Reliability Problems: General Principles and Applications in Mechanics of Solid and Structures* (ed. F. Casciati), 89–136. New York: Springer Verlag Wien.

Bocchini, P. and Deodatis, G. (2008). Critical review and latest developments of a class of simulation algorithms for strongly non-Gaussian random fields. *Probab. Eng. Mech.* 23: 393–407.

Bourinet, J.M., Deheeger, F., and Lemaire, M. (2011). Assessing small failure probabilities by combined subset simulation and support vector machines. *Struct. Saf.* 33 (6): 343–353.

Box, G.E.P. and Cox, D.R. (1964). An analysis of transformations. *J. R. Stat. Soc. Ser. B* 26 (2): 211–252.

Box, G.E.P. and Muller, M.E. (1958). A note on the generation of random normal deviates. *Ann. Math. Stat.* 29 (2): 610–611.

Breitung, K. (1984). Asymptotic approximation for multi-normal integrals. *J. Eng. Mech.* 10 (3): 357–366.

Bucher, C.G. (1988). Adaptive sampling: an iterative fast Monte-Carlo procedure. *Struct. Saf.* 5 (2): 119–126.

Bucher, C.G. and Bourgund, U. (1990). A fast and efficient response surface approach for structural reliability problems. *Struct. Saf.* 7 (1): 57–66.

Cai, G.Q. and Elishakoff, I. (1994). Refined second-order reliability analysis. *Struct. Saf.* 14: 267–276.

Cai, C.H., Lu, Z.H., Xu, J., and Zhao, Y.G. (2019). Efficient algorithm for evaluation of statistical moments of performance functions. *J. Eng. Mech.* 145 (1): 06018007.

Cai, C.H., Lu, Z.H., and Zhao, Y.G. (2020). Moment method with Box-Cox transformation for structural reliability. *J. Eng. Mech.* 146 (8): 04020086.

Chakraborty, S. and Chowdhury, R. (2016). Assessment of polynomial correlated function expansion for high-fidelity structural reliability analysis. *Struct. Saf.* 59: 9–19.

Chang, Y. and Mori, Y.A. (2013). Study on the relaxed linear programming bounds method for system reliability. *Struct. Saf.* 41: 64–72.

Chang, C.H., Tung, Y.K., and Yang, J.C. (1994). Monte Carlo simulation for correlated variables with marginal distributions. *J. Hydraul. Eng.* 120 (3): 313–331.

Chen, X. and Tung, Y.K. (2003). Investigation of polynomial transformation. *Struct. Saf.* 25 (4): 423–445.

Chen, J.B. and Wan, Z.Q. (2019). A compatible probabilistic framework for quantification of simultaneous aleatory and epistemic uncertainty of basic parameters of structures by synthesizing the change of measure and change of random variables. *Struct. Saf.* 78: 76–87.

Chen, J., Kong, F., and Peng, Y. (2017). A stochastic harmonic function representation for non-stationary stochastic processes. *Mech. Syst. Signal Process.* 96: 31–44.

Choi, M. and Sweetman, B. (2010). The Hermite moment model for highly skewed response with application to tension leg. *J. Offshore Mech. Arct. Eng.* 132 (2): 021602.

Choi, S.K., Grandhi, S.K., and Canfield, R.A. (2007). *Reliability-Based Structural Design*. London: Springer.

Clough, R.W. and Penzien, J. (2003). *Dynamics of Structures*. Computers and Structures, Inc.

Coleman, J.J. (1959). Reliability of aircraft structures in resisting chance failure. *Oper. Res.* 7 (5): 639–945.

Cornell, C.A. (1967). Bounds on the reliability of structural systems. *J. Struct. Div.* 93 (1): 171–200.

Cornell, C.A. (1969). *Structural safety specification based on second-moment reliability*. Symposium: International Association of Bridge and Structural Engineering, London.

Cramer, H. (1966). On the intersections between the trajectories of a normal stationary stochastic process and a high level. *Ark. Mat.* 6 (4–5): 337–349.

Cramer, H. and Leadbetter, M.R. (1967). *Stationary and Related Stochastic Process*. Hoboken, NJ.: Wiley.

Crandall, S.H. (1970). First-crossing probabilities of the linear oscillator. *J. Sound Vib.* 12 (3): 285–299.

Daghigh, M. and Makouie, S.H. (2003). Application of β-unzipping method in offshore system reliability analysis under changing load pattern. Proceedings of the 22nd International Conference on *Offshore Mechanics and Arctic Engineering*, 1–7.

David, H.A. (1981). *Order Statistics*, 2e. New York: Wiley.

Deng, J. and Pandey, M.D. (2008). Estimation of the maximum entropy quantile function using fractional probability weighted moments. *Struct. Saf.* 30 (4): 307–319.

Deodatis, G. (1996). Simulation of ergodic multivariate stochastic process. *J. Eng. Mech.* 122: 778–787.

Deodatis, G. and Micaletti, R. (2001). Simulation of highly skewed non-Guassian stochastic processes. *J. Eng. Mech.* 127 (12): 1284–1295.

Der Kiureghian, A. (1989). Measures of structural safety under imperfect states of knowledge. *J. Eng. Mech.* 115 (5): 1119–1140.

Der Kiureghian, A. (2008). Analysis of structural reliability under parameter uncertainties. *Probab. Eng. Mech.* 23: 351–358.

Der Kiureghian, A. and Dakessian, T. (1998). Multiple design points in first- and second-order reliability. *Struct. Saf.* 20 (1): 37–50.

Der Kiureghian, A. and De Stefano, M. (1991). Efficient algorithm for second-order reliability analysis. *J. Eng. Mech.* 117 (12): 2904–2923.

Der Kiureghian, A. and Ditlevsen, O. (2009). Aleatory or epistemic? Does it matter? *Struct. Saf.* 31: 105–112.

Der Kiureghian, A. and Liu, P.L. (1986). Structural reliability under incomplete probability information. *J. Eng. Mech.* 112 (1): 85–104.

Der Kiureghian, A., Lin, H.Z., and Hwang, S.J. (1987). Second-order reliability approximations. *J. Eng. Mech.* 113 (8): 1208–1225.

Ditlevsen, O. (1979a). Generalized second moment reliability index. *J. Struct. Mech.* 7 (4): 435–451.

Ditlevsen, O. (1979b). Narrow reliability bounds for structural systems. *J. Struct. Mech.* 7 (4): 453–472.

Ditlevsen, O. (1981). Principle of normal tail approximation. *J. Eng. Mech.* 107: 1191–1208.

Ditlevsen, O. (1986). Duration of visit to critical set by Gaussian process. *Probab. Eng. Mech.* 1 (2): 82–93.

Ditlevsen, O. and Bjerager, P. (1986). Methods of structural system reliability. *Struct. Saf.* 3 (3): 195–229.

Ditlevsen, O. and Madsen, H.O. (1996). *Structural Reliability Methods*. New York: Wiley.

Du, W., Luo, Y., and Wang, Y. (2019). Time-variant reliability analysis using the parallel subset simulation. *Reliab. Eng. Syst. Saf.* 182: 250–257.

Ellingwood, B., Galambos, T.V., MacGregor, J.C., Cornell, C.A. (1980). *Development of a probability based load criterion for American National Standard A58*, National Bureau of Standards Publication 577, Washington, DC.

Ellingwood, B., MacGregor, G., and Galambos, T.V. (1982). Probability based load criteria: load factor and load combinations. *J. Eng. Mech.* 108 (5): 978–997.

Enright, M.P. and Frangopol, D.M. (1998). Failure time prediction of deteriorating fail-safe structures. *J. Struct. Eng.* 124 (12): 1448–1457.

Fan, W.L., Wei, J.H., Ang, A.H.-S., and Li, Z.L. (2016). Adaptive estimation of statistical moments of the responses of random systems. *Probab. Eng. Mech.* 43: 50–67.

Fan, W.L., Ang, A.H.-S., and Li, Z.L. (2017). Reliability assessment of deteriorating structures using Bayesian updated probability density evolution method (PDEM). *Struct. Saf.* 65: 60–73.

Faravelli, L. (1989). Response surface approach for reliability analysis. *J. Eng. Mech.* 115 (12): 2763–2781.

Feng, Y.S. (1988). Enumerating significant failure modes of a structural system by using criterion methods. *Comput. Struct.* 30 (5): 1153–1157.

Feng, Y.S. (1989). A method for computing structural system reliability with high accuracy. *Comput. Struct.* 33 (1): 1–5.

Fiessler, B. (1979). *Das programm system FORM zur Berechnung der versagenswahrscheinlichkeit von komponenten von tragsystemen*, Merichte zur zuverlassigkeits theorie der bauwerke, No. 43, Technical University, Munich.

Fiessler, B., Neumann, H.-J., and Rackwitz, R. (1979). Quadratic limit states in structural reliability. *J. Eng. Mech. Div.* 105 (4): 661–676.

Filippou, F.C., Popov, E.P., and Bertero, V.V. (1983). *Effect of bond deterioration on histeretic behavior of reinforced concrete joints*, Berkeley, CA: Earthquake Engineering Research Center, No. EERC Rep. 83–19.

Fisher, R.A. and Cornish, E.A. (1960). The percentile points of distributions having known cumulants. *Technometrics* 2: 209–225.

Fleishman, A. (1978). A method for simulating non-normal distributions. *Psychometrika* 43 (4): 521–532.

Freudenthal, A.M. (1956). Safety and the probability of structural failure. *Trans. Am. Soc. Civ. Eng.* 121 (1): 1337–1397.

Galambos, T.V. and Ellingwood, B. (1982). Probability based load criteria: assesment of current design practice. *J. Eng. Mech.* 108 (5): 957–977.

Gao, S., Zheng, X.Y., and Huang, Y. (2017). Hybrid C- and L-moment–based Hermite transformation models for non-Gaussian processes. *J. Eng. Mech.* 144 (2): 04017179.

Gentleman, J.F. (1975). Algorithm S 88: Generation of all NCR combinations by simulating nested Fortran DO loops. *J. R. Stat. Soc.* 24 (3): 374–376.

Ghazizadeh, S., Barbato, M., and Tubaldi, E. (2011). New analytical solution of the first passage reliability problem for linear oscillator. *J. Eng. Mech.* 138 (6): 695–706.

Gorman, M.R. (1980). Reliability of structural systems, Ph.D. Thesis, Case Western Reserve University, Cleveland, OH: 320–332.

Greenwood, J.A., Landwehr, J.M., Matalas, N.C., and Wallis, J.R. (1979). Probability weighted moments: definition and relation to parameters of several distributions expressible in inverse form. *Water Resour. Res.* 15 (5): 1049–1054.

Grigoriu, M. (1983). Approximate analysis of complex reliability problems. *Struct. Saf.* 1 (4): 277–288.

Grigoriu, M. (1984). Crossings of non-Gaussian translation process. *J. Eng. Mech.* 110 (4): 610–620.

Grigoriu, M. (1995). *Applied Non-Gaussian Processes*. Englewood Cliffs, NJ: PTR Prentice-Hall.

Grigoriu, M. (1998). Simulation of stationary non-Gaussian translation processes. *J. Eng. Mech.* 124 (2): 121–126.

Grimmelt, M.J. and Schueller, G.I. (1982). Benchmark study on methods to determine collapse failure probabilities of redundant structures. *Struct. Saf.* 1 (2): 93–106.

Gumbel, E. (1958). *Statistics of Extremes*. New York: Columbia University Press.

Gurley, K.R. (1997). *Modeling and simulation of non-Gaussian processes*. Ph.D. thesis, Dept. of Civil Engineering and Geological Sciences, Univ. of Notre Dame, Notre Dame, IN.

Gurley, K.R. and Kareem, A. (1997). Analysis, interpretation, modeling and simulation of unsteady wind and pressure data. *J. Wind Eng. Ind. Aerodyn.* 69–71: 657–669.

Gurley, K.R. and Kareem, A. (1998a). A conditional simulation of non-normal velocity/pressure fields. *J. Wind Eng. Ind. Aerodyn.* 77–78: 39–51.

Gurley, K.R. and Kareem, A. (1998b). Simulation of correlated non-Gaussian pressure fields. *Meccanica* 33: 309–317.

Gurley, K.R., Kareem, A., and Tognarelli, M.A. (1996). Simulation of a class of non-normal random processes. *Int. J. Non Linear Mech.* 31 (5): 601–617.

Gurley, K.R., Tognarelli, M.A., and Kareem, A. (1997). Analysis and simulation tools for wind engineering. *Probab. Eng. Mech.* 12 (1): 9–31.

Harold, L., Nelson, J., and Granger, C.W.J. (1979). Experience with using the Box-Cox transformation when forecasting economic time series. *J. Econom.* 10: 57–69.

Harr, M.E. (1989). Probabilistic estimates for multivariate analysis. *Appl. Math. Modell.* 13 (5): 313–318.

Hasofer, A.M. and Lind, N.C. (1974). Exact and invariant second-moment code format. *J. Eng. Mech. Div.* 100 (1): 111–121.

Hastings, W.K. (1970). Monte Carlo sampling methods using Markov chains and their applications. *Biometrika* 57 (1): 97–109.

He, J. (2009). Numerical calculation for first excursion probabilities of linear systems. *Probab. Eng. Mech.* 24 (3): 418–425.

He, J. (2015). Approximate method for estimating extreme value responses of nonlinear stochastic dynamic systems. *J. Eng. Mech.* 141 (7): 04015009.

He, J. and Zhao, Y.G. (2007). First passage times of stationary non-Gaussian structural responses. *Comput. Struct.* 85 (7): 431–436.

He, J., Gao, S.B., and Gong, J. (2014). A sparse grid stochastic collocation method for structural reliability analysis. *Struct. Saf.* 51: 29–34.

Ho, T.C.E., Surry, D., Morrish, D., and Kopp, G.A. (2005). The UWO contribution to the NIST aerodynamic database for wind loads on low buildings: part 1. Archiving format and basic aerodynamic data. *J. Wind Eng. Ind. Aerodyn.* 93 (1): 1–30.

Hohenbichler, M. and Rackwitz, R. (1981). Non-normal dependent vectors in structural safety. *J. Eng. Mech.* 107 (6): 1227–1238.

Hohenbichler, M. and Rackwitz, R. (1988). Improvement of second order reliability estimates by importance sampling. *J. Eng. Mech., ASCE* 114 (12): 2195–2199.

Hong, H.P. (1996). Point-estimate moment-based reliability analysis. *Civ. Eng. Syst.* 13 (4): 281–294.

Hong, H.P. (1998). An efficient point estimate method for probabilistic analysis. *Reliab. Eng. Syst. Saf.* 59 (3): 261–267.

Hong, H.P. (2000). Assessment of reliability of aging reinforced concrete structures. *J. Struct. Eng.* 126 (12): 1458–1465.

Hong, H.P. and Lind, N.C. (1996). Approximation reliability analysis using normal polynomial and simulation results. *Struct. Saf.* 18 (4): 329–339.

Hosking, J.R.M. (1990). L-moments: analysis and estimation of distributions using linear combinations of statistics. *J. R. Stat. Soc. B.* 52 (1): 105–124.

Hosking, J.R.M. and Wallis, J.R. (1997). *Regional Frequency Analysis: An Approach Based on L-Moments.* Cambridge, UK: Cambridge University Press.

Hurtado, J.E. (2007). Filtered importance sampling with support vector margin: a powerful method for structural reliability analysis. *Struct. Saf.* 29 (1): 2–15.

Ibrahim, R.A. (1987). Structural dynamics with parameter uncertainties. *Appl. Mech. Rev.* 40 (3): 309–328.

Impollonia, N., Muscolino, G., and Ricciardi, G. (1998). *Improved approach in the dynamics of structural systems with mechanical uncertainties*. Presented at Proceedings of the 3rd International conference on Computational Stochastic Mechanics, Santorini, Greece, June 14–17.

Inverardi, P. and Tagliani, A. (2003). Maximum entropy density estimation from fractional moments. *Commun. Stat. Theory Methods* 32 (2): 327–345.

ISO 2394 (2015). *General Principles on Reliability for Structures*. Switzerland: International Standardization Organization.

Israel, M.S. (1986). Probabilistic basis of partial resistance factors for use in concrete design. Master's thesis, The Johns Hopkins University, Baltimore, MD.

Jensen, J.J. (1994). Dynamic amplification of offshore steel platform responses due to non-Gaussian wave loads. *Mar. Struct.* 7 (1): 91–105.

Ji, X.W., Huang, G.Q., Zhang, X.X., and Kopp, G.A. (2018). Vulnerability analysis of steel roofing cladding: influence of wind directionality. *Eng. Struct.* 156: 587–597.

JIS-G-3101 (1976). *Rolled Steels for General Structure*. Tokyo: Japanese Industrial Standard (in Japanese).

JIS-G-3192 (1977). *Dimensions, Mass and Permissible Variations of Hot Rolled Steel Plates, Sheets and Strip*. Tokyo: Japanese Industrial Standard (in Japanese).

Johnson, N.L. and Kotz, S. (1969). *Distributions in Statistics, Discrete Distributions*. New York: Houghton Mifflin.

Johnson, N.L. and Kotz, S. (1970a). *Distributions in Statistics, Continuous Univariate Distributions-1*. New York: Houghton Mifflin.

Johnson, N.L. and Kotz, S. (1970b). *Distributions in Statistics, Continuous Univariate Distributions-2*. New York: Houghton Mifflin.

Johnson, N.L. and Kotz, S. (1972). *Distributions in Statistics, Continuous Multivariate Distributions-2*. New York: Houghton Mifflin.

Kanda, J. (1981). *A new extreme value distribution with lower and upper limits for earthquake motions and wind speeds*. Proceeding 31st Japan National Congress for Applied Mechanics, 351–360.

Kanda, J. (1994). Application of an empirical extreme value distribution to load models. *J. Res. Nat. Inst. Stand. Technol.* 99 (4): 413–420.

Knuth, D.E. (1969). *The Art of Computer Programming: Seminumerical Algorithms*, vol. 2. MA: Addison-Wesley Professional.

Knutson, T.R., McBride, J.L., Chan, J. et al. (2010). Tropical cyclones and climate change. *Nat. Geosci.* 3: 157–163.

Kobayashi, S. (1977). *Differential geometry of curves and surfaces*. Tokyo: Syoukabou Co.

Koyluoglu, H.U. and Nielsen, S.R.K. (1994). New approximations for SORM integrals. *Struct. Saf.* 13 (4): 235–246.

Krishnaiah, P.R. (1981). *Analysis of Variance*. Elsevier Science & Technology Books.

Kuschel, N., Pieracci, A. and Rackwitz, R. (1998). *Multiple β-points in structural reliability*, in Proceeding of 3rd International Conference on Computational Stochastic Mechanics, Santorini, Greece: 14–17.

Lancaster, H.O. (1957). Some properties of bivariate normal distribution considered in the form of a contingency table. *Biometrika* 44: 289–292.

Langley, R.S. (1988). A first passage approximation for normal stationary random processes. *J. Sound Vib.* 122 (2): 261–275.

Lee, Y. and Song, J. (2011). Risk analysis of fatigue-induced sequential failures by branch-and-bound method employing system reliability bounds. *J. Eng. Mech.* 137 (12): 807–821.

Leu, L.J. and Yang, S.S. (2000). *System reliability analysis of steel frames considering the effect of spread of plasticity*. 8th ASCE Specialty Conference on Probabilistic Mechanics and Structural Reliability, 19–22.

Li, K.S. (1992). Point-estimate method for calculating statistical moments. *J. Eng. Mech.* 118 (7): 1506–1511.

Li, H.S. and Cao, Z.J. (2016). MatLab codes of subset simulation for reliability analysis and structural optimization. *Struct. Multidiscip. Optim.* 54 (2): 391–410.

Li, J. and Chen, J.B. (2009). *Stochastic Dynamics of Structures*. Singapore: Wiley.

Li, H.N. and Zhao, Y.G. (2005). Investigation and analysis of Chuetsu earthquake damage in Niigata Province of Japan. *J. Nat. Disasters* 14 (1): 165–174. (in Chinese).

Li, G.Q., Cao, H., Li, Q.S., and Huo, D. (1993). *Structural Dynamic Reliability Theory and Its Application*. Beijing: Seismological Press (in Chinese).

Li, J., Chen, J.B., and Fan, W.L. (2007). The equivalent extreme-value event and evaluation of the structural system reliability. *Struct. Saf.* 29 (2): 112–131.

Li, Q.W., Wang, C., and Ellingwood, B.R. (2015). Time-dependent reliability of aging structures in the presence of non-stationary loads and degradation. *Struct. Saf.* 52 (13): 132–141.

Li, Q.W., Wang, C., and Zhang, H. (2016). A probabilistic framework for hurricane damage assessment considering non-stationarity and correlation in hurricane actions. *Struct. Saf.* 59: 108–117.

Lin, Y.K. (1967). *Probabilistic Theory of Structural Dynamics*. Malabar, FL: Krieger.

Lin, Y.K. (1970). First-excursion failure of randomly excited structures. *AIAA J.* 8 (4): 720–725.

Lind, N.C. (1979). Optimal reliability analysis by fast convolution. *J. Eng. Mech. Div.* 105 (3): 447–452.

Liu, P.L. and Der Kiureghian, A. (1986). Multivariate distribution models with prescribed marginals and covariances. *Probab. Eng. Mech.* 1 (2): 105–112.

Livesley, R.K. (1975). *Matrix Methods of Structural Analysis*, 2e. New York: Pergamon Press.

Low, B.K. (2007). Reliability analysis of rock slopes involving correlated nonnormals. *Int. J. Rock Mech. Min. Sci.* 44 (6): 922–935.

Lu, Z.Z., Song, S.F., and Yue, Z.F. (2008). Reliability sensitivity method by line sampling. *Struct. Saf.* 30 (6): 517–532.

Lu, Z.H., Zhao, Y.G., and Ang, A.H.-S. (2010). Estimation of load and resistance factors based on the fourth moment method. *Struct. Eng. Mech.* 36 (1): 19–36.

Lu, Z.H., Wang, X.F., and Zhao, Y.G. (2014). High-order moment method for structural reliability with unknown distribution. In: *Presented at Proceedings of the 17th working conference of the IFIP WG7.5 on Reliability and Optimization of Structural Systems* (IFIP 2014), Huangshan, China, July, 3–7.

Lu, Z.H., Cai, C.H., and Zhao, Y.G. (2017a). Structural reliability analysis including correlated random variables based on third-moment transformation. *J. Struct. Eng.* 143 (8): 04017067.

Lu, Z.H., Hu, D.Z., and Zhao, Y.G. (2017b). Second-order fourth-moment method for structural reliability. *J. Eng. Mech.* 143 (4). 06016010.

Lu, Z.H., Leng, Y., Dong, Y. et al. (2019). Fast integration algorithms for time-dependent structural reliability analysis considering correlated variables. *Struct. Saf.* 78: 23–32.

Lu, Z.H., Cai, C.H., Zhao, Y.G. et al. (2020). Normalization of correlated random variables in structural reliability analysis using fourth-moment transformation. *Struct. Saf.* 82: 101888.

Lutes, L.D. and Sarkani, S. (2004). *Random Vibrations: Analysis of Structural and Mechanical Systems*. Amsterdam: Elsevier.

MacGregor, J.G. (1988). *Reinforced Concrete Mechanics and Design*. Englewood Cliffs, NJ: Prentice-Hall.

MacKenzie, C.A. and Winterstein, S.R. (2011). *Comparing L-moments and conventional moments to model current speeds in the North Sea*. In: T. Doolen and E. Van Aken, eds., Proceedings of the 2011 Industrial Engineering Research Conference, IERC 2011, Reno, NV.

Madsen, H.O. and Tvedt, L. (1990). Methods for time-dependent reliability and sensitivity analysis. *J. Eng. Mech.* 116 (10): 2118–2135.

Madsen, H.O., Krenk, S., and Lind, N.C. (1986). *Methods of Structural Safety*. NJ: Prentice-Hall: Englewood Cliffs.

Mahadevan, S. and Raghothamachar, P. (2000). Adaptive simulation for system reliability analysis of large structures. *Comput. Struct.* 77 (6): 725–734.

Mander, J.B., Priestley, M.J., and Park, R. (1988). Theoretical stress-strain model for confined concrete. *J. Struct. Eng.* 114 (8): 1804–1826.

Masters, F. and Gurley, K.R. (2003). Non-Gaussian simulation: cumulative distribution function map-based spectral correction. *J. Eng. Mech.* 129 (12): 1418–1428.

Mehta, V.K. and Dasgupta, B. (2012). A constrained optimization algorithm based on the simplex search method. *Eng. Optim.* 44 (5): 537–550.

Melchers, R.E. (1987). *Structural Reliability, Analysis and Prediction*. New York: Wiley.

Melchers, R.E. (1994). Structural system reliability assessment using directional simulation. *Struct. Saf.* 16 (1–2): 23–37.

Melchers, R.E. (1999). *Structural Reliability; Analysis and Prediction*, 2e. West Sussex, UK: Wiley.

Melchers, R.E. and Beck, A.T. (2018). *Structural Reliability Analysis and Prediction*, 3e. Hoboken, NJ: Wiley.

Melchers, R.E. and Tang, L.K. (1984). Dominant failure modes in stochastic structural systems. *Struct. Saf.* 2 (2): 128–143.

Metropolis, N., Rosenbluth, A.W., Rosenbluth, M.N. et al. (1953). Equation of state calculations by fast computing machines. *J. Chem. Phys.* 21 (6): 1087–1092.

Miao, F. and Ghosn, M. (2011). Modified subset simulation method for reliability analysis of structural systems. *Struct. Saf.* 33 (4–5): 251–260.

Mirza, S.A. and MacGregor, J.G. (1982). Safety and limit states design for reinforced concrete. *Can. J. Civ. Eng.* 3 (3): 484–513.

Mori, Y. (2002). Practical method for load and resistance factors for use in limit state design. *J. Struct. Constr. Eng.* 559: 39–46. (in Japanese).

Mori, Y. and Ellingwood, B.R. (1993a). Time-dependent system reliability analysis by adaptive importance sampling. *Struct. Saf.* 12 (1): 59–73.

Mori, Y. and Ellingwood, B.R. (1993b). Reliability-based service-life assessment of aging concrete structures. *J. Struct. Eng.* 119 (5): 1600–1621.

Mori, Y., Tajiri, H., and Mizutani, Y. (2017). *Practical method for load and resistance factors using shifted lognormal approximation*, Safety, Reliability, Risk, Resilience and Sustainability of Structures and Infrastructure, 2017.8: 664–673.

Mori, Y., Tajiri, H. and Higashi, Y. (2019). *A study on practical method for load and resistance factors using shifted lognormal approximation*, Proc. 9th Japan Conf. on Structural Safety and Reliability, Vol. 9, 8 (CD-ROM) (in Japanese).

Moses, F. (1982). System reliability developments in structural engineering. *Struct. Saf.* 1: 3–13.

Murotsu, Y., Okada, H., Taguchi, K. et al. (1984). Automatic generation of stochastically dominant failure modes of frame structures. *Struct. Saf.* 2 (1): 17–25.

Murphy, S.A. and Van Der Vaart, A.W. (2000). On profile likelihood. *J. Am. Stat. Assoc.* 95 (450): 449–485.

Naess, A. (1987). The response statistics of nonlinear, second-order transformations to Gaussian loads. *J. Sound Vib.* 115 (1): 103–129.

Naess, A. (1990). Approximation first passage and extremes of narrow-band Gaussian and non-Gaussian random vibrations. *J. Sound Vib.* 138 (3): 365–380.

Naess, A. and Gaidai, O. (2009). Estimation of extreme values from sampled time series. *Struct. Saf.* 31 (4): 325–334.

Naess, A., Gaidai, O., and Batsevych, O. (2010). Prediction of extreme response statistic of narrow-band vibrations. *J. Eng. Mech.* 136 (3): 290–298.

Nafday, A.M. (1987). Failure mode identification for structural frames. *J. Struct. Eng.* 113 (7): 1415–1432.

Nataf, A. (1962). Determination des Distributions dont les Marges sont Donnees. *C.R. Acad. Sci.* 22: 42–43.

Nelder, J.A. and Mead, R. (1965). A simplex method for function minimization. *Comput. J.* 7 (4): 308–313.

Nie, J.S. and Ellingwood, B.R. (2005). Finite element-based structural reliability assessment using efficient directional simulation. *J. Eng. Mech.* 131 (3): 259–267.

Nowak, A.S. and Collins, K.R. (2000). *Reliability of Structures*. NewYork: McGraw-Hill.

Nowak, A.S., Lutomirska, M., and Sheikh Ibrahim, F.I. (2010). The development of live load for long span bridges. *Bridge Struct.* 6 (1–2): 73–79.

Ochi, M. (1986). Non-Gaussian random processes in ocean engineering. *Probab. Eng. Mech.* 1 (1): 28–39.

Ono, T. and Idota, H. (1986). Development of high-order moment standardization method into structural design and its efficiency. *J. Struct. Constr. Eng.* 365: 40–47.

Ono, T., Idota, H., and Kawahara, H. (1986). A statistical study on the resistance of steel column and beam using higher order moments. *J. Struct. Constr. Eng.* 370: 19–27. (in Japanese).

Ono, T., Idota, H., and Dozuka, A. (1990). Reliability evaluation of structural systems using higher-order moment standardization technique. *J. Struct. Constr. Eng.* 418: 71–79. (in Japanese).

Osborne, J.W. (2010). Improving your data transformations: applying the Box-Cox transformation. *Pract. Assess. Res. Eval.* 15 (12): 1–9.

Oswald, G.F. and Schueller, G.I. (1984). Reliability of deteriorating structures. *Eng. Fract. Mech.* 20 (3): 479–488.

Papaioannou, I., Papadimitriou, C., and Straub, D. (2016). Sequential importance sampling for structural reliability analysis. *Struct. Saf.* 62: 66–75.

Papoulis, A. (1984). *Probability, Random Variables and Stochastic Processes*. New York: McGraw-Hill.

Parkinson, D.B. (1978). First-order reliability analysis employing translation systems. *Eng. Struct.* 1 (1): 31–40.

Patil, G.P., Boswell, M.T., Ratnaparkhi, M.V., and Joshi, S.W. (1985a). *Dictionary and Classified Bibliography of Statistical Distributions in Scientific Work, Vol. 1, Discrete Models*. MD: International Co-Operative Publishing House, Burtonsville.

Patil, G.P., Boswell, M.T., and Ratnaparkhi, M.V. (1985b). *Dictionary and Classified Bibliography of Statistical Distributions in Scientific Work, Vol. 2, Univariate Continuous Models*. Burtonsville, MD: International Co-Operative Publishing House.

Patil, G.P., Boswell, M.T., Ratnaparkhi, M.V., and Roux, J.J.J. (1985c). *Dictionary and Classified Bibliography of Statistical Distributions in Scientific Work, Vol. 3, Multivariate Models*. Burtonsville, MD: International Co-Operative Publishing House.

Pearson, K. (1895). Contributions to the mathematical theory of evolution. *Philos. Trans. R. Soc. London* 186 (4): 343–414.

Pearson, E.S., Johnson, N.L., and Burr, I.W. (1979). Comparison of the percentage points of distributions with the same first four moments, chosen from eight different systems of frequency curves. *Commun. Stat. Simul. Comput.* B8 (3): 191–229.

Penmetsa, R.C. and Grandhi, R.V. (2002). Structural system reliability quantification using multipoint function. *AIAA J.* 40 (12): 2526–2531.

Pierre, L.M.S., Kopp, G.A., Surry, D., and Ho, T.C.E. (2005). The UWO contribution to the NIST aerodynamic database for wind loads on low buildings: part 2. Comparison of data with wind load provisions. *J. Wind Eng. Ind. Aerodyn.* 93 (1): 31–59.

Piric, K. (2015). Reliability analysis method-based on determination of the performance function's PDF using the univariate dimension reduction method. *Struct. Saf.* 57: 18–25.

Puig, B. and Akian, J. (2004). Non-Gaussian simulation using Hermite polynomials expansion and maximum entropy principle. *Probab. Eng. Mech.* 19 (4): 293–305.

Rackwitz, R. and Fiessler, B. (1978). Structural reliability under combined load sequence. *Comput. Struct.* 9 (5): 489–494.

Rahman, S. and Xu, H. (2004). A univariate dimension-reduction method for multi-dimensional integration in stochastic mechanics. *Probab. Eng. Mech.* 19 (4): 393–408.

Rajashekhar, M.R. and Ellingwood, B.R. (1993). A new look at the response surface approach for reliability analysis. *Struct. Saf.* 12 (3): 205–220.

Ramberg, J. and Schmeiser, B. (1974). An approximate method for generating asymmetric random variables. *Commun. ACM* 17 (2): 78–82.

Ramberg, J.S., Dudewicz, E.J., Tadikamalla, P.R., and Mykytka, E.F. (1979). A probability distribution and its uses in fitting data. *Technometrics* 21 (2): 201–214.

Rice, S.O. (1944). Mathematical analysis of random noise. *Bell Syst. Tech. J.* 23 (3): 282–332.

Rice, S.O. (1945). Mathematical analysis of random noise. Part III: statistical properties of random noise currents. *Bell Syst. Tech. J.* 24 (1): 46–156.

Rosenblatt, M. (1952). Remarks on a multivariate transformation. *Ann. Math. Stat.* 23 (3): 470–472.

Rosenblueth, E. (1975). Point estimates for probability moments. *Proc. Natl. Acad. Sci. U.S.A.* 72 (10): 3812–3814.

Rubinstein, R.Y. (1981). *Simulation and Monte Carlo Method*. New York: Wiley.

Rzhanitzyn, R. (1957). It is necessary to improve the standards of design of building structures, in Allan, D.E. (Transl.). A statistical method of design of building structures, Technical translation No. 1368. National Research Council of Canada, Ottwa.

Sakamoto, J. and Mori, Y. (1995). *Probability analysis method by fast Fourier transform: accuracy, efficiency, and applicability of the method.* Proc. of Asian-Pacific Symposium on Structural Reliability and its Applications, Nov., Tokyo, 32–39.

Sakamoto, J. and Mori, Y. (1997). Probability analysis method using fast Fourier transform and its application. *Struct. Saf.* 19 (1): 21–36.

Sankaran, N. (1959). On eccentric chi-square distribution. *Biometrika* 46: 235–237.

Sankaran, N. (1963). Approximations to the non-central chi-square distribution. *Biometrika* 50: 199–204.

Sarpkaya, T. and Issacson, M. (1981). *Mechanics of Wave Forces on Offshore Structures.* New York.: Van Nostrand Reinhold.

Shields, M.D. and Deodatis, G. (2013). A simple and efficient methodology to approximate a general non-Gaussian stationary stochastic vector process by a translation process with applications in wind velocity simulation. *Probab. Eng. Mech.* 31: 19–29.

Shields, M.D., Deodatis, G., and Bocchini, P. (2011). A simple and efficient methodology to approximate a general non-Gaussian stationary stochastic process by a translation process. *Probab. Eng. Mech.* 26: 511–519.

Shinozuka, M. (1983). Basic analysis of structural safety. *J. Struct. Eng.* 109 (3): 721–740.

Shinozuka, M. and Deodatis, G. (1991). Simulation of stochastic processes by spectral representation. *Appl. Mech. Rev.* 44 (4): 191–204.

Shinozuka, M. and Jan, C.M. (1972). Digital simulation of random processes and its applications. *J. Sound Vib.* 25 (1): 111–128.

Shooman, M.L. (1968). *Probabilistic Reliability: An Engineering Approach.* New York: McGraw-Hill.

Shueller, G.I. and Shinozuka, M. (1987). *Stochastic Methods in Structural Dynamics.* Boston, MA.: Martinus Nijhoff Publishers.

Singh, R. and Lee, P. (1993). Frequency response of linear systems with parameter uncertainties. *J. Sound Vib.* 168 (1): 71–92.

Slifker, J.F. and Shapiro, S.S. (1980). The Johnson system: selection and parameter estimation. *Technometrics* 22 (2): 239–246.

Sobol, I.M. (1993). Sensitivity estimates for nonlinear mathematical models. *Math. Modell. Comput. Exp.* 1 (4): 407–414.

Song, B.F. (1992). A numerical integration method in affine space and a method with high accuracy for computing structural system reliability. *Comput. Struct.* 42 (2): 255–262.

Song, J. and Der Kiureghian, A. (2003). Bounds on system reliability by linear programming. *J. Eng. Mech.* 129 (6): 627–636.

Song, S.F., Lu, Z., and Qiao, H. (2009). Subset simulation for structural reliability sensitivity analysis. *Reliab. Eng. Syst. Saf.* 94 (2): 658–665.

Sorensen, J.D. (2004). *Structural Reliability Theory and Risk Analysis.* Aalborg, Denmark: Aalborg Univ.

Spacone, E., Filippou, F.C., and Taucer, F.F. (1996). Fibre beam–column model for non-linear analysis of RC frames: part I formulation. *Earthquake Eng. Struct. Dyn.* 25 (7): 711–725.

Stuart, A. and Ord, J.K. (1987). *Kendall's Advanced Theory of Statistics*. London.: Charles Griffin & Company Ltd.

Sundar, V.S. and Shields, M.D. (2016). Surrogate-enhanced stochastic search algorithms to identify implicitly defined functions for reliability analysis. *Struct. Saf.* 62: 1–11.

Takada, T. (2001). Discussion on LRFD in AIJ-WG of limit state design; private communication.

Thoft-Christensen, P. (1990). Consequence modified β-unzipping of plastic structures. *Struct. Saf.* 7 (2): 191–198.

Thoft-Christensen, P. and Murotsu, Y. (1986). *Application of Structural Systems Reliability Theory*. Berlin: Springer.

Tichy, M. (1994). First-order third-moment reliability method. *Struct. Saf.* 16 (3): 189–200.

Tung, Y.K. (1999). Polynomial normal transform in uncertainty analysis. In: *ICASP 8, Application of Statistics and Probability* (eds. R.E. Melchers and M.G. Stewart), 167–174. Netherlands: A. A Balkema Publishers.

Tvedt, L. (1983). Two second-order approximations to the failure probability. Veritas Rep. RDIV/20-004083. Det norske Veritas: Oslo.

Tvedt, L. (1990). Distribution of quadratic forms in the normal space–application to structural reliability. *J. Eng. Mech., ASCE* 116 (6): 1183–1197.

Ugata, T. (2000). Reliability analysis considering skewness of distribution–simple evaluation of load and resistance factors. *J. Struct. Constr. Eng.* 529: 43–50.

Ugata, T. and Moriyama, K. (1996). Simple method of evaluating the failure probability of s structure considering the skewness of distribution. *Nucl. Eng. Des.* 160: 307–319.

Val, D.V. (2007). Deterioration of strength of RC beams due to corrosion and its influence on beam reliability. *J. Struct. Eng.* 133 (9): 1297–1306.

Vanmarcke, E.H. (1975). On the distribution of the first passage time for normal stationary process. *J. Appl. Mech.* 42: 215–220.

Wang, L., Ma, Y.F., Zhang, J.R. et al. (2015). Uncertainty quantification and structural reliability estimation considering inspection data scarcity. *J. Risk Uncertainty Eng. Syst. Part A: Civ. Eng.* 1: 04015004.

Wang, C., Li, Q.W., and Ellingwood, B.R. (2016). Time-dependent reliability of aging structures: an approximate approach. *Struct. Infrastruct. Eng.* 12 (12): 1566–1572.

Wang, C., Zhang, H., and Li, Q.W. (2017). Time-dependent reliability assessment of aging series systems subjected to non-stationary loads. *Struct. Infrastruct. Eng.* 13 (12): 1–10.

Wen, Y.K. and Chen, H.-C. (1987). On fast integration for time variant structural reliability. *Probab. Eng. Mech.* 1987, 2 (3): 156–162.

Winterstein, S.R. (1988). Nonlinear vibration models for extremes and fatigue. *J. Eng. Mech.* 114 (10): 1772–1790.

Winterstein, S.R. and Kashef, T. (2000). Moment-based load and response models with wind engineering applications. *J. Solar Energy Eng.* 122 (3): 122–128.

Winterstein, S.R. and Mackenzie, C.A. (2011). Extremes of nonlinear vibration: comparing models based on moments, L-moments, and maximum entropy. *J. Offshore Mech. Arct. Eng.* 135 (2): 185–195.

Wolfram, S. (2003). *The Mathematica Book*, 5e. Wolfram Media, Inc.

Wu, Y.T. and Burnside, O.H. (1990). *Computational methods for probability of instability calculations*. Presented at AIAA/ASME/ASCE/AHS/ASC 31th Structures, Structural Dynamics and Materials Conference, A90-2928311-39.

Xu, L. and Cheng, G.D. (2003). Discussion on: moment methods for structural reliability. *Struct. Saf.* 25 (2): 193–199.

Xu, J. and Dang, C. (2019). A novel fractional moments-based maximum entropy method for high-dimensional reliability analysis. *Appl. Math. Modell.* 75: 749–768.

Xu, J. and Kong, F. (2019). Adaptive scaled unscented transformation for highly efficient structural reliability analysis by maximum entropy method. *Struct. Saf.* 76: 123–134.

Xu, J. and Lu, Z.H. (2017). Evaluation of moments of performance functions based on efficient cubature formulation. *J. Eng. Mech.* 143 (8): 06017007.

Xu, H. and Rahman, S. (2004). A generalized dimension-reduction method for multidimensional integration in stochastic mechanics. *Int. J. Numer. Methods Eng.* 61 (12): 1992–2019.

Xu, J., Dang, C., and Kong, F. (2017). Efficient reliability analysis of structures with the rotational quasi-symmetric point and the maximum entropy methods. *Mech. Syst. Signal Process.* 95 (3): 58–76.

Yamazaki, F. and Shinozuka, M. (1988). Digital generation of non-Gaussian stochastic fields. *J. Eng. Mech.* 114 (7): 1183–1197.

Yang, L. and Gurley, K.R. (2015). Efficient stationary multivariate non-Gaussian simulation based on a Hermite PDF model. *Probab. Eng. Mech.* 42: 31–41.

Yang, L., Liu, J., and Yu, B. (2014). Adaptive dynamic bounding method for reliability analysis of structural systems. *China Civ. Eng. J.* 47 (4): 38–46. (in Chinese).

Yao, H.J. and Wen, Y.K. (1996). Response surface method for time-variant reliability analysis. *J. Struct. Eng.* 122 (2): 193–201.

Yuan, S. (2007). *Programming Structural Mechanics*. Beijing, China: Higher Education Press (in Chinese).

Zhang, R.X. and Mahadevan, S. (2003). Bayesian methodology for reliability model acceptance. *Reliab. Eng. Syst. Saf.* 80: 95–103.

Zhang, X.F. and Pandey, M.D. (2013). Structural reliability analysis based on the concepts of entropy, fractional moment and dimensional reduction method. *Struct. Saf.* 43: 28–40.

Zhang, T. and Yang, B. (2017). Box-Cox transformation in big data. *Technometrics* 59 (2): 189–201.

Zhao, Y.G. and Ang, A.H.-S. (2002). Three-parameter Gamma distribution and its significance in structure. *Int. Comput. Struct. Eng.* 2 (1): 1–10.

Zhao, Y.G. and Ang, A.H.-S. (2003). System reliability assessment by method of moments. *J. Struct. Eng.* 129 (10): 1341–1349.

Zhao, Y.G. and Ang, A.H.-S. (2012). On the first-order third-moment reliability method. *Struct. Infrastruct. Eng.* 8 (5): 517–527.

Zhao, Y.G. and Jiang, J.R. (1992). An advanced first order second moment method. *Earthquake Eng. Eng. Vib.* 12: 49–57. (in Chinese).

Zhao, Y.G. and Jiang, J.R. (1995). A structural reliability analysis method based on genetic algorithm. *Earthquake Eng. Eng. Vib.* 15 (3): 916–924. (in Chinese).

Zhao, Y.G. and Lu, Z.H. (2006). Load and resistance factors estimation without using distributions of random variables. *J. Asian Archit. Build. Eng.* 5 (2): 325–332.

Zhao, Y.G. and Lu, Z.H. (2007a). Fourth-moment standardization for structural reliability assessment. *J. Struct. Eng.* 133 (7): 916–924.

Zhao, Y.G. and Lu, Z.H. (2007b). The first three moments for some commonly used performance function. *J. Struct. Constr. Eng.* 613: 31–37.

Zhao, Y.G. and Lu, Z.H. (2007c). Applicable range of the fourth-moment method for structural reliability. *J. Asian Archit. Build. Eng.* 6 (1): 151–158.

Zhao, Y.G. and Lu, Z.H. (2008). Cubic normal distribution and its significance in structural reliability. *Struct. Eng. Mech.* 28 (3): 263–280.

Zhao, Y.G. and Lu, Z.H. (2011). Estimation of load and resistance factors using the third-moment method based on the 3P-lognormal distribution. *Front. Archit. Civ. Eng. China* 5 (3): 315.

Zhao, Y.G. and Ono, T. (1998). System reliability evaluation of ductile frame structure. *J. Struct. Eng.* 124 (6): 678–685.

Zhao, Y.G. and Ono, T. (1999a). A general procedure for first/second-order reliability method (FORM/SORM). *Struct. Saf.* 21 (2): 95–112.

Zhao, Y.G. and Ono, T. (1999b). New approximations for SORM: part 1. *J. Eng. Mech.* 125 (1): 79–85.

Zhao, Y.G. and Ono, T. (1999c). New approximations for SORM: part 2. *J. Eng. Mech.* 125 (1): 86–93.

Zhao, Y.G. and Ono, T. (2000a). New point estimates for probabilistic moments. *J. Eng. Mech.* 126 (4): 433–436.

Zhao, Y.G. and Ono, T. (2000b). Third-moment standardization for structural reliability analysis. *J. Struct. Eng.* 126 (6): 724–732.

Zhao, Y.G. and Ono, T. (2001). Moment methods for structural reliability. *Struct. Saf.* 23 (1): 47–75.

Zhao, Y.G. and Ono, T. (2004). On the problems of the fourth moment method. *Struct. Saf.* 26 (3): 343–347.

Zhao, Y.G. and Sun, J.J. (1991). On dynamic sensitivity analysis of structures. *Earthquake Eng. Eng. Vib.* 11 (4): 28–38. (in Chinese).

Zhao, Y.G., Ono, T., and Idota, H. (1999). Response uncertainty and time-variant reliability analysis for hysteretic MDF structures. *Earthquake Eng. Struct. Dyn.* 28 (10): 1187–1213.

Zhao, Y.G., Ono, T., Idota, H., and Hirano, T. (2001). A three-parameter distribution used for structural reliability evaluation. *J. Struct. Constr. Eng.* 546: 31–38.

Zhao, Y.G., Ono, T., and Kato, M.A. (2002a). Second-order third-moment reliability method. *J. Struct. Eng.* 128 (8): 1087–1090.

Zhao, Y.G., Ono, T., and Kiyoshi, I. (2002b). Monte Carlo simulation using moments of random variables. *J. Asian Archit. Build. Eng.* 1 (1): 13–20.

Zhao, Y.G., Lu, Z.H., and Ono, T. (2006a). A simple third-moment method for structural reliability. *J. Asian Archit. Build. Eng.* 5 (1): 129–136.

Zhao, Y.G., Lu, Z.H., and Ono, T. (2006b). 4P-Lambda distribution and its applications to structural reliability assessment. *J. Struct. Constr. Eng.* 604: 47–54.

Zhao, Y.G., Zhong, W.Q., and Ang, A.H.-S. (2007). Estimating joint failure probability of series structural systems. *J. Eng. Mech.* 133 (5): 588–596.

Zhao, Y.G., Lu, Z.H., and Zhong, W.Q. (2014). Time variant reliability analysis with consideration of parameter uncertainties. *Struct. Infrastruct. Eng.* 10 (10): 1276–1284.

Zhao, Y.G., Zhang, X.Y., and Lu, Z.H. (2018a). Complete monotonic expression of the fourth-moment normal transformation for structural reliability. *Comput. Struct.* 196: 186–199.

Zhao, Y.G., Zhang, X.Y., and Lu, Z.H. (2018b). A flexible distribution and its application in reliability engineering. *Reliab. Eng. Syst. Saf.* 176: 1–12.

Zhao, Y.G., Li, P.P., and Lu, Z.H. (2018c). Efficient evaluation of structural reliability under imperfect knowledge about probability distributions. *Reliab. Eng. Syst. Saf.* 175: 160–170.

Zhao, Y.G., Zhang, L.W., Lu, Z.H., and He, J. (2019). First passage probability assessment of stationary non-Gaussian process using the third-order polynomial transformation. *Adv. Struct. Eng.* 22 (1): 187–201.

Zhao, Y.G., Tong, M.N., Lu, Z.H., and Xu, J. (2020). Monotonic expression of polynomial normal transformation based on the first four L-moments. *J. Eng. Mech.* 146 (7): 06020003.

Zimmerman, J.J., Ellis, J.H., and Corotis, R.B. (1993). Stochastic optimization models for structural reliability analysis. *J. Struct. Eng.* 119 (1): 223–239.

Zong, Z. and Lam, K.Y. (1998). Estimation of complicated distribution using B-spline functions. *Struct. Saf.* 20 (4): 341–355.

Zwillinger, D. (2018). *CRC Standard Mathematical Tables and Formulae*. CRC Press.

Index